Standard error of estimate

$$S_{Y \cdot X} = S_Y \sqrt{(1 - r_{YX}^2)}$$

Standard error of the mean

$$\sigma_{\overline{X}} = \frac{\sigma_X}{\sqrt{N}}$$

Standard score of a sample mean in a population

$$Z_{\overline{X}} = \frac{\overline{X} - \mu}{\sigma_{\overline{X}}}$$

Limits of the α percent confidence interval for μ. Use positive Z for upper limit, negative Z for lower limit

$$CI = \pm Z_{\alpha/2}(\sigma_{\overline{X}}) + \overline{X}$$

Definition of the unbiased estimate of σ^2. Take the square root to estimate the standard deviation.

$$\hat{S}_X^2 = \frac{\sum(X_i - \overline{X})^2}{N - 1} = \frac{SS}{N - 1}$$

Sample estimate of the standard error of the mean

$$\hat{S}_{\overline{X}} = \frac{\hat{S}_X}{\sqrt{N}}$$

Formula for Student's t

$$t_{df} = \frac{\overline{X} - K}{\hat{S}_{\overline{X}}}$$

Confidence interval for μ from sample statistics

$$CI = \pm t_{\alpha/2}\hat{S}_{\overline{X}} + \overline{X}$$

Standard error of the difference between means

$$\sigma_{\overline{X}_1 - \overline{X}_2} = \sqrt{\frac{\sigma_1^2}{N_1} + \frac{\sigma_2^2}{N_2}}$$

Critical ratio test for a two-sample hypothesis

$$Z_{\overline{X}_1 - \overline{X}_2} = \frac{(\overline{X}_1 - \overline{X}_2) - (\mu_1 - \mu_2)}{\sigma_{(\overline{X}_1 - \overline{X}_2)}}$$

Formula for the pooled estimate

$$\hat{S}_p^2 = \frac{(N_1 - 1)\hat{S}_1^2 + (N_2 - 1)\hat{S}_2^2}{(N_1 + N_2 - 2)}$$

Estimate of standard error of the difference

$$\hat{S}_{\overline{X}_1 - \overline{X}_2} = \sqrt{\frac{\hat{S}_p^2}{N_1} + \frac{\hat{S}_p^2}{N_2}}$$

t-test for two-sample hypotheses

$$t_{df} = \frac{(\overline{X}_1 - \overline{X}_2) - (\mu_1 - \mu_2)}{\hat{S}_{\overline{X}_1 - \overline{X}_2}}$$

Computing the direct difference

$$D_i = X_{i1} - X_{i2}$$

Standard deviation of direct differences

$$\hat{S}_D = \sqrt{\frac{N\sum D_i^2 - (\sum D_i)^2}{N(N - 1)}}$$

Standard error of the mean of differences

$$\hat{S}_{\overline{D}} = \frac{\hat{S}_D}{\sqrt{N}}$$

t-test for hypotheses about differences

$$t_{df} = \frac{\overline{D} - K}{\hat{S}_{\overline{D}}}$$

Determining variance accounted for in a t-test

$$r_{pb}^2 = \frac{t^2}{t^2 + df}$$

Effect size of a difference between means

$$d = \frac{(\overline{X}_1 - \overline{X}_2) - (\mu_1 - \mu_2)}{\hat{S}_p}$$

Effect size for a one-sample hypothesis

$$d = \frac{\overline{X} - K}{\hat{S}_X}$$

Basic Statistics for
the Behavioral Sciences

Basic Statistics for the Behavioral Sciences

Robert M. Thorndike
Western Washington University

Dale L. Dinnel
Western Washington University

Upper Saddle River, New Jersey
Columbus, Ohio

**To ELT, RLT, and EST, and with
special thanks to Jacob Cohen.**

Library of Congress Cataloging-in-Publication Data

Thorndike, Robert M.
  Basic statistics for the behavioral sciences / by Robert M. Thorndike and Dale L. Dinnel.
      p. cm.
  Includes bibliographical references and index.
  ISBN 0-13-861055-X
  1. Psychometrics.   2. Education—Statistics.   I. Dinnel, Dale L.   II. Title.

BF39.T476 2001
150′.1′5195—dc21

00-062461

Vice President and Publisher: Jeffery W. Johnston
Executive Editor: Kevin M. Davis
Development Editor: Gianna M. Marsella
Editorial Assistant: Christina Kalisch
Production Editor: Mary Harlan
Design Coordinator: Diane C. Lorenzo
Cover Design: Jason Moore
Cover Art: SuperStock
Illustrations: The PRD Group, Inc.
Production Coordination: Betsy Keefer
Production Manager: Laura Messerly
Director of Marketing: Kevin Flanagan
Marketing Manager: Amy June
Marketing Services Manager: Krista Groshong

This book was set in Garamond by The PRD Group, Inc. It was printed and bound by Courier/Kendallville, Inc. The cover was printed by The Lehigh Press, Inc.

SPSS is a registered trademark of SPSS Inc. Copyright 2000 SPSS Inc.

10 9 8 7 6 5 4 3 2 1
ISBN: 0-13-861055-X

Preface

A Unique Approach to Statistics

There are dozens of books available that introduce students to statistics for education and psychology. In view of this array of choices, why have we gone to the labor of preparing a new one? Simply stated, we were not satisfied with the available choices. Between us, we have been teaching statistics for more than 50 years, and we have used many of the books that are currently available. There are a number of things we feel our students should understand about statistics that, unfortunately, we have not found sufficiently covered in these texts.

In contrast, we have tried to write a book that provides students with a clear understanding of statistics as a tool of research. To do this, we focus on helping students understand how and why particular statistics have been developed, with special emphasis on helping students see what questions a specific statistic can answer. This focus and our persistent effort to present statistics within the context of the research process distinguish our text from others.

An Emphasis on Understanding What Questions Individual Statistics Can Answer

Most introductory statistics books do not bother to explain to students where individual statistical indices come from, the logic behind their development, what their relationship is to other statistical indices, under what circumstances they are most useful, and what insights they can give about a set of data. We feel these questions are critically important.

For example, the concept of least squares is first introduced in Chapter 4, as a logical outgrowth of the concepts of mean and variance. Then in Chapter 7, the discussion of prediction is developed using the mean as a least-squares predictor and the variance as an index of error. Students can see the logic of linear regression building before their eyes and understand it as a straightforward expansion of ideas they have already mastered. This development of least-squares prediction is used later in Chapter 16 to show how the analysis of variance relates to the comparison of means and to linear models. It is also used to link correlation with measures of effect size in Chapter 13.

One major emphasis of this approach, which we believe sets this book apart from many of the alternatives, is a focus on understanding the logic and relatedness of all statistical methods. We do not slight accuracy in computation, but we place a high value on helping students understand what they are doing, why they are doing it, and how what they are doing relates to their research question.

An Emphasis on Statistics in the Context of Research

Another significant element that sets our book apart is an emphasis on the research applications of statistics. For most psychologists and educators, statistical methods are not ends in themselves but tools to answer research questions. Throughout this book, we keep the research questions to be answered constantly before the reader. Each chapter begins by posing a set of research questions, and then is developed with the goal of answering those questions.

The relatedness of different statistical methods and the focus on answering practical research questions are brought together in *two hypothetical research studies* that are introduced early in the book and used throughout the remaining chapters to illustrate the application of the statistical methods discussed. The studies are woven throughout the text and provide rich and multiple opportunities for students to see and apply statistical methods in realistic contexts. By applying all of the different methods to data from two particular studies, we also emphasize that good study design allows the investigator to answer a number of related questions.

The research focus of the book is further illustrated by our continuing emphasis on the value of sound research design and measurement. We insist that the source of the data and the methods of data collection are of primary importance and that no amount of elegant statistical analysis can extract meaning from a bad study.

An Accessible Text That Supports Student Comprehension

We are both well aware of the high level of anxiety that statistics can sometimes produce. As a result, we have made a particular effort to provide a book that is readable and accessible, but not "dumbed down." Rather than avoid difficult concepts, we do our best to provide clear, thoughtful explanations and straightforward verbal and mathematical definitions.

In addition, students are given many opportunities to practice and test their new knowledge to make sure their understanding is complete. For example, after each topic is discussed, a *Now You Do It* feature presents students with a practical example of the issue and asks them to work through a related problem. Suggestions and guidance are given, and at the end of the feature the solution is provided. Such examples scaffold a student's learning as concepts are reexplained, procedures are reviewed, and statistical operations are placed in real-life contexts.

To further help students achieve the goal of acquiring both a computational and a conceptual understanding of statistics in the context of research applications, each chapter ends with an extensive set of *exercises*. These problems challenge students to recall information, to calculate answers by hand and with statistical software, such as SPSS, and to use critical thinking skills to solve realistic research-based statistical problems. Because statistics works in a kind of cumulative fashion where new understandings are built upon old understandings, students are strongly encouraged to work through the problems at the end of each chapter before moving ahead. The cumulative nature of statistics is further reinforced by ongoing examples in the problem sets, which allow students to see how one step in the analysis leads to another.

Critical Coverage of the Role of Computers and Technology in Statistics

Computers have become a large part of data analysis, and we have included a discussion of doing each kind of analysis using SPSS. We have not, however, simply made this a cookbook of computer procedures. The underlying logic is developed for each statistic, and traditional hand-computation methods are also illustrated.

After the logic and computations have been shown, a box called *How the Computer Does It* guides the student through the steps to achieve the same result with the computer package. Where the program makes specific assumptions, these are pointed out, and students are shown how to change the default options if this is possible. Shortcomings and restrictions that the program imposes are also discussed in these boxes, a few of which are playfully titled *The Computer Doesn't Do It*.

In addition, two technology-based supplements will help students gain confidence using the computer to assist in their statistical endeavors: The CD-ROM at the back of the book provides electronic versions of the data sets for the hypothetical studies and the end-of-chapter exercises, and the Companion Website provides online resources and opportunities for additional practice for students.

Unique Content and Coverage of Special Topics

There are a number of points in data analysis where there are alternative ways to analyze data or where there is controversy about the best approach. These topics are highlighted in boxes called *Special Issues in Research*. Students' statistical competence will not suffer if they skip over these boxes, but their understanding of and appreciation for the research literature will be enhanced by exposure to these issues.

Another topic that we have found inadequately treated in many contemporary statistics books is the logic of testing the null hypothesis. We have been careful to provide a clear explanation of this logic and have provided examples of its correct use and some common mistakes associated with it. In addition, we discuss measures of effect size and variance accounted for as viable alternatives to null hypothesis testing. Some of the criticisms of traditional methods are also discussed.

Finally, this textbook extends its unique coverage with an emphasis that the logic of sampling distributions and hypothesis testing applies to all statistics, not just the mean, as is the case with many other books. We present *chi-square* in its correct form, as a test statistic for hypotheses about variances, and show how it can be used to place confidence intervals on variances. The F distribution also is introduced first as a test of the null hypothesis that two population variances are equal. Likewise, we illustrate the correct way to test hypotheses about correlation coefficients and to form confidence intervals for ρ using Fisher's Z transformation. In each case, the question that a researcher would be trying to answer, and how the particular statistic answers it, are kept at the center of the discussion.

Supplements

We believe that students learn best when they have ample opportunities to practice and apply their new knowledge and understandings in authentic contexts, and that the quality of instruction improves when teachers have access to a wide variety of instructional resources. Toward those ends, we have created a number of useful supplements for both teachers and students.

Data Sets CD-ROM
Packaged free with every copy of the textbook, the data sets CD-ROM houses the complete data packages for the two hypothetical studies featured throughout the text, as well as four additional data sets appearing in the end-of-chapter exercises. Students may use these data sets with popular statistical software packages such as SPSS, Systat, and MiniTab.

Student Study Guide
The Student Study Guide provides students with additional opportunities to review chapter content, practice statistical procedures, and apply statistical concepts using realistic problems and exercises.

Companion Website
The Companion Website, located at *http://www.prenhall.com/thorndike,* provides a wealth of online resources to individuals using this text. Students can use the Companion Website to review chapter content, take online quizzes, do research on the Internet or gain additional experience with statistics using suggested Websites, and engage in online discussions with their peers. PowerPoint slides can be also downloaded from this site. Using the Syllabus Manager tool, professors can create and maintain the class syllabus online, allowing students electronic access to the syllabus at any time.

Instructor's Manual
The Instructor's Manual includes transparency masters and the print version of the testbank. In addition, it includes all answers to the even-numbered exercises from the text.

Computerized Testbank Software
The computerized testbank software gives instructors electronic access to the test questions printed in the Instructor's Manual, allowing them to create and customize exams on their computer. The software can also perform a number of statistical operations to help professors manage their courses and gain insight into their students' progress and performance. Computerized testbank software is available in both Macintosh and PC/Windows versions.

Transparency Package

Transparencies are available to instructors in two ways: as transparency masters located in the Instructor's Manual or as PowerPoint slides available for download from the Companion Website.

Acknowledgments

The production of this book extended over many years, and many people contributed to its features. Our students and colleagues gave us valuable feedback on much of the general outline and approach. The editorial staff at Merrill/Prentice Hall, and Barbara Lyons in particular, were invaluable in helping us focus our thinking and writing. Western Washington University generously provided release time so we could work with fewer interruptions.

We are also very grateful for the comments from these reviewers, who responded with thought and care to the many drafts of this manuscript: Robert Barcikowski, Ohio University; Yong Dai, Louisiana State University, Shreveport; Mary I. Dereshiwsky, Northern Arizona University; Thomas D. Dertinger, University of Richmond and George Washington University; Jim C. Fortune, Virginia Tech; Thomas Frantz, SUNY Buffalo; Michael Gayle, SUNY New Paltz; Neal Grandgenett, University of Nebraska, Omaha; Carl Huberty, University of Georgia; Robert L. Kennedy, University of Arkansas, Little Rock; Alejandro Lazarte, Auburn University; Dale Shaw, University of Northern Colorado; and Michael E. Walker, The Ohio State University.

Discover the Companion Website Accompanying This Book

The Prentice Hall Companion Website: A Virtual Learning Environment

Technology is a constantly growing and changing aspect of our field that is creating a need for content and resources. To address this emerging need, Prentice Hall has developed an online learning environment for students and professors alike—Companion Websites—to support our textbooks.

In creating a Companion Website, our goal is to build on and enhance what the textbook already offers. For this reason, the content for each user-friendly website is organized by chapter and provides the professor and student with a variety of meaningful resources. Common features of a Companion Website include:

For the Professor

Every Companion Website integrates **Syllabus Manager**™, an online syllabus creation and management utility.

- **Syllabus Manager**™, provides you, the instructor, with an easy, step-by-step process to create and revise syllabi, with direct links into Companion Website and other online content without having to learn HTML.
- Students may logon to your syllabus during any study session. All they need to know is the web address for the Companion Website and the password you've assigned to your syllabus.
- After you have created a syllabus using **Syllabus Manager**™, students may enter the syllabus for their course section from any point in the Companion Website.
- Clicking on a date, the student is shown the list of activities for the assignment. The activities for each assignment are linked directly to actual content, saving time for students.
- Adding assignments consists of clicking on the desired due date, then filling in the details of the assignment—name of the assignment, instructions, and whether or not it is a one-time or repeating assignment.
- In addition, links to other activities can be created easily. If the activity is online, a URL can be entered in the space provided, and it will be linked automatically in the final syllabus.
- Your completed syllabus is hosted on our servers, allowing convenient updates from any computer on the Internet. Changes you make to your syllabus are immediately available to your students at their next logon.

For the Student

- **Chapter Objectives**—outline key concepts from the text.
- **Interactive Self-quizzes**—complete with automatic grading that provides immediate feedback for students.

After students submit their answers for the interactive self-quizzes, the Companion Website **Results Reporter** computes a percentage grade, provides a graphic representation of how many questions were answered correctly and incorrectly, and gives a question by question analysis of the quiz. Students are given the option to send their quiz to up to four email addresses (professor, teaching assistant, study partner, etc.).

- **Message Board**—serves as a virtual bulletin board to post—or respond to—questions or comments to/from a national audience.
- **Chat**—real-time chat with anyone who is using the text anywhere in the country—ideal for discussion and study groups, class projects, etc.
- **Web Destinations**—links to WWW sites that relate to chapter content.

To take advantage of these and other resources, please visit the *Basic Statistics for the Behavioral Sciences* Companion Website at

www.prenhall.com/thorndike

Brief Contents

Contents

Chapter 3
Measures of Central Tendency or Location   58

Chapter 4
Measures of Variability   82

Part III
Logical Considerations in the Design and Analysis
of Research

Chapter 11
Sampling Distributions and Statistical Estimation
of the Mean   260

Chapter 12
Statistical Hypotheses, Decisions, and Error   282

Part IV
Inferential Statistics in Action   309

Chapter 13
Testing Hypotheses About Means   310

Chapter 14
Inferences About Variances and Correlations   346

Basic Statistics for
the Behavioral Sciences

Part I

Describing Single Variables

Chapter 1

The Nature and Purpose of Data: Foundations for Research

CHAPTER OBJECTIVES

After studying this chapter, you should be able to:

❏ State the difference between a statistic and a parameter.

❏ State four purposes researchers have for collecting information.

❏ List the properties of good information and research data.

❏ State the differences between direct and indirect methods of observation and identify examples of each.

❏ State why sampling of content, individuals, times, and locations is important for generalizing research findings.

❏ Describe the four basic types of studies researchers use, and their advantages and disadvantages.

❏ Describe the differences between discrete and continuous variables and identify examples of each.

❏ Define the four levels of scales and state research contexts in which each can be used.

❏ Identify independent, dependent, status and control variables when designing or reading research studies.

One hundred years ago psychologists and educators believed that learning to read Latin or Greek trained the mind and made learning other academic subjects easier. This theory, called the doctrine of formal discipline, was reflected in American educational practice through the early decades of the 20th century. Beginning around 1901, however, psychologists and educators began to test (and ultimately rejected) this theory. In so doing, they laid the groundwork for dealing scientifically with information in the fields of education and psychology. They began to collect information systematically and compare their observations to predictions made on the basis of educational theories. What types of information do you think they would have needed to obtain to test the doctrine of formal discipline? If you were faced with a problem like this, how would you collect and analyze the necessary information to make a sound decision?

Introduction: The Place of Information in Science

All sciences advance knowledge in the same basic way, by gathering information about the world and developing explanations to account for what is consistent in that information.

This is a book about information in education and psychology: how to collect it, how to process it, and how to interpret it according to scientific methods. In other words, we will discuss some of the broad principles of information collection and processing used in education and psychology; then we will apply these principles to see how psychologists and educators detect regularities in their information and reach conclusions.

As we think about the place of information in science, it is important to be mindful that people often equate perceptions with information. We all know from our daily lives, however, that some "facts" of perception are true, while other "facts" of perception turn out not to be true, to be only partially true, or to be true in some situations but not in others. For example, perception tells us that the sun travels westward across the sky, and for thousands of years this was the accepted wisdom, although this "fact" of perception is false. Similarly, educators once believed that all children who experienced trouble learning to read were mentally retarded, but research has shown this "truth" to be false in many cases. One role of the scientific method in expanding human understanding of our universe, including the fields of education and psychology, is to provide a set of procedures for testing the truth of our perceptions and beliefs. Strategies and techniques for separating good information from bad, and for identifying what conclusions can appropriately be drawn from the available information, occupy a large portion of this book. Understanding these strategies and techniques will help you become a more intelligent collector, processor and interpreter of information, both in your college work and afterward.

In addition to being a book about information in science, this is also a book about language. All languages are made up of words and rules for combining these words to communicate ideas. Educational researchers and behavioral and social scientists have a common language called **statistics** which they use to communicate the results of their studies. Statistics is a branch of applied mathematics used to summarize quantities of information and help investigators draw sound conclusions. Although sociologists, psychologists, educators, and economists may speak different dialects of the language of statistics and may investigate very different subjects, they all use the same basic terms and rules to communicate their findings. The dialect of statistics that we will introduce in this book is that used by educators and psychologists.

We are all bombarded daily by information communicated in the language of statistics. For example, suppose that in last year's collegiate softball championship the most valuable player had a batting average of .400 with seven extra-base hits. This information in its original form would have listed the batter's performance each time at bat, about 30 separate pieces of information. The single number that characterizes her performance is the result of applying statistical methods: dividing the number of hits (in this case 12) by the number of times at bat (30) produces the player's batting average (.400). In this case, the use of statistics to summarize information has simply saved space in your newspaper. As we shall see, making information manageable by summarizing it is one of the many ways statistical methods serve us all.

Athletic information is just one example of statistics in everyday life. Each year most states give batteries of achievement tests to pupils in selected grades and report the results in the form of school and district averages, which are an indicator of how well or poorly the state's education system is doing. Economists provide statistical information on employment rates and average income. Articles presenting the results of medical research report in statistical form the incidence of drug and alcohol use on college campuses and the changing incidence of AIDS and other conditions. The list is almost endless. When you watch the news or read the newspaper tonight, see how many examples you can identify of information that uses statistics.

Some Vocabulary of Statistics

We defined statistics as a branch of applied mathematics. Scientists use the word in singular form as well. To a scientist a **statistic** is the number resulting from the application of the mathematical procedures of statistics to a body of information from a sample. A batter's batting average is a statistic. The number of traffic deaths in your state attributable to drinking and driving in a year is a statistic, as is the average score earned by fourth graders on a reading achievement test. The plural form, **statistics,** may refer to more than one statistic (for example, a team's batting statistics or the number of drinking-related fatalities

Statistics *is a branch of applied mathematics used to summarize quantities of information and help investigators draw sound conclusions.*

A **statistic** *is the number resulting from the application of the mathematical procedures of statistics to a body of information from a sample.*

in each of the 50 states) or it may refer to the branch of applied mathematics called statistics. In the latter case, statistics is a singular word.

We also use the word statistic in everyday language, where it has a different meaning. For example, television ads warn us: "Don't become a statistic. Don't drive after drinking." In this case, the word statistic refers to a piece of information to which the mathematical procedures of statistics might be (but have not been) applied. This is not the meaning the word has for scientists.

Three other terms that we will encounter often are **population, sample,** and **parameter.** Each will be defined in more detail in later chapters, but we will introduce them here in a preliminary way.

*A **population** is all the people or objects to which we wish to apply our conclusions.*

A **population** is the set of all the people or objects to which we wish to generalize the conclusions we reach from applying statistical methods. For example, the registrar at your college might make a survey of student attitudes toward the institution's policy on allowing students to drop classes. He or she would probably want to apply the conclusions reached from the survey to all students at the institution; the entire student body, therefore, would comprise the population. Similarly, when the state department of education tests all children in selected grades, the objective is to be able to draw conclusions about the entire educational system. Therefore, all pupils in grades K–12 would constitute the population.

*A **sample** is a subgroup of a population.*

A **sample** is a subgroup of a population. Usually, it is not feasible to gather information from the entire population, so we observe a sample from the population, then generalize the results we find in the sample to the population as a whole. For example, rather than sending surveys to all currently enrolled students, the registrar might send surveys to a randomly selected ten percent of the student body, intending to generalize the opinions expressed by these students, the sample, to the whole population. Rather than testing all pupils in all grades, a state department of education tests students at selected levels, who constitute a sample, with the objective of generalizing to the entire school population.

*A **parameter** is a numerical value that describes a characteristic of a population.*

The terms **parameter** and **statistic** go hand in hand with the terms *population* and *sample*. A **parameter** is a numerical value that describes a characteristic of a population. For example, the proportion of students at your college who approve a more liberal policy for dropping a class is a parameter of that population. A **statistic,** as we saw earlier, is a numerical value that describes a characteristic of a sample. The proportion of students in the sample who approve a more liberal drop policy is a statistic.

One of the major uses of the discipline of statistics is to help us make appropriate and accurate generalizations about populations based on the information obtained from samples drawn from those populations. For example, using statistical techniques, the major news services are able to predict accurately who has won an election before very many votes have been counted. Politicians and businesses test new ideas and products by conducting polls or market surveys of a sample of the population. Medical researchers decide whether a new drug will have the desired effect by giving it to a sample.

▶ *Now You Do It*

We encounter many situations in our daily lives where samples are used to represent populations. For example, the amount a TV network can charge for advertising during a show such as "The Oprah Winfrey Show" or the Super Bowl depends on how many people are assumed to be watching. TV viewing patterns are estimated by the Nielsen company, primarily using monitors attached to the TV sets in a few thousand "representative" households. The monitors record when the sets are on, and what channel they are tuned to. The results recorded by these monitors are used to estimate viewing patterns of the American public as a whole. How accurate are such estimates? What factors might prevent the Nielsen company from getting accurate results? Consider what kinds of errors or distortions would be likely to occur in each of the following situations.

1. Monitors are put only in homes with young children.
2. A large number of people in the sample work the swing (4 PM to midnight) or graveyard (midnight to 8 AM) shift.
3. Most of the monitors are in homes close to either New York or Los Angeles.
4. Monitors are put on the TV set in the living room, but the family watches programs on another set while in bed.

5. Some people leave their TV sets on to entertain their pets while they are away at work.

What other common situations can you think of where a sample is used to represent a population? Under what circumstances might distortions or errors occur? Has the food service at your university distributed a questionnaire about preferences? What would happen if you and ten friends filled out 50 questionnaires each?

Answers
1. We would get an overestimate of the amount of time people spend watching cartoons and children's shows, and an underestimate of the amount of evening primetime viewing.
2. We would overestimate the amount of daytime TV people watch.
3. We would learn little about what middle America and residents in rural areas watch.
4. We would underestimate the amount of primetime and late night TV people watch.
5. We would substantially overestimate human viewing time.

Why Do Researchers Collect Information?

For years, government agencies have been gathering vast quantities of information for the apparent purpose of filling up computer disks. At least, that is the way it seems to many people! We are led to ask the most fundamental question: Why has the information been collected? We cannot answer for the government, but researchers usually collect information for one of four reasons, all of which involve increasing the amount and accuracy of our knowledge of the world around us.

Description.   The most basic purpose for gathering information is *description*. If we want to describe a baseball player's performance, for example, we can collect information about the number of times at bat, number of hits, number of fielding chances, number of errors, and so on, almost to infinity in the case of baseball. The primary information, the record of the player's performance on each occasion, includes so much detail, however, that it does not permit us to see trends. A similar problem arises when teachers keep track of the progress of their students. The records for a single class in a single subject can be extensive, but confusing in their detail. Likewise, an industrial psychologist who is monitoring the performance of a group of new employees to evaluate a personnel selection procedure will need to present his or her findings in summary form so management can see the trends and appraise the results. In cases like these the application of statistics enables us to describe what has happened, or what is presently happening, in summary form so that we can comprehend the information and convey it efficiently to other people. The procedures presented in Part I of this book focus on description and are called descriptive statistics. **Descriptive statistics** are procedures for summarizing a body of information into one or a few numbers.

Descriptive statistics summarize a body of information into one or a few numbers.

Evaluation.   Usually we wish to go beyond mere description. Perhaps the most frequent use of statistical procedures is to evaluate a situation or outcome. Educators use statistical methods to evaluate the effect of different teaching methods on students. Psychologists use statistical methods to evaluate success rates from different types of therapy, and business leaders use evaluations of production and sales statistics to guide corporate decisions. An **evaluation** is a judgment about the quality or goodness of an outcome.

An evaluation is a judgment about quality or goodness.

The methods of statistics help people to make value judgments, but the values themselves are imposed by people. Two people with dissimilar value systems may make quite different judgments or draw quite different conclusions from the same information. For example, test scores showing that a new reading program, which is more expensive than the old one, has raised reading achievement in your school district by five percent might be interpreted favorably by a group of parents who are concerned about the progress their children are making in reading, but the same information might be seen by a taxpayer watchdog group as showing that the program was not having an effect worth its cost. Since most of the statistical procedures used for evaluation are the same as those used for description, we will not treat the two topics separately, but merely note that evaluation involves attaching value to an outcome based on a comparison of descriptive statistics.

Suppose you were to hear the following statistic: In the past year 640,000 acres of old-growth forest were cut in the western states. By itself, this statistic does not carry any value judgments; it is neutral. However, it is likely that different people, representing assorted interest groups, might use this number to support their own positions and agendas. Some might find it large or small, good or bad. How might you view this statistic and use it with your constituency if you were:

1. The manager of a plywood mill who is faced with laying off workers because you cannot get enough timber to keep the plant running full time.
2. A conservationist worried about diminishing habitat for bald eagles.
3. A logger who has been laid off because the quota of trees for your area has been reached.
4. A commercial fisher who is worried about deterioration of spawning grounds for salmon.

How might you counter the arguments of each of these people?

Prediction. There are a number of situations in which it is desirable to determine the relationship between two sets of information. For example, is there a relationship between spelling achievement and reading ability? Do people who have more education also have higher lifetime income? In one sense, this is a descriptive problem: We can describe two sets of information in terms of the relationship between them. However, methods have been developed in statistics that allow us to go beyond merely describing existing relationships. If we assume that a relationship will continue, we can use this information to **predict** what is likely to happen.

Statistical prediction is forecasting future events or performance based on statistical analysis of information. For example, your institution is using the statistics of prediction when it forecasts student performance in college (often in the form of admissions decisions) on the basis of admissions test scores and high-school records. Businesses use statistics to forecast the acceptability of prospective employees and even the probable success of new products and advertising campaigns. Political candidates and pollsters forecast the outcomes of elections. Such forecasts are not always correct, but the use of predictive statistics can substantially reduce the number and size of mistakes these individuals and groups might otherwise make. Part II of this book is devoted to using statistics to describe relationships and to make predictions.

Research. In addition to description, evaluation, and prediction, statistical methods may be used for **research**. Doing research means going beyond the given information to reach conclusions about truth, the laws of nature, or some other abstract principle. For example, the information at hand may show that breast-fed babies are happier than babies fed formula. A research approach, going beyond the given information, would ask if this result is true only for the babies who were studied or whether it is likely to be true for all babies. In other words, in a research application we ask whether the conclusions we have reached from the information at hand can be generalized to future observations. A tentative statement of what we believe to be the true state of affairs is called a **hypothesis**. By collecting data, we can test whether our hypothesis, for example, that breast-fed babies are generally happier than formula-fed babies, is likely to be true. The mathematical and logical principles involved in this type of generalization are known as **inferential statistics** and are discussed in Parts III and IV of this book.

Prediction is forecasting future events or performance based on statistical analysis.

Research involves going beyond the given information to reach conclusions about truth, the laws of nature, or some other abstract principle.

A hypothesis is a tentative statement of what we believe to be true.

Inferential statistics is a set of procedures used to test hypotheses and to help reach generalizations beyond the set of observations at hand.

Desirable Properties of Information

One of the most famous photographs in American political history shows a smiling Harry Truman holding a newspaper with a huge headline proclaiming "Dewey defeats Truman" for president. Obviously, the newspaper editor who wrote this headline had reached his conclusion on the basis of poor information. Information can be used in many ways, but the quality of a description or conclusion is only as good as the original information. Good decisions cannot be made from poor information. (However, it is perfectly possible to make a poor decision from the best information.) There are two characteristics that information must have if it is to be useful for scientific or social purposes. The first, most fundamental requirement is that each single piece of information must itself be **accurate**. The second is that the information must be **representative**.

Accuracy

In many cases, accuracy of information may reasonably be taken for granted. For example, there is relatively little question about the accuracy of the number of days a child has been absent from school in a year: The number is publicly verifiable, and, with the possible exception of clerical errors and days when the child was late for school and mistakenly counted absent, not open to question. Congressional voting records, age and sex of children, age and occupation of parents, test scores, and many other pieces of information are assumed to be correct. However, the accuracy or correctness of information may be compromised in various ways. Two of the most common factors that have a bearing on the accuracy of information are the source of the information and the level of inference or abstraction that is required to interpret it.

*Information is **accurate** if the measurement results in the correct numerical value.*

    Source of Information. Suppose that we are collecting parental age and occupation information from some school children. These are verifiable and knowable pieces of information. Likewise, a child's statement of his or her father's age or mother's occupation can be accurately recorded. However, children's statements about their parents' ages or occupations may nonetheless be inaccurate indicators of the parents' true ages or occupations; the children may simply not have the correct information to give us. The problem here is the potentially inaccurate data due to the source of the information. Ensuring that we have an accurate source of information is a serious challenge. In one doctoral study, for example, trained classroom observers showed less than perfect agreement regarding the biological sex of the teachers they were observing! This problem could have been avoided if the investigator had asked the teachers themselves, who would have given accurate information. (However, inaccuracy could also have crept in because of lack of care in recording their observations, or from errors in processing the reports.) Such situations lead us to wonder about the accuracy of other, less verifiable information that the classroom observers reported.

In collecting information, we should generally select the source that is most likely to give us information that is accurate or correct.

    Level of Inference. Does your score on an exam always reflect exactly how well you have mastered the course material? The number of test items you got right is accurately knowable. However, most teachers assume that your exam score gives information beyond the items themselves, providing indirect information about your learning or mastery of the course content. Similarly, the clinical psychologist who uses a questionnaire to get information about a client's personal functioning is less interested in specific responses to specific items than in the information these responses convey about the person in general. In such cases—and whenever we use one piece of information as an indicator of some unobserved characteristic—there is a genuine risk that the information we have does not accurately represent the characteristic we wish to know about. This risk is so serious that it is treated at length in books on research design and on psychological measurement and testing under the heading of **validity**. Accurate information allows valid inferences. You can find a more extensive treatment of these topics in the references at the end of this chapter.

    Validity problems are greatest when a high level of inference is required to relate what is actually observed (responses to test questions or other overt behavior) to what we are interested in studying (general traits of the person). In simple cases such as stated age or occupation, the measurement of such physical properties as height or distance, or the price of a stock on the New York Stock Exchange, no inference is required because the measurement procedure defines the characteristic. This type of information is said to be obtained by **direct observation** or **direct measurement. Direct measurement** occurs when the measurement procedure itself defines the characteristic.

Direct measurement occurs when the measurement procedure itself defines the characteristic.

    In the preceding examples, the teacher and the clinical psychologist both wish to infer the existence of a trait underlying the observed behavior. Their interest centers not on the observations themselves but on an unobservable characteristic or trait that is *assumed* to be related to the observations. The trait may or may not exist, and if it does exist, the observations may or may not be related to it. There is at best an *indirect* relationship between the information in our observations and the use or interpretation that we make of it. Therefore, the accuracy of information based on **indirect observation,** or **indirect**

Indirect measurement occurs when the measuring operation does not completely define the trait being measured.

measurement, tends to be relatively low. **Indirect measurement** occurs when the measuring operation does not completely define the trait being measured. Your level of understanding of statistics, for example, may be indicated by your score on a midterm exam, but your understanding is not defined by the test score. Most of the measurements made by psychologists and educators are indirect measurements because the traits of interest are assumed to be more than just what is observed. Even the measurement of a trait that seems straightforward, such as reading achievement or job performance, is a finite sample of behavior, and the educator or psychologist is interested in skills or behaviors beyond those that can be observed.

Information gathered via indirect measurement is not without meaning. Many of our actions in daily life require that we infer meaning from indirect or incomplete information, such as the smile on a classmate's face or the way someone is carrying a bowl of soup toward us. However, in science there are dangers in being too confident that our indirect measurements are correct, and these dangers are much greater than in cases where there is a direct relationship between the observations and the trait of interest. We should not refrain from studying traits that are not completely defined by how they are measured. However, our conclusions must be quite tentative because of the extended chains of inference that relate them to our observations.

▶ *Now You Do It*

Can you think of situations in daily life where one type of behavior is used as an indirect indicator of another? Do you subscribe to a ski magazine (you can substitute a fashion magazine, a computer games magazine, or a periodical relating to any other subject) and receive advertising in the mail from ski resorts or ski equipment manufacturers? Why are catalog and retail sales businesses interested in magazine subscription lists? They assume that if you read about something such as skiing (or fashion, or computers), you have an above average interest in that subject and are more likely than people in general to spend money on related activities. They expect to get more sales for their advertising dollar by targeting more likely prospects.

The same principle is applied in science; one behavior is taken as an indicator of another. Think back on places where you have encountered indirect measurement, perhaps in a journal article you read recently or in research that one of your professors has described in a lecture, where indirect measurement was involved. What inferences were being made about the traits?

Representativeness

Information is **representative** if it comes from a sample of people who are like the population and it covers all aspects of the traits or situations about which we wish to know.

In addition to accuracy, both in the source and the nature of the information, it is desirable to have information that is **representative**. If we wish to reach conclusions that extend beyond the available information (that is, the information is a sample from a population of information), the information we have must adequately represent the population. Suppose, for example, that we wish to study attitudes toward changing sex roles in American society. For convenience, we stand outside the dining hall and ask those students who will speak to us a series of prepared questions about changing sex roles. How representative is this information? About what and whom can we reach conclusions?

We can generalize our findings only to other people who are like those who are in our sample.

Sampling of People. One restriction on our ability to generalize our findings may arise from a limited sampling of people. In the preceding example, our information does not include people who could not speak to us because they do not go to college, do not go to our college, or do not go to the dining hall; it also does not include those people who would not answer our questions. We can only generalize our results to other people who are like those in our sample. This principle of generalizing our results only to people (or animals, or locations, and so forth) that are like the ones we observed is so important that polling organizations and commercial testing agencies go to great lengths to assure that their samples are representative. For example, when a standardized test is being prepared for use in schools, the publisher may try the test out on a sample of up to 250,000 students to assure that the interpretation of scores is justified for students across the country.

Psychology has often, and justly, been criticized for being the science of the behavior of the white rat and the college sophomore because these have often been the subjects used in psychology experiments. Unless rats and college sophomores are representative of larger universes of organisms, conclusions based on their behavior may not generalize well to other animals or people. Issues of sampling and generalization are discussed in more detail in Chapter 9.

Sampling of Content. In addition to making sure that the individuals we study are representative of the population we are interested in, we must make certain that our measures sample effectively the content related to the trait of interest. With regard to our questionnaire about changing sex roles, for example, we should ask: Do the questions in our survey cover all facets of attitudes toward changing sex roles? Each individual question may accurately reflect a feature of the trait, but a good information-gathering procedure will reveal all aspects of the trait. Unless our survey is comprehensive in its coverage of the facets of attitudes about sex roles, we cannot generalize the information beyond the types of questions asked.

The same problem arises with achievement tests. The questions on a mathematics achievement test do not cover every possible bit of mathematics knowledge; they sample from the population of possible bits of knowledge. When gathering indirect information, we can seldom cover the topic of interest exhaustively. Rather, a good test or questionnaire will have a representative sample of the possible observations so that we can generalize from the particular items to all items relating to the trait.

A good test or questionnaire will contain a representative sample of items relating to the trait of interest.

Sampling of Times and Locations. There are many other characteristics of the measurement situation that can affect the representativeness of our results. In addition to our concerns regarding who and what are being measured, we need to be cautious about when and where the observations are made.

Some characteristics fluctuate during the day. For example, your sense of well-being or alertness, and therefore your responses to an attitude questionnaire or your performance on a test, may be affected by how recently you have eaten or whether you have just gotten up. Are you a "morning person," or do you feel you work better later in the day? Many students avoid early morning classes because they "can't wake up." Some find themselves sleepy after lunch. If all of our observations are made at the same time of day or at the same point in the academic term, this may introduce systematic distortions. To assure that the results of our questionnaire accurately reflect the attitudes of the student body, we might have to collect responses at several times of day. Similarly, to avoid systematic distortions, testing companies administer tests in the fall and in the spring so future test scores can be interpreted relative to tests administered at similar times of the year.

The surroundings in which observations are made can also affect those observations. Students might express somewhat different attitudes if they were stopped on the way to class, or were responding in a busy dining hall, rather than if they were approached in the quiet and privacy of their rooms. In order to generalize our results over a range of occasions or contexts, we may have to draw our sample from different locations.

Obviously, careful planning is necessary to ensure that the information we obtain adequately represents the topic, the individuals and the situations about which we wish to know. No amount of elegant analysis can compensate for deficiencies in the basic information-collecting process. The principles of analysis and interpretation described in the chapters that follow can be applied to any set of numbers. In order for the results of an analysis to have meaning, the numbers that go into the analysis must themselves have meaning; unless the numbers are accurate and representative, their meaning is questionable at best.

If the surroundings or time of day can affect measures of the trait of interest, a good study will sample from various surroundings and times of day.

▶ *Now You Do It*

Suppose that you are interested in studying how much college students know about AIDS. You have constructed a five-item test to measure their knowledge, and you plan to administer the test to groups of students. To obtain participants, you put up posters in the dorms announcing testing sessions to be held on Saturday in the library and ask for students to volunteer.

There are, however, problems with the plan of this study. Can you identify what some of them are? How will your ability to generalize the results to all students be restricted, and what might you do to improve the generalizability of the results?

To begin to answer these questions, first, look at the test itself. Will you be able to obtain an adequate sample of the domain of knowledge with only five questions? Probably not. Therefore, you will only be able to generalize your finding about knowledge to the topics covered. Perhaps you should develop a longer test.

Depending on the type of institution you attend, students who live in dorms may differ in systematic ways from those who have apartments or other living arrangements. Therefore, if very many students live in facilities other than dorms, your sample may not represent all students. Advertising more widely may help.

Doing your testing on a Saturday may also affect your results in systematic ways. Is there a football game that day? Do large numbers of students tend to go home for the weekend, or go skiing, or go to a large nearby city? If yours is an urban campus where many students live off campus, those who would be available on a Saturday may differ from students in general. Providing several testing sessions on alternate days and times will allow a wider variety of students to participate.

Finally, there is always the problem in psychological and educational research that those who are willing to take part in a study may be different from other students. Because we are usually limited to using volunteers we seldom have a way to check on this problem, but we need to make our study as attractive as possible to assure that our volunteers come from all segments of the student population.

Types of Studies

We have said that science uses information from observations of the world to describe and understand that world. Four broad strategies are used in education and psychology to collect information. Some are more appropriate in certain situations and each has advantages and disadvantages. Statistics can be of use in any study but is more essential in some types of studies than in others.

Case Studies

*A **case study** is an intensive investigation of a single individual.*

An intensive study of a single individual is called a **case study** or, sometimes, a single-subject study. Investigations of this type are common in clinical psychology where a single client may be studied carefully and over a long period of time. An intensive study of a single industrial plant or classroom also qualifies as a case study. Sigmund Freud developed many aspects of his theory of personality through case studies, as did the psychologists who studied people with multiple personalities. Alfred Binet and Jean Piaget both made intensive studies of their own children in developing their theories of cognitive development.

We get a richly detailed description of one individual from a case study, but the findings may not apply to anyone else. Case studies are particularly useful for developing insights that may lead to other studies with greater generalizability. They can also provide an understanding of rare phenomena, where we may be able to find only one person who possesses the traits of interest. For example, Arthur Jensen (1990) studied a woman from India named Shakuntala Devi, who could perform amazing feats of mental arithmetic, to test a hypothesis about intelligence. He tested her elementary information processing speed to see if this could help explain her unusual ability and found that she was unexceptional in this regard; her ability was due to some other extraordinary capacity.

Field Studies

*In a **field study**, observations are made in natural environments.*

We often wish to know how individuals function in their normal environments. A **field study** can help us gather this information. In a field study, observations are made of people, animals, or social systems, and care is taken to ensure that the process of observation does not affect those being observed. Anthropologists conduct field studies when they observe cultures in operation, and ethologists do the same when they watch hives of bees or troops of baboons. An educational psychologist may gather information about classroom interactions between elementary school pupils and their teacher, or a clinician may study the behavior of schizophrenic patients in a psychiatric ward. Because the observations are made in the natural environment, field studies are also called *naturalistic observations*. Probably the most famous example of this type of study is Jane Goodall's observations of the behavior of chimpanzees. In that case, she almost joined the troop!

A special variation on the field study occurs when an investigator sets up a particular situation in a public place and then observes how people respond. There are a number of well-known studies in social psychology that have used this methodology. For example, in studies of bystander apathy a confederate of the investigator may fake a heart attack in a shopping center. The data for the study are the reactions of the people nearby to the plight of the person in distress: Who helped, and how? Who didn't? Although the investigator manipulates the situation, the people are not aware that they are being studied, and therefore behave as they would if the investigator were not present.

Both single case and field studies can result in quantitative information. The most common form of quantification is counts of the number of times particular classes of events occurred. For example, an observer in a psychiatric ward might record the number of violent and nonviolent patient interactions, an ethologist might record the number of threat behaviors emitted by the dominant animal in the baboon troop, or an educational researcher might record the number of gender-exclusive statements made by a teacher. However, quantification is often of secondary concern in studies of this type. Their unique advantage is the richness of the descriptive information they provide. Results are often given in the form of narrative descriptions of the behavior rather than in numerical form, and statistical procedures other than counting are seldom applied. Studies like these are therefore sometimes referred to as qualitative research.

Case studies and field studies, also known as qualitative research, often report their results using narrative descriptions rather than statistical analysis.

Survey Studies

A very common form of research in the behavioral and social sciences is the **survey study**. A survey study attempts to determine people's status or level on specific characteristics by asking them questions or otherwise measuring them. You certainly have encountered survey studies in the form of opinion polls and attitude surveys that fill newspapers, particularly during election campaigns. Studies of this type attempt to describe people in terms of their responses to various controlled stimuli such as questions about political preferences or drug use. The survey method is more general than this, however. When a school district assesses all third-grade children with a test of reading comprehension, for example, it is conducting a survey study because the objective is to determine the current status of third graders on the trait of reading comprehension. Our questionnaire about attitudes toward changing sex roles would also qualify as a survey study because our purpose is to determine where people stand on the issue, in other words, their status on the trait. Much of the research in education and in clinical and counseling psychology involves survey study methods.

Survey studies determine the status of respondents on the traits of interest.

Survey studies are more likely than case or field studies to produce results in the form of numbers or scores that signify an amount of something. This occurs for two reasons.

Table 1-1 Characteristics of Types of Studies		
Type	Advantages	Restrictions
Case study	Data rich in detail. Can study unusual cases.	Findings may not apply to others. Results seldom quantitative.
Field study	Situation is lifelike; generalizes to "real world."	Often difficult, time-consuming and expensive.
Survey study	Results in quantitative form. Efficient.	Limited to expressing relationships between traits.
Experimental study	Environment tightly controlled. Possibility of causal inference.	May lack relationship to "real world."

First, the characteristics typically studied by this method must be observed indirectly, using a large number of responses to assure representativeness. The observation is reported as the number of items answered correctly (as in an ability test) or the number of choices of a particular kind that were made (as in an attitude questionnaire). Second, studies of this type often are concerned with characterizing a large number of people or generalizing to people who did not take part in the study. In such cases some sort of numerical representation of the observations is needed in order to describe them.

Experimental Studies

*In an **experiment,** one feature is manipulated or allowed to change, and the change in some other characteristic is measured; everything else is held constant.*

In an **experimental study,** or **experiment,** a single feature of a situation is manipulated or allowed to change, and the results of the change are observed; everything else is held constant. The hallmark of the experiment is *control* of all factors in the observation situation except those we specifically wish to study. In formal experiments, such as those conducted by chemists, the scientist holds all aspects of the environment constant and then changes one feature, such as the temperature of a gas, to observe the effect of this change on another feature, such as the pressure of the gas on the walls of the container. Since everything else remains constant, the change in pressure can be attributed to the change in temperature. The advantage of the experiment is that we can be more confident in this kind of study than in any other that there is a causal relationship between our manipulation (changing temperature) and the outcome (change in pressure) because everything else has been held constant.

True experiments are difficult to conduct in many areas of concern to the social and behavioral sciences. Psychologists can control most of the factors affecting the behavior of rats learning to run mazes, but it is virtually impossible to satisfy the requirements for control when studying the effects of one program of reading instruction or one form of therapy compared to another. Where it is possible to exercise control, it is often at the expense of making the situation highly artificial, thus reducing our ability to generalize the findings to non-experimental situations. We will have more to say about the requirements for a good experimental study in Part III of this book.

Measurement: Giving Meaning to Numbers

Measurement is the assignment of numbers to objects or events according to a set of rules.

Information must be in the form of numbers if we are to apply mathematics to it. If information has been gathered in some other form, for example, as verbal descriptions, videotapes of events, or essay responses to test questions, these must be converted into numerical form before they can be subjected to statistical analysis. Whether the observations are obtained in numerical form or converted to numbers at a later stage, the numbers are being used to represent something else, an attribute of the person or thing being observed. They take on meaning in addition to, or beyond, their numerical quality. The process of attaching meaning to numbers is known as **measurement**. One widely accepted definition of measurement is the assignment of numbers to objects or events according to a set of rules. We are all familiar with many types of measurements that affect our lives daily. Measurements of distance, weight, and time are so familiar that we hardly stop to consider where they come from. But suppose we say that Jane's weight is 50; without some other reference, the number is meaningless. What is missing is the set of rules that led to the assignment of the number 50 to Jane. Our impression of Jane would be quite different if the number 50 represented kilograms rather than pounds.

Measurement and the Origin of Data

Numbers that have taken on meaning as the result of measurement are called **data**. Note that the word *data* is a plural form. The singular form, datum, refers to a single measurement or observation. Therefore, whenever we refer to a collection of observations we will say that the data *are* distributed in some form or that the data *have* some property.

Data are numbers that have taken on meaning as a result of measurement.

The set of rules for assigning numbers to objects or events is called a **scale,** and unless the scale is known, a number imparts little or no information and cannot be interpreted correctly. In many areas of the physical sciences and of our everyday lives, the rules for assigning numbers to objects or events are so familiar and well developed that mere mention of the scale of measurement is sufficient to provide the necessary meaning. For example, if we know that Jane's 50 refers to pounds, we have a fairly accurate idea of her weight. Likewise, if the 50 refers to kilograms, we have an equally good, but different, idea. If we know that Jane weighs 50 kilograms (about 110 pounds), we know that if we put Jane on one end of a balance beam and a 100-pound sack of flour on the other, the beam will tip toward Jane. However, is Jane heavy or light? If we are told that Jane is six feet tall we get one impression, but if she is eight years old we get quite a different impression. The same number conveys different information depending on other aspects of the context, that is, other pieces of information. Similarly, most of our common scales require additional specification when used for educational or psychological research.

A *scale* is a set of rules for making measurements.

There are many situations in education and psychology where it is necessary to use scales that are quite different from the scales used for weight and length. In such cases, it may be necessary to specify in some detail what rules were used for assigning numbers to the objects or events that have been observed. Suppose, for example, we wish to "measure" the gender of teachers in school classrooms and we say that teacher X is a 1. You do not know whether X is a male or female. If we now say that the scale is that males receive a zero and females receive a 1, you know that X is female. Confusion or disagreement about the scale is one major source of inaccuracy in the information used by investigators in the behavioral sciences. It may, for example, explain the lack of agreement among the classroom observers regarding teachers' sex described earlier. The mathematical processes of the statistical analysis do not know or care what the data mean. Unless investigators remember the meaning (they don't always!), the interpretation of the analysis and the conclusions reached will be useless.

> ### ◗ *Now You Do It*
>
> Suppose someone told you that your score on the Thorndike Precarious Prognosticator of Performance (the T3P; psychologists and educators love acronyms) was 300. What additional information would you need to make sense of this score? Would you want to know how others your age scored? What types of questions were on the test? What the highest and lowest possible scores were? You would get quite a different impression if you were told that geniuses scored 250 than if you were told that average performance was 500.
>
> Perhaps you can think of other situations where you would need clear specification of the scale to interpret the results of measurement correctly. What is the number of calories you need to maintain your weight? Do you think you would be satisfied with a score of 20 on your first statistics exam? What additional information would you need in each case before you could decide?

Importance of Valid and Reliable Measures

There are two general concerns that all scientists share regarding their measurements. Their first and most central concern is assuring that the measurement process results in numbers that actually reflect the trait or property of interest. This is a concern with the **validity** of the measurement. A measurement is valid to the extent that differences in the numbers assigned to different individuals correspond to true differences in the trait we wish to measure. A ruler can provide a valid measure of your height, but it is useless as

Validity concerns the extent to which differences in the numbers assigned to different individuals correspond to true differences in the trait.

a way to measure your self-confidence. In order to measure this latter trait we would probably ask you to respond to a series of statements about how you think you would act or feel in various situations. If the numbers resulting from the measuring procedure are valid, then people who have higher levels of self-confidence would receive higher scores on the set of items. Likewise, the tests you take in your classes are valid only to the extent that the differences in test scores between you and your classmates correspond to differences in your actual mastery of the course material and objectives.

Scientists wish to ensure not only that a measuring procedure is valid, but also that it is free of random variation. For example, one of your authors (RMT) got a new electronic bathroom scale not too long ago. The first time he weighed himself the reading seemed a little high, so he stepped off the scale and tried it again. This time the reading was 7 pounds lower! He tried the scale several more times, and each time the reading was different. Although his weight had not changed over the 10 minutes he spent experimenting with the scale, the readings showed random variation around the value his old spring scale had given consistently. The next day he took the fancy electronic scale back because it did not give reliable information. The degree to which a measuring instrument yields consistent results is known as its **reliability**. A reliable measuring device will give you the same score each time you are measured unless you have actually changed your status on the trait.

Issues of reliability and validity of measurement are critical for psychologists and educators because of the indirect nature of their measurements and the inherent variability in human behavior. Unless the variables we study are measured reliably and validly, we cannot draw conclusions with any confidence. In fact, with unreliable or invalid measures, any conclusions we draw may be completely wrong. Reliability and validity are usually covered in books on testing, but because these are such important properties of our data, we have included a brief discussion of the related issues in Appendix B.

> *A **reliable** measure is one that is free from random variation.*

▶ *Now You Do It*

1. Ordinarily, we assume the measures we encounter are reliable and valid. For example, what would happen if the gas gauge in your car was unreliable, sometimes reading too high, sometimes too low? The gauge might read half full when you ran out of gas, or you might stop at a filling station with the gauge on empty and be able to add only three gallons.
2. Have you ever experienced an unreliable or invalid measure? How about a test in one of your classes where there were so many questions that you didn't have time to finish them all and marked the last 10 randomly? How reliable would your score on those items be? Is it possible for a measure to be reliable but not valid?
3. You can measure your height reliably, but would this be a good measure of your knowledge of statistics or your self-confidence? How else might one get a reliable measure that is not valid? Why is it impossible for a measure to be valid but not reliable?

Discrete and Continuous Traits

When we consider numbers in the abstract sense of the real number system usually used in arithmetic and algebra, the numbers are pure numbers and nothing more. We can add them, subtract them, multiply and divide them, find squares and square roots, and, in general, perform any mathematical operation. (Some types of number systems do not permit all of the above operations. The integer system, for example, does not always allow division because the result of the operation may not be an integer, and even the real number system does not permit finding the square root of a negative number.)

Discrete Scales and Traits. Numbers that result from measurement, and therefore have meaning beyond their "numberness," can be restricted in various ways. For example, if you play softball, it is reasonable to interpret your batting average of .400 as meaning that you got 400 hits per 1000 times at bat (or 4 out of 10, or 40 out of 100), but it is not reasonable to say that you got 4/10ths of a hit each time at bat. Viewed from a slightly different perspective, which we will discuss later, your batting average can be interpreted as implying a probability of .40 that you will get a hit on any given opportunity.

The problem with the second interpretation, that you got 4/10ths of a hit each time at bat, is that each hit occurs as a whole event and cannot be subdivided. Events or characteristics that occur in an all-or-none fashion are called **discrete events** or **characteristics,** and the scales used to measure them are called **discrete scales**. Children in a family or in a classroom represent discrete events, and family or classroom size are discrete characteristics. The same is true of the number of fans at a sporting event, the number of credits a student has earned toward a degree, and many other situations that involve counting how many times something has occurred. Measures of any of these events represent discrete scales.

Discrete scales are like the integer number system. Except for certain specific applications, such as the first interpretation above (batting averages) and the third interpretation (probability of getting a hit), fractional values are not meaningful in a scale like this. Counting the number of times something happens or the number of objects of a particular kind is the most common way to produce a discrete scale. Consider, for example, the fate of the unfortunate third child in the average American family (in which there are 2.3 children) and it will be clear why we must be careful when we interpret data of this kind.

Continuous Scales and Traits. When fractional values are permissible and meaningful, we say that we have a **continuous scale**. (Strictly speaking, it is the characteristic being measured that is discrete or continuous, not the scale.) It is reasonable to think of a student's knowledge of statistics as being a continuous trait because we can imagine that no two students have exactly the same amount of the trait and that differences in knowledge or ability can occur in very small pieces. We consider height and weight to be continuous traits, as are time, temperature, and degrees of emotions like stress, happiness, and test anxiety.

The permissibility and meaningfulness of fractional values depend on (1) how we conceptualize the trait and (2) the level of development of the measurement scale. If the characteristic of interest is a physical property (for example, length, height, width), fractional scale values probably make sense, and the scales available (for example, meter sticks) are sufficiently well-developed that we may reasonably permit and interpret fractional values. We use test and questionnaire items in education and psychology in much the same way. That is, each item a student passes (or attitude statement he endorses) is interpreted as an increment on the trait, so the total number of items passed (or endorsed) is seen as placing the person at a level on a continuous trait. The test or questionnaire is treated like a ruler.

Most of the measurements that confront us in our daily lives involve traits that we consider to be continuous and scales that are well-developed, so we are accustomed to thinking in terms of continuous measurements. However, while some traits, such as length, weight, and time, may be continuous, *all* procedures used to measure them result in discrete scales at some level of refinement. For example, if RMT's electronic bathroom scale only reports weight in pounds, the scale is discrete at the level of pounds even though his actual weight might be measurable to an accuracy of a fraction of an ounce with a more refined scale. In psychology and education the point where scales become discrete often is at the level of the individual item. For every characteristic, there is a point beyond which our current technology does not permit subdivisions, even though they might be possible in theory. We will discuss some consequences of this situation in Chapter 2.

> ▶ *Now You Do It*

Consider each of the following situations. Which characteristics and scales are discrete and which are continuous?

1. Your grandmother's recipe for angel food cake calls for 12 eggs and 4 tablespoons of sugar.
2. The number of books in your home.
3. The number of bathrooms in your house.
4. Gas mileage.

Answers
1. The eggs are discrete (although a scale for "Eggbeaters," an egg substitute, would not be), but the sugar is continuous because fractions of a tablespoon are possible.

Discrete events or *charac-teristics* occur in an all-or-none fashion.

Discrete scales report how many times a discrete event has happened.

A *continuous trait* is one in which the differences between individuals can be arbitrarily small.

A *continuous scale* is a scale used to measure a continuous trait.

2. Discrete scale, although the underlying trait the scale may be intended to measure, such as literacy, could be continuous.
3. Again, discrete, but the trait of affluence this question is often meant to measure is continuous.
4. Continuous. Both the amount of gas your car used and the number of miles you traveled are continuous variables, so their ratio is continuous.

Types of Scales

As we have suggested, the numerical meaning that numbers possess as a result of measurement varies and depends on the measurement procedures employed. Different types of measurement procedures produce different types of scales. We will focus here on the classification of scales into four levels as proposed by Stevens (1951). Other ways of classifying measurement scales have been suggested; these are usually based on the type of transformation the scale permits without loss of meaning. We will give a brief review of one alternative classification system. A more thorough treatment of it can be found in Hays (1994).

Nominal Scales. We saw earlier that it was possible to form a scale of gender by specifying that if an individual is male, he is assigned the numerical symbol 0 and if the individual is female, she is assigned the numerical symbol 1. We normally think of numbers as indicating *quantity;* in other words, the larger the number, the more we have of something. But what quantity is expressed by our scale for gender? Our gender scale is nothing more than a rule that substitutes the number zero for the word *male* and the number 1 for the word *female*. We could reverse the scale, calling males 1 and females zero, without affecting the amount or quality of information, or, in fact, we could use any pair of numbers we choose. A scale that substitutes numbers for words or names is called a **nominal scale**. Nominal scales are not uncommon in education and psychology. Numerical labels are frequently used to designate gender or occupational categories or groups in a study. For example, you might assign numbers to represent different schools in your district, different classrooms in a study of reading achievement, or different types of therapy in a study of the effectiveness of counseling. In each case the same number is assigned to people who would otherwise receive the same verbal label (female, securities analyst, student in Johnson High School, member of Mr. Smith's class, client in rational emotive therapy).

Consider again the nominal scale for gender. Would it make any sense to say that the average gender of study participants in an experiment was .40? If males are zeros and females are 1's, a value of .40 is not a meaningful value on the scale for two reasons: First, the scale, like all nominal scales, is discrete. Values between the named categories are not meaningful. Second, there is no information about the amount of anything. The operation $0 + 1 = 1$ is not meaningful in this context. The only things we can do with a nominal scale are count the number of times an event or observation in each category occurred and express this frequency as a proportion of the total number of observations. That is, the value of .40 in the foregoing example would tell us that 40 percent of the group are females. Nominal scales are considered the lowest level of scale because they require only recognition of categories.

Ordinal Scales. Many times it is possible to identify an order in the categories that make up a scale; that is, the different numerical values indicate relatively more or less of the characteristic of interest. Such scales are called **ordinal scales**. Examples of this second level of measurement are not difficult to find in education and psychology. They often take the form of rank orders of preferred activities or preferred individuals. For example, a teacher might ask each child in his class to put several play activities in the order in which he or she would like to do them. Another common use of ordinal scales is to express judgments of the relative merit of workers, artistic productions, and so forth. Perhaps the most widely used ordinal scales are those for measuring attitudes and values. You certainly have encountered situations where you have been asked to indicate how strongly you agree or disagree with a statement or series of statements by ascribing a numerical value to your feelings. All of these scales order the objects or events or attitudes according to

A **nominal scale** substitutes numbers for names or other labels designating individuals.

In **ordinal scales** the order of the numbers assigned to people corresponds to their order on the trait. The numbers assigned to individuals must preserve their order on the trait.

relative amounts of the trait of interest. One of the rules for assigning numbers in an ordinal scale is that the order of the numbers must preserve the order of the people on the trait.

Data from ordinal scales carry more meaning than data from nominal scales, but such data still cannot be treated like ordinary numbers. The numbers on an ordinal scale convey information about relative position, but not about amounts of a trait in some absolute sense. Specifically, they do not convey information about amounts of differences. For example, there is no guarantee that the difference between the person with the highest score on a test and the person scoring second is of the same magnitude as the difference between the people in second and third place. In fact, when we order people according to a characteristic, such as height, where better measuring procedures exist, we typically find much greater differences between people at the extremes of the ordering than we do among those in the middle. In such cases, equal differences in ordinal score correspond to unequal differences in the amount of the trait. Therefore, special procedures are necessary for analyzing ordinal data. These are discussed in Chapter 15.

Interval Scales. Another fairly common type of measurement results in numbers that convey information not only about order but also about magnitudes of differences. **Interval scales** are scales on which equal numerical differences imply equal differences in underlying characteristics. Most measurements of human abilities are of this general type. Scale values carry information about order (that is, relative position) and about *amount* (that is, level) of the characteristic in question. The differences, or intervals, between numbers carry a constant meaning and it is appropriate to employ more sophisticated mathematical procedures with the numbers, such as addition and subtraction.

Interval scales are scales on which equal numerical differences imply equal differences in underlying characteristics.

A familiar example of an interval scale is the Celsius scale of temperature, on which water freezes at 0° and boils at 100° under standard conditions. These two reference points define the unit for the scale; the unit is 1/100th of the difference between the freezing point and boiling point of water. Each unit between these two points is equal (by definition), and we may add more units of the same size above 100 and below zero. Equal numerical differences at any point on the scale reflect equal changes in the characteristic—temperature—being measured.

Perhaps the most common scales treated as interval scales in education and psychology are those for measuring abilities. For example, when your professor gives you a multiple choice test, your score probably is reported as the number of questions you answered correctly. Each item is treated as equal to every other item, and the ability increment needed to answer item 7 is treated as being the same as the ability increment needed to answer item 20. Each item you answer correctly is assumed to require one more "unit" of statistics knowledge, so the difference between a score of 15 and one of 20 is treated as equal to the difference between a score of 25 and one of 30.

SPECIAL ISSUES IN RESEARCH

Restrictions on the Meaning of Numbers

It is important to be aware that numbers do not remember where they come from, and, ordinarily, they will raise no objections, no matter what we do to them. We can add and subtract them, multiply and divide them, take logs and exponentiate, and even nominal data will not object. But **data** are both more **and** less than numbers. They are more than numbers because they have been given additional meaning by the process of measurement; they represent those objects or events to which they have been assigned. Data are also less than numbers because the meaning or nature of the objects or events they represent, the scale of measurement, may not logically permit various numerical operations. Average gender is one example of such a problem; average rank is another example. Suppose George ranks third in verbal ability and fifth in quantitative ability in a group of 50 students. His average rank is fourth, but if all 50 students were ranked in terms of total performance, George's consistency would probably earn him a rank of first or second. The numbers representing the ranks do not themselves object to being averaged, but in most cases, the order that results from averaging does not make sense and does not correspond to the order that would result from a more appropriate procedure.

We shall return to the meaning of numbers from time to time. For the present, it is necessary only to emphasize the intimate relationship between the way the data were collected and how many attributes of numbers they possess. Scales below the interval level are quite restricted in the attributes of numberness that they possess.

Interval scales have some restrictions; in particular, the definition of zero on the interval scale is arbitrary. The Celsius scale for temperature uses the freezing point of water as an arbitrary zero point. Any other point could have been chosen for zero (such as the freezing point of a saturated salt water solution which Fahrenheit chose for his scale) without affecting calculations very much and without changing the usefulness of the scale at all. Likewise, the zero point on a statistics test is arbitrary and depends on how "hard" the professor makes the test. In fact, most scales for measuring human abilities, such as the one for the Wechsler Adult Intelligence Scale, use an arbitrary reference point near the center of the set of scores and define the scale in equal units above and below this anchor. Zero on these scales does not mean zero intelligence or ability, an issue that we will consider in more detail in Chapter 5.

Ratio Scales. The major consequence of the arbitrary zero point on interval scales is that statements involving ratios of scale values are not meaningful. For example, stating that 20°C is twice as hot (a ratio of 2 to 1) as 10°C is not meaningful; neither is the statement that 50°C is half as hot as 100°C. In order for ratio statements to make sense, the zero point on the scale must be defined so that it means exactly none of the trait. A zero beyond which values are not possible is often called an *absolute zero* or a *rational zero*. A scale that has an absolute zero and equal intervals is called a **ratio scale**.

> *A **ratio scale** has equal intervals and an absolute zero point.*

Ratio scales provide the most sophisticated measurement and are found in the most well-developed sciences. For example, length and mass are measured on ratio scales, as is temperature on the Kelvin scale. Generally, the numbers resulting from measurements using ratio scales can be treated like the ordinary numbers to which we are accustomed.

A single basic scale can be either an interval scale or a ratio scale, depending on how it is used. The measurement of time provides an example. The time 2 AM, or two hours after midnight, is an interval scale statement, but the fact that a statistics exam lasted two hours is a ratio scale statement. Because we figure time through the day from an arbitrary zero point defined as when the sun is exactly on the opposite side of the earth from our position (midnight), we can talk meaningfully about events that have negative time values (they occurred before zero) and we have an interval scale of time. However, if we are considering a measure of time to reflect a duration, like the two hours for the statistics exam, then we have a ratio scale because a duration of time of less than zero is not meaningful. Thus, the same scale could represent interval or ratio information depending on the context.

Each level of scale possesses all of the properties of meaning and numberness of the scales below it, and something more. These properties are summarized in Table 1-2. Because each level of scale builds on its predecessors, each can be converted into a scale of a lower level. This is a desirable and necessary feature for some types of interpretations, and for some statistical procedures.

▶ *Now You Do It*

Researchers need to be aware of the level of scaling their measurements represent so they can conduct the most appropriate and informative analyses. Assume you have collected some data using the methods described below. What level of scaling does each represent?

1. The number of hours people in a sensory deprivation experiment spend sleeping.
2. Ratings of teacher effectiveness made by parents on a five-step scale from poor to excellent, filled out after a parent-teacher conference.
3. The brand of computer that you own.
4. Scores on the Stanford-Binet Intelligence Scale.

What are some of the other common measurements that we meet in daily life? On what level of scale is each measured?

Answers
1. This is a ratio scale because it has a rational zero and negative values are not possible.

Table 1-2	Four Major Levels of Scales and Their Properties
Scale	**Properties**
Nominal	Numbers are names of categories of a variable.
Ordinal	Numbers represent the order of categories or individuals.
Interval	Equal differences between numbers represent equal amounts of the trait. Both negative and positive values are possible because the location of zero is arbitrary.
Ratio	Zero means exactly none of the trait. Values on both sides of zero in the same scale are not possible.

2. This is an ordinal scale. We cannot assume that the differences between steps are equal, but they do have an order.

3. This is nominal. There is no order to the brands themselves, even though we might create an ordinal scale of rated quality and order the brands by these ratings.

4. Interval. Zero intelligence is not meaningful, but scores spread out in equal units above and below an arbitrary reference point.

SPECIAL ISSUES IN RESEARCH
Scales and Permitted Transformations

Another way to think about differences among scales is the types of transformations or changes we can carry out on them without changing the information they contain. Transforming scales can be useful in statistical analysis because it can give us values that are easier to work with and whose meaning can be conveyed more easily. We will encounter several common transformations in Chapter 5.

A nominal scale gives us information about group membership, but nothing else. Therefore, not even the order of the numbers is meaningful, and we can change the numbers in any way so long as group identity is preserved. This means that the only thing we cannot do is to give two groups the same number. For example, in our scale of gender, where 0 = male and 1 = female, we could add 10 to both scale values, or we could multiply both scale values by 20. We could even add 10 to the male value and multiply the female value by 500. The only thing we could not do would be to make a change such as adding 2 for the males and 1 for the females because then both groups would have the same scale value, which would destroy group uniqueness.

With ordinal scales, any change that preserves the order *and* uniqueness of the original categories is permitted. We could, for example, square each scale value, or multiply each by the same positive constant, or add the same value to each score. We could not, however, multiply each score by a negative number, which would reverse the direction of the scale, and hence the relationship between the order of the entities and the order of the numbers. Transformations that preserve the order of the scores are called *monotonic* (*single direction*) *transformations.*

Interval scales require that, in addition to uniqueness and order, the information about distances on the scale be preserved. Some transformations that are permitted for ordinal scales, such as squaring scale values, change the distances (the distance between 2 and 3 is 1 unit, but the distance between 2^2 and 3^2 is 5 units) and so cannot be used with interval scales. In fact, the only transformations permitted for interval scales are those involving adding or subtracting a constant from every score and those that involve multiplying or dividing every score by a positive constant because only these transformations maintain distance and order relationships. Scale modifications of this kind are called *linear transformations.*

Ratio scales require that the value of zero be maintained in addition to the properties of uniqueness, order, and distance. Since adding or subtracting a constant from zero changes its value, the only transformations that are permitted with ratio scales are those of multiplying or dividing each scale value by a positive constant. Thus, we can see that the more "numberness" a scale possesses, the more restricted we are in how we can change the scale. (See Michell [1990] for an alternative view of the nature of psychological scales.)

Summary of Permitted Transformations

Scale	Transformation Must Preserve
Nominal	Uniqueness
Ordinal	Uniqueness and Order
Interval	Uniqueness, Order and Distance
Ratio	Uniqueness, Order, Distance, and Zero Point

More Meaning in Numbers

We have seen that numbers take on meaning as the result of measurement using a particular scale. But numbers can take on meaning beyond the properties of their measurement scale and come to represent specific kinds of relationships. This additional meaning is conferred by the way in which the data are collected; it takes on special importance when the results of measurement are used in science. The meaning given to numbers by the way they are collected is frequently discussed under the heading of **research design**. Research design involves deciding what the characteristics of the measurement situation are. Who should be measured, and under what conditions?

Research design involves deciding what the characteristics of the measurement situation are.

The first issue we must face in research design is whether more than one value of a trait should occur in the set of data with which we are concerned. Should we include in our study both males and females, people of different ages, people who faint at the sight of blood and others who do not? For example, we might measure a large group of people on a trait such as test anxiety and select for inclusion in our study only those people with a particular score on the trait. If only one value or type of the trait occurs, the trait is said to be a **constant**. Characteristics that are the same for all participants in the study are constants. If, for example, a sample includes only 25-year-old males, then age and gender are constants. In conducting a true experiment, it is necessary to keep almost all aspects of the situation constant; this is called **experimental control**.

A constant is a characteristic on which everyone in the study is the same.

The characteristics whose influence we wish to study must be different for different people; because their values vary from person to person, they are called **variables**. Studies in psychology and education generally involve describing single variables (e.g., counting the ways in which they occur and the frequency of different values), specifying relationships among variables, and determining how changes in one variable affect others. Examples of common variables in psychological and educational research include age, gender, cognitive ability, or score on an attitude questionnaire, personality inventory or memory task. Note that a characteristic that is a variable in one study can be a constant in another.

A variable is a characteristic that has more than one value in a set of data.

Independent and Dependent Variables

Field and survey studies in the behavioral and social sciences generally involve at least two variables. In some cases, the investigator is interested in making observations in a natural setting and seeing which variables seem to go together. In other studies, naturally occurring categories of one variable (for example, sex or age) are chosen and some other variable (such as mathematical ability or aggression) is measured to determine whether there is a systematic relationship between the first variable and the second. Because people can differ in many ways other than the ones we are studying, the kinds of conclusions we can draw from field studies and surveys often are limited to statements that two variables seem to go together or be related. For this reason studies of this kind sometimes are called **correlational studies**. The fact that we lack control over potentially important variables means we cannot conclude that changes in one variable cause changes in another, only that they tend to occur together.

A correlational study is one in which we examine the relationship between two or more naturally occurring variables.

As we saw earlier, in a true **experimental study** the investigator imposes the value of one variable (for example, type of teaching or drug dose) on the participants and measures a second variable (learning, reaction time, pupillary dilation, and so forth). The variable that the investigator imposes on the participants is called the **independent variable** and the measured outcome is called the **dependent variable**. Studies that involve imposition of an independent variable by the investigator are the only ones that permit us to infer that the change in one variable (the independent variable) causes a change in the other (dependent) variable, and then only when we have controlled all the other possible variables.

An experimental study examines the relationship between a manipulated independent variable and a dependent variable.

A manipulated independent variable is one in which the investigator assigns participants to particular levels of the variable.

It is important to distinguish among the different types of independent variables because the type of variable we are using affects meaning in the data, which in turn determines the type of conclusions we can draw. There are two types of independent variables, manipulated variables and status variables. A true experimental study involves a **manipulated independent variable** and a **dependent variable**. The independent variable is completely under the investigator's control; the investigator *assigns* research participants to levels of the independent variable. For example, a statistics instructor might wish to

A dependent variable is a measurement of the behavior that is influenced by the independent variable.

study whether using structured lecture notes improves students' comprehension of material. The variable is a manipulated independent variable if some classes are given structured notes (one level) and others are not (a second level), because students are assigned to study conditions by the investigator. The dependent variable *depends upon* (or is influenced by) the manipulation of the independent variable. It is the measured outcome (statistics comprehension in the case described here).

Many, perhaps most, variables that are of interest to social and behavioral scientists cannot be brought under the investigator's control in the sense of a true experiment. Personality and ability characteristics, gender, attitudes, needs, and a host of other variables can be controlled *by the selection of individuals for study*, but we are practically and (often) ethically prevented from imposing values for these variables on our research participants. For example, forcing a randomly selected group of people to smoke two packs of cigarettes a day for 50 years and comparing them with a similarly selected group who never experienced tobacco smoke would help us answer the question of whether there is a causal relationship between smoking and various diseases, but the study would be neither practical nor ethical. For this reason, many research questions in education and psychology must rely on weaker designs and the weaker inferences they permit.

Variables that participants bring to a study as part of their identity are called **status variables** because they reflect the individual's status on some characteristic. Gender, age, cognitive abilities, personality characteristics, family income and social status are some examples of status variables that are of interest to educational and psychological researchers. People may be selected for a study based on their level on some status variable, but because status variables cannot be imposed on the research participants, we cannot infer that they bear causal relationships with dependent variables; we can infer relationship, but not causation.

It is fairly common to refer to status variables as independent variables, and we shall adopt that convention in the remainder of this book, but you should be aware that we can draw different conclusions depending on whether an independent variable is a status variable or a manipulated variable. Studies in which status variables are treated as independent variables are often called **quasi-experiments**.

Status variables are characteristics that research participants bring to the study. They are descriptive of who the individual is.

Quasi-experiments are studies in which the independent variable is a status variable.

▶ *Now You Do It*

To test your comprehension of some of the key topics in this chapter, we have included below a description of a hypothetical study that might have been conducted by a team of psychologists. After you read the description of the study, you will find a series of questions, along with the answers. Try to answer the questions yourself before looking at the answers.

A team of clinical psychologists who treat many phobic clients believes that fear of dentists is highly focalized; it is different from fear of medical practices generally. In order to investigate this topic, they plan to conduct a study of the effect of dental fear using a word recognition task. Because their practice includes a wide variety of people to whom they wish to be able to generalize their findings, they advertise in the newspaper for people to participate in their study. The volunteers for the study provide information on their age and gender; then they take two tests, one to measure their verbal ability and a second to assess their general perceptual speed. They also complete a questionnaire that measures their fear of dentists and dental situations.

When all this information is collected, each person is assigned randomly to one of three testing conditions in which they view a series of 10 words flashed on a computer screen for a fraction of a second. After each word is flashed on the screen, the participant types into the computer the word that he or she thinks was displayed. People in one group see 10 words that are neutral in content. People in the second group receive 10 general medical terms, and people in the third group see 10 terms relating to dentistry and dental practice. The number of correctly identified words of each kind is recorded. There are 20 people in each condition: 10 men and 10 women.

Questions
1. Which variable is the dependent variable?
2. Which variable is a manipulated independent variable?

3. Which variables are status variables?
4. Which of the status variables would the investigators be most likely to consider an independent variable?
5. What level of scale does each of the following variables represent?
   Gender
   Age
   Perceptual Speed Score

Answers

1. Number of correctly identified words is the measured outcome or dependent variable.
2. The only variable to which people are assigned is the kind of words they see (neutral, medical or dental).
3. Age, gender, verbal ability, perceptual speed and dental fear score are all status variables.
4. Dental fear score.
5. Gender is nominal, age is ratio, and perceptual speed would ordinarily be treated as interval.

Two Hypothetical Studies

The methods of data analysis that are described in the later chapters of this book provide many ways to ask questions of data. A single data set may be used to answer a large number of questions, depending on the variables that have been measured. Appendices C and D contain data from two hypothetical studies which are introduced below. There are enough different variables in each set to enable us to apply most of the methods of analysis described in this book to each set. In many chapters we will present a detailed application of the analyses to the first data set. We will also provide the answers that you should get by applying the same methods to the second data set. If you follow our computational procedures through the first example, then apply the same steps to the second set of data, you should get the same results we did. We will also show, on a chapter-by-chapter basis, how to analyze the data using the Student SPSS software available with this book. You may want to enter one or both sets of data into this or another statistical analysis computer program to see how the program works, what kinds of output the program provides and how its results may differ from hand calculations. Small differences between programs often occur, particularly with some descriptive statistics and graphs, because of the assumptions and options selected by the person who wrote the program.

First Hypothetical Study

All students in the behavioral and social sciences at Big Time University (BTU) are required to take a course in statistics from the Statistics for Analyzing Data (SAD) Department. The SAD Department has three faculty: the department chair, Alice B. Numbercruncher; and two associates, Wil B. Quick (called slick Willie by his students) and Macrow Hardy, the department computer whiz. One day at a faculty meeting Dr. Hardy observed that some students who seemed to know the course material, according to their homework assignments and in-class performance, did not do well at all on formal exams. She suggested that the department undertake a study to see whether particular students suffered from test anxiety and whether a change in the way lab sections were taught might alleviate the problem. The three faculty members agreed that each would collect some data on the students in their section of the course and that each would then try out one of three different ways to teach the lab section.

Assume that you are an advanced student at BTU majoring in SAD and that you have been asked to assist in the design of the study and the analysis of the data. Think about what variables you might want to measure and how you would go about measuring them. What level of scale does each variable represent? Also, think about questions you might have about the data and how you would attempt to answer them. We will introduce the study the faculty designed in the next chapter and explore ways to extract the information the data have to give us throughout the remainder of the book.

Second Hypothetical Study

A recent news release by the Political Organization of Women (POW) asserted that a conservative men's religious group called the Searchers has as one of its objectives to reverse many of the gains made by women in the past 20 years. Professor Victor Cross, a social psychologist, and his class decide to examine whether there is a relationship between affiliation with this men's religious organization and attitudes toward equality for women. At the next public rally of the Searchers, Professor Cross and his students advertise for men to participate in their study. The questionnaire they hand out requests the following demographic data: age, years of education, annual income, whether they are Roman Catholic or Protestant, and whether they are formally enrolled members of the Searchers. Sixty volunteers are selected and asked to stop by the Psychology Department at their convenience to fill out an additional questionnaire and complete an attitude inventory. The questionnaire measures their degree of religious involvement using questions about time spent on church and related activities. The attitude survey is designed to assess their beliefs about political and economic equality between men and women. There are three forms of the survey: one neutral and factual, one in which the language is aggressively pro-feminist, and one that is strongly conservative about gender equity issues. Which form of the survey a man gets is determined at random. What are some of the questions that you might want to answer using these data?

SUMMARY

Statistics is a branch of applied mathematics that provides a language for behavioral and social science communication. Information is collected for the purposes of *description, evaluation, prediction,* and *research.* Useful information must be *accurate,* in the sense that observations must be correctly recorded, and *representative,* in that the observations must reflect variations in the events of interest. *Data* are pieces of information that have been put in numerical form. The process of creating data (that is, putting information into numerical form) is known as *measurement.* A *scale* is the set of rules used in assigning numbers to objects or events. Scales may be *nominal, ordinal, interval,* or *ratio,* depending on the quantitative meaning that the numbers convey.

An *experimental control,* or constant, is a trait that has the same value for all individuals observed. Traits that take on more than one value are known as *variables;* if they are under the investigator's control, they are called *independent variables.* The responses made by study participants are measures of the *dependent variable.* The kinds of inferences we can make about the relationships between independent and dependent variables depend on whether the independent variables are *status variables* or *manipulated variables.*

REFERENCES

American Psychological Association. (1982). *Ethical principles in the conduct of research with human participants.* Washington, DC: Author.

American Psychological Association. (1992). Ethical principles of psychologists and code of conduct. *American Psychologist, 47,* 1597–1611.

Dooley, D. (1995). *Social research methods* (3rd ed.). Upper Saddle River, NJ: Prentice Hall.

Hays, W. L. (1994). *Statistics* (5th ed.). San Antonio, TX: Harcourt.

Jensen, A. R. (1990). Speed of information processing in a calculating prodigy. *Intelligence, 14,* 259–274.

Michell, J. (1990) *An introduction to the logic of psychological measurement.* Hillsdale, NJ: Erlbaum.

Rosnow, R. L., & Rosenthal, R. (1996). *Beginning behavioral research* (2nd ed.). Upper Saddle River, NJ: Prentice Hall.

Stevens, S. S. (1951). *Handbook of experimental psychology.* New York: Wiley.

Thorndike, R. M. (1997). *Measurement and evaluation in psychology and education* (6th ed.). Upper Saddle River, NJ: Merrill/Prentice Hall.

EXERCISES

1. From a statistical perspective, define a population and a sample.
2. What is the difference between a statistic and a parameter?

3. In a recent state election, a political initiative was introduced to set limits on the number of years an elected state official could hold a particular political position. An exit poll conducted by a newspaper of 852 voters indi-

cated that 51% of the voters voted in favor of the initiative. When the final results of the election were released, only 45% of the voters supported the initiative.

  a. What is the population?
  b. What is the sample?
  c. What is the parameter?
  d. What is the statistic?
  e. What explanations can you provide for the difference in the exit poll outcome and the election outcome?

4. List the four reasons for collecting data. How do they differ? From your major area of study, provide two examples of how data are used. Describe two examples of how data are used in newspapers or magazines.

5. Discuss the desirable properties of information and research data that make them useful.

6. Define direct measurement. Define indirect measurement. How are these two forms of measurement different? Provide at least four examples of each type of measurement.

7. If we are to reach conclusions that go beyond the available information (i.e., the sample from the population), the information we have must adequately represent the population. Discuss why sampling of people, content, time and location is important for generalizing research findings.

8. Describe the four basic types of research studies. What are the advantages and the disadvantages of each type of research study?

9. Explain why it is important to have measurement scales that are reliable and valid.

10. Describe the difference between a discrete scale and a continuous scale. Provide at least four examples of each type of scale.

11. In your own words, describe the four types of scales. Provide at least two examples of each scale.

12. For each of the following, state whether the measurement represents a discrete or continuous variable and the type of scale (i.e., nominal, ordinal, interval, ratio) involved:

  a. Time it takes a rat to run a maze.
  b. The number of errors or false turns a rat makes in running a maze.
  c. Placing the rat in category 1 if it ran the maze faster than the average time and category 2 if it ran the maze slower than the average time.
  d. Intelligence as measured by an IQ test.
  e. The political party affiliation of individuals who responded to a survey.
  f. A rank ordering of 10 qualities of your ideal mate.
  g. A participant's reaction time measured from the onset of a stimulus on a computer screen until a response is typed on the computer.
  h. A measure of depression on a 21-item questionnaire on which participants rate each item on a scale of 0–6.
  i. The number of recall errors of participants after they viewed a videotape of a crime and were provided some misleading information in a discussion after the videotape.
  j. Arranging a series of photographs of people in order of attractiveness.
  k. Categorizing people according to religious preference.

13. In research terminology there are different types of variables. Define the terms dependent variable, manipulated independent variable, status variable, and control variable and provide an example of each.

14. For each of the following, identify the dependent variable and the independent variable. For each independent variable, indicate whether it is a manipulated independent variable or a status variable. In addition, indicate the variables that are controlled in each example.

  a. Participants are asked to rate the expertise of the counselor in a videotaped counseling session on a 7-point Likert scale. Participants are shown videotape in which the counselor uses one of three counseling approaches: client-centered, Gestalt, or rational-emotive. In reality the counseling videotape is a scripted tape in which actors are portraying the "counselor" and the "client." The counselor and the client are the same people in each tape. In addition, the counselor and the client are wearing the same attire in each of the tapes. Three counseling psychologists, each of whom was an expert in one of the three approaches, wrote the scripts for the videotapes. Furthermore, the problem the client presented the counselor was the same in each tape, and each counseling session was 30 minutes long. The videotapes were all made in a counselor's office.

  b. A psychometrician wants to determine if male participants have higher scores on a test of spatial reasoning than female participants. The participants are all college students who are 18–20 years of age.

  c. An educational psychologist wishes to determine if participants who read an advance organizer prior to reading a prose passage will recall more information in the passage than participants who merely read the prose passage. Participants are given 25 minutes to recall as much of the passage as they can remember in a free recall format.

  d. A clinical psychologist is interested in determining whether a deep muscle relaxation technique will reduce heart rate more than a relaxation technique utilizing an imaginal fantasy. Participants are given 5 minutes to allow their heart rate to stabilize before baseline heart rates are taken. Participants are directed through the relaxation technique via a tape-recording that lasts for 20 minutes. At the end of that time, a second measure of heart rate is taken.

  e. A sports psychologist is interested in determining if a combination of mental and physical practice is more effective in increasing the number of free throws made compared to physical practice only. Participants are 12-year-old females who are just learning the art of free-throw shooting. Participants shoot 10 free throws in one session for 10 sessions. All participants shoot at the same goal using the same basketball. The number of free throws made in the first session is subtracted from the number of free throws in the last session to provide a measure of change.

  f. A human factors psychologist is interested in determining if one interior arrangement of an automobile will result in better scores on a driving simulation test than a second arrangement that is more widely used. All participants use the same 20-minute videotape with the driving simulation machine.

  g. A developmental psychologist is interested in comparing scores on a measure of self-esteem for three age categories: 10, 14, and 18. The number of males and females in each age group is the same.

15. The members of the faculty in the Mathematics Department in the Floodgate School District are concerned about their students' level of understanding of algebraic concepts. At a department meeting, faculty discussed several factors related to student understanding of algebra. During the discussion, several faculty members indicated that a recent movement toward teaching algebra using a metacognitive approach had resulted in some preliminary studies that demonstrated that students not only were learning algebra better than when a traditional approach was used but were also more positive in their attitudes toward algebra. Unlike previous studies with this age group, these findings were consistent for both males and females. However, there were also some studies that did not find these effects. Furthermore, differences in biological sex were present for both algebra achievement and attitudes toward algebra. The algebra teachers at Gatekeeper High School had received training at a summer institute that focused on teaching algebra using a metacognitive approach, and the algebra teachers at Lockedgate High School had not attended the summer institute. Assume that you are an educational psychologist serving as a consultant to the mathematics faculty in the Floodgate School District and have been asked to assist in the design of the study and the analysis of the data. What variables would you want to measure and how would you go about measuring them? What level of scale does each variable represent? What questions would you have about the data and how would you attempt to answer them?

16. For many years the supply of blood has not kept pace with the demand. Currently, the United States is at a crisis point with regard to this issue. Health psychologists have been interested in recruitment strategies that might increase both the number of first-time donors as well as the number of repeat donors. The research in this area has indicated that middle-aged men are more likely to donate blood than any other group. In general, men are more likely to donate blood than women. Several different recruitment methods have been used in the past with varying degrees of success. In addition, researchers have investigated the reasons why people donate or fail to donate blood. Several personality factors have emerged as salient factors related to blood donation, including dimensions of empathy, perspective-taking, empathic concern, altruism, and personal distress. In addition, certain fears, such as the fear of needles and the fear of blood, have also been shown to be related to blood donation. Furthermore, the amount and quality of information about the need for blood donations, the process of donating, the risk factors involved in blood donation, and other informational items may be salient factors in blood donation. For those individuals who donated, health psychologists are interested in determining symp-

tomatic reactions, such as fainting, before they donated as well as symptomatic reactions after they donated. The researchers are also interested in participants' intent to donate in the future as well as the reasons for deferral for those who attempted to donate but were not accepted. Assume that you are a health psychologist involved in blood donor research. What variables would you want to measure and how would you go about measuring them? What level of scale does each variable represent? What questions would you have about the data and how would you attempt to answer them?

17. A person's conception of self, such as self-esteem, is said to be a function of the cultural values, either individualistic or collectivistic, to which he or she has been exposed. In some cultures the role of the individual is emphasized more than the role of significant others (e.g., individualistic cultures), such as family or peer groups, whereas in other cultures the reverse is true (e.g., collectivistic cultures). Thus, self-esteem measures with an individualistic orientation and self-esteem measures with a collectivistic orientation may be useful. In addition, the gender roles that operate within each culture have been shown to impact a person's conceptualization of self. Furthermore, one important conceptualization of self is how an individual achieves in different situations and the motivation that drives that achievement. This achievement motivation factor has also been shown to be a function of both gender and culture. In addition, how an individual construes himself or herself as independent or interdependent has been shown to be a function of both gender and cultural values. Assume that you are a social psychologist involved in research on the conceptualization of self. What variables would you want to measure and how would you go about measuring them? What level of scale does each variable represent? What questions would you have about the data and how would you attempt to answer them?

18. Achievement in mathematics and selection of potential careers that utilize mathematics have often been attributed to factors such as prior achievement in mathematics, perceptions of mothers' attitudes toward mathematics, perceptions of fathers' attitudes toward mathematics, perceptions of mathematics teachers' attitudes toward mathematics, mathematics anxiety, etc. The factors listed above may be differentiated on the basis of gender. In addition, there may be a developmental difference in the factors listed above depending on the age or grade level of the individuals involved. You are an educational psychologist who has been hired as a consultant by the local school district to investigate these factors. What variables would you want to measure and how would you go about measuring them? What level of scale does each variable represent? What questions would you have about the data and how would you attempt to answer them?

Chapter 2

Summarizing Data in Frequency Distributions and Graphs

CHAPTER OBJECTIVES

After studying this chapter, you should be able to:

❑ Describe the four kinds of frequency distributions researchers use to organize their data.

❑ List and explain the principles for preparing a grouped distribution.

❑ Prepare an appropriate frequency distribution from a set of data.

❑ Plot points in the Cartesian coordinate system.

❑ Recognize and give the primary features of the graphs introduced in the chapter.

❑ Prepare an appropriate graph of a set of data.

THE BTU TEST ANXIETY STUDY

At the end of the last chapter, the SAD faculty of Big Time U were pondering a study of the influence that anxiety might have on statistics students' test performance and of how the problem might be alleviated. Your assignment was to suggest some variables that the faculty should consider in their study and to propose some questions that the study might answer. You will now be introduced to the study that the faculty actually designed and conducted. As you read about it, think about these questions: How many of the variables are similar to ones you would have chosen? If you included other variables, what questions would your variables address that are not covered by the ones that are included, and why are those questions important? How would you collect data on those variables in an efficient and easily usable way?

The faculty came up with a list of 12 variables that they felt were reasonably related to statistics test anxiety and test performance. They could be collected within the budget (one of the realities of research) and in usable form. Most could be gathered with a simple questionnaire or obtained from the university's records. The rest could be collected as the study progressed. Table 2-1 shows the variables and how they were collected.

On the first day of class, all students were told about the study and asked whether they would be willing to participate. In line with the guidelines for research with human participants, they were told that taking part was completely voluntary, that they could refuse without affecting their course grade or standing, that they would be allowed to withdraw at any time and that at the end of the term they would be able to find out the results of the study. In addition, all data were coded to remove individual identification prior to analysis, and participation status and data were kept confidential. There were 340 students enrolled in the course, and 68 of them agreed to take part in the study.

Table 2-1	Twelve Variables Collected in the BTU Test Anxiety Study
Age	from a student self-report questionnaire recorded to the nearest year
Sex	self-report, female = 0, male = 1
Major	self-report, psychology = 1, education = 2, all others grouped as 3
High school GPA	from college admissions records, recorded to 2 decimal places
College GPA	from college records
Number of math courses taken	from college records
Math aptitude test score	from college admissions records
Math anxiety	from a questionnaire administered on the first day of class
Homework completed	number of assignments completed, from class records
Homework grade	average homework grade, from class records
Course section	from class records
Final exam score	from class records

First Steps in Organization

As suggested in Chapter 1, careful planning in the selection of variables is very important for determining the types of conclusions that can be drawn from a study. Good planning in data collection will also facilitate handling the data to be analyzed. For example, it may be possible to arrange for automatic recording of data by an electronic recorder or computer, or it may be possible to prepare a questionnaire so that responses can be read by optical scanning equipment. However, information does not come to an investigator in neat and orderly packages very often. Even with the most careful planning and sophisticated equipment, the data, as provided by the real world, come in a disorganized and incomprehensible mass of numbers. This can be confirmed by a glance at the complete data for all 68 students in the BTU study, found in Appendix C. Part of the same data is shown in Table 2-2.

Table 2-2 presents two types of information. First, there is a series of numbers in order from 1 to 68; this is a variable measured on a nominal scale. Each number identifies a particular individual from whom information was obtained. The second set of numbers represents the information itself, the data. As we have learned, the qualities of these numbers depend upon the procedures used to collect them. Although these data are the

Table 2-2   Final Exam Scores for 68 Students in the BTU Statistics Course

Student ID	Score on Final	Student ID	Score on Final	Student ID	Score on Final
1	37	24	34	47	25
2	47	25	61	48	43
3	19	26	44	49	40
4	37	27	55	50	29
5	33	28	19	51	55
6	48	29	50	52	32
7	20	30	23	53	28
8	44	31	48	54	59
9	44	32	34	55	41
10	34	33	13	56	54
11	38	34	65	57	29
12	14	35	25	58	16
13	4	36	38	59	30
14	33	37	46	60	30
15	35	38	25	61	18
16	37	39	34	62	42
17	8	40	7	63	14
18	37	41	47	64	43
19	37	42	25	65	39
20	31	43	20	66	56
21	22	44	22	67	39
22	49	45	21	68	29
23	28	46	45		

scores on the final exam of our 68 students who have just completed the introductory statistics course at BTU, they might also represent the scores obtained by 68 second-grade school children on a test of reading achievement, or the number of minutes 68 monkeys took to solve a puzzle problem. Whatever the source, it is not unusual for researchers to obtain data in this general form—values of a variable or variables of research interest paired with some form of participant identification.

Reasons for Summarizing Data

What can we say about these data? What do they tell us about what happened in our study? If very few individuals were involved, we could perhaps take in the whole picture at a single glance and describe our data without further processing. However, whenever there are more than a few pieces of information, or we wish to progress beyond mere description to evaluation, prediction, or research, it is necessary to put the data into a different form. The form we choose will depend upon how concise we wish our description to be and the types of further conclusions we wish to draw.

Processing data almost always involves trade-offs. We must discard some types of information or lose some detail in the data in order to gain a clearer picture of what remains. (This is a little like saying that to be able to see the entire forest, we must move far enough away so that we can no longer identify the individual trees.) In this chapter, we will consider some of the first and most basic stages in the process of summarizing data. The frequency distributions and graphs we will introduce involve minimal information loss at a considerable gain in perspective.

Frequency Distributions

*A **frequency distribution** is a listing, in order, of each value of the variable that occurred, along with the number of times that value occurred.*

The first step in getting to know the data usually involves sorting them by numerical value and counting the number of times each value occurs. Looking through the data in Table 2-2, we note that individual 13 has the lowest score, 4, and the highest score, 65, belongs to individual 34. The scores for the other 66 people lie between these two extremes. If we list all possible scores from the lowest to the highest, together with the number of people who obtained each score, the result is a **frequency distribution.** A frequency distribution allows us to tell how many times (how frequently) each of the possible scores in this range occurs.

SPECIAL ISSUES IN RESEARCH

Why Do the Computations by Hand?

Computers and pocket calculators have taken over most of the work of statistical analysis. It is unlikely that you will ever have to perform by hand many of the operations covered in this book, once you have learned the basic procedures. For this reason, some books skip parts of the simple, old-fashioned material we cover in this chapter. However, computers do not know where the data came from; the numbers have no meaning for the calculator. Only the mind of the human operator is capable of interpreting the results of an analysis in terms of the meaning of the original data. For this reason, and because some of the useful procedures we describe here are not available in most standard computer packages, we will provide a detailed account of some basics.

The computer tends to separate the investigator from her/his data by doing most of the work and making it easy to go directly to a very abstract summary of the data. This has a triple disadvantage. First, you may never become aware of some levels of detail in your data. Studying frequency distributions, graphs, and simple descriptive statistics can give you a feeling for what your data mean that is hidden by more abstract analyses. Second, operating only with abstract statistical indices can cause you to forget where the numbers came from and to reach inappropriate conclusions. It is important that investigators, particularly those just entering a field of inquiry, have available and use data-summarizing procedures that do not discard too much detail. Finally, every computer program is constructed with certain default conditions. Unless you tell the program otherwise, the data will be analyzed in a way that was decided by the computer programmer, not by you, the person analyzing the data. In order to make good decisions about what values the program should use for your data, rather than just passively accepting the programmer's judgment, you need to know the principles on which the program operates, the assumptions it is using, and whether these are right for your situation. For all these reasons, we need to get behind the computer screen and interact with our data in a close and personal way.

Raw Frequency Distributions

The task of making a frequency distribution is simple and not too time consuming if done correctly. Here, on a step-by-step basis, is how to make a frequency distribution like that shown in Table 2-3, using the final exam scores of our 68 statistics students.

1. Find the highest and lowest scores.
2. Write down all possible scores from highest to lowest in decreasing order in a column labeled X. (Note that the direction in which you list the scores is arbitrary. Many computer programs give you the option of top-down or bottom-up organization. We recommend using decreasing order because it makes some of our later graphing tasks a little easier and it seems more appropriate to have the spatial organization mirror the numeric one.)
3. Go through the list of individuals, and make a tally mark next to the score each person obtained. (Person 1 gets a tally mark next to 37, person 2 next to 47, person 3 next to 19, and so forth.)
4. Count the number of tallies for each score value and write that number in a column labeled f.

Notice that in Table 2-3 the column for score values is labeled X and the column containing the number of tally marks is labeled f. The letter X is widely used in statistics to designate a variable, in this case final exam score. The letter f is generally used to label frequencies, or the number of times each score value occurs. These two columns, the values of the variable X *and the frequency, f,* with which each occurs, constitute the *frequency distribution.* (The column of tally marks is an intermediate step shown here for convenience, but it would not be included in the final table.)

*The letter **X** stands for the possible values of the variable.*

*The letter **f** stands for the frequency with which each value of the variable occurs.*

Table 2-3 Raw Frequency Distribution of Statistics Final Exam Scores for 68 Students					
X	Tallies	f	X	Tallies	f
65	/	1	34	////	4
64		0	33	//	2
63		0	32	/	1
62		0	31	/	1
61	/	1	30	//	2
60		0	29	///	3
59	/	1	28	//	2
58		0	27		0
57		0	26		0
56	/	1	25	////	4
55	//	2	24		0
54	/	1	23	/	1
53		0	22	//	2
52		0	21	/	1
51		0	20	//	2
50	/	1	19	//	2
49	/	1	18	/	1
48	//	2	17		0
47	//	2	16	/	1
46	/	1	15		0
45	/	1	14	//	2
44	///	3	13	/	1
43	//	2	12		0
42	/	1	11		0
41	/	1	10		0
40	/	1	9		0
39	//	2	8	/	1
38	//	2	7	/	1
37	/////	5	6		0
36		0	5		0
35	/	1	4	/	1

Preparing a frequency distribution involves minimal loss of information or detail in the data. In this first stage of data analysis, what is lost is the identification of particular individuals as sources of information. Each separate score appears in the table, but the numbers identifying particular people have been discarded. We shall see as we go along that descriptive statistics involves discarding detail in successive steps in order to develop a broader perspective on the information we wish to analyze. Every gain from condensing and simplifying the data must usually be paid for with a loss of detail. While unavoidable, it is important not to discard the detailed information too quickly. In later chapters we will encounter single indices that can summarize any number of scores, but going directly to these indices without examining a frequency distribution can leave us unaware of important features of the data.

A frequency distribution such as that in Table 2-3 is known as a raw frequency distribution. Every score value is listed, even if its frequency is zero. In many situations, this type of frequency distribution does not provide much simplification; this is the case in Table 2-3, where we have substituted 62 separate score values for the 68 original scores. Many of the possible values of X have zero frequency, and the scores are so widely spread that it is hard to discern a pattern to them. In short, there is still too much detail in the data. One way to condense the distribution further is simply to drop values of X for which f is zero as most computer programs do. However, when preparing the frequency distribution yourself, you cannot take this step until after the tallies have been made because you will not know which scores have $f = 0$.

HOW THE COMPUTER DOES IT

Most computer programs produce frequency distributions that are slightly different from the one shown in Table 2-3. The difference is simply that they omit the values with zero frequency; a computer program would yield 41 distinct values for our example instead of 62, dropping the 21 scores with zero frequency. To ask our SPSS program to produce a frequency distribution, we need to open the file containing the data we wish to analyze. Clicking on the STATISTICS option produces a window that looks like this.

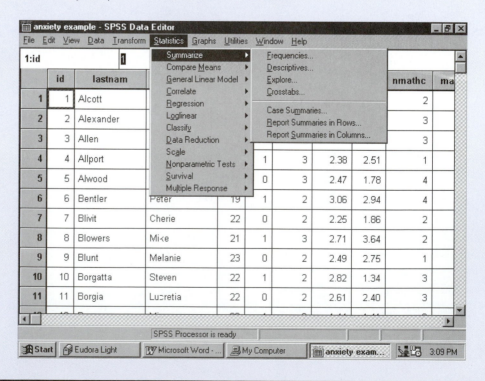

Selecting the FREQUENCIES procedure from within the SUMMARIZE group and inserting FINAL as the variable to be analyzed will produce the following result. The output includes three additional columns with percentage information which we will discuss shortly.

Score on Stat Final

Valid	Frequency	Percent	Valid Percent	Cumulative Percent
4	1	1.5	1.5	1.5
7	1	1.5	1.5	2.9
8	1	1.5	1.5	4.4
13	1	1.5	1.5	5.9
14	2	2.9	2.9	8.8
16	1	1.5	1.5	10.3
18	1	1.5	1.5	11.8
19	2	2.9	2.9	14.7
20	2	2.9	2.9	17.6
21	1	1.5	1.5	19.1
22	2	2.9	2.9	22.1
23	1	1.5	1.5	23.5
25	4	5.9	5.9	29.4
28	2	2.9	2.9	32.4
29	3	4.4	4.4	36.8
30	2	2.9	2.9	39.7
31	1	1.5	1.5	41.2
32	1	1.5	1.5	42.6
33	2	2.9	2.9	45.6
34	4	5.9	5.9	51.5
35	1	1.5	1.5	52.9
37	5	7.4	7.4	60.3
38	2	2.9	2.9	63.2
39	2	2.9	2.9	66.2
40	1	1.5	1.5	67.6
41	1	1.5	1.5	69.1
42	1	1.5	1.5	70.6
43	2	2.9	2.9	73.5
44	3	4.4	4.4	77.9
45	1	1.5	1.5	79.4
46	1	1.5	1.5	80.9
47	2	2.9	2.9	83.8
48	2	2.9	2.9	86.8
49	1	1.5	1.5	88.2
50	1	1.5	1.5	89.7
54	1	1.5	1.5	91.2
55	2	2.9	2.9	94.1
56	1	1.5	1.5	95.6
59	1	1.5	1.5	97.1
61	1	1.5	1.5	98.5
65	1	1.5	1.5	100.0
Total	68	100.0	100.0	
Total	68	100.0		

Grouped Frequency Distributions

The next stage in simplifying our data is to prepare a **grouped frequency distribution** by discarding information about the number of times each *particular* score occurred and grouping together similar scores into intervals. This step eliminates values of *X* that have zero frequency and reduces the number of categories. Instead of taking single values of *X*, we now use *intervals* and count the frequency of scores in each interval.

*A **grouped frequency distribution** uses a coarser measurement scale by ignoring small differences in scores and grouping similar scores together in intervals.*

Limits of Intervals. Measuring continuous variables always involves approximation because the measurement scale has categories covering a range on the trait, and each observation is assigned the label of the category within which it falls. For example, Mike Blowers and Melanie Blunt received the same score, 44, on the statistics final. A single obtainable score, such as Mike and Melanie's scores of 44, represents the **midpoint**—the center—of an interval on the measurement scale of a continuous trait. It is the score given to all individuals who fall within the interval, and it is halfway between the limits of the interval. The interval around each obtainable value on the scale is called a **real interval.** The extreme trait values of a real interval are called the **real limits** of the interval; they represent the points on the trait continuum where the obtainable scale values change. The real limits of each interval are always exactly halfway between the midpoint of that interval and the midpoints of the next lower and next higher intervals. For example, on the scale of statistics knowledge, the value 45 represents a real interval 1 unit wide with a *lower real limit* at 44.5000 units and an *upper real limit* at 45.5000 units. The upper real limit of one interval is the lower real limit of the next. The trait value 44.5000 is the point where the reported scale value changes from 44 to 45, and the value 45.5000 is where the score changes from 45 to 46.

Preparing a grouped frequency distribution entails recognizing another kind of interval as well, the **score interval.** In contrast to the real interval, which reflects the continuous nature of the underlying trait, the score interval (sometimes called an apparent interval) reflects a range of values on the measurement scale itself, or a grouping of adjacent obtainable scores. For example, if we group together the test scores 44, 45, and 46, we have a score interval running from 44 to 46. Its limits, which are themselves obtainable scale values, are known as **score limits** (also called the apparent limits). The intervals and limits that are used to prepare a grouped frequency distribution are score intervals and score limits. The relationships between the underlying continuum, the score scale, and the grouped frequency scale are shown in Figure 2-1.

Preparing a Grouped Frequency Distribution

In preparing a grouped frequency distribution, it is important to select the intervals carefully. While there are no hard and fast rules, the following principles provide sound guidance and will usually produce a satisfactory result:

1. *It is desirable to have each interval contain an odd number of score values.* The reason for this is that if the grouped frequency distribution is to be used as a starting point for later computations, the middle value of the interval, its midpoint, will be used to represent all scores in the interval. When there are an even number of score values, the midpoint will be halfway between two scores and thus a decimal fraction.

A **midpoint** is the center of an interval of measurement for a continuous trait. It is the score given to all individuals who fall within the interval, and it is halfway between the limits of the interval.

A **real interval** is an interval on a continuous scale.

The **real limits** of an interval are the points where one interval changes into another one. They are the points on a continuous trait where the reported scale values change. They are halfway between two adjacent midpoints or scale scores.

A **score interval** is a grouping of adjacent obtainable scores.

Score limits are the most extreme score values that are included in any given score interval.

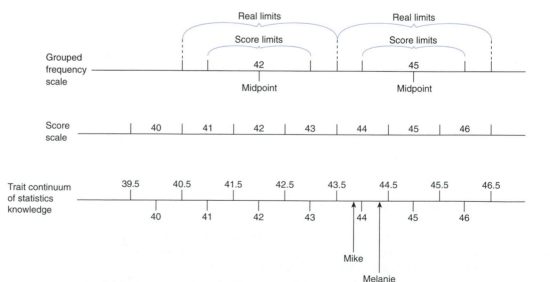

Figure 2-1 Measurement of a continuous variable involves forming score intervals. Each interval has a midpoint that is the obtainable score value and real limits. In a grouped frequency distribution several score intervals are combined.

Using an odd number of scores will result in the midpoint being a whole number. We can make an exception to this guideline for intervals of 10 or a multiple of 10.

2. As a general rule, *there should be between 10 and 20 intervals in a grouped frequency distribution*. The reason for this is that using fewer than 10 intervals often loses too much detail and conceals important information, while having more than 20 intervals may present unnecessary and confusing detail. If there are few observations, you should use fewer intervals, perhaps even as few as 5 or 6; if there are many observations, use more, up to about 20.

3. *All intervals should be the same width*. There are two reasons for this principle. First, because grouping the data introduces some approximation to the original information, it is important that the reduction in precision be uniform throughout the scale. Otherwise some scores will be distorted more than others. Second, when unequal intervals are used, the shape of the underlying distribution can be drastically altered.

   There is one exception to the rule of equal interval widths. When a distribution has a few very extreme scores it is sometimes necessary to use open-ended intervals at one or both ends of the distribution to avoid making the intervals in the center too wide, thereby losing too much detail about the majority of the cases. However, open-ended intervals should be used only as a last resort because these intervals have no defined midpoints, and therefore any scores that fall in them cannot be used for further computations. One alternative is to place an arbitrary midpoint one-half interval beyond the limits of the defined intervals. For example, if you are using an interval width of 5 and the highest defined interval has the real limits 97.5 and 102.5, you could reasonably use an arbitrary midpoint of 105 for an open-ended interval to include all values of 103 or higher.

4. *Intervals should be selected so that they do not include impossible values*. If we are preparing a grouped frequency distribution for a set of test scores where the lowest possible score is zero and the highest possible score is 100, then the bottom interval should not extend below zero and the top interval should not extend above 100.

5. *Make the interval midpoints or one of the score limits a multiple of the interval width*. After we have decided on an interval size, the next question is where to start the intervals. One convenient rule of thumb is to choose an interval that has as its midpoint or as its lower or upper score limit a value that is a multiple of the interval width. Thus if we choose an interval of 3, the midpoint of our bottom interval could be 3, 6, 9, and so on, or if we choose 5, the lowest interval would have a lower limit of 5.

With the above principles in mind, we can prepare a grouped frequency distribution for the statistics knowledge test data in Table 2-2. First, we must decide on the number of intervals. The best way to do this is to subtract the lowest obtained value on the scale (4) from the highest (65) and divide the resulting value (61) by 10 and by 20. (Some authors suggest dividing by interval widths of 3, 5, 7, and so on, until a result between 10 and 20 is obtained. Either method achieves the same result.) Dividing by 10 shows what interval width will yield 10 intervals, and dividing by 20 shows what interval width will yield 20 intervals. This operation reveals that an interval width of 6 will result in about 10 intervals and an interval width of 3 will yield about 20 intervals. Four and five are also possible choices for interval width, but four does not have our first desirable property, an odd number of scores in the interval.

The grouped frequency distributions that would result from using the above procedures on our final exam data are given in Table 2-4. Using an interval width of 3 results in 22 intervals, while an interval width of 5 yields 13 intervals. Both of these grouped frequency distributions are close enough to meeting the criteria listed above that the choice between them must be dictated by some other factor.

Perhaps the most important factor that would determine our choice between 3 and 5 for interval width is that larger intervals lose more information. We noted earlier that measuring continuous traits always involves some approximation. In the process of preparing a grouped frequency distribution, we are decreasing the precision of our measurement (perhaps it would be better to say that we are using coarser approximations) in the hope

Table 2-4 Grouped Frequency Distributions of Scores on the Statistics Final Exam for Intervals of 3 and 5

Interval of 3			Interval of 5		
Score Interval	Midpoint X	Frequency f	Score Interval	Midpoint X	Frequency f
65–67	66	1	63–67	65	1
62–64	63	0	58–62	60	2
59–61	60	2	53–57	55	4
56–58	57	1	48–52	50	4
53–55	54	3	43–47	45	9
50–52	51	1	38–42	40	7
47–49	48	5	33–37	35	12
44–46	45	5	28–32	30	9
41–43	42	4	23–27	25	5
38–40	39	5	18–22	20	8
35–37	36	6	13–17	15	4
32–34	33	7	8–12	10	1
29–31	30	6	3–7	5	2
26–28	27	2			
23–25	24	5			
20–22	21	5			
17–19	18	3			
14–16	15	3			
11–13	12	1			
8–10	9	1			
5–7	6	1			
2–4	3	1			

of getting a better overall picture of the data. The loss of precision occurs because we use the midpoint of each interval to represent all of the scores in that interval. In Table 2-4, the score limits of each interval are given in the column labeled **Score Interval,** and the midpoint is given in the next column, labeled X; as before, X designates the values of the variable. In a grouped frequency distribution, the midpoints of the intervals are the only values of X that we retain. All observations that fall in an interval are given the midpoint of that interval as their value for X, regardless of the value they had in the original frequency distribution.

The three lowest scoring individuals in Table 2-3 had scores of 4, 7, and 8, respectively. In a grouped frequency distribution with an interval of 5, the scores of 4 and 7 are represented by an X-value of 5, which is the midpoint of the lowest interval, and the score of 8 is represented by an X-value of 10, the midpoint of the next interval. None of these midpoint values is a score that one of these students actually earned. We have exchanged precise knowledge of their scores for a simplification of the data. Note that all three people are in separate categories when the interval width is 3, but none of the midpoints corresponds to one of the scores actually earned.

For many years grouped frequency distributions were an important aid in computing the descriptive statistics we discuss in later chapters. As computers have taken over most of the labor of data analysis, the need for grouping the data as an aid to computation has largely disappeared. However, there are two reasons why it is still essential that you understand the principles for grouping data. First, the data sometimes come to us in grouped form. For example, a survey may collect information about respondents' ages by decade. When data such as these are analyzed, even by computer, the effects of grouping are present, and a decision must be made at the planning stage about how to form the groups. Second, preparing graphs of data, as we show later in this chapter, almost always involves grouping to reduce the number of intervals. This is necessary because the limits of visual presentation require that we use a relatively small number of categories to present the information.

To test your understanding of the principles for making a grouped frequency distribution, let's try making one for the high school grade point average (HSGPA) variable included in the BTU test anxiety data set in Appendix C. Because this variable is in a very different measurement scale than the final exam score is, we must proceed with caution.

Begin by reviewing the five steps we used to prepare the grouped frequency distribution in Table 2-4. Next, think about the variable itself. What are the units? Looking at the data in Appendix C, we can see that each "score"—each GPA—is reported to two decimal places. The "unit" is .01. Now decide on the number of intervals. The highest score is 4.00, and the lowest score is 1.04, so the range is from 1.04 to 4.00 or 2.96 grade units. Dividing by 10 and by 20 produces possible intervals of about .30 and .15. Two quite reasonable values that seem to meet most of our criteria are .15 and .25. A quarter of a grade unit has a nice "ring" to it, so let's use that one. As you begin to construct a frequency distribution, you will soon see that some of our criteria for forming intervals rule out certain options. For example, we cannot have interval *midpoints* that are multiples of .25, and have intervals that do not include impossible values (the top interval would include GPAs above 4.00). Perhaps the best solution, at least the one that is most consistent with all five principles we have learned, is to use upper score limits that are multiples of .25 and to start with an interval from 1.01 to 1.25. The midpoint of this interval, 1.13, then represents all three scores in the interval. You should get the following results from this decision about the intervals:

Interval	Midpoint (X)	Frequency (f)
3.76–4.00	3.88	2
3.51–3.75	3.63	4
3.26–3.50	3.38	3
3.01–3.25	3.13	5
2.76–3.00	2.88	6
2.51–2.75	2.63	9
2.26–2.50	2.38	12
2.01–2.25	2.13	7
1.76–2.00	1.88	4
1.51–1.75	1.63	10
1.26–1.50	1.38	3
1.01–1.25	1.13	3

Cumulative Frequency Distributions

Another useful way to organize and summarize information is in the form of a **cumulative frequency distribution.** A cumulative frequency distribution shows how many people received scores at or below each point in the distribution. It is often of benefit to know how many scores fall at or below a given scale value, and the cumulative frequency distribution provides this information.

To make a cumulative frequency distribution, start with either a raw or grouped frequency distribution. The cumulative frequency distribution is easily prepared from either of these starting points by listing values of X in descending order and counting the number of scores that fall at or below each value of X. In practice, we simply add the frequency in each interval to the sum of the frequencies in the intervals below it. The process is illustrated in Table 2-5 for the two grouped frequency distributions in Table 2-4. Notice that

1. the values of X are now the upper score limits of the intervals,
2. the frequencies are the same as those in Table 2-4, and
3. the cumulative frequency (**cf**) in each interval is equal to the sum of the cumulative frequency in the interval below and the frequency in the interval itself.

For example, in the columns with an interval width of 3, the cumulative frequency for an X of 25 includes the 15 people whose scores are below 23 (the cf for 22) and the 5 people whose scores were in the interval 23–25. The same procedure is followed throughout the distribution: The cf for a score of 37 includes the 35 people who scored at or below 34

*A **cumulative frequency distribution** shows how many people received scores at or below each point in the distribution.*

	Interval of 3			Interval of 5	
Upper Limit X	Frequency f	Cumulative Frequency cf	Upper Limit X	Frequency f	Cumulative Frequency cf
67	1	68	67	1	68
64	0	67	62	2	67
61	2	67	57	4	65
58	1	65	52	4	61
55	3	64	47	9	57
52	1	61	42	7	48
49	5	60	37	12	41
46	5	55	32	9	29
43	4	50	27	5	20
40	5	46	22	8	15
37	6	41	17	4	7
34	7	35	12	1	3
31	6	28	7	2	2
28	2	22			
25	5	20			
22	5	15			
19	3	10			
16	3	7			
13	1	4			
10	1	3			
7	1	2			
4	1	1			

(its *cf*) plus the 6 people who fall in the interval bounded by 37. Cumulative frequency distributions are primarily useful for computing various measures of relative position such as percentiles and percentile ranks which we introduce in Chapter 5.

Relative Frequency Distributions

All of the frequency distributions we have discussed so far have one common shortcoming: They are all specific to a given total number of observations and cannot be generalized to larger or smaller groups. The way to overcome the problem is to report the frequency for each score or interval as a *proportion* of the total number of observations. Any frequency distribution, whether raw, grouped, or cumulative, can be presented in terms of the relative

---

SPECIAL ISSUES IN RESEARCH

Which Limits to Use

We have illustrated the cumulative frequency distribution using the upper score limits of the intervals because in a grouped frequency distribution these are the points below which all people in a given interval fall. There are two alternative procedures that you may encounter, each of which has its own justification.

The first alternative uses the values of the midpoints rather than the upper score limits of the intervals as the values of X. The rationale for this choice is that the midpoints are the score values that are used to represent all observations in an interval. If we were to make this change, it would alter

the values in the X column, but not the cumulative frequencies.

The second alternative is to use the midpoints of the intervals and assume that the scores in each interval are evenly spaced throughout the interval, so half of the scores in any interval would fall at or below its midpoint. Adopting this approach is somewhat more complicated because it requires that we add half of the cases falling in an interval to the sum of the frequencies for the lower intervals. It also means that in some cases the cumulative frequencies defined in this way will not be whole numbers.

Table 2-6   Relative Frequency and Relative Cumulative Frequency Distributions for Final Exam Scores Using an Interval of Five

Midpoint X	Frequency f	Relative Frequency rf	Upper Limit	Cumulative Frequency cf	Relative Cumulative Frequency rcf
65	1	.015	67	68	1.000
60	2	.029	62	67	.985
55	4	.059	57	65	.956
50	4	.059	52	61	.897
45	9	.132	47	57	.838
40	7	.103	42	48	.706
35	12	.176	37	41	.603
30	9	.132	32	29	.426
25	5	.074	27	20	.294
20	8	.116	22	15	.221
15	4	.059	17	7	.103
10	1	.015	12	3	.044
5	2	.029	7	2	.029

frequency of each score by the simple process of dividing each frequency or cumulative frequency by the total number of cases observed. A **relative frequency distribution** shows the proportion of the total group who received each score or fell in each interval.

Table 2-6 presents the relative frequencies and relative cumulative frequencies for the final exam scores in Table 2-5 using an interval width of 5. The values in the relative frequency (*rf*) column are obtained by dividing each frequency by 68, while the relative cumulative frequencies (*rcf*) are the *cf* values, also divided by 68.

The primary advantage of relative frequency distributions (and *rcf* distributions) is that they are independent of the particular number of individuals observed. A frequency of 5 in a group of 50 scores has the same meaning that a frequency of 10 has in a group of 100, or 75 in 750. This means that we can directly compare two relative frequency distributions, regardless of the sizes of the groups on which they are based. The importance of this will become more apparent as we consider graphs.

*A **relative frequency distribution** shows the proportion of the total group who received each score or fell in each interval.*

HOW THE COMPUTER DOES IT

Most computer programs, including SPSS, are not set up to create grouped frequency distributions, but they can read data that have been converted into a grouped form. They group the data when preparing graphs, as we shall see shortly, but the thinking of the programmer is, Why throw away information when I have a computer? Therefore, you cannot get results like those in Tables 2-5 and 2-6 from most programs except as a byproduct of graphing. However, if you look back at the SPSS-produced frequency distribution we showed earlier, you will see that the program automatically produces a relative frequency distribution (column 3—Percent) and a relative cumulative frequency distribution (column 5—Cumulative Percent). (The column labeled "Valid Percent" takes into account situations where data are missing for some individuals. It is the relative frequency, counting only people with valid data. The relative cumulative frequency is also for people with valid data.)

◗ *Now You Do It*

Is there anything you need to adjust to prepare a cumulative or relative frequency distribution for the high school GPA data we worked with earlier? Not really; once you have chosen a satisfactory set of intervals and determined the interval midpoints and upper score limits, a few minor changes will give the desired result.

Cumulative Frequency			Relative Frequency		
Upper Limit	f	cf	Midpoint	f	rf
4.00	2	68	3.88	2	.029
3.75	4	66	3.63	4	.059
3.50	3	62	3.38	3	.044
3.25	5	59	3.13	5	.074
3.00	6	54	2.88	6	.088
2.75	9	48	2.63	9	.132
2.50	12	39	2.38	12	.176
2.25	7	27	2.13	7	.103
2.00	4	20	1.88	4	.059
1.75	10	16	1.63	10	.147
1.50	3	6	1.38	3	.044
1.25	3	3	1.13	3	.044

Graphs: Pictures of Frequency Distributions

A picture is worth a thousand words, or so the saying goes. Certainly, for some forms of communication, pictures are more useful or persuasive than words. The same principle holds true for some types and purposes of communication involving quantitative information. Pictures that convey quantitative information are called **graphs.** In this section, we discuss the advantages and disadvantages of several types of graphs and how to prepare the ones that are widely used in education and psychology.

Graphs are pictures that convey quantitative information.

Stem-and-Leaf Plots

Our first example really is a grouped frequency distribution constructed in a visual way, which loses minimal information. It is called a **stem-and-leaf plot** because the basis for grouping forms a "stem" and the individual observations are listed as "leaves" sprouting from the stem.

A **stem-and-leaf plot** is a graph of a frequency distribution in which the stem is the first part of each score and the leaf is the last digit.

A stem-and-leaf plot for the data from Table 2-2 is shown in Table 2-7. In this presentation the stems are the multiples of 10 (00, 10, 20, and so on) with their final 0 dropped. The leaves are the last digits of the individual observations. Thus, the first person, with a score of 37, has a stem value of 3 and a leaf of 7; the second person (score of 47) has a stem of 4 and a leaf of 7. The other 66 people are listed in the same way, each with his or her stem value determining which row they are in and their leaf being listed in that row.

There are two important features of the stem-and-leaf plot. First, no information except individual identification is lost, so there is no loss of precision. Second, making the plot itself produces a picture of the data, showing that most of the students earned scores between 20 and 49, with fewer people scoring at the extremes. Notice also that it is conventional to list the leaves on each stem in order of magnitude rather than by the order they appear in the original data listing. Also, possible score values that do not occur are not listed, so there is no gap in the picture. Stem-and-leaf plots are useful for presenting small-to-moderate-sized data sets involving small-to-moderate-sized numbers and not too

Table 2-7 Stem-and-Leaf Plot of the Data from Table 2-2	
Stem	**Leaves**
6	15
5	045569
4	012334445677889
3	00123344445777778899
2	001223555588999
1	3446899
0	478

many distinct values. They retain too much detail for more than about 100 cases or scales that would result in more than about 15 stems.

▶ *Now You Do It*

Are there any other variables from the BTU test anxiety study for which we might prepare a stem-and-leaf plot? If we select high school GPA, what would we use for stem values? There are only four values for the first digit, and then all the leaves would be two-digit decimals. If we use the first two digits (e.g., 2.8, 2.9, and so on) we would have 29 stems with few leaves on any one stem. A stem-and-leaf plot of this variable does not look promising.

Several other variables show similar problems, either not enough stems to show detail or too many stems to show overall pattern. A stem-and-leaf plot is not for every distribution. There is one variable, however, that should produce a good plot, math aptitude test score. See if you can construct a proper stem-and-leaf plot for these data. When you are finished, the results should look like this.

3	1356789
4	0111223344444555556778
5	0011344566666777888
6	00111223444556799
7	112479

Pie Charts and Bar Graphs

Everyone who reads newspapers or magazines is familiar with the **pie graph,** or **pie chart,** which is widely used to illustrate the division of a whole into its parts. Government agencies often use pie charts to illustrate sources of tax revenue or areas of expenditure. It is called a pie graph because it shows how the revenue pie, for example, is sliced up into various parts or sources.

In psychology and education, pie charts are most often used to show the distribution of discrete status variables. For example, we might wish to show the distribution of the number of mathematics courses taken by the students participating in our test anxiety study. The pie chart, with one slice for each of the six values the variable has, would look as shown in Figure 2-2. Each slice is proportional to the number of people who have taken that number of courses.

A second type of graph that is very similar to the pie chart in its application is the **bar graph,** or **bar chart,** which is used to show the frequencies of various nominal categories. For example, schools and colleges often report enrollment information on gender, numbers of freshmen, sophomores, juniors and seniors, or the racial (numbers of students representing various minority groups) composition of the institution in the form of a bar graph. The lengths of the bars in the graph indicate the relative or absolute frequencies of the various categories of the variable. The bar graph for our number of math courses taken variable is shown in Figure 2-3. The information in the bar graph is exactly the same as that in the

*A **pie chart,** or **pie graph,** shows the division of a whole into its parts.*

*A **bar graph** uses bars of different lengths to show the frequency for each of several discrete categories.*

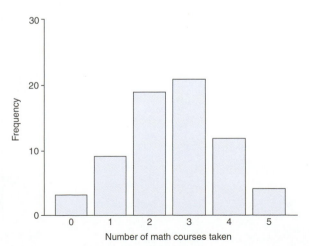

Figure 2-3 Bar graph showing the number of individuals in each category of number of math courses taken. The abscissa is a nominal scale.

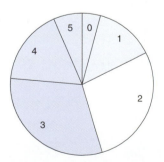

Figure 2-2 Typical pie graph showing the division of a whole into its parts for the number of math courses taken by students at BTU.

pie graph. Note that in the bar graph the bars are separated; they do not touch each other. This indicates the discrete scale nature of the variable being graphed.

The most frequent use of graphs in education and psychology is to illustrate frequency distributions for continuous variables. Two types of graphs, the **histogram** and the **frequency polygon,** are used with raw, grouped and relative frequency distributions, while the **cumulative frequency curve** is the graph of a cumulative frequency distribution. Because these three types of graphs are widely used, and because they form the graphic base for many of our later topics, it is important to consider their principal features, preparation, and interpretation in some detail.

The Coordinate System

Many statistical relationships can be graphed using a space of two dimensions. To do this we use a system for labeling the dimensions of our space and locating points within the space that is known as the Cartesian coordinate system. It is named for René Descartes, the seventeenth-century French philosopher and mathematician who developed it.

We begin by defining two **axes.** These are lines at right angles to each other; one horizontal, the other vertical. Mathematicians have given the name **abscissa** to the horizontal dimension of a two-dimensional space, while the vertical dimension is generally called the **ordinate.** The abscissa is also sometimes referred to as the *base line.* As shown in Figure 2-4, the abscissa divides the space in half horizontally, and the ordinate divides it in half vertically, resulting in four equal sections, or **quadrants.** The ordinate and abscissa are the axes of the space, and the point where they cross is called the **origin.**

The axes are divided into equal units outward from the origin. On the abscissa positive values are to the right of the origin and negative values are to the left. The ordinate is labeled with positive values above the origin and negative values below it. The values of both axes are zero at the origin. This numerical labeling of the axes is shown in Figure 2-4.

The advantage of labeling the axes in this way is that a single value expresses both direction and distance from the origin along either axis. Using two axes labeled in this way permits us to locate any point in the space simply by giving the abscissa and ordinate values. These values are generally given in pairs with the value for the abscissa given first. A pair of values given in this way defines the location of a point in the space; these values are known as the **coordinates** of the point. In Figure 2-5, three points with coordinates $(2, 4)$, $(-1, 3)$, and $(-3, -1)$ are shown.

Each axis can be labeled to represent the different values that some variable may possess. When we wish to express a frequency distribution in the form of a graph, the usual procedure is to label the abscissa with the values of the measured variable (X) and the ordinate with the possible frequencies (f). This convention is quite arbitrary, and later,

*The **abscissa** is the horizontal axis in the Cartesian coordinate system.*

*The **ordinate** is the vertical axis in the Cartesian coordinate system.*

*The **origin** is the point where the ordinate and abscissa cross.*

*The **coordinates** of a point are its values on the abscissa and the ordinate.*

*By convention, the variable is called **X,** and its values are put on the abscissa. The frequency is designated **f,** and its values are put on the ordinate.*

Figure 2-4 The two-variable coordinate system showing the four quadrants formed by the ordinate and the abscissa.

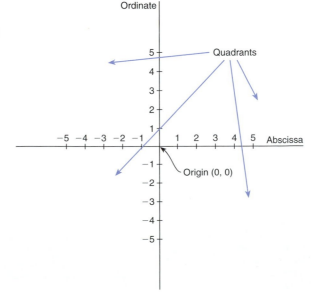

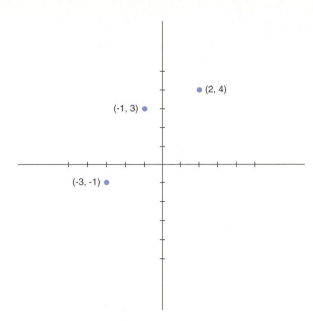

Figure 2-5 Three points plotted in a two-coordinate system. The first value given is for the abscissa.

we shall find that other types of graphs in this coordinate system with the axes differently defined can be quite useful as well.

Because most of the data that psychologists and educators collect come in the form of positive numbers, we will find ourselves using the first, or upper right, quadrant more often than the others. All of the graphs presented in this chapter, for example, fall in the first quadrant. However, we shall see that some transformations of data that help compare scores on different variables, and some more advanced applications require us to use two or all four quadrants.

Histograms

Graphs of frequency distributions may label the abscissa either with the real limits of the intervals or with the midpoints of the intervals. A **histogram** is a graph of a frequency distribution in which the abscissa *usually* is labeled with the real limits of the intervals, and frequency is shown by the height of a bar. (Many computer packages label the midpoints rather than the real limits, and it does not make a lot of difference which convention is used. When preparing a graph by hand, it is easier to label the real limits because they must be used to draw the bars anyway.) Each individual may be represented by a square. The squares are stacked on top of each other until everyone is accounted for. Figure 2-6 shows a histogram for the grouped frequency distribution with an interval width of 5 derived from Table 2-4. The numbers in the squares correspond to individual identification numbers from Table 2-2 and show how a histogram is constructed from the data of the individuals. Ordinarily, the squares representing the individual scores would not be shown, and the histogram would appear as it does in Figure 2-7. Histograms differ from bar graphs in that they are applied to data from a continuous scale, which is why the bars of the histogram touch each other.

To prepare a histogram, we must first prepare a raw or grouped frequency distribution, then lay out the axes. We label the ordinate *f* for frequency and mark off enough values to include the largest frequency from our frequency distribution. We label the abscissa with the real limits or midpoints of the score categories of our measured variable. If the values are all fairly large, it is customary to chop out a section of the abscissa and indicate by a broken line that some space has been omitted. (Many graphing programs solve this problem by simply starting the abscissa with the lowest value at the origin. See the graph in How the Computer Does It on page 44.) Finally, we draw in a bar over each interval so that the top of the bar is even with the frequency in the interval.

The important thing to remember about histograms is that they indicate the number of people (or objects or events) that fall in each *interval* on a continuous variable. Even a histogram of a raw frequency distribution reflects the fact that our measurement is

*A **histogram** is a graph of a frequency distribution in which the abscissa usually is labeled with the real limits of the intervals, and frequency is shown by the height of a bar.*

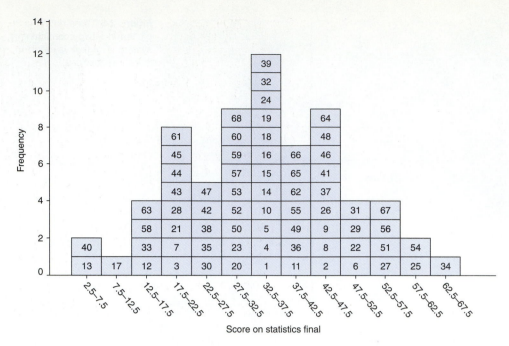

Figure 2-6 Histogram of scores on the statistics final at BTU. When making a histogram, each person is viewed as a square over the value of her/his score interval. Numbers indicate particular individuals.

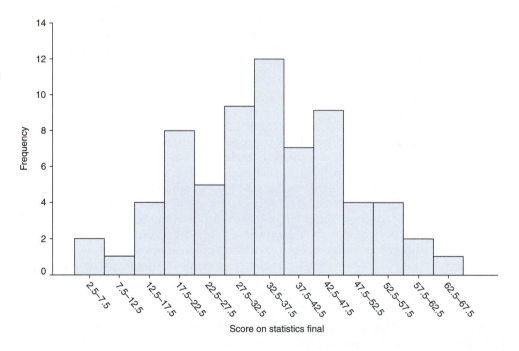

Figure 2-7 The squares are not shown in the final histogram. Each bar starts at the lower real limit of its interval and extends to the upper real limit of that interval.

approximate and we are really only sure that the observations fall between two limits, the limits of our intervals. The height of the bar indicates the number of observations in the interval.

▶ *Now You Do It*

Earlier, you prepared a frequency distribution for the high school GPA variable of the BTU Test Anxiety Study. How would you go about making a histogram of these data? Should you use the same intervals? In general, the principles that lead to a good grouped frequency distribution will also produce a good histogram. The 12 intervals that we used for the frequency distribution will work well and produce the following figure.

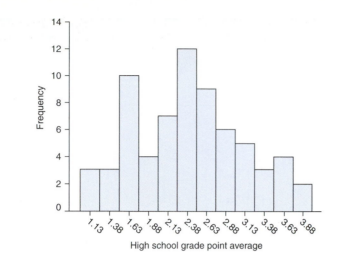

High school grade point average

Frequency Polygons

Instead of focusing on the limits of the intervals on the abscissa, we could direct our attention to the assigned score values, the midpoints. When the abscissa is labeled with the assigned score values, either the raw scores of the raw frequency distribution or the interval midpoints of the grouped frequency distribution, and the limits of the intervals are ignored, the resulting graph is called a **frequency polygon.**

*The **frequency polygon** shows the number of people or observations that are being represented by each of the listed numerical values, usually interval midpoints.*

HOW THE COMPUTER DOES IT

SPSS has the ability to produce a wide variety of graphs, including histograms. The graphing options are easy to use, but you may have to edit the graph to make it look like you want it to. The first thing to do is click on the Graphs drop down menu and select Histogram. You should get a screen that looks like this.

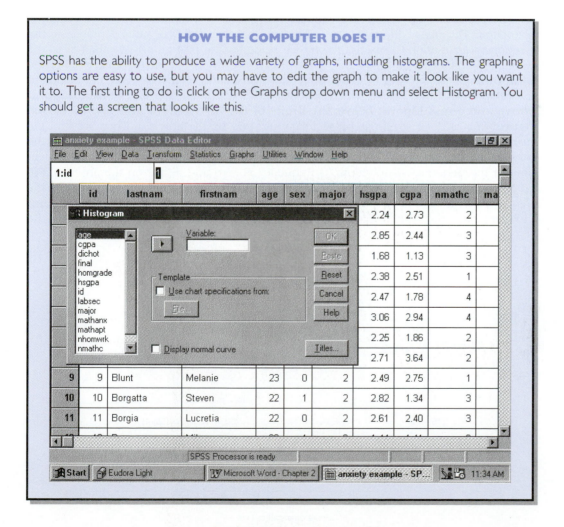

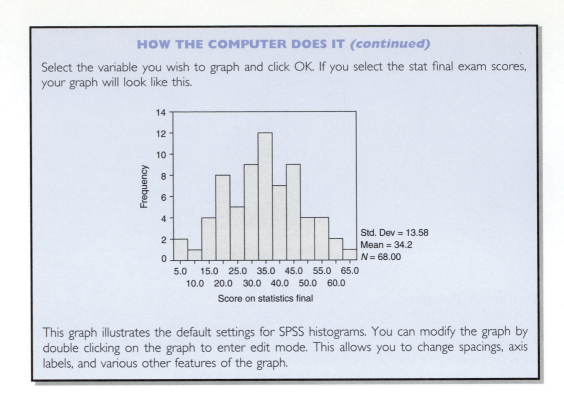

HOW THE COMPUTER DOES IT *(continued)*

Select the variable you wish to graph and click OK. If you select the stat final exam scores, your graph will look like this.

This graph illustrates the default settings for SPSS histograms. You can modify the graph by double clicking on the graph to enter edit mode. This allows you to change spacings, axis labels, and various other features of the graph.

The procedure for making a frequency polygon is very similar to that for making a histogram. The major difference is in labeling the abscissa. We start by preparing the frequency distribution (or using the one we made for the histogram). We then label the ordinate *f* and mark off frequency values as before. Next, we mark off and label the **midpoints** of the intervals from our frequency distribution on the abscissa. In addition to the midpoints of intervals in which observations actually fall, **we label one midpoint above and one midpoint below the limits of the distribution and form a closed figure.** This is done so the graph will start and end on the abscissa. Then we find the **point** that has as its coordinates the value of the midpoint of the first interval and the frequency of that midpoint. Because the first midpoint is for the interval below the lower limit of the frequency distribution, its frequency is zero and it falls on the abscissa. For each of the other midpoints, we find the point that has the interval midpoint and the frequency as its coordinates. These points are plotted and labeled with their coordinates in Figure 2-8. (Ordinarily, we include the points themselves but not the coordinates in the

Figure 2-8 In preparing a frequency polygon, the first step is to plot the points that have the midpoints of the intervals and the frequencies as their coordinates. For example, the point (15,4) shows that 4 people are in the interval that has 15 as its midpoint.

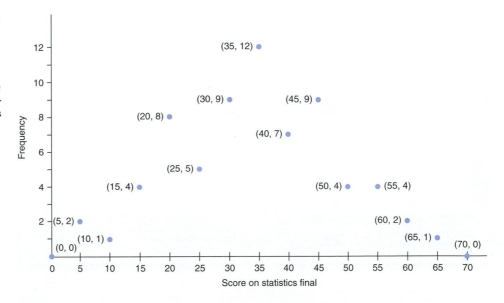

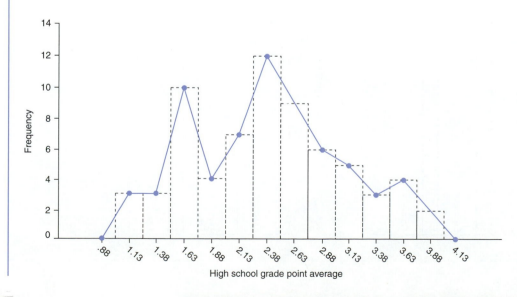

Figure 2-9 The frequency polygon is completed by connecting the points with straight lines.

frequency polygon.) After all of the points have been located, they are connected with a line, as shown in Figure 2-9. This is what the frequency polygon would look like for the final exam scores of our statistics students.

A frequency polygon and a histogram convey exactly the same information because they are both prepared from the same frequency distribution. The difference is one of focus or emphasis. The frequency polygon emphasizes the numerical values that are used to represent observations (the midpoints), while the histogram emphasizes score intervals. Figure 2-10 illustrates the similarities and differences between them. Note that each point for the frequency polygon falls in the middle of the top of the histogram bar for that interval.

▶ *Now You Do It*

Earlier in this chapter you prepared a frequency distribution and a histogram for the high school GPA variable from the BTU Test Anxiety Study. Now use the same data to prepare a frequency polygon. Is there any reason to change the intervals you are using? What values should be placed on the abscissa? Try to make your frequency polygon on the same scale as the histogram so you can compare them. If you have done everything correctly, your frequency polygon should look like this. The histogram is shown in dashed lines so you can compare the two.

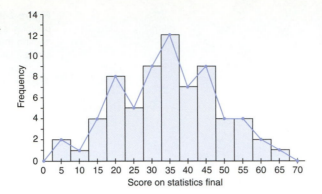

Figure 2-10 The frequency polygon and histogram convey the same information about a set of data.

SPECIAL ISSUES IN RESEARCH

Proportions for Graphs

In preparing a histogram or frequency polygon it is customary to mark off the scales on the two axes in such a way that the final graph will be about three inches wide for every two inches of height. There is no mathematical or statistical reason for this rule of thumb, but it does produce a graph that has a more pleasing and balanced appearance. Also, the way one chooses to scale the ordinate and abscissa can have an important impact on the impression the graph makes. If you think about the graphs you see in advertisements, the importance of scale quickly becomes apparent. A graph showing the reliability of different brands of cars can imply a large difference in favor of one manufacturer by showing only the top part of each bar

on an expanded scale, whereas the graph of the entire picture would give the impression of very little difference. Consider, for example, the two bar graphs shown in the figure below comparing the number of majors in psychology, education, and other departments who are enrolled in the SAD introductory statistics course at BTU. The height of the education bar is the same in both graphs, but which one do you think the chairperson of the education department might present to argue for his/her department having greater control over the course content? What other situations have you encountered recently where the scale of a graph was manipulated to create a particular impression?

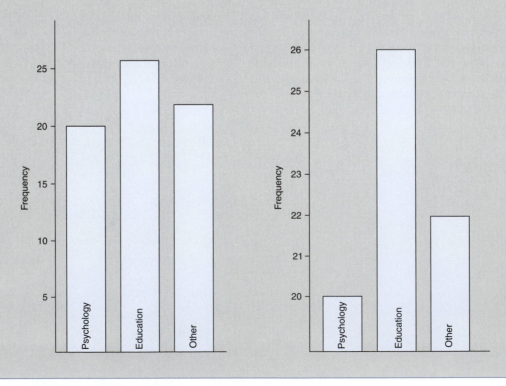

The Cumulative Frequency Curve

It is also possible to make a graph of the cumulative frequency distribution. Such a graph is called a **cumulative frequency curve.** A cumulative frequency curve shows how many people scored below a given value in the cumulative frequency distribution.

The procedures for preparing a cumulative frequency curve differ only slightly from those for preparing a frequency polygon. First, the axes are laid out and labeled. The abscissa is labeled with the upper real limits (or interval midpoints if that is how the cumulative frequency distribution was constructed), and the ordinate is labeled with cumulative frequency values. Then for each upper limit a point is located that has as its coordinates the values of the upper limit and the cumulative frequency for that limit. Finally, the points are connected with a line. Figure 2-11 shows the cumulative frequency curve for the distribution of scores in Table 2-5 using an interval width of 5.

There are several important characteristics of the cumulative frequency curve.

1. Like the frequency polygon, it begins with the point (either upper limit or midpoint) of the first interval below the distribution. This means it will start on the abscissa with a *cf* of zero.
2. The cumulative frequency curve never drops back toward the abscissa. While the line of the frequency polygon tends to rise to its highest point near the middle of the distribution and then drops back to the abscissa, the line for the cumulative frequency curve *never* drops back at all.
3. The curve rises most rapidly where the frequency is greatest (that is, the values corresponding to the high points of the frequency polygon) and is horizontal in areas where the frequency is zero. Notice that the curve in Figure 2-11 rises most rapidly over the values between 27.5 and 47.5. It is almost horizontal where the change in cumulative frequency is low.

The curve in Figure 2-11 has a shape that is very common among cumulative frequency curves. It has the general shape of an *S,* rising slowly at first, then more rapidly, and finally slowly again.

Graphs of Relative Frequency Distributions

Earlier in this chapter, we discussed frequency distributions, in which the relative frequencies were the *proportions* of observations falling in each interval (interval frequencies divided by *N*). It is possible and often desirable to prepare graphs of these distributions. A relative frequency polygon or histogram is constructed from a relative frequency distribu-

*A **cumulative frequency curve** shows how many people scored below a given value in the cumulative frequency distribution.*

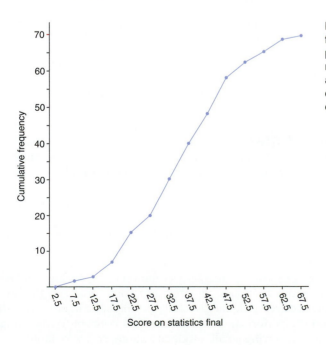

Figure 2-11 The cumulative frequency curve is formed by plotting points whose coordinates are the score values and their cumulative frequencies. These points are then connected with straight lines.

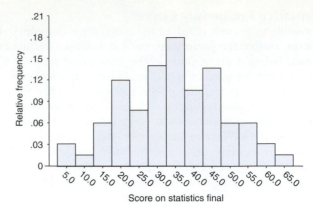

Figure 2-12 In a relative frequency histogram the relative frequency *f/N* of each score is plotted instead of the frequency.

tion in exactly the same way as graphs of other frequency distributions, except that the ordinate is labeled and marked off in proportions. Figure 2-12 shows the relative frequency histogram for the relative frequency distribution in Table 2-6; the shape and size of this figure are exactly the same as those of Figure 2-7. The only difference between them is that the height of the histogram in Figure 2-7 reflects the actual frequency, or number of cases, whereas the height in Figure 2-12 reflects the relative frequency, or proportion, of cases.

You will recall that the area under a histogram bar expresses the number of individuals who received a score in that interval. As in Figure 2-6, the bars are made up of squares representing people or observations. The total area of the bars is 68 units in this example, corresponding to the 68 individuals in the study. Similarly, the total area under the frequency polygon for the same data (Figure 2-9) is equivalent to the total number of observations. Also, for both of these graphs, the area of the graph corresponding to a particular interval is equal to the frequency in that interval.

In the case of the relative frequency histogram or polygon, the total area under the graph is 100 percent of the group, or a proportion of 1.00. Just as the sum of the frequencies in the intervals, or of the areas of the histogram bars, equals the total frequency, so the sum of the proportions in the relative frequency distribution or of the areas of its histogram bars equals the total proportion, or 1.00. In later chapters we will make extensive use of histograms and polygons of relative frequencies, which are sometimes called **unit distributions.** All such distributions have the following properties:

1. The height of the curve above the abscissa indicates relative frequency or proportion.
2. The area under the curve is unity (1.00).

Because relative frequencies have the great advantage of being independent of the number of cases that have been observed and can, therefore, be used to illustrate general principles rather than particular examples, relative frequency graphs are the ones to use when comparing distributions based on quite different sample sizes.

*A **unit distribution** is one in which frequencies are expressed as proportions and the proportions sum to 1.00.*

Line Graphs and Box-and-Whisker Plots

There are times when we may wish to show a trend in some variable over time or across several groups. For example, we often see graphs of population trends or of the number of people who have had a particular disease in different years or decades. School districts may show patterns of rising or falling test scores over time or by building. In cases like these we really are using graphs to show the relationship between two variables, because both the values of the measured variable and the points in time or the groups take on different values. One variable, usually the measured variable, is shown on the ordinate, and the other, time or group, is shown on the abscissa.

The application of line graphs that is particularly common in psychology and education is one in which different groups form one variable and some outcome measure is the other variable. For example, a developmental psychologist might be interested in the effect of magnitude of reward on the performance of children in a concept-formation task. The groups might be identified as receiving small, moderate, large, and very large rewards for

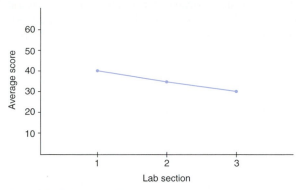

Figure 2-13 A line graph shows a trend in group averages.

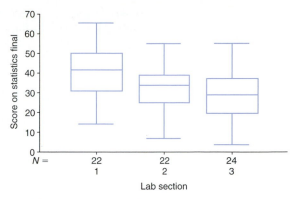

Figure 2-14 Box-and-whisker plots show the center and spread of scores for several groups in one graph.

performance, and the outcome measure or dependent variable might be the number of trials or the time taken to form a concept. Similarly, the faculty of the SAD Department might want to graph the final exam performance of the students in the three different lab sections. The graph of these results would ordinarily be a plot of the average performance of the three groups (average number of items correct) and might result in a graph such as that shown in Figure 2-13.

A slightly different graph can show both the relationship between group membership and average score on the dependent variable, and the spread of performance in each group around that average. It is called a **box-and-whisker plot,** or just a **box plot** (see Figure 2-14). The heavy line marks the median of each group's distribution (see Chapter 3); the box includes the middle 50 percent of cases; and the thin lines, or whiskers, extend to the highest and lowest scores. An ordinary line graph such as Figure 2-13 just shows the mean value for each group (see Chapter 3) connected with a line. The box plot gives this information and more.

*A **box plot** or **box-and-whisker plot** shows both the average score of each group on the dependent variable and the spread of each group around this value.*

Choosing the Best Frequency Distribution or Graph

What is the correct or best way to summarize a set of data? There really is no one answer to this question. The method you choose must depend on your purpose and on the type of data you wish to summarize. A graph is often the best presentation of a set of data to achieve high impact, but it has the disadvantage that it does not permit further quantitative analysis. If you choose to use a graph, which one you use will depend on what you want to emphasize about your data. A pie chart shows the same basic information as a relative frequency histogram—the proportion of the total set of data that is of each kind—but the pie chart emphasizes breaking the whole down into parts or proportions and is appropriate for discrete data. The relative frequency histogram emphasizes the frequency of occurrence of particular scores or intervals on a continuous scale and the spread of scores across a range of values. Likewise, the histogram and frequency polygon provide identical information, but one highlights the intervals while the other focuses attention on the midpoints. Effective communication is the only guide.

Frequency distributions pose a similar problem. Should we group the data? What intervals should we use? The answers to these questions depend on what we plan to do with the data and what computing equipment we have for the analysis. If there are a very large number of cases and we have a computer to do the analysis, we may want to prepare a raw frequency distribution and analyze the data in their original form. For example, district and statewide test results are based on enough cases that there will not be gaps or discontinuities in the data, and computers will be available. In fact, the data will almost certainly come on a computer disk in cases such as these.

On the other hand, if we have a small number of cases that scatter widely over the score scale, or if our computing equipment is very limited, we may wish to prepare a grouped frequency distribution and analyze the data in this coarser form. In that case, our choice of interval width and midpoints is arbitrary, but some values will generally give more satisfactory results than others. We suggested earlier in this chapter that having between 10 and 20 intervals, using intervals with an odd number of values so that our midpoints will be whole numbers, and selecting midpoints that are multiples of the interval width usually give good results, but these are just guidelines.

What Frequency Distributions Show

Inspecting the frequency distribution can reveal several important aspects of the data. First, inspection can give us a general idea of where the center of the data is, and what the most typical values are. Tables 2-3 and 2-4 show that most students earned scores on the statistics final exam between about 20 and 50 with smaller numbers falling outside this range. We will deal with more exact measures of where the center of the distribution is in Chapter 3.

The second piece of information that we can glean from inspecting the frequency distribution is an idea of how widely the scores spread out from their center. Our scores cover a range from a low of 4 to a high of 65. Whether this is a large amount of variation or not depends on the nature of the variable that we are measuring. For example, the variable of high school GPA shows much less spread, only from 1.04 to 4.00. We can also see whether there are individuals who are very extreme on the trait. Such cases, falling far from the main body of the distribution, are called *outliers*. Outliers can have a disproportionately large effect on some of the statistics we will be discussing in Chapters 3 and 4.

Finally, the frequency distribution, and the graphs of distributions to an even greater degree, show us the shape of the distribution. Do the scores tend to pile up at one end, or, as in the case of the statistics final exam scores, does the frequency tend to drop off more or less equally on both sides from a high point in the middle? Some statistical indices assume that the distribution has a particular shape. If this assumption is not met, it can lead to incorrect conclusions. Therefore, it is important to look for abnormalities in the shape of the distribution before using such measures. Indices that reveal the shape of the distribution will be covered in Chapter 4.

SUMMARY

When data are first collected, they are generally in the form of a list of individuals and their assigned value on some measurable variable. It is usually desirable to simplify and summarize the data by preparing a *frequency distribution*. Observations are grouped by their numerical values, and identification of individual observations is discarded. When there are more than 20 distinct values of the variable, it is often desirable to condense the scores further into a *grouped frequency distribution* with 10 to 20 score *intervals. The number of score values in each interval should be odd, so that the midpoints* of the intervals are whole score values.

Frequency distributions show the number of observations occurring in each category. *Cumulative frequency distributions* show the number of scores that are less than or equal to particular points on the score scale. *Relative frequency distributions* and *relative cumulative frequency distributions* reflect the proportion of observations in each interval and the proportion of scores at or below points on the score scale respectively.

Several kinds of graphs are useful aids for understanding and communicating about data. *Pie charts* and *bar graphs* are used to depict frequencies or proportions for discrete variables. Graphs of frequency distributions for continuous variables generally take the form of *histograms,* where columns are used to show frequencies in an interval, and *frequency polygons,* which show the frequencies of score values or midpoints of intervals. *Cumulative frequency curves* show the number or proportion of cases falling at or below each score value; these curves are often S-shaped. *Stem-and-leaf plots* also show the shape of the frequency distribution, and *box plots* and *line graphs* are used to show trends over time or over groups.

FREQUENCY DISTRIBUTIONS AND GRAPHS FOR THE SEARCHERS STUDY DATA

You may wish to test out your mastery of the principles we have been discussing by applying them to a completely different set of data. There are several variables from the study that Dr.

Table 2-8 Frequency Distribution of Ages of Participants in the Searchers Study			
Interval	X	f	cf
59–61	60	2	60
56–58	57	4	58
53–55	54	5	54
50–52	51	3	49
47–49	48	5	46
44–46	45	8	41
41–43	42	3	33
38–40	39	4	30
35–37	36	4	26
32–34	33	2	22
29–31	30	6	20
26–28	27	6	14
23–25	24	4	8
20–22	21	4	4

Table 2-9 Frequency Distribution of Income for Participants in the Searchers Study			
Interval	X	f	cf
58,500–61,499	60,000	1	60
55,500–58,499	57,000	4	59
52,500–55,499	54,000	5	55
49,500–52,499	51,000	9	50
46,500–49,499	48,000	6	41
43,500–46,499	45,000	11	35
40,500–43,499	42,000	11	24
37,500–40,499	39,000	4	13
34,500–37,499	36,000	5	9
31,500–34,499	33,000	1	4
28,500–31,499	30,000	2	3
25,500–28,499	27,000	1	1

Cross and his students conducted (described at the end of Chapter 1) that we might wish to summarize in graphs or frequency distributions. We know from the study design that there are 30 men who are members of the Searchers and 30 who are not; that there are 30 Roman Catholics and 30 Protestants; and that there are 20 men who took each form of the attitude survey; but we should examine the distributions of the age, income, attitude toward equality, and degree of religious involvement variables.

Looking through the age data, we find that two men, Earl Anderson and Cal Knudson, are 59 and Karl Baltch is 20. Therefore, the distance covered by the age variable is 59–20 or 39 years. Dividing this value by 10 gives us 3.9, so intervals of three will meet our guideline. Intervals of 5 would yield only about 8 intervals, which may be too few. Let's assume we use an interval of three and midpoints that are multiples of three. The results you should get for the frequency distribution and cumulative frequency distribution are given in Table 2-8.

The income data present a slightly different problem. Here, the scores cover incomes from a low of 27,731 for Moam Chompsky to a high of 58,682 for Sly Van Donnette. This span covers 30,951 values. We could try various intervals, but one easy approximation is to ignore the last three places and look only at the 30. This is equivalent to saying we will use an interval that is a multiple of 1000. Dividing 30 by 3 yields 10. Therefore, if we use intervals of 3000, we should get a usable result, as is shown in Table 2-9.

The attitude toward equality scores range from a low of 24 for Bill Battle to a high of 60 for Earl Anderson, a distance of 36 units. If you are in favor of an interval width of three, we agree with you, and although there is no real reason to prefer one alternative over another, having the interval midpoints be multiples of three seems reasonable. This choice produces the distribution in Table 2-10.

Finally, the degree of religious involvement scores run from a low of 6, for Sandy McCallum, to a high of 31 shared by Carl Simpson and Sly Van Donnette. Because these scores cover only 25 scale points, we might not ordinarily group them. However, let's prepare a grouped frequency distribution for practice. Using an interval width of 3 and midpoints that are multiples of the interval width should give you the frequency distribution given in Table 2-11.

We encountered stem-and-leaf plots as a way to show the frequency distribution in a more graphical form. Table 2-12 contains the stem-and-leaf plot for the age data. The stem-and-leaf plot for attitude toward equality scores is shown in Table 2-13. Note how the plot shows clearly the high concentration of scores in the 40s. The income data do not lend themselves to a stem-and-leaf plot because the numbers are so large and widely spread. One alternative would be to consider only the first three digits, but then we would have either 4 stems with two-digit leaves or 32 stems with one-digit leaves. A standard grouped frequency distribution probably is better. Because the other two variables have such restricted ranges of values we would not ordinarily construct stem-and-leaf plots for them either.

Table 2-10 Frequency Distribution of Attitude Toward Equality Scores of Participants in the Searchers Study			
Interval	X	f	cf
59–61	60	1	60
56–58	57	0	59
53–55	54	3	59
50–52	51	1	56
47–49	48	8	55
44–46	45	12	47
41–43	42	8	35
38–40	39	9	27
35–37	36	2	18
32–34	33	5	16
29–31	30	3	11
26–28	27	6	8
23–25	24	2	2

Table 2-11 Frequency Distribution for Degree of Religious Involvement Scores for Participants in the Searchers Study			
Interval	X	f	cf
29–31	30	3	60
26–28	27	3	57
23–25	24	4	54
20–22	21	7	50
17–19	18	19	43
14–16	15	16	24
11–13	12	4	8
8–10	9	3	4
5–7	6	1	1

Table 2-12 Stem-and-Leaf Plot of Age for the Searchers Study	
Stem	Leaves
5	00233445667799
4	001234455566678899
3	001134556799
2	0112344567777899

Table 2-13 Stem-and-Leaf Plot of Attitude Toward Equality Scores	
Stem	Leaves
6	0
5	2455
4	0001111222344445555556678888899
3	112234457888999
2	456778889

As we saw earlier in this chapter, a histogram or frequency polygon is simply a graph of the relationship between scores on the variable and the frequency of those scores. We lay out the values of the variable on the abscissa and the values of frequency on the ordinate. In a grouped frequency distribution each bar represents an interval, but when the scores have not been grouped,

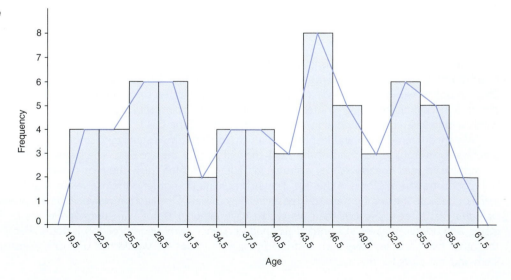

Figure 2-15 Histogram of age scores for data from the Searchers Study using an interval width of 3. Frequency polygon is shown in solid color lines.

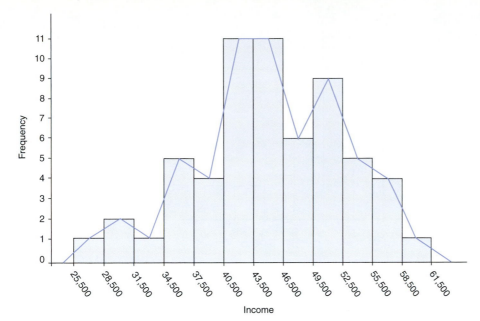

Figure 2-16 Histogram of income scores for data from the Searchers Study using an interval width of 3000. Frequency polygon is shown in solid color lines.

each bar corresponds to a single score. You should also construct frequency polygons or histograms of the distributions for each of these four variables. Histograms, with frequency polygons overlaid, are shown in Figures 2-15 through 2-18.

If you use SPSS to graph these data, the program may select a set of intervals different from the ones you might choose. If this happens, you will have to experiment with the program to get the representation you want. It is important that you make an active decision about the characteristics of your graph rather than accept passively what the computer program gives you.

The cumulative frequency curves are prepared from the scores and the values in the cf column. We have used the upper real limits of the intervals. These curves are plotted in Figures 2-19 through 2-22.

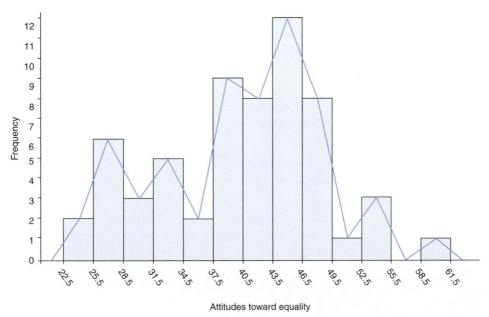

Figure 2-17 Histogram of attitude toward equality scores for data from the Searchers Study using an interval width of 3. Frequency polygon is shown in solid color lines.

Figure 2-18 Histogram of degree of religious involvement scores for data from the Searchers Study using an interval width of 3. Frequency polygon is shown in solid color lines.

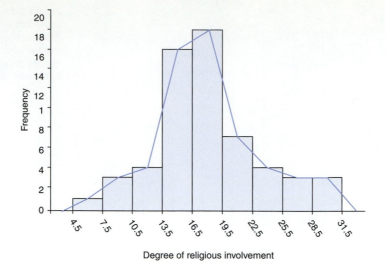

Figure 2-19 Cumulative frequency curve of age for data from the Searchers Study using an interval width of 3.

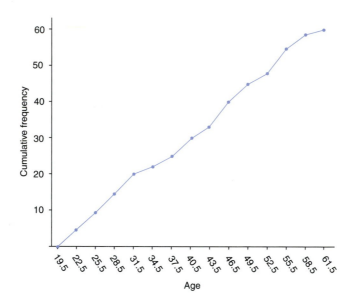

Figure 2-20 Cumulative frequency curve of income for data from the Searchers Study using an interval width of 3000.

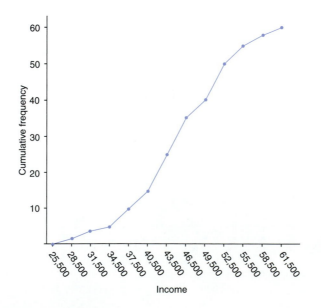

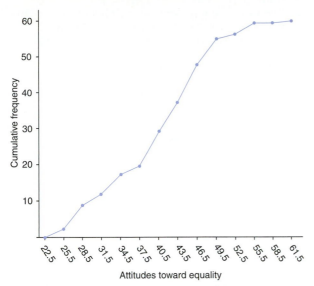

Figure 2-21 Cumulative frequency curve of attitude toward equality for data from the Searchers Study using an interval width of 3.

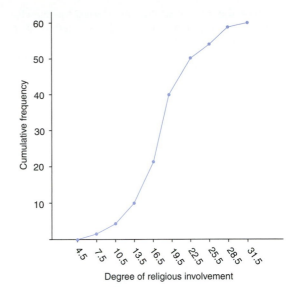

Figure 2-22 Cumulative frequency curve of degree of religious involvement for data from the Searchers Study using an interval width of 3.

REFERENCES

Henry, G. T. (1994). *Graphing data: Techniques for display and analysis*. Applied Social Research Methods (Vol. 36). Thousand Oaks, CA: Sage.

Jacoby, W. G. (1997). *Statistical graphics for univariate and bivariate data*. Thousand Oaks, CA: Sage.

Tufte, E. R. (1997). *Visual and statistical thinking: Displays of evidence for making decisions*. Cheshire, CT: Author.

Wallgren, A., Wallgren, B., Persson, R., Jorner, U., & Haaland, J. (1996). *Graphing statistics and data*. Thousand Oaks, CA: Sage.

EXERCISES

1. In collecting data, it is sometimes so disorganized and incomprehensible that we cannot arrive at any conclusions about the data. As a result, researchers often organize and summarize their data for better understanding. One way of organizing and summarizing data with minimal loss of information is to construct some type of frequency distribution. Describe each of the four types of frequency distributions.

2. A survey of the models of cars in a parking lot at a major shopping center was conducted for an advertising firm. A row of the parking lot was randomly selected and the models of cars in the row were recorded as follows:

Chevrolet	Mazda	Ford	Oldsmobile	Pontiac
Dodge	Chevrolet	Toyota	Chevrolet	Ford
Honda	Honda	Nissan	Toyota	Volvo
Toyota	Buick	Cadillac	Chevrolet	Ford
Honda	Toyota	Ford	Chevrolet	Nissan
Ford	Honda	Chevrolet	Honda	Toyota
Dodge	Honda	Chevrolet	Toyota	Volvo
Oldsmobile	Ford	Toyota	Mercedes Benz	Honda

   a. Although the data listed above are qualitative, summarize these observations in a frequency table.

   b. What types of graphs are appropriate for these data? Construct an appropriate graph for the data.

   c. Although the selection of the row was random, what concerns do you have about the representativeness of the sample?

3. For the following sets of data, prepare raw frequency distributions listing the scores from highest to lowest, tally marks for the appropriate scores, frequency of each score, and cumulative frequency of each score:

   a. The following represents scores on the first statistics quiz: 20, 20, 19, 19, 19, 18, 18, 18, 18, 16, 16, 16, 15, 15, 14, 14, 14, 14, 13

   b. The following represents scores on an algebra test: 22, 10, 27, 10, 12, 27, 16, 28, 11, 23, 19, 27, 20, 23, 20, 16, 13, 21, 11, 12, 14, 18, 22, 23, 13, 16, 31, 45, 18, 24, 27, 49, 38, 22, 31, 35, 41, 47, 29, 25, 24, 27, 22, 18, 17, 34, 33, 41, 31, 33, 28, 28, 29, 34, 19, 25, 36, 25, 33, 29, 28, 22

   c. The following represents scores on a measure of spatial visualization: 31, 32, 30, 18, 19, 30, 23, 22, 24, 15

4. In your own words, describe the principles for preparing a grouped frequency distribution.

5. For each of the following descriptions of data sets, find the appropriate interval width(s):

   a. A data set contains 56 scores that range from a lowest score of 23 to a highest score of 63. The lowest possible score is 20 and the highest possible score is 65.

   b. A data set contains 119 scores that range from a lowest score of 24 to a highest score of 96. The lowest possible score is 10 and the highest possible score is 100.

   c. A data set contains 38 scores that range from a lowest score of 68 to a highest score of 115. The lowest possible score is 60 and the highest possible score is 120.

   d. A data set contains 88 scores that range from a lowest score of 2 to a highest score of 48. The lowest possible score is 0 and the highest possible score is 50.

6. Suppose the interval width you selected for Exercise 5(d) was 3. What intervals should you select so that you do not violate any of the other principles?

7. For the scores below, construct a grouped frequency table. The following data represent scores on measures of test anxiety:

Score	Tally
79	/
78	/
77	//
76	//
75	///
74	//
73	
72	/
71	///
70	///// /
69	///// ///
68	////
67	/////
66	////
65	///
64	////
63	//
62	/////
61	///
60	/
59	//
58	
57	//
56	///
55	/
54	/
53	
52	/
51	/

Construct a frequency histogram from the information in the grouped frequency distribution table.

8. Use the grouped frequency table that you constructed in Exercise 7 to construct a table that contains both cumulative frequency distribution and relative frequency distribution information. Construct a cumulative frequency polygon and a relative frequency polygon using information in the table.

9. What data set(s) from Exercise 3 could be appropriately represented by a grouped frequency table? Defend your answer. Construct a grouped frequency table for the appropriate data set(s) from Exercise 3.

10. In a Cartesian coordinate system, on which axis is the abscissa and on which axis is the ordinate? Plot the following points in a Cartesian coordinate system: (0, 4), (−2, 3), (5, 6), (3, 5), (−4, −8), (−3, 0), (7, 2), (−6, −2), (2, −8), (−2, 8)

11. For the following data sets, construct a stem-and-leaf display, a grouped frequency table, a frequency polygon, and a histogram:
    a. The following represents scores on the feminine scale of the Bem Sex Role Inventory: 77, 58, 64, 73, 87, 71, 91, 64, 72, 84, 65, 78, 72, 75, 55, 78, 43, 54, 55, 79, 69, 76, 79, 97, 67, 69, 90, 54, 57, 58, 97, 43, 51
    b. The following represents scores on the masculine scale of the Bem Sex Role Inventory: 75, 80, 103, 65, 79, 88, 71, 86, 77, 78, 49, 86, 76, 96, 76, 92, 87, 87, 80, 91, 79, 92, 91, 65, 78, 91, 88, 89, 105, 76, 67, 80, 102, 79, 57, 77, 74

    c. The following represents scores on a measure of the cognitive level: 74, 50, 66, 65, 67, 54, 54, 78, 52, 72, 53, 62, 81, 68, 54, 62, 74, 85, 42, 33, 38, 37, 86, 32, 85, 31
    d. The following represents scores on a measure of assertiveness: 75, 37, 13, 60, 49, 76, 88, 12, 70, 89, 61, 53, 93, 45, 52, 99, 17, 59, 40, 63, 94, 59, 86, 75, 39, 80, 45, 65, 38

12. From the grouped frequency tables that you constructed in Exercise 11(a) through (d), construct a table that contains the cumulative frequency distribution and relative frequency distribution. Construct a cumulative frequency curve.

13. On a survey, individuals in the United States were asked to indicate the race they considered themselves to be. The data were coded as follows:

1 = Aleut	2 = American Indian
3 = Asian American	4 = Black/African American
5 = Eskimo	6 = Hispanic/Latino
7 = Pacific Islander	8 = White/European American
9 = Other race	

Construct a frequency table for the data below. Then construct an appropriately labeled bar graph and pie chart.
8, 6, 7, 4, 4, 3, 3, 2, 8, 3, 5, 8, 6, 6, 8, 9, 1, 8, 3, 4, 6, 7, 8, 8, 4, 2, 3, 6, 8, 7, 8, 6, 9, 5, 8, 8, 6, 6, 4, 6, 8, 4, 3, 8, 8, 9, 8, 6, 8, 8
What can you conclude from the graphs?

14. a. On a survey, individuals were asked to indicate the political party with which they were registered. The data were coded as follows:

1 = Republican Party	2 = Democratic Party
3 = Libertarian Party	4 = Independent Party
5 = No Party Affiliation	6 = Other

Construct a frequency table for the data below. Then construct an appropriately labeled bar graph and pie chart.
1, 2, 2, 2, 1, 1, 3, 2, 1, 2, 5, 4, 2, 3, 4, 6, 2, 1, 1, 3, 5, 2, 1, 1, 2, 4, 4, 2, 2, 2, 1, 2, 3, 6, 4, 6, 1, 2, 2, 1
What can you conclude from the graphs?

15. a. Use the SPSS program and SPSS data bank for the data set named math.sav to obtain a frequency table for the post-instruction score on the mathematics subtest of the Iowa Test of Educational Development (ited9).
    b. From the frequency table, construct a grouped cumulative frequency table. Construct a grouped cumulative frequency curve from this table.
    c. Using the SPSS program, construct a grouped frequency histogram.

16. a. Use the SPSS program and SPSS data bank for the data set named self.sav to obtain a frequency table for Socially-Oriented Achievement Motivation (soam) for Japanese participants only (coded 1 under the variable name of citizen), and then obtain a frequency table for Socially-Oriented Achievement Motivation (soam) for U.S. participants only (coded 2 under the variable name of citizen).
    b. From the frequency tables for each group, construct a grouped cumulative frequency table for each group. For each group, construct a grouped cumulative frequency curve from this table.

c. Using the SPSS program, construct a grouped frequency histogram for each group.

17. a. Use the SPSS program and SPSS data bank for the data set named career.sav to obtain a frequency table for level of mathematics anxiety (anxiety).
   b. From the frequency table in part (a), construct a grouped cumulative frequency table. Construct a grouped cumulative frequency curve from this table.
   c. Using the SPSS program, construct a grouped frequency histogram and pie chart for level of mathematics anxiety that matches the table in part (b).
   d. Use the SPSS program and SPSS data bank to construct a bar graph and a pie chart for career interest area (interest) from the data set career.sav.
   e. Use the SPSS program and SPSS data set career.sav to construct a box plot of level of mathematics anxiety

(anxiety) by the category of grade level (gradelvl) and a box plot of level of mathematics anxiety (anxiety) by the category of career interest area (interest).

18. a. Use the SPSS program and SPSS data bank for the data set named donate.sav to obtain a frequency table for empathic concern (empathy).
   b. From the frequency table, construct a grouped cumulative frequency table. Construct a grouped cumulative frequency curve from this table.
   c. Using the SPSS program, construct a grouped frequency histogram.
   d. Use the SPSS program and SPSS data bank to construct a box plot of empathic concern (empathy) by the answer to the question, "Did you actually donate blood?" (donate).

Chapter 3

Measures of Central Tendency or Location

CHAPTER OBJECTIVES

After studying this chapter, you should be able to:

❏ Explain why *researchers* use numerical indices to describe the central tendency of a set of data.

❏ List the desirable properties of an index of location or central tendency.

❏ Define the mode, median (both definitions) and mean.

❏ List the advantages and disadvantages of each measure.

❏ Given a research problem, identify which measure of central tendency or location is most appropriate to convey the findings.

❏ Determine the mode, median (both definitions) and mean for a set of raw data or a frequency distribution.

❏ Define and identify skewed and multimodal frequency distributions.

❏ State the factors that should influence a researcher's choice of an index of central tendency or location.

Think back to the data from our 68 test-anxious statistics students in the SAD Department at BTU. Can you think of any ways that would be useful to summarize the information about their performance on the final exam, or their status on other variables? How might you describe one of the frequency distributions or graphs derived from the raw data with a simple summary index?

Introduction

We saw in Chapter 2 that frequency distributions and graphs bring order to a set of data and organize it so we can gain an overall picture of the information. These methods work best when it is important to lose very little detail or to communicate a quick overall picture of the data. No detail is lost with a raw frequency distribution or a stem-and-leaf plot, only individual identification. Grouping the scores into intervals introduces some inaccuracies, but the pattern in the data may become clearer.

To further manipulate and perform analyses on the data in order to summarize them, we must go back to the original scores. Our objective is twofold. First, we wish to develop efficient ways to describe the data and communicate their properties; second, these descriptive tools should permit further analysis and comparison. In addition, we would like to lose as little detail in the information as possible in the process of summarizing and describing.

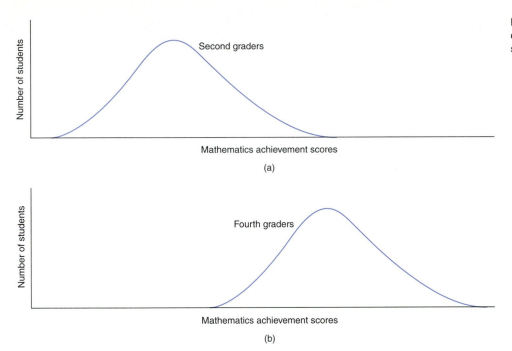

(a)

(b)

Two features of a set of data are essential in its description. The first is the location of the data on the scale of measurement. For example, how would the distribution of scores on a test of mathematics achievement of a group of second graders differ from that of fourth graders? The shapes of the two distributions may be the same, but the scores of one group cluster in one part of the scale, the scores of the other in a different part of the scale. Figure 3-1 shows graphs of two distributions that are the same in every respect except their location on the scale. We will find it convenient to represent location on the scale with a number showing where the center of the distribution falls. This numerical index of location is called a **measure of central tendency.**

A measure of central tendency allows a researcher to describe where a set of data is located on a scale or, often more importantly, to compare two or more sets of data in terms of their location. Suppose, for example, you wanted to know whether students taught to read using phonics become, on average, better spellers than students taught to read by another method. How would you present your results? You would get the clearest answer if you compared the spelling scores for the two groups. Measures of central tendency would give the best comparison.

A measure of central tendency is a number showing where a distribution of scores falls on a scale of measurement.

▶ *Now You Do It*

What are some other variables on which you might expect groups to differ in location? How about the number of trials it takes for a monkey to learn a discrimination task? We might expect monkeys who receive a reward for every correct solution to learn in fewer trials than monkeys who are rewarded on only every fifth trial.

Have you read recently, either in your coursework or in the newspaper, about groups that differ in their status on some trait; for example, the differences between ethnic groups in academic test scores or differences between men and women in aggressiveness? What measures of location were reported?

The second feature of the data that we must describe is the degree of spread, or variability, among observations. When we talk about "degree of spread," or "variation," in scores, we are really asking how many different scores occurred; also, did people show large individual differences, or did most people receive about the same score? In Figure 3-2, the two frequency distributions are centered at the same location on the scale but differ in their variability, or spread, around their center. As is apparent in the figure, children of a given age tend to show more variability, or spread, in their reading achievement than

Figure 3-2 Distributions with the same center but different variability (for reading and math achievement test scores for fourth graders).

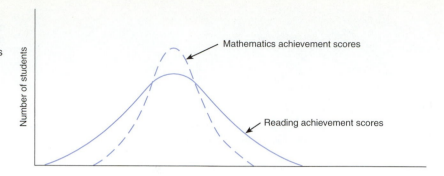

in their mathematics achievement (perhaps because reading is less dependent on specific instruction). Similarly, we would expect to see greater variability in the incomes of real estate salespeople than school teachers, probably because income is more dependent on luck and effort for the former than the latter. In this chapter, we will discuss indices of location or central tendency; in Chapter 4 we will present measures of the spread, or variability, among scores.

Desirable Properties of an Index of Central Tendency

There are three measures of the location, or central tendency, of a distribution that psychologists and educators have found particularly useful; they are the **mode,** the **median,** and the **mean**. Each of these indices serves certain purposes more or less well than the others. In addition, each can be evaluated in terms of four properties that are generally desirable in an index of location or central tendency; these are **uniqueness, stability, representativeness,** and an **ability to be used in further computations and analyses**.

*An index if **unique** if only one value can occur for a given set of data.*

An index is **unique** if, for a given set of data, one and only one numerical value is possible. For example, we would not want to have to say that the center of our distribution of math anxiety scores is at both point A and point B. It may seem obvious that there can be only one center in a set of observations, but this depends on how we define the center.

*A **stable** index does not change much from one sample to another.*

An index is **stable** if it does not change much when different groups of similar individuals are observed. For example, scores from different first-grade classes in a school district or classes of statistics students from different years should have about the same location on a scale of achievement unless there is some reason to expect a difference. Small changes in the data from group to group should not produce large changes in our measure of central tendency.

*An index of central tendency is **representative** if it is close to a large number of the observations in the distribution.*

An index of central tendency is **representative** if it is close to a large number of the observations in the distribution. In a sense, measures of central tendency would seem to be representative by definition; however, we shall see that they satisfy this requirement in different ways and to different degrees, depending partly on the properties of the index and partly on the shape of the frequency distribution.

Finally, complete analysis of a set of data, extracting all of the information that the numbers contain, often requires going beyond just a measure of central tendency. It is desirable that an index permit additional analyses to allow more of the information to be recovered from the data. We will see in future chapters how indices of central tendency allow us to answer a variety of questions.

Measures of Central Tendency for Raw Data

The Mode

*The **mode** is the most frequently occurring score in a distribution.*

The **mode** is the simplest measure of location, or central tendency, of a distribution. The mode is the numerical value of the score that occurs most frequently in a distribution; it is the score that the greatest number of people earned. From Table 3-1 (Table 2-3 reprinted here) it is easy to see that the mode of our set of statistics final exam scores is 37. This score occurred five times; this figure is called the **modal frequency**. The modal frequency tells us how many observations fell at the mode. Note also that the mode corresponds to

*The **modal frequency** is the number of scores falling at the mode.*

X	Tallies	f	X	Tallies	f
65	/	1	34	////	4
64		0	33	//	2
63		0	32	/	1
62		0	31	/	1
61	/	1	30	//	2
60		0	29	///	3
59	/	1	28	//	2
58		0	27		0
57		0	26		0
56	/	1	25	////	4
55	//	2	24		0
54	/	1	23	/	1
53		0	22	//	2
52		0	21	/	1
51		0	20	//	2
50	/	1	19	//	2
49	/	1	18	/	1
48	//	2	17		0
47	//	2	16	/	1
46	/	1	15		0
45	/	1	14	//	2
44	///	3	13	/	1
43	//	2	12		0
42	/	1	11		0
41	/	1	10		0
40	/	1	9		0
39	//	2	8	/	1
38	//	2	7	/	1
37	/////	5	6		0
36		0	5		0
35	/	1	4	/	1

Table 3-1   Raw Frequency Distribution of Statistics Final Exam Scores for 68 Students From Table 2-3

the highest point in the graph of a frequency distribution because the height of the graph shows frequency.

The mode is used most often when the data are on a nominal scale, such as the lab section and college major variables from the SAD Department study. It is generally the only measure of central tendency that makes sense when the data do not possess quantitative information. You might emphasize the mode when you must select one and only one value of the variable. Suppose, for example, you ran a college bookstore but could stock only one style of college logo sweatshirt. You would satisfy the greatest number of students by stocking the modal style (modal in the sense that it would be the one chosen by the greatest number of students). If you, as a teacher, have time to play only one word game with your first-grade class, you would please the greatest number of children by selecting the most popular game, the one the largest number of children would choose. Because such situations seldom occur in the behavioral sciences, other indices are more frequently used.

Looking at Table 3-1 shows that one advantage of making a frequency distribution is that the mode immediately becomes obvious. Another clear advantage is that it is easy to determine the mode as a measure of central tendency; no computation is involved. Thus, the mode has two characteristics that make it desirable as an index of central tendency: (1) It satisfies our criterion that it is representative in the sense that *more observations with the modal value occurred than with any other,* and (2) it is easy to obtain.

Unfortunately, the mode does not fare very well on our other criteria. Because the mode is based only on a count of the number of scores at each value, only the observations that are at the mode actually determine its value. In this sense, it is not a very representative index; for the final exam scores in the BTU test anxiety study only 5 of 68 observations were used to determine its value.

Figure 3-3 Some distributions have more than one mode.

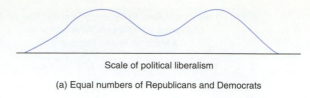

Scale of political liberalism

(a) Equal numbers of Republicans and Democrats

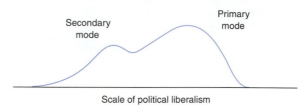

Secondary mode

Primary mode

Scale of political liberalism

(b) Democrats outnumber Republicans

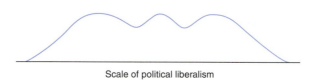

Scale of political liberalism

(c) Republicans, Democrats, and middle-of-the road Independents

A **bimodal distribution** is one in which two scores both satisfy the definition of a mode.

A second problem with the mode is that it may not yield a unique value. Suppose that four people instead of five had received a score of 37 in our example, so that there are three different score values, 25, 34, and 37, each with a frequency of four. Since there is no way to choose between them, there is no single mode. This problem occurs fairly frequently in raw, ungrouped frequency distributions because often there are only a few cases at each score value. When two scores involve the same number of cases, the distribution is called **bimodal**. If there are two bumps in the distribution and one bump is higher than the other, we speak of the higher one (the one with the greater frequency) as the **primary mode** and the one with the lesser frequency as the **secondary mode**. The scores in Table 3-1 show a primary mode and two secondary modes.

When several scores occur with the same frequency, or the distribution has several high points, the distribution is called **multimodal**. Figure 3-3 shows three possible multimodal frequency distributions. The first graph might reflect the distributions of scores on a scale of political liberalism when equal numbers of Democrats and Republicans are represented in the group. The second graph, with a primary mode and a secondary mode, shows what percent of the group were Democrats, and the third could occur if we interviewed equal numbers of liberals, moderates, and conservatives. A frequency distribution that has only

A **unimodal** frequency distribution has only one mode.

one mode is said to be **unimodal.** The graphs in Figures 3-1 and 3-2 are of unimodal distributions. Most distributions of psychological and educational data are unimodal.

▶ *Now You Do it*

Multimodal distributions occur most frequently when the distribution includes individuals from two or more distinct populations. In fact, the existence of more than one mode is usually a clear indication that this is the case and if you encounter such a situation it should prompt you to identify what populations those might be. Primary and secondary modes most often occur when there are more individuals from one population than the other included in the sample. What would you expect the distribution of scores on a measure of willingness to take physical risks to look like if the sample included equal numbers of skiers and golfers? What would the distribution of math anxiety scores look like if it included some students who were retaking a statistics course after failing, some students taking it for the first time, and some students who were mathematics majors?

In the first case, we would probably expect to find a bimodal distribution because skiing is a sport that, by its nature, involves physical risk; in fact, the thrill of the risk is often given by skiers as a reason they like the sport. Golf, on the other hand, involves little physical risk. In the second case, we would expect the three groups to differ quite markedly in their levels of anxiety, and therefore to probably produce a distribution with three modes.

Because it is based only on frequency, the mode presents two additional disadvantages (see Table 3-2). First, it tends to be unstable from one group to another because relatively small shifts in frequency can cause large changes in the mode. Second, the mode is not derived mathematically from the scores themselves; it represents only a count, so it cannot be used in any further analyses. For these reasons, the mode is seldom satisfactory as the only index of central tendency. It is sometimes reported in conjunction with one of the other indices, but except for cases where the data are clearly nominal, it is seldom reported alone.

The Median

A second way of defining the center of a distribution is to use the point that cuts the distribution exactly in half; this value is called the **median**. The median is the point below which exactly 50 percent of the distribution falls.

*The **median** is the point which has 50 percent of the distribution below it.*

In Chapter 1 we saw that some scales are continuous and others are discrete. On a continuous scale there are (theoretically) infinitely many possible values. Adopting this position about continuous scales has some implications for how we view the median. On the other hand, if we consider the units of the measurement scale to represent the finest possible gradations of the trait so that all people who receive the same score are considered exactly equal (that is, we treat the measurement scale as discrete), this has different implications for the median.

For discrete scales, the median can be obtained by counting off cases from the frequency distribution. However, for continuous scales, it is necessary to compute a value for the median. In either case, the first thing to do is to determine how many observations there are in 50 percent of the distribution.

The Median as the Middle Score. In the discrete scale case, a person's score is assumed to provide a precise position and all people who receive the same score are exactly alike; there is no way, even theoretically, to differentiate between them. When this assumption is made, as it usually is for computer programs that provide descriptive statistics, the median is defined as whatever score the middle person received, regardless of how many people got the same score. For a discrete scale, the **median** is the score earned by the middle person (odd number of scores) or is the average of the scores for the middle two people (even number of scores).

*For a discrete scale, the **median** is the score earned by the middle person (odd number of scores) or it is the average of the scores for the middle two people (even number of scores).*

Suppose, for example, that we showed 11 male statistics students a list of ten statements related to mathematics anxiety, statements such as "Every time I see a number, I freeze up and cannot think," and ask them to mark the statements that are true of them. The results might be like the scores in Table 3-3. If we think of ordering these 11 people by their scores, the order would be

(3), (6,9), (2,4,8,10), (1,7,11), (5)

where the parentheses indicate individuals who received the same score and are therefore equal on the scale (but not on the underlying trait, as we discussed in Chapter 2, Figure 2-1). Starting at the bottom, the sixth person in order (student number 8) is the middle individual. There are 5 people below him and 5 above. Defining the median as the score

Table 3-2	Properties of the Mode	
Definition	**Advantage**	**Disadvantages**
The score that occurs most frequently	1. Easy to obtain	1. Not necessarily representative 2. May not yield a unique value 3. Unstable from one group to another 4. Cannot be used in further analyses

			Frequency Distribution	
Individual	Score	X	f	cf
1	6	7	1	11
2	5	6	3	10
3	3	5	4	7
4	5	4	2	3
5	7	3	1	1
6	4			
7	6			
8	5			
9	4			
10	5			
11	6			

Table 3-3 Scores of 11 Male Statistics Students on a 10-Item Math Anxiety Scale

Remember that X stands for a variable, in this case, the number of anxious statements checked, f stands for frequency, and cf stands for cumulative frequency.

earned by the middle person, we conclude that individual number 8 is the middle person and his score of 5 is the median.

When there are an even number of cases, the median is defined as the average of the two middle scores. For example, to find the median of the distribution of final exam scores in Table 3-1, where there are 68 scores, we would have 34 people below the median and 34 above. The 34th and 35th people both have scores of 34, so this would be the median, but if one had scored 34 and the other 35, the median would have been 34.5, the average of the two scores.

The Median as a Point on a Continuous Score Scale. If the underlying trait is considered to be continuous, the score of any individual is only an approximate indication of his or her status on the trait. The **median** is then defined as the value on the score scale that cuts the distribution exactly in half. If we consider anxiety to be a continuous trait, we might represent the actual amount of anxiety of each person as a position along a line. The true state of affairs for the data from Table 3-3 might be as shown in Figure 3-4, where each individual's identification number has been placed above an X at his position on the line. Individuals 10, 2, 8, and 4 all received a score of 5 because their locations are all between the real limits of the score interval 5, but they do not share exactly equal amounts of the trait. If we could determine their exact locations on the trait they would be ordered as shown in Figure 3-4. The same is true for students in the other intervals.

To find the median of this set of scores, we must first find 50 percent of N (.50N). Because N is 11, the value we need is 5.5; that is, 5.5 of the 11 people fall below the median and 5.5 fall above it. The median cuts the middle person in half. If N had been an even number, the median would fall halfway between two people.

On a continuous scale the **median** *is defined as the value on the score scale that cuts the distribution exactly in half.*

Figure 3-4 Distribution of scores for 11 male students on a measure of mathematics anxiety. Scores are distributed evenly in each interval.

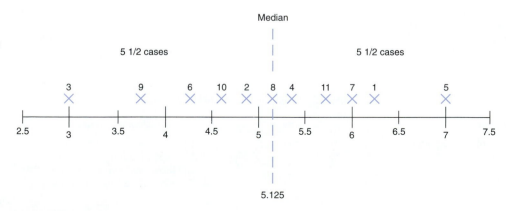

To find the median we must find the point that corresponds to the 5.5th person in the distribution. To do this, we assume that the individual scores are distributed evenly throughout each interval. Thus, the two people in score interval 4 in Figure 3-4 each occupy one-half of the interval; the four people in interval 5 each occupy one-fourth of the interval; and the three people in score interval 6 each occupy one-third of that interval.

Now we can find the median; we know that it is the point that splits the sixth person in half. The sixth person is individual number 8, the third person in the real interval 4.5–5.5. We can see from Figure 3-4 that this person occupies the part of the score scale from 5.0 to 5.25. Since we must split the portion of the score scale that person 8 occupies in half, the median must be 5.125.

Computing Procedures. Of course we would never use this graphic procedure to find a median in a real situation. However, the principle underlying the computation of the median follows these steps exactly: First, we must find $.50N$, then locate this point on the score scale as a distance above the *lower real limit* of the interval containing the point we seek. The formula that is usually used to find the median from a frequency distribution is

$$Mdn = LL_j + \frac{.5(N) - cf_{j-1}}{f_j} \qquad (3.1)$$

This formula looks much more complicated than it actually is. If we dissect it and examine its parts in light of what we already know, it will not be so formidable. First, the subscripts j and $j - 1$ are only place indicators in the frequency distribution; they are used to specify particular scores. N is the total number of individuals, the symbol LL means the lower real limit of a score, and f and cf refer to the frequency and cumulative frequency columns of the distribution in Table 3-3. The formula is a sentence that tells us how to combine these values in order to compute the median.

The first thing to do is compute $.50N$. This gives us the number of people who fall below the median. Next, we must find the interval that contains the $.50(N)$th individual; we do this by looking in the cf column. The first score where the value in the cf column equals or exceeds $.50(N)$ is the score that contains the median. Call this score j (score j is the score 5, with real limits 4.5 and 5.5, in our example). The median math test anxiety score for these 11 students lies somewhere in this interval. Now we can find all the other values.

LL_j is the lower real limit of score j ($LL_j = 4.5$).
cf_{j-1} is the value from the cf column for the score (row of the table) immediately below (1 less than) the score j ($cf_{j-1} = 3$).
f_j is the value from the f column for score j ($f_j = 4$).

By substituting the values from Table 3-3, formula 3.1 now reads

$$Mdn = 4.5 + \frac{.50(11) - 3}{4}$$

$$Mdn = 4.5 + [(5.5 - 3)/4]$$
$$= 4.5 + [2.5/4]$$
$$= 4.5 + .625 = 5.125.$$

There is a direct relationship between what we did in Figure 3-4 and the computations of formula 3.1. In each case, we are, in effect, counting off people until we get to the one we need. In formula 3.1, we locate LL_j as our starting point and as an anchor to the score scale itself. The term $.50(N)$ tells us the total number of individuals we need; cf_{j-1} tells us how many we already have by the time we get to LL_j; and the difference $[.50(N) - cf_{j-1}]$ tells us how many observations we need from score j itself. Dividing this value by f_j (the number of students in the interval) tells us what *proportion of the interval from 4.5 to 5.5* we need (remember that we assumed the people are spread evenly throughout the interval).

As another example, let us take the data in Table 3-4; they are the scores obtained by all 340 students who are taking the introductory statistics course this term in the SAD Department of BTU on a 10-item quiz. (Our 68 students are a sample from this population.) To find the median test score, we first compute $.50N = 170$. The first interval where cf is greater than or equal to 170 is the score interval 7, so this is interval j. We may then

Table 3-4 Frequency Distribution of Scores on a Statistics Quiz			
Quiz Score	f	cf	
10	8	340	
9	26	332	
8	84	306	
7	132	222	score *j*
6	67	90	
5	15	23	
4	6	8	
3	2	2	

compute the median.

$$Mdn = 6.5 + \frac{170 - 90}{132}$$

$$= 6.5 + 80/132$$
$$= 6.5 + 0.606 = 7.106.$$

What would the median be for these scores if we used our first definition of the median? The middle person is one of the people who scored 7, so 7 would be the median. As is often the case, the two definitions give us a slightly different answer. The difference is seldom large enough to be of consequence when we are dealing with raw data, but, as we shall see shortly, when the data have been grouped, particularly when the intervals are large, the two procedures can yield quite different values. If we are comparing groups on the basis of their medians, we might conclude that two groups are quite similar using one definition and rather different using the other.

▶ *Now You Do It*

Let's use the final exam scores listed in Table 3-1 to test your understanding of how to compute a median. Since there are 68 students in the group, we must find the point below which 34 of them fall. Which score is the score *j*? How many students earned scores below this score, and how many earned this score? Since cf_{34} is 35, 34 is score *j*. Thirty-one students got scores of 33 or below, and four students earned scores of 34. Applying formula 3.1, you should get

$$Mdn = 33.5 + \frac{34 - 31}{4} = 33.5 + 3/4 = 34.25$$

The median has some definite advantages over the mode as an index of central tendency (see Table 3-5).

1. It tends to be more stable from group to group than the mode because many more scores must change for the median to shift very much.

Table 3-5 Properties of the Median		
Definition	Advantage	Disadvantages
1. The middle score OR 2. The point on the score scale below which half of the distribution falls	More stable from group to group than the mode	1. Not necessarily representative 2. Cannot be used in further calculations because it is not computed from the scores themselves but from their order

2. The median must be a single, unique value for a given distribution because there can be only one point that cuts the distribution exactly in half. However, the value you get will depend on which definition of the median you use. As we shall see shortly, when the median is computed from grouped data, the choices made in selecting intervals will also affect the median.
3. Its computation is not difficult.
4. It is representative in the sense that it splits the distribution in half.

However, the median does have some drawbacks.

1. It may be in a score interval that actually contains very few scores, or even none; this is most likely to happen in the case of a seriously bimodal distribution.
2. The median is not actually computed from the score values themselves, but from the order of the individuals. It is either midway between the middle two ordered

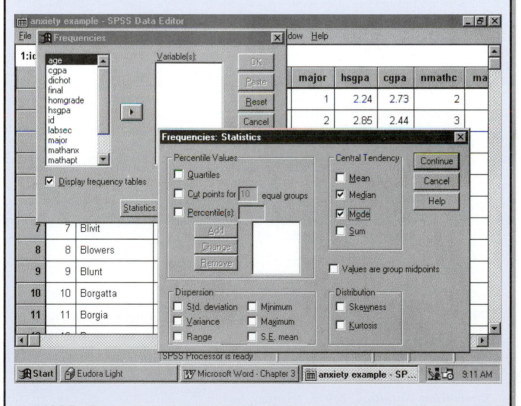

observations (if there are an even number of cases) or at the middle of the middle observation if N is odd. The only thing that ties the median to the score scale is the lower limit of one score interval.

3. No scores are used in its computation, so it cannot be used in further analyses except in a few restricted instances.

The Mean

When people are asked to give a value that is most typical of a group of observations (for example, what single value would you use to represent your academic performance last term?), the most common answer is the average. The term used in statistics to refer to this concept is the **mean** or, more properly, the **arithmetic mean,** because there are other types of means, such as geometric and harmonic means, that are used for special purposes. We shall restrict our concern to the arithmetic mean and use the single term mean to refer to it.

*The **mean** is the sum of a set of scores divided by the number of scores that have been summed.*

Most people know how to find an average: Simply add up the values being considered, and divide that sum by the number of values added. For example, if your institution reports grades on a 12-point scale with 0 representing F and 12 representing A, and your grades last term were 12 (A), 10 (B+), 7 (C+), and 9 (B) respectively, then to find your mean grade, we would find the sum

$$\text{Sum} = 12 + 10 + 7 + 9 = 38$$

and divide it by the number of things that have been added. (This assumes all courses carried the same number of credits.)

$$\text{Mean} = 38/4 = 9.5.$$

This basic series of steps is always followed in finding a mean; however, in some cases it will be convenient to express the steps in a slightly different way.

Summation Notation. Before venturing further, we should become familiar with what statisticians call summation notation. Data analysis frequently involves adding up a long list of numbers, and a compact way of expressing this process is very useful. There is a special symbol, called the **summation operator,** which looks like an M on its side; this symbol, Σ, is the capital Greek letter sigma. It has a particular meaning, just as the symbol + has a particular meaning. Σ means find the sum of the values that follow it.

*The symbol Σ is the **summation operator** and means "find the sum of what follows."*

The symbol Σ is used in conjunction with the name of some variable; ΣX instructs you to find the sum of the values of the variable X. If X represents your grades in four courses last term, then the expression ΣX means find $12 + 10 + 7 + 9$. If X is the score on a final exam in statistics and the values of X are those given in Table 2-2, then ΣX instructs you to find the sum of those 68 scores.

Subscripts. It is often necessary to specify the particular observations that are to be summed. This is accomplished by using **subscripts,** values that appear half a line down from the variable name. (In the formula for the median, j and $j - 1$ appeared as subscripts.) In Table 2-2, each individual's score was paired with a number that uniquely identified that person. X stands for the test score, and the individual number paired with X specifies the particular test score to which we are referring. The identification number is placed below the letter as a subscript; thus, X_{10} refers to the score on X of individual 10 (Steven Borgatta), and the value of X_{10} from Table 2-2 is 34. Likewise, X_{15} refers to the score of 35 obtained by Jim Cattell, X_{17} has the value 8, and X_{40} is 7. Subscripts are very useful because they allow us to keep track of and refer to long lists of data in a simple shorthand form. (Note that subscripts are also essential in working with computers because they refer to memory locations in the computer. In our work, however, they refer to locations in the data table.)

Subscripts are particularly helpful when used in a general form; that is, when a letter is used as a variable subscript. The symbol X_i refers to any of several values of X—those values that could be substituted for i. When subscripts are used in this way, it is often necessary to specify the permissible values of i, and this is done in either of two ways. The most common way is to specify the limits of the subscript on the summation sign.

$$\sum_{i=8}^{15} X_i$$

means to sum those values of the variable X that have subscripts from 8 to 15. In terms of our statistics final exam scores from Table 2-2,

$$\sum_{i=1}^{6} X_i = 37 + 47 + 19 + 37 + 33 + 48 = 221$$

and

$$\sum_{i=8}^{15} X_i = 44 + 44 + 34 + 38 + 14 + 4 + 33 + 35 = 246$$

An alternative way of specifying subscript limits is to use parentheses following the variable. For example,

$$\sum X_i \qquad (i = 1, 10).$$

In general, the first time an equation is presented and when there might be ambiguity about limits, we give the limits of each summation. When it is clear what subscript values are to be included, only the summation operator, $\sum$, and the subscript on the variable will be given. Thus, $\sum X_i$ means that summation is of the values of X for all i observations. That is, the limits $(i = 1, N)$, where N is the total number of cases, are assumed.

For some topics we discuss in later chapters it will be necessary to keep track of both the group we are referring to and the individual within the group. We will adopt the general convention that the subscript i refers to the score of an individual and the subscript j refers to a group or interval. Thus, in our discussion of the median we referred to the score interval that contained the median as j, and in our formula for the mean we will use the subscript i because the reference is to the scores of individual subjects.

Computing Procedures for the Mean. Using summation notation (the summation operator and subscripts) allows us to write a very compact expression for finding a mean. We use the symbol $\overline{X}$ (X-bar) to stand for the mean of the values of X. Using summation notation, we may then say that

$$\overline{X} = \frac{\sum_{i=1}^{N} X_i}{N} \qquad (3.2)$$

We will use this general formula many times in the remainder of this book, each time changing slightly what we mean by X. Applying this short mathematical instruction to the data in Table 2-2, we find that the sum of those 68 numbers is 2328 and therefore the

Table 3-6 Properties of the Mean		
Definition	**Advantages**	**Disadvantage**
The sum of scores divided by the number of scores	1. A single unique value 2. Representative in that every score enters into its computation 3. Most stable from group to group 4. May be used in further computations	Affected by a few extreme scores because every score enters into its computation

mean is (2328)/68 or 34.24. In words, formula 3.2 says that the mean of the X's equals the sum of the X's, going from the first to the Nth value of the subscript i, divided by the number of X's that have been summed.

The mean has certain definite advantages over the mode and the median as a measure of central tendency (see Table 3-6).

1. It shares with the median the feature of being a unique value for any given set of data.
2. Unlike the median, every single score enters directly into the computation of the mean. In this sense, it is more representative of all the data than the median is.
3. The mean is generally less subject to fluctuation from one group to another; it is more stable.
4. Perhaps most important, the mean is a mathematical function of all the scores and, therefore, may be used in further computations. It is permissible and useful to add means, subtract means, and treat them just like the original scores. Most of the techniques of data analysis that will occupy our attention in later chapters are based on computations involving the mean.

The mean does have one significant disadvantage compared to the median. Because every score enters into the computation of the mean, the mean is more affected by extreme or atypical scores. For example, if the top two scores among our 68 students had been 100's instead of 61 and 65, the median would not change, but the mean would be 35.32, a change of over a point as a result of changing only two scores.

Measures of Central Tendency for Frequency Distributions

We saw in Chapter 2 that it is often useful to arrange our data in frequency distributions or grouped frequency distributions (that is, to list the scores or score intervals and the number of people scoring in each) so that the properties of the data can be more easily seen. Grouping the data introduces some loss of detail in the frequency distribution because all of the scores in an interval are represented by the midpoint of the interval, but we may be willing to pay this price in exchange for the added understanding that grouping provides.

Various ways of finding indices of central tendency from frequency distributions have been developed. Originally, these procedures evolved to make computations involving large sets of data easier and less time-consuming. Modern computers have eliminated the computational drudgery, but learning how to handle grouped data will help you in three ways. First, data may come to you already grouped into intervals. Suppose you see some figures on educational expenditures and achievement or on income and crime rates in an article you want to use in a term paper and the analysis doesn't seem to be quite right, or only the frequency distribution is given and you want to use the median or mean. Or you are attending a PTA meeting where some parents are arguing against full inclusion of students with disabilities at your child's school, and the parents present the mean for grouped data to support their position. You might want to make your own analyses on the spot, perhaps computing a median, to argue for a conclusion different from that of the parent group. In either of these situations you may not have immediate access to a computer, so simplified procedures would be helpful. Finally, if you see the procedures

Table 3-7 Grouped Frequency Distributions of Statistics Final
Exam Scores for an Interval of 5 from Table 2-4

Score Interval	Midpoint X	Frequency f	Cumulative Frequency cf
63–67	65	1	68
58–62	60	2	67
53–57	55	4	65
48–52	50	4	61
43–47	45	9	57
38–42	40	7	48
33–37	35	12	41
28–32	30	9	29
23–27	25	5	20
18–22	20	8	15
13–17	15	4	7
8–12	10	1	3
3–7	5	2	2

for computing the mean applied in a similar but not identical context, it may help you develop a more thorough understanding of those procedures. The procedures for finding a mode, median, or mean remain the same for data in a frequency distribution, whether grouped or not; it is only the details of the application that change slightly. We use the grouped frequency distribution of final exam scores for an interval of 5 from Table 2-4, repeated here as Table 3-7.

Mode

The **mode** for a grouped frequency distribution is the midpoint of the interval with the highest frequency. The **modal class,** or **modal interval,** is the interval that has the highest frequency. For example, in Table 3-7 where we grouped our statistics final exam scores in intervals of 5, the mode is 35, the midpoint of the interval with a frequency of 12, and the modal class is 33–37. Note that it makes a difference how the intervals have been selected: If an interval width of 3 is used for these data, the mode is 33, with a modal class of 32–34. Choosing a different place to start our intervals can have a substantial effect as well. Also, as was the case for raw data, a small change in frequency can move the mode. Changing two people a few points could change the mode by as much as 10 points.

*The midpoint of the interval with the highest frequency is the **mode** of a grouped frequency distribution.*

*The **modal class** or **modal interval** is the interval that has the highest frequency.*

Median

Computing the median from a grouped frequency distribution is almost identical to computing it from the raw frequency distribution when the median is defined as the middle point on the score scale. The only difference is that we must adjust for the width of the interval. The formula is

$$Mdn = LL_j + \left[\frac{.5(N) - cf_{j-1}}{f_j} \right](J) \qquad (3.3)$$

where each symbol has the same meaning that it had in equation 3.1, and the symbol J here stands for the width of an interval. Using the grouped frequency distribution with an interval width of 5 in Table 3-7, we first calculate $.50N = 34$. The first interval where the cf equals or exceeds 34 is the interval 33–37. The **lower real limit** of this interval is 32.5; cf_{j-1} is 29; $f_j = 12$ and the interval width (J) is 5. Therefore, the expression given in formula 3.3 for finding the median may be rewritten for our statistics final exam scores as

$$Mdn = 32.5 + \left[\frac{34 - 29}{12} \right](5)$$

$$= 32.5 + [5/12](5) = 32.5 + (.42)5$$
$$= 32.5 + 2.1 = 34.6$$

The parts of this formula operate in exactly the same way that they did for raw data. The lower real limit LL_j locates us on the score scale, and *the expression in brackets tells* us what proportion of the interval we need to make up the five cases required for half of the group. Because we have assumed that the 12 observations in the interval are spread evenly throughout the interval, taking that proportion of the interval is equivalent to taking the 5 people. Notice that if we had not multiplied by *J*, our computations would have resulted in a value of 32.92, which is quite in error. We multiply by the interval width to account for the fact that the interval is 5 score units wide. Notice also that formula 3.1 is a special case of formula 3.3, where $J = 1$. Thus, formula 3.3 is the proper general formula for the median and may be used either for raw or grouped frequency distributions.

As was the case with the mode, the median computed from a grouped frequency distribution depends to some degree on the choices that were made about the intervals to be used. If we were to compute the median for an interval width of 3 we would get a slightly different answer. In that case, using the distribution in Table 2-4, our computations would go as follows:

$$Mdn = 31.5 + \left[\frac{34 - 28}{7} \right] (3)$$
$$= 31.5 + [6/7](3) = 31.5 + 2.57$$
$$= 34.07$$

We noted earlier that there are two ways to define the median, as the score earned by the middle person and as the point on the score scale below which 50 percent of the distribution falls. In many computer programs the median is defined as the score obtained by the middle person in the distribution. When this definition is used for raw data, the difference between the two ways of defining and determining the median will never be more than $\frac{1}{2}$ score point and is rarely of importance. However, for grouped data the difference can be substantial because this definition leads to using the midpoint of the interval containing the middle person as the median. Under such circumstances the median reported could be as much as half the interval width in error (relative to the raw data value), and the value reported could divide the group into quite unequal halves. As we suggested earlier, this might cause a researcher to draw quite different conclusions about the similarities or differences between groups and might influence our conclusions from a study. It is important to examine the output from the computer to determine which definition is being used. SPSS for Windows, for example, allows the user to indicate that data values are interval midpoints. Selecting this option in the program invokes use of formula 3.3, which makes it possible to determine the median correctly for either definition.

▶ *Now You Do It*

In Chapter 2 we asked you to prepare a grouped frequency distribution for the high school GPA scores for our 68 statistics students. In the example, we suggested using an interval width of .25. Go back and take a look at that frequency distribution (p. 35). How should we go about finding the mode and median for these scores?

The mode is easy. The interval with the highest frequency (12) is the one from 2.26 to 2.50, so the modal class or modal interval is the interval 2.26–2.50. The mode is the midpoint of this interval, or 2.38.

The median is a little trickier. What are the steps you must take? How many students' scores fall below the median? What is the interval width—what value shall we use for *J*? With these questions in mind, we proceed just as before, but carefully.

With 68 students, 34 of them fall below the median. The interval containing the median is therefore the interval 2.26–2.50 (it is the first one where the cumulative frequency exceeds 34), but what is its lower real limit? Since these data are reported in units of .01, the real limits of this interval are 2.255 and 2.505, one-half unit beyond the score limits, and the interval width, *J*, is .25. Therefore, the correct values to put into formula 3.3 are

$$Mdn = 2.255 + \left[\frac{34 - 27}{12} \right] (.25)$$
$$= 2.255 + (7/12)(.25)$$
$$= 2.255 + (.58)(.25)$$
$$= 2.255 + .145 = 2.40$$

Notice that if we had forgotten to multiply by the interval width, we would have gotten a value of 2.84, which would place us in the interval 2.75–3.00, more than a whole interval in error.

Mean

You may have noticed that we did not mention in our earlier discussion of the mean the procedures for computing the mean from a frequency distribution. The reason is twofold. First, the computations summarized in formula 3.2 are widely familiar, so most people already know how to use this formula even though they may never have seen it. We did not need unnecessary complexity at that point; dealing with summation notation and subscripts was enough. Second, computation of the mean from a frequency distribution uses exactly the same procedure and formula for either a raw or a grouped frequency distribution. Therefore, we delayed introducing the procedures for a raw score frequency distribution until it was time to handle both.

Let us look once again at the data from Table 3-7, the grouped frequency distribution of final exam scores with an interval width of 5. There are two observations in the interval 3–7 which are represented by the midpoint value, 5; there is one observation in the interval 8–12, which is represented by the midpoint value, 10; and so on for each interval. The procedure for finding a mean is to add together each of the values of X. From the *grouped frequency distribution,* we have observed a value of $X = 5$ for the two people in the lowest interval, a value of $X = 10$ for the one person in the second interval, a value of $X = 15$ four times, $X = 20$ eight times, and so on. One way to find the mean would be to find

$$5 + 5 + 10 + 15 + 15 + 15 + 15 + 20 + \cdots + 60 + 60 + 65$$

However,

$$5 + 5 = 2(5) = 10$$

$$15 + 15 + 15 + 15 = 4(15) = 60$$

the sum of eight scores of 20 equals $8(20) = 160$ and so on. In fact, for each interval, the sum of the observations in the interval is equal to the midpoint of the interval (the numerical value of X_j) multiplied by the number of observations in the interval (the frequency f_j). Carrying out this operation and placing the products in a column labeled $f_j X_j$ (f_j times X_j) gives us the results shown in Table 3-8. (Note that we use the subscript j because our summation is over intervals.)

Summing the values in the f column across the 13 intervals gives us

$$\sum_{j=1}^{13} f_j = N = 68,$$

Table 3-8 Computing the Sum of Scores for Grouped Data			
Interval	X_j	f_j	$f_j X_j$
63–67	65	1	65
58–62	60	2	120
53–57	55	4	220
48–52	50	4	200
43–47	45	9	405
38–42	40	7	280
33–37	35	12	420
28–32	30	9	270
23–27	25	5	125
18–22	20	8	160
13–17	15	4	60
8–12	10	1	10
3–7	5	2	10
Sums		68	2345

the number of scores that have been summed. That is, the sum of the interval frequencies is equal to the total number of cases. Summing the values in the f_jX_j column yields

$$\sum_{j=1}^{13} f_jX_j = \sum_{i=1}^{68} X_i$$

where the X_j are the 13 interval midpoints and the X_i are the scores of the 68 students *as represented by the interval midpoints*. That is,

$$\sum_{i=1}^{68} X_i = 5 + 5 + 10 + 15 + 15 + 15 + 15 + 20 + \cdots + 65$$

and

$$\sum_{j=1}^{13} f_jX_j = 2(5) + 1(10) + 4(15) + \cdots + 2(60) + 1(65)$$

Either way of performing the operation yields the same sum of scores, 2345, and the same mean of $2345/68 = 34.49$. Notice that while this value is different from the value of $2328/68 = 34.24$ that we obtained earlier from the raw data, the difference is small and is due to grouping and using interval midpoints to replace exact scores on X, not to multiplication. If we had computed an f_jX_j for each score value listed in Table 3-1 and summed across all score values, we would have obtained the same 2328 that we got using $\sum X_i$.

The general equation for computing a mean from any frequency distribution is

$$\overline{X} = \frac{\displaystyle\sum_{j=1}^{J} f_jX_j}{\displaystyle\sum_{j=1}^{J} f_j} \tag{3.4}$$

Here, J is used to indicate the number of intervals, so the summation takes place across all score intervals. Those scores with zero frequency are multiplied by zero and thus have no effect on the total. Formulas 3.2 and 3.4 are equivalent. Just as with the median, the formula for raw data is a special case of the formula for grouped data where the interval width is 1.

As we saw with the median, the choice of interval width and where to start the intervals can make a difference in the value we obtain for the mean computed from grouped data. If we had chosen to use an interval width of 3 with our statistics final exam scores from Table 2-4 we would have found a sum of

$$\sum_{j=1}^{22} f_jX_j = 1(3) + 1(6) + 1(9) + \cdots + 2(60) + 0(63) + 1(66)$$
$$= 2319$$

and a mean of $2319/68 = 34.10$. In general, using narrower intervals will produce a result closer to that obtained from the raw data.

The formulas for finding the mode, median, and mean for raw data and for grouped data are presented side-by-side in Table 3-9. You should study these formulas and be sure you can see the similarities and differences.

Table 3-9 Formulas for Raw Scores and Grouped Data		
Statistic	**Raw Score Formula**	**Frequency Distribution Formula**
Mode	Score with the highest frequency	Midpoint of interval with highest frequency
Median	Score of middle person or $LL_j + \left[\dfrac{.5(N - cf_{j-1})}{f_j}\right]$	$LL_j + \left[\dfrac{.5(N - cf_{j-1})}{f_j}\right](J)$
Mean	$\overline{X} = \dfrac{\displaystyle\sum_{i=1}^{N} X_i}{N}$	$\overline{X} = \dfrac{\displaystyle\sum_{j=1}^{J} f_jX_j}{\displaystyle\sum_{j=1}^{J} f_j}$

HOW THE COMPUTER DOES IT

Most pocket calculators have a special function that allows you to compute the mean for data in a frequency distribution. The process usually involves entering the score (either the raw score or interval midpoint), storing this value, and then entering the frequency. The calculator automatically accumulates $\sum f_j X_j$ and $\sum f_j$ in its memory. When you have entered each value and its frequency you either retrieve the two sums and perform the division, or the calculator may have a key (often labeled m or μ [Greek letter mu]) that computes the mean automatically. You should check the instructions for your calculator to see how it performs this function.

Computer programs ordinarily will not accept data in the form of a frequency distribution. Most programs are set up to read one or more data records for each individual. Therefore, if the data come to you in the form of a frequency distribution, it often is easier to do the calculations with a calculator, rather than enter dozens or hundreds of individual scores. However, if your data file contains scores as interval midpoints (as in the example we gave earlier of computing the median as the midpoint of the score distribution) the program will give you results exactly equal to those you would obtain with formula 3.4. Demographic data on ages, family incomes, and similar variables are the ones most likely to result in individual data points being recorded as interval midpoints.

▶ *Now You Do It*

Let's use the data from Table 3-4 to check your understanding of how to compute the mean from a frequency distribution. In that table (page 66) we found a median of 7.106 for the scores of 340 students in a SAD introductory statistics course weekly quiz. All we have in the table are the score values and their frequencies. What are the values we will use for X_j and f_j? Can you tell already what value will appear in the denominator?

To find the mean we could add together the 8 scores of 10, the 26 scores of 9, and so forth, or enter the 340 individual scores into a computer, but you will find it much easier to use formula 3.4 and a calculator. If you write down the individual products in the fX column, you would get

X	f	fX
10	8	80
9	26	234
8	84	672
7	132	924
6	67	402
5	15	75
4	6	24
3	2	6
Sums	340	2417

The mean quiz score, then, is 2417/340 = 7.109. On average, the class earned just over 7 out of 10 points. Whether this is good, average or poor performance, of course, depends on the nature of the quiz questions and the instructor's objective in giving the quiz, but clearly the average student has not achieved complete mastery of the material. How are you doing in mastering the mean?

The Shape of Frequency Distributions

We discussed frequency distributions and their graphs in Chapter 2. We bring them up again here because graphs can sometimes help develop an understanding of the properties of the mode, median, and mean, and because the relationship between these three measures of location can tell us something about the shape of the frequency distribution.

Figure 3-5 The shape of the distribution affects the positions of measures of central tendency. Skewed distributions are asymmetric.

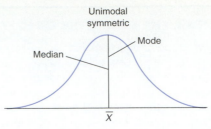

(a) Distribution of general cognitive ability scores

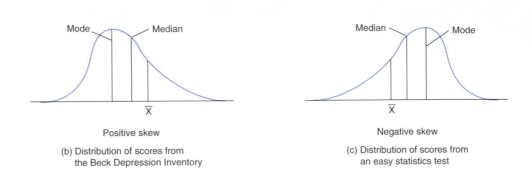

(b) Distribution of scores from the Beck Depression Inventory

(c) Distribution of scores from an easy statistics test

First, we must master a little vocabulary. Figure 3-5 shows three frequency curves illustrating different shapes that a frequency distribution may possess. The first of these is called a **unimodal symmetric distribution**. There is a single mode which is exactly in the middle of the distribution, and the frequency in each score interval drops off at the same rate on both sides of the mode. In any unimodal symmetric distribution, the mean and median will have the same value as the mode. Distributions with this general shape are quite common in work with psychological tests, anthropometric measures, and educational-achievement test scores. For example, general cognitive ability as measured by tests like the Stanford-Binet or the Wechsler Scales often shows a unimodal symmetric distribution, as do scores on achievement tests such as the Iowa Tests of Basic Skills. Many of the procedures discussed in Parts II and III of this book are based on the assumption that the data have this type of distribution.

*A **skewed** frequency distribution has more observations at one end of the scale than at the other.*

The other two curves in Figure 3-5 represent frequency distributions that are **skewed**. A frequency distribution is skewed when there are more observations at one end of the scale than at the other. For example, you would get a skewed distribution if you gave an achievement test that is too hard (scores pile up at the bottom) or too easy (scores pile up at the top) for your students. (The distribution of quiz scores in the previous example looks unimodal symmetric, indicating that it was neither too hard nor too easy for these students.) Such variables as personal income and reaction time often show skewed distributions because there is a lower limit to the scale but no upper limit. An example from clinical psychology is the Beck Depression Inventory, an instrument designed to assess clinical depression in psychiatric patients. This inventory is often used in research on the effects of depression in normal groups such as college students, but the responses you can choose from are so extreme that few college students select a response other than zero, resulting in a strongly skewed distribution of scores.

*A **positively skewed distribution** occurs when most scores bunch up at the low end of the distribution. The mean will be highest, the median next, and the mode lowest.*

*A **negatively skewed distribution** occurs when most scores bunch up at the high end of the distribution. The mean will be lowest, the median next, and the mode highest.*

Bunching up the scores at one end and the existence of extreme scores at the other moves the mode toward one end of the distribution and the mean toward the other. When the mean has a numerical value that is higher than the median and mode (Figure 3-5b), the resulting distribution is **positively skewed**. For example, we would expect this to happen if we have given students a test that is too hard for them. Positive skew occurs when most scores bunch up at the low end of the distribution. The mean will be highest, the median next, and the mode lowest. A distribution has **negative skew** (Figure 3-5c) when the mean is lower than the median and mode (the test is too easy). Negative skew occurs when most scores bunch up at the high end of the distribution. The mean will be lowest, the median next, and the mode highest. We discuss quantitative indices of skew in Chapter 4.

Choosing an Index

There are three factors to consider when choosing which index of location, or central tendency, to use. The first is the level of numerical meaning of the numbers—what type of scale do the numbers represent? If the data represent a nominal scale, the mode is usually the only index that makes sense. It is mathematically possible to compute a median or a mean for any set of numbers; the question is whether the resulting value has a meaningful interpretation. For example, it would not be useful to compute the mean lab section for our BTU statistics students because there is no meaningful interpretation of the result, but the mean of the biological gender variable (because it is dichotomous and scored 0 or 1) would tell us the proportion of our students who are female.

For data measured on an ordinal scale, either the mode or the median may be useful, depending on other considerations, but the mean should not be used. For example, colleges often request rank in high school class as part of their admissions information. But if Big Time U wanted to compare itself to other schools in the state in terms of the high school ranks of its first-year students, the appropriate statistic would be the median rank, not the mean rank. (Note that this example introduces an additional complication; low numbers represent higher performance. Therefore, the school attracting the more able students would be the one with the lower median. In cases like this it is very important to remember which end of the scale is up.) When the measurement scale has an interval or ratio level of meaning (for example, when the scores represent ability measures or reaction times), any of the indices we have discussed may be used.

The second factor to consider is the use to which the index will be put. There are some situations where the only thing that counts is identifying the most frequent event. The mode is the score that is most likely to occur. In other cases, getting an equal number of observations on each side of the center may be important, and the median is the index to use. If we select a person at random, he or she is just as likely to score below the median as above it. However, when the scaling properties of our measurements permit it, we usually find that the mean is the index that satisfies the widest variety of uses.

The final factor to consider is the shape of the frequency distribution. This is closely related to use, because only when the distribution is not symmetrical will the values for the median and mean be different. In fact, a difference between the mean and median indicates asymmetry. (The mode does not always help here because it is possible to have a bimodal symmetric distribution.) Remember that the mean will be the highest measure of central tendency for a positively skewed distribution and the lowest for a negatively skewed distribution because it is more influenced by extreme scores. Thus, if our use of the index would be benefited by a higher value, we might choose the mean in the case of a positively skewed curve and the median if the skew is negative. For example, suppose the school board in your district is having a hearing on teachers' salaries, and the school principals are paid twice as much as the average teacher. Would you want to include the principals with the teachers for the purpose of establishing district salaries? Your answer to this question would depend on whether you were trying to show that teachers are underpaid or that the district is spending a relatively large amount for salaries. This is just one example of how statistics can sometimes be used to support either side of an argument; for others, see *Flaws and Fallacies in Statistical Thinking* (Campbell, 1974), *How to Lie with Statistics* (Huff, 1954) or *Statistical Illusions* (Huck & Sandler, 1984).

▶ *Now You Do It*

1. Suppose you are interested in "locus of control" as a personality variable. You have recruited 100 study participants and want to divide them into two equal groups, those who are "above average" and those who are "below average." What would be the best way to define "average"?

2. You and a friend have a brown paper bag of jelly beans to share. The bag contains 20 red ones, 15 green ones, and 25 white ones. To make eating the jelly beans more interesting, you decide to draw one bean at a time and try to guess the color of the bean you will draw. If you guess correctly, you get to eat the bean, but if you are incorrect, you must put the bean back. What color should you guess on your first try, and why?

3. Godpater's Cookie Company claims that their average oatmeal-raisin cookie contains seven raisins. You don't believe them and decide to draw a sample of 25 cookies, count the raisins, and compare your results with their claim.

What measure should you use to describe your findings?

Answers
1. The median, because it divides the group in half.
2. White, because it is the mode, the most frequently occurring color.
3. The mean, because it will use information from every cookie, is the most stable, and the data are on a ratio scale.

SUMMARY

There are many occasions when it is desirable to use a numerical index to summarize a set of data, because frequency distributions and graphs are inefficient and do not lend themselves to further analysis. Location and spread are two features that should be included in the summary of a frequency distribution. Three indices of location or central tendency are commonly used in the behavioral and social sciences. They are the *mode,* or most frequently occurring score; the *median,* which is the value on the score scale that has exactly 50 percent of the observations below and 50 percent above; and the *mean,* or arithmetic average of the scores. The median and mean can be computed from raw or grouped frequency distributions. The best index to use depends on the type of scale the data represent, the shape of the frequency distribution, and the use the index is to serve. The properties of the measures of central tendency are summarized in Table 3-10.

Table 3-10 Summary of Measures of Central Tendency		
Scale of Measurement	Index of Central Tendency	Properties
Nominal	Mode	Numbers contain no quantitative information other than frequency.
Ordinal	Mode	Gives most frequent value. Appropriate for discrete data.
	Median	Point above and below which half of cases lie. Indicates middle position. No information about distance. More stable than mode.
Interval	Mode	Same as ordinal
	Median	Same as ordinal
	Mean	Uses information about distances between observations. More stable than the median.

MEASURES OF CENTRAL TENDENCY FOR THE SEARCHERS STUDY DATA

In Chapter 2 we constructed the frequency distributions and graphs for the variables age, attitude toward equality, income, and degree of religious involvement. Now we can use this information to find measures of central tendency for these variables. Because there are no missing data (all subjects have scores on all variables) the sample size, N, is 60 throughout our computations. The measures of central tendency and the quantities used to compute them are given in Table 3-11. Definition 1 provides values of the median using the score of the middle individual. The values of the median under definition 2 have been determined using equation 3.3.

There are two points of particular interest in these results. First, the modal age is very different for the grouped age data than for the raw scores. The raw score mode of 27 is really quite far from the middle of the set of scores, but it falls there because, probably by chance, four men in the sample were of this age. The median and mean are both about 40, suggesting that the mode is out of place due to chance effects. Second, there is no mode for the raw scores on income because each value occurs just once. When the data cover a large range of values, no mode may exist. Finally, there are two modal values for the grouped income data, 42,000 and 45,000, so there is no unique mode for these scores.

The next question is, what do these measures of central tendency tell us about this sample? We can tell from the average age that these are not typical college students but an older group

Table 3-11 Quantities Used to Determine the Mean, Median, and Mode for the Searchers Data from Raw Scores and Grouped Frequency Distributions

Age

	Raw Scores	Interval of 3
LL_j	39.5	37.5
cf_{j-1}	28	26
f_j	2	4
Median—Definition 1	40.50	40.50
Median—Definition 2	40.50	40.50
ΣfX	2383	2385
Mean	39.72	39.75
Modal Interval		44–46
Mode	27	45

Attitude Toward Equality for Women

	Raw Scores	Interval of 3
LL_j	40.5	40.5
cf_{j-1}	27	27
f_j	4	8
Median—Definition 1	41	42
Median—Definition 2	41.25	41.62
ΣfX	2424	2421
Mean	40.40	40.35
Modal Interval		44–46
Mode	45	45

Income

	Raw Scores	Interval of 3
LL_j	45,129.5	43,500
cf_{j-1}	29	24
f_j	1	11
Median—Definition 1	45,293.5	45,000
Median—Definition 2	45,130.5	45,136
ΣfX	2,715,252	2,718,000
Mean	45,254.20	45,300
Modal Interval		40,500–43,499
		43,500–46,499
Mode	None	42,000 or 45,000

Degree of Religious Involvement

	Raw Scores	Interval of 3
LL_j	16.5	16.5
cf_{j-1}	24	24
f_j	6	19
Median—Definition 1	17.5	18
Median—Definition 2	17.5	17.45
ΣfX	1086	1077
Mean	18.10	17.95
Modal Interval		17–19
Mode	14	18

of men, and we can tell from their incomes that they are near the American average on this characteristic. We know these things from what we know about the distributions of these variables in the general population. But what about their attitudes toward equality for women and their degree of religious involvement? We really cannot tell much about these crucial variables because we do not have clear information about their population distributions. This is one of the problems with many variables in psychology; their meaning depends a lot on the context of their use. Many scores in psychology get much of their meaning from how they are distributed in the population. We address these issues in Chapter 5.

REFERENCES

Campbell, S. K. (1974). *Flaws and fallacies in statistical thinking*. Upper Saddle River, NJ: Prentice Hall.

Hays, William (1995). *Statistics* (5th ed.). San Antonio, TX: Harcourt.

Huck, S. W., & Sandler, H. M. (1984) *Statistical illusions: Problems*. New York: Harper & Row.

Huck, S. W., & Sandler, H. M. (1984). *Statistical illusions: Solutions*. New York: Harper & Row.

Huff, D. (1954). *How to lie with statistics*. New York: Norton.

Kirk, R. E. (1990). *Statistics: An introduction* (3rd ed.). Fort Worth, TX: Holt, Rinehart & Winston.

EXERCISES

1. Why do researchers use numerical indices to describe the central tendency or index of location of a set of data?

2. Describe the desirable properties of an index of location or a measure of central tendency.

3. What is the definition of the mode of a set of data? What are the advantages and disadvantages of using the mode as an index of location or a measure of central tendency?

4. What are the definitions of the median of a set of data? What are the advantages and disadvantages of using the median as an index of location or a measure of central tendency?

5. What is the definition of the mean of a set of data? What are the advantages and disadvantages of using the mean as an index of location or a measure of central tendency?

6. Find the mean, median, and mode for the data listed below. In computing the median use both approaches (i.e., the discrete scale approach and the continuous scale approach). Do these two approaches yield the same median? If not, account for any differences.
   a. The following represents scores on the first statistics quiz: 20, 20, 19, 19, 19, 18, 18, 18, 18, 16, 16, 16, 15, 15, 14, 14, 14, 14, 13
   b. The following represents scores on an algebra test: 22, 10, 27, 10, 12, 27, 16, 28, 11, 23, 19, 27, 20, 23, 20, 16, 13, 21, 11, 12, 14, 18, 22, 23, 13, 16, 31, 45, 18, 24, 27, 49, 38, 22, 31, 35, 41, 47, 29, 25, 24, 27, 22, 18, 17, 34, 33, 41, 31, 33, 28, 28, 29, 34, 19, 25, 36, 25, 33, 29, 28, 22
   c. The following represents scores on a measure of spatial visualization: 31, 32, 30, 18, 19, 30, 23, 22, 24, 15

7. The data sets in Exercise 11 from Chapter 2 are presented below. Find the mean, median, and mode for each data set using the raw data. In addition, find the mean, median, and mode of these data sets using the grouped frequency distributions that you constructed in Exercise 11 from Chapter 2. Account for any differences that may occur in each of these measures.
   a. The following represents scores on the feminine scale of the Bem Sex Role Inventory: 77, 58, 64, 73, 87, 71, 91, 64, 72, 84, 65, 78, 72, 75, 55, 78, 43, 54, 55, 79, 69, 76, 79, 97, 67, 69, 90, 54, 57, 58, 97, 43, 51
   b. The following represents scores on the masculine scale of the Bem Sex Role Inventory: 75, 80, 103, 65, 79, 88, 71, 86, 77, 78, 49, 86, 76, 96, 76, 92, 87, 87, 80, 91, 79, 92, 91, 65, 78, 91, 88, 89, 105, 76, 67, 80, 102, 79, 57, 77, 74
   c. The following represents scores on a measure of the cognitive level: 74, 50, 66, 65, 67, 54, 54, 78, 52, 72, 53, 62, 81, 68, 54, 62, 74, 85, 42, 33, 38, 37, 86, 32, 85, 31

   d. The following represents scores on a measure of assertiveness: 75, 37, 13, 60, 49, 76, 88, 12, 70, 89, 61, 53, 93, 45, 52, 99, 17, 59, 40, 63, 94, 59, 86, 75, 39, 80, 45, 65, 38

8. Suppose the raw frequency distribution of scores on a measure of generalized anxiety is as follows: 55, 95, 43, 74, 46, 49, 93, 77, 64, 59, 82, 37, 48, 33, 75, 64, 50, 50, 72, 64, 74, 75, 50, 43, 60, 75, 40, 65, 79, 54, 52, 55, 55, 47, 65, 92, 61, 44, 50, 69, 35, 40, 48, 8, 66, 6, 12, 33, 25, 38, 26, 40, 42, 31, 26, 25, 34, 33, 88, 30
   a. Determine the mean, median and mode from the raw data.
   b. Prepare a grouped frequency distribution using about 10 intervals.
   c. Find the mean, median, and mode for the grouped data in part (b).
   d. Prepare a grouped frequency distribution using about 20 intervals.
   e. Find the mean, median, and mode for the grouped data in part (d).
   f. What differences exist in the mean, median, and mode for parts (a), (c), and (e)? Account for any differences that exist in each measure.

9. What do we mean when we say that a distribution of scores is symmetric? What is the relationship of the three measures of central tendency when the distribution of scores is symmetric?

10. What do we mean when we say that a distribution of scores is positively skewed? What is the relationship of the three measures of central tendency when the distribution of scores is positively skewed?

11. What do we mean when we say that a distribution of scores is negatively skewed? What is the relationship of the three measures of central tendency when the distribution of scores is negatively skewed?

12. Suppose we have the following distribution of scores on a test: 5, 3, 2, 6, 6, 7, 8, 6, 3, 8, 4, 5, 7, 9, 7, 6, 10, 8, 9, 6, 4, 5, 7, 4, 5. Without graphing the scores in any manner, describe what the distribution of scores would look like and provide a rationale for your answer.

13. For each of the following descriptions, indicate which measure of central tendency would be the most appropriate. Support your answer.
   a. A researcher is interested in test anxiety scores of clients who participated in an anxiety reduction workshop. The distribution of these anxiety scores is unimodal and symmetric.
   b. A first grade teacher is interested in determining which of several reinforcers seems to be the most potent based on the frequency of selection in a two-week test period. The reinforcers include candy, stickers, Pokemon cards, extra time to play games on the

computer, being selected helper of the day, and extra time to play with Legos.

c. Students from different schools participate in a district-wide speech contest. Students are rank ordered on 15 different speech categories. Schools are then compared on the ranks of their students in all of these categories, and an overall winner is selected.

d. Forty-five rats are timed on a maze completion task. The results are highly variable. Two rats completed the maze in exceptionally fast times whereas two other rats took more than 10 times longer than the next slowest rat. The rest of the rats seemed to cluster around two different time intervals.

e. Blood phobic patients are presented with 10 words from four different categories: words that are highly related to blood, words that are highly related to the medical profession but not highly related to blood, words that are highly pleasant, and words that are unpleasant but not related to blood. A baseline measure of galvanic skin response is taken. The galvanic skin response is then measured in response to the words presented in random order. The researcher is interested in the category of words that led to the most frequent change in galvanic skin response.

14. Use the grouped frequency table you constructed for Exercise 15 in Chapter 2 to complete part (a) below:

a. Calculate the mean, median, and mode for the post-instruction score on the mathematics subtest of the Iowa Test of Educational Development (ited9).

b. Use the SPSS program and SPSS data (math.sav) bank provided with your textbook to find the mean, median, and mode for the post-instruction score on the mathematics subtest of the Iowa Test of Educational Development in part (a).

c. What differences exist in the mean, median, and mode for parts (a) and (b)? Account for any differences that exist in each measure.

15. Use the grouped frequency table you constructed for Exercise 16 in Chapter 2 to complete parts (a) and (b) below:

a. Calculate the mean, median, and mode for Japanese participants only (citizen = 1) for Socially-Oriented Achievement Motivation (soam).

b. Calculate the mean, median, and mode for American participants only (citizen = 2) for Socially-Oriented Achievement Motivation (soam).

c. Use the SPSS program and SPSS data (self.sav) bank provided with your textbook to find the mean, median, and mode for Socially-Oriented Achievement Motivation (soam) for Japanese participants only (citizen = 1).

d. Use the SPSS program and SPSS data (self.sav) bank provided with your textbook to find the mean, median, and mode for Socially-Oriented Achievement Motivation (soam) for American participants only (citizen = 2).

e. Compare the values for parts (a) and (c) and the values for parts (b) and (d). What differences exist in the mean, median, and mode for each culture? Account for any differences that exist in each measure.

16. Use the grouped frequency table you constructed for Exercise 17 in Chapter 2 to complete part (a) below:

a. Calculate the mean, median, and mode for anxiety about doing mathematics (anxiety).

b. Use the SPSS program and SPSS data (career.sav) bank provided with your textbook to find the mean, median, and mode for anxiety about doing mathematics (anxiety).

c. What differences exist in the mean, median, and mode for parts (a) and (b)? Account for any differences that exist in each measure.

17. Use the grouped frequency table you constructed for Exercise 18 in Chapter 2 to answer part (a) below:

a. Calculate the mean, median, and mode for empathic concern (empathy).

b. Use the SPSS program and SPSS data (donate.sav) bank provided with your textbook to find the mean, median and mode for empathic concern (empathy).

c. What differences exist in the mean, median, and mode for parts (a) and (b)? Account for any differences that exist in each measure.

18. Construct histograms of the following distributions:
a. a negatively skewed distribution
b. a bimodal symmetric distribution
c. a symmetric distribution with no mode
d. a positively skewed distribution
e. a unimodal symmetric distribution

19. For each of the distributions in Exercise 18, what is the relationship between the mean, the median, and the mode?

20. Discuss the factors that influence a researcher's choice of an index of central tendency or location.

Chapter 4

Measures of Variability

CHAPTER OBJECTIVES

After studying this chapter, you should be able to:

❑ Explain why researchers use numerical indices of the variation or spread in their data.

❑ Define the terms range, interquartile range, semi-interquartile range, standard deviation, and variance, and state the properties of each index.

❑ Identify which measure of spread is most appropriate for a given set of data.

❑ Determine the range, interquartile range, Q, S, and S^2 for a set of raw data.

❑ Determine the same statistics for data in a frequency distribution or grouped frequency distribution.

❑ Define the sum of squares.

❑ State the principle of least squares.

❑ Define and identify positively and negatively skewed distributions.

❑ Define and identify platykurtic, mesokurtic, and leptokurtic distributions.

❑ Properly interpret the values of the indices of skewness and kurtosis.

We have seen in Chapter 3 how the SAD faculty at Big Time U could use measures of central tendency to describe the location of a distribution of scores, such as those on the final course exam. But location is not the only way that two frequency distributions can differ. They can also differ in their spread; that is, in how much of the measurement scale the distribution covers. How might you describe the amount of spread in a distribution? As you think about that question, remember the properties that we would like our descriptive measures to have: stability, uniqueness, representativeness, and the ability to be used in future analyses.

Introduction

In Chapter 3, we described three indices of the central tendency of a set of measurements. Using one or more of these statistics, it is possible to specify where on the measurement scale of a variable, such as scores on our statistics final exam at Big Time U, the GPA's of the students, or their mathematics anxiety inventory scores, the center of a distribution falls. This is extremely important information, and a vast body of literature has been developed in psychology and education based on comparisons of means or medians.

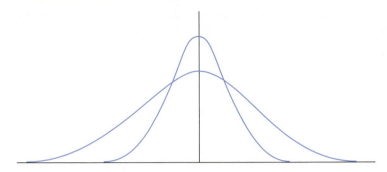

However, the location of the distribution is not the only way that groups of measures or sets of scores may differ. In this chapter, we will discuss the other major way that two or more sets of measurements may differ from each other: their spread over the scale of measurement, which is known as their **variability.** Variability is a reflection of the individual differences between the people or events we are studying.

*Measures of **variability** describe how large the differences between individuals are on a trait.*

The two frequency distributions shown in Figure 4-1 have the same mean, median, and mode. They are located in the same place on the measurement scale, and they are both symmetric, but they are clearly different from each other. In one distribution the scores cluster tightly around the mean with many scores at or near the mean. A distribution such as this one might be expected on a mathematics achievement test for fifth graders or for the yearly earnings of assembly-line workers in an automobile factory. The other distribution shows relatively fewer observations occurring at any single score value (although scores near the mean still have higher frequencies than those farther away), and the scores spread out across a much wider range of scale values. We might expect a distribution like this for the mathematics achievement scores of a group of fourth through sixth graders or the annual incomes of real estate salespeople.

The reasons for these differences in spread of scores lie (1) in the characteristics of the groups and (2) in the nature of the trait. Our schoolchildren differ more in age and experience in the second case than the first, so we should expect them to show greater variability in mathematics achievement than would the more homogeneous group. With regard to the second example, assembly line workers usually are paid a salary based on hours worked and seniority, so they will show little variation in income. But people who sell real estate are paid by commission depending on how much they sell. Sales can show huge differences from person to person, so we might expect data like these to show great variability.

▶ *Now You Do It*

What other examples can you think of where you would expect to find differences in the amount of spread shown by different groups? How about the high school GPA's of freshmen at an elite private college compared to the GPA's of freshmen at a large public university? Or the amount of depression found among elderly people in a nursing home compared to the general population? Are there situations in your own experience where you have encountered highly homogeneous or highly heterogeneous groups of people? Are there certain traits you can think of where people show more individual differences than others?

Students admitted to elite private colleges usually differ from students at large public universities both in the average level of high school GPA (higher) and in variability (they are more homogeneous). This does not mean that there are not students at public universities who are as bright as the best at private colleges, but that the range of talent is much wider at public institutions. Likewise, we might find some people in nursing homes who are quite happy, but most would probably score at the depressed end of a scale of mood. What other traits did you come up with? How about the effects of development? Don't people generally come to show greater individual differences in most traits as they reach adulthood than they did as children?

To give a complete description of any frequency distribution we must specify its location, its spread, and its general shape. Just as we found it useful to summarize the location of our distribution of final exam scores or mathematics anxiety scores or GPA's with a single numerical index, we shall find it convenient and informative to have a single index that reflects the spread of scores in the distribution. As was the case for central tendency, we have a choice of three such indices.

Nominal Scales

Before discussing measures of variability, we again must consider nominally scaled measurements. Remember that nominal measurement does not carry information about more or less of the characteristic. For example, using the nominal scale 1 = psychology, 2 = education, 3 = other for college major, increasing numerical values do not mean more or less of any trait. The mode is the only meaningful index of central tendency for such measurements, and it does not indicate location, only the category with greatest frequency. Since the categories of such a scale have no fixed or meaningful order, the concept of the center of the distribution is meaningless. With the exception mentioned in Chapter 3 of a dichotomous 0,1 variable where the mean conveys the proportion of cases scoring 1, the only summary statement we can make is that category X_j (for example, if $j = 2$, X_j means education major) is the most frequently occurring category.

The same properties of nominal scales that prevent interpreting the mode as an index of location also prevent us from developing any index of variability or spread for a set of nominal measurements. The very idea of spread implies a set of categories with meaningful directional relations to each other. At the very least, the categories of the scale must have a fixed order for even the most crude interpretations of spread. To see that this is the case, remember that the categories on a nominal scale could be designated by letters or words rather than numbers without losing information. The interpretation of the mode would not be altered, but direction and distance on the scale are not meaningful.

The only way that we can approach the concept of variability for data on a nominal scale is to indicate the number of different categories into which our observations are divided. Since the numbers or labels attached to the categories may be completely arbitrary, statistics such as the range are not appropriate. Thus, if we were to collect information about the number of people who major in education, psychology, and other subjects, we might summarize our findings by saying that we observed three categories of college major, but not that the range of scores was three scale units. In the discussions to follow, we shall assume that measurements have been made at the level of an ordinal, interval, or ratio scale. (Kirk [1999] suggests an index based on the number of categories and the evenness of frequencies as a measure of variability for nominal variables.)

The Range

Raw Data

The simplest index of the spread of a set of raw scores is called the range. It is found by the formula

$$Range = (H - L) + 1 \tag{4.1}$$

That is, find the difference between the highest score and the lowest score and add 1 score unit to that difference. This makes the range extend from the lower real limit of the lowest score to the upper real limit of the highest score. To compute the range for the set of statistics final exam scores in Table 3-1, we find

$$H = 65$$
$$L = 4$$
$$Range = (H - L) + 1 = (65 - 4) + 1 = 61 + 1 = 62,$$

which yields a range equal to the number of values of X in the table.

There are two interpretations of the range that are useful in many contexts. First, for data measured on an integer scale, the range specifies the number of unique score values that are covered by a frequency distribution. In our example, there are 62 possible scores from the lowest observed score to the highest observed score. There are several score

The **range** is the distance from the lower real limit of the lowest obtained score to the upper real limit of the highest obtained score.

values that have zero frequency, but these are still unique and obtainable score values within the range of scores.

The second interpretation is that the range is the distance on the measurement scale between the lower real limit of the lowest score interval and the upper real limit of the highest score interval. The upper real limit of the interval 65 on the underlying continuous scale is 65.5, and the lower real limit of interval 4 is 3.5. We may then state the range as

$$65.5 - 3.5 = 62,$$

which is the same value that we obtained using formula 4.1. Since this second interpretation of the range implies a distance on the measurement scale, it requires that the data have at least interval scale properties.

It is meaningful to use the range when describing variability among the members of a statistics class on a scale of height, on a scale of time to complete an examination, or on the scale of the number of points earned on that examination. Each of these measurements can be thought of as representing at least an interval scale. On the other hand, rank in class is an ordinal scale, so the range could not be interpreted as distance between highest and lowest ranks in the class. Note, however, that the first interpretation of the range as the number of unique score values, which in the case of class rankings is equivalent to the number of individuals in the group, is still appropriate.

The range is a statistic that illustrates clearly a point first made in Chapter 1 and to which we return periodically: Numbers by themselves have no scale properties; they can be manipulated by any of the procedures of mathematics, and the results will be meaningful as *mathematical results*. However, the scientific conclusions that we can draw from any analysis of a set of data are restricted to the meaning that was given to the numbers by the procedures used to generate them. When a statistic has more than one interpretation, we may be restricted in the interpretations we can make by the scale properties or other aspects of the data. This is quite clearly the case for the range, and we shall see that the principle applies elsewhere as well.

To make matters more complex, some statistics texts and most computer programs define the range as the difference between the highest and lowest scores, $H - L$. This is a perfectly acceptable definition and is useful as an index of spread, but it does not permit either of the interpretations described here. We have chosen to use the definition in equation 4.1, which Kirk (1999) calls the **inclusive range,** because it permits the two interpretations given above and it is just as useful as the slightly simpler definition for comparing two distributions.

As a measure of the spread of a set of scores, the range has the advantage of being quickly and easily computed; its meaning is also easy to understand. However, certain features of the range make it less desirable than some other indices of variability. First, the range uses very little of the information in a set of measures. It is based on only two scores at the extremes of the distribution and, for this reason, is quite unstable. It can show wide fluctuations from one group to another group because of the chance presence of a single extreme individual. Second, while the range yields a unique value for any particular set of data, it cannot be used in any additional analyses and does not permit uses or interpretations beyond description of the data at hand.

Grouped Frequency Distributions

The range for data from a grouped frequency distribution can be found by subtracting the midpoint of the lowest interval from the midpoint of the highest interval and adding one interval width, that is,

$$Range = (MP_H - MP_L) + J. \tag{4.2}$$

The range defined in this way runs from the lower real limit of the lowest interval to the upper real limit of the highest interval. For example, if we wish to compute the range for the statistics final exam scores using an interval width of five (the data in Table 2-4), we would get

$$Range = (65 - 5) + 5 = 65$$

The range defined in this way will tend to be slightly higher than for the raw scores that produced the grouped frequency distribution unless scores go out to the limits of both

Table 4-1 Properties of the Range	
Advantages	**Disadvantages**
Easy to compute	Depends only on two scores
Provides a unique value	Unstable from group to group
Easy to understand	Not used in further computations

the top and bottom intervals, but the alternative, defining the range as the difference between the upper and lower midpoints, will be necessarily too low. Notice that the range, as we have defined it here, is equal to the distance between the lower real limit of the lowest interval and the upper real limit of the highest interval, just as it was for raw scores.

$$Range = 67.5 - 2.5 = 65$$

The properties of the range are summarized in Table 4-1.

▶ *Now You Do It*

Let's take a look at the high school GPA data again. How should we go about finding the range of scores, first for the raw data, then for the data grouped as in Chapter 3?

Once again, it is important that we keep the units and intervals in mind. What is the unit for HSGPA? What is the interval width?

HSGPA is in units of .01. Equation 4.1 tells us to subtract the lowest score, 1.04, from the highest score, 4.00, and add one unit to the difference. Therefore,

$$Range_{GPA} = (4.00 - 1.04) + .01 = 2.97$$

This value tells us that the distance from the lower real limit of the lowest score to the upper real limit of the highest score is 2.97. Because the unit is .01, there are 297 possible scores in this range.

For the grouped data, the range is the difference between the extreme midpoints plus an interval. Because the interval width is 0.25, the range is

$$Range_{GPA} = (3.88 - 1.13) + .25 = 3.00$$

which is very close to the raw data value and much closer than the result we would obtain from the simple difference between midpoints. It is always the case that the results we get from grouped data estimate the values from raw data, but here the estimate from equation 4.2 is clearly superior because it is much closer to the values obtained with the raw scores.

HOW THE COMPUTER DOES IT

We can use either the FREQUENCIES procedure or the DESCRIPTIVES procedure from the STATISTICS menu of SPSS to find the range. Since you probably will be using FREQUENCIES to get the frequency distribution anyway, you can get the range printed out at the same time. It is included as one of the optional statistics. Computer programs generally do not use the definition of the range we advocate. Rather, they use the simple (highest − lowest) definition. For raw data, this is not a large problem, but for grouped data, as we have seen, it can result in inaccurate estimates. SPSS, for example, uses the highest midpoint minus the lowest midpoint for the range, even when the data are identified as interval midpoints. To have the program print the range, simply click on the box next to the range.

---

*A **quartile** is a point in the score scale that cuts off 25% of the distribution.*

*The **first quartile (Q_1)** cuts off the bottom 25% of the distribution.*

The Interquartile Range

A second way to represent the variability, or spread, in a set of scores is to locate points on the score scale that include some specified percent of the frequency distribution. Two points are of particular interest in this regard: the value that cuts off the bottom 25 percent of the distribution, known as the **first quartile** or Q_1; and the value that cuts off the bottom 75 percent (or top 25 percent) of the distribution, known as the **third quartile** or Q_3. Note that Q_1 and Q_3 bracket or include between them the middle 50 percent of the distribution. For the set of 68 final exam scores of the BTU statistics class, Q_1 would separate the bottom 17 students (25% of 68) from the rest of the group, and Q_3 would separate the top 17 students. The middle 34 students (50%) fall between these two points.

The **third quartile (Q_3)** cuts off the bottom 75% (or top 25%) of the distribution.

The word quartile, of course, means quarter and refers to quarters of the frequency distribution. The first quartile marks off the bottom 25 percent, the second quartile (Q_2, also known as the median) marks off the next 25 percent (or the bottom 50 percent), and the third quartile marks off the next 25 percent (or separates the top 25 percent from the bottom 75 percent). There are four quarters in the distribution but only three quartiles. The **interquartile range** is the distance or number of values on the score scale that are needed to include the middle 50 percent of the scores; that is, the distance between Q_3 and Q_1. The **semi-interquartile range,** often given the symbol Q, is half of the interquartile range. In terms of formulas

The **interquartile range** is the distance between the first and third quartiles. It includes the middle 50% of scores.

$$Interquartile\ Range = Q_3 - Q_1 \qquad (4.3)$$

and

$$Q = (Q_3 - Q_1)/2$$

Computations for Raw Data

The computations needed to find Q_1 and Q_3 are almost identical to those for finding the median. First, we must prepare the frequency distribution and cumulative frequency distribution. We may then use the following formulas to compute Q_1 and Q_3.

$$\left.\begin{array}{l} Q_1 = LL_{.25} + \dfrac{.25N - cf_{25-1}}{f_{.25}} \\[2mm] Q_3 = LL_{.75} + \dfrac{.75N - cf_{75-1}}{f_{.75}} \end{array}\right\} \qquad (4.4)$$

In each of these formulas, the terms have meanings very similar to those in formula 3.1. As with the median, the first step is to determine how many individuals are needed to make up 25% ($.25N$) and 75% ($.75N$) of the sample respectively. These values are used to find the scores containing the persons we need. Next, we find the lowest score whose cumulative frequency equals or exceeds the required value, and the lower real limit of this score gives us the value of LL. The required value of cf (from score $j - 1$) gives us the cumulative frequency up to the score containing the desired point, and f is the frequency of that score.

We can use our statistics quiz data from Table 3-4 (reproduced here as Table 4-2) to illustrate the computation of quartiles for raw data. To determine Q_1, we must find $.25N$ [$.25(340) = 85$], locate the interval that includes this 85th individual, and find the necessary values from the table. Using the rule from Chapter 3 that the interval we seek is the first one where the cumulative frequency equals or exceeds $.25N$, we select the score interval 5.5–6.5, which has the midpoint 6. The 85th person is one of the 67 people who received a score of 6. The lower limit of this interval is 5.5, so we compute Q_1 to be

$$Q_1 = 5.5 + \frac{85 - 23}{67}$$
$$= 5.5 + (62/67)$$
$$= 5.5 + (.93)$$
$$= 6.43$$

Computation of Q_3 proceeds in exactly the same way with the necessary adjustments.

$$.75N = .75(340) = 255$$
$$LL_{.75} = 7.5$$
$$Q_3 = 7.5 + \frac{255 - 222}{84}$$
$$= 7.5 + (33/84)$$
$$= 7.5 + (.39)$$
$$= 7.89$$

Table 4-2 Frequency Distribution of Scores on a Statistics Quiz

X	f	cf	
10	8	340	
9	26	332	
8	84	306	interval .75
7	132	222	
6	67	90	interval .25
5	15	23	
4	6	8	
3	2	2	

The interquartile range is the difference between Q_3 and Q_1, so for these quiz scores it is

$$Q_3 - Q_1 = 7.89 - 6.43$$
$$= 1.46.$$

It takes 1.46 units to cover the middle 50 percent of this group. The semi-interquartile range is half this amount, or .73.

▶ *Now You Do It*

Let's use the raw scores from the statistics final exam data (Table 2-3) as a second example of computing quartiles. What are the steps to follow? First find .25N and .75N; then identify the scores and extract the necessary f and cf values.

We need 17 people for Q_1 and 51 people for Q_3. Sixteen people have scores of 23 or below, there are no scores of 24, and 4 people scored 25. Therefore, the person we need is one of the four people in the 24.5–25.5 score interval. Likewise, the 51st person is one of the three people in the interval 43.5–44.5. Applying equation 4.4, you should get values of 24.75 for Q_1 and 43.83 for Q_3. The difference between them, the interquartile range, is 19.08, and Q is 9.54.

Computations for Grouped Data

The interquartile range and semi-interquartile range can also be computed from a grouped frequency distribution. The only thing we need to do is make the same modifications to formula 4.4 that we made to the formula for the median, then multiply the value we obtain

---

SPECIAL ISSUES IN RESEARCH

An Alternative Way to Define the Quartiles

In Chapter 3 we pointed out that the median can be defined as whatever score is earned by the middle person in the distribution. The same principle can be applied to the quartiles. Using this definition, Q_1 becomes the score earned by the .25Nth person and Q_3 is the score earned by the .75Nth person. For the data in the example we have been considering, these would be the scores for the 85th and 255th people re-spectively, or 6 and 8. The interquartile range would therefore be 2 using this definition, one-third larger than the value we obtained. We shall see in Chapter 5, where we discuss the general case of percents of a distribution, that the difference between the two ways of obtaining these points can yield some very strange results.

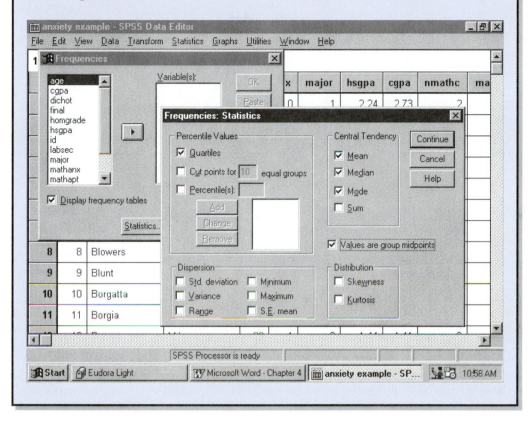

from the fraction by the interval width. This gives us

$$Q_1 = LL_j + \left[\frac{.25(N) - cf_{j-1}}{f_j} \right] (J)$$
(4.5)

$$Q_3 = LL_j + \left[\frac{.75(N) - cf_{j-1}}{f_j} \right] (J)$$

We can use our statistics final exam data in Table 3-7 (repeated here as Table 4-3) for an interval width of 5 to illustrate the computations. To find Q_1, we first find

$$.25N = .25(68) = 17.$$

Inspecting the cf column shows that the interval 23–27 contains the 17th person, and its lower limit is 22.5. Then

$$Q_1 = 22.5 + \left[\frac{17 - 15}{5} \right] (5)$$

$$= 22.5 + [2/5](5)$$
$$= 22.5 + 2.0$$
$$= 24.5$$

Table 4-3	Grouped Frequency Distributions of Statistics Final Exam Scores for an Interval of 5 from Table 3-7		
Score Interval	Midpoint X	Frequency f	Cumulative Frequency cf
63–67	65	1	68
58–62	60	2	67
53–57	55	4	65
48–52	50	4	61
43–47	45	9	57
38–42	40	7	48
33–37	35	12	41
28–32	30	9	29
23–27	25	5	20
18–22	20	8	15
13–17	15	4	7
8–12	10	1	3
3–7	5	2	2

Using the same general procedure to find Q_3, we get

$$.75N = .75(68) = 51$$

$$LL_{.75} = 42.5$$

$$Q_3 = 42.5 + \left[\frac{51 - 48}{9}\right](5)$$

$$= 42.5 + [3/9](5)$$
$$= 42.5 + 1.7$$
$$= 44.2,$$

so the interquartile range is

$$Q_3 - Q_1 = 44.2 - 24.5 = 19.7,$$

and the semi-interquartile range is $Q = 19.7/2 = 9.85$.

Meaningful interpretation of the quartiles requires the values of the measurement scale to have a fixed order that is related to the trait of interest. However, because the meaning of each quartile value is determined by the number of observed cases falling below that point rather than by how far away those cases are, equal units on the score scale need not be assumed. The data are treated as though they were obtained by using an ordinal scale. Only the observations in the immediate vicinity of the quartile points take part in the calculations.

To illustrate the consequences of this point, consider the two small frequency distributions in Table 4-4. Each has ten cases and is symmetric around the same median value of 50. The range for distribution A is $(100 - 0) + 1 = 101$, while the range for distribution B is only $(53 - 47) + 1 = 7$. The range indicates that the two distributions are very different

Table 4-4	Two Small Frequency Distributions				
	A			B	
X	f	cf	X	f	cf
100	1	10	53	1	10
52	0	9	52	0	9
51	2	9	51	2	9
50	4	7	50	4	7
49	2	3	49	2	3
48	0	1	48	0	1
0	1	1	47	1	1

in spread. Computing the interquartile ranges for the two distributions yields the following results; for distribution A:

$$.25N = .25(10) = 2.5 \qquad\qquad .75N = .75(10) = 7.5$$
$$Q_1 = 48.5 + [(2.5 - 1)/2] \qquad\qquad Q_3 = 50.5 + [(7.5 - 7)/2]$$
$$Q_1 = 49.25 \qquad\qquad Q_3 = 50.75$$
$$\text{Interquartile range} = 1.5$$

For distribution B:

$$.25N = .25(10) = 2.5 \qquad\qquad .75N = .75(10) = 7.5$$
$$Q_1 = 48.5 + [(2.5 - 1)/2] \qquad\qquad Q_3 = 50.5 + [(7.5 - 7)/2]$$
$$Q_1 = 49.25 \qquad\qquad Q_3 = 50.75$$
$$\text{Interquartile range} = 1.5$$

The interquartile ranges for the two distributions are identical! The reason for this result is that extreme cases do not enter into the computation of quartiles; in fact, information from only two score intervals is used to compute any quartile.

▶ **Now You Do It**

Let's check up on your comprehension of the interquartile range using the high school GPA data. These data are very "thin" or "sparse"; that is, there are 68 people spread over 297 score values. Very few score values have more than one person and there are many values that no one obtained, so we would almost certainly use grouped data.

Let's go back to the frequency distribution you prepared for these data in Chapter 2. We suggested that you use an interval of .25 and begin with an interval of 1.01–1.25. If you did that, you should have the following distribution.

Interval	Midpoint X	Frequency f	Cum Freq cf
3.76–4.00	3.88	2	68
3.51–3.75	3.63	4	66
3.26–3.50	3.38	3	62
3.01–3.25	3.13	5	59
2.76–3.00	2.88	6	54
2.51–2.75	2.63	9	48
2.26–2.50	2.38	12	39
2.01–2.25	2.13	7	27
1.76–2.00	1.88	4	20
1.51–1.75	1.63	10	16
1.26–1.50	1.38	3	6
1.01–1.25	1.13	3	3

What are the questions you must answer before you start? We need to know how many people fall below each quartile and how wide the intervals are. The answers are $(.25)68 = 17$ people below Q_1 and $(.75)68 = 51$ people below Q_3 with an interval of .25.

What are the intervals we should use? The first cf greater than 17 is 20, so Q_1 is in the interval 1.76–2.00; the first cf greater than 51 is 54, so Q_3 falls in the interval 2.76–3.00. Applying formula 4.5 should give you

$$Q_1 = 1.8175 \text{ or } 1.82 \text{ and}$$
$$Q_3 = 2.88$$

The interquartile range is $2.88 - 1.82 = 1.06$ and $Q = .53$.

Properties of the Interquartile Range

The insensitivity of the interquartile range to the presence of extreme scores is perhaps its most appealing feature (see Table 4-5). We have seen that the range, to the contrary, can be greatly affected by a single extreme score. It can therefore be quite unrepresentative of the whole distribution and very unstable from group to group. The interquartile range, based on the middle 50 percent of the distribution, is much more stable from one group to another and uses more of the information from the entire distribution. Of course, both the range and the interquartile range provide unique values for a given distribution (unlike the problem of multiple modes in expressing central tendency).

Box-and-Whisker Plots Revisited

You were introduced to the box-and-whisker plot, or box plot, as one form of graph in Chapter 2. As you learned, graphs of this kind show the locations of the centers of a series of groups and the spread of each group around its center. We are now ready to explain exactly what these values are.

The box plot from Chapter 2 is reproduced here as Figure 4-2. The heavy line in the middle of each box is the median. The top of each box corresponds to Q_3, and the bottom line is Q_1, so the box covers the interquartile range. The whiskers extending above and below the box go to the upper and lower limits of the distribution, so the distance from the tip of one whisker to the other covers the range. A plot like this makes it easy to compare two or more distributions at the same time in terms of their locations and spread.

The Variance and Standard Deviation

The concepts of variance and standard deviation, our next focus, are at the heart of most of the remaining topics in this book. Therefore, it is extremely important to develop a thorough understanding of these concepts.

Conceptual Formulation

First, let us consider the meaning of the concept of variability and our purpose in using statistics. We noted in Chapter 1 that two reasons for collecting data were to describe groups of individuals or phenomena and to predict future events or performances. In most situations, it makes sense to say that we would like our predictions or descriptions to be as close to reality as possible. We can, of course, be completely accurate within the limits of our measurement procedures by reporting every observation. However, for reasons discussed in Chapter 3, we usually find it desirable to report a single value to represent the center or location of the set of scores. We described three statistics that can be used

Table 4-5 Properties of the Interquartile Range	
Advantages	**Disadvantage**
Not affected by extreme scores	Not useful in other calculations
Provides a unique index	

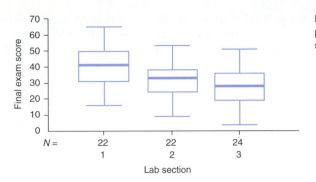

Figure 4-2 Box-and-whisker plot of final exam scores for students by lab section.

for this purpose, the mode, the median, and the mean, and have mentioned some of the advantages and limitations of each.

Any time we use a single value to represent several numerically different scores, that value will be a better representation of some scores than others in the sense that it will be closer to some scores than others. For example, the median for each set of scores in Table 4-4 is 50. If we use 50 to describe these sets of scores, it will be exactly right for four of the ten individuals in each set and in error by some amount for the remaining six. One way to think about variability is in terms of the errors that we incur in using a measure of central tendency to describe a set of observations; this is probably the most common way that psychologists and educators use the term.

It is common in engineering and business applications of statistics to think of variability in another way, as the difference between a targeted or intended outcome and what actually occurred. Suppose, for example, that students in a chemistry class are instructed to use *exactly* 0.25 grams of substance X in an experiment. Each sample is supposed to weigh 0.2500 grams; however, since no measuring process is exactly perfect, not all of the samples will weigh that exact amount. There will be errors in measurement resulting in some overweight samples and some that are underweight. We might use the same sort of index to describe these errors in measurement that we would use to describe the errors in description that result from using the median or mean to represent all cases; in each case, the objective is to describe the amount of variation around some point in the distribution.

It is not really necessary to think of variation in terms of error. In the measuring example, where there is a target measurement, error is an appropriate concept. However, when describing a group of people, we should not think that because people differ from one another, some are right and others are in error. A century ago, it was popular in psychology to regard the mean as nature's ideal, and data values differing from the mean were viewed as nature's errors. One of the objectives in designing research studies was to reduce or eliminate such errors. While behavioral science has outgrown the idea that the mean is right and variation from the mean is undesirable, the original terms error and deviation continue in the vernacular of statistics. Therefore, when we use the terms error and deviation in what follows, we do so from tradition and for convenience of exposition and do not imply any positive or negative affective value in being at the mean or differing from the mean.

One of the major objectives of psychological and educational research is to understand and account for the differences among people on various traits such as personality, ability, achievement, and attitudes. The SAD faculty at Big Time U, for example, want to understand how differences in test anxiety and other student characteristics may relate to performance on their statistics exams. When we make predictions about people, as when ability measures are used to forecast academic performance, we are really making statements about individual differences, or variability. We are trying to use differences among people on one measure to account for or predict differences among those people on another measure. Therefore, we may also view measures of variation as aids in describing the nature and extent of individual differences.

Deviations from the Mean

Suppose we choose to use the mean score on the statistics final to describe the performance of our group of BTU statistics students. There are several reasons beyond those already mentioned in Chapter 3 for making this choice. We shall discuss some of them shortly, but for now, we may decide to use the mean more or less arbitrarily. Given that we use the mean exam score of 34.24 to represent the group, our representation of each individual will be in error, or inaccurate, by some amount. This will always be true unless the mean happens to be an obtainable score. For each individual, we can represent the magnitude of this inaccuracy by a **deviation score,** which is the difference between an individual's observed score and the mean of the distribution of scores. For example, Louisa Alcott earned a score of 37 on the final, so her deviation score is $37 - 34.24 = 2.76$. A deviation score is ordinarily represented by a lower case letter; thus, if Louisa's observed score is represented as X_i, her deviation score would be represented as x_i and defined as

$$x_i = X_i - \overline{X} \tag{4.6}$$

which expresses the score in terms of its distance from the mean of the scores.

Deviation scores have two important properties to which we must pay careful attention. First, deviation scores are **signed numbers;** that is, some of them are positive and some of them are negative (about half of each). We *must* keep the sign of a deviation score because the magnitude tells us how far from the mean the score is, and the sign tells us in which direction to go. For example, Harold Lange got a score of 25 on the final, making his deviation score -9.24. If we disregard the sign, and simply call his score 9.24, he has a higher score than Louisa does, which is not true to the original data. Neglecting the sign destroys the relationship between the deviation scores and the original data, and the shape of the frequency distribution changes dramatically. By thinking about this for a minute, you can see that the result is chaos. If we keep the sign, then by simply reversing our process and adding the mean to the deviation scores, we can recover the original raw scores. (For example, $-9.24 + 34.24 = 25$, Harold's original score.) When the sign is lost, we have also lost the ability to return to the original scores. We can no longer distinguish between values that were below the mean and those that were above, so the frequency distribution is cut at the mean and the bottom portion is, in effect, folded over onto the top portion.

The second important property of deviation scores is that for any group of N observations, such as our 68 final exam scores in statistics, the sum of the deviations from the mean of the group is zero; that is,

$$\sum_{i=1}^{N} x_i = 0$$

(For most of the remaining formulas in this chapter the summation is over all cases, so the limits $i = 1, N$ will be omitted. The exception is for computing with frequency distributions; when we discuss them, the summation limits will be used again.) To see that the sum of deviation scores must be zero, we use the definition of deviation scores from equation 4.6.

$$\sum x_i = \sum (X_i - \overline{X})$$
$$= \sum X_i - \sum \overline{X}$$

Because $\overline{X}$ is a constant (that is, it has only one value for a given distribution), summing it N times is the same as multiplying it by N. Using this point and substituting the definition of the mean from equation 3.2, we get

$$\sum x_i = \sum X_i - N \frac{\sum X_i}{N}$$

which gives us

$$\sum x_i = \sum X_i - \sum X_i = 0$$

This will always be true for the sum of deviations of any set of scores around their mean. It also makes sense intuitively for $\sum x_i$ to be zero because the mean is approximately in the center of the distribution (indeed, it is exactly at the center in terms of deviation scores), and roughly half the scores are positive and half negative. They should average out,

Table 4-6 Comparison of Raw Scores and Deviation Scores		
	Sum	Mean
Raw Scores	ΣX	$\overline{X}$
Deviation Scores	0	0

therefore, to about zero. The fact that $\Sigma x_i = 0$ tells us the mean of a set of deviation scores must also be zero. (See Table 4-6.)

The Variance

Since a set of deviation scores must sum to zero and their mean must be zero regardless of the shape of the distribution, the deviation scores themselves do not help us much in describing the spread of a set of scores. Two methods have been used to resolve the problem. The first is to find the mean of the absolute values of the deviation scores; that is, chop off the signs, treat all numbers as positive, add them up, and divide by N. The resulting statistic, known as the *average deviation*, was popular in the years before World War II, but it has fallen out of use because, like the range and interquartile range, it cannot be used in further computations, and, as we saw earlier, disregarding the sign of deviation scores destroys their relationship with the original data.

A much more satisfactory solution is to square each deviation score (multiply it by itself) and find the *mean of these squared deviations*. This operation results in a statistic called the **variance,** which, for reasons that will become apparent shortly, is given the symbol S^2. The variance of variable X is the mean of squared deviations from the mean of X.

The **variance** is the mean of squared deviation scores.

$$S_X^2 = \frac{\Sigma x_i^2}{N} \tag{4.7}$$

Notice that we are using the general operations for finding a mean (summing and dividing by the number of things summed) but that the operations are being applied to something other than observed scores. We are finding the mean of squared deviations from the mean, so the resulting statistic is sometimes called a mean-squared deviation. Since deviation scores can be expressed in terms of raw scores ($x_i = X_i - \overline{X}$), the variance can also be written

$$S_X^2 = \frac{\Sigma(X_i - \overline{X})^2}{N} \tag{4.8}$$

where the operation of squaring is performed before summation. The application of this formula to the set of 11 anxiety inventory scores is shown in Table 4-7. Looking at the deviation scores in column two, you can see that they sum, within rounding error, to zero.

The Standard Deviation

The variance has proven to be an extremely useful and important statistic. Most of the statistical techniques used by educators, psychologists, and many other behavioral and social scientists are based on the concepts of the mean and variance. We shall use these concepts extensively in the chapters that follow. However, the variance in its present form has a drawback for some descriptive purposes. Because it is computed from squared deviation scores, its numerical value will generally exceed the total range of scores in the distribution. It is not on the same scale as the original scores but on the square of that scale. We can correct this problem by taking the positive square root of the variance, thereby undoing the effects of squaring. The resulting statistic, which is in the same scale as the original measurements, is called the **standard deviation.**

The **standard deviation** is the positive square root of the variance.

The standard deviation, which is sometimes referred to as the [square] root-mean-squared deviation to express the procedures for computing it, is given the symbol S (sometimes SD is used as the symbol for the standard deviation). Obviously, S^2 is used

Table 4-7 Computing the Variance and Standard Deviation for the Math Anxiety Scores of 11 Male Students		
X	x	x^2
6	.91	.8281
5	−.09	.0081
3	−2.09	4.3681
5	−.09	.0081
7	1.91	3.6481
4	−1.09	1.1881
6	.91	.8281
5	−.09	.0081
4	−1.09	1.1881
5	−.09	.0081
6	.91	.8281
Sums 56	.01	12.9091

$$\overline{X} = 5.09 \qquad S^2 = 1.17$$
$$S = 1.08$$

for the variance because the variance is the square of the standard deviation. (More properly, S is defined as the positive square root of the variance.) The definition of S in the form of an equation is

$$S_X = \sqrt{\frac{\sum x_i^2}{N}}$$

or

$$S_X = \sqrt{\frac{\sum (X_i - \overline{X})^2}{N}}$$

(4.9)

The two equations are equivalent and give identical numerical results if used to compute the standard deviation of the scores of N individuals on variable X.

We will use the anxiety scores of our 11 male students from Table 3-3, repeated here as Table 4-7, to illustrate how to compute the variance and standard deviation from the definitional formulas 4.8 and 4.9 and to show how the variance and standard deviation summarize the spread of individuals around their mean. The 11 raw scores are listed in the first column of the table. You should use formula 3.2 to show that the mean of these scores is 5.09. The deviation of each person from that mean is shown in column 2, labeled x, and the squares of the deviation scores, x^2, are shown in column 3. The sums of the three columns are shown in the bottom row.

Formula 4.8 tells us that the variance is the mean of squared deviation scores, or the mean of the values in column 3. Therefore, the variance of these 11 scores around their mean is

$$12.9091/11 = 1.1736 \text{ or } 1.17.$$

The standard deviation is the square root of this value, or 1.08.

One of the properties of the standard deviation is that, unlike the interquartile range, it uses information from all of the scores. Earlier in this chapter (Table 4-4) we saw that two quite different distributions could produce the same interquartile range. We now look at how these two sets of scores behave when computing the standard deviation.

Using the ten scores from part A of Table 4-4, the steps for computing the variance and standard deviation are shown in columns 1–3 of Table 4-8. Each of the ten scores is listed in column 1; its deviation from the group's mean of 50 is given in the second column; and the squared deviations are given in column 3. The same information for the scores from part B of Table 4-4 is given in columns 4–6, and the two S's and S^2's are computed below. Although the interquartile ranges for these two distributions are identical, the

Table 4-8 Extreme Scores Have a Large Effect on the Variance and Standard Deviation

Part A Scores			Part B Scores		
X_i	x_i	x_i^2	X_i	x_i	x_i^2
0	−50	2500	47	−3	9
49	−1	1	49	−1	1
49	−1	1	49	−1	1
50	0	0	50	0	0
50	0	0	50	0	0
50	0	0	50	0	0
50	0	0	50	0	0
51	1	1	51	1	1
51	1	1	51	1	1
100	50	2500	53	3	9

Calculations

$\sum X_i = 500, \quad \sum x_i = 0$

$\overline{X} = 50, \quad \sum x_i^2 = 5004$

$S_X^2 = \dfrac{\sum x_i^2}{N} = \dfrac{5004}{10} = 500.4$

$S_x = \sqrt{S_X^2} = \sqrt{\dfrac{\sum x_i^2}{N}}$

$\quad = \sqrt{500.4} = 22.37$

$\sum X_i = 500, \quad \sum x_i = 0$

$\overline{X} = 50, \quad \sum x_i^2 = 22$

$S_X^2 = \dfrac{\sum x_i^2}{N} = \dfrac{22}{10} = 2.2$

$S_x = \sqrt{2.2} = 1.48$

variances and standard deviations are markedly different, reflecting the influence of the two extreme scores in distribution A. Like the mean, S and S^2 are affected by the presence of extreme scores, while the interquartile range, like the median, is not. We will examine other properties of S and S^2 after we explore simpler ways to compute these two indices.

Computing S and S^2 from Raw Data

Formulas 4.8 and 4.9 are known as definitional formulas for the variance and standard deviation, and they express in a fairly direct way the mathematical meaning of these two statistics. However, we would rarely use these formulas to compute a variance or standard deviation with data from an actual study because the mean is seldom a whole number, and when it is not, computations using these formulas can be extremely tedious and rounding error is introduced. With a little algebraic sleight of hand, it is possible to convert these equations into ones that look more complicated but are really much easier to use. (Computers and pocket calculators generally use the computational formulas that follow because they are more accurate, require fewer steps than the definitional formulas, and provide quantities that can be used in calculations for other statistics.)

There are two useful computational formulas for the variance. Each of them involves finding the quantity

$$\sum_{i=1}^{N} X_i^2$$

which is found by squaring each individual's score and taking the sum of these squares. For the math anxiety scores in Table 4-7 this means finding $6^2 + 5^2 + \cdots + 5^2 + 6^2 = 298$. If the mean has already been computed or will be needed later, it may be convenient to use the formula

$$S_X^2 = \frac{\sum X_i^2}{N} - \overline{X}^2 \qquad\qquad (4.10)$$

That is, we find the sum of squared raw scores, divide this quantity by N and subtract the square of the mean from the result. This series of operations gives us the variance. The standard deviation, of course, is the square root of the variance.

Table 4-9 Computing the Variance and Standard Deviation for the Math Anxiety Scores of 11 Male Students Using Raw Score Formulas

X	X²	Formula 4.10
6	36	$S_X^2 = \dfrac{\sum X_i^2}{N} - \overline{X}^2$
5	25	
3	9	$= (298/11) - 5.0909^2$
5	25	$= 27.0909 - 25.9174$
		$= 1.1735$
7	49	**Formula 4.12**
4	16	$S_X^2 = \dfrac{N\sum X_i^2 - (\sum X_i)^2}{N^2}$
6	36	
5	25	$= \dfrac{11(298) - 56^2}{11^2}$
4	16	$= (3278 - 3136)/121$
5	25	$= 142/121$
6	36	$= 1.1736$
Sums 56	298	$S = 1.0833$

However, using the mean can still involve decimals and may introduce a very small rounding error. An alternate formula for the variance that requires only one division step is

$$S_X^2 = \frac{\sum X_i^2 - \dfrac{(\sum X_i)^2}{N}}{N} \qquad (4.11)$$

or its equivalent

$$S_X^2 = \frac{N\sum X_i^2 - (\sum X_i)^2}{N^2} \qquad (4.12)$$

Both equations substitute the formula for the mean in the second part of the equation, then manipulate that expression to simplify calculations.

These formulas are particularly convenient to use with calculators that can accumulate sums of scores and sums of squared scores in a single step. In Table 4-9, we illustrate the steps in calculating S_X and S_X^2 using formulas 4.10 and 4.12 with the math anxiety scores of our 11 male students from Table 4-7. These calculations show that the equations for S and S^2 all give exactly the same results as the definitional formula (4.9) (within a rounding error determined by the accuracy with which the mean has been calculated). The difference of .0001 between the two results is due to rounding of the mean. If the mean is a whole number, the formulas agree exactly, but if, as is usually the case, it is a decimal, formula 4.12 gives slightly more accurate results. Notice, also, that each quantity needed in formula 4.11 or 4.12 is simply the sum of the values in one of the columns.

▶ *Now You Do It*

Now, let's take a slightly larger example. How much variability is there in scores on the statistics final exam? The data from Table 2-2 are repeated here in Table 4-10. Here, because there are 68 scores, we would probably want to use a computer, but let's go through the process with a calculator so you can see the steps. Every calculator is different, so you will have to check the manual for yours to get the correct sequence.

First, what quantities do we need? Clearly, we need the sum of scores, but we have that from calculating the mean in Chapter 3. There we found the sum of scores to be 2328 and the mean was 34.24. We also need N, but you know that is 68. The only other

Table 4-10 Final Exam Scores for 68 Students in the BTU Statistics Course

Student ID	Score on Final	Student ID	Score on Final	Student ID	Score on Final
1	37	24	34	47	25
2	47	25	61	48	43
3	19	26	44	49	40
4	37	27	55	50	29
5	33	28	19	51	55
6	48	29	50	52	32
7	20	30	23	53	28
8	44	31	48	54	59
9	44	32	34	55	41
10	34	33	13	56	54
11	38	34	65	57	29
12	14	35	25	58	16
13	4	36	38	59	30
14	33	37	46	60	30
15	35	38	25	61	18
16	37	39	34	62	42
17	8	40	7	63	14
18	37	41	47	64	43
19	36	42	25	65	39
20	31	43	20	66	56
21	22	44	22	67	39
22	49	45	21	68	29
23	28	46	45		

quantity we need is the sum of the squared scores, ΣX_i^2. That is, we need to find $37^2 + 47^2 + \cdot\cdot\cdot + 39^2 + 29^2$. If you square each score in the table and add them up, you should get 92,056. Putting these values in formula 4.12 yields

$$S^2 = \frac{68(92,056) - (2328)^2}{68^2} = \frac{6259808 - 5419584}{4624}$$

$$= 840224/4624$$

$$= 181.71$$

$$S = \sqrt{181.71} = 13.48$$

Formula 4.11 produces identical results, and formula 4.10 gives us $S^2 = 181.39$ and $S = 13.47$. The difference, of course, is due to rounding the mean to two decimal places. (If you used a computer to find the variance and standard deviation of final exam scores or if you pressed the key for the variance on your calculator, you probably got slightly different values from the ones given here. The computer probably gave you $S^2 = 184.42$ and $S = 13.58$. We will address the reason for these differences shortly.)

Computing S and S^2 from a Frequency Distribution

When our data are in the form of a frequency distribution, the computations of the standard deviation and variance require a slight revision in our approach. This revision is exactly the same as the one we encountered when we computed the mean from a frequency distribution or grouped frequency distribution. Recalling that when dealing with data in a frequency distribution, we multiply each score value by its frequency, we can rewrite equation 4.7 as

$$S_X^2 = \frac{\sum_{j=1}^{J} f_j x_j^2}{N} \tag{4.13}$$

In this equation, there are J separate score values, each with its own frequency f_j. Of course, we can substitute $(X_j - \overline{X})^2$, the definition of a deviation score, for x_j^2, and N is equal to the sum of the individual interval frequencies or Σf_j. The details of how this

Table 4-11 Computing the Variance and Standard Deviation for Math Anxiety Scores of 11 Male Students from a Frequency Distribution

X	f	x	x²	fx²
7	1	1.91	3.6481	3.6481
6	3	.91	.8281	2.4843
5	4	−.09	.0081	.0324
4	2	−1.09	1.1881	2.3762
3	1	−2.09	4.3681	4.3681

$$\Sigma fx^2 = 12.9091$$

$$S^2 = \frac{12.9091}{11} = 1.17$$

$$S = \sqrt{1.17} = 1.08$$

formula works are illustrated in Table 4-11 using the data from Table 4-7. Note that we get exactly the same sum of squared deviation scores that we got for these data in Table 4-7.

We would not ordinarily use equation 4.13 to compute S^2 for the same reason that we would not use equation 4.7. Instead, we note that the following equalities hold for data in a frequency distribution:

$$\sum_{j=1}^{J} f_j X_j = \sum_{i=1}^{N} X_i$$

$$\sum_{j=1}^{J} f_j X_j^2 = \sum_{i=1}^{N} X_i^2$$

$$\sum_{j=1}^{J} f_j = N$$

and, by substituting these terms into equations 4.10 and 4.12, we obtain

$$S_X^2 = \frac{\sum_{j=1}^{J} f_j X_j^2}{\sum_{j=1}^{J} f_j} - \overline{X}^2 \tag{4.14}$$

and

$$S_X^2 = \frac{\sum_{j=1}^{J} f_j \left(\sum_{j=1}^{J} f_j X_j^2 \right) - \left(\sum_{j=1}^{J} f_j X_j \right)^2}{\left(\sum_{j=1}^{J} f_j \right)^2} \tag{4.15}$$

These equations, particularly 4.15, look formidable indeed, but they really only represent terms computed by summing the columns of the frequency distribution. The standard deviation is found by taking the square root of the result from either formula.

We will use the frequency distribution of statistics quiz scores for the 340 students in the SAD course at BTU (from Table 3-4, which has been reproduced here as Table 4-12) to illustrate the use of equations 4.14 and 4.15. This example demonstrates the advantages of using frequency distributions and the appropriate equations for computing S^2 or S when the number of observations is fairly large. Rather than squaring and summing each of 340 scores, we can carry out all computations on only the eight score values and their frequencies. Even with the aid of an advanced modern calculator, the computations are easier, and the likelihood of error is less when working with the frequency distribution.

Using formula 4.14 in Table 4-12, we take the sum from the column labeled $f_j X_j^2$, which is 17,651, and divide it by the sum from the column labeled f_j. We then subtract the square of the mean (7.1088^2). The resulting variance is 1.3797. Using formula 4.15 with the same

Table 4-12 Computing S^2 for Statistics Quiz Scores from the Frequency Distribution of Raw Scores

X_j	f_j	$f_j X_j$	X_j^2	$f_j X_j^2$
10	8	80	100	800
9	26	234	81	2,106
8	84	672	64	5,376
7	132	924	49	6,468
6	67	402	36	2,412
5	15	75	25	375
4	6	24	16	96
3	2	6	9	18
$\Sigma f_j = 340$		$\Sigma f_j X_j = 2{,}417$		$\Sigma f_j X_j^2 = 17{,}651$

Calculations

$\overline{X} = 7.1088$

Using formula 4.14 $\qquad S_X^2 = \dfrac{\sum\limits_{j=1}^{J} f_j X_j^2}{\sum\limits_{j=1}^{J} f_j} - \overline{X}^2$

$S_x^2 = \dfrac{17{,}651}{340} - (7.1088)^2 = 51.9147 - 50.5350$

$\qquad = 1.3797 \qquad S_x = 1.175$

Using formula 4.15 $\qquad S_X^2 = \dfrac{\sum\limits_{j=1}^{J} f_j \left(\sum\limits_{j=1}^{J} f_j X_j^2\right) - \left(\sum\limits_{j=1}^{J} f_j X_j\right)^2}{\left(\sum\limits_{j=1}^{J} f_j\right)^2}$

$S_x^2 = \dfrac{340(17{,}651) - (2{,}417)^2}{(340)^2} = \dfrac{6{,}001{,}340 - 5{,}841{,}889}{115{,}600}$

$\qquad = \dfrac{159{,}451}{115{,}600} = 1.3793 \qquad S_x = 1.174$

data, we multiply the sum of the $f_j X_j^2$ column by the sum from the f_j column. We subtract the square of the sum from the $f_j X_j$ column from this product. Finally, we divide that result by the square of the sum from the f_j column to obtain 1.3793. The difference between the two variances is due to rounding error in computing the mean in formula 4.14.

Any of the equations 4.13–4.15 can also be used when the data are presented in the form of a grouped frequency distribution. The only adjustment that we have to make in our thinking is to remember that we are using midpoints of the intervals as the score values. Applying formula 4.15 to grouped frequency data is illustrated in Table 4-13 for our data from Table 3-7 with an interval of 5. For the sake of comparison, we note that the standard deviation of this set of scores from raw data was $S = 13.48$. The difference between this value and the value of 13.43 found in Table 4-13 is due to inaccuracies introduced by grouping.

Scores are grouped for convenience and in order to obtain a better view of the distribution with the knowledge that grouping may have small and unpredictable effects on the statistics computed from the data. However, the wide availability of modern computational technology has made it possible and relatively painless to have the best of both worlds. Outside the statistics classroom, few people are likely to encounter a situation where a large amount of data must be analyzed by a device as primitive as a pocket calculator. Even the smallest personal computer will have sufficient memory for a sophisticated statistics program and the original data and can produce a frequency distribution and a mean and standard deviation computed from the raw data. This enables the user to have the advantageous overview provided by grouping without introducing grouping or rounding errors into the mean and standard deviation. The only case where you may have no alternative but to use the methods for calculating statistics from a frequency distribution as described here is when the data come to you already grouped.

Table 4-13 Calculation of S^2 and S from the Grouped Frequency Distributions of Statistics Final Exam Scores for an Interval of 5 from Table 3-7

Score Interval	Midpoint X	f	fX	X²	fX²
63–67	65	1	65	4225	4225
58–62	60	2	120	3600	7200
53–57	55	4	220	3025	12100
48–52	50	4	200	2500	10000
43–47	45	9	405	2025	18225
38–42	40	7	280	1600	11200
33–37	35	12	420	1225	14700
28–32	30	9	270	900	8100
23–27	25	5	125	625	3125
18–22	20	8	160	400	3200
13–17	15	4	60	225	900
8–12	10	1	10	100	100
3–7	5	2	10	25	50

$$\Sigma f_j = 68 \qquad \Sigma f_j X_j = 2345 \qquad \Sigma f_j X_j^2 = 93125$$

$$S^2 = \frac{68(93125) - (2345)^2}{68^2} = 180.25$$

$$S = 13.43$$

▶ *Now You Do It*

Let's apply the techniques you have learned in this section to the GPA data for the SAD statistics students, first as raw scores, then from the grouped frequency distribution we prepared in Chapter 2. What three quantities do we need?

The sum of the raw GPA's is 165.43, and the sum of the squared GPA's is 437.50. N, of course, is 68, so we can find the variance to be

$$S^2 = \frac{68(437.50) - 165.43^2}{68^2} = \frac{29750 - 27367.08}{4624}$$

$$= 2382.92/4624 = .515$$

$$S = .718$$

If you did this analysis on a computer, you probably got a slightly different result. The computer would have given you .524 for the variance and .724 for the standard deviation. The reason for this difference is explained later in the section on bias and statistical estimation.

The grouped frequency distribution gives slightly different results. Using the grouped distribution for these data that we prepared in Chapter 2, ΣfX^2 is 428.0742, and ΣfX is 163.84. These quantities give us the following:

$$S^2 = \frac{68(428.0742) - 163.84^2}{68^2} = \frac{29109.05 - 26843.55}{4624}$$

$$= 2265.5/4624 = .490$$

$$S = .700$$

Properties of the Standard Deviation

For most situations where a statistic reflecting the spread of a set of measures is needed, the standard deviation is the best statistic to use. It is more stable from one group to another than either the range or the interquartile range; it is more representative of the group in the sense that each score is included in its computation, and it provides a single index that is unique for the particular set of scores. (See Table 4-14.) However, its primary virtue lies in the fact that the standard deviation or its square, the variance, can be used in further computations and in the development of additional statistical concepts. In prepa-

Table 4-14 Properties of the Standard Deviation	
Advantages	**Disadvantage**
Most stable measure of spread	Affected by extreme scores
Most representative	
Unique	
Is used in further computations	

ration for our later discussions, we must now focus on two terms that are closely related to the standard deviation.

Sum of Squares

The term **sum of squares** (usually given the symbol SS) appears frequently in statistics; it always refers to a sum of squared deviations, usually the squared deviations from a mean. We have already used the sum of squares in developing the concept of the variance. The expressions $\sum x_i^2$ and $\sum(X_i - \overline{X})^2$ both refer to the sum of squared deviations from the group's mean, where the summation is across all members of the group. The variance may then be defined as the sum of squares divided by the number of observations.

One reason for focusing attention on the sum of squares is that there are occasions when we need to combine the variances of two or more groups. Sums of squares can be added together for different groups, while variances and standard deviations generally cannot. A combined variance for two or more groups is called a **pooled variance.** To find the pooled variance, we add together the sums of squares and divide the total sum of squares by the total number of cases. For example, if we know that two classes took the same final exam in statistics, that the standard deviation in one class with 20 students was 5, and the standard deviation in the other class with 25 students was 10, what is the pooled variance? To find the answer, we must work backwards. Given $S_1 = 5$ and $S_2 = 10$, we find $S_1^2 = 25$ and $S_2^2 = 100$. The variance is defined as the sum of squares divided by N. Therefore, we can find the sum of squares (SS) by multiplication

> The **sum of squares** is the sum of the squared deviations of a set of scores around their mean.

> A **pooled variance** is a variance that results from combining the sums of squares from two or more groups. It is found by dividing the pooled sum of squares by the total number of cases.

$$SS = N(S^2).$$

In our example, $SS_1 = 20(25) = 500$ and $SS_2 = 25(100) = 2500$. We find the total sum of squares for the combined group (SS_P) by addition

$$SS_P = SS_1 + SS_2$$
$$= 500 + 2500 = 3000.$$

The resulting sum of squares has been obtained by summing 45 deviations, 20 in the first class and 25 in the second, so to find the pooled variance, we divide the pooled sum of squares by the total number of deviations that were summed.

$$S_P^2 = \frac{SS_P}{N_T} = \frac{SS_1 + SS_2}{N_1 + N_2}$$
$$S_P^2 = 3000/45 = 66.67$$

The simple average of the two variances would be

$$\frac{25 + 100}{2} = 62.5,$$

which is not the correct answer.

The pooled standard deviation is $S_P = 8.16$. Notice that this value differs from what would be obtained by averaging the two original standard deviations to get 7.5. We can add sums of squares to obtain pooled variances and standard deviations, but we cannot average the variances or standard deviations directly.

Least Squares

The other important concept related to the variance and standard deviation is the concept of least squares. Here we tie together the mean as the index that is generally most descriptive of the scores in a distribution and the standard deviation as an index of the accuracy of that description.

Least squares is not a statistic or a numerical value; it is a property that a statistic may possess or a criterion that a statistic may or may not fulfill. *It is a principle that states that we should select as a descriptive value for a set of data the statistic that has the smallest sum of squares.* Thus, if we follow the principle of least squares in selecting a single value to describe the final exam scores of our 68 statistics students, we must select the value for which the sum of squared deviations from that selected value

$$\sum_{i=1}^{N} (X_i - A)^2$$

(where A might be anything) is a minimum.

It can be shown that for any group of scores *the mean of the group satisfies the principle of least squares.* The quantity $\sum(X_i - \overline{X})^2$ will be smaller than the sum of squared deviations from any other value. It is the value in the center of the distribution and the one most descriptive of each score in the sense that the sum of deviations from the mean is a minimum value. The variance and standard deviation are indices of how well the mean fulfills this descriptive function. *The fact that the mean satisfies the principle of least squares is a strong argument in favor of using the mean as an index of central tendency.* We will employ the principle of least squares again in our discussion of correlation and regression.

Bias and Statistical Estimation (Why Formula 4.12 and Your Computer Disagree)

Earlier we said that one of the purposes of statistics is to estimate the parameter in a population from results obtained in a sample. Whenever we compute a statistic, it can be viewed as an estimate of the corresponding population parameter. Thus, the sample mean is an estimate of the mean in the population from which the sample was drawn, and the same is true of each of the other indices we have discussed so far.

We also noted in Chapter 3 several desirable properties that a statistic should have. These included uniqueness, stability, representativeness, and the ability to be used in further computations. At this stage we can add one more property to the list. We would like our statistics to be **unbiased.**

A statistic is said to be **biased** if it is systematically either larger or smaller than its corresponding parameter. All of the measures of central tendency discussed in Chapter 3 are unbiased. What this means is that if we were to take a large number of samples from a population and compute the mean (or median or mode) in each, the mean of these means (or medians or modes) would be very close to the population mean (or median or mode). The same is true of the range and interquartile range as estimates of their respective parameters. This is not true, however, of the variance and standard deviation. Both S and S^2 are **negatively biased,** which means that the sample values tend to be systematically smaller than the population values. We will discuss the reason for this bias in detail in Chapter 12.

HOW THE COMPUTER DOES IT

The variance that we defined in equation 4.8 is the correct variance for describing the sample at hand. That is, it is the proper descriptive value for the sample. The same is true for the standard deviation as the square root of this variance. In Chapter 12, when we are interested in drawing inferences about populations, we will need to understand why the sample variance is a biased estimate and how we introduce a correction for that bias. Until then, we will use the descriptive sample variance in our examples and discussion. However, most computer programs and calculators automatically correct for the bias in the variance and standard deviation and do not give you the option of obtaining the proper descriptive value. The correction is simple; we merely substitute $(N - 1)$ for N in the denominator of equation 4.8. Any sample variance can be converted to an unbiased estimate of its population variance by multiplying the sample variance by $N/(N - 1)$, and any unbiased estimate can be converted into the descriptive sample variance by multiplying by $(N - 1)/N$. (Note that these conversions are applied to the variances, not the standard deviations.) Therefore, if the results you get differ from the ones we provide, first check to see whether your software computed the unbiased estimate.

Table 4-15 Comparison of the Range, Interquartile Range, and Standard Deviation	
Advantages	**Disadvantages**
Range	
Easy to compute	Depends only on two scores
Provides a unique value	Unstable from group to group
Easy to understand	Not used in further computations
Interquartile Range	
Not affected by extreme scores	Not useful in other calculations
More stable than the range	
Provides a unique index	
Standard Deviation	
Most stable measure of spread	Affected by extreme scores
Most representative	
Unique	
Is used in further computations	

Choosing an Index of Variation

We now have three general ways to describe the variation or spread in a set of data. As was the case for measures of location, we should choose the index that (1) matches the properties of our data in terms of scale and (2) is most useful. (See Table 4-15.)

For nominal data the only measure that makes any sense is a count of the number of categories that have occurred. This is one of the interpretations of the range, so the range is the only index we can use.

With ordinal data we have both the range and the interquartile range available. Each emphasizes a different aspect of variation. The range is sensitive to extreme cases and shows the limits of variation in our data. If we want to focus attention on this aspect of the distribution, then the range is an appropriate statistic for the job. On the other hand, because it considers only the extremes, it is unrepresentative of variation in the middle of the distribution, and it is unstable from group to group.

The interquartile range and Q describe the variation in the middle half of the group. They depend on this half of the distribution for their result, but are totally unaffected by extreme scores. They are more stable than the range because they use information from more of the cases. For ordinal data the best solution probably is to report both of the available statistics, the range and interquartile range.

When the data have the properties of an interval scale, the standard deviation is almost always the best index to use. Because it is based on every score in the distribution, it is more stable than other measures, and it can be used in further procedures. Almost all of the remaining topics in this book will make use of the standard deviation or the variance. The only time when the standard deviation can be problematic is when the distribution is seriously asymmetric. With a highly skewed distribution a small number of scores in one tail of the distribution will have extremely large deviations and will inflate the variance and standard deviation.

Statistics of Skewness and Kurtosis

The effect of the shape of the frequency distribution on measures of location was introduced in Chapter 3. There we noted that there could be more extreme scores at one end of the distribution than the other and that this often occurs when the scale is curtailed in one direction or when the measure is not appropriate for the group. For example, reaction time cannot be less than zero, but it can be very long. Test scores cannot go below zero or above perfect, but they can bunch up at one end or the other if the test is too easy or too hard. The result of such occurrences is a skewed distribution, and the skew can be positive or negative depending on the direction in which the extreme scores lie.

A second way in which distributions can vary is in their degree of "peakedness." This property is shown in Figure 4-3 and is called **kurtosis.** Distributions that are low and flat, such as the one in Figure 4-3(a), are called *platykurtic,* while those that climb to a sharp

Kurtosis *refers to the degree of flatness or peakedness in a distribution.*

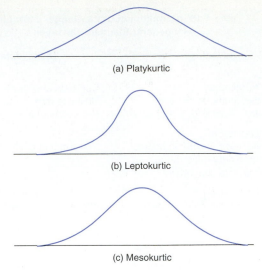

Figure 4-3 Three distributions differing in kurtosis: (a) platykurtic, (b) leptokurtic, and (c) mesokurtic.

(a) Platykurtic

(b) Leptokurtic

(c) Mesokurtic

peak [Figure 4-3(b)] are said to be *leptokurtic*. Distributions of moderate peakedness [Figure 4-3(c)] are *mesokurtic*. Two distributions can have the same variance and differ in their degree of kurtosis. The complete description of a frequency distribution requires that we report the deviation from symmetry, the skewness, and the deviation from middle peakedness, the kurtosis. Knowledge of skewness and kurtosis can also help us decide which statistical methods are appropriate for a particular set of data. Some of the more advanced procedures require that the data have a symmetric mesokurtic shape known as the normal distribution.

Measures of Skewness

We noted in Chapter 3 that one of the effects of skewness is to move the mean away from the median in the direction of the skew. One way to describe the degree of skewness is therefore to compare the median and the mean. *Pearson's index of skewness,* which uses this relationship to quantify the degree of skew, is computed as

$$P_S = \frac{3(\overline{X} - Mdn)}{S_X} \tag{4.16}$$

This index will be positive for positively skewed distributions, zero for symmetric distributions, and negative for negatively skewed distributions. Larger numbers indicate more severe skewness.

An alternative measure of skew is based only on deviations from the mean. The fundamental definition is

$$Skew = \frac{\sum (X_i - \overline{X})^3}{N(S_X^3)} \tag{4.17}$$

That is, we find the deviation of each person from the mean, raise those deviations to the third power, sum the results, and divide by N times the cube of the standard deviation.

Applying equations 4.16 and 4.17 to the mathematics aptitude scores for the 68 statistics students at Big Time U gives us the following values for skewness:

$$P_S = \frac{3(53.53 - 54.67)}{11.44} = -.299$$

$$Skew = .135$$

Both of these values are very small, indicating that the distribution is approximately symmetric. The fact that the signs are in opposite directions is a function of the differences between the formulas. Equation 4.17 is the more precise index, and whenever there is more than minimal skew, the two indices will agree quite closely.

Many computer programs provide this latter index of skewness as part of their descriptive statistics package, and a little reflection will reveal its properties. Remember that about half of the deviations from the mean are negative and that the sum of the deviations is zero. When we cube these deviations, the negatives remain negative, and cubing gives relatively greater weight to the larger deviations. A symmetric distribution will have an equal number of large deviations in both directions. If the distribution is negatively skewed, there will be more large negative deviations than positive ones. Conversely, a positively skewed distribution will have more large positive deviations. Therefore, this skewness index is similar to Pearson's index in that it will be zero for symmetric distributions and otherwise its sign will indicate the direction of skew. Larger values mean a greater degree of skewness. The index given by equation 4.17 has the advantage that it can be used in further tests that will be described in a later chapter.

A Measure of Kurtosis

The most commonly used index of kurtosis is similar to the second skewness index in that it is based on deviations from the mean. The definition of the kurtosis index is

$$K = \frac{\sum (X_i - \overline{X})^4}{N(S_X^4)} - 3 \qquad (4.18)$$

The numerator is the sum of the fourth powers of the deviations, and this sum is divided by N times the fourth power of the standard deviation (which is also the square of the variance). The value 3 is subtracted from the entire statistic to make K have a value of 0 for the particular mesokurtic distribution called the normal distribution (which we will meet in Chapter 6). Negative values of K indicate a flat or platykurtic distribution, while positive values reveal a leptokurtic or highly peaked distribution. Larger deviations from zero indicate more extreme kurtosis. The result of applying equation 4.18 to the mathematics aptitude data yields a value of $-.795$, implying a slightly platykurtic distribution.

HOW THE COMPUTER DOES IT

You can obtain all of the statistics described in this chapter simply by checking the appropriate boxes in the Statistics submenu of the FREQUENCIES menu from the STATISTICS-SUMMARIZE procedure. None of these summary statistics is provided automatically, so you must select the ones you want. We have selected all of them. The screen will look like this.

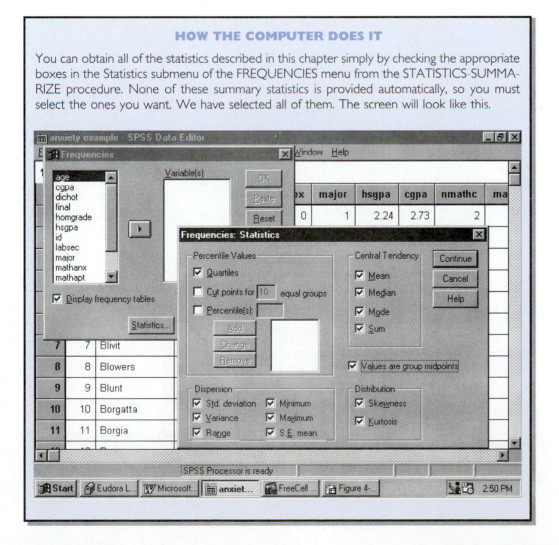

The SPSS default output comes in a long string that is not suited for direct tabular presentation. You will have to experiment with editing the output to get the look you want. One option looks like this. The footnotes show that we treated the data as though they were interval midpoints.

Statistics

			Score on Stat Final
N	Valid	Statistic	68
	Missing	Statistic	0
Mean		Statistic	34.24
		Std. Error	1.65
Median		Statistic	34.40[a]
Mode		Statistic	37
Std. Deviation		Statistic	13.58
Variance		Statistic	184.42
Skewness		Statistic	−.016
		Std. Error	.291
Kurtosis		Statistic	−.383
		Std. Error	.574
Range		Statistic	61
Minimum		Statistic	4
Maximum		Statistic	65
Sum		Statistic	2328
Percentiles	25.00	Statistic	24.20[b]
	50.00	Statistic	34.40
	75.00	Statistic	43.80

[a]Calculated from grouped data.
[b]Percentiles are calculated from grouped data.

As an alternative, you can use the DESCRIPTIVES menu from the STATISTICS-SUMMA-RIZE procedure. The same basic statistics are provided, except the median and quartiles, but several are computed automatically. The output from either program also provides values called *standard errors* for the mean and for the skewness and kurtosis indices. We will explain what these values mean in a later chapter. They have to do with the amount of variation you can expect in the value of the statistic from sample to sample. Remember, the computer will probably give you the estimate of the population variance and standard deviation rather than the values that are descriptive of your particular sample, but you can get the descriptive variance by multiplying the value from the program by $(N − 1)/N$.

SUMMARY

There are three widely used indices of the spread or variability of a set of scores. The *range* is the distance on the scale from the lower limit of the lowest observed score category to the upper limit of the highest observed score category. The range also equals the number of score values covered by the distribution. The range requires an interval scale of measurement for the first interpretation.

The *interquartile range* is the number of measurement units needed to encompass the middle 50 percent of the distribution. The *first quartile* (Q_1) identifies the lowest quarter of the distribution, the *third quartile* (Q_3) cuts off the top quarter of the distribution, and the interquartile range is $Q_3 − Q_1$. The interquartile range is an appropriate index of variation for data that have been obtained by ordinal scale measurement.

The most generally useful and widely used indices of variability are the *standard deviation* and its square, the *variance*. The standard deviation may be used any time it is appropriate to compute a mean. It is computed from the *sum of squares* of deviations from the mean. The mean satisfies the *principle of least squares,* which states that the statistic having the smallest sum of squares should be selected as a descriptive value for a set of data, and the variance is an index of the variation around the mean.

The description of a distribution is completed by computing an index of *skewness,* usually using the cubed deviations, and an index of *kurtosis* obtained from the fourth powers of the deviations.

MEASURES OF VARIATION OR SPREAD FOR THE SEARCHERS DATA

Professor Cross and his students wish to describe the amount of variation that their participants showed on several of the measures that were collected. You can check your understanding of the topics in this chapter by computing each of the indices we have discussed. Table 4-16 provides the quantities needed to compute the range, interquartile range, variance, and standard deviation for the data from Dr. Cross's study of the Searchers. For each variable, Q_1 and Q_3 were computed

Table 4-16 Quantities Used to Determine the Quartiles, Interquartile Range, Variance, and Standard Deviation for the Searchers Study Variables

Age

		Raw Scores	$J = 3$
Range		$(59 - 20) + 1 = 40$	$(60 - 21) + 3 = 42$
Q_1	$.25N = 15$		
	LL_j	28.5	28.5
	cf_{j-1}	14	14
	f_j	2	6
	Definition 1	29.0	29.0
	Definition 2	29.0	30.0
Q_3	$.75N = 45$		
	LL_j	48.5	46.5
	cf_{j-1}	44	41
	f_j	2	5
	Definition 1	49.0	48.9
	Definition 2	49.0	48.0
Interquartile Range		20.0	19.9
Sum of Scores		2383	2385
Sum of Squared Scores		102,573	102,879
S^2	Descriptive	132.14	134.59
	Estimate	134.38	136.87
S	Descriptive	11.50	11.60
	Estimate	11.59	11.70

Income

		Raw Scores	$J = 3000$
Range		30,952	36,000
Q_1	$.25N = 15$		
	LL_j	40,997.5	40,500
	cf_{j-1}	14	13
	f_j	1	11
	Definition 1	40,998.5	41,045.5
	Definition 2	40,998	42,000
Q_3	$.75N = 45$		
	LL_j	50,581.5	49,500
	cf_{j-1}	44	41
	f_j	1	9
	Definition 1	50,842.5	50,833.3
	Definition 2	50,842	51,000
Interquartile Range		9844	9787.8
Sum of Scores		2,715,252	2,718,000
Sum of Squared Scores		125,995,538,700	126,270,000,000
S^2	Descriptive	51,983,028.12	52,410,000
	Estimate	52,864,096.39	53,298,305
S	Descriptive	7209.93	7239.48
	Estimate	7270.77	7300.57

Table 4-16 *Continued*

Attitude Toward Equality

		Raw Scores	J = 3
Range		(60 − 24) + 1 = 37	(60 − 24) + 3 = 39
Q_1	.25N = 15		
	LL_j	33.5	31.5
	cf_{j-1}	14	11
	f_j	2	5
	Definition 1	34.0	33.9
	Definition 2	34.0	33.0
Q_3	.75N = 45		
	LL_j	44.5	43.5
	cf_{j-1}	39	35
	f_j	6	12
	Definition 1	45.5	46.0
	Definition 2	45.0	45.0
Interquartile Range		11.5	12.1
Sum of Scores		2424	2421
Sum of Squared Scores		101,980	101,745
S^2	Descriptive	67.51	67.62
	Estimate	68.65	68.77
S	Descriptive	8.22	8.22
	Estimate	8.29	8.29

Degree of Religious Involvement

		Raw Scores	J = 3
Range		(31 − 6) + 1 = 26	(30 − 6) + 3 = 27
Q_1	.25N = 15		
	LL_j	13.5	13.5
	cf_{j-1}	8	8
	f_j	8	16
	Definition 1	14.38	14.81
	Definition 2	14.0	15.0
Q_3	.75N = 45		
	LL_j	20.5	19.5
	cf_{j-1}	44	43
	f_j	3	7
	Definition 1	20.83	20.36
	Definition 2	21	21
Interquartile Range		6.45	5.55
Sum of Scores		1086	1077
Sum of Squared Scores		21,284	20,889
S^2	Descriptive	27.12	25.95
	Estimate	27.58	26.39
S	Descriptive	5.21	5.09
	Estimate	5.25	5.14

using formula 4.5, and the variance and standard deviations are sample descriptive values. The values you would need to compute the variance and standard deviation from a frequency distribution are also given. Note that when the data have been grouped into intervals, the grouping has a small effect, but when the frequency distribution has not been grouped the results are the same as with raw data. In addition, the values for the quartiles using the alternate definition are given, as are the variance and standard deviation that estimate their respective population parameters.

REFERENCE

Kirk, R. E. (1999). Statistics: An introduction (4th ed.). Orlando, FL: Harcourt Brace.

EXERCISES

1. In an effort to summarize data, researchers rely on measures of central tendency. However, this may limit researchers' interpretations of the information too severely. Explain why researchers need numerical indices of the variation or spread of their data to more completely characterize the information.

2. Suppose that you have the following set of scores on a measure of heart rate in response to a fearful stimulus such as a rattlesnake: 92, 88, 64, 110, 128, 77, 75, 93, 101, 91, 95, 112, 99, 85, 79, 68, 101, 105, 108, 96. What is the range of the data above? What are the two interpretations of this range? Suppose you have a second group of 20 participants to whom you show the same fearful stimulus that results in the following heart rate measures: 76, 72, 78, 68, 69, 70, 66, 74, 128, 79, 65, 72, 73, 71, 64, 78, 79, 70, 77, 72. What is the range of this second set of scores? Compare the ranges of the two distributions of scores and examine the scores in each distribution. What problem(s) do you see with using the range as a measure of variability?

3. Calculate the range in two ways: using the raw data as presented below and using the grouped frequency table you constructed for the data in Exercise 11 in Chapter 2.
   a. Use the scores on the feminine scale of the Bem Sex Role Inventory listed in Chapter 2, Exercise 11a.
   b. Use the scores on the masculine scale of the Bem Sex Role Inventory listed in Chapter 2, Exercise 11b.
   c. Use the scores on the measure of the cognitive level listed in Chapter 2, Exercise 11c.
   d. Use the scores on the measure of assertiveness listed in Chapter 2, Exercise 11d.

4. What is the interquartile range? The concept of interquartile range relies on knowing that the distribution is divided into four equal parts in terms of percent of scores. How many points are required to divide any distribution into four equal parts? What names do we give to these points and what do they represent? What is the semi-interquartile range? What is an alternate way of conceptualizing the semi-interquartile range?

5. Calculate the interquartile range in two ways: using the raw data as presented below and using the grouped cumulative and relative frequency table you constructed for the data in Exercise 12 in Chapter 2.
   a. Use the scores on the feminine scale of the Bem Sex Role Inventory listed in Chapter 2, Exercise 11a.
   b. Use the scores on the masculine scale of the Bem Sex Role Inventory listed in Chapter 2, Exercise 11b.
   c. Use the scores on the measure of the cognitive level listed in Chapter 2, Exercise 11c.
   d. Use the scores on the measure of assertiveness listed in Chapter 2, Exercise 11d.
   For each of the data sets, what are the semi-interquartile ranges for both the raw score method and the grouped frequency method?

6. Suppose you have a group of 20 participants to whom you show the same fearful stimulus, a syringe, that results in the following heart rate measures: 76, 72, 78, 68, 69, 70, 66, 74, 128, 79, 65, 72, 73, 71, 64, 78, 79, 70, 77, 72. What are the interquartile range and the range of this set of scores? Compare the interquartile range and the range of the distribution of scores and examine the scores in the distribution. Given this comparison, what can you conclude about these two measures of variability?

7. In Chapter 2 you were introduced to the box plot without much further elaboration. Construct a box plot of the data presented in Exercise 2. Suppose the two groups represent clients from two different rattlesnake phobia treatment approaches. Furthermore, suppose the two groups were initially no different on the measure of fear response, heart rate. What can you conclude about the groups based on the graphs?

8. For the box plots below, compare the different distributions in terms of their location and spread.
   a. According to gender role socialization theories, females should score higher than males on empathic concern. What can you conclude from the index of location and spread of the data? Justify your answer.

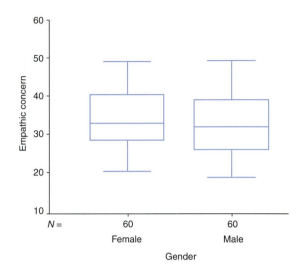

   b. The graph below may require some explanation. Forty rats were trained to run a maze for five weeks. At the end of the five-week training, the rats did not differ in their running speed and were randomly assigned to receive one of four types of water: water with no additives (control), water with 2% aluminum sulfate, water with 4% aluminum sulfate, and water with 8% aluminum sulfate. Aluminum sulfate is hypothesized to cause memory problems if it is ingested on a frequent basis. Based on the graph what can you conclude? Justify your answer.

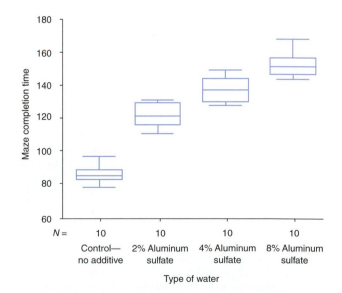

9. Calculate the variance in two ways: using the conceptual formula (4.8) and using the computational formula (4.12). Use the mean that you calculated for the data in Exercise 7 in Chapter 3 and round off all numbers to

two decimal places when doing the calculations. Once you have the variance, compute the standard deviations. For each set of data, do you get the same value for the variance using the two different formulas? If not, please explain why.

   a. Use the scores on the feminine scale of the Bem Sex Role Inventory listed in Chapter 2, Exercise 11a.

   b. Use the scores on the masculine scale of the Bem Sex Role Inventory listed in Chapter 2, Exercise 11b.

   c. Use the scores on the measure of the cognitive level listed in Chapter 2, Exercise 11c.

   d. Use the scores on the measure of assertiveness listed in Chapter 2, Exercise 11d.

10. Using the grouped frequency tables that you constructed in Exercise 11 from Chapter 2 and the grouped frequency formula for variance (4.15), compute the variances for each of the data sets in Exercise 9. Do these values differ from those you obtained in Exercise 9? If yes, please explain why.

11. In Exercise 9 you used a sum of squares when you calculated the variance using the conceptual formula (4.8). What does the concept of sum of squares entail? Why do we need to square these deviation values? What is the principle of least squares as it applies to the sum of squares?

12. The standard deviations and variances that you calculated in Exercises 9 and 10 are negatively biased. Explain what this means.

13. For each of the following data sets, identify which measure(s) of variation is (are) most appropriate and provide a rationale for your decision.

   a. A researcher is interested in test anxiety scores of clients who participated in an anxiety reduction workshop. The distribution of these anxiety scores is unimodal and symmetric.

   b. A first grade teacher is interested in determining which of several reinforcers seem to be the most potent based on the frequency of selection in a two-week test period. The reinforcers include candy, stickers, Pokemon cards, extra time to play games on the computer, being selected helper of the day, and extra time to play with Legos.

   c. Students from different schools participate in a district-wide speech contest. Students are rank ordered on 15 different speech categories. Schools are then compared on the ranks of their students in all of these categories, and an overall winner is selected.

   d. Forty-five rats are timed on a maze completion task. The results are highly variable. Two rats completed the maze in exceptionally fast times whereas two different rats took more than 10 times longer than the next slowest rat. The rest of the rats times were normally distributed.

   e. Blood phobic patients are presented with 10 words from four different categories: words that are highly related to blood, words that are highly related to the medical profession but not highly related to blood, words that are highly pleasant, and words that are unpleasant but not related to blood. A baseline measure of galvanic skin response is taken. The galvanic skin response is then measured in response to the words presented in random order. The researcher is interested in the category of words that led to the most frequent change in galvanic skin response.

14. Frequency distributions are typically described by their shapes. One way that a distribution is classified is by its symmetry or skewness. A second way that a distribution is classified is by its kurtosis. Explain the meaning of skewness and kurtosis.

15. When classifying frequency distributions in terms of skewness, the distributions usually fall into one of two categories: positively skewed or negatively skewed. What does it mean to have a frequency distribution that is positively skewed? What does it mean to have a frequency distribution that is negatively skewed? Draw a histogram that represents data that are positively skewed and a histogram that represents data that are negatively skewed.

16. When classifying frequency distributions in terms of kurtosis, the distributions fall into one of three categories: platykurtic, mesokurtic, or leptokurtic. Define each of these categories and draw a histogram that illustrates each category.

17. For the following sets of scores, calculate the measure of skewness using formula 4.17 and the measure of kurtosis using formula 4.18. Based on these values, what can you conclude about the shape of the frequency distribution?

   a. Scores on a statistics quiz: 8, 14, 13, 9, 11, 12, 10, 11, 12, 13, 9, 10, 11, 10, 12, 11

   b. Scores on an anxiety measure: 23, 21, 25, 23, 22, 24, 26, 21, 23, 22, 23, 25, 20, 23, 24, 22, 23, 24, 23, 23

   c. Scores on a measure of creativity: 62, 68, 63, 60, 62, 65, 66, 61, 62, 63, 61, 67, 65, 64, 62, 60, 64, 62, 63, 62, 61

   d. Scores on a science achievement test: 81, 80, 76, 80, 79, 78, 77, 81, 80, 79, 78, 80, 81, 79, 80

18. Using the responses from Exercise 3 in Chapter 2 and Exercise 6 in Chapter 3, determine the interquartile range, variance, standard deviation, index of skewness, and index of kurtosis for each of the following sets of data.

   a. Use the scores on the first statistics quiz from Chapter 2, Exercise 3a.

   b. Use only the following scores on the algebra test from Chapter 2, Exercise 3b: 22, 10, 27, 10, 12, 27, 16, 28, 11, 23, 19, 27, 20, 23, 20, 16, 13, 21, 11, 12, 14, 18, 22, 23, 13, 16, 31

   c. Use the scores on the measure of spatial visualization from Chapter 2, Exercise 3c.

19. For the following sets of data, calculate the measure of skewness using both formula 4.16 and formula 4.17. What do these values indicate about the skewness of the distribution of scores for each set of data? In addition, calculate the measure of kurtosis using formula 4.18 for each set of data. What does this value indicate about the kurtosis of the distribution of scores for each set of data? Note: You have already done many of the calculations necessary for using these formulas. Check the results of Exercise 11 in Chapter 2, Exercise 7 in Chapter 3, and Exercise 5 in this chapter.

   a. Use the scores on the feminine scale of the Bem Sex Role Inventory listed in Chapter 2, Exercise 11a.

   b. Use the scores on the masculine scale of the Bem Sex Role Inventory listed in Chapter 2, Exercise 11b.

   c. Use the scores on the measure of the cognitive level listed in Chapter 2, Exercise 11c.

   d. Use the scores on the measure of assertiveness listed in Chapter 2, Exercise 11d.

20. Use the SPSS program and SPSS data (math.sav) bank to complete the following:
   a. Find the interquartile range, range, standard deviation, variance, skewness, kurtosis, and quartiles for each of the variables that follow: middle school grade point average (msgpa), middle school mathematics grade point average (mathgpa), pre-instruction attitude toward mathematics (preatt), post-instruction attitude toward mathematics (postatt), pre-instruction score on a novel algebra problem solving test (preprbso), post-instruction score on a novel algebra problem solving test (postpbso), pre-instruction score on a general problem solving test (pregenps), post-instruction score on a general problem solving test (postgnps), pre-instruction score on the mathematics subtest of the Iowa Test of Educational Development (ited8), post-instruction score on the mathematics subtest of the Iowa Test of Educational Development (ited9), and grade point average in algebra for the academic year (algpa).
   b. Using the index of skewness and the index of kurtosis, describe the characteristics of the distribution of scores for each variable.
21. Use the SPSS program and SPSS data (self.sav) bank to complete the following:
   a. Find the range, standard deviation, variance, skewness, kurtosis, and quartiles for Japanese participants only (citizen = 1) for the following variables: Collectivism Scale Score (collect), Individually-Oriented Achievement Motivation (ioam), Socially-Oriented Achievement Motivation (soam), Individualistic Self-Esteem (indse), Collectivistic Self-Esteem Scale (cses), Independent Self-Construal (indsc), and Interdependent Self-Construal (intsc).
   b. Using the index of skewness and the index of kurtosis, describe the characteristics of the distribution of scores for each variable.
22. Use the SPSS program and SPSS data (self.sav) bank to complete the following:

   a. Find the range, standard deviation, variance, skewness, kurtosis, and quartiles for American participants only (citizen = 2) for the following variables: Collectivism Scale Score (collect), Individually-Oriented Achievement Motivation (ioam), Socially-Oriented Achievement Motivation (soam), Individualistic Self-Esteem (indse), Collectivistic Self-Esteem Scale (cses), Independent Self-Construal (indsc), and Interdependent Self-Construal (intsc).
   b. Using the index of skewness and the index of kurtosis, describe the characteristics of the distribution of scores for each variable.
23. Use the SPSS program and SPSS data (career.sav) bank to complete the following:
   a. Find the range, standard deviation, variance, skewness, kurtosis, quartiles for the following variables: confidence in doing mathematics (confmath), perception of mother's attitude toward mathematics (mother), perception of father's attitude toward mathematics (father), success at doing mathematics (succesat), perception of current teacher's attitude toward mathematics (teacher), perception of mathematics as a male domain (maledom), perception of the usefulness of mathematics (useful), anxiety about doing mathematics (anxiety), and perception of mathematics as a fun activity (mathfun).
   b. Using the index of skewness and the index of kurtosis, describe the characteristics of the distribution of scores for each variable.
24. Use the SPSS program and SPSS data (donate.sav) bank to complete the following:
   a. Find the range, standard deviation, variance, skewness, kurtosis, and quartiles for the following variables: perspective taking (perspect), fantasy (fantasy), empathic concern (empathy), personal distress (distres), and altruism (altruism).
   b. Using the index of skewness and the index of kurtosis, describe the characteristics of the distribution of scores for each variable.

Chapter 5

Describing the Individual

CHAPTER OBJECTIVES

After studying this chapter, you should be able to:

❏ Identify research situations where direct and indirect measurements are appropriate.

❏ Identify whether a scale represents absolute or relative measurement.

❏ State the importance of norm groups for interpreting what scores mean.

❏ Identify appropriate norm groups for different uses of measurement.

❏ Compute any percentile or percentile rank, given the frequency distribution.

❏ State the properties of percentiles and percentile ranks and identify which would be more appropriate for a particular research situation.

❏ Apply any linear transformation to a set of data.

❏ State the properties of Z-scores and identify situations where it is appropriate to use them.

❏ Compute measures of central tendency and variability for transformed scores.

Have you ever wondered why intelligence test scores have the values they do, or why Scholastic Achievement Test scores are given in hundreds? Where do the scales these tests use come from, and how do they get their meaning? Psychological and educational measurements differ in some fundamental ways from measures in other sciences. There are no inches of the mind or pounds of educational achievement. Reaching agreement on how to measure the variables of interest to educators and psychologists is one of the greatest challenges researchers in these disciplines face. In this chapter we will discuss some of their solutions to the problem of measuring human characteristics.

Introduction

Now that we have become familiar with ways to summarize a set of measurements and to describe a distribution by location and spread, we can examine in more detail some of the fundamental properties of the measurement operation itself, the process by which measurements become meaningful data. We have seen that measurement in the behavioral and social sciences is almost always indirect. We must infer an individual's status on a trait from measures of manifestations of the trait rather than by observing the trait itself. For example, a positive, or "agree," response to the statement "I like most of the people I meet" is not, in itself, the trait of friendliness; rather, it is a response that we would expect from someone who possesses a lot of the trait, and so we infer a high amount of the trait for people who choose this and similar statements. Likewise, being able to solve an analogies problem such as "Bird is to Air as Fish is to _?_," is not itself intelligence, but the result of applying intelligence to the stated problem. Measurements of social relationship structures, personality, academic achievement, and many other characteristics involve inferences of this kind.

The indirect nature of our measurements presents the greatest problem faced by researchers in the fields of education and psychology. They have been unable to reach general agreement about what constitutes a satisfactory measure of many traits of interest. The practical consequences are serious; two scientists doing research on the effect of having teaching assistants in public-school classrooms, for example, may reach different conclusions simply because of differences in the measures they used and in the way they collected their data. Until researchers in psychology, education, and related disciplines are able to develop direct measures of traits of interest or to reach widespread agreement on indirect measures, it is unlikely that we will overcome the problem of inconsistent results from studies, or that we will discover the kinds of universal regularities found in sciences such as physics or chemistry.

Nevertheless, we are not doomed to perpetual ignorance regarding social and behavioral phenomena. There _are_ ways to classify measurement operations and to manipulate numbers obtained from measurements that can help us draw useful conclusions about the status of an individual. In general, these procedures, such as those for measuring cognitive ability and academic achievement, have been developed to their highest degree for educational settings. However, they provide ways for describing observations in quantitative terms and for placing individual observations in a meaningful context that have proven useful in other settings as well.

▶ _Now You Do It_

What kinds of variables depend on indirect measurement, and how might we develop ways to measure them? How should a first-grade teacher, for example, measure a child's level of good citizenship or a psychologist assess a client's social competence?

The first step is to decide what behaviors are indicators of the characteristic. What do children who are good citizens do? Do they

Share toys?
Push each other on the playground?
Wait until others have finished speaking?

What other relevant characteristics can you think of? Once we have developed a list of observable behaviors, we can begin to measure citizenship.

How about social competence? What do socially competent people do? Can you see why it is particularly difficult to come up with ways to measure personality characteristics and why much more progress has been made with the measurement of abilities? Even though the task is difficult, a scientific approach to studying behavior necessitates measuring the properties of people that we wish to study.

Frames of Reference

In Chapter 1, we defined measurement as _the assignment of numbers to objects according to a set of rules_. The set of rules was called a _scale_. Rules and their scales can take various forms. In direct measurement, the rules involve applying the scale directly to the object and reading off the value of interest without further inference. For example, if we wish to know Johnny's height, we can ask him to stand next to a yardstick and read off the appropriate value. Similarly, when he steps on a scale, we can read off his weight directly. In indirect measurement, we cannot measure the value of interest, _per se_. Indirect measures get their meaning from the frame of reference in which they occur. Johnny's arithmetic or spelling test scores, for example, cannot be interpreted in isolation from other information, such as the scores of his classmates. There are several ways to define a frame of reference for measurements in the social and behavioral sciences.

Absolute and Relative Measurement

First, we may differentiate between **absolute measures** and **relative measures.** With an absolute measurement, the numbers convey the desired information without further manipulation; the frame of reference is provided by the definition of the measuring operation. Direct measurements, such as temperature and length, often represent an absolute reference frame. For example, the number of words that a student can type per minute represents absolute measurement; we need not go beyond the observation itself to obtain useful information about the individual's performance. Knowing that the keyboard operator we just hired can type an average of 75 words per minute, we can expect that our new employee will complete the typing of a 7500-word manuscript before lunch. Other examples of absolute measurement include the number of errors that a monkey makes in a discrimination learning task, and a subject's decision reaction time in a cognitive processing experiment.

A relative measurement gets its meaning from comparison with other measures. For example, we may note that Mary spelled 15 words correctly on a spelling test, but we can't answer the question, Is Mary a good speller? Rather, this single question opens up a whole range of complex issues and answers. We must respond, "Is Mary a good speller compared to what, or to whom?" What do we mean by the term "good speller"? We might draw one conclusion if Mary spelled 15 words correctly out of 15 and an entirely different conclusion if she had been given 100 words to spell. Our conclusion would be further qualified by Mary's age and by the difficulty of the words she was given. We might want to know whether Mary is a good speller relative to her classmates or relative to students her age across the country. Mary's teacher would also want to know whether Mary's spelling had improved, which implies comparing her performance today with that at an earlier point in time.

Whether a measurement is in an absolute or relative frame of reference depends upon how it is to be used. Johnny's height represents an absolute measurement, but when this measurement is used to decide whether he is short or tall, the question requires an answer in a relative frame of reference. Most research on behavior depends on relative measurement; that is, it involves some form of comparison with other measures, either measures of other traits or measures of other individuals. This is most obvious in educational settings, but it is equally true in psychological and sociological research. We draw the conclusion that a rat has learned to run through a maze by comparing its running time or number of errors on one trial, or block of trials, with its performance on a previous trial or trials. We conclude that an educational film has changed children's attitudes about smoking by comparing the statements or choices they made before and after seeing the film. Even conclusions about the percentage of alpha waves in an electroencephalogram take on most of their meaning only when compared to the behavior of others, or to the behavior of the same person under different conditions.

In this chapter, we will focus on the aspects of relative measurement that describe a single individual's behavior in relation to that of others. The procedures that we will be discussing are widely used (and misused) in public education, industry, government, and

---

SPECIAL ISSUES IN RESEARCH

Frames of Reference

There are three relative reference frames that we commonly encounter in educational and psychological research. The first and most common, which we will consider in more detail shortly, is the **normative** reference frame. The normative reference frame compares the measurement of a person to the scores of other people on the same scale.

The second reference frame is called **ipsative** measurement (from the Latin *ipse,* meaning self). In ipsative measurement a person's score is compared to his or her own scores on other measurements. For example, personality and interest inventories often use this frame of reference because they ask questions about a person's relative preferences for various activities.

The third commonly used frame of reference, found mainly in education, is called the **criterion** frame of reference. The score takes its meaning from comparison with a domain of content. For example, a third grade spelling test might sample 20 words from a list of 100 that the students are supposed to know. A student's score is interpreted relative to the domain of 100 words. For a more detailed discussion of frames of reference, see the books on psychological and educational measurement listed at the end of the chapter.

---

clinical psychological practice. You have experienced the application of the principles of relative measurement yourself throughout your academic career, from elementary school to college; they provided the basis for inferring meaning from the standardized tests you have taken and from many other academic assessment procedures that represent measurements of cognitive ability within a particular frame of reference.

Defining a Standard

The definition of some type of standard is fundamental to the measurement process. We have seen that in direct measurement there is widespread agreement about the standard to which a particular entity is compared. In absolute measurement as well, the measuring instrument itself provides the standard for comparison, and a meaningful numerical value is available directly. A relative measurement does not provide a meaningful numerical value until the initial result has been compared to some standard other than the measuring instrument itself. That standard is typically based on the scores earned by other people on the same measurement instrument. Most of the decisions that are made about selecting students for special school programs and employees in industry, for example, depend upon relative measurements; that is, the comparison of scores earned by one individual with those of others. Such questions as "Is Georgia the best person for this job?" clearly imply comparing Georgia's test scores and other qualifications to those of other potential employees. The question "Is Johnny ready to enter second grade?" also suggests comparison with a standard that is defined by the behavior of others; in this case, average beginning second graders. Tasks are placed at the second-grade level in part because they are things that most children who are about seven years old and who have completed two years of school can do. A standard that involves comparing a measurement of one person to measurements of other people is called a **normative standard.**

A **normative standard** involves comparing one person's score on a measurement with the scores of other people.

Norm Groups and Standardization

Normative measurement expresses an individual's status on some trait in comparison with the scores of other people. The scores of people define the standard, and the people who are used to provide the standard form a **norm group.** *The scores of the members of the norm group* **and the procedures used to collect those scores** *become the standard or scale for measurement of an individual.* (For more information on norm groups and measurement, see Thorndike, 1997, Anastasi and Urbina, 1997, or other books on psychological and educational measurement.)

A **norm group** is a group of people who provide the scores with which an individual's score is compared.

The fundamental principle underlying the use of normative measurement is that the person we are measuring must be compared to a norm group that she or he might belong to. It probably will not be very useful, for example, to compare the behavior of a ten-year-old on a task with that of a group of high-school seniors. The proper standard (except in a few rare cases) would be the behavior of other ten-year-olds. (The principle of using an appropriate norm group is so important that major commercial producers of educational and psychological tests spend millions of dollars on identifying and measuring norm groups, in order to provide adequate normative comparisons to a wide variety of potential test users.)

Each individual is a member of many norm groups, some more and others less appropriate for scientific or educational uses. Your instructors probably use you and your classmates, possibly in conjunction with their recollections of former classes, as a norm group for assigning grades. Your academic achievement in high school was judged by your rank in your class or your grades (both normative comparisons), but you probably also took a standardized academic aptitude/achievement test for admission to college. In the latter case, the norm group was a combination of all the other students who took the test at the time you did and hundreds of thousands of others who took this or a similar test before you. If you apply to graduate school, your performance on several measures will be compared to that of other applicants. In each case, your potential is assessed by comparing you with a group of similar individuals; this is the essence of the normative frame of reference.

But what if the building where you take a test is right next to a railroad switching yard with boxcars crashing into each other unpredictably throughout the testing session? This question points out a factor—the uniformity of conditions under which the measurements are made—that is important in all scientific measurements but that is particularly acute in normative comparisons. Defining a set of uniform measurement conditions is known as **standardization.** A standardized measure, in particular, a standardized test, is one for which detailed instructions have been prepared to cover all aspects of the measurement procedure. The standardized tests given throughout the country every year are administered under carefully specified conditions, including verbatim instructions, precise time limits, and, as far as possible, quiet and well-lighted surroundings.

Standardization is the process of defining a uniform set of measurement conditions, including instructions for making the measurement, time limits, and so forth.

When the standard is defined under specified conditions, it is necessary to make other measurements under very similar conditions. Measurements using the same questionnaire or set of test items, but given under different conditions, become measurements with a different scale. Failure to recognize the need to control the conditions surrounding data collection can render the information useless or, even worse, misleading.

The need for standardization affects many other data collection situations beyond the typical standardized test. Opinion pollsters must be careful to present their questions in the same way every time, or a change in procedure on the part of interviewers may be mistaken for a change in people's attitudes. Experimental psychologists studying avoidance behavior in mice must be certain that changes in heat, light, noise, and other conditions do not occur between observations; otherwise, they cannot draw proper conclusions from their data. Control of variables that are not intended to be parts of the measurement operation is a critical feature of good research in all areas of science, but it is particularly important where the organisms being observed are as complex as humans.

The problems involved in getting good measurements, particularly of human characteristics, are so complex and so important that their study comprises an entire subarea in psychology and education. We will introduce some of the issues in a later chapter, but books on measurement theory and experimental design such as those listed at the end of this chapter are also an important resource for anyone studying human behavior or conducting psychological or educational research.

Indices of Relative Position

The norm group provides the standard for relative, or norm-referenced, measurement. The individual's status on a trait such as verbal reasoning, as measured by the SAT, is expressed as the relative position of her or his score in the frequency distribution of scores in the norm group. When we are using relative measures, our attention is focused not on the directly observed raw score—the number of questions answered correctly on the SAT, for example—but on the relative position of that score in the distribution of scores obtained by an appropriate reference group.

You already have learned some ways of expressing relative position by using the mean or median of the distribution or the quartiles. Comparing Johnny's score with the mean of the norm group, we may say that he is average, above average, or below average; or that Sally's score of 48 on an arithmetic test places her in the top 25 percent of her class (it exceeds Q_3). (Notice that in the second case, the norm group is Sally's class; this is frequently the case for relative measures in education.) However, it is often desirable to express relative position with greater precision than merely stating that a score is above or below the mean or is in some quarter of the distribution. There are several ways to make more precise statements about a person's relative position. The first two that we will discuss are ordinal measures very similar to quartiles; the others make use of the mean and standard deviation.

Percentiles

A percentile specifies the point in a distribution below which a specified proportion of the scores fall.

In Chapters 3 and 4, we described the median and the quartiles. These are special cases of the general concept of a **percentile,** which refers to the point in a distribution below which some specified proportion of the scores fall. In computing the median for the final exam scores of the students at BTU, we found the point below which 50 percent of the distribution fell (the 50th percentile); to locate the first and third quartiles (Q_1 and Q_3), we sought the 25th and 75th percentiles, respectively. In fact, the median (Q_2) and the first

Table 5-1 Scores and Percentile Ranks of 340 Students on a 25-Point Quiz			
X	f	cf	PR$_j$
24	3	340	99
23	0	337	99
22	5	337	98
21	16	332	95
20	5	316	92
19	29	311	87
18	13	282	81
17	24	269	76
16	43	245	66
15	57	202	51
14	40	145	37
13	15	105	29
12	28	90	22
11	20	62	15
10	12	42	11
9	8	30	8
8	8	22	5
7	5	14	3
6	5	9	2
5	3	4	1
4	1	1	.1

(Q_1) and third (Q_3) quartiles are simply points corresponding to special percentiles chosen because they divide the distribution into four equal parts. The same general procedure that we used to find these points may be used to calculate any percentile of the distribution.

The 340 students in the statistics course at BTU were given a 25-point quiz. The results were as shown in the first two columns of Table 5-1. (The first three columns contain the score values, frequencies, and cumulative frequencies. We will explain the last column shortly.) The instructor has decided that the top 12 percent of the class (and ties) will get As, the next 25 percent will get Bs, the next 40 percent are Cs, the next 18 percent Ds, and the bottom 5 percent earn grades of F. In order to assign the grades, the instructor must find certain percentiles. If the top 12 percent are to get As, the instructor must find the 88th percentile (the top 12 percent exceed the scores of the bottom 88 percent); next, the instructor must locate the 63rd percentile. (The top 12 percent plus the next 25 percent make the top 37 percent. Sixty-three percent of the class fall below the top 37 percent.) To identify the 40 percent receiving Cs, the 18 percent receiving Ds, and the 5 percent receiving Fs, the instructor must find the 23rd and the 5th percentiles.

The general form of the equation that the instructor would use to find the value on the score scale (X_p) that corresponds to percentile p is

$$X_p = LL_j + \left(\frac{pN - cf_{j-1}}{f_j} \right)(J) \tag{5.1}$$

Each of the terms in this equation has the same meaning that it had for finding the median. The only difference is that we have substituted pN (p times N) for $N/2$. The letter p stands for the desired proportion of the distribution; if we are looking for the 88th percentile, the value of p is .88. To find the 67th percentile, it is .67.

X_p is the point on the score scale that cuts off the desired proportion of the distribution; to find X_{88} (the score corresponding to the 88th percentile), the computations proceed as follows:

$$pN = 0.88(340) = 299.2.$$

This tells us that we are looking for the 299.2th person (299.2 is 88% of 340), and by reference to the *cf* column of Table 5-1, we find that this individual is one of the 29 who scored 19 on the quiz because this is the score value where the cumulative frequency first exceeds 299.2; therefore, $LL_j = 18.5$ (the lower real limit of the interval that includes 19), and we take the other necessary values from the table and plug them into equation 5.1.

$$X_{.88} = 18.5 + \left(\frac{299.2 - 282}{29}\right)(1)$$

$$= 18.5 + (17.2/29)$$
$$= 18.5 + .59 = 19.09$$

The scale value 19.09 is the point that separates the top 12 percent of the distribution from the bottom 88 percent. Because the instructor included ties (in fact, although they differ in their knowledge of the subject, the instructor's quiz lacked sufficient precision to detect those differences), all 29 people who scored 19, plus the 29 who got higher scores, would receive grades of A.

Next, the instructor must find the 63rd percentile to determine who receives a B. Using the same procedure, we find

$$pN = .63(340) = 214.2$$

$$X_{.63} = 15.5 + \left(\frac{214.2 - 202}{43}\right)(1) = 15.78$$

The instructor will give Bs to all students who scored 16 or above.

▶ *Now You Do It*

You should now take a moment to calculate for yourself, using the same procedure, $X_{.23}$ (= 12.08) and $X_{.05}$ (= 7.88). The minimum score for a C is 12, and the minimum score for a D is 8.

Formula 5.1 can be used to find any percentile for any distribution. If a grouped frequency distribution is being used, it is important to remember to use the lower real limit of the *interval* and to multiply the fraction in the right-hand part of the formula by the interval width (J) before adding the result to LL_j (as we did for the median and quartiles); otherwise, computations always follow the steps outlined. (Even though the data were not grouped, we carried the interval width of 1 through both computations above to make this point.)

▶ *Now You Do It*

Now let's check your understanding by using two examples from the BTU statistics anxiety study. Find the 35th percentile and then the 60th percentile for final exam scores from the grouped frequency distribution with an interval width of 5. The last place we encountered these data was in Table 4-3.

Applying equation 5.1 to the data in Table 4-3, we find that we need 23.8 people to make up 35 percent of the group and 40.8 people to make up 60 percent. We therefore get

$$X_{.35} = 27.5 + \left(\frac{23.8 - 20}{9}\right)(5) = 29.61$$

$$X_{.60} = 32.5 + \left(\frac{40.8 - 29}{12}\right)(5) = 37.42$$

Percentile Ranks

Percentiles do not give us a relative measurement scale directly because they are not linked to obtainable values in the original score scale. For example, when we find the 88th percentile, the answer (19.09 in the case of our BTU statistics quiz) usually is not a score someone could actually earn. To obtain a scale that relates the original scores directly to the cumulative frequency distribution of a norm group, we must find out *what percent of*

We have previously discussed the fact that the computer does not deal with grouped data unless the scores are put in that way or we tell the program that the scores are interval midpoints. Therefore, when we ask the computer to analyze a set of scores, the results can be quite different from what we might expect. For example, if we ask SPSS to find the 35th and 50th percentiles for the data on student age in the BTU statistics class, and don't tell it to consider the scores as grouped data, we are told that both the 35th and 50th percentiles are 21.0! However, if we tell the program that the data represent group midpoints, the corresponding percentiles are 20.61 and 21.26, which are the same values we would get using equation 5.1. This happens because, even with percentiles, the program uses the score value obtained by the person in question unless we make it do otherwise. Such a practice is not a big problem when the scores spread over a large range with a small percentage of the scores at any particular value, as is typically the case with final exam scores.

To compute selected percentiles using SPSS, open the STATISTICS menu and go to the FREQUENCIES submenu. There you will find a button labeled STATISTICS which will present you with a menu that looks like this:

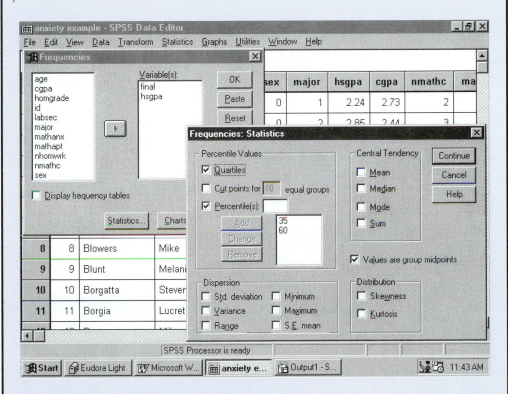

Click on the Percentile(s) button and type the desired percentile values into the box. Click "Add" after each value is entered. Be sure to click the "Midpoints" box to get values calculated by equation 5.1.

One of the features of most computer programs is that they will not compute fractional percentiles. That is, although we could ourselves compute the 26.48th percentile using equation 5.1, the program will only compute the 100 whole percentiles.

the distribution falls below each **observed score** value. The percent of the distribution that falls below an observed score is the **percentile rank** of that score or *PR*.

There are important differences between percentiles and percentile ranks. It is possible to specify any percentile and compute a point on the continuous score scale below which that proportion of the distribution falls. Regardless of the number of scores in a distribution, it is always possible (except for the trivial case where the variance is zero) to determine an infinite number of percentile points. Percentile ranks, on the other hand, focus on the

The **percentile rank** of an observed score shows the proportion of the reference group that earned scores below that score.

▶ Comparison of Percentiles and Percentile Ranks

Percentiles	**Percentile Ranks**
Start from specification of a particular percent of the distribution.	Start from a specified obtainable score value.
Compute a point on the score scale.	Compute a proportion of the score distribution.
Can have an infinite number of values.	Can have only as many values as there are obtainable scores.

discrete scale of obtainable score values (see comparison above). For example, we could compute the 36.21835th percentile for the data in Table 5-1, but there are only 21 score values in the distribution, and, therefore, only 21 computable percentile ranks for the data.

Computing percentile ranks is quite straightforward; we must find the number of individuals whose scores fall below a given observed score value and divide that number by the total number of individuals. Remember from our discussion in Chapter 2 that the observed score value is at the middle of each score interval and that we assume that the people whose scores place them in the interval are actually spread throughout the interval.

HOW THE COMPUTER DOES IT

SPSS creates a new variable to compute percentile ranks. To obtain percentile ranks, click on the TRANSFORM menu and select Rank Cases. Select the variable for which you wish percentile ranks; then click on the Rank Types button. The screen should look like this:

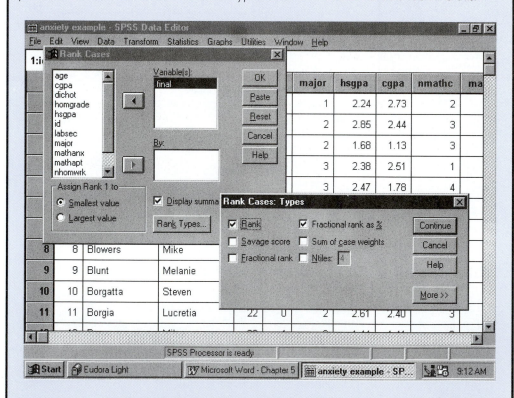

Select Fractional Rank as %. The program will produce two new variables, each of which is named with a variation on the name of the variable being converted to *PRs*. The new variables will be added at the end of your data file. In the example above, we are finding percentile ranks for "final." The two new variables are called "rfinal" and "pfinal." Rfinal contains each person's score converted to a rank within the group, and pfinal is the percentile rank of each person's score. The default creates variables with four digits (two decimal places), so to obtain results such as those in Table 5-1 you would need to round the scores reported by the program.

Therefore, to find the number of people in an interval who score below the midpoint of the interval, we must divide the frequency in the interval by 2. The number of people who fall below the particular score value is the number of people who are below the interval entirely plus those people who are in the interval but below its midpoint.

The general formula for computing PR_j is

$$PR_j = \frac{cf_{j-1} + \frac{1}{2}f_j}{N} \tag{5.2}$$

In this formula, PR_j is the percentile rank of observed score X_j, cf_{j-1} is the cumulative frequency for the next lower score value, f_j is the frequency with which the score X_j occurs, and N, of course, is the total number of individuals. To find the PR for a score of ten on the quiz in Table 5-1, we proceed as follows:

$$PR_{10} = \frac{30 + (.5)(12)}{340}$$
$$= 36/340$$
$$= .1059$$

Percentile ranks are almost always rounded to two places with the decimal point omitted, so this result would be read as a percentile rank of 11 (some books show formula 5.2 multiplied by 100).

In the fourth column of Table 5-1, the percentile rank of each score is given. Note that PR_{23} is not the same as PR_{22} even though score value 23 has zero frequency. This occurs because only half of the five people who scored 22 are counted as being below 22 but all five of them fall below 23. Because we use only half of the cases in the extreme score categories and include only score values within the observed score range, percentile ranks of 0 and 100 are not possible. With small sets of scores such as this one, percentile ranks are reported to two digits. When very large groups have been measured (as is often the case with commercial standardized tests) and it is desirable to make fine distinctions at the extremes of the distribution, percentile ranks such as the value of .1 that we have given for a score of 4 may be reported. In general, percentile ranks are not used with grouped data because percentile ranks refer to specific score values rather than intervals, although SPSS will compute PRs for grouped data and some books show the formula.

▶ *Now You Do It*

In Chapter 3 we used the quiz scores of all 340 statistics students at BTU to compute the median. The table below shows the frequency distribution. What percentile rank is associated with each score? You should get the values listed at the bottom of the table.

Statistics Quiz Scores for 340 Students at BTU

X	f	cf
10	8	340
9	26	332
8	84	306
7	132	222
6	67	90
5	15	23
4	6	8
3	2	2

The percentile ranks to two places, starting with 3, are as follows: 0.3, 1, 5, 17, 46, 78, 94, 99. Note that the PRs change quite slowly where there are few cases (at the bottom) and quickly where frequencies are large. Also, there are only 8 possible PRs for these data, but we could compute as many percentiles as we wish. You could, for example, compute the 19.8th percentile and the 19.9th percentile, and they would be different, but there are only 8 PRs.

Linear Transformations

One of the disadvantages of percentiles and percentile ranks as normative frames of reference is that they provide only ordinal information and therefore in general should not be used as scores in more advanced analyses where interval level information is assumed. For example, percentile ranks are reported for students who took the SAT, but when information from the SAT is used to predict college performance, an interval scale form of score (such as the standard scores discussed in this chapter) is used so that we can employ the power of statistical prediction as described in Chapter 7. There are many occasions in psychology and education when we would like to be able to perform further analyses that require interval level data. For example, researchers studying the effects of different educational methods often want to compare the means of samples exposed to different teaching materials. A developmental psychologist might want to ask whether the variability in a trait such as cognitive ability changes with age. Such analyses require development of a normative expression for the scores that retains their original metric qualities.

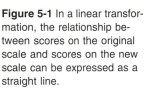

*A **linear transformation** is a rescaling of scores that can be described by a straight line.*

The class of data rescalings that retains the interval properties of the data is called a **linear transformation.** We use the term "linear transformation" because a graph of the relationship between the original data and the rescaled data is a straight line. If, for example, we show the scale of the original data on the abscissa and place the new scale on the ordinate, we get a graph such as that in Figure 5-1. Each value on the original scale corresponds to exactly one value on the new scale, and these paired values fall on a straight line. The value P on the original scale translates to P′ on the new scale, Q translates to Q′, R to R′, and so on.

The Linear Equation

One of the advantages of a linear transformation is that it is possible to write a simple equation for a straight line to show the relationship between the original scale and the new scale. The equation for a straight line requires two pieces of information: the rate at which the line rises or falls, called the **slope,** or **B**, and the overall elevation of the line above or below the origin, called the **intercept,** or **A**. The intercept is the point where the line crosses the ordinate.

*The **slope (B)** is the rate at which a line rises (positive) or falls (negative).*

*The **intercept (A)** shows the height of the line above or below the origin. It is the ordinate value of the line when the abscissa value is zero.*

The general form of the equation for a linear transformation from the original scale value X into the paired value X′ on the new scale may be written as

$$X' = BX + A \tag{5.3}$$

The **slope,** B, is the rate at which the line rises or falls. In formal terms, *the slope is the number of units of change in the scale of the ordinate per unit of change in the scale of the abscissa*. It therefore expresses the relative magnitudes of the units on the two scales.

Figure 5-1 In a linear transformation, the relationship between scores on the original scale and scores on the new scale can be expressed as a straight line.

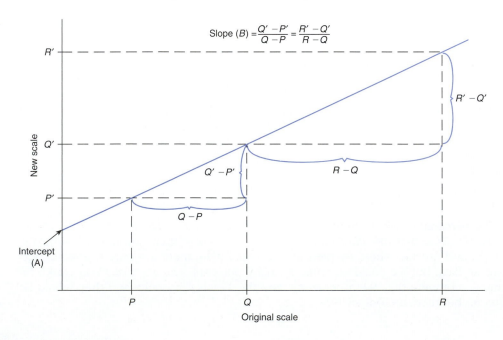

If a unit on the new scale is half as large as a unit on the original scale, the slope will be 2 because two units on the new scale correspond to one unit on the original scale. Similarly, if we want to convert from a scale of feet to a scale of inches, it takes 12 times as many inches to express a person's height as it does feet, and the slope for the transformation is 12. Notice that if we want to reverse the process, going from inches to feet, the slope is 1/12 because 1/12 of a unit on the new scale corresponds to one unit on the original scale.

One familiar use of a linear relationship is the conversion from the Fahrenheit temperature scale to the Celsius, or centigrade, scale. Placing the Fahrenheit scale on the ordinate and the Celsius scale on the abscissa, the equation of the linear relationship between the two scales is

$$F = 9/5C + 32.$$

This equation tells us that a unit on the Celsius scale is 9/5 as large as a unit on the Fahrenheit scale, or phrased differently, it takes 9 units on the Fahrenheit scale to equal 5 units on the Celsius scale. The intercept value of 32 tells us that a value of 0 on the Celsius scale corresponds to a value of 32 on the Fahrenheit scale. Using this equation, it is a simple task to compute the value on the F-scale for any temperature on the C-scale. For example, a value of 157 on the C-scale translates to

$$F = 9/5(157) + 32 = 282.6 + 32 = 314.6.$$

The slope can be either positive or negative. A positive slope indicates a rising line; a negative slope reveals a descending line. We will encounter both positive and negative linear relationships in Chapters 7 and 8, when we discuss relationships between variables. However, when we are concerned with linear transformations the slope is always positive. This means that the order of original scale values is retained in the transformed scale. Referring to Figure 5-1, if Q is greater than ($>$) P and $R > Q$, then $R' > Q' > P'$. Also, the relative magnitudes of differences remain the same in the new scale as they were in the original. If the difference ($Q - P$) is equal to ($R - Q$), then ($Q' - P'$) = ($R' - Q'$). Likewise, if the difference ($Q - P$) is less than ($<$) ($R - Q$), then ($Q' - P'$) < ($R' - Q'$). That differences remain proportional means that linear transformations can only be applied to variables that have been measured on an interval or ratio scale because it is only in these scales that the magnitude of a difference has meaning.

Any linear transformation can be reversed by simply applying the reverse of the original steps in reverse order. For example, we can go from a value on the Fahrenheit scale to one on the Celsius scale by first subtracting 32, then dividing by 9/5. (Originally, to convert from the Celsius scale to the Fahrenheit scale, we first multiplied the degrees Celsius by 9/5, then added 32.)

A linear transformation can be reversed by applying the reverse steps in the reverse order.

▶ *Now You Do It*

There are a number of linear transformations that we encounter regularly. We have mentioned inches to feet and Fahrenheit to Celsius. What other common transformations do you encounter? Suppose you were to take a vacation in Canada, where they use the metric system. The posted speed limit is 100, but your car speedometer is calibrated in miles per hour. How fast should you be going? Suppose you stop to buy gasoline. Your car's gas tank will hold 12 gallons. About how many liters should you expect to buy, and how much will the gas cost in U.S. dollars? Answering each of these questions requires a linear transformation. What answers do you come up with?

There are .62 miles to the kilometer. The slope is .62, so you should be going about 62 mph.

Since a liter is a little larger than a quart, you will probably need about 45 liters to fill your tank. The answer to the last question depends on the exchange rate and the price of gas.

Recentering the Distribution
(Adding or Subtracting a Constant)

Sometimes it is desirable to perform the simplest of all linear transformations on a scale: adding or subtracting a constant. There are two reasons why we may wish to perform a transformation of this kind. The first is to get rid of some troublesome property of the scores, such as minus signs. If a distribution of scores includes some negative values, this can be both a nuisance and a potential source of calculation errors. Negative values can be removed from the data by the simple expedient of adding a constant to all scores. If, for example, we wanted to study the effects of biofeedback on anxiety test scores, the data would be in the form of changes from an initial baseline reading, and anyone who showed a drop in anxiety would have a negative score. The problem of negative values could be eliminated by adding a constant to every score to make them all positive.

The second reason that we may want to add or subtract a constant from every score is that it may be a necessary intermediate step to an important result. For example, to compute deviation scores, we subtracted a constant, the mean, from every score in the distribution; this had the effect of moving the center of the distribution (considering the mean to be the center) to zero and making roughly half of the scores negative.

Adding or subtracting a constant has the effect of changing the magnitudes of the scores without changing the numerical values of the distances between score points (the units). Suppose, for example, that two students received test scores of 150 and 170 on a statistics final. If we subtract a constant of 100 from both scores, they become 50 and 70, but the difference between them is still 20. We could subtract 150 from both scores (getting 0 and 20) or 200 from both (getting −50 and −30), but the difference between the two scores remains 20 units. Likewise, adding any constant to both scores would leave their difference unchanged.

Adding or subtracting a constant, *A*, from every member of a set of scores recenters the distribution at a new point *A* units away without changing the shape or spread of the distribution. When a constant is added or subtracted for computational facility, the value of the constant is not crucial and may be chosen for convenience. One common practice is to subtract from every score a value near the lowest score; this means that all scores will be positive numbers but smaller than the original values (we ignore from here on the case where the measurement results in negative numbers on the original scale; in that case, we might wish to add a constant). Of course, if the constant is being subtracted as part of a specific computation, such as to obtain a standard deviation, we must subtract that particular constant. In Figure 5-2, the effects of subtracting the lowest score and the mean are illustrated for a distribution running from 75 to 150 with a mean of 100. Note that neither the positive skew of the distribution nor its range changes; all that happens is that the distribution, otherwise unaltered, slides up or down the scale to center on a new point equal to the original center minus the constant. In particular, the spread of scores, their variability, does not change.

Effects of Recentering on Measures of Central Tendency

Recentering the distribution has a simple effect on the measures of central tendency: Subtracting a constant from each score has the effect of subtracting that same constant

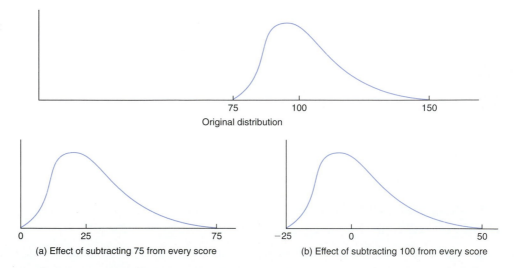

Figure 5-2 An original distribution transformed (a) so that its lowest value is zero and (b) so that its mean is zero.

from the original values of the mode, median, and mean. For example, consider the following small data set.

10, 20, 20, 30

The mean of these numbers is 20, as are the median and mode. If we subtract 10 from every number, the transformed scores are

0, 10, 10, 20

The mean is now 10, as are the mode and median. Each has changed by an amount equal to the value we subtracted from the raw scores. In general, if we subtract a constant, A, from every score in a distribution, the mean of the new distribution is $\overline{X}' = \overline{X} - A$, and the mode and median are changed by the same amount.

Changing Units of Measurement (Multiplying or Dividing by a Constant)

Adding or subtracting a constant from each score involves only the intercept portion of a linear transformation; the size of the scale units remains the same. It is sometimes useful to change the size of the units on the measurement scale by using a slope other than 1.0. For example, all standardized tests, such as personality and ability scales, use transformations that change the size of the units. This may be done either to ease computational labors or, as in the case of test scores, to reach scale units that have a special meaning. In either case, the transformation may involve both adding or subtracting one constant and multiplying or dividing by another. Each constant may be chosen arbitrarily, but in fact some constants will prove more useful than others.

Central Tendency of Transformed Scores

The effect of multiplying or dividing each score by a constant on the mode, median, and mean is to multiply or divide each statistic by that same constant. For example, if the four scores

10, 20, 20, 30

are each divided by 5, they become 2, 4, 4, 6. The original scores had a mean of 20, but the new scores have a mean of 4. If a distribution has a mean of 200 and we divide every score in the distribution by 5, the mean of the transformed scores will be 200/5 or 40. In general, if the original scores have a mean of $\overline{X}_O$, the mean of scores that have been transformed by multiplying by K will be $\overline{X}_N = K \cdot \overline{X}_O$.

To obtain the values of the statistics for the data in their original form, it is necessary to perform the reverse of the transformation. If a transformation involved dividing each score by 5, the mean, median, or mode of the transformed scores can be converted back to the original scale by multiplying the mean, median, or mode of the transformed scores by 5.

Variability of Transformed Scores

When either or both portions of a linear transformation are applied to a set of raw scores, the transformation has an effect on the measures of central tendency that is the same as the transformation that was applied to the data. Adding a constant, A, to all scores increases the mean or median of those scores by A, and multiplying all scores by a constant, B, produces a new distribution with a mean or median equal to the original mean or median multiplied by B. On the other hand, the effects of transformations on the range and interquartile range as well as on variance and standard deviation are a little less straightforward.

Transformations that change the size of the units affect the range, interquartile range, and standard deviation in the same way. Dividing each score by a constant has the effect of dividing the range, interquartile range, and standard deviation by that same constant; this will be true regardless of whether the distribution has previously been recentered by subtracting a constant. Likewise, multiplying the scores by a constant will multiply the range, interquartile range, and standard deviation by the same constant. To use our small 4-score example again, the simple range ($H - L$) of the scores

$$10, 20, 20, 30$$

is 20, the interquartile range is 10, and the standard deviation is 7.07. If we divide each score by 5, the set of scores

$$2, 4, 4, 6$$

has a simple range of 4, an interquartile range of 2, and a standard deviation of 1.414. In each case, we can find the statistic of the transformed scores by simply applying the transformation to the original statistic. Likewise, we can recover the value of the original statistic by simply applying the reverse of the transformation to the statistic computed from the transformed scores.

In the case of the variance, we must proceed in a slightly different way. Because the variance involves squared distances, the effect of the transformation is squared. Therefore, the square of the transformation must be applied to the original variance to obtain the proper transformed value. If we multiply two scores by a constant, the square of the distance between them is multiplied by the squared constant. For example, if we multiply the numbers 2 and 4 by 2, we get 4 and 8. The squared distance between the first pair of numbers is

$$(4 - 2)^2 = 2^2 = 4,$$

but the squared distance between the second pair is

$$(8 - 4)^2 = 4^2 = 16.$$

The constant we multiplied the original numbers by was 2, so the difference between the new squares is the original squared distance (4) multiplied by the squared constant ($2^2 = 4$), or 16.

A comparison of the effects of different transformations is shown in Table 5-2. Here the math aptitude scores of the first six students from the statistics class at BTU are listed in column 1. Column 2 contains the scores transformed by subtracting 45 from each score. Note that the sum of squares, variance, and standard deviation at the bottom of the table are unchanged. Column 3 lists the scores after they have been divided by 5. Although ΣX and ΣX^2 do not look much different from the second column, the sum of squares and variance are much smaller. In fact, if we multiply the sum of squares and variance from

▶ EFFECTS OF ADDITIVE AND MULTIPLICATIVE TRANSFORMATIONS

Statistic	Action	Effect
Mean	Add or Subtract	Adds or subtracts the same constant
	Multiply or Divide	Multiplies or divides by the same constant
Median	Add or Subtract	Adds or subtracts the same constant
	Multiply or Divide	Multiplies or divides by the same constant
Mode	Add or Subtract	Adds or subtracts the same constant
	Multiply or Divide	Multiplies or divides by the same constant
Range	Add or Subtract	No effect
	Multiply or Divide	Multiplies or divides by the same constant
Interquartile Range	Add or Subtract	No effect
	Multiply or Divide	Multiplies or divides by the same constant
Standard Deviation	Add or Subtract	No effect
	Multiply or Divide	Multiplies or divides by the same constant
Variance	Add or Subtract	No effect
	Multiply or Divide	Multiplies or divides by the square of the constant

Table 5-2 Results of Applying Additive and Multiplicative Transformations to Math Aptitude Scores for Six Students

Student	Raw Scores	Transformed Scores		
		Subtract 45	Divide by 5	Both
Alcott	60	15	12	3
Alexander	56	11	11.2	2.2
Allen	46	1	9.2	.2
Allport	44	−1	8.8	−.2
Allwood	60	15	12	3
Bentler	61	16	12.2	3.2
ΣX	327	57	65.4	11.4
ΣX^2	18,109	829	724.36	33.16
SS	287.5	287.5	11.5	11.5
S^2	47.9	47.9	1.92	1.92
S	6.9	6.9	1.38	1.38

column 3 by 25 (the square of the transformation) we get the exact values from columns 1 and 2. Likewise, multiplying the $S = 1.38$ from column 3 by 5 (the transformation) produces 6.9, the standard deviation of the untransformed scores. Column 4 shows the results of the two-part transformation of first subtracting 45, then dividing by 5. The values of ΣX and ΣX^2 are much smaller than those in column 3, but the sum of squares, variance, and standard deviation have not changed, confirming that only the multiplicative part of a transformation affects the variance.

▶ *Now You Do It*

Have you ever encountered situations where you had to deal with large numbers and wished you could make them more manageable? Suppose you are interested in population trends for states and you obtain data from the Census Bureau reported in thousands. How would you want to handle the data? For example, you are told that the number of deaths from heart disease for each of the last five years is 257,000, 335,000, 264,000, 372,000 and 298,000. Try computing the mean, variance, and standard deviation of these data in their present form; then try it after dividing each score by 1000.

For the original scores you should get a value for ΣX of 1,526,000, a mean of 305,200, a value for ΣX^2 of 475,158,000,000, a variance of 1,884,560,000, and a standard deviation of 43,411.5. But if you first divide each score by 1000, you get a sum of 1,526 and a mean of 305.2. ΣX^2 is 475,158, the variance is 1,884.56, and the standard deviation is 43.4115. These numbers are quite a bit smaller and much easier to deal with, but we can go back to the original scale simply by multiplying the mean and standard deviation by 1000. We get ΣX^2 and the variance by multiplying their transformed values by 1000^2 or 1,000,000.

Standard Scores

There is one particular linear transformation that has proven to be extremely useful in the behavioral and social sciences. It is so common that it has been given a special name, the **standard score transformation,** and a special symbol, the letter **Z**. Scores that have been transformed in this way are known as **standard scores,** or **Z-scores.** Z-scores form the basis for the scales used by most standardized tests, such as the SAT and the Minnesota Multiphasic Personality Inventory. If you want to really understand the meaning of your SAT scores, you must master Z-scores.

The standard score transformation involves recentering the distribution at zero and changing the size of the unit to equal the standard deviation of the original distribution. This is accomplished by first subtracting the mean from each score X_i to obtain the deviation score x_i and then dividing each resulting deviation score by the standard deviation of the

Standard scores, or Z-scores, have a mean of zero and a standard deviation of 1.0.

We are not usually worried about large numbers when we have a computer to do the work, but if the data come to you already transformed and you want to put them back into their original scale, the computer can help. All statistical packages include ways to transform data. SPSS uses the menu called TRANSFORM, which will perform a wide variety of transformations on your data. If you wish to use a simple transformation such as the one we applied in Table 5-2, you would use the COMPUTE subprogram. In it, you can create a new transformed variable or replace the original variable with the transformed one. The transformation we used in Table 5-2 would look like this:

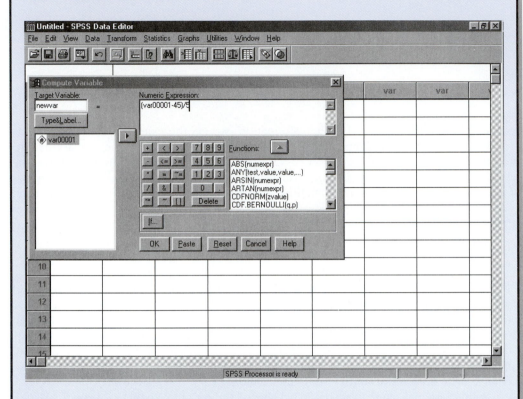

You will find it useful to explore the ways your statistics program can transform the data. Remember, as we discussed in Chapter 4, most programs do not compute the sample variance and standard deviation. Rather, they will give you the unbiased estimate of the variance. The difference can be quite large, as you will see if you ask the computer to analyze the heart disease data from our example in the previous Now You Do It.

original distribution. This may be symbolized

$$Z_i = \left(\frac{X_i - \overline{X}}{S_X}\right) = \frac{x_i}{S_X} \tag{5.4}$$

where the standard score for each individual (Z_i) is found by subtracting the mean ($\overline{X}$) from the observed score (X_i) and dividing by the standard deviation (S_X). For example, in a distribution with a mean of 23.7 and a standard deviation of 4.6, the Z-score corresponding to a raw score of 20 would be

$$Z_i = \frac{20 - 23.7}{4.6} = -.804$$

The raw score of 20 is .804 standard score units *below* the mean.

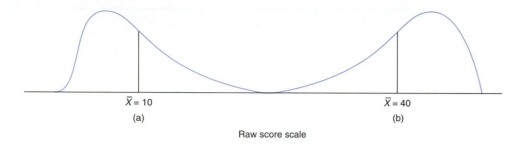

$\bar{X} = 10$ $\bar{X} = 40$

(a)   (b)

Raw score scale

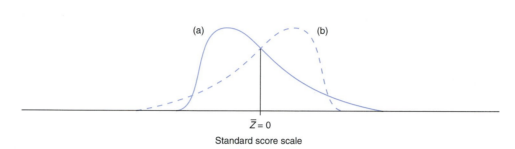

(a)   (b)

$\bar{Z} = 0$

Standard score scale

Figure 5-3 When two distributions are transformed to Z-scores, their means become zero, but the shapes of the distributions are unchanged.

The reason the standard score transformation is so useful is that it provides a common base for all normative frames of reference. Suppose, for example, someone asked you if you were taller than you are heavy. How could you answer? You could respond by using standard scores. Standard scores can take very different distributions such as height and weight, convert them to a comparable metric and then express each observation in terms of a distance in standard score units above or below a standard point.

There are two features of standard scores that we must always keep in mind. The first is that standard scores are expressed relative to the mean and standard deviation of the particular group on the particular trait being measured. Two groups may differ dramatically in their level or variability on some trait (for example, gender differences in height), but when each distribution has been converted to the Z-score metric using its own mean and standard deviation, the two new distributions will have the same mean (0) and standard deviation (1). The average man has a height Z-score of zero (he is at the mean for men), as does the average woman (she is at the mean for women), but the man will be taller because his score is from a different distribution. Likewise, two traits, such as reading and mathematics achievement, may differ in the amount of variability they show on a raw score scale, but their Z-score distributions will always have the same standard deviation, 1.00. The critical factor in a comparison is that, to be comparable, the standard scores must have been developed on the same norm group or equivalent norm groups.

The second important feature of Z-scores is that the transformation (or any linear transformation) does not change the shape of the original distribution. If the original scores had a markedly skewed distribution or a platykurtic distribution, the Z-scores will also. Because Z-scores are often used with a special distribution called the normal distribution (discussed in Chapter 6), people sometimes assume that the transformation to Z-scores makes the distribution normal in shape. This is not the case. No linear transformation changes the shape of a distribution. Special procedures, some of which are discussed in Chapter 6, are needed to change the shape of the distribution. The effects of the standard score transformation on two different distributions are shown in Figure 5-3. The two distributions start out with very different means (10 and 40) and quite different shapes (one is positively skewed, the other is negatively skewed). After transformation to Z-scores, both distributions have means of 0, but their shapes are still different.

▶ *Properties of Z-Scores*

1. Z-scores have a mean of zero.
2. Z-scores have a standard deviation of 1.0.
3. The distribution of Z-scores has the same shape as the distribution of original scores.

▶ *Now You Do It*

Suppose you have collected some data on intelligence and shoe size. Frodo, an inquisitive fellow and a participant in your study, asks you whether his feet are bigger than his brain. How would you answer his question?

Although the question seems silly, we can actually give Frodo a rational answer if we phrase the question in terms of Z-scores. Let's say the mean and standard deviation of intelligence scores in your data are 101 and 14.5, and the mean and standard deviation of shoe size are 10.3 and 1.7. Frodo's scores were 105 and 11.5, so his Z-scores would be +.28 for intelligence and +.71 for shoe size. You might therefore reasonably conclude that his feet are bigger than his brain within the normative frame of reference of your study.

Standard Scores and Normative References

Standard scores and linear transformations of them are the most widely used alternative to percentile ranks for creating a relative measurement scale. After an appropriate norm group has been measured, the mean and standard deviation of the distribution are found, and every score in the scale can be expressed as a Z-score indicating its direction from the mean of the norm group and its distance in standard deviation units. This series of computations results in a new set of scores, about half of which are negative and half are positive. For our class of 340 BTU statistics students (Table 5-1), the mean quiz score was 14.72, with a standard deviation of 3.71. Z-scores on this quiz are shown in Table 5-3. The

	Table 5-3	Percentile Ranks, Z-Scores, and Transformed Standard Scores of 340 Students on a 25-Point Quiz			
X	**f**	**cf**	**PR$_j$**	**Z**	**IQ**
24	3	340	99	+2.50	138
23	0	337	99	+2.23	133
22	5	337	98	+1.96	129
21	16	332	95	+1.69	125
20	5	316	92	+1.42	121
19	29	311	87	+1.15	117
18	13	282	81	+ .88	113
17	24	269	76	+ .61	109
16	43	245	66	+ .35	105
15	57	202	51	+ .08	101
14	40	145	37	− .19	97
13	15	105	29	− .46	93
12	28	90	22	− .73	89
11	20	62	15	−1.00	85
10	12	42	11	−1.27	81
9	8	30	8	−1.54	77
8	8	22	5	−1.81	73
7	5	14	3	−2.08	69
6	5	9	2	−2.35	65
5	3	4	1	−2.61	61
4	1	1	.1	−2.88	57

first four columns of this table are the same as those in Table 5-1, and the fifth column lists the Z-score that is equivalent to each observed score. The last column lists a transformation of these Z-scores explained below.

Transformations of Z-Scores. Standard scores are seldom left in their original form when used to describe a person's performance; they are almost always subjected to some type of additional linear transformation for final presentation. There are two reasons for this: (1) Z-scores involve small decimal numbers. Since it is inconvenient to work with numbers like .57 as scores, a transformation is used to get rid of the decimal point. (2) About half of the scores in any Z-score distribution are always negative, which is undesirable both computationally and socially. Who wants to think of themselves as a negative quantity? Therefore, the scores are transformed to remove the negative sign.

Several types of linear transformations of Z-scores are commonly used in education and psychology. All involve multiplying the Z-score by an arbitrary standard deviation to remove the decimal point and adding an arbitrary mean to make all scores positive. Perhaps the most familiar is the so-called IQ scale. This scale uses an arbitrary mean of 100 and an arbitrary standard deviation of 15 (or, for one test, 16). The linear transformation has the general form

$$X_{A_i} = (S_A)Z_i + \overline{X}_A \tag{5.5}$$

The score on the new arbitrary scale (X_{A_i}) is equal to the standard score (Z_i) multiplied by an arbitrary standard deviation (S_A) and added to an arbitrary mean ($\overline{X}_A$); for the IQ scale, the equation is

$$IQ_i = Z_i(15) + 100.$$

In this metric, a Z-score of +1 corresponds to an IQ score of 115, and a Z-score of −1 is equal to an IQ score of 85. In column 6 of Table 5-3 the quiz scores have been transformed into a scale with a mean of 100 and a standard deviation of 15. Notice that we have gotten rid of the decimals and minus signs. The mean of scores on the new scale is 100, and the standard deviation is 15. Like any linear transformation, this one can be reversed to find the Z-score.

$$Z_i = \frac{X_{A_i} - \overline{X}_A}{S_A}$$

One of the advantages of arbitrary linear transformations when applied to standard scores is that we can move freely back and forth between different scales because they are all linear transformations of the same base, Z-scores. Several measures of personality use a transformation with an arbitrary mean of 50 and standard deviation of 10; some ability tests in government and industry are transformed to a mean of 100 and $S_A = 20$. The Scholastic Aptitude Tests (SAT) use a scale with a mean of 500 and a standard deviation of 100. Each has been developed on a norm group and conveys the same general information about the position of an individual's score as some number of standard deviations above or below the mean score of that norm group. Any one of these scales could replace any other without gain or loss of information. That is, the College Board could just as well report SAT scores on a scale with a mean of 100 and a standard deviation of 15 or a scale with a mean of 50 and a standard deviation of 10 instead of the scale they do use. They are all equivalent (except that they use different norm groups), and they are all equivalent to Z-scores. If we know the mean and standard deviation of the transformation, we can convert from one metric to another (for example, go from the SAT scale to the IQ scale or vice versa) by applying formula 5.5 or its reverse.

There is a danger inherent in this linear equivalence of scales, however. It does not mean that we can compare a person's scores on two different instruments directly. For this to be possible, the two tests must have used the same norm group, or special efforts must have been made to make the scales comparable. For example, it is NOT appropriate to say that an SAT score of 600 is equivalent to a Wechsler IQ score of 115. Both scores are one standard deviation above the mean, but the norm groups are quite different. The norm group for the Wechsler scale uses a nationally representative sample while the SAT norms are based on prospective college students, a much more select sample. This issue is discussed in greater detail in books on psychological and educational measurement.

The Wechsler Adult Intelligence Scale reports scores for cognitive ability on a scale with a mean of 100 and a standard deviation of 15. Suppose you wanted to convert Wechsler scores to a scale with a mean of 50 and a standard deviation of 10. What would the following Wechsler scores be in the new scale?

55, 85, 105, 130

The first step is to convert the Wechsler scores to Z-scores. Using $\overline{X} = 100$ and $S_X = 15$ you should get Z-scores of

$-3.0, -1.0, +.33,$ and $+2.0$

Applying the second transformation to the Z-scores gives us

$NewScore = 10Z_i + 50$

so the four transformed scores would be

20, 40, 53.3, and 70

Z-Scores and Percentile Ranks. Let us now compare the information provided by Z-scores or their transformations with the information that we get from percentile ranks. The percentile rank scale is based upon the number of individuals at each score point. If there are more individuals at one place in the distribution than in another, there will be a more rapid change in the percentile rank; the difference in *PR* between two adjacent raw scores will not be the same everywhere in the distribution. In contrast, a standard score scale is based on a single unit of distance on the raw score scale that depends only on the standard deviation, with the zero point determined by the mean. The difference in Z-score between any two adjacent raw scores will be the same everywhere in the distribution, regardless of frequency.

We can illustrate this difference using the data in Table 5-3. The difference of one raw score unit is equal to .27 units on the Z-score scale at all points in the distribution. The change in raw score from 6 to 7 is .27 Z-units and so is the change from 18 to 19. However, the change of one raw score point from 6 to 7 results in a *PR* change of one point, while the raw score change from 18 to 19 means a *PR* change of six points because of the difference in frequency.

These two ways of expressing normative information are very different because the percentile rank scale changes the shape of the frequency distribution (it is not a linear transformation), while the standard score scale does not change the shape of the distribution. If we were to prepare three histograms using the two normative scales and the raw score scale on the abscissas, we would get the results shown in Figure 5-4. Histograms (a) and (b) are identical in shape, showing a moderate negative skew; histogram (c), representing the *PR* scale, is a rectangular distribution showing equal frequency in every interval; this occurs because every interval of 10 percent contains 34 observations (in our example) without regard to the number of raw score points covered, and this results in bars of equal height. In fact, any set of data will yield a rectangular distribution when converted to percentile ranks.

Another way to think about the difference between scales based on linear transformations of the raw scores (that is, Z-scores and their transformations) and the percentile rank scale is to think of the raw score scale and linear transformations as rigid and the *PR* scale as elastic. The raw score scale and its linear transformations retain the size of their units regardless of the shape of the distribution, while the *PR* scale stretches its units in some areas and contracts them in others, depending on the shape of the distribution, to achieve a uniform frequency. We will use this property of the *PR* scale in Chapter 6 to perform a very useful nonlinear transformation.

Interpreting Normative Information

There are several factors that we must keep in mind when using normative information. First and foremost, the information gets its meaning from comparison with some norm group; it can be of no higher quality than the basis for comparison permits. If the norm

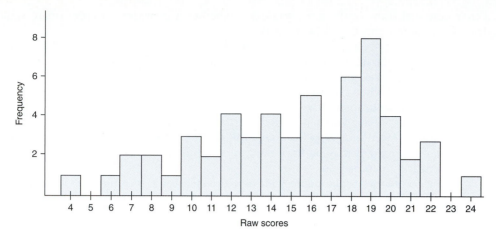

Figure 5-4 Effects of using two different transformations of a score scale.

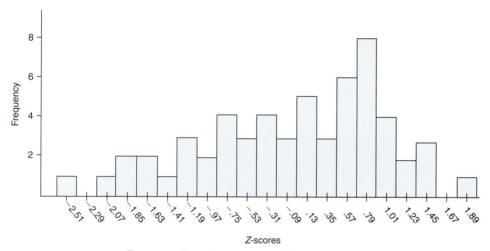

Z-score transformation leaves shape of the distribution unchanged

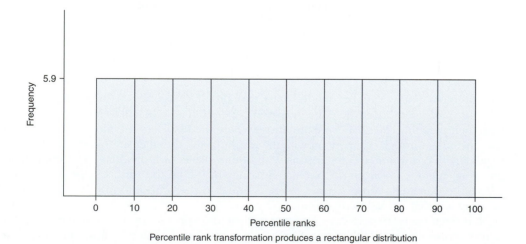

Percentile rank transformation produces a rectangular distribution

group was poorly defined or the measurements of the norm group were made carelessly, no amount of effort on the part of a subsequent user of the scale will ensure high-quality information. The data must be collected from the norm group under uniform conditions that are described in sufficient detail to allow other users of the scale to duplicate the measurement conditions. Also, the norm group must be described clearly enough to enable

Chapter 5: Describing the Individual **135**

other potential users of the scale to decide whether it provides an appropriate standard for the measurements they wish to make.

Some ways of scaling change the shape of the frequency distribution while others do not. Specifically, the information from a percentile rank form of scale is ordinal information and produces a rectangular distribution to achieve uniform frequency. It is important to keep this in mind when choosing which type of scale to use and what analyses to perform on the data in order to answer the questions that prompted the study. If ease of interpretation is particularly important, you may wish to use the percentile rank scale because the meaning of the results is more straightforward with this scale. On the other hand, if further analyses are to be performed, it may be important to retain the properties of the original units by using a linear transformation.

A very common mistake made with normative scales is placing too much faith in the precision of the measurements. Standard scores can be calculated to any number of decimal places, thus giving the impression of precise, high-quality information. In some areas such as business or economic applications, it is quite possible that these numerical values are correct. For example, we may be able to determine the income of a major corporation down to the last penny, and the dollar values of employee benefits, cost of raw materials, and so on, may be known exactly. However, in many areas in education and psychology, particularly where test scores are concerned, the measurements are not well-defined, and reporting scores or statistics with many decimal values gives a false sense of precision. Therefore, there is seldom reason to use more than one or two decimal places in standard score scales in psychology, sociology, or education because the quality of the original information seldom justifies more than this level of precision. For this reason, the publishers of commercial tests often warn potential users of the hazards inherent in assuming too much precision.

There are a great many other cautions to be observed when interpreting normative scale information for individuals or specific applications. They are thoroughly treated in books that focus on measurement, such as those listed at the end of this chapter. For our purposes, it is sufficient to note that for many problems in the social and behavioral sciences, two types of standards can be used in defining a scale to give meaning to the numbers we wish to analyze. One is the data collection procedure itself. Sociologists and psychologists often administer questionnaires and analyze the numbers without reference to any standard other than the verbal content of the questions. The behavior is quantified in terms of the number of errors made, the numerical value the person selects, or some other aspects of the data collection situation. The primary alternative to this procedure, and the main subject of this chapter, is using a group's responses as a standard to give meaning to those of an individual. Neither standard is inherently superior to the other, but it is essential to give careful thought to the way that the numerical results will get their meaning before the data are collected.

SUMMARY

Measurement scales can be divided into two broad categories, relative and absolute. *Absolute scales* are those where the measuring operation is sufficient to invest the numbers with their meaning. *Relative scales* are those where the results of the measuring operation gain meaning through comparison with the results from other measurements. Most relative scales involve *normative* comparisons with other objects or individuals. The comparison group is called a *norm group*. Normative scales may take the form of *percentile ranks,* where the percentile rank of a raw score is equal to the proportion of the scores in the norm group that fall below that point.

It is often useful to perform a *linear transformation* on a set of data to make interpretation of scores easier. All linear transformations may be represented by a straight line when the scale before transformation is placed on one axis of a graph and the scale after transformation is placed on the other. A linear transformation may involve *recentering* the distribution (adding or subtracting a constant), *changing the scale unit* (multiplying or dividing by a constant), or both. Recentering changes the value of the mode, mean, and median but not the interquartile range or standard deviation. Changing the scale unit by multiplying or dividing by a constant will affect both the indices of central tendency and the indices of variability. The range, interquartile range, and standard deviation will be multiplied or divided by the constant, while the variance will be multiplied or divided by the square of the constant used for rescaling.

A particularly useful linear transformation is the *standard score* or *Z-score transformation,* where the mean is subtracted from each score and the deviation score is divided by the standard deviation. All distributions that have had a *Z*-score transformation have a mean of zero and a standard deviation of 1. No linear transformation affects the shape of the distribution.

ANALYZING THE SEARCHERS STUDY DATA

Now try using one of the variables from the Searchers study to clarify further the use of percentiles, percentile ranks, and *Z*-scores. The frequency distribution for the degree of religious involvement scale appears in Table 5-4. Suppose we decide that we will call the top 15 percent of respondents "highly involved," the remainder of the top third "moderately involved," the middle third "neutral," the next 18 percent "moderately uninvolved," and the bottom 15 percent "definitely uninvolved." This breakdown is not unlike what is often done in analyzing research data. That is, groups are formed from what is a continuous variable in the original data.

Table 5-4 Frequency Distribution of Degree of Religious Involvement Scores for the Searchers Study

Score	Frequency	Cumulative Frequency	Score	Frequency	Cumulative Frequency
21	2	60	18	6	36
30	1	58	17	6	30
29	0	57	16	5	24
28	1	57	15	3	19
27	1	56	14	8	16
26	1	55	13	3	8
25	2	54	12	1	5
24	2	52	11	0	4
23	0	50	10	2	4
22	3	50	9	1	2
21	3	47	8	0	1
20	1	44	7	0	1
19	7	43	6	1	1

In order to form the groups described above, we must find the 15th, 33rd, 67th, and 85th percentiles. Table 5-5 contains the values you would need to compute these percentiles and the resulting points on the score scale. Compute the values yourself before checking them against the table.

After you have computed the desired percentiles, calculate the percentile ranks and *Z*-scores for each of the raw scores in Table 5-4. The mean for these data is 18.10 and the standard deviation of the sample is 5.21. The raw scores, percentile ranks, and *Z*-scores are given in Table 5-6. Again, compute the values for yourself before consulting the table. Why are there larger positive *Z*-scores than negative ones?

Table 5-5 Calculating Percentiles for Degree of Religious Involvement Scores for the Searchers Study

Percentile	PN	LL_i	cf_{j-1}	f_j	X_p
85	51	23.5	50	2	24.0
67	40	18.5	36	7	19.07
33	20	15.5	19	5	15.70
15	9	13.5	8	8	13.62

Table 5-6 Percentile Ranks and *Z*-scores for Degree of Religious Involvement Scores for the Searchers Study

Score	Percentile Rank	Z-Score	Score	Percentile Rank	Z-Score
31	98.3	+2.48	18	55.0	−0.02
30	95.8	+2.28	17	45.0	−0.21
29	95.0	+2.09	16	35.8	−0.40
28	94.2	+1.90	15	29.2	−0.60
27	92.5	+1.71	14	20.0	−0.79
26	90.8	+1.52	13	10.8	−0.98
25	88.3	+1.32	12	7.5	−1.17
24	85.0	+1.13	11	6.7	−1.36
23	83.3	+0.94	10	5.0	−1.55
22	80.8	+0.75	9	2.5	−1.75
21	75.8	+0.56	8	1.6	−1.94
20	72.5	+0.36	7	1.6	−2.13
19	65.8	+0.17	6	0.8	−2.32

REFERENCES

Anastasi, A., & Urbina, S. (1997). *Psychological testing* (7th ed.). Upper Saddle River, NJ: Prentice-Hall.

Cohen, R. J., Swerdlik, M. E., & Smith, D. K. (1992). *Psychological testing and assessment* (2nd ed.). Mountain View, CA: Mayfield.

Cronbach, L. J. (1990). *Essentials of psychological testing* (5th ed.). New York: Harper & Row.

Gregory, R. J. (1996). *Psychological testing: History, principles, and applications* (2nd ed.). Needham Heights, MA: Allyn & Bacon.

Linn, R. L. (1989). *Educational measurement* (3rd ed.). New York: American Council on Education/Macmillan.

Nunnally, J. C., & Bernstein, I. H. (1994). *Psychometric theory* (3rd ed.). New York: McGraw-Hill.

Rogers, T. B. (1995). *The psychological testing enterprise.* Pacific Grove, CA: Brooks/Cole.

Thorndike, R. M. (1997). *Measurement and evaluation in psychology and education* (6th ed.). Upper Saddle River, NJ: Prentice-Hall.

EXERCISES

1. Researchers in educational settings have found a strong relationship between academic achievement and self-esteem. However, self-esteem requires indirect measurement. What behaviors are indicators of positive self-esteem? How would you measure these behaviors?

2. An individual is classified as depressed based on the frequency of occurrence of a large number of behavioral indicators. Thus, depression seems to be an indirectly measured construct. What behaviors are indicators of depression? How would you measure these behaviors?

3. Define direct and indirect measurement and provide at least three examples of each. For each example of indirect measurement, indicate what some of the behavioral indicators are for that measure and how you would measure these behavioral indicators.

4. Describe the difference between absolute and relative measurement. Provide at least three examples of each type of measurement.

5. Classify each of the following as absolute measurement or relative measurement.
   a. The number of free throws made out of 100 attempts to determine if different types of practice are more effective.
   b. Rating of anxiety in different situations involving spiders as a measure of fear of spiders.
   c. The score on a series of problem solving tasks as an indicator of the effectiveness of problem solving instructional techniques.
   d. The proportion of time a child selects a sticker as a reward.

   e. A measure of creativity as indicated by a score on the Torrance Creativity Test.

6. Why are group norms important in the interpretation of scores of the measure of some construct or behavior?

7. For each of the following measurement situations, identify the appropriate norm group.
   a. Some of the requirements for becoming a licensed psychologist in the state of Washington include completing the requirements for a Ph.D., having 2,000 hours of supervised counseling and achieving at least the criterion score on a test about psychology.
   b. Teachers in some states must take a competency test before they are granted a teaching certificate.
   c. As a part of the supporting documents for applying to law school, applicants must take the LSAT exam.
   d. Applicants for a job at a local burger joint must complete a basic competency test.
   e. As a part of the requirements for entering a high school program for the gifted and talented, students must take an individually administered intelligence test.

8. Suppose the following scores represent performance on a quiz. The instructor wants to assign grades such that the upper 15% receive As, the next 25% receive Bs, the next 35% receive Cs, the next 20% receive Ds, and the bottom 5% receive Fs. Find the percentiles associated with each letter grade. 20, 20, 19, 19, 19, 19, 18, 18, 18, 18, 18, 16, 16, 16, 15, 15, 14, 14, 11

9. Suppose the scores below represent measures of daily stress. A researcher wants to categorize the scores based

on the following system: the upper 10% are highly stressed, the next 20% moderately stressed, the next 40% are somewhat stressed, the next 20% are slightly stressed, and the bottom 10% are not stressed. Find the percentiles associated with each classification. 22, 17, 27, 17, 15, 27, 16, 25, 21, 23, 19, 27, 20, 23, 18, 22, 23, 24, 16, 25, 24, 27, 28, 19, 21, 23, 18, 17, 15, 16, 25, 28, 29, 22, 21, 20, 21, 24, 25, 19, 17, 16, 22, 24, 27, 18, 21, 24, 20, 24

10. Suppose the following grouped frequency distribution represents test anxiety measures. The upper 15% of the individuals form the highly anxious group, the next 25% form the moderately anxious group, the next 40% form the somewhat anxious group, and the bottom 20% form the non-anxious group. Find the percentiles associated with each classification.

Interval	Frequency	Cumulative Frequency
60–64	8	155
55–59	11	147
50–54	15	136
45–49	23	121
40–44	36	98
35–39	26	62
30–34	19	36
25–29	13	17
20–24	4	4

11. Use the quiz scores in Exercise 8 to calculate the percentile ranks of the following quiz scores: 12, 14, 17, 18

12. Use the stress scores in Exercise 9 to calculate the percentile ranks of the following stress scores: 16, 20, 24, 26, 28

13. Discuss the difference between percentiles and percentile ranks. Provide at least two situations in which a percentile would be more appropriate and at least two situations in which a percentile rank would be more appropriate.

14. Use the SPSS data (career.sav) bank and the SPSS program to find the 15th, 28th, 56th, 73rd, and 89th percentiles for the following variables: confidence in mathematics (confmath), success at doing mathematics (succesat), perception of the usefulness of mathematics (useful), anxiety about doing mathematics (anxiety), and perception of mathematics as a fun activity (mathfun).

15. Calculate the mean, variance, and standard deviation for the quiz scores in Exercise 8.
   a. Calculate what the mean, variance and standard deviation would be if the instructor decided to add 5 points to each score.
   b. Calculate what the mean, variance, and standard deviation would be if the instuctor decided to subtract 11 points from each score.
   c. Calculate what the ~~ructo~~ ~~ariance~~, and standard deviation would b~~ , variance~~ ~~r~~ decided to multiply each sco~~structor~~ ~~deci~~
   d. Cal~~ ~~ and standard devi-~~ ~~led to divide

e. Given the results of parts (a) through (d) what rules can you state regarding transformations and their effects on the mean, variance, and standard deviation?
f. Convert each of the scores in Exercise 8 to standard scores or Z-scores. What are the mean and standard deviation of this set of standard scores? Draw a frequency distribution of the raw scores and the standard scores. How do the shapes of the frequency distributions compare?
g. Suppose that we do not want to represent the scores in Exercise 8 as standard scores because some of them will be negative values. We decide to change the scores to the IQ scale (i.e., $\overline{X} = 100$ and $S = 15$). What would the IQ scale be for each score? What will be the mean and standard deviation of this set of IQ scale scores? How will this transformation affect the shape of the frequency distribution?

16. Researchers administered the neuroticism subscale of the Eysenck Personality Inventory and the measure of social phobia.

Participant	Neuroticism	Social Phobia
1	6	11
2	4	12
3	19	22
4	8	10
5	5	12
6	9	17
7	9	15
8	6	13
9	15	15
10	14	23
11	10	15
12	11	17
13	20	26
14	19	18
15	20	26
16	16	22

What are the mean, median, and mode of each sample? For each sample calculate the range, the interquartile range, the semi-interquartile range, the variance, and the standard deviation. Find the Z-scores that correspond to each of the raw scores. Calculate the mean and standard deviation for the Z-scores for each sample. Compare the values between the two samples. Do you notice anything unusual?

17. In general, what are the properties of Z-scores? Identify situations in which it is appropriate to use Z-scores.

18. Suppose you are told that raw scores on a measure of quantitative ability are transformed using the equation $X' = 50X + 100$. If the mean and standard deviation of the raw scores were 14.4 and 2.80, respectively, what would be the mean and standard deviation of the transformed scores?

19. Suppose you are told that raw scores on a measure of quantitative ability are transformed using the equation $X' = 50X - 10$. If the mean and standard deviation of the transformed scores were 468 and 125, respectively, what would be the mean and standard deviation of the raw scores?

Part II

Describing Relationships

Chapter 6

Probability and the Normal Distribution

CHAPTER OBJECTIVES

After studying this chapter, you should be able to:

❏ Define the three basic kinds of probability and recognize applications of each.

❏ Compute empirical probabilities.

❏ Apply the addition and multiplication laws of probability to compute theoretical probabilities.

❏ Define the binomial distribution.

❏ Determine the probability of a binomial event.

❏ Define the normal distribution.

❏ Use the table of the normal distribution to determine the probability of an outcome for discrete or continuous events.

❏ Compute confidence intervals using the normal distribution.

Suppose you are the human resources director for a hospital, or perhaps you are a clinical or educational psychologist studying personality or academic achievement. Or you may be a teacher who has received standardized testing reports on the students in your class. In all of these cases and a multitude of others, it is necessary to work with and to analyze data.

Until now, we have focused on methods for making a set of observations on a single variable and describing their distribution. We have been content to summarize the data as they occurred without attempting to impose any theory or explanation on them. For example, we found the means and standard deviations of the variables collected by the SAD faculty from their 68 students at Big Time U. This is sufficient if our objective is merely to describe rather than to predict and understand. However, as HR director you may be asked to develop a program to predict how well future employees will perform their tasks, or if you are a psychologist you may be trying to find an explanation for a personality disorder. As a teacher, you may wish to determine whether a curricular change improves your students' performances on a standardized test. When we have broader goals such as these, we must go beyond the data we have collected to make predictions about events we have not yet observed (and may never have an opportunity to observe completely). Explaining or understanding a phenomenon such as academic achievement or a personality disorder implies that we are able to predict its occurrence from a knowledge of other events and forecast its consequences in the form of other future outcomes. In this chapter we take a brief excursion into probability theory, which will help us to answer questions of prediction and explanation that are so important to a scientific understanding of psychology and education.

Introduction

The ultimate objective of scientific research is to understand the phenomena under study. Since this implies going beyond the limits of any finite set of observations, it is necessary for all scientists to assume that the entities or events they observed are like those they might have observed but did not. A chemist studying the properties of a complex chemical makes the assumption that the molecules of the chemical that are available for study are not different from all other molecules of the same substance. The physicist working on problems relating to subatomic particles or the propagation of light makes the same assumption. For research in the natural sciences, this assumption is not difficult to justify and is often left unstated; the researcher can generalize the results obtained in a study of substance *X* or process *Y* to other examples of the same substance or process.

Unfortunately, it is a much more difficult and complex matter to generalize from observed to unobserved entities or events in other branches of science. Anyone who has depended on a weather forecast in planning a day's activities knows that such predictions are often (some might say usually) in error. Although meteorology is a natural science and its phenomena follow natural laws, the sheer number of events that function in causal relationships to each other make exact predictions impossible with current technology. Consequently, weather forecasts are presented in the form of probabilities. We are told that there is a 40 percent chance of rain today, for example, and we have learned to adjust our behavior accordingly.

Although the situation is difficult in meteorology, it is often more challenging in the social and behavioral sciences. The organisms being studied often have minds of their own and behave in inconsistent ways. There is much more variability among individuals, even when the individuals are microscopic organisms, than there is from one molecule of a chemical or photon of light to another. There is also a vast array of forces, both internal and external, acting on any organism at any time, and individual organisms themselves change over time.

Such problems with variability and uncertainty have made it necessary for researchers in psychology and education to adopt methods of data analysis that take them into account when making predictions and generalizations. In this chapter, we will discuss the principles of chance and probability, the same admissions of uncertainty found in weather forecasting, and show how they produce a particular frequency distribution, the normal distribution, that has far-reaching significance for the statistical analysis of research data.

▶ *Now You Do It*

Issues of probability are reflected in many facets of our everyday lives, not just in weather forecasting and statistics. In what other circumstances is probability a factor? Think about an insurance company. How would they have use for a knowledge of probability? In what ways would knowing probabilities affect rates for insurance?

Insurance companies are perhaps the biggest users of information about probability. They use information on accident rates in different age groups to set automobile insurance rates, information on survival rates to various ages to determine how much to charge for life insurance, and so on. The insurance company is gambling that you will drive safely enough and live long enough, and pay enough in premiums, so that when they have to pay out on a claim, they will have made sufficient income from your premiums to cover what they have to pay out. If their probabilities are wrong (or at least, wrong very often) they will go out of business.

Can you identify other areas in which probability plays an important part in your life?

Types of Probability

Probability is a concept that is familiar to almost everyone in one way or another. We have already mentioned its application to the weather forecast in the evening news, and to the past (and probable future) batting performance of baseball players. Most people have an intuitive feel for what probability means when it comes to making guesses in the face of uncertainty. There are, in fact, three somewhat different forms of probability, and we need to understand each of them in order to grasp the application of probability to scientific research.

Subjective Probability

What can best be described as your gut feeling is called **subjective probability.** Subjective probability is the name we give to hunches and beliefs that are based on unquantified past experience. This type of probability estimate guides much of our day-to-day living. For example, when you are driving down the street, you may see another car stopped in a driveway or a person standing on the street corner. How you proceed depends on your predictions about the future behavior of the other driver or the pedestrian. You have a set of expectations derived from past experience, and you use these expectations as the basis for estimating what the other person will do. You attach a subjective probability to each of the other person's possible behaviors, such as the pedestrian's stepping into the street or the driver's pulling out in front of you, and you then act in terms of your subjective probabilities by slowing down or maintaining your speed.

Although subjective probabilities govern many aspects of our daily behavior, they lack sufficient precision to be useful beyond the earliest stages of scientific inquiry. Subjective probability, our educated guess about what we will find in a study, may guide our initial investigation, but we need more formal definitions to make scientific decisions. Statistics, as an aid in science, uses two other forms of probability—theoretical probability and empirical probability—to aid in the prediction of the outcomes of scientific studies and the interpretation of their results.

Theoretical Probability

Theoretical probability is a branch of mathematics that deals with probability in the abstract. Working in this realm, we define a theoretical object, such as an ideal coin or card deck, that fits a mathematical abstraction and deduce how the object will behave, given our assumptions and definitions. In the world of theoretical probability things are true because we say they are. Theoretical probability "experiments" are thought experiments, which do not result in the actual collection of data. For example, we will discuss a number of theoretical probability experiments involving coin tosses. It is not necessary, or even particularly useful, to toss any real coins in developing our laws of probability because the laws hold exactly only in the abstract. Instead, we define an "ideal coin" that has the precise properties we assign to it, and we study the "behavior" of this coin under the conditions of interest. This is a little like doing research in collaboration with the Queen of Hearts whom Alice meets in Wonderland, where the world behaves just as she says it will.

Empirical Probability

The form of probability that is based upon carefully recorded past experience or observation is called **empirical probability.** Empirical probability is just like theoretical probability except that the "data" come from experience rather than from definition. Thus, instead of defining an ideal coin that has a probability of coming up heads of 50 percent (.50) on each toss, we might observe a real coin and determine the probability of its coming up heads from a number of observations. As we shall see in the chapters ahead, many of the applications of statistics in psychology and education involve a comparison between empirical probabilities derived from data and theoretical probabilities derived from expectations based upon a theory.

Distribution of Chance Events

Assume that a Las Vegas casino operator wants to introduce a new gambling game where the players flip silver dollars, trying to match a coin flipped by an employee of the casino. The casino owner wants to know what outcomes can be expected if the new game is introduced, so she hires a statistician who makes a prediction based on the logic and the computation of probabilities. Logical processes and applications of probability theory lead to her prediction.

First, the statistician assumes that if a single coin is tossed, it is equally likely to come up heads or tails and it will never stay on edge. We can state this assumption in the language of empirical probability: in a very large number of tosses (K), the coin will come up heads half the time ($K/2$) and tails half the time. We now define the probability that our coin will come up heads (P_H) as the number of times that a head occurred (N_H) divided by the number of tosses (N)

$$P_H = \frac{N_H}{N} \qquad\qquad (6.1)$$

In our example,

$$P_H = \frac{K/2}{K} = \frac{1}{2} = .50$$

The probability that the coin will come up tails is found in the same way and is also equal to .50. We note that in this example only two outcomes are possible: Either the coin comes up heads or it comes up tails. Since something happens on every toss, the probability of something happening is 1.0; therefore, we can find the probability of either outcome once we know the probability of the other:

$$P_H = 1.0 - P_T \text{ or}$$
$$P_T = 1.0 - P_H$$

The coin toss is an example of what is called a **binary event.** In this case, where there are only two possible outcomes, the two probabilities must add up to 1.0. The general rule is that the sum of the probabilities of all possible outcomes must add up to 1.0, which includes the probability that nothing will happen if that is a possible outcome. If we roll a die and count the number of spots, there are six possible outcomes, so this is not a binary event; however, the probabilities of the possible outcomes must still add to 1.0.

The statistician has now provided the casino owner with the information that the probability that a coin will come up heads is .50 and is equal to the probability that it will come up tails (assuming that the casino doesn't tamper with the coins, but that's another story). But the game the casino owner has in mind involves players matching coins with a person who works for the casino, known as "the house": Match the house and the player wins; fail to match and the house wins. What may we now expect to find?

The house flips a coin that may come up heads or tails and the player does the same; there are now two binary events and four possible outcomes as shown here:

Event	House	Player	Winner
1	H	H	player
2	H	T	house
3	T	H	house
4	T	T	player

There are two ways for the player to win and two ways for the house to win. Looking at this table, it seems reasonable that the probability of each unique outcome (*HH, HT,*

A **binary event** is one that can occur in one of two forms.

SPECIAL ISSUES IN RESEARCH

Theoretical Probability in ESP Research

There are certain research situations of interest to psychologists where theoretical probability comes into play quite directly. Perhaps the most widely known is the study of extra-sensory perception, or ESP. Can people receive messages by means other than through the usual five senses? How might you set about to test someone's claim that they could receive such messages?

A very simple ESP experiment might have a person try to guess whether a coin tossed by someone else came up heads or tails. We could have the coin tosser (call him the sender)

in one room and the person who claims to have ESP (call her the receiver) in another room isolated from the first one. The sender tosses a coin and concentrates on the outcome, and the receiver tries to correctly identify whether the coin came up heads or tails. Most ESP experiments use a more complex situation such as a deck of cards in which there are four or five different symbols used, but we will stick with the simpler binary experiment for now because it is easier to follow the reasoning and the mathematics.

TH, TT) is 1/4, the probability that the house will win is 2/4 = 1/2, and the probability that the player will win (house lose) is also 1/2.

Here intuition serves us well. However, when the number of events taking place becomes large or the two probabilities are not .50, the calculations become more complicated. Suppose, for example, that we have two people playing against the house (three coins being tossed) and each player bets $1; there are now eight possible outcomes as shown here, all equally likely:

Event	House	Player 1	Player 2	Winners
1	H	H	H	1 and 2
2	H	H	T	house and 1
3	H	T	H	house and 2
4	H	T	T	house
5	T	H	H	house
6	T	H	T	house and 2
7	T	T	H	house and 1
8	T	T	T	1 and 2

On events 1 and 8, both players win and the house loses $2, while on events 4 and 5, the house wins from both players. The other four toss combinations result in a split where one player loses, the other player wins, and the house breaks even. It is still relatively easy to see that the probability of any single outcome is now 1/8, that the house will win two dollars 2/8ths or 1/4th of the time, lose two dollars 1/4th of the time, and come out even 4/8ths or 1/2 of the time, and that the sum of these probabilities is 1.0, but things are becoming complex rather quickly. For example, adding only one more player increases the number of possible outcomes to 16.

▶ *Now You Do It*

Suppose you are running the ESP experiment we discussed in Special Issues in Research on page 145. Let us assume that ESP does not exist, which is exactly the same as assuming that the casino operator's coins are fair. Let's call an outcome where the sender and receiver match each other a right answer [R] and an outcome where they disagree a wrong answer [W]. What are the possible outcomes on the first three trials of our experiment?

The answer is that we should expect outcomes exactly like those the statistician described for the casino. That is, the receiver might match the sender on all three trials, or she might match the sender on any two of the trials, any one of the trials, or none of the trials. We might list the outcomes this way:

Trial 1	Trial 2	Trial 3	Outcome
R	R	R	3
R	R	W	2
R	W	R	2
R	W	W	1
W	R	R	2
W	R	W	1
W	W	R	1
W	W	W	0

Notice that we once again have eight possible outcomes. We would expect each outcome to be equally likely if ESP does not exist, so we might say that the probability of the receiver matching the sender on all three trials is 1/8 or .125. The probability that the receiver will fail to match the sender on all three trials is also .125.

Assumptions of Probability Theory

Up to this point we have left implicit three basic assumptions about the nature of our events. The first is that **the list of outcomes is exhaustive;** it includes all possible results.

*A list of outcomes is **exhaustive** if it includes all possible results.*

This assumption permits us to find any one probability by subtracting the others from 1.0. For example, if we know that the probability of drawing a club from a deck of cards is .25, we also know that the probability of a card other than a club is $(1 - .25)$ or .75 because "club" and "not club" form an exhaustive list of the possible outcomes. The second assumption is that **the outcomes are mutually exclusive;** that is, only one of them can happen at a time. It is not possible, for example, for a toss of three coins to result in three heads and three tails at the same time. While the truth of this assumption seems obvious, there are situations where it does not apply, and when this happens it is necessary to use some different rules. The third assumption is that **each toss is independent of all the others.** In other words, what happens to one coin does not affect what happens to another, and what happens on one trial does not influence later trials. Phrased in terms of probabilities, this means that P_H for player 2 is .50 regardless of what happens to the coins of player 1 and the house, and that the probability of each particular outcome—for example, *HTH*—is 1/8 on each trial regardless of what happened on the last trial.

Two Laws of Probability

When the conditions of exhaustiveness, mutual exclusivity and independence are satisfied, two simple laws of probability can help us to calculate the expected outcomes. The first is the **addition law** of probability, also known as the **either/or law.**

Likewise, the probability that **either A or B or C** will occur is equal to the *sum* of their separate probabilities. A general statement of this law is that the probability that any one of *several* events will occur is equal to the sum of their separate probabilities.

The second law of probability is the **multiplication law,** or the **and law** of probability. In its general form, this law states that the probability that *each of several* independent events will occur is equal to the *product* of their separate probabilities.

Our first example, where one player is attempting to match the house, can be used to illustrate the application of these two laws. Remembering that $P_H = .5$ and $P_T = .5$ for each coin separately, we can use the multiplication law to find that the probability of a head on coin 1 *and* a head on coin 2 is

$$P_{H_1 H_2} = P_{H_1} \cdot P_{H_2} = .50 \times .50 = .25 = 1/4.$$

We can use the same computations to show that each of the other outcomes also has a probability of 1/4. This is the same probability for this event that we found when we listed the possible events.

The addition law can be used to compute the probability that the house will win. The house wins in case 2 or 3 (*HT* or *TH*), each of which has a probability of .25 (found from applying the multiplication law). Therefore, the probability that the house will win is

$$P_{Housewins} = P_{HT or TH} = P_{HT} + P_{TH} = .25 + .25 = .50$$

which obviously corresponds to our previous expectation. This also confirms the probability we found previously.

For the more complex problem of three coins, the same principles apply. The multiplication law tells us that the probability of three heads or of any other specific outcome, such as *HTH* or *TTH*, is $1/2 \cdot 1/2 \cdot 1/2 = 1/8$ [or $(1/2)^3$ or $.5^3$]. The addition law reveals that the probability that the house will win two dollars is $1/8 + 1/8 = 1/4$ ($P_{HTT or THH} = P_{HTT} + P_{THH}$). Likewise, the probability that both players will win and the house lose two dollars is

$$P_{HHH or TTT} = P_{HHH} + P_{TTT} = .125 + .125 = .25.$$

The probability that the house will break even (house and one player win) is $1/8 + 1/8 + 1/8 + 1/8 = 1/2$. If we know the probability of each simple event, and if our assumptions of independence, mutual exclusiveness, and exhaustiveness hold, we can use the addition and multiplication laws to calculate the probability of any complex event.

◗ Two Laws of Probability

Addition Law	The probability that *either* event A *or* event B will occur is equal to the *sum* of their separate probabilities ($P_{A or B} = P_A + P_B$).
Multiplication Law	The probability that *both* event A *and* event B will occur is equal to the *product* of their separate probabilities ($P_{A and B} = P_A \cdot P_B$).

Let's now see how the laws of probability apply to the outcomes we would expect to find for our ESP experiment. Suppose we conduct 4 trials with Sender A and Receiver B. What are the possible outcomes, and how probable is each one? Can you apply the two laws of probability to find the answer?

First, let's find the probability of any given single outcome. Because there are four trials, we are looking at the probability of each of four specific events, for example, matches on trials 1 and 3 and failures on trials 2 and 4. This is equivalent to the sequence R–W–R–W or R–and–W–and–R–and–W. The presence of "and" clues us that we must use the multiplication law. Therefore, the probability of any specific sequence is

$$(.5)(.5)(.5)(.5) = .5^4 = .0625.$$

Each of the 16 possible outcomes has a probability of .0625 or 1/16 of being the one that occurs.

You should now satisfy yourself that there are 16 possible unique outcomes of this experiment, one of which is

$$R \text{ and } R \text{ and } R \text{ and } R.$$

When you are finished, you should be able to count up 1 outcome with 4 matches, 4 outcomes with 3 matches, 6 outcomes with 2 matches, 4 outcomes with 1 match, and 1 outcome with no matches.

The probaility of getting a member of the class of outcomes "1 match" involves getting either the first *or* the second *or* the third *or* the fourth instance of this class, so we will apply the addition rule to find the probability. Since one match can occur in any of four ways, the addition law tells us that the answer is the sum of the individual probabilities or

$$.0625 + .0625 + .0625 + .0625 = 4(.0625) = .25.$$

There is a 25 percent chance that we will observe exactly one match. The case of two matches can occur in any of six ways, so the probability of this outcome is .375.

Unequal Probabilities

The case of three binary events of equal probability is a simple example of a more general and complex problem. Suppose, for example, that organized crime gained control of the casino and substituted loaded coins for those in use. As any good gambler knows, it doesn't take much of an edge for the house to win in the long run. Imagine that the new coins have a probability of coming up heads of .6 rather than .5. We can use the addition and multiplication laws to compute the probabilities of the various outcomes as follows:

For each coin	$P_H = .60$	$P_T = (1 - .6) = .40$
For two coins	$P_{HH} = (.60)(.60)$	$P_{TT} = (.40)(.40)$
	$= .36$	$= .16$
	$P_{TH} = P_{HT} = (.60)(.40) = .24$	
For three coins	$P_{HHH} = (.60)(.60)(.60) = .216$	
	$P_{TTT} = (.40)(.40)(.40) = .064$	
	$P_{HHT} = (.60)(.60)(.40) = .144$	
	$P_{HTT} = (.60)(.40)(.40) = .096$	
	$P_{HTH} = P_{HHT} = P_{THH}$	$P_{THT} = P_{TTH} = P_{HTT}$

For the two-coin case, we see that the probability that the player will win by matching the house (either *HH* or *TT*) is .36 + .16 = .52 rather than the .5 that we had earlier. Of course, the probability that the house will win can be found to be 1.0 − .52 = .48 or

$$(.4 \cdot .6) + (.4 \cdot .6) = .24 + .24 = .48.$$

The same thing happens with two players (three coins). The probability that both will match the house (either *HHH* or *TTT*) is .216 + .064 = .280 instead of the .25 we found earlier, while the chance that the house will win big (*HTT* or *THH*) is .096 + .144 = .240. The odds of a split, which can occur in any of four ways, are .096 + .096 + .144 + .144 = .48. (Note that again the probabilities of all classes of outcomes sum to 1.0.) Clearly, if the house leaves the game defined in this way (match the house and you win), the odds are in favor of the players, and the house will lose in the long run. Although it may seem surprising to use gambling examples and coin tosses to discuss statistics, gambling actually did provide a major early impetus for developing the mathematics of probability.

▶ *Now You Do It*

At this point, let's see how the principles we have been discussing apply to our research on ESP. Suppose we have set up a study in which the sender draws a card from a standard card deck and the task of the receiver is to determine whether the card is red or black. What is the chance of a match on any single trial? The answer, of course, is .50 because there are two red suits and two black suits (or 26 red cards and 26 black ones). Therefore, the probabilities we determined for tosses of fair coins can be applied directly to the ESP experiment. What are the probabilities of the five possible outcomes that could occur on the first four trials?

The probability of four successive matches by the receiver is $.50^4$ or .0625. If we observed a result like this, we might be inclined to believe in ESP (we discuss the rules for making this kind of decision in a research setting in Part III). You should be able to show that the probabilities of the other outcomes are

$$3 \text{ matches (4 ways to happen)} = .25$$
$$2 \text{ matches (6 ways to happen)} = .375$$
$$1 \text{ match (4 ways to happen)} = .25$$
$$0 \text{ matches (1 way to happen)} = .0625$$

Of course, these probabilities sum to 1.00.

Now suppose we ask the sender to send the suit of the card and the receiver to state whether it is a club, diamond, heart, or spade. What is the probability that there will be a match on a given trial *if ESP does not exist?* The answer, of course, is 1/4 or .25. What are the probabilities of the five possible outcomes from four trials? You should be able to show that they are

$$4 \text{ matches (1 way to happen)}$$
$$1(.25)(.25)(.25)(.25) = .0039$$

$$3 \text{ matches (4 ways to happen)}$$
$$4(.25)(.25)(.25)(.75) = 4(.0117) = .0469$$

$$2 \text{ matches (6 ways to happen)}$$
$$6(.25)(.25)(.75)(.75) = 6(.0352) = .2112 \text{ (.2109 if you do not round)}$$

$$1 \text{ match (4 ways to happen)}$$
$$4(.25)(.75)(.75)(.75) = 4(.1055) = .4220 \text{ (.4219 if you do not round)}$$

$$0 \text{ matches (1 way to happen)}$$
$$1(.75)(.75)(.75)(.75) = .3164$$

The Binomial Distribution

The probabilities of various outcomes that can occur in only one of two ways, such as trying to match a sender in an ESP experiment, form a distribution known as a **binomial distribution.** A binomial distribution describes the probability of each class of outcomes for groups of binary events. For example, when there are two binary trials such as tossing a coin twice, the binomial distribution contains the probabilities of the three possible outcomes (0, 1, or 2 heads). For four trials, the distribution contains the probabilities of the five possible outcomes.

The properties of the binomial distribution can be illustrated with one last coin-toss example. Imagine that we shake up ten coins in a shaker, then dump them out on a table, and count the number of coins that come up heads. Assuming that $P_H = .5$ for all coins, what is the most probable outcome? (If you answered five *H* and five *T*, you are right.) The next question, what is the probability of this outcome, is much more difficult to answer.

*A **binomial distribution** is a distribution of probabilities of events that can occur in one of only two ways.*

The probability of any particular outcome can be calculated by our multiplication rule. A particular outcome is one where we specify exactly which coins come up heads, for example, that coins 1–5 are heads and 6–10 are tails. For another particular outcome, we could specify that coins 1–4 and 10 are heads, and that coins 5–9 are tails, or that odd-numbered coins are heads and even-numbered coins are tails. Each of these examples results in the general outcome of five heads, but each is itself the conjunction of ten different elementary binary events. In other words, the outcome of heads on 1–5 and tails on 6–10 can occur only if we get heads on each of the first five tosses and tails on each of the last five tosses. Therefore, according to the multiplication rule, the probability of any *particular* outcome is computed as the product of the ten individual probabilities, or $.5^{10}$. Our handy pocket calculator reveals that 1/2 raised to the tenth power is 1/1024 or .0009765625, a little less than one in a thousand.

We can calculate the total number of possible outcomes by multiplication as well. To find the total number of outcomes, we obtain the product of the number of possible outcomes for each entity (coin in this case). Since there are two possible outcomes for each of the ten coins, we obtain $2_1 \cdot 2_2 \cdot 2_3 \cdot \cdots \cdot 2_{10} = 2^{10} = 1024$ possible particular outcomes, each of which has the same probability of occurring.

When there are more than two possible outcomes for each element, the number of possibilities grows much faster. In the case of dice, for example, there are six faces, or six possible outcomes. Two dice yield 36 possible unique outcomes ($6 \cdot 6$), three dice give 216 ($6 \cdot 6 \cdot 6$), and four dice produce 1296 (6^4) possibilities. In general, the total number of possible unique outcomes is the number of alternatives per element (two in the case of coins, six for dice) raised to the power of the number of elements. Thus, for N coins the number of possible unique outcomes is 2^N, and for N dice it is 6^N.

The total number of possible unique outcomes is the number of alternatives per element raised to the power of the number of elements.

We still haven't answered our original question, however: What is the probability of obtaining five heads when ten coins are shaken up and dumped on the table? There are several different configurations of coins that would give us five heads, and our problem at this point is to figure out how many such configurations there are.

There are two ways to solve the problem. The first is to do what we did with three coins; list all particular outcomes and count the number of these that give us five heads. Unfortunately, with 1024 particular outcomes, this would be quite an unpleasant task. The alternative is to compute what mathematicians call the number of combinations; this is done with the formula

$$_NC_r = \frac{N!}{r!\,(N-r)!} \tag{6.2}$$

N = the total number of objects (coins),
r = the number of outcomes of a particular kind (heads),
$(N-r)$ = the number of outcomes of the other kind (tails or not-heads), and
$!$ = means factorial.

A factorial is a successive product where the number is reduced by one at each stage; for example,

$$3! = 3 \cdot 2 \cdot 1 = 6;$$
$$5! = 5 \cdot 4 \cdot 3 \cdot 2 \cdot 1 = 120;$$

and, in general,

$$N! = N \cdot (N-1) \cdot (N-2) \cdot \cdots \cdot 2 \cdot 1.$$

In our problem, we want to find the number of combinations of ten coins that will give us five heads, so we use formula 6.2 to find

$$_{10}C_5 = \frac{10!}{5!\,(10-5)!} = \frac{10 \cdot 9 \cdot 8 \cdot 7 \cdot 6 \cdot 5 \cdot 4 \cdot 3 \cdot 2 \cdot 1}{(5 \cdot 4 \cdot 3 \cdot 2 \cdot 1) \cdot (5 \cdot 4 \cdot 3 \cdot 2 \cdot 1)}$$

Canceling where possible, we get

$$= \frac{9 \cdot 2 \cdot 7 \cdot 2}{1} = 252.$$

X	f	cf	P	Z
10	1	1024	.00098	+3.16
9	10	1023	.0098	+2.53
8	45	1013	.0439	+1.90
7	120	968	.1172	+1.26
6	210	848	.2051	+.63
5	252	638	.2461	0
4	210	386	.2051	−.63
3	120	176	.1172	−1.26
2	45	56	.0439	−1.90
1	10	11	.0098	−2.53
0	1	1	.00098	−3.16

Table 6-1 Possible Outcomes from Tossing Ten Coins

This tells us that of the 1024 particular events, there are 252 that give us five heads. Therefore, from formula 6.1, the probability that we get five heads from tossing ten coins is

$$252/1024 = .246.$$

Just to complete the picture of what can happen with ten coins, we can calculate the number of ways each of the various other outcomes can occur. We find each of these results using formula 6.2 and enter them into Table 6-1 as a frequency distribution. Notice that the distribution is symmetric around 5.

▶ *Now You Do It*

Now try your hand at determining the number of combinations using formula 6.2 by verifying the frequencies in Table 6-1. For example, the number of ways eight heads can happen is

$$_{10}C_8 = \frac{10!}{8!(10-8)!} = \frac{10 \cdot 9 \cdot 8 \cdot 7 \cdot 6 \cdot 5 \cdot 4 \cdot 3 \cdot 2 \cdot 1}{(8 \cdot 7 \cdot 6 \cdot 5 \cdot 4 \cdot 3 \cdot 2 \cdot 1)(2 \cdot 1)}$$
$$= \frac{10 \cdot 9}{2 \cdot 1} = 45$$

The other values you should get are listed in the table.

Earlier we discussed the possibility of using biased coins, and we saw what that would do to the probabilities of some simple outcomes. How would biased coins affect the outcomes of our 10-coin example? The answer requires that we go back and examine how we calculated the probability of each unique event.

As we have seen, when either of two outcomes is equally likely, all possible unique outcomes have the same probability. However, this changes when one outcome is more likely than the other. If we call the probability of a coin coming up heads P and the probability of not heads (tails) Q, we know from the multiplication law that a particular combination of heads and tails can be expressed as a product of Ps and Qs. For example, the combination $HTHHT$ has the probability $P \cdot Q \cdot P \cdot P \cdot Q$. Grouping together terms that are equivalent, we can rewrite this product as $P \cdot P \cdot P \cdot Q \cdot Q = P^3Q^2$. If $P = .60$ and $Q = .40$, the probability of this combination (or any particular combination of 3 heads and 2 tails) is

$$(.60)(.60)(.60)(.40)(.40) = .60^3 \cdot .40^2 = .03456,$$

which can be written symbolically as P^3Q^2.

We can combine these symbols with the ones we used earlier to come up with a general expression for the probability that a particular number of heads and tails will occur for any number of coin tosses. If there are N coins, each of which has a probability of coming up heads of P (and a probability of tails of Q), the probability of any specific combination resulting in exactly r heads on a single trial is

$$P^r Q^{N-r} \tag{6.3}$$

In the case of N binary events, each of which has a probability of coming up one way of P, the probability of any particular combination resulting in exactly r events of the first kind is $P^r Q^{N-r}$.

For the particular example of 15 heads from a toss of 20 coins, this would be $P^{15} Q^5$.

The probability given by formula 6.3 is for each specific outcome. To find the probability of a class of outcomes such as "15 heads in 20 coins," we need to combine the probability of each single event from formula 6.3 with the number of ways that event can happen from formula 6.2. The resulting formula gives us the probability of the class of 15 heads in 20 coins. In general, the probability of exactly r events of a chosen kind from a set of N events is given by the formula

$$P_{r\, of\, N} = \frac{N!}{r!\,(N-r)!}\,(P^r Q^{N-r}) \tag{6.4}$$

Working out equation 6.4 for $P_{15\, of\, 20}$ with $P = .60$ and $Q = .40$, we would obtain

$$P_{15\, of\, 20} = \frac{20!}{15!\,(5)!}\,(.6^{15}\,.4^5)$$

$$= \frac{(20)(19)(18)(17)(16)}{(5)(4)(3)(2)(1)}\,(.000470)(.01024)$$

$$= 15{,}504\,(.000004815)$$

$$= .07465$$

The probability is about .075 that a toss of these 20 coins, each with $P_H = .60$, would produce an outcome of exactly 15 heads.

Formula 6.4 is known as the **binomial probability** formula. It is applied to situations where we are looking for the probability of a particular number of discrete events and the outcomes are defined in such a way that there are only two kinds of outcomes (such as heads/tails or successes/failures). It is not widely used in educational and psychological research, but it provides a good first look at probability because the examples are tangible. (Two places where the binomial distribution is applied are ESP experiments such as the ones described earlier and studies of guessing on educational tests. If, for example, you thought two students were cheating, you could look at their wrong answers to see whether they chose the same wrong answer more frequently than would be expected from random guessing.)

▶ Now You Do It

Let's check your understanding of binomial probability using a slightly different kind of problem. Suppose you believe that any ape can learn statistics. Your classmates doubt your claim and challenge you to prove it by running a study of statistics performance in a chimpanzee who has been exposed to the same series of lectures that you have had in your class. The instructor agrees to have the lectures taped, and you are going to show them to Washoe, the famous chimp who learned sign language, to see whether she can learn statistics. You and your classmates agree to test Washoe's mastery of statistics using a multiple choice test where each item has five alternatives. Washoe will take the test after watching each videotaped lecture twice. What should we expect if your classmates are right?

The first thing we must do is determine Washoe's probability of success on each item if she knows nothing about statistics. What are the values of P and Q? With five alternative answers, the chance that Washoe will get an item right by chance is 1/5. If we call the probability of a correct answer P, then $P = .20$. Since $Q = 1 - P$, $Q = .80$.

Let's assume we are going to give Washoe a 12-item test over a set of four lectures. What should we expect to find? How many outcomes are possible? What is the most probable score? What is the probability of each possible outcome?

To answer these questions we must compute the binomial probabilities. There are 13 possible outcomes ranging from Washoe's missing every item at one extreme, to her answering every item correctly at the other. Once we have listed these outcomes, we must calculate how many ways each one can occur, using formula 6.2. Then we can use formula 6.3 to calculate the probability of any single event from a given class. For example, there are 495 ways that Washoe could get exactly four items right and eight wrong. Likewise, there also are 495 ways to get eight items right and four wrong. The probability of any specific example of four items correct is $(.2^4)(.8^8)$ or .000268, and the probability of any example of eight items correct and four wrong is $(.2^8)(.8^4)$ or .00000105. Finally, the probability of any class of outcomes is the product of the probability of one example times the number of ways it can happen. Thus, the probability of four right and eight wrong is $(.000268)(495)=.1329$, and the probability of eight right and four wrong is $(.00000105)(495)=.000519$. These two outcomes have different probabilities because Washoe is much more likely to miss an item than to get one right if she does not know statistics. If *you* know statistics, you should be able to get the following probabilities for Washoe's test scores.

Right	$\dfrac{N!}{r!\,(N-r)!}$	$P_r Q^{N-r}$	Probability
0	1	.0687	.0687
1	12	.0172	.2062
2	66	.0043	.2835
3	220	.00107	.2362
4	495	.000268	.1329
5	792	.000067	.0532
6	924	.0000168	.0155
7	792	.0000042	.0033
8	495	.00000105	.000519
9	220	.00000026	.000058
10	66	.000000066	.0000043
11	12	.000000016	.0000002
12	1	.000000004	.000000004

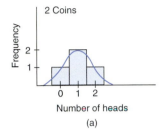

(a)

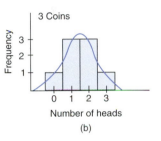

(b)

The Normal Distribution

There are many occurrences in the world of human experience that seem to be governed more or less by the random processes of chance. A variety of games (cards, dice, and so on) use chance events for entertainment, but the pattern of chance events that has been observed in games also seems to fit a variety of naturally occurring phenomena that are of scientific interest. For example, in the early nineteenth century studies of the distributions of human physical characteristics such as height and weight led scientists to postulate that the mean of each distribution of human characteristics was Nature's (or God's) ideal, and the variability of people around that mean was due to chance factors in their development. In fact, the model provided by the mathematics of probability fit the observed data so well that people spoke of the *normal law of error,* and the frequency distribution that fit these measurements came to be called the **normal distribution.**[1]

We can begin our study of the normal distribution by looking again at the frequency distributions of chance events. Histograms for the cases in which two coins, three coins, and four coins are tossed in succession are shown in Figure 6-1. Coin tosses and their outcomes are discrete events, but if we assume that the abscissa represents a continuous variable, we can draw a smooth curve through each of the figures. Although the curves have somewhat different shapes, they are all symmetric and unimodal; thus the mean,

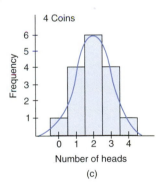

(c)

Figure 6-1 Histograms of expected outcomes from tossing two, three, and four coins. The normal distributions have been drawn in over the histograms.

---

[1]The normal distribution is a specific mathematical function that was first published about 200 years ago by the German mathematician Karl Friedrich Gauss. This equation,

$$b = \frac{N}{\sigma\sqrt{2\pi}}\, e^{-(X-\mu)^2/2\sigma^2}$$

gives the height (*b*) of a theoretical frequency polygon for any possible score in the distribution when the distribution has a skewness of zero (from formula 4.17) and kurtosis of zero (from formula 4.18). The Greek letters μ and σ stand for the mean and standard deviation of a theoretical population.

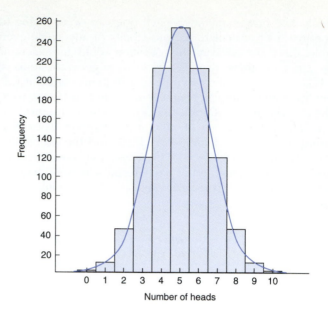

Figure 6-2 Histogram of expected outcomes from tossing ten coins. The normal distribution has been drawn in over the histogram.

median, and mode coincide. Each of the curves is also shaped somewhat like a bell. Each is a normal curve. A curve of this shape is also commonly called a bell curve, or a Gaussian curve, after its discoverer, Karl Friedrich Gauss.

The three smooth curves in Figure 6-1 are all normal curves. They are not exactly the same shape because they have different means and standard deviations, and each represents a different number of events. For example, the curve for two coins represents four possible outcomes, while the curve for four coins includes the 16 unique outcomes that are possible with four coins. Also, each theoretical curve only approximates the histogram from which it was drawn. However, as the number of score categories (coin tosses in this case) increases, the similarity between the histogram and the normal curve increases. Figure 6-2 contains the histogram that we would expect for ten coins (the frequencies come from Table 6-1); superimposed on the histogram is the normal curve that approximates the frequency polygon we would get for ten coins. Clearly, by the time we have the 11 score categories that can occur for ten events, the similarity between the frequency distribution of chance events and the normal distribution is apparent.

There are many bell-shaped curves that are normal curves. In fact, there is a possible normal curve for every possible combination of mean and standard deviation. It would be very useful if we could find a way to reduce all of these curves to a single curve; then we would have to worry about only one curve. There is, in fact, a way to reduce all of these curves, or theoretical frequency distributions, to the same scale, and we have already discussed the necessary procedures at some length. We can use a linear transformation to put all of the distributions on a single common scale.

The most obvious transformation is the standard score or Z-score transformation, which converts each frequency distribution to a mean of zero and a standard deviation of one. The distributions for two, three, and four coins are converted to Z-scores in Table 6-2. In turn, the frequency polygons for two, three, four, and ten coins are plotted in standard score form in parts (a) to (d) of Figure 6-3, along with the normal distribution. (Refer to Table 6-1 for the probabilities and Z-scores of the ten-coin problem.)

It is clear from Figure 6-3 that as the number of categories increases from 3 to 4 to 5 to 11 (2, 3, 4, and 10 coins), the difference between the coin-toss frequency polygon and the normal curve decreases. If we were to plot an example for 25 categories, it would be virtually impossible to see the difference between the normal curve and the frequency polygon of binomial chance events. This increasing similarity between the normal curve and the frequency distribution for chance events has enabled statisticians and scientists to use the normal curve, which has known mathematical properties, as a model of many scientific phenomena. Perhaps the most well-known recent example is the distribution of scores on tests of cognitive ability, which tend to form a normal distribution, and which gave rise to the book title *The Bell Curve*.

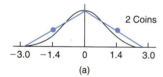

2 Coins

(a)

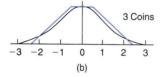

3 Coins

(b)

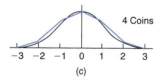

4 Coins

(c)

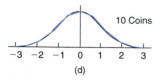

10 Coins

(d)

Figure 6-3 Standard score frequency polygons for expected outcomes from coin tosses. The normal distributions have been drawn in over the frequency polygons.

Table 6-2	Frequency Distributions, Probabilities, and Z-Scores for Expected Outcomes								
	2 Coins			**3 Coins**			**4 Coins**		
Heads	f	P	Z	f	P	Z	f	P	Z
0	1	.25	−1.414	1	.125	−1.732	1	.0625	−2.0
1	2	.5	0	3	.375	−.577	4	.25	−1.0
2	1	.25	+1.414	3	.375	+.577	6	.375	0
3				1	.125	+1.732	4	.25	+1.0
4							1	.0625	+2.0

Calculations

$\overline{X} = 1.0$		$\overline{X} = 1.5$	$\overline{X} = 2$
$S = .707$		$S = .866$	$S = 1$

The Standard Normal Distribution and Probability

Look back to Figures 6-1 and 6-2; in these two figures, the frequencies of the various outcomes of our coin tosses are shown as bars in a histogram. Each bar is one unit wide and as many units high as the expected frequency of that particular outcome; that is, if we were to toss two coins four times, we should get zero heads once, one head twice, and two heads once on the average over a large number of trials. Another way to look at it is that on a large number of tosses of two coins, we would expect to get two heads 25 percent of the time, one head 50 percent of the time, and no heads on the remaining 25 percent of tosses.

The essential point here is that *we can equate the areas of the bars in our histograms with the relative frequencies or probabilities of each of our outcomes.* Figure 6-4 shows the four histograms we would expect to get from 1000 repetitions of our two-, three-, four-, and ten-coin tosses. The total area under each histogram is 1000 units. The area of each

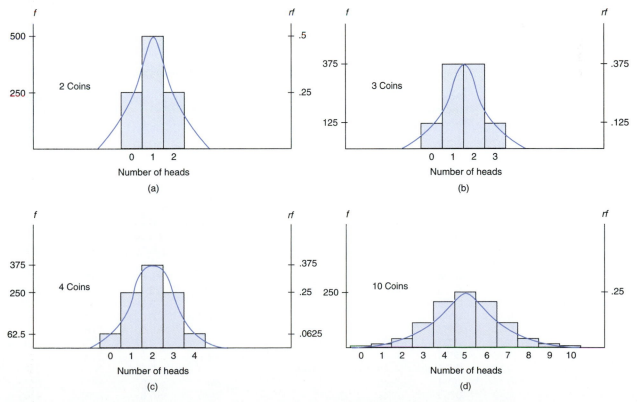

Figure 6-4 Histograms of frequencies (ordinate on the left side) and relative frequencies (ordinate on the right side) of the expected results from tossing two, three, four, and ten coins 1000 times. The normal distributions have been drawn in over the histograms.

bar can be divided by the total area under the histogram, which converts each bar to a relative frequency or proportion of the total, and this, by equation 6.1, is the probability of that event. For example, the area of the bar for one head in the relative frequency histogram for the case of four coins in Figure 6-4 is .25, which means that the probability of this outcome is .25. As we discussed in Chapter 2, converting to relative frequencies does not change the shape of the frequency distribution. This point is emphasized in Figure 6-4 by the use of two scales for the ordinate. The scale on the left of the histogram is in the metric of number of cases (f), and the scale on the right is in relative frequency (rf). By expressing each histogram in relative frequencies, we convert the total area under each curve to 1.0, and the area associated with each outcome is its probability. We refer to these as **unit histograms** because they enclose one unit of area; all relative frequency histograms or polygons are unit histograms or polygons, enclosing a total area of 1.0.

We have seen that it is possible to convert all normal distributions to one with a mean of zero and a standard deviation of one by transforming them to standard score form. It is also possible to convert normal distributions to unit area by expressing them in relative frequencies. These transformations have the very substantial advantage of reducing all normal distributions to a single distribution with precisely known mathematical properties. This single curve, or distribution, is known as the **standard normal distribution.** Some of the properties of the standard normal distribution are presented in Appendix Table A-1.

The most widely used feature of the standard normal distribution is that there is a known relationship between Z-scores, area, and probability. The primary reference point in the distribution is the mean. Appendix Table A-1, a part of which is shown here as Table 6-3, lists the area under the curve between the mean and any Z-score. All normal distributions converted to Z-score form have the same proportion of their area between the mean and a particular Z-score, so the areas listed in Appendix Table A-1 characterize all normal distributions.

We can summarize the importance of the standard normal distribution quite simply. Many traits of interest to psychologists and educators have frequency distributions that very closely approximate the normal distribution. Relative frequency or probability can be expressed as the area of a portion of the normal distribution. Therefore, if some trait such as cognitive ability has a normal distribution, we can tell what the probability of an observation in a particular range of the trait is by computing Z-scores and determining the

When expressed as relative frequency, **the area of a histogram bar is equal to the probability that the given event will occur.**

*The **standard normal distribution** is a normal distribution in which area corresponds to probability or relative frequency.*

	Area Between Z and $\overline{X}$	Area Beyond Z		Area Between Z and $\overline{X}$	Area Beyond Z		Area Between Z and $\overline{X}$	Area Beyond Z
Z			Z			Z		
0.00	.0000	.5000	0.21	.0832	.4168	0.42	.1628	.3372
0.01	.0040	.4960	0.22	.0871	.4129	0.43	.1664	.3336
0.02	.0080	.4920	0.23	.0910	.4090	0.44	.1700	.3300
0.03	.0120	.4880	0.24	.0948	.4052	0.45	.1736	.3264
0.04	.0160	.4840	0.25	.0987	.4013	0.46	.1772	.3228
0.05	.0199	.4801	0.26	.1026	.3974	0.47	.1808	.3192
0.06	.0239	.4761	0.27	.1064	.3936	0.48	.1844	.3156
0.07	.0279	.4721	0.28	.1103	.3897	0.49	.1879	.3121
0.08	.0319	.4681	0.29	.1141	.3859	0.50	.1915	.3085
0.09	.0359	.4641	0.30	.1179	.3821	0.51	.1950	.3050
0.10	.0398	.4602	0.31	.1217	.3783	0.52	.1985	.3015
0.11	.0438	.4562	0.32	.1255	.3745	0.53	.2019	.2981
0.12	.0478	.4522	0.33	.1293	.3707	0.54	.2054	.2946
0.13	.0517	.4483	0.34	.1331	.3669	0.55	.2088	.2912
0.14	.0557	.4443	0.35	.1368	.3632	0.56	.2123	.2877
0.15	.0596	.4404	0.36	.1406	.3594	0.57	.2157	.2843
0.16	.0636	.4364	0.37	.1443	.3557	0.58	.2190	.2810
0.17	.0675	.4325	0.38	.1480	.3520	0.59	.2224	.2776
0.18	.0714	.4286	0.39	.1517	.3483	0.60	.2257	.2743
0.19	.0753	.4247	0.40	.1554	.3446	0.61	.2291	.2709
0.20	.0793	.4207	0.41	.1591	.3409	0.62	.2324	.2676

Table 6-3   Sample of Appendix Table A-1, the Normal Distribution

area of the distribution within that range. For example, what is the probability that a randomly selected student will have a Wechsler IQ score between 90 and 110? What proportion of students get SAT scores above 560? We can use the table of the normal distribution to answer questions such as these.

Using the Normal Curve Table

Appendix Table A-1 has three major columns, repeated across the table to conserve space. The first lists all possible positive Z-scores to two decimal places from zero to 2.75. Beyond this point there are so few cases that entries in the table go in increments of .05, then .10. The second column lists the area between the Z-score given in the first column and the mean; the third column gives the area beyond the given Z-score.

Because the normal distribution is symmetric (see Figure 6-4), it is necessary to present only the top half of the curve in the table. The area between the mean and a particular Z-score below the mean is the same as it is for that Z-score above the mean. The only difference, but an important one for some applications, is that the sign of a Z below the mean is negative. The sign of Z tells you in which direction to go from the mean, and the magnitude of Z tells you how far to go.

Four Questions About Area and Probability

There are four questions about area or probability in relation to the normal distribution that we commonly ask in educational and psychological research. All four are questions about the probability of an event or score occurring in a specified range of Z-scores.

Question 1. First, what portion of the area of a normal distribution falls between a particular Z-score and the mean; that is, what is the probability of a score in this range? The answer to this question can be read directly from the table; for example, what portion of the area falls between $Z = +.90$ and the mean? Referring to the table, we find that the entry in column 2, "Area Between Z and $\overline{X}$," for a Z of .90 is .3159; this means that 31.59 percent of the area under the curve is included in this region. Another way to view this result is to say that we would expect 3,159 out of 10,000 observations to fall in that portion of the normal distribution between the mean and a Z-score of +.90.

Question 2. The second question is, what portion of the area of the normal distribution falls beyond a given Z-score? The answer to this question may be obtained either directly from the column labeled "Area Beyond Z" in Appendix Table A-1 or by subtracting the area between Z and the mean from .5000. Because the normal distribution is a unimodal symmetric distribution, the mean, median, and mode coincide; therefore, one-half of the distribution, or 50 percent, is above the mean. For our Z of .90 the first method (the entry for .90 in column 3 of Appendix Table A-1) yields .1841 and the second method does also (.5000 − .3159 = .1841).

Question 3. The third question that we may ask is, what portion of the area under the normal curve lies between two Z-scores? The answer to this question involves adding or subtracting values from the table. For example, we find the area between $Z = +.50$ and $Z = +1.50$ by looking up the tabled values .1915 ($Z = .50$) and .4332 ($Z = 1.50$); the area between them is the difference, .2417.

There is one important thing to watch for when determining the area between two Z-scores: If they have the same sign, they are on the same side of the mean, and the area is found by subtracting the smaller area from the larger. If the Z-scores have different signs, they are on opposite sides of the mean, and the separate areas are added. This principle is illustrated in Figure 6-5. If the Z-scores in the previous example had been −1.5 and +.5, the solution would have been .4332 + .1915 = .6247 as shown in the right half of the figure.

Question 4. Our final question is the converse of the last one: what portion of the normal distribution lies beyond a pair of Z-scores? For example, what proportion of the distribution lies outside the range from −1.5 to +.5? In this case, if the Z-scores have opposite signs, which they almost always do for applications of this kind, the answer is given by the sum of the two areas from column 3 of Appendix Table A-1. For example,

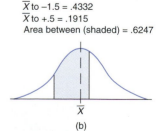

$\overline{X}$ to +1.5 = .4332
$\overline{X}$ to +.5 = .1915
Area between (shaded) = .2417

(a)

$\overline{X}$ to −1.5 = .4332
$\overline{X}$ to +.5 = .1915
Area between (shaded) = .6247

(b)

Figure 6-5 Proportion of the normal distribution between two Z-scores.

the area below $Z = -1.5$ is .0668, and the area above $Z = +.5$ is .3085, so the answer to the question is .3753 or 37.53 percent. Because the total area under the curve is 1.0, the answer to this question can also be found as 1.0 minus the area between the two Z-scores in the right of Figure 6-5. In the unusual situation where we wish to find the area outside two Z-scores that have the same sign, we first find the area between the Z-scores as we would for question 3; then we subtract that area from 1.0.

We use areas under the normal curve for several important purposes. The most common use of the normal curve is in making statements about probabilities; also, the relationship between area and probability is used in forecasting what will happen. For example, the normal curve can be used to estimate the probability that a sample mean will fall in a particular region. As we will see in Chapters 13 and 14, the normal curve also lets us make probability statements about other statistics. The normal curve is used because it is simple and widely available, and it seems to fit reasonably well many of the probability problems that scientists encounter in their research.

▶ *Summary of Questions About the Normal Distribution*

Question	Answer
1. What is the area between the given Z-score and the mean?	Enter Appendix Table A-1 with the value of Z and take out the value in column 2.
2. What is the area beyond the given Z-score?	Enter Appendix Table A-1 with the value of Z and take out the value in column 3, or subtract the value in column 2 from .5000.
3. What is the area between two Z-scores?	**Case 1:** Both Zs have the same sign. Find the areas between the mean and each Z-score; then subtract the smaller area from the larger.
	Case 2: Zs have opposite sign. Find the areas between the mean and each Z-score; then add the areas.
4. What is the area beyond two Z-scores?	**Case 1:** Zs have opposite sign. Find the area beyond each Z from Appendix Table A-1, column 3; then add the two areas.
	Case 2: Zs have the same sign. Find the area between the Zs as in Question 3, case 1; then subtract that area from 1.00.

▶ *Now You Do It*

The Thorndike Precarious Prognosticator of Statistical Perfection (TPPSP) is a measure of statistics knowledge that has been normed for students in introductory statistics. The faculty in the SAD Department at Big Time U have decided to administer the test at the end of the term to see how their students stand relative to the national norm group (remember from Chapter 5 that norm groups are used to provide a reference frame for test scores). The TPPSP has a normal distribution with a mean of 150 and a standard deviation of 20. Assuming that the SAD students at BTU are just like the norm group, what proportion of them should earn scores below 160? What proportion should receive scores below 125? If the faculty selected a student at random, what is the probability that her score would be between 145 and 155? If a second student is selected at random, what is the probability that his score will be more than 1.5 standard deviations from the mean?

Answers
The answers are .6915, .1056, .1974, and .1336.

Relationship Between the Binomial and Normal Distributions

Earlier in this chapter we presented the binomial distribution as the distribution of discrete dichotomous random events. We saw that as the number of classes of events increased, the binomial distribution came to look like the normal distribution. The apparent similarity is quite real, but to explore it we need to be able to express binomial events in the language of standard scores. This requires that we know the mean and standard deviation of a binomial variable. For example, what are the mean and standard deviation of the distribution of tosses of six coins or of matches to the sender on 12 ESP trials?

Remembering that P is the probability of the event of interest and N is the number of elementary events, the mean of a binomial distribution of N events of probability P is

$$\overline{X}_{binomial} = N \cdot P \tag{6.5}$$

The standard deviation can be found from the formula

$$S_{binomial} = \sqrt{N \cdot P \cdot Q} = \sqrt{N \cdot P(1 - P)} \tag{6.6}$$

Using these two formulas we can find the Z-score for any binomial outcome and explore the normal distribution as a convenient approximation to the binomial distribution. For example, the mean and standard deviation of the distribution of the number of heads expected from tosses of four coins, in which the probability of a head on each toss is .5, are

$$\overline{X} = (4)(.5) = 2$$
$$S = \sqrt{4(.5)(.5)} = 1$$

We can illustrate the use of the normal distribution approximation to the binomial distribution in calculating probabilities with our coin-toss example for ten coins. The exact probabilities for each of the 11 possible outcomes, given to four places, are reproduced in Table 6-4; the Z-scores are also given for the various outcomes. However, these Z's are of little use to us because they represent midpoints of the intervals that would exist if the coin tosses represented a truly continuous variable. Recall from our discussion of the histogram in Chapter 2 that the score values represent the midpoints of score intervals when a discrete scale is being used to measure a continuous trait. We must therefore view the bar for the discrete category six heads as covering the interval that has real limits of

X	P	Midpoint Z	Limit Z	Area Z to $\overline{X}$	Area in Interval
			3.48	.4998	
10	.00098	3.16			.0020
			2.85	.4978	
9	.0098	2.53			.0110
			2.22	.4868	
8	.0439	1.90			.0439
			1.58	.4429	
7	.1172	1.26			.1140
			.95	.3289	
6	.2051	.63			.2034
			.32	.1255	
5	.2461	0			.2510
			−.32	.1255	
4	.2051	−.63			.2034
			−.95	.3289	
3	.1172	−1.26			.1140
			−1.58	.4429	
2	.0439	−1.90			.0439
			−2.22	.4868	
1	.0098	−2.53			.0110
			−2.85	.4978	
0	.00098	−3.16			.0020
			−3.48	.4998	

Table 6-4 Using the Normal Distribution to Estimate Probabilities

5.5 and 6.5, and it is the *Z*-scores for these points that we have to use with Appendix Table A-1. The *Z*'s of these interval limits are listed in the fourth column of Table 6-4 under "Limit *Z*." The next column of the table contains the portions of the normal distribution between these *Z*-scores and the mean. In the final column are the portions of the normal distribution that lie in each interval, found by answering questions of type 3 as outlined previously. The entries in the final column are the estimates that we would make of the probabilities of each of the possible outcomes of tossing ten coins using the normal distribution approximation rather than calculating the exact binomial probabilities using formula 6.4. For example, the probability of exactly 9 heads, from column 2 of Table 6-4, is .0098. Using the normal distribution, the probability of this outcome is found by subtracting .4868 (the area between the mean and the *Z*-score for 8.5 heads) from .4978 (the area between the mean and the *Z*-score for 9.5 heads). The result is .0110.

The most striking feature of the approximate probabilities in the last column is that they are so close to the exact values in column 2 of the table. The largest discrepancy, which occurs for the scores of 3 and 7, is only .0032. The normal probability distribution provides a very close fit to the data for ten coins; as the number of coins or different outcome categories increases, the fit gets better. From this we can see that whenever a measurement procedure results in a scale with more than about ten categories and the distribution is reasonably close to a bell shape, the normal distribution can be used as an acceptable substitute for the more cumbersome binomial distribution. These conditions are met in the case of many studies in educational and psychological research. It should also be clear, however, that if the distribution is not symmetrical, we cannot use the normal distribution.

Percentile Ranks, Probability Intervals, and the Normal Distribution

The relationship between standard scores in the normal distribution and the probability of various outcomes is widely used as a basis for forecasting continuous traits. Scores on achievement tests and personality and interest inventories, for example, are often assumed to fall in a normal distribution; that is, if we were to administer an appropriate test of reading achievement to all high-school students in our state, we might expect the frequency distribution of scores to be very nearly normal in shape. Although we know that the scores actually come from a scale with discrete units in the form of test items, there are enough different scale values so that we can treat the resulting measurements as forming a continuous scale. Suppose that students' scores on the reading test ranged from 10 to 150 with a mean of 80, a standard deviation of 20, and a frequency distribution in the shape of the normal curve. It would be conventional to say that *the scores are normally distributed with mean 80 and standard deviation 20.*

We can use Appendix Table A-1 and information about the frequency distribution of test scores to make some general descriptive probability statements about people's performances on the test. Because there is a consistent relationship between percentile ranks and *Z*-scores in normal distributions, we can move back and forth between *Z*-scores, percentile ranks, and probability.

Z-Scores and Percentile Ranks

Z-scores can be converted into percentile ranks by finding what proportion of the normal distribution falls below the desired Z-score.

First, we can translate *Z*-scores directly into percentile ranks. This is accomplished by determining the portion of the distribution that falls below a given *Z*-score from Appendix Table A-1. For negative *Z*'s, the percentile rank is simply the area of the distribution beyond the *Z*-score. Thus, a person with a *Z*-score of -1.28 has a percentile rank of 10 because the area beyond a *Z* of -1.28 is .1003. Positive *Z*-scores are above the mean, so their percentile ranks are found by adding .50 to the area between the mean and *Z*. A *Z*-score of $+.60$ corresponds to a percentile rank of 73 (.50 below the mean and .2257 between the mean and a *Z* of $+.60$). Of course, both of these *Z*-scores can be transformed back into the original scale or into any other desired scale using formula 5.5. The *Z* of -1.28 becomes a raw score of 54 in our distribution, with a mean of 80 and a standard deviation of 20 [$20(-1.28) + 80 = 54$] or, stated the other way, a raw score of 54 has a percentile rank of 10 in a normal distribution that has a mean of 80 and a standard deviation of 20.

Now You Do It

Suppose you are a counselor in a local high school. The school has recently administered a battery of achievement tests to all students in the eleventh grade, and you are responsible for interpreting the scores for students and their parents. You know that percentile ranks are easier for people to understand, but the test scores have been reported as Z-scores. The testing company has told you that the scores are normally distributed, so you decide to convert the Z-scores into percentile ranks. What are the percentile ranks for the following five Z-scores? (answers below)

$$-1.63 \qquad -.53 \qquad .44 \qquad .98 \qquad 2.60$$

Now suppose the testing company has reported scores in a scale with a mean of 500 and a standard deviation of 100. Five students received scores of 337, 447, 544, 598, and 760. How can you relate scores on such a scale to percentile ranks?

Do you remember the idea of linear transformation that we covered in Chapter 5? That gives us the solution. If we transform each score into a Z-score, we can use the table of the normal distribution (Appendix Table A-1) to relate these scores to the percentile rank scale. The Z-transformed scores are -1.63, $-.53$, $+.44$, $+.98$, and $+2.60$ (the same ones we had above). Therefore, the percentile ranks are as follows:

Z	Percentile Rank
-1.63	05
$-.53$	30
$+.44$	67
$+.98$	84
$+2.60$	99.5

Z-Scores and Probability

The second widespread use of the normal distribution and Appendix Table A-1 is in making statements about the probability of outcomes that fall within particular ranges of scores. Suppose that we write down on separate slips of paper the scores obtained by each of the students who took our reading test (*mean* = 80, S = 20) and place the slips in a large drum. If we mix the slips well and reach into the drum, what is the probability that we will draw a score that will be less than 75? What is the probability that the score we draw is between 50 and 110? We can use the normal distribution to answer these questions.

The solutions to these problems are found by (1) converting the raw scores to Z-scores and (2) using Appendix Table A-1 to determine the areas of the normal distribution. In the first situation, we have

$$Z_{75} = (75 - 80)/20 = -.25.$$

Since we want to know the probability of a score below 75, the answer is given by the area beyond a Z of $-.25$, as shown in Figure 6-6. This is found to be .4013, so we can say that the probability of drawing a score of less than 75 is .40. Note that a more precise answer is obtained if we use a raw score of 74.5, the lower limit of the score interval; in this case, $Z_{74.5} = -.275$, and the probability of a score below this point is .39. However, it is almost always sufficiently accurate to use the raw scores themselves when the distribution has a range exceeding about 25.

The answer to the second problem is also found by converting the raw scores to Z's; the difference is that we use the area between the two scores to find the desired probability. We find $Z_{50} = -1.50$ and $Z_{110} = +1.50$, as shown in Figure 6-7. From Appendix Table A-1, the area between each Z-score and the mean is .4332, so the area between the two Z's is .8664, and the probability of obtaining a raw score between 50 and 110 is .87. Again, a more exact procedure would be to use the real limits of the score intervals or 49.5 and 110.5, which yields an area between the two scores of .8726, but in this case, our .87 probability does not change.

Probability Intervals

Sometimes it is also useful to reverse the question from the previous section and ask, "Between what two scores does some specified percent of the distribution fall?" For

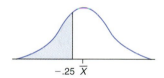

Figure 6-6 The probability of a score below 75 can be found by finding the proportion of the normal distribution falling below a Z-score of $-.25$.

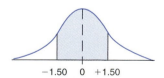

Figure 6-7 The probability of a score falling between raw scores of 50 and 110 can be found by finding the proportion of the normal distribution falling between Z-scores of $+1.50$ and -1.50.

example, we might want to know what score range on the reading test includes the middle 50 percent of the students in our district. When the question is asked in this way, we have what we might call a **probability interval** or **PI**. A probability interval is that portion of the frequency distribution that is centered on the mean and contains the specified percent of the distribution. We might ask for the 50 percent PI, or the 75 percent, 90 percent, or 95 percent probability interval; whatever probability interval we are seeking, the answer will be expressed in units of the original raw score scale.

We use Appendix Table A-1 in reverse to find probability intervals. Because a probability interval is symmetric around the mean, we divide the desired percent in half and look up that value in the table. To find the 50 percent PI, we use column 2 to look up the Z that has 25 percent of the distribution falling between itself and the mean; this value is about .675, so we know that 50 percent of the distribution falls between $Z = -.675$ and $Z = +.675$, and the 50 percent probability interval for the distribution of reading scores we have been discussing

$$(\overline{X} = 80, S = 20)$$

is from 66.5 [$(-.675 \cdot 20) + 80$] to 93.5 [$(+.675 \cdot 20) + 80$] in terms of the raw score scale. We find the 90 percent PI in the same way, except that we now need 45 percent of the distribution on each side of the mean. Referring to Appendix Table A-1 we find the necessary Z to be 1.645; therefore, the 90 percent probability interval is from -1.645 to $+1.645$ in Z-scores or from 47.1 to 112.9 in raw scores. Note that for the special cases of the 90 percent, 95 percent, 98 percent, and 99 percent PIs the values of Z are given to three decimal places. This is because these probabilities are the most widely used ones, both for probability intervals and for topics in hypothesis testing discussed in Chapter 13.

▶ *Finding Probability Intervals*

We may summarize the general procedure for finding probability intervals as follows:

1. A probability interval is always symmetric and always stated in terms of some percentage or level of probability (P). Typical PIs are 90%, 95%, and 99%. Probability intervals usually are centered on the mean of a distribution or on a predicted score for a person, in which case they are called prediction intervals.
2. Once the probability value is stated, we go to the table of the normal distribution and look up the Z-score for which the area between the mean and Z is 1/2 the value specified (call it Z_{PI}).
3. The upper limit of the probability interval is found to be

$$PI_{Upper} = + Z_{PI}(S) + central\ value$$

where the central value is a mean or predicted score.
4. The lower limit of the probability interval uses the negative of the Z-score

$$PI_{Lower} = - Z_{PI}(S) + central\ value$$

5. The P percent PI is the area or range between PI_{Lower} and PI_{Upper}.

▶ *Now You Do It*

To check yourself, make sure you can determine the 95 percent raw score probability interval for the distribution of scores that has a mean of 80 and a standard deviation of 20. The answer you get should be 40.8 to 119.2. Remember that you need 47.5 percent on each side of the mean, so the Z is 1.96.

SUMMARY

The concept of *probability* was introduced with examples of binary events; for example, the probability of event A, obtaining three heads on a toss of five coins, is equal to the number of ways that event A can happen divided by the total number of things that can happen. The probability of event A may also be viewed as the number of times that we can expect A to occur over a long series of trials. In this sense, probability is the relative frequency of event A in the distribution of all possible outcomes.

The *binomial distribution* describes the probability distribution of classes of dichotomous events. This distribution uses the addition and multiplication laws of probability for events that are *mutually exclusive, exhaustive,* and *independent.* When the probability of one class of binary outcomes (e.g., heads on a coin) is equal to .50, the probability distribution is symmetric around the point $N \cdot P$ (equation 6.5) with a standard deviation of $\sqrt{N \cdot P \cdot Q}$ (equation 6.6).

The *normal distribution* is a precisely defined mathematical function that closely approximates the distribution of many chance events. The frequency polygon of the normal distribution is called the *normal curve* and is a *unimodal symmetric* curve. When a normal distribution is expressed in terms of the relative frequencies of Z-scores, a single curve can be used to fit any distribution. This curve is called the *unit normal curve* because the area enclosed by the curve is 1.0. The area between the mean of this curve and any Z-score can be found in Appendix Table A-1. The values in this table can be used to find the area of the curve between any two Z-scores.

The area under the portion of the unit normal curve that lies between two Z-scores can be viewed as the probability that an event with a value in that range will occur, which makes it possible to state consistent relationships between Z-scores in a normal distribution and the percentile ranks of those Z-scores. When we ask the question, "What range includes a specified portion of the distribution?"—usually 90%, 95%, or 99%—the range is called a *probability interval* or *PI.*

EXERCISES

1. Define the three basic kinds of probability and provide an original example of each.
2. For each of the following, identify which of the three basic types of probability is represented.
   a. What is the probability of rolling two dice and obtaining a total score of 7?
   b. What is the probability that your favorite uncle, whom you seldom see, will tell you how much you have changed since the last time he saw you?
   c. What is the probability of drawing a red bead on the first draw, a white bead on the second draw, and a blue bead on the third draw from a container containing 50 red beads, 38 white beads, and 76 blue beads?
   d. What is the probability that Smokey will win the prestigious Jockey Cup based on the last six races of each horse that is entered inthe race?
   e. What is the probability of the driver of the car in front of you who is driving slowly making a right-hand turn without signaling the turn?
   f. Based on the performance of the students at Lerner High School, what is the probability that students will score above the national average on the National Math Test this year?
   g. Given the rates of cancer in Atomic City in the last ten years, what is the probability that a person will be diagnosed with cancer in his or her lifetime?
   h. What is the probability of drawing an ace from a normal deck of playing cards?
3. Two players enter a contest in which they try to match the color of the bead drawn by the host of the contest. Each player and the host will draw a bead from their own covered jar. In each covered jar there are 50 red beads, 50 green beads, and 50 purple beads. If a player matches the host, she or he receives $2. If a player fails to match the host, she or he pays the host $1. List all of the possible outcomes. Is the game fair? Why or why not?

4. What are the three assumptions of probability theory?
5. State the two laws of probability.
6. Using the answers to Exercise 3, what is the probability that the host and the two players will all draw a bead of the same color? What is the probability that the host and one and only one player will draw a bead of the same color? What is the probability that the host will draw a bead that is a different color from those of the two players?
7. Suppose that you are using a deck of playing cards that, unknown to you, is missing 8 cards: 2 hearts, 4 diamonds, 1 spade, and 1 club. Thus, the deck of playing cards contains 44 cards. Note: A normal deck of playing cards contains 13 cards in each category (heart, diamond, spade, club).
   a. What is the probability of selecting a red card (hearts or diamonds)? What is the probability of selecting a black card (spades or clubs)?
   b. What is the probability of selecting two red cards if the card is replaced after the first drawing and the deck of cards is reshuffled before selecting the second card? What is the probability of selecting two black cards if the card is replaced after the first drawing and the deck of cards is reshuffled before selecting the second card?
   c. What is the probability of selecting two different colors of cards if the first card is replaced and the deck of cards is reshuffled before the second card is drawn?
   d. Find the probabilities of all possible combinations of colors of cards if three cards are drawn and a card is replaced and the deck reshuffled after each drawing. Note that the order in which the cards are drawn makes a difference (e.g., *rrb* is different than *rbr*). What is the sum of all of these probabilities?

8. Suppose that we are interested in studying mathematics anxiety reported by second-year algebra students at a local high school. According to demographic information collected prior to the measures of mathematics anxiety, the subject pool includes 313 males and 387 females; there are 179 freshmen, 202 sophomores, 217 juniors, and 102 seniors. If we select participants randomly, what are the probabilities that the first person selected will be
   a. female?
   b. a junior?
   c. a freshman or a senior?
   d. a senior male?
   e. a freshman female?
   f. What conditions must be assumed to be true before the probabilities in (c), (d), and (e) are correct?

9. A psychologist is interested in studying concept formation in young children. A computer is programmed to randomly present stimuli that vary in shape (square and circle) and color (red and blue). The program is written such that a square is five times more likely to appear than a circle, and the object is ten times more likely to be red than blue. What is the probability that the first figure presented on the screen will be
   a. a circle?
   b. a red object?
   c. a blue square?
   d. a red circle?
   e. Is it appropriate to use the addition law of probability to determine the probability that the first figure presented on the screen is blue or a square? Why or why not?

10. As a part of their testing program, Testmore High School requires its students to take the California Achievement Tests in the ninth and eleventh grades. The Language subtest contains 50 questions, each of which is a five-alternative multiple choice item. Since language is not your strength and you are tired since this is the last of six subtests that comprise the California Achievement Tests, you decide to randomly mark your answer sheet.
    a. What is the probability of answering any one item (e.g., Item 21) correctly?
    b. What is the probability of answering the first and the last items correctly?
    c. What score are you most likely to receive?

11. How many possible outcomes are there for each of the following if the order in which the cards are drawn makes a difference (e.g., *rrb* is different from *rbr*):
    a. the color combinations of the card for 6 draws from a standard deck of playing cards given that each card is replaced and the deck of cards is shuffled after each draw.
    b. the face value of the card for 5 draws from a standard deck of playing cards given that each card is replaced and the deck of cards is shuffled after each draw.
    c. the suit of the card for 8 draws from a standard deck of playing cards given that each card is replaced and the deck of cards is shuffled after each draw.
    d. the value of the face of a die that is rolled 7 times.
    e. the color combinations of an equal number of beads that are red, white, green, blue, and purple after 4 draws, replacing each bead after a draw.

12. What is a binomial distribution? Provide at least three examples of a binomial distribution.

13. Calculate the probability of the following events using formula 6.2:
    a. drawing 4 red cards out of 6 attempts from a standard deck of playing cards given that each card is replaced and the deck of cards is shuffled after each draw.
    b. obtaining 2 heads on 8 tosses of an unbiased coin.
    c. drawing no red beads after 10 attempts with replacement after each draw from a jar containing an equal number of red beads and blue beads.

14. Calculate the probability of the following events using formula 6.4.
    a. In a given deck of standard playing cards, 5 diamonds, 3 hearts, 2 clubs, and 2 spades are missing. What is the probability of drawing 4 red cards out of 6 attempts from this deck of playing cards given that each card is replaced and the deck of cards is shuffled after each draw?
    b. Suppose a coin is biased so that it comes up heads 75% of the times it is tossed. What is the probability of obtaining 2 heads on 8 tosses of this biased coin?
    c. A jar contains four times as many blue beads as red beads. What is the probability of drawing no red beads after 10 attempts with replacement after each draw from the jar?

15. Find the probability of obtaining exactly three 2's if an ordinary die is tossed 5 times.

16. A basketball player hits 80% of her free throws in game situations. What is the probability that she will make exactly 3 of her next 8 free shots in a basketball game?

17. What is the normal distribution?

18. Find the proportion of the area under the normal distribution that lies
    a. below $Z = .50$
    b. above $Z = .50$
    c. below $Z = -1.50$
    d. above $Z = -.75$
    e. below $Z = 2.25$
    f. below $Z = 1.64$
    g. between $Z = .25$ and $Z = .75$
    h. between $Z = 0$ and $Z = 1.75$
    i. between $Z = -1.96$ and $Z = 1.96$

19. Suppose you are given a 10-question multiple-choice statistics quiz with four response alternatives for each question. However, you have neglected to study the material that the quiz covers so you have to make a random guess for each question. List the possible outcomes of the quiz scores. Find the probability of each possible outcome. What is the most probable score?

20. Scores on the Scholastic Aptitude Test (SAT) are normally distributed with a mean of 500 and a standard deviation of 100.
    a. What is the probability of obtaining a score greater than 640?
    b. What is the probability of obtaining a score less than 390?
    c. What is the probability of obtaining a score between 725 and 800?
    d. What is the probability of obtaining a score either less than 375 or greater than 650?

21. Scores on standard IQ measures usually have a mean of 100 and a standard deviation of 15. Many schools use, among other things, IQ scores in making the determination of exceptionality.
    a. In general, individuals with IQ scores below 70 are designated for further consideration as "educably men-

tally retarded." What is the probability of a person selected at random receiving this designation?

b. Individuals whose IQ scores lie between 90 and 110 are considered to be in the "normal" IQ range. What is the probability of a person selected at random receiving the "normal" designation?

c. Individuals whose IQ scores are above 130 are designated "gifted." What is the probability of a person selected at random receiving this designation?

22. For each of the following, calculate the mean and the standard deviation.

a. the number of red cards that can be drawn from a standard deck of playing cards when 5 cards are drawn if once a card is drawn it is replaced and the deck of cards is reshuffled.

b. the number of green beads that can be drawn from a jar containing an equal number of red beads and green beads. When 8 beads are drawn and, once a bead is drawn, it is replaced and the jar of beads is shaken.

c. the number of tails in 10 tosses of an unbiased coin.

d. the number of red cards that can be drawn from a deck of cards that is missing 5 hearts, 3 diamonds, 3 spades, and 1 club when 7 cards are drawn, if once a card is drawn, it is replaced and the deck of cards is reshuffled.

e. the number of green beads that can be draw from a jar that contains four times as many green beads as red beads when 4 beads are drawn and, once a bead is drawn, it is replaced and the jar of beads is shaken.

f. the number of tails in 6 tosses of a biased coin that will result in heads 70% of the time.

23. A psychology instructor was asked to teach two sections of a social psychology class, one for students in humanities and one for students in business administration. The same final exam was administered to each section, with the resulting distributions being normal. Humanities students had a mean of 81 and a standard deviation of 9 on the final exam while the business administration students had a mean of 69 and a standard deviation of 15.

a. Jennifer, a humanities student, earned a score of 77 on the final exam. What is her percentile rank in the humanities section? If she had been a business administration student and achieved the same score, what would her percentile rank be?

b. Martin is a business administration student who received a score on the final exam that corresponded to a percentile rank of 33. What would his percentile rank be if he were a student in the humanities?

c. Alex scored at the 55th percentile in the business administration section while Chris scored at the 43rd percentile in the humanities section. Who had the higher standing on the exam relative to her or his class?

24. Suppose that a self-concept measure has a mean score of 54 and a standard deviation of 5.61. What self-concept scores correspond to the following probability intervals:

a. 90 percent?

b. 95 percent?

c. 99 percent?

Chapter 7

Prediction

CHAPTER OBJECTIVES

After studying this chapter, you should be able to:

❏ Identify applications of different types of predictions in research settings.

❏ Define the different types of decisions that people make using data and identify instances of each.

❏ Identify predictor and criterion variables in research settings.

❏ Define the standard error of estimate and use it to specify prediction intervals around group means.

❏ State the assumptions underlying linear regression and identify instances where they might be violated.

❏ Given a set of data, compute the slope and intercept of the regression line.

❏ Given a regression equation, compute the predicted score for an individual.

❏ Prepare a scatter plot of a set of scores and draw in the regression line.

❏ Compute a prediction interval around a predicted score.

In all likelihood, your college or university asked you to submit your high school transcript and scores on a standardized test as part of your application for admission. Have you ever wondered why they did that, or how the information was used? If you enlisted in one of the military services, you had to take a battery of tests. How do the military services use the test information? Almost every facet of modern life involves prediction of future performance, ranging from academic or military performance to employee performance, to the performance of stocks. How does one go about making the most accurate predictions possible?

Introduction

Our lives are affected in many ways by statistical prediction. Schools use tests to predict academic success and to assign students to classes and programs; colleges and universities use tests and grades to select the students who will be admitted based on predictions of success. Tests are widely used in business, industry, government service, and the military to predict who the best employees will be or to place people in the jobs they are most likely to perform satisfactorily. Clinical psychologists use test scores to choose a treatment program that is most likely to be successful with a client. The list easily could be extended.

Questions about the propriety of such uses of prediction have generated widespread controversy in the news media. Articles in magazines and newspapers have criticized many of the instruments that are widely used for prediction. Television programs and specials have broadcast much the same message. Some states have passed laws seriously restricting the use of tests for predicting educational achievement, and Congress and government agencies have developed regulations and guidelines concerning how prediction information may be used in educational and occupational selection. Clearly, there are practical as well as educational reasons for developing an understanding of what statistical prediction is and how it can be used.

Types of Prediction

Prediction generally involves forecasting. To make a forecast, we use information available at the present time to answer basic questions: What is the most likely outcome at some future time? Which outcome has the highest probability of occurrence? Or, more generally, what is the probability for each of several possible outcomes? For example, if you have participated in a career counseling program, you may have taken a vocational interest inventory that asked you questions about things you like to do. Your pattern of answers would have allowed a counselor to identify occupations that you would have a high probability of liking and others that are almost certainly not for you. Such information could be useful in helping you select your college major and even to think about your future professional focus.

Any sound prediction strategy seeks to best use the available information in order to determine the probability of each of several possible outcomes. However, a very important point to keep in mind, and one that is often overlooked, is that different individuals and groups have different purposes in making predictions and different valuations of the outcomes. You, as an applicant to college, are primarily interested in the accuracy of the predictions made about you, but the admissions officer of the university is concerned with the accuracy of predictions for the incoming class as a whole, and the affirmative action officer is concerned with predictions about the performance of minority students. Alternative prediction procedures may be appropriate for these different purposes and values, and it is important to understand the issues involved in making predictions.

The first distinction we must make is between what might be called **subjective prediction** and **statistical prediction.** A subjective prediction is one that is based on informal past experience, much like the subjective probability discussed in Chapter 6. For example, we predict whether it is going to rain today through our informal observations of the sky, cloud patterns, and wind. We predict what some other person will do in a given situation on the basis of our previous experience with the person in similar situations. Such predictions, although often quite accurate, are generally categorical. We might say, without qualification, "It is going to rain," or "Joe will lose his temper when he sees the scrape on his car." Statistical prediction, on the other hand, resembles empirical probability; it uses the results of formal prior observations such as test scores and grades to make forecasts about future events such as how well you will perform on a job or in graduate school. For example, as a school counselor, you might use quantitative information in the form of test scores to predict what Sara's grade point average will be at Big Time U. Statistical prediction is the focus of this chapter.

There are two kinds of statistical prediction: **point prediction** and **interval prediction.** In point prediction a specific outcome or particular value is forecast. For example, on the basis of your overall grade point average, your likely score on the final exam in your statistics course could be predicted. In contrast, interval prediction results in a statement about a range of values that may occur. For example, a college, in deciding to admit a student, is predicting that his or her academic performance will fall within a range that is considered to be satisfactory.

Predictions and Decisions

A prediction usually results in a decision to take an action of some kind. We can identify four categories of decisions, which are defined by the nature of the action and whose interest is being served.

Subjective prediction is based on an informal appraisal of past experience.

Statistical prediction uses the results of formal prior observations to make quantitative forecasts about the future.

In *point prediction* a specific outcome is forecast.

Interval prediction results in a statement about a range of values that may occur.

Selection Decisions. **Selection decisions** are decisions that involve picking out some individuals from a larger group. The individuals may be selected in, as when a college accepts some applications but not others, or they may be selected out, or identified for exclusion, as when the military rejects some potential enlistees on the basis of low test scores. In either case, there is an organization—for example, a school or college, a company, the U.S. military—that is doing the selecting. Usually, the organization has limited resources and wishes to realize the highest return from its prediction decisions. A college selects the students it believes most likely to obtain degrees or those most likely to earn the best grades. A company tries to select people who will be good workers and remain with the company for a long time.

Selection decisions have specific characteristics: (1) They are made by the organization for the benefit of the organization. (2) The organization usually makes many selection decisions, some of which will be correct and others incorrect. The organization seeks to maximize the number of correct decisions, and therefore, of necessity, measures success in terms of overall outcomes rather than individual cases. The procedures described in later sections of this chapter are those widely used to obtain the best selection decisions.

Classification Decisions. A second type of decision is the **classification decision.** This is also a decision made by an organization, but it differs from the selection decision in that no one is excluded. Instead, each individual is assigned to a category, treatment, or job. The objective is to obtain maximum performance across the average of all categories in the organization. Military-duty assignments are classification decisions. Some recruits receive electronics training, others are trained to be cooks and bakers, and others are assigned to various combat specialties. The distribution depends on how many individuals are needed in each area and which recruits are likely to make the best contribution within specific areas. Procedures for making and evaluating classification decisions are beyond our scope, but are treated in books on decision theory.

Placement Decisions. The **placement decision** is the third type of prediction decision; in this case, the focus is on benefit to the individual. A placement decision involves assigning each individual to the category that suits her/him best. For example, school personnel use test scores and other information to assign children to the learning environments that should produce the greatest learning success for each child. Placement decisions are made both by the organization and by the individual and often involve such subjective factors as preferences and role perceptions in addition to more objective information about past performance and probable success. The ultimate accuracy of the decision may be difficult or impossible to assess because the definition of success usually is a subjective feeling of well-being. Very few placement decisions will be made in the lifetime of a specific individual, so each decision is very important.

Personal Decisions. The fourth class of decisions may be called **individual,** or **personal, decisions.** These are the choices that have to be made at various stages in a person's life, such as which college to attend, what major to pursue, whom to marry, and so on. Most decisions of this type are subjective, although they may be based on some quantitative information. For example, you may have quantitative information in the form of test scores to help in selecting a college, and this information may lead to point or interval predictions of success. However, you will make the actual decision by combining the quantitative information with your personal values and aspirations. The mix will necessarily be subjective and cannot take advantage of the principles of accurate quantitative prediction developed below.

The different types of decisions require different kinds of information. In this chapter we will discuss how quantitative information in the form of test scores, grades, and similar types of information can be used to make the most accurate predictions possible. Throughout this discussion we will have as our goal to minimize the errors that are made, realizing that it is not possible to forecast the future without error.

▶ *Now You Do It*

Can you identify what kind of decision is being made in each case below? Try to explain your answers.

1. Caleb has to decide whether to write a term paper on the psychoanalytic theory of Freud or the educational theory of John Dewey.
2. Janelle has been accepted for graduate study at Harvard.
3. Oscar has been diagnosed with lung cancer.
4. Liz has been assigned to infantry training.

After you have identified the type of decision in each of the above examples, try to give an example yourself of each kind of prediction decision. Ask a friend to identify the decision types in your examples.

Answers
1. Caleb's decision is a personal one.
2. Janelle's decision to apply to Harvard was a personal one, and Harvard's decision to accept her was a selection decision.
3. Oscar's diagnosis is a placement decision.
4. Liz's assignment is a classification decision.

The Principle of Least Squares—Again

In general, although predictions are made for individuals, they are based on the past behavior of a group of similar people. We seldom have a long and detailed history of one person's behavior, and we therefore must rely on the past behavior of a group of similar people under similar conditions. The basic principle on which prediction operates is quite simple. We have an individual, call him Ishmael, for whom a prediction is to be made. By collecting some current or past information about Ishmael—for example, his age, education, and score on the SAT—we identify him as a member of a group of similar individuals whom we have encountered in the past. If Ishmael is an applicant for admission to our university (or perhaps employment on our whaling ship) and we wish to predict his grades (or job performance), we can base our prediction on the performance of previous applicants who were like him in age, education, SAT score, and so forth. Our prediction is that Ishmael will behave in his grades (or job performance) like the group that he most resembles on the variables we have available.

The information that we collect in order to identify groups to which Ishmael belongs is called **predictor information;** Ishmael's age, education, and test score are predictor information. The specific variables—age, education, and SAT score—are known as **predictor variables.** The behavior that we wish to predict is called the **criterion variable;** in Ishmael's case, grade average (or job performance) is the criterion variable.

The Prediction Process
Our study of Big Time U statistics students can give us a chance to take a closer look at the prediction process. Suppose, for example, that we wish to predict scores on the next weekly quiz in the statistics course. The information that we have as a predictor is the number of homework assignments each class member has completed prior to taking the quiz. We also know the performance of some students from last year's class both on the homework assignments and on the quiz. The data from last year are given in Table 7-1.

A little quick calculation reveals that last year's class taken as a whole had a mean score on the quiz of 6, and the standard deviation was 1.37. If we wished to choose a single score to describe the performance of last year's class, the score to choose would be 6. Why? Because 6 is the mean of the distribution, and the mean, as we know from our discussion in Chapter 4, is the center of the distribution *in a least-squares sense*. That is, the mean is the point or score that is closest to every score; it is the score with the smallest sum of squared differences between it and the other scores in the distribution.

Ahab (a friend of Ishmael) is a member of this year's class. What is our prediction—our best guess—of what his score on the quiz will be? If we have no further information about Ahab, our best guess is that his score will be 6. We can make this prediction by viewing Ahab, at least potentially, as a member of a group of similar individuals we have encountered in the past—last year's statistics class—whose mean score was 6. We know that our prediction of a score of 6 is likely to be in error, but we have no way of knowing whether our prediction is too high or too low. However, in choosing the mean, we have selected

*A **predictor variable** is a variable that is used for making predictions.*

*The **criterion variable** is the behavior that we wish to predict.*

	Table 7-1		Quiz Scores and Homework Assignments for 30 Students		
Individual	Homework Completed	Quiz Score	Individual	Homework Completed	Quiz Score
1	1	5	16	3	7
2	2	7	17	2	8
3	1	6	18	2	5
4	3	5	19	1	4
5	3	9	20	2	6
6	1	5	21	2	7
7	2	4	22	3	8
8	2	6	23	1	5
9	1	4	24	3	7
10	1	7	25	2	5
11	3	6	26	1	6
12	3	8	27	3	6
13	2	6	28	3	7
14	3	7	29	1	5
15	1	3	30	2	6

Least-squares prediction involves making predictions for which the sum of squared errors of prediction is minimized.

the value where the average error is zero, and the sum of squared errors, which is reflected by the standard deviation, is minimized. This is a case of **least-squares prediction.**

▶ *Now You Do It*

Suppose we actually use 6 as our prediction for each person. The errors of prediction for our 30 people then become −1, 1, 0, −1, 3, −1, . . . , 1, −1, 0. Convince yourself that these errors sum to zero and that the sum of squared errors is 56. Now suppose we had used a value of 7 as our prediction. What would the errors look like? If you subtract 7 from every score in Table 7-1 and find the sum and sum of squares of the errors, you should obtain the values −30 and 86. Obviously, our average error and the sum of squared errors have both increased. Now try 5. You should get +30 and 86. There is no value other than 6 that will give you as small a sum of squares as 6 will. Therefore, 6 is our least-squares prediction for these 30 people.

Adding Predictor Information

Suppose that we actually have some additional information about Ahab: we know that he handed in all three homework assignments. This is predictor information; that is, information that allows us to put Ahab in a subgroup of people who are like him in some way. If we sort the people from last year's class into groups defined by the number of homework assignments completed, Ahab can be identified as a member of the group that completed all three assignments.

Our figures show that a third of the class falls in each of three groups. Computing the mean quiz score for each group reveals that those who handed in only one assignment earned an average of 5 points on the quiz, while the other two groups had means of 6 and 7, respectively. Since Ahab's group has a mean of 7, this should be our new prediction for him because *this is the mean of the subgroup of which he is a member.* The standard deviation of scores for this subgroup is 1.1 instead of the 1.37 we found for the total class, indicating that the average error is smaller than it was for the entire group. Clearly, using our predictor information has reduced the size of our errors.

The consequences of using the subgroup means to reduce our errors are shown in Table 7-2. If we admit ignorance of group membership and use the overall mean, our sums of squared errors for each of the three groups are 22, 12, and 22, respectively, for a total of 56. Using the subgroup means, the sum of squared errors in each group is 12, for a total of 36. We are still making errors, but there are fewer and smaller errors when we use the information about subgroup membership and subgroup means. Note that when

Table 7-2 Errors Using the Overall Mean Compared to Errors Using Subgroup Means

	Group 1			Group 2			Group 3	
Score	Overall Mean (6)	Subgroup Mean (5)	Score	Overall Mean (6)	Subgroup Mean (6)	Score	Overall Mean (6)	Subgroup Mean (7)
5	−1	0	7	1	1	5	−1	−2
6	0	1	4	−2	−2	9	3	2
5	−1	0	6	0	0	6	0	−1
4	−2	−1	6	0	0	8	2	1
7	1	2	8	2	2	7	1	0
3	−3	−2	5	−1	−1	7	1	0
4	−2	−1	6	0	0	8	2	1
5	−1	0	7	1	1	7	1	0
6	0	1	5	−1	−1	6	0	−1
5	−1	0	6	0	0	7	1	0
Sum of Errors	−10	0		0	0		10	0
Sum of Squared Errors	22	12		12	12		22	12

we talk about the total amount of error in our predictions, we are referring to the pooled sum-of-squares, 36 in this case. Each group's sum-of-squares is computed around its own mean, and the sum-of-squares for the entire class is the sum of the individual group sums-of-squares, 12 + 12 + 12 in our example.

Predictor information is useful if it reduces our errors; in fact, we generally reserve the term predictor for information that does reduce our errors. In the example above, knowing the number of homework assignments completed reduced our errors; therefore, we would say that this variable predicts quiz performance. On the other hand, the color of Ahab's socks is probably not useful in reducing our error in guessing his quiz score, so sock color would not be considered a predictor variable.

A graphical representation of the use of a predictor variable is illustrated in Figure 7-1 for the homework-quiz score example from Table 7-1. The numbers correspond to the particular individuals in the table. The vertical dimension is quiz score, and the abscissa represents the number of homework assignments completed. It is standard practice when preparing a graph of a predictor variable and a criterion variable to place the predictor on the abscissa and the criterion on the ordinate. The frequency distribution on the axis labeled quiz score is the distribution for all 30 members of the class. The three smaller frequency distributions contain the quiz scores of the people who handed in one, two, and three homework assignments, respectively; the three smaller distributions can be summed to yield the larger one.

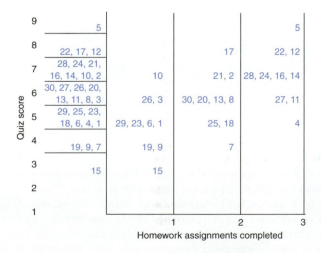

Figure 7-1 Frequency distributions of quiz scores for total group and for subgroups based on homework completed.

Let's confirm your understanding of how to show the relationship between a categorical predictor variable and a criterion variable using the BTU data for lab section and final exam. Like college major and number of homework assignments completed, lab section is a categorical variable. The data for all 12 variables in the BTU study are listed in Appendix C.

The first step is to find the overall mean and standard deviation and the subgroup means and standard deviations. Whether you do the computations by hand or use a computer program such as SPSS you should get the following results. The values in parentheses are the standard deviations from the computer output. Remember that the computer uses $N - 1$ in the denominator.

Lab Section

1	Mean	40.68	
	N	22	
	Std. Deviation	13.57	(13.89)
2	Mean	32.59	
	N	22	
	Std. Deviation	11.73	(12.01)
3	Mean	29.83	
	N	24	
	Std. Deviation	12.64	(12.91)
Total	Mean	34.24	
	N	68	
	Std. Deviation	13.48	(13.58)

Here the differences between groups are somewhat larger than they were for college major, and the standard deviations within groups are somewhat smaller. It appears likely that knowing a student's lab section will reduce our errors in predicting scores on the final because knowing a student's lab section will allow us to make quite different predictions. We would predict a score of 40.68 for students in section 1, 32.59 for students in section 2, and 29.83 for those in section 3. The line graph that you prepare to show the relationship should look approximately like this.

Comparison of final exam scores by lab section

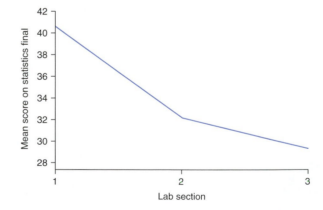

Errors in Prediction

The Standard Error of Estimate

The mean criterion score of the group to which the person belongs is always a least-squares prediction for a given amount of information. If no predictor information is available, the overall mean of the criterion variable is the best we can do, and the overall standard deviation is an index of the amount of error we are likely to make on the average. For example, when we did not know how many homework assignments Ahab had completed, our best guess for his quiz score was 6, and the standard deviation was 1.37. When there is a predictor variable, such as the number of homework assignments completed, we can use it to form subgroups. The mean of each subgroup is a least-squares prediction for the members of that subgroup, and the standard deviation of the subgroup-members' scores around their mean is the index of error. Thus, when we found out that Ahab had completed all three homework assignments, we could use the mean of his subgroup, 7, as a prediction for him, and the standard deviation of scores within his group, 1.1, was our measure of error. This standard deviation of observed scores from their predicted values, or the standard deviation of errors of prediction, is called the **standard error of estimate** and is given the symbol S_{Y-X}.

*The **standard error of estimate** is the standard deviation of observed scores around their predicted values, or the standard deviation of errors of prediction.*

HOW THE COMPUTER DOES IT

All statistics packages have a number of different ways to help you determine and understand the relationships between predictor variables and criterion variables. As described in Chapter 2, graphs can be an aid to discovering the meaning in a set of data. In this chapter we have been discussing how to use information from one variable to help predict a person's status on another variable. The two graphs shown here, one a box-and-whisker plot, the other a line graph, illustrate how graphs can reveal the different predictions we would make for students in the statistics class at Big Time U based on their majors. The box-and-whisker plot shows both group medians (which, for symmetrical score distributions, are equal to the group means) and the spread of scores around those predictions. The line graph shows the different predictions we would make for the members of each group, but not the spread of scores.

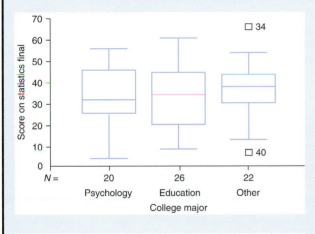

Box-and-whisker plot of scores on the statistics final by college major

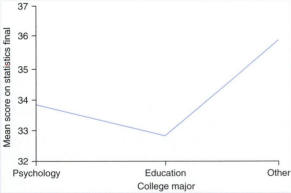

Comparison of means of final exam for different college majors

An alternative way to express the relationship between college major and final exam score is to compute—then graph—the means of the groups. This can be done using the MEANS subcommand of the COMPARE MEANS procedure on the STATISTICS menu. By selecting final as the dependent variable and major as the independent variable, as shown here, you will get the results shown below.

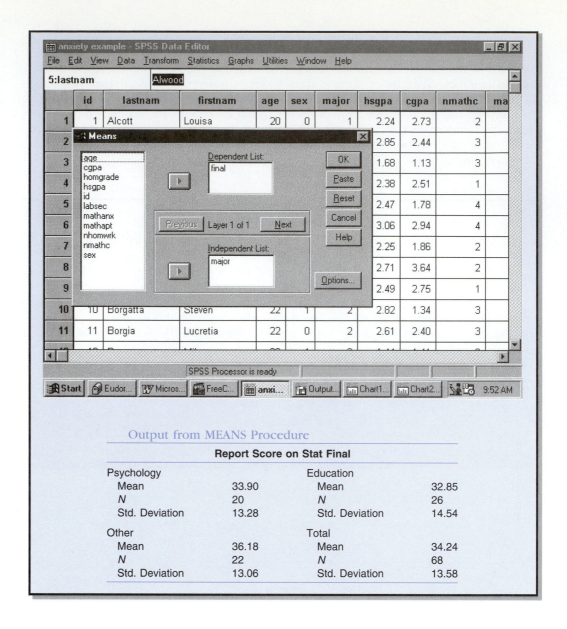

Output from MEANS Procedure

Report Score on Stat Final

Psychology		Education	
Mean	33.90	Mean	32.85
N	20	N	26
Std. Deviation	13.28	Std. Deviation	14.54
Other		Total	
Mean	36.18	Mean	34.24
N	22	N	68
Std. Deviation	13.06	Std. Deviation	13.58

The mean of the smallest group with which an individual can be identified is the best prediction of her/his score on the criterion variable. However, for a particular individual this prediction may not be as accurate as the overall mean. For example, if Ahab had earned a score of 6 on the quiz, our prediction would have been more accurate if we had ignored the number of homework assignments entirely. We would have predicted 7 based on the number of homework assignments he completed, but our prediction would have been 6 without this predictor. Nevertheless, *over a series of predictions, the least-squares predictions made by using the subgroup means will result in the smallest average error.* Because it minimizes the overall error, this strategy is used in selection and placement decisions.

Probability and Error

It is often useful to know how large we may expect our errors to be in making predictions. In other words, we may want to ask what the probability is that our prediction will be in error by some specified amount. For example, what is the probability that Ahab will get a score below 6 on the quiz? We can use the standard error of estimate (because it is the standard deviation of observed scores around their predicted values) and the relationship between probability and the normal distribution to answer this question.

Ahab handed in three homework assignments; therefore, our best prediction of his quiz score is 7 because this is the mean of his group. The standard error of estimate, which is the standard deviation of observed scores around this predicted score, is the standard deviation we should use to compute Z. The standard error of estimate is 1.1, and the lower limit of the score category 6 is 5.5; so we calculate the Z-score as

$$Z_{5.5} = \frac{5.5 - 7}{1.1} = \frac{-1.5}{1.1} = -1.36$$

Looking up the area beyond a Z of -1.36 in Appendix Table A-1, we find the value .0869, which we round to .09; that is, our calculations tell us that there is a 9 percent chance (or about 1 in 11) that Ahab will actually get a score below 6 on the quiz. Notice that this conforms very closely to our frequency distribution where one of the ten people who handed in all three homework assignments actually received a quiz score of less than 6. Using exactly the same procedure, we can also determine the probability that someone who handed in only one homework assignment will get a score above 6. The predicted score for people in this group is 5, the standard error of estimate is 1.1, and we would need to use the value 6.5 in computing the Z-score.

$$Z_{6.5} = \frac{6.5 - 5}{1.1} = \frac{1.5}{1.1} = +1.36$$

Notice that once again we must be careful to use the real limits of the score interval in our computations because of the small number of values in the distribution; otherwise, our probability estimates will not be as accurate. If we had forgotten or ignored the fact that a score value of 6 is used to represent an interval on the continuous trait from 5.5 to 6.5 and had used 6 instead of the interval limit, 5.5, we would have found a Z-score of $-.91$ for Ahab. The area of the normal curve that falls beyond this point is .1814 or .18, which is just double our previous answer. Had we made this mistake, we would have overestimated the probability of a score below 6. Of course, as we saw with the normal distribution approximation to the binomial distribution in Chapter 6, as the number of score categories in the distribution increases, the importance of this distinction is diminished because the differences between the probabilities become small.

▶ *Now You Do It*

To make sure you know how to find the probabilities associated with scores, let's determine the probability of a score of 5 or lower in the overall group, then in each of the subgroups. How shall we proceed?

We know that the mean for the complete group is 6.0 with a standard deviation of 1.37. A score of *5 or lower* includes the interval 4.5–5.5, so the limit of our range is 5.5. Converting this limit of 5.5 into a Z-score yields $Z = -.36$. The probability that we would observe a Z-score of $-.36$ or lower is the area beyond the Z. Looking in Appendix Table A-1, the area beyond a Z of $-.36$ is .3594, so the probability is .36 that a randomly selected person in this class will earn a score of 5 or lower.

Working with the subgroups, we must proceed a little differently in each case. For example, the score of 5 is exactly at the mean of group 1, but because this interval runs from 4.5 to 5.5, we must determine the probability of a score below 5.5. The Z for 5.5 in this group is $(5.5 - 5)/1.1 = +.45$ (remember that within a group we use the standard error of estimate of 1.1 rather than the standard deviation for the entire sample of 1.37). The area below a Z of $+.45$ is .6736, so the probability of a score of 5 or less in this group is .67. For the second group with a mean of 6, the Z of 5.5 is $-.45$, so the probability is .33 that a member of this group will earn a score of 5 or less. Finally, in the third group a score of 5.5 has a Z-score of $(5.5 - 7)/1.1 = -1.36$, yielding a probability of only .09 that a member of this group will score that low.

Pooling the Standard Errors of Estimate. When the standard deviations within the groups are approximately equal, it is customary to use the **pooled standard deviation** as a standard error of estimate for all groups. The primary reason to pool is that the resulting standard error of estimate, because it is based on a larger number of cases than the group statistics, will be more stable than the standard errors of estimate of the individual groups.

The pooling proceeds in exactly the same way that you learned in Chapter 4. First square each group standard deviation to obtain the variance; then multiply the variance by its sample size (N) to yield the sum-of-squares within the group. Next, add the within-group sums-of-squares to obtain SS_p. Finally, divide SS_p by the total number of cases.

We can illustrate the process of pooling the standard errors of estimate using the data from our BTU statistics class. In "Now You Do It" on page 172 you were asked to find the standard deviations for the different lab sections. The values you found there are repeated here and are used to compute the pooled sum-of-squares of 10912.70. Dividing this quantity by 68 yields the pooled variance, and its square root, 12.67, is the pooled standard error of estimate.

Lab Section	SD	Variance	N	SS
1	13.57	184.14	22	4051.19
2	11.73	137.59	22	3027.04
3	12.64	159.77	24	3834.47
Pooled Values	12.67	160.48	68	10912.70

Notice that the pooled standard error of estimate (12.67) is smaller than the standard deviation of 13.48 that we found for the total group. Because the pooled standard error of estimate is an index of our error of prediction after we have used the information in our predictor variable, this difference reveals how much our error of prediction is reduced by using the group means instead of the overall mean.

Prediction Intervals

The most widespread use of the standard error of estimate is to place probability intervals on our predictions. When applied to predictions, these probability intervals are called **prediction intervals.** A probability interval, as we saw in Chapter 6, is a range of values selected so that we have some specified level of confidence or assurance that the interval includes the unknown value we are estimating or predicting. Confidence is stated in terms of probability; a 95 percent prediction interval is one where the probability is .95 that the interval includes the unknown actual score of the person for whom the prediction is being made. In our example where students had turned in different numbers of homework assignments, 9 percent of the distribution of quiz scores of people who had completed three homework assignments would be expected to fall below the score category 6; likewise, 9 percent would be expected to score above 8 ($Z_{8.5} = +1.36$). Therefore, 82 percent will fall in the range 5.5 to 8.5. The score values 6 and 8 are the score limits of the 82 percent prediction interval, and category limits 5.5 and 8.5 are the real limits. We can also say that we expect about 82 percent of the people who have handed in three homework assignments to earn scores between 6 and 8. In addition, the probability is about 82 percent that Ahab will earn a quiz score between 6 and 8.

Prediction intervals are extremely useful when making predictions for individuals because they allow us to make approximate predictions (interval predictions) with a specified probability of being wrong. Because prediction intervals yield predictions over a range of possible criterion variable scores, they emphasize the approximate nature of our predictions.

▶ *Now You Do It*

Let's use the final exam data from the different lab sections at Big Time U to check your understanding of computing prediction intervals. We found that the pooled standard error of estimate was 12.67 (remember that when the group standard deviations are approximately equal we use the pooled value because it is more stable). The mean for lab section 1 was 40.68. What is the probability that a randomly selected student from that section will have a final exam score above 45? What interval will include the middle 90 percent of student scores?

The first question asks what proportion of the distribution falls above a score of 45 in a distribution with a mean of 40.68 and a standard deviation of 12.67. Using the real limit of the score, 45.5, gives us a Z-score of $+.38$. Looking in Appendix Table A-1 we find that the area beyond a Z of .38 is .3520. Therefore, the chances are about 35 percent that a student selected at random from section 1 will have a score above 45.

The 90 percent prediction interval requires that we find the interval around 40.68 that includes 90 percent of the distribution. If we divide the interval evenly, putting 45 percent on each side (5 percent in each tail), we can find the Z that cuts off the top 5 percent of the distribution. This value is 1.645. The prediction interval extends 1.645 Z units above and below the mean. Converting the Z's back into the metric of raw scores we obtain

$$Upper\ Limit = +1.645(12.67) + 40.68 = 61.52$$
$$Lower\ Limit = -1.645(12.67) + 40.68 = 19.84$$

The interval from 19.84 to 61.52 includes 90 percent of the distribution. Therefore, our interval prediction for someone in lab section 1 is that they will get a score between these values.

Linear Relationships

So far, we have been discussing situations where the predictor variable is a categorical variable such as college major or number of homework assignments completed. However, many useful predictor variables are continuous variables such as ability test score, reaction time, or age. We now explore how to express the relationship between two variables that are both continuous.

Two variables are said to be related or to have a relationship when the value of one variable changes as the value of the other is changed. There are always two or more variables involved, and the relationship may be simple or complex. Our discussion at this point will be restricted to the case of two variables and to the simplest form of relationship.

A relationship between two variables is known as a **bivariate relationship;** one of the variables is given the label X and the other Y. When prediction is involved, it is customary to use X for the predictor variable and Y for the criterion variable. In our previous example, the number of homework assignments would be the X-variable and the quiz score would be Y. In cases where prediction is not involved, it makes little difference which variable is called X, but it is conventional when graphing the relationships to put the X-variable on the abscissa and the Y-variable on the ordinate. Y is then treated as the criterion variable even if no predictions are to be made.

The relationship between X and Y is expressed in the form of an equation. Since Y is the variable to be predicted, the equation has the general form

$$Y = f(X),$$

where the right-hand term, which is read "function of X," can assume any form needed. Some possible forms for $f(X)$ are

$$X^2 + 2X + 96 \text{ and}$$
$$17X^3 - 386X^2 + 0.0037X + 23.$$

By far the most common form of relationship used in psychology and education is the **linear relationship,** which means that the relationship can be described by a straight line. The equation for a linear relationship is

$$Y = BX + A \tag{7.1}$$

B is the rate at which the line rises or falls (slope), and A is the value of Y when X is zero (intercept) (see Chapter 5, equation 5.3). The difference between equation 7.1 and equation 5.3 is that one (equation 5.3) transforms one scale into another in a one-for-one relationship, while the other (equation 7.1) shows the best prediction of one variable (Y) from another variable (X).

Graphing Relationships

The Scatter Plot

A set of observations never follows a straight line perfectly; the graph of a set of observations on two variables is called a **scatter plot.** It is formed by plotting a point to represent each

A **bivariate relationship** is a relationship between two variables.

When plotting a bivariate relationship it is customary to put the predicted or criterion variable on the Y-axis and the predictor variable on the X-axis.

A **linear relationship** is one that can be described by the equation for a straight line.

A **scatter plot** is a graph showing the relationship between two continuous variables. Each person is represented by a point which has as its coordinates the person's scores on the two variables.

Table 7-3 Scores of 20 Students on Two Quizzes		
Individual	Quiz 1 (X)	Quiz 2 (Y)
1	4	3
2	4	4
3	6	5
4	4	2
5	5	4
6	5	3
7	5	4
8	4	4
9	2	3
10	6	4
11	3	2
12	3	1
13	4	2
14	3	3
15	2	2
16	2	1
17	3	3
18	4	3
19	3	4
20	5	6

person. Every member of the group is observed on both variables, and we follow essentially the same procedure in making a scatter plot that we followed in preparing a frequency distribution in Chapter 2, with the difference that we have to consider both variables simultaneously.

The steps in making a scatter plot are outlined in Figure 7-2 using the set of scores in Table 7-3. In Figure 7-2(a), the coordinate axes are laid out with the proper numerical labels, since the data show scores up to 6 on each quiz. If data on one or both variables are to be grouped, the axes should reflect this; in other words, each axis is labeled just as it would be in preparing an ordinary frequency distribution.

The most common form for a scatter plot is that shown in Figure 7-2(b). In this figure, the X- and Y-axes are shown with tallies used to indicate the number of individuals who received each combination of scores. Here, two people (numbers 1 and 18) got the score combination $X = 4$, $Y = 3$, so there are two tallies in this square. Each of the other squares has a number of tallies equal to the number of people who received that combination of scores.

When both variables have approximately normal distributions and the relationship between them can be described by a straight line, the shape of the scatter plot is elliptical with more tallies clustered in the center of the ellipse. This general feature of the scatter plot is shown in Figure 7-2(c).

▶ Now You Do It

Once again, let's use the BTU data to check your ability to prepare a scatter plot. Let's look at the relationship between scores on the math anxiety test and final exam. If you are doing it by hand, remember that you have to decide what scale you are going to use to mark off the axes. How should you decide?

The rules for a good scale for a scatter plot are essentially the same as those for preparing a univariate frequency plot; it is preferable to have between 10 and 20 intervals. If you are using computer software, you may have to adjust the default settings the program uses to achieve the results you want. For example, we had to change the labeling on the axes to get an appropriate number of categories. Your results should look approximately like this:

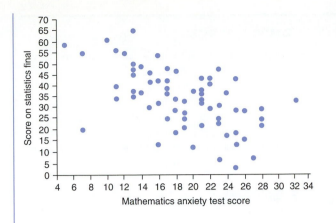

Scatter plot of relationship between math anxiety and final exam scores

If you used a computer to make the plot, the grid lines are probably not included.

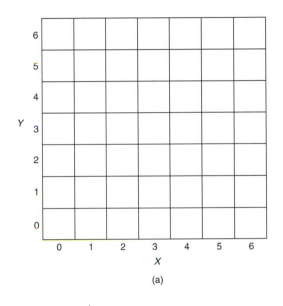

(a)

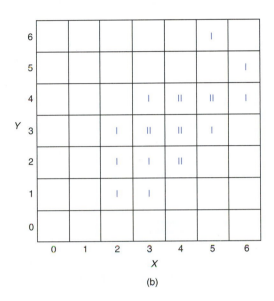

(b)

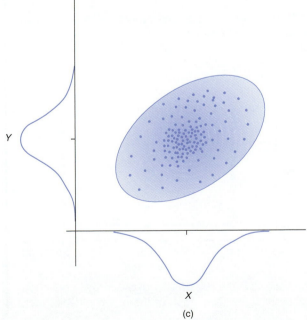

(c)

Figure 7-2 Preparation of a bivariate (two-variable) scatter plot where X = quiz 1 and Y = quiz 2.

HOW THE COMPUTER DOES IT

All statistical packages have graphing procedures for preparing scatter plots of various sorts. For example, SPSS can prepare a simple scatter plot, or it can produce an overlay plot where the members of different subgroups are identified by symbols such as circles and squares. It can also produce simple scatter plots of several groups arranged for comparison in a format called a matrix. If there are three variables, the program will prepare a three-dimensional plot. Within the GRAPH menu, select the SCATTER procedure and the SIMPLE subcommand. You must then DEFINE the variables you want to include in the plot. Here, we have included score on the statistics final for our BTU data as the predicted variable and lab section as the predictor variable.

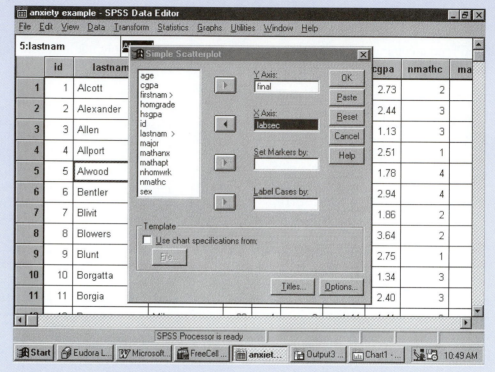

The results look like this. Notice that each group forms a column. The result is somewhat like a box-and-whisker plot, but with the individual scores shown. Notice that the points are more densely packed in the middle portion of each column.

Scatter plot of final exam scores by lab section

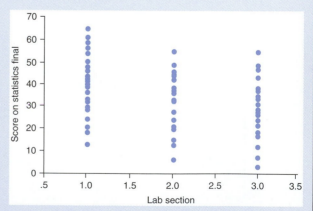

It is more common to use a scatter plot to show the relationship between two continuous variables such as college GPA and final exam score. The graph of these two variables looks like this:

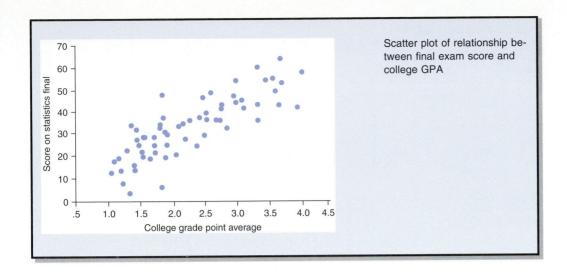

Scatter plot of relationship between final exam score and college GPA

Linear Prediction and Description

We saw in Chapter 4 that the mean provides a least-squares description of a group of observations, and our discussion in the first part of this chapter extended the concept of least squares to prediction. Actually, the distinction between the two is one of purpose, since the same set of observations can be used for either prediction or description. We measure two variables for a group of people, and if we want to describe the degree of relationship between the two variables, the same indices are computed that would be used for prediction. When prediction is the objective, we describe the relationship between the two variables in a sample we already have in hand. Then, by assuming that the relationship is the same for other, as yet unmeasured people, we use the description of the first sample to make forecasts for future cases. For example, it is common for colleges and universities to use high school grades to predict success in college. High school grade point average is the predictor variable, and college grade point average, usually after the first term, is the criterion. Each year a college collects high school grade point information from incoming students and college grade point information from the same students after the first term. The relationship between the two variables is a description of this group, the current freshman class. The relationship can also be used to make college grade point average predictions for next year's freshman class on the basis of those students' high school grade point information alone. Prediction always requires one sample in which both the predictor and criterion variables are available in order to determine the appropriate equation describing the relationship between the two. Then, predictions can be made for future samples using this equation.

Assumptions for Linear Relationships

Observations of the real world seldom, if ever, form a perfect linear relationship; instead, they form an ellipse like the scatter plot in Figure 7-2(c). For each value of X, several different values of Y are observed. In a perfect linear relationship only one value of Y would occur for each value of X, and the scatter plot would form a straight line.

The data in Tables 7-1 and 7-3 are typical of the kind of data obtained in social and behavioral science research. In the first case, the X-variable is categorical, and it is possible to use subgroup means on the Y-variable as least-squares predictors of Y-scores. In the second example, however, we can reasonably consider both of the quiz scores to be continuous variables. That is, there are no explicit subgroups on the X-variable. This feature is illustrated in Figure 7-2(c), where both variables are shown as having normal distributions. When both variables are continuous, it is necessary to introduce some assumptions in order to obtain least-squares predictions.

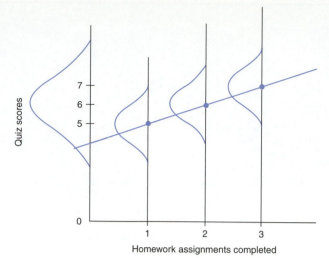

Figure 7-3 In a linear relationship the means of the subgroups all fall on a straight line.

Quiz scores

7
6
5
0

1   2   3

Homework assignments completed

---

*The **linearity assumption** says that the means of Y-scores for all values of X fall on a straight line.*

*A **linear relationship** is one where a straight line provides a least-squares prediction of Y from X.*

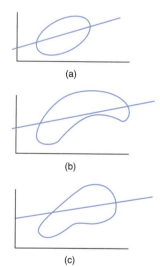

(a)

(b)

(c)

Figure 7-4 When the scatter plot forms a regular ellipse [part (a)], the linearity assumption will be met. In irregularly shaped scatter plots the Y-score means will not fall on a straight line [parts (b) and (c)].

*The **normality assumption** states that at any value of X the Y-scores are normally distributed around their mean.*

Linearity Assumption. Our first and most important assumption is called the **linearity assumption.** It states that the relationship between X and Y can be described by a straight line. Another way to say this is that we assume it is possible to find a straight line that passes through the Y-score means of all possible groupings on the X-variable. The straight line so determined provides a least-squares prediction of Y for any value of X. It does so because it passes through the value that the mean of Y-scores *would have* if we were to compute it for all individuals who have that value of X.

This principle is illustrated in Figure 7-3, which reproduces in schematic form the data from Figure 7-1. The overall distribution of quiz scores has a mean of 6 and a standard deviation of 1.37. Each of the smaller distributions represents the scores of the people handing in 1, 2, or 3 homework assignments. Notice that the means all fall on a straight line. Thus, the value of the Y-coordinate for the line when $X = 1$ is the mean of that group, the value of Y when $X = 2$ is the mean of that group, and so on.

It is important to understand that there is a straight line that provides a least-squares prediction of Y from X. This is what is meant in statistics by the term **linear relationship.** If the data satisfy the linearity assumption, then it is possible to find a straight line of the form

$$Y = BX + A$$

such that for every value of X, the value Y is a least-squares predictor of Y-scores for the individuals who have that value of X. The linear relationship is a straight line that is fitted to the observed data so that it satisfies the principle of least squares.

The scatter plot of a set of data that fulfills the linearity assumption will form a simple ellipse such as the one in Figure 7-4(a). Each part of the scatter plot falls around the line in a symmetric way. When the linearity assumption is not met, the scatter plot may resemble a banana, as is shown in Figure 7-4(b). In this case, the means of some possible groups do not fall on a straight line, and the scatter plot is not symmetric around the line. In psychological and educational research the data usually fulfill the linearity assumption, but it is always a good idea to check for failures of the assumption by preparing a scatter plot. Serious violations of the assumption will be revealed by asymmetrical scatter plots.

Normality Assumption. Our second important assumption when we make predictions is that the scores of the individuals about whom we are making predictions have an approximately normal distribution. If they do, the Y-scores for the people who score at any given value of X will form a normal distribution. This principle is illustrated in Figure 7-5. The Y-scores for people whose X-score was X_1 form a normal distribution, as do the Y-scores of the people at X_2. This implies that the errors of prediction are due to chance, because, as we have seen, chance events tend to follow the normal curve. Stating this another way, we assume that the errors of prediction are random events that form a normal distribution around the line. This does not mean that the individual differences among the

---

people in any group are due to errors, only that we have no way of knowing for whom our predictions will be in error.

Homogeneity of Variance Assumption. The third assumption is that the *variance of observations around any point on the line is the same at every point on the line*. The standard deviation of *Y*-scores for those people who received a score of X_1 in Figure 7-5 is the same as that of people who earned a value of X_2 on the *X*-variable. Another way of saying this is that the variances of errors of prediction are **homogeneous,** or that the standard error of estimate is the same everywhere. Figure 7-4(c) illustrates a scatter plot of data that fail to satisfy the homogeneity of variance assumption.

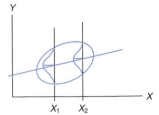

Figure 7-5 When the three assumptions of linearity, normality, and homogeneous variances are met, the mean of any slice through the scatter plot will fall on the regression line, and the standard deviation of scores around their predicted values will be the same for all slices.

Making Predictions

In using the number of homework assignments completed to predict test scores, we computed group *Y*-score means and used these means as our least-squares predictors of *Y* for the several values of *X* (groups). We have now seen that for continuous predictor variables, it is possible to find a line, called the **regression line,** that will serve the same function but will yield a predicted value of *Y* for any value of *X* without the need to form groups.

Predicted values of *Y* are determined for any value of *X* by using the equation for the regression line, which is called the **regression equation.** When predictions are to be made, the regression equation is generally written

$$\hat{Y}_i = B_{Y \cdot X} X_i + A_{Y \cdot X} \tag{7.2}$$

where the caret ($\hat{\ }$) over the *Y* indicates a predicted or estimated score. The symbol indicates that the relationship between *X* and *Y* is not exact and that the observed data points spread out around the line. The subscript *i* on $\hat{Y}$ and *X* indicates that the prediction is for an individual, and the subscript $Y \cdot X$ on *B* and *A* means that *Y*-scores are being predicted from *X*-scores. The values of $\hat{Y}$ all fall on the regression line, which means that each $\hat{Y}$ is assumed (from the linearity assumption) to be the mean of *Y*-scores for a particular value of *X*.

If we know the slope (*B*) of the line and its intercept (*A*), we can use equation 7.2 to compute a $\hat{Y}$ for any value of *X*. The data plotted in Figure 7-6 are listed in Table 7-4, and we will use these data to illustrate the computations to find *B* and *A*. By procedures we will discuss shortly, the slope of the regression line is found to be +.45 and its intercept

*The assumption of **homogeneity of variance** means the variance of scores around any point on the line is the same as at every other point on the line.*

*The **regression line** is the line that provides least-squares predictions of Y-scores for any value of X.*

*The equation of the regression line is called the **regression equation.***

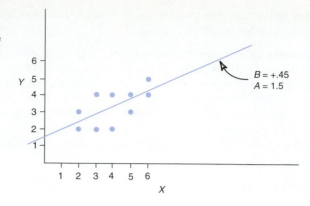

Figure 7-6 Scatter plot of 10 pairs of scores from Table 7-4 for which the slope of the regression line is +.45 and the intercept is +1.5.

is +1.5. Placing these values in equation 7.2 gives us the equation for the regression line:

$$\hat{Y}_i = +.45X_i + 1.5$$

By inserting any value of X_i in this equation, we can compute the predicted value, $\hat{Y}$, for the people with that X-score; for example, if $X_i = 2$

$$\hat{Y}_i = .45(2) + 1.5$$
$$= .9 + 1.5 = 2.4.$$

Similar computations for each of the other observed values of X yield $\hat{Y}$s of 2.85 for $X = 3$; 3.3 for $X = 4$; 3.75 for $X = 5$; and 4.2 for $X = 6$.

Once we have predictions for each person, we can compute the errors of prediction. They can then be used to compute a standard error of estimate, which is the standard deviation of the distribution of distances of data points from the regression line. Column 4 of Table 7-4 contains the predicted values $\hat{Y}_i$. The errors of prediction $(Y_i - \hat{Y}_i)$ are given in column 5, and the squares of these errors appear in the last column. Note that the sum of errors (but not the sum of squares) is exactly zero. This is the same thing that we observed regarding the sum of deviations from the mean in Chapter 4 and illustrates the fact that the regression line is exactly like a mean that runs all the way through the scatter plot. At the end of Table 7-4, ΣY, ΣY^2, and S_y are given. The value of S_y is, of course, an index of the average amount of error that we would incur by using $\overline{Y}$ as our prediction for each individual. Also included in the table are $\Sigma(Y_i - \hat{Y}_i) (= 0)$, $\Sigma(Y_i - \hat{Y}_i)^2$, and the standard error of estimate.

Table 7-4   Computing the Standard Error of Estimate

Individual	X	Y_i	$\hat{Y}_i$	$(Y_i - \hat{Y}_i)$	$(Y_i - \hat{Y}_i)^2$
1	2	3	2.4	.6	.36
2	4	4	3.3	.7	.49
3	6	5	4.2	.8	.64
4	4	2	3.3	−1.3	1.69
5	5	4	3.75	.25	.0625
6	5	3	3.75	−.75	.5625
7	3	4	2.85	1.15	1.3225
8	6	4	4.2	−.20	.04
9	2	2	2.4	−.40	.16
10	3	2	2.85	−.85	.7225

Calculations

$\Sigma Y_i = 33$

$\Sigma Y_i^2 = 119$

$S_y = 1.005$

$\Sigma(Y_i - \hat{Y}_i) = 0$

$\Sigma(Y_i - \hat{Y}_i)^2 = 6.05$

$S_{Y \cdot X} = .778$

Linear Least Squares

We have said that the regression line is like a continuous mean of Y-scores for each value of X. However, the data in our example show that the value on the line, $\hat{Y}_i$, is not equal to the mean of Y-scores for any observed value of X. At $X = 2$, $\overline{Y}$ is 2.5 , but $\hat{Y}_i = 2.4$; similar discrepancies occur at each value of X. The data do not fulfill our assumption of linearity precisely. The linearity assumption, like all other assumptions that we make about theoretical properties in statistics, is never exactly true for any set of observations. The question we must ask is whether the departure from our assumptions is too great to warrant making those assumptions. In fact, the experience of over 95 years of research in education and psychology indicates that there are very few occasions when educational and psychological research data cannot be analyzed more fruitfully with the assumption of a simple linear relationship than with any other kind. The simplest way to test this assumption is to prepare a bivariate scatter plot and look at it. If the plot has an approximately elliptical shape, we know that we are on safe ground with our assumption of linearity.

▶ *Now You Do It*

Why do you suppose psychologists and educational researchers find so few relationships that are not linear? What does a nonlinear relationship imply? Can you think of any situations where you might expect to find a nonlinear relationship?

What does a nonlinear relationship look like? One example is the "banana-shaped" relationship in Figure 7-4(b) where low levels of trait Y occur with low levels of trait X, high levels of Y occur with moderate levels of X, and moderate levels of Y occur with high levels of X. The reverse, where high levels of Y occur with low levels of X, low levels of Y occur with moderate levels of X, and moderate levels of Y occur with high levels of X, is another possibility. But what traits might we expect to give relationships like these?

The answer is that we would expect very few such relationships, and it is nature that has taught us to expect few such events. Psychologists have studied the relationship between motivation and performance and have found that very high levels of motivation can produce reduced levels of performance, resulting in a curvilinear relationship. Educators have found a similar relationship between anxiety (which we might view as a type of motivation) and measures of achievement. Moderate anxiety improves achievement, but high levels of anxiety can produce low achievement scores.

Finding the Slope of the Regression Line

Slope coefficients (B's) and intercepts (A's), like all statistics, are computed from the observed data using some relatively simple (but fearsome looking) formulas. First, remember that we defined the regression line as the least-squares line, which means that the line we seek is the one for which errors are minimized. The errors are $(Y_i - \hat{Y}_i)$, and the least-squares criterion states that the line we seek is the one that yields the smallest possible value of $\Sigma(Y_i - \hat{Y}_i)^2$.

The value of $B_{Y \cdot X}$ that yields the least-squares prediction of Y-scores from X-scores is defined as

$$B_{Y \cdot X} = \frac{\Sigma(Y_i - \hat{Y})(X_i - \overline{X})}{\Sigma(X_i - \overline{X})^2} \tag{7.3}$$

That is, we take each person's deviation from the mean of the Y's and multiply it by the deviation of their X-score from the mean of the X's. This quantity is called a *cross product*. We next find the sum of the cross products and divide this sum by the sum-of-squares of the X-scores. This is called the definitional formula for $B_{Y \cdot X}$. (The denominator of equation 7.3 contains the sum-of-squares of the predictor variable. If Y is used to predict X, SS_Y would be used in the denominator.)

As we saw in the case of the variance, definitional equations are not convenient to use when computing the actual values of statistics. The same is true of the slope coefficient.

Table 7-5 Computing the Slope from Deviation Scores and Raw Scores

X	Y	$X - \bar{X}$	$Y - \bar{Y}$	XY	$(X - \bar{X})(Y - \bar{Y})$	$(X - \bar{X})^2$
2	3	−2	−.3	6	.6	4
4	4	0	.7	16	0	0
6	5	2	1.7	30	3.4	4
4	2	0	−1.3	8	0	0
5	4	1	.7	20	.7	1
5	3	1	−.3	15	−.3	1
3	4	−1	.7	12	−.7	1
6	4	2	.7	24	1.4	4
2	2	−2	−1.3	4	2.6	4
3	2	−1	−1.3	6	1.3	1

$\Sigma X = 40$ $\Sigma Y = 33$ Sums 141 9.0 20.0
$\bar{X} = 4.0$ $\bar{Y} = 3.3$
$\Sigma X^2 = 180$ $\Sigma Y^2 = 119$

$$B_{Y \cdot X} = \frac{\Sigma(Y - \bar{Y})(X - \bar{X})}{\Sigma(X - \bar{X})^2} \qquad B_{Y \cdot X} = \frac{N(\Sigma XY) - (\Sigma X)(\Sigma Y)}{N\Sigma X^2 - (\Sigma X)^2}$$

$$= \frac{9}{20} \qquad\qquad = \frac{10(141) - (40)(33)}{10(180) - (40)^2}$$

$$= .45 \qquad\qquad = .45$$

By algebraic manipulation of equation 7.3, we can obtain a computational formula for $B_{Y \cdot X}$ that uses only raw scores.

$$B_{Y \cdot X} = \frac{N(\Sigma X_i Y_i) - (\Sigma X_i)(\Sigma Y_i)}{N(\Sigma X_i^2) - (\Sigma X_i)^2} \tag{7.4}$$

The numerator contains N times the sum of cross products of the raw scores (each person's score on Y multiplied by their score on X and summed across people) and the product of the sums of raw scores for the two variables. The denominator of formula 7.4 is simply the numerator of the formula for the variance of X.

Since we would probably already have ΣY, ΣX, and ΣX^2 from computing means and standard deviations, ΣXY is the only new quantity to find. The computational form has the double advantage of being easier to use and less prone to rounding errors than the definitional one, even though it does look more complex. An alternate and equivalent formula that uses the means is

$$B_{Y \cdot X} = \frac{\Sigma X_i Y_i - N(\overline{XY})}{\Sigma X_i^2 - N\bar{X}^2} \tag{7.5}$$

The raw data from Table 7-4 are reproduced in Table 7-5 to provide an example of computing $B_{Y \cdot X}$ using either the definitional (7.3) or the raw score (7.4) formula. In this example, the numbers are simple enough so that the slope of +.45 works out quite easily either way. However, it is generally the case that the computational formula 7.4 or its equivalent will be easier to use and give a more accurate answer because they involve less rounding.

Formula 7.3 simply requires that we divide the sum of the $(X - \bar{X})(Y - \bar{Y})$ column by the sum of the $(X - \bar{X})^2$ column. Formula 7.4 seems more complicated with this small number of scores, but is really easier to use when N is reasonably large. For the numerator, we multiply the sum of the XY column by N and subtract from this product the product of the sum of the X column by the sum of the Y column. The computations for the denominator are the same as for the numerator of a variance.

Computing the Intercept

The complete regression equation requires not only the slope, B, but also the intercept, A. That is,

$$\hat{Y}_i = BX_i + A$$

where B is obtained from formula 7.4. The value of A is, of course, the value of $\hat{Y}_i$ where the regression line crosses the Y-axis (when the value of X_i is zero). For a given set of data, we can then find the intercept by the formula:

$$A = \bar{Y} - B\bar{X} \qquad (7.6)$$

The intercept can easily be computed for the data in Table 7-5. With $\bar{X} = 4.0$, $\bar{Y} = 3.3$, and $B_{Y \cdot X} = +.45$, we get

$$A_{Y \cdot X} = 3.3 - (+.45)(4)$$
$$= 3.3 - 1.8 = 1.5.$$

▶ *Now You Do It*

Let's use the relationship between math anxiety score and final exam score to practice procedures for computing B and A. If you use SPSS, you should get the following output by selecting final as the dependent variable and math anxiety as the independent variable.

Coefficients

Model	Unstandardized Coefficients		Standardized Coefficients		
	B	Std. Error	Beta	t	Sig.
1 (Constant)	58.902	4.845		12.158	.000
Math Anxiety Test Score	−1.321	.249	−.547	−5.315	.000

ªDependent Variable: Score on Stat Final

The information you need is given in the column labeled B. "Constant" refers to the intercept. Therefore, the regression equation is $\hat{Y} = -1.32X + 58.90$. The negative value for B means that higher values of anxiety predict lower values of test performance. In the event you chose to look at a scatter plot, you should have gotten something like this:

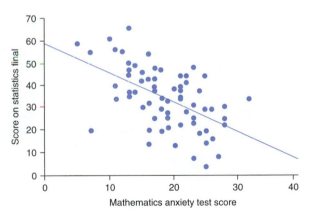

Scatter plot of final exam scores and math anxiety test scores

Interpreting the Regression Equation

The regression equation is used in making least-squares predictions. It gives one basic piece of information about the nature of the relationship between the two variables: the direction or sign of the relationship. The algebraic sign (+ or −) of the slope coefficient tells us whether the value of $\hat{Y}$ gets larger or smaller as the value of the predictor is increased. A positive sign on B means that the line rises as we go from left to right, as shown in part (a) of Figure 7-7; we call this a positive relationship between the two variables. A negative sign on $B_{Y \cdot X}$ tells us that the value of $\hat{Y}$ decreases as the value of X increases; this situation, shown in part (b) of the figure, is called a negative relationship. In the two examples we have been using to illustrate regression, the one between GPA and final exam score showed a positive relationship ($B = +14.15$), and the one between math anxiety and final exam score showed a negative relationship ($B = -1.32$). (We show you how to make the computer do these computations in the box on pp. 188 and 189.)

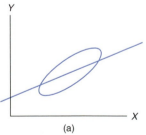

(a)

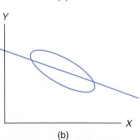

(b)

Figure 7-7 The sign of a regression coefficient shows a positive slope (a) or a negative slope (b).

HOW THE COMPUTER DOES IT

There are two topics that we need to cover about computer applications at this point: (1) putting the regression line into the scatter plot, and (2) computing the regression equation. We will tackle the plotting problem first, then the regression equation.

The process of putting the regression line into the scatter plot using SPSS requires that you first prepare the plot of the two variables, then edit the resulting graph. We have already covered how to make the basic graph. Once the output with the graph is displayed, double-click on the graph to invoke the graph editor. Next, select OPTIONS from the CHART menu. A screen like the following will be superimposed on the chart:

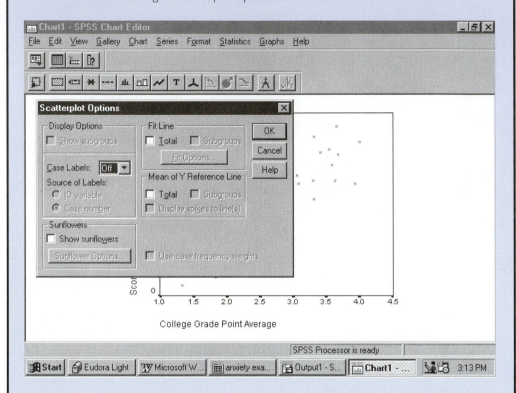

Click on the Fit Line Total box, and the regression line will be placed on the scatter plot and look like this for the relationship between the statistics final exam scores and the college GPAs. The plot shown here is not exactly as it is displayed by SPSS. We have edited the plot a little in order to adjust labels and to show the full range of the X-variable to illustrate the location of the intercept.

Scatter plot of relationship between final exam score and college GPA

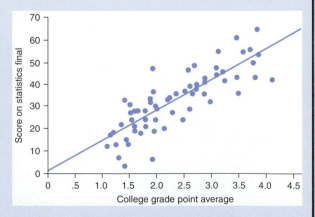

Our second topic is how the computer calculates the regression equation. For most statistics programs, bivariate linear regression is a subtopic within the broader subject of

multiple regression. This is particularly true of SPSS, which includes a wide array of options that need not concern us and that only clutter the output for our purposes. What you must learn to do is go to the part of the output that you need and ignore the statistics that don't apply to your problem. To get a regression analysis of the data we are considering here, click on the STATISTICS menu and select the REGRESSION-LINEAR option. You should get a screen that looks like this:

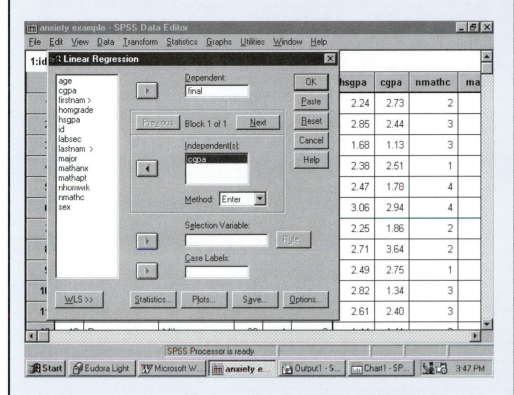

Click on the "final" variable and the DEPENDENT arrow, then on the "cgpa" variable and the INDEPENDENT arrow. Then click OK. One section of the output will look like this:

Coefficients

Model	Unstandardized Coefficients		Standardized Coefficients		
	B	Std. Error	Beta	t	Sig.
1 (Constant)	2.125	2.815		.755	.453
College Grade Point Average	14.152	1.172	.830	12.078	.000

[a]Dependent Variable: Score on Stat Final

In the column labeled *B* the value in the (Constant) row is *A* (2.125 in this case), and the value of 14.152 is *B*. Therefore, for these data, the regression equation is

$$\hat{Y}_i = 14.152X_i + 2.125$$

Disregard the other values.

B conveys information of limited interpretive value in most educational and psychological applications. The sign indicates the direction of the relationship, but the numerical value of B does not indicate the strength of the relationship. The B of $+14.15$ for predicting final exam score from college GPA does not necessarily imply a larger or stronger relationship than the B of -1.32 for the math anxiety test.

The value of B is heavily influenced by the relative magnitudes of the variances of the two variables. Suppose, for example, that the numerator of formula 7.5 found for a given set of data is 5000. This value, and the degree of association it represents, will be the same regardless of which variable is the predictor. Now suppose that the sum of squares for one variable (X) is 50,000 and for the other (Y) it is 500. Then, the value of $B_{Y \cdot X}$ for predicting Y from X would be $+0.10$, while the value of $B_{X \cdot Y}$ for predicting X from Y would be $+10.0$. One slope is 100 times larger than the other, but the degree of relationship between the two variables is the same regardless of which is considered the predictor. This difference is due solely to the unequal variances.

The reason that the magnitudes of regression coefficients seldom have direct meaning in most psychological and educational applications is that most of the variables we study have arbitrary units. This is not always so, but scores on personality and ability tests, for example, are on a completely arbitrary scale whenever they use a normative frame of reference. An exception to this general rule is a case like that of number of homework assignments handed in and quiz score (neither of which is norm referenced) from earlier in this chapter. There, the B coefficient was 1, and we could conclude that handing in one more homework assignment resulted, on the average, in earning one additional point on the quiz.

A positive relationship between two variables is no better or worse than a negative one, and its meaning depends entirely on the meaning of the variables themselves. There are some situations where the direction of measurement (which end of the scale is up) may be more or less arbitrary. For example, a scale that measures the trait political liberalism (high score is liberal) might show a negative relationship with chronological age. The same scale, rescored as a measure of political conservatism (high score is conservative), would then be positively related to age but the B would have the same numerical value. Similarly, it is common to find a negative association between reaction time and conventional measures of cognitive ability. The negative sign indicates a positive relationship between quick reactions and cognitive ability because short reaction times (small numbers) mean more of the trait of quickness. The most important thing to remember in interpreting regression slopes is which end of the original scale is up.

Likewise, $A_{Y \cdot X}$ has little interpretive significance by itself. There are situations where it is useful to compare intercepts and slopes for different groups, but the intercept for a single group tells us nothing beyond what is needed to compute the most accurate predictions possible for this set of data.

Prediction Intervals from the Regression Line

We defined the standard error of estimate as the standard deviation of errors of prediction (that is, the square root of the average squared distance of the data points) from the regression line. It may be used just like any other standard deviation to compute Z-scores because under the assumptions that we have made about the scatter plot, a point on the regression line is always the mean of Y-scores for people with a particular value of X. When we can assume a linear relationship with homogeneous variances and normally distributed errors of prediction, we can use the normal curve and its associated probabilities just as we did in our earlier example of homework and quiz scores. For any value of X_i, we can compute a $\hat{Y}_i$ and a prediction interval around the predicted value.

Once we have $\hat{Y}_i$, there are two ways to set the prediction interval. The first way is to ask how likely it is that the prediction will be in error by more than K points. The second is to specify some probability (P) of error and to set our prediction interval so that the probability of an individual's actual score falling outside the interval is only P.

Suppose that Dr. Numbercruncher is giving her statistics class a final exam. She has been teaching the course for several (her students claim it is 75) years, and each year she computes the regression equation for the relationship between midterm and final test scores (statistics professors are prone to do that sort of thing). This year, she tells her class that on the basis of years of experience she has concluded that the appropriate regression slope is 0.84, the intercept is 37, the standard error of estimate is 4.15, and the minimum score on the final to get at least a C is 60.

Jezebel is worried about her performance; she got a score of 34 on the midterm and would like to know what her chances are on the final. What can we predict from the information given? First, we compute Jezebel's predicted final score. Using Dr. Number-cruncher's regression equation, we get

$$\hat{Y}_i = +.84(34) + 37$$
$$= 65.56$$

as Jezebel's predicted score. The value 65.56 is the mean score we would expect on the final for people who scored 34 on the midterm, and it is our least-squares prediction for Jezebel. Since the critical score for a C is 60, we find Z for *a score below 60* in a distribution with a mean of 65.56 and a standard deviation (the standard error of estimate) of 4.15 (note that we are using 59.5, the lower real limit of the score category 60).

$$Z_{59.5} = \frac{59.5 - 65.56}{4.15} = \frac{-6.06}{4.15} = -1.46$$

Looking in the table of the normal curve, we find that the probability of a Z beyond -1.46 is .0721, so we conclude that there is only about one chance in 14 that Jezebel will get a D or less. (Note that Jezebel might be able to alter this probability with a little extra studying.)

This example does not involve a true prediction interval, although it does involve a kind of problem that is frequently encountered. We might rephrase the question: What is the probability that a student will obtain a score more than five points above or below his or her predicted score? Again using the real limits of the raw scores, we find the Z of $+5.5$ or -5.5 to be

$$Z = \frac{\pm 5.5}{4.15} = \pm 1.33.$$

We use Appendix Table A-1 to find that the probability of a score more than five points above the predicted one is .0918, and the same is true for the lower side. Therefore, we can say that the probability is about .18 (or 18 percent) that our prediction will be in error by more than five points for any given individual. Looking at it from the other point of view (the glass is half full rather than half empty), we can conclude that 82 percent of our predictions are likely to be in error by five points or less. The crucial problem, for which there is no adequate solution at the present time, is how to identify the people for whom the error will be large.

The second and more conventional way of expressing a prediction interval is to specify a probability and work backward to the score limits. For example, Jezebel still has a predicted score of 65.56, but she would like to know a range that she may be 90 percent certain will include her score. She is willing to accept a 5 percent chance of being too low and a 5 percent chance of being too high.

The solution to this problem, as described in Chapter 6, is to look up the Z beyond which 5 percent of the normal distribution falls; from Appendix Table A-1, we find this value to be 1.645. Five percent of the distribution falls above $Z = +1.645$ and another 5 percent falls below $Z = -1.645$. We convert these Z-scores to deviations by multiplying by $S_{Y \cdot X}$ (4.15) to get ± 6.83. That is, 90 percent of our errors will be less than 6.83 units, or Jezebel with her predicted score of 65.56 can be 90 percent certain that the interval from $(65.56 - 6.83) = 58.73$ to $(65.56 + 6.83) = 72.39$ includes the score that she will actually earn. In other words, the 90 percent prediction interval for Jezebel's score is 58.73 to 72.39.

▶ *Now You Do It*

You may find it helpful at this time to review our discussion of Z-scores, probability, and the normal distribution in Chapter 6. You should be able to see how the same principles are at work in probability intervals and prediction intervals. Of course, if our assumptions of homogeneous variances and normal distributions for the errors of prediction from the regression line are not met, the prediction interval and its associated probability will not be correct. However, for most practical work, the discrepancy would have to be fairly large for the inaccuracies to be important.

Let's check your understanding of prediction using the regression equation and prediction intervals with another problem from Dr. Numbercruncher's class. Absalom is a classmate of Jezebel's who got a score of 52 on the midterm. What is his predicted score, and what is the 95 percent prediction interval?

Recalling that the regression equation is $\hat{Y} = (+.84)X + 37$, Absalom's predicted score on the final is 80.68. Looking in Appendix Table A-1, we find the Z that encloses 95 percent of the distribution (2.5 percent in each tail) to be ± 1.96. Therefore, the 95 percent prediction interval for Absalom's score is from $(-1.96)(4.15) + 80.68 = 72.5$ to $(+1.96)(4.15) + 80.68 = 88.8$. In terms of raw scores, Absalom is likely to score between 73 and 88.

SUMMARY

Predictions occur in many facets of our lives. Prediction decisions may involve selection, classification, placement, and personal decisions. We usually find that applying the **least-squares principle** to the relationship between a **predictor (X)** and a **criterion (Y)** results in the best overall prediction.

When the predictor variable is a discrete group-membership variable, the subgroup mean on the criterion variable is the least-squares prediction for each individual in the group. The **standard error of estimate ($S_{Y \cdot X}$)** is the standard deviation of errors of prediction. It can be used to compute a prediction interval around the prediction.

We generally assume that the relationship between two continuous variables is best described by a straight line. The scores of individuals are represented in a **scatter plot** that is usually elliptical in shape. Predictions are made from a straight line which has the equation $\hat{Y} = B_{Y \cdot X}X + A$. If the assumption of a **linear relationship** is met, the value of $\hat{Y}$ is a least-squares prediction of Y for every value of X, and the standard error of estimate is the standard deviation of deviations from the line. Using $S_{Y \cdot X}$, it is possible to determine a **prediction interval** for any prediction or to state a probability that the criterion score of a person with a given predictor score will be above or below a particular value. The slope of the regression line indicates the direction of a relationship between two variables but gives no information about its strength or about the accuracy of predictions made from the regression line.

REGRESSION ANALYSIS OF THE SEARCHERS STUDY DATA

The data that Dr. Cross and his students collected provide us an opportunity to review what we have learned about regression analysis with a new set of data. There are several variables that might attract our attention, but the one we are likely to want to predict is scores on the attitudes toward equality scale. What variables do you think might predict attitude scores?

One variable we might expect to be related to attitudes is age. We might have a hunch that older men are likely to hold more conservative attitudes. We might also expect to find some relationship between attitudes and income, although here the direction seems less certain. Degree of religious involvement and form of the attitude questionnaire are other variables that might predict attitude scores. Now, make a scatter plot of each of these relationships and compare your results with those given below. Next, compute the regression equations to be sure you can get the results we found. When you analyze the results for the attitude questionnaire, be careful.

Finally, and before you look at the answers, what do you think will happen to the regression line of the relationship between income and attitude if we predict income from attitude? Compare the two scatter plots that result from SPSS. Why do they look similar when the regression equations are so different? The answer lies in the standard deviations.

The results you should have gotten are shown below. Be sure you can pick out the slope and intercept coefficients from the output if you used SPSS.

We can also use the regression equations we found for these variables to make predictions about what a person's attitude score will be. For example, based on his age, we would predict an attitude score of 45.98 for Sandy Champagne. This prediction is in error by only .98; however, if we had used his religious involvement score our prediction would have been 42.23, an error of 2.77. For John Gibbs, on the other hand, our prediction from age would be 30.38, an error of almost 9 points, while our prediction from his involvement would be 37.82. In the real world of prediction we have no way of knowing which variable will be the best predictor for a given person or for which people our predictions will be most accurate. All we know for sure is that, across the group as a whole, least-squares procedures will yield the smallest average error.

Age and Attitude Results

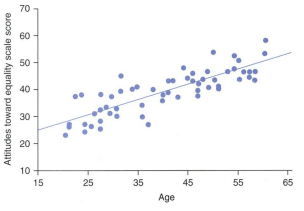

Scatter plot of age and attitudes toward equality scale score

Regression Equation	Standard Error of Estimate
$\hat{Y} = .600X + 16.58$	4.47

Questionnaire Form and Attitude Results

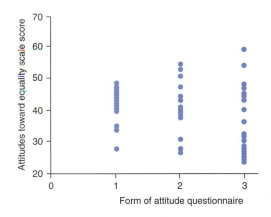

Scatter plot of questionnaire form and attitudes toward equality scale score

Attitudes Toward Equality scale score
Neutral
    Mean   41.95
    N   20
    Std. Deviation   5.53
Profeminist
    Mean   41.40
    N   20
    Std. Deviation   7.65
Conservative
    Mean   37.85
    N   20
    Std. Deviation 10.66
Total
    Mean   40.40
    N   60
    Std. Deviation   8.29

Note that in this case the plot shows the relationship to be nonlinear. When this happens, it often is best, particularly with a categorical (in this case nominal) variable, to compute subgroup means. Whenever the predictor variable is nominal, the order of the subgroups is arbitrary and the idea of a regression line does not make sense.

Results for Degree of Religious Involvement and Attitude

Scatter plot of degree of religious involvement and attitudes toward equality scale score

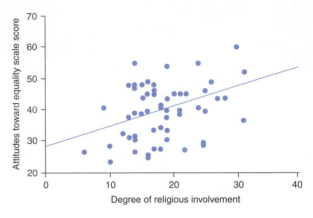

Regression Equation

$$\hat{Y} = .63X + 29.00$$

Standard Error of Estimate

7.54

Results for Predicting Attitude from Income

Scatter plot of income and attitudes toward equality scale score

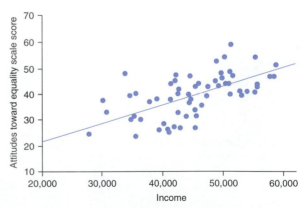

Regression Equation

$$\hat{Y} = .000735X + 7.13$$

Standard Error of Estimate

6.38

Results for Predicting Income from Attitude

Scatter plot of income and attitudes toward equality scale score

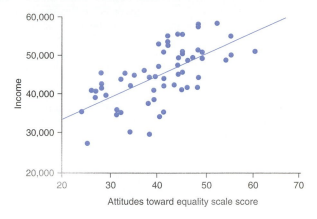

Regression Equation

$$\hat{Y} = 566.19X + 22380$$

Standard Error of Estimate

5602.52

REFERENCES

Darlington, R. B. (1990). *Regression and linear models*. New York: McGraw-Hill.

Edwards, A. L. (1984). *An introduction to linear regression and correlation* (2nd ed.). New York: Freeman.

Fox, J. (1997). *Applied regression analysis, linear models, and related methods*. Thousand Oaks, CA: Sage.

Hamilton, L. C. (1992). *Regression with graphics: A second course in applied statistics*. Pacific Grove, CA: Brooks/Cole.

Ott, R. L. (1993). *An introduction to statistical methods and data analysis* (4th ed.). Belmont, CA: Duxbury.

Pedhazur, E. J. (1982). *Multiple regression in behavioral research: Explanation and prediction* (2nd ed.). New York: Holt, Rinehart and Winston.

Younger, M. S. (1985). *A first course in linear regression* (2nd ed.). Boston: Duxbury.

EXERCISES

1. Define the four types of decisions; then provide two original examples of each type of decision.
2. For each of the following situations, state whether the decision is selection, classification, placement, or individual.
   a. An adolescent must decide whether or not to join a gang.
   b. An academic advisor helps a student decide what courses to take to fulfill general requirements.
   c. A juvenile counselor assists in making job assignments for juvenile offenders.
   d. A personnel director considers applicants to interview for a position within a company.
   e. A school psychologist examines students' records to determine the appropriate classes.
   f. A social worker determines the most appropriate foster home for an abused child.
3. For each of the following, identify the predictor and criterion variables.
   a. A researcher wants to determine a person's tendency to procrastinate by taking measures of perfectionism, self-esteem, locus of control, and level of pessimism.
   b. A researcher wants to determine a person's ability to achieve in algebra by taking measures of formal operational thinking, spatial ability, and understanding of basic mathematics.
   c. A researcher wants to determine a person's level of social phobia by taking measures of self-confidence, neuroticism, and embarrassability.
   d. A teacher wants to predict a child's reading ability by taking measures of the number of books at home, the education level of the parents, the number of hours a parent reads to the child, and the child's vocabulary.
4. Students were asked to practice a physical skill during the course of the week and record the amount of time they spent practicing. At the end of the week, they were given a skill test. The data are presented below:

Student	Hours of Practice	Skill Test Score	Student	Hours of Practice	Skill Test Score
1	2	16	19	3	18
2	3	19	20	1	9
3	2	13	21	1	8
4	1	10	22	3	17
5	3	17	23	2	15
6	1	7	24	2	15
7	1	9	25	3	17
8	2	15	26	1	9
9	1	8	27	2	14
10	3	16	28	3	18
11	2	16	29	1	11
12	1	10	30	2	14
13	3	18	31	2	17
14	1	9	32	3	19
15	3	20	33	2	15
16	1	8	34	2	16
17	3	18	35	1	10
18	2	14	36	3	19

a. Without considering the number of hours of practice, what is the best prediction for the score on the skill test? How much error is involved in this prediction?

b. What is the best prediction for the score on the skill test for those who practiced one hour? two hours? three hours? How much error is involved in each prediction (i.e., What is the standard error of estimate for each group?)?

5. Using the data in Exercise 4(b), answer the following.

a. Suppose Alex practiced the physical skill for one hour during the week. What is the probability that he will score above 10 on the skill test?

b. Suppose Andrea practiced the physical skill for two hours during the week. What is the probability that she will score below 13 on the skill test?

c. Suppose Antionne practiced the physical skill for three hours during the week. What is the probability that he will score above 18 on the skill test?

6. For the data listed below, construct a scatter plot of the data using attitude about science as the predictor variable and the final exam score in science as the criterion variable.

Student	Attitude About Science	Science Final Exam Score	Student	Attitude About Science	Science Final Exam Score
1	50	70	11	55	80
2	40	60	12	65	90
3	40	65	13	70	95
4	60	85	14	50	68
5	65	85	15	55	74
6	58	78	16	62	88
7	48	68	17	55	78
8	50	73	18	68	94
9	60	84	19	45	64
10	60	82	20	55	70

7. Suppose the slope of the regression line for the data in Exercise 6 is 1.15 and the Y-intercept is 13.72.

a. What is the equation of the regression line for the data in Exercise 6?

b. Using the equation from part (a) and the data from Exercise 6, what is the predicted science final exam score for each student?

c. What is the standard error of estimate if we use the regression equation in part (a) for the data in Exercise 6?

8. Suppose the following information represents the average score on weekly history quizzes and the grade on the final exam for 10 students:

Student	Weekly Quizzes Average	History Final Exam Score	Student	Weekly Quizzes Average	History Final Exam Score
1	8	42	6	5	33
2	7	39	7	3	23
3	7	40	8	6	36
4	4	28	9	6	36
5	7	43	10	7	40

a. Calculate the slope of the regression line.

b. Calculate the Y-intercept of the regression line.

c. Use the regression equation to calculate the predicted final exam score of each student given her or his weekly quiz average.

d. What is the standard error of estimate for these data?

9. Suppose that Randrick must get at least a C− (34 or higher) on the final exam in history to assure that he will pass the course. Using the regression equation from Exercise 8 and the fact that Randrick's weekly quiz average is 6, answer the following:

a. What is the probability that Randrick will pass the course with at least a C−?

b. What is the probability that Randrick will obtain a score that is more than 3 points above or 3 points below the predicted value?

c. What is the range of scores that Randrick can be 95 percent certain will contain his score?

10. An educational psychologist is interested in high school students' attitudes toward their English teacher in an elective class that requires a formal level of thinking. Since the class is an elective class, students from any high school grade can enroll in the course. The educational psychologist reasons that students in the course who are not at the formal level of thinking will be frustrated, and this frustration will be reflected in their attitudes toward the English teacher. In general, cognitive development is thought to become increasingly more formal as the level of education increases. The following data are reported in pairs that indicate grade in high school (i.e., freshman = 1, sophomore = 2, etc.) and attitude toward the English teacher:

3,36 3,37 2,36 2,37 1,53 1,45 4,45 4,42 1,54
2,39 3,34 4,39 4,58 3,29 2,42 1,42 2,24 3,42
4,47 1,37 2,45 1,50 4,49 3,43 4,56 2,38 3,30
1,56 2,30 1,39 4,46 3,44 3,22 1,39 4,47 2,35
3,33 4,51 1,45 2,44

a. If we assume that grade in high school (i.e., freshman, sophomore, etc.) is a discrete variable, explain how the principle of least squares can be used to predict attitude scores for students who were not a part of this study.

b. Provide your least-squares prediction for each grade in high school.

c. What is the 90% confidence interval for sophomores?

d. Is it appropriate to use a regression line for the data provided above? Explain why or why not.

11. A researcher was interested in how well students' attitudes toward mathematics at the beginning of the school year predicted final exam scores in a ninth grade algebra course. He obtained the following results:

Student	Attitude Toward Mathematics (X)	Final Exam Score (Y)
1	59	114
2	58	114
3	45	100
4	52	100
5	46	72
6	45	102
7	60	120
8	36	70
9	44	64
10	49	76
11	43	84
12	41	68
13	60	120
14	46	76
15	49	94
16	42	80

Student	Attitude Toward Mathematics (X)	Final Exam Score (Y)
17	43	52
18	33	62
19	23	76
20	42	80
21	46	84
22	60	108
23	43	74
24	51	84
25	51	66
26	51	102
27	32	68
28	37	74
29	42	56
30	44	93

a. For the information provided in the table, construct a scatter plot with the attitude toward mathematics scores on the X-axis and the final exam scores on the Y-axis. Once the scatter plot has been constructed, sketch where you estimate the best fitting regression line should be placed.

b. Calculate the slope and intercept of the regression line. Write the regression equation.

c. At the beginning of the next school year, students were administered the Attitude Toward Mathematics scale. What will be the predicted final exam scores of three students whose Attitude Toward Mathematics scores were 26, 43, and 53?

d. What is the probability of a student getting a final exam score of 70 or below if that student has a score of 43 on the Attitude Toward Mathematics scale?

e. What is the probability of a student getting a final exam score of 100 or above if that student has a score of 26 on the Attitude Toward Mathematics scale?

12. Social phobia is defined in such a way that neuroticism appears to be a component of social phobia. Clinical researchers administered the Neuroticism subscale of the Eysenck Personality Inventory (NEPI) and the measure of social phobia (SP) to determine if NEPI would serve as an effective predictor variable of SP. A random sample of the data is presented below:

Participant	Neuroticism (X)	Social Phobia (Y)
1	6	11
2	4	12
3	19	22
4	8	10
5	5	12
6	9	17
7	9	15
8	6	13
9	15	15
10	14	23
11	10	15
12	11	17
13	20	26
14	19	18
15	20	26
16	16	22
17	9	14
18	6	14
19	17	20
20	14	23

a. For the scores listed above, calculate the slope of the regression line, the intercept of the regression line, and the standard error of measurement.

b. What is the regression equation?

13. A sports psychologist was interested in doing research on the relationship between anxiety and performance on a motor skill (e.g., free throw shooting). The following results were recorded for 15 participants:

Participant	Anxiety (X)	Number of Free Throws Made (Y)
1	56	8
2	41	11
3	32	15
4	51	10
5	24	16
6	35	13
7	21	18
8	49	14
9	44	12
10	33	9
11	29	13
12	48	7
13	59	10
14	39	12
15	38	15

a. For the information provided in the table above, construct a scatter plot with the anxiety scores on the X-axis and the number of free throws made on the Y-axis. Once the scatter plot has been constructed, sketch where you estimate the best fitting regression line should be.

b. Compute the equation of the regression line for predicting the number of free throws made from the anxiety score.

c. Given that Julie has an anxiety score of 46, how many free throws do you predict she will make?

d. What is the standard error of estimate for the data listed above?

e. What is the 95% confidence interval for Julie's predicted free throws made?

14. Use the SPSS program and SPSS data bank to find the scatter plot for each of the following. In addition, insert the regression line that best fits the data. [Note: For the variable postatt, the higher the score, the more negative the attitude toward math.]

a. The predictor variable (independent variable) post-instruction math attitude (postatt) and the criterion variable (dependent variable) final algebra grade point average (algpa) from the data file math.sav.

b. The predictor variable (independent variable) post-instruction math attitude (postatt) and the criterion variable (dependent variable) mathematics score on the Iowa Test of Educational Development for 9th Graders (ited9) from the data file math.sav.

c. The predictor variable (independent variable) confidence in doing mathematics (confmath) and the criterion variable (dependent variable) final grade in mathematics (fingrd) from the data file career.sav.

d. The predictor variable (independent variable) empathic concern (empathy) and the criterion variable (dependent variable) altruism (altruism) from the data file donate.sav.

15. Use the SPSS program and SPSS data bank to find the regression analysis and equation for each of the following.
   a. The predictor variable (independent variable) post-instruction math attitude (postatt) and the criterion variable (dependent variable) final algebra grade point average (algpa) from the data file math.sav.
   b. The predictor variable (independent variable) post-instruction math attitude (postatt) and the criterion variable (dependent variable) mathematics score on the Iowa Test of Educational Development for 9th Graders (ited9) from the data file math.sav.
   c. The predictor variable (independent variable) confidence in doing mathematics (confmath) and the criterion variable (dependent variable) final grade in mathematics (fingrd) from the data file career.sav.
   d. The predictor variable (independent variable) empathic concern (empathy) and the criterion variable (dependent variable) altruism (altruism) from the data file donate.sav.

Chapter 8

Accuracy of Prediction: Correlation

CHAPTER OBJECTIVES

After studying this chapter, you should be able to:

❏ Given data from a study with continuous variables,

- Compute the correlation coefficient.

- Determine B from r.

- Compute the standard error of estimate.

- Determine the predicted standard score and predicted raw score given the individual's raw or standard score on the predictor variable.

- Determine the prediction interval for a predicted score.

- Interpret r and r^2 in terms of the given study.

❏ Given a scatter plot, state the approximate value of the correlation coefficient.

Educational and psychological researchers often speak of two variables as being related. Academic aptitude scores are related to academic achievement scores, scores on a vocational interest measure are related to stated job satisfaction, and amount of study time is related to academic performance in statistics courses. Relationships between variables are so ubiquitous that there is a saying that when a butterfly flaps its wings in China, it eventually rains in New York. Every variable in the universe is related to every other variable in the universe, but some relationships are much stronger and more direct than others. What are the conditions that allow relationships to exist? Can a constant characteristic be related to something? (If you think about this carefully, you will realize that something that shows no variation cannot be related to anything else.) You know from your own experience that some variables predict future events much more accurately than others. How can we express the strength of a predictive relationship? How, for example, might Dr. Numbercruncher express the strength of the relationship between math anxiety scores and performance on the statistics final exam for her class at Big Time U?

Introduction

In the 1880s the English scientist Francis Galton (Galton, 1886[1]) was studying patterns of heredity between parents and children with regard to a wide variety of characteristics,

---

[1]Galton, F. (1886). Family likeness in stature. *Proceedings of the Royal Society of London, 40*, 42–73. The derivation of equation 8.1 was provided by Karl Pearson, one of the foremost English statisticians of the late 19th and early 20th centuries, in 1896.

including height, weight, and academic achievement. He noticed similarities between generations, for example, that taller than average parents tended to have taller than average children, and shorter than average parents usually had shorter than average children. As part of his research, he found he needed a way to express the degree of similarity between parents and their children on these variables. His answer to this problem has provided psychological and educational researchers with one of their most valuable research tools. He developed an index he called the coefficient of co-relation, which we now call the correlation coefficient. In this chapter we will examine how the correlation coefficient is developed and how it relates to the accuracy with which we can predict scores on one variable from scores on another.

Definition of the Correlation Coefficient

Galton used the positions of the two members of a pair of scores relative to the scores of other individuals on the same variables to develop the correlation coefficient. For example, in Galton's studies, a father's height and his son's height would make up a pair of scores. Similarly, the scores of BTU student Patty Alexander on the mathematics aptitude test and the mathematics anxiety scale would make up a pair of scores. In either case, we would like to know whether there is a tendency for high scores on one variable of the pair to go together with high scores for the other member of the pair. Do fathers who are above average in height tend to have sons who are also above average in height? Do people who are above average on math anxiety tend also to be above average on math aptitude; or, do they tend to be below average? In other words, is there a systematic pattern relating the position of one score to the position of the other?

In Chapter 7 we saw that the scatter plot provides one way to describe the relationship between two variables. We plotted one variable, called Y, on the ordinate, and a second variable, called X, on the abscissa. We will use the data for the 20 pairs of quiz scores from Table 7-3, reprinted in Table 8-1, to illustrate how we can develop an index of the strength of this relationship. The scatter plot is given in Figure 8-1.

We can see from Figure 8-1 that there is a generally positive relationship between the two sets of scores. People who scored higher on the first quiz also tended to score higher

TABLE 8-1	Raw Scores and Standard Scores for 20 Students on Two Quizzes			
Student	Quiz 1	Quiz 2	Z_1	Z_2
1	3	4	−.66	.64
2	4	4	.17	.64
3	6	5	1.82	1.44
4	4	2	.17	−.96
5	5	4	.99	.64
6	5	3	.99	−.16
7	5	4	.99	.64
8	4	4	.17	.64
9	2	3	−1.49	−.16
10	6	4	1.82	.64
11	3	2	−.66	−.96
12	3	1	−.66	−1.76
13	4	2	.17	−.96
14	3	3	−.66	−.16
15	2	2	−1.49	−.96
16	2	1	−1.49	−1.76
17	3	3	−.66	−.16
18	4	3	.17	−.16
19	3	4	−.66	.64
20	5	6	.99	2.24

$$\sum_{i=1}^{20} Z_{1_i} Z_{2_i} = 12.43$$

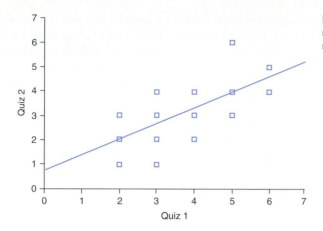

Figure 8-1 Scatter plot of scores for quiz 1 and quiz 2 showing the regression line.

on the second. Our question is, how strong is the relationship? Galton used *Z*-scores to provide an answer. He defined his index of co-relation as the mean of the cross products of *Z*-scores, and he gave his index the label *r*. The correlation between the two variables, *X* and *Y*, is therefore given by the equation

$$r_{YX} = \frac{\Sigma_i Z_{X_i} Z_{Y_i}}{N}$$

(8.1)

Using this formula, the correlation between *X* and *Y* is found by (1) multiplying each person's *Z*-score on *X* by his or her *Z*-score on *Y*, (2) summing these cross products across all people, and (3) dividing the result by *N*. Equation 8.1 is the definitional formula for the **correlation coefficient.**

The *Z*-scores corresponding to the raw scores on the two quizzes are shown in columns 4 and 5 of Table 8-1. These *Z*-scores are plotted in Figure 8-2, where the means are also shown. You can see that there is a general tendency for people who scored above the mean on quiz 1 to have also scored above the mean on quiz 2. Now, let us think about what information *Z*-scores contain.

We know that *Z*-scores are signed numbers and that the sign indicates whether the score is above or below the mean. When we multiply two *Z*-scores together, the sign of the product depends on whether the two *Z*-scores have the same sign. If both *Z*-scores are positive, the product is positive, and if both are negative, the product is also positive. However, if the two *Z*-scores have opposite signs, the product is negative. Therefore, *if there is a tendency for both Z-scores in a pair to be on the same side of the mean, either both above the mean or both below it, the sum of their products will be positive.* Not all individuals will have both of their *Z*-scores with the same sign, but most will. This tendency

*The **correlation coefficient** is defined as the mean of the cross products of Z-scores.*

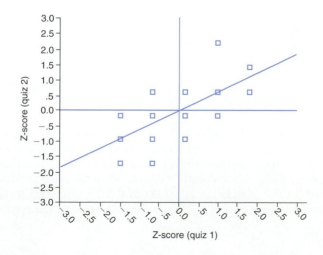

Figure 8-2 Scatter plot of *Z*-scores from Figure 8-1.

is what we mean by a **positive relationship.** Conversely, if there is a tendency for a person's Z-score on one variable to be positive when the other is negative, the sum of the products will be negative. This is what we mean by a **negative relationship.** The correlation coefficient will be a positive number if the relationship is positive, and it will be a negative number if the relationship is negative. The sign of the correlation coefficient gives the same information about direction of relationship that is conveyed by the sign of B, the slope of the regression line. For any given set of data, r and B will always have the same sign.

The sum of the cross products of the Z-scores on our two quizzes is given at the bottom of Table 8-1. Its value is $+12.43$. If we divide this quantity by 20, we find that the correlation between the two quizzes is $+.622$. The fact that the correlation is positive is consistent with our observations from the scatter plot.

Let's take a very simple set of pairs of scores and explore how the pattern of scores affects the correlation. We have two variables, X and Y, each of which can take any value between 1 and 5. Suppose we obtain the following results.

X	Y
1	1
2	2
3	3
4	4
5	5

Each person received exactly the same score on both variables. If you make a scatter plot of these data, the set of points will all fall on a straight line like this.

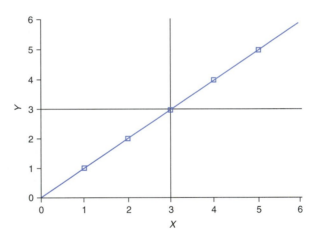

Now compute the Z-scores for each raw score and the products of those Z-scores. You should get the following values.

Z_X	Z_Y	$Z_X Z_Y$
-1.414	-1.414	2.00
$-.707$	$-.707$	.50
0	0	0
.707	.707	.50
1.414	1.414	2.00

The sum of cross products of the Z-scores ($\Sigma Z_X Z_Y$) is 5.00, and, according to equation 8.1, the correlation is $+1.00$. Each person's relative position on Y is identical to his or her relative position on X. What happens if we reverse the values for Z_Y, so that all of the positive scores become negative and vice versa? Of course, all of the cross products become negative, their sum is -5.00, the correlation is -1.00, and the scatter plot looks like this. In this case, each person's relative position on Y is the inverse of his or her relative position on X.

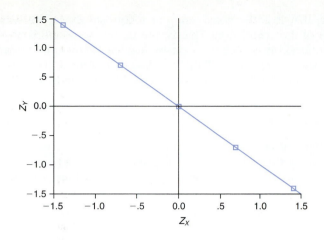

Now consider what happens if only some of the scores are switched, as follows. What happens to the correlation coefficients?

Z_X	Z_Y	$Z_X Z_Y$	Z_X	Z_Y	$Z_X Z_Y$
1.414	1.414	2.00	1.414	−1.414	−2.00
.707	−.707	−.50	.707	.707	.50
0	0	0	0	0	0
−.707	.707	−.50	−.707	−.707	.50
−1.414	−1.414	2.00	−1.414	1.414	−2.00
	$\sum Z_X Z_Y$	3.00		$\sum Z_X Z_Y$	−3.00

In both cases the correlation is less than perfect ($r = +.60$ and $r = -.60$) and, if you make a scatter plot, the points do not fall on a single straight line.

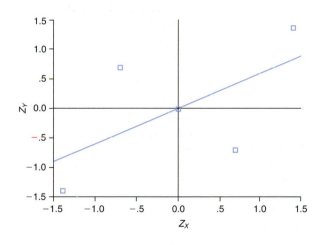

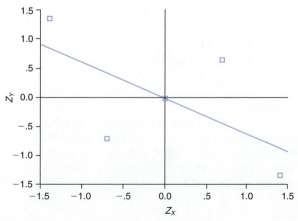

Notice, however, that it is the consistency or inconsistency in the extreme values that controls the sign of the correlation. That is, the farther someone is above or below the mean, the more effect consistency or inconsistency has on the correlation. We can expect people who are near the mean to fall more or less randomly above or below it, but those who are at the extremes of their distributions should keep their relative positions.

▶ *Now You Do It*

Now try your hand at computing a correlation. There are 12 pairs of test scores listed here. We have done the work of computing the Z-scores for you. Use equation 8.1 to compute the correlation. Also, prepare a scatter plot for these scores.

Test 1	Test 2	$Z_{test\ 1}$	$Z_{test\ 2}$
11	22	−1.33	.52
14	20	.38	−.10
15	17	.96	−1.03
15	14	.96	−1.96
12	23	−.76	.83
13	23	−.19	.83
10	25	−1.91	1.45
12	22	−.76	.52
14	18	.38	−.72
16	17	1.53	−1.03
15	24	.96	1.14
13	19	−.19	−.41

You should get $\sum_i Z_{X_i} Z_{Y_i} = -8.23$ and a correlation of −.686. The scatter plot should look like this. Notice that the scores tend to run from the upper left of the plot to the lower right.

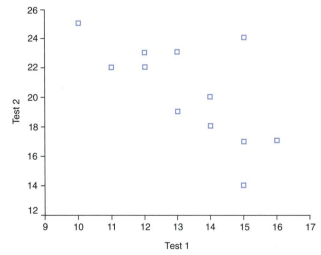

Computational Formulas for the Correlation Coefficient

Deviation Score Formula

In Chapter 4 we saw that it is possible to develop formulas that are easier to use in computation than the definitional formulas are. This is particularly true for the correlation coefficient. The definitional formula requires that we first compute Z-scores, and Z-scores have decimals and minus signs, both of which make calculations more difficult and more error prone. Starting with equation 8.1,

$$r_{YX} = \frac{\sum_i Z_i Z_i}{N}$$

we can substitute for the Z's to get

$$r_{YX} = \frac{\Sigma \left(\frac{x_i}{S_X}\right)\left(\frac{y_i}{S_Y}\right)}{N}$$

$$r_{YX} = \frac{\Sigma x_i y_i}{N S_Y S_X} \tag{8.2}$$

In this equation x_i and y_i are deviation scores. Equation 8.2 is known as the deviation score formula for the correlation coefficient.

Raw Score Formula

We can move from equation 8.2 to a raw score formula for r by substituting raw-score equivalents for each of its terms and by performing some algebraic sleight of hand to collect and simplify terms. These changes lead us to the following raw score equation for r (all summations are from 1 to N):

$$r_{YX} = \frac{N[\Sigma(X_i Y_i)] - (\Sigma X_i)(\Sigma Y_i)}{\sqrt{N\Sigma X_i^2 - (\Sigma X_i)^2} \sqrt{N\Sigma Y_i^2 - (\Sigma Y_i)^2}} \tag{8.3}$$

While equation 8.3 looks imposing, this formula is not difficult to use, and because it involves no decimals or negative numbers and much less rounding of numbers, it gives better results from raw data than either equations 8.1 or 8.2. We have encountered all of the terms in the equation before. The numerator is identical to the numerator for B in equation 7.4, and the denominator terms come from the raw score formulas for the standard deviations of X and Y. Formula 8.3 tells us to

1. Multiply each person's raw score on Y by his or her raw score on X and sum the resulting cross products to get $\Sigma X_i Y_i$.
2. Multiply the sum of cross products by N.
3. Find the sum of the Xs and the sum of the Ys and multiply these two sums together.
4. Subtract the quantity in step 3 from the quantity in step 2. This gives the value in the numerator.
5. Square each person's score on X, sum the squared scores, and multiply the result by N.
6. Find the sum of the X-scores, square it, and subtract the result from the quantity in step 5.
7. Take the positive square root of the quantity in step 6.
8. Repeat steps 5–7 for the Y-scores.
9. Find the product of the quantity in step 7 and the quantity in step 8.
10. Divide the quantity in step 4 by the quantity in step 9.

We will use the 20 pairs of quiz scores from Table 8-1 to illustrate the use of all three equations for computing r. The raw scores are in the X and Y columns of Table 8-2. These are followed by the deviation scores x and y and the standard scores. Then the cross products $X_i Y_i$, $x_i y_i$, and $Z_{X_i} Z_{Y_i}$ are given in the last three columns. Finally, the necessary sums and sums of squares are shown at the bottom of the table. The sums of cross products are given at the bottom of the cross products columns. Cross products are computed and summation takes place for all N individuals. S_x and S_y, computed by equation 4.12, are 1.21 and 1.25 respectively.

Selecting the appropriate values from the table, we find that equation 8.1 yields an r of

$$r_{YX} = \frac{\Sigma Z_{X_i} Z_{Y_i}}{N} = \frac{12.43}{20} = +.622$$

This can be compared to the value that we get from applying equation 8.2,

$$r_{YX} = \frac{\Sigma x_i y_i}{N S_X S_Y} = \frac{18.80}{20(1.21)(1.25)} = \frac{18.80}{30.25} = +.621$$

TABLE 8-2 Quantities Needed to Compute the Correlation Coefficient for 20 Quiz Scores Using the Raw Score, Deviation Score, and Definitional Formulas

	X	Y	x	y	Z_x	Z_y	XY	xy	$Z_X Z_Y$
1	3	4	−.80	.80	−.66	.64	12.00	−.64	−.42
2	4	4	.20	.80	.17	.64	16.00	.16	.11
3	6	5	2.20	1.80	1.82	1.44	30.00	3.96	2.62
4	4	2	.20	−1.20	.17	−.96	8.00	−.24	−.16
5	5	4	1.20	.80	.99	.64	20.00	.96	.63
6	5	3	1.20	−.20	.99	−.16	15.00	−.24	−.16
7	5	4	1.20	.80	.99	.64	20.00	.96	.63
8	4	4	.20	.80	.17	.64	16.00	.16	.11
9	2	3	−1.80	−.20	−1.49	−.16	6.00	.36	.24
10	6	4	2.20	.80	1.82	.64	24.00	1.76	1.16
11	3	2	−.80	−1.20	−.66	−.96	6.00	.96	.63
12	3	1	−.80	−2.20	−.66	−1.76	3.00	1.76	1.16
13	4	2	.20	−1.20	.17	−.96	8.00	−.24	−.16
14	3	3	−.80	−.20	−.66	−.16	9.00	.16	.11
15	2	2	−1.80	−1.20	−1.49	−.96	4.00	2.16	1.43
16	2	1	−1.80	−2.20	−1.49	−1.76	2.00	3.96	2.62
17	3	3	−.80	−.20	−.66	−.16	9.00	.16	.11
18	4	3	.20	−.20	.17	−.16	12.00	−.04	−.03
19	3	4	−.80	.80	−.66	.64	12.00	−.64	−.42
20	5	6	.1.20	2.80	.99	2.24	30.00	3.36	2.22
Sums	76	64	0.00	0.00	0.00	0.00	262	18.80	12.43
Sum of Squares	318	236							

Finally, using equation 8.3, we find r to be

$$r_{YX} = \frac{N[\Sigma(X_i Y_i)] - (\Sigma X_i)(\Sigma Y_i)}{\sqrt{N\Sigma X_i^2 - (\Sigma X_i)^2}\sqrt{N\Sigma Y_i^2 - (\Sigma Y_i)^2}}$$

$$= \frac{20(262) - 76(64)}{\sqrt{20(318) - (76)^2}\sqrt{20(236) - (64)^2}} = \frac{376}{\sqrt{584}\sqrt{624}}$$

$$= +.623.$$

The three values are virtually identical; the differences are due to errors introduced by rounding in the calculation of the deviation scores and Z-scores. That is, the value +.623 is the most accurate because rounding errors can only occur in the square root and division steps of the computation.

Of the three examples, the amount of computation appears to be greatest in the third. However, the raw score equation (8.3) uses only columns 1, 2, and 7 from Table 8-2, with only the cross product column requiring any computation. Either of the other two formulas requires five columns of the table, three of which involve computation. Therefore, even if the other two formulas did not involve rounding error, they require enough additional intermediate computation to make them more time-consuming and cumbersome than equation 8.3.

◆ *Now You Do It*

Computing correlations by hand can be quite a chore, and you may never do it yourself again, but it is often worthwhile to go through the labor of such computations once to see how the pieces are generated and how they fit together. Why don't you choose one of the correlations involving college GPA (X), final exam score (Y), and math anxiety (Z) from the data gathered at Big Time U to test yourself. The quantities you should get are

$$\Sigma XY = 5883.43 \qquad \Sigma XZ = 2745.53 \qquad \Sigma YZ = 40,675$$
$$\Sigma X = 154.29 \qquad \Sigma Y = 2328 \qquad \Sigma Z = 1270$$
$$\Sigma X^2 = 392.55 \qquad \Sigma Y^2 = 92,056 \qquad \Sigma Z^2 = 25,842$$

Putting these values into formula 8.3 correctly should give you the correlations $r_{XY} = +.830$, $r_{XZ} = .453$, $r_{YZ} = -.547$. Notice that the correlation between math anxiety and the other two variables is negative. What does this mean? Is it something you might expect?

HOW THE COMPUTER DOES IT

All statistics packages have programs to compute correlation coefficients of various kinds. The one we have been discussing is known as the Pearson product moment correlation, abbreviated PMC, or sometimes PPMC. It is the appropriate coefficient to use when both variables are measured on an interval or ratio scale and have approximately normal distributions. To obtain a PMC from SPSS, simply select the Correlate procedure from the STATISTICS menu and click on "Bivariate." You should get a screen that looks like this:

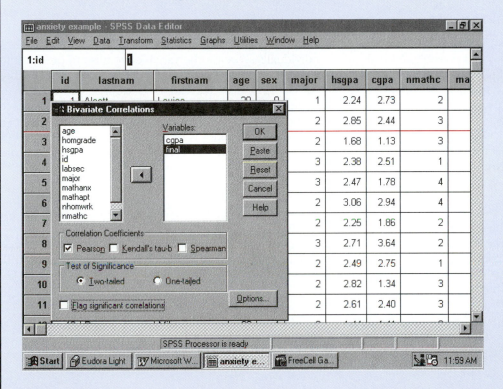

Select the variables you wish to correlate and move them into the Variables box. The Options button allows you to select output of means and standard deviations as well as correlations. Then click OK.

You can find the correlations between several variables at the same time using a computer. The output usually is arranged in a special table called a **correlation matrix.** Each row of the matrix refers to one of the variables you have selected, as does each column. Thus, to find the correlation between any pair of variables, simply go to the cell that has one variable as its row and the other variable as its column. The matrix of correlations between college GPA, math anxiety, and statistics final exam score looks like this. Note that the values in the diagonal of the matrix are 1.00 because each variable correlates perfectly with itself. Also, the correlations below the diagonal mirror the ones above because the correlation of X with Y is the same as the correlation of Y with X.

	CGPA	Final	Anxiety
College Grade Point Average	1.000	.830	−.453
Score on Stat Final	.830	1.000	−.547
Math Anxiety Test Score	−.453	−.547	1.000

The Correlation Coefficient as an Index of Strength of Prediction

The correlation coefficient is perhaps most widely used to tell us how much of the variance in one variable, for example, score on the statistics final at BTU, can be predicted from another variable, such as math anxiety score. To see how this happens, we must return to our discussion of the regression line from Chapter 7. There we saw that the regression line provides a least-squares prediction of scores on one variable, Y, from scores on another variable, X. The relationship has the general form

$$\hat{Y}_i = BX_i + A$$

where $\hat{Y}_i$ is the predicted Y-score for person i. The values of B and A are determined so that the sum of squared errors of prediction,

$$\sum_{i=1}^{N} (Y_i - \hat{Y}_i)^2$$

is minimized.

It is possible to break up an individual's score on the variable we wish to predict into three parts. Look at the scatter plot of our 20 pairs of quiz scores in Figure 8-3. The Y-score of the last individual (Y_{20}, call him George) is 6. We can view George's score of 6 as composed of a portion due to $\overline{Y}$, the mean of all Y's; a portion due to the difference between his predicted score ($\hat{Y}_{20}$) and $\overline{Y}$ ($\hat{Y}_{20} - \overline{Y}$); and a portion that is the difference between his observed score and his predicted score ($Y_{20} - \hat{Y}_{20}$). The last part is the error remaining after we have made our best prediction. We can now express George's score symbolically as

$$Y_{20} = \overline{Y} + (\hat{Y}_{20} - \overline{Y}) + (Y_{20} - \hat{Y}_{20})$$

George had a score of 5 on quiz 1. The slope and intercept of the regression line for these data are $B = +.644$ and $A = +.753$. Therefore, his predicted score on quiz 2 is 3.97. The mean of all 20 scores for this group on quiz 2 is 3.2. Therefore, we can view George's score of 6 on quiz 2 as composed of an overall mean effect of 3.2, plus the deviation of his predicted score from the overall mean of $+.77$, plus an error of prediction of $+2.03$. That is,

$$6 = 3.2 + (3.97 - 3.2) + (6 - 3.97)$$
$$6 = 3.2 + (.77) + (2.03)$$

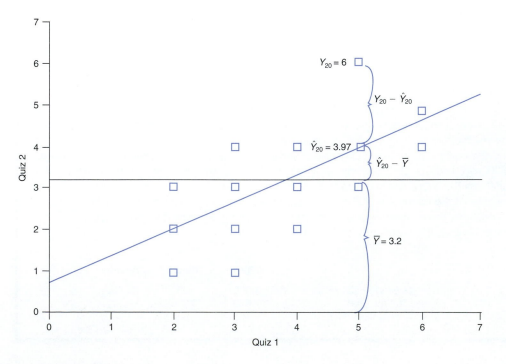

Figure 8-3 Scatter plot of 20 quiz scores showing portions of a score for individual 20. The observed score (Y_i) is composed of the overall mean ($\overline{Y}$), the deviation of the predicted score ($\hat{Y}_i$) from $\overline{Y}$, and the deviation of the observed score (Y_i) from $\hat{Y}_i$.

Each score in Figure 8-3 can viewed in the same way. For example, the third person (call her Carol) received a score on quiz 2 of 5, and, because her score on quiz 1 was 6, her predicted score is 4.62. Therefore, Carol's score of 5 can be seen as composed of the overall mean (3.2) plus the difference between her predicted score and $\overline{Y}$ (4.62 − 3.2) plus the remaining error (5 − 4.62).

$$Y_3 = 3.2 + (4.62 - 3.2) + (5 - 4.62)$$
$$Y_3 = 3.2 + (+1.42) + (+.38)$$

Remember from our discussion in Chapter 7 that the mean is our least-squares prediction for a person if we know nothing about her or him. Therefore, George's deviation from the overall mean (+2.80) is the error we would make in guessing his score on quiz 2 if we did not know his score on quiz 1. This original error can be broken down into two pieces, a piece which is knowable from knowing the regression line, and a piece which remains in error even after we have used the information from the regression line. Subtracting $\overline{Y}$ from both sides of the equation to express George's score as a deviation from the overall mean yields

$$(Y_{20} - \overline{Y}) = (\hat{Y}_{20} - \overline{Y}) + (Y_{20} - \hat{Y}_{20}) \qquad (8.4)$$
$$(6 - 3.2) = (3.97 - 3.2) + (6 - 3.97)$$
$$2.80 = (.77) + (2.03)$$

This shows us that George's deviation from the mean ($Y_{20} - \overline{Y} = 2.80$) is composed of a portion that is due to the difference between his predicted score ($\hat{Y}_{20}$) and $\overline{Y}$ (.77) and a portion that is due to error of prediction (2.03). Carol's deviation of +1.8 from $\overline{Y}$ is composed of a portion that is predictable using the regression line (4.62 − 3.2 = 1.42) and a remaining error of .38. Table 8-3 lists the quiz 2 scores for each of the 20 people in Table 8-1, along with their deviations from the mean ($Y - \overline{Y}$), their predicted score ($\hat{Y}$), the deviation of their predicted score from the mean ($\hat{Y} - \overline{Y}$), and the deviation of their observed score from their predicted score ($Y - \hat{Y}$).

	TABLE 8-3	**Components of the Sums of Squares**			
	Quiz 2	$Y - \overline{Y}$	$\hat{Y}$	$\hat{Y} - \overline{Y}$	$Y - \hat{Y}$
1	4	.80	2.69	−.51	1.31
2	4	.80	3.33	.13	.67
3	5	1.80	4.62	1.42	.38
4	2	−1.20	3.33	.13	−1.33
5	4	.80	3.97	.77	.03
6	3	−.20	3.97	.77	−.97
7	4	.80	3.97	.77	.03
8	4	.80	3.33	.13	.67
9	3	−.20	2.04	−1.16	.96
10	4	.80	4.62	1.42	−.62
11	2	−1.20	2.69	−.51	−.69
12	1	−2.20	2.69	−.51	−1.69
13	2	−1.20	3.33	.13	−1.33
14	3	−.20	2.69	−.51	.31
15	2	−1.20	2.04	−1.16	−.04
16	1	−2.20	2.04	−1.16	−1.04
17	3	−.20	2.69	−.51	.31
18	3	−.20	3.33	.13	−.33
19	4	.80	2.69	−.51	1.31
20	6	2.80	3.97	.77	2.03
Sum	64	0.00	64	0.00	0.00
Sum of Squares		31.20		12.10	19.10
Variance		1.56		.60	.96
Standard Deviation		1.25		.77	.98

Let's use the relationship between the math anxiety test score and the score on the statistics final at Big Time U to review the parts of a score. The mean final exam score was 34.24, and in Chapter 7 we found that the regression equation for predicting final exam score from math anxiety test score was

$$\hat{Y} = -1.32X + 58.90$$

Louisa Alcott and Sue Embretson both had scores of 14 on the math anxiety test. Louisa's score on the final was 37, and Sue earned a 44. What is each student's predicted score? What are their deviations from the mean? How much does the error of prediction change when we add the math anxiety test as a predictor? You should get the following answers:

	Y	$\hat{Y}$	$Y - \bar{Y}$	$\hat{Y} - \bar{Y}$	$Y - \hat{Y}$
Louisa	37	40.42	2.76	6.18	−3.42
Sue	44	40.42	9.76	6.18	+3.58

Notice that each student receives the same predicted score because their scores on the math anxiety scale were the same. They also have the same effect due to prediction $(\hat{Y} - \bar{Y})$. It is in the remaining error $(Y - \hat{Y})$ that they differ.

To see how breaking each person's deviation from the mean into two parts, one that is predictable from X and one that remains error of prediction, allows us to express the strength of a relationship, we must convert equation 8.4 into variance-like terms. It can be shown that by squaring equation 8.4 and summing across the N individuals, we obtain

$$\sum_{i=1}^{N} (Y_i - \bar{Y})^2 = \sum_{i=1}^{N} (\hat{Y}_i - \bar{Y})^2 + \sum_{i=1}^{N} (Y_i - \hat{Y}_i)^2 \qquad (8.5)$$

The term on the left of the equals sign is the sum of squared deviations of observed scores in this distribution from the mean. We discussed this sum of squares in Chapter 4. It is called the **total sum of squares**, or SS_T, and for these data it has a numerical value of 31.20.

> *The **total sum of squares** is the sum of squared deviations of observed scores from the overall mean.*

Both of the other terms are also sums of squared deviations. The first is the sum of squared deviations of the predicted scores from the mean. This term contains the individual differences in Y that are predicatable from X. It is called the **sum of squares due to prediction**, or SS_P, and has a value in these data of 12.10. The other term is the sum of squared deviations of observed scores from their predicted values. This term contains the uncertainty about a person's score that remains after we have used the information in X. It is called the **sum of squares due to error**, or SS_E, and has the value 19.10 in these data. Notice that

> *The **sum of squares due to prediction** is the sum of squared deviations of the predicted scores from the overall mean.*

$$SS_T = SS_P + SS_E$$
$$31.2 = 12.1 + 19.1$$

> *The **sum of squares due to error** is the sum of squared deviations of the observed scores from their predicted scores.*

Dividing the total sum of squares into a piece that results from predictions and a piece that contains the remaining error is called partitioning the sum of squares.

Dividing the sums of squares on both sides of equation 8.5 by N, we have an equation with three variance terms.

$$\frac{\sum (Y_i - \bar{Y})^2}{N} = \frac{\sum (\hat{Y} - \bar{Y})^2}{N} + \frac{\sum (Y_i - \hat{Y}_i)^2}{N} \qquad (8.6)$$

> *The **predictable variance**, $S_{\hat{Y}}^2$ is the variance in Y that is predictable from knowing a person's score on X using the least-squares regression equation.*

This equation shows that, like SS_T, the total variance of Y-scores $\sum (Y_i - \bar{Y})^2/N$ can be divided into a portion of variance that is due to prediction from the regression line, called the **predictable variance**, $\sum (\hat{Y}_i - \bar{Y})^2/N$, and a portion that remains error after we have used our predictor information, called the **variance error of estimate**, $\sum (Y_i - \hat{Y}_i)^2/N$. For the set of quiz scores in Table 8-3 we get 1.56 for the overall variance, .60 for the predictable variance, and .96 for the error variance. Writing equation 8.6 in terms of the variances, we get

$$S_Y^2 = S_{\hat{Y}}^2 + S_{Y \cdot X}^2 \tag{8.7}$$

$$1.56 = .60 + .96$$

In this equation $S_{Y \cdot X}^2$ is the same variance-error-of-estimate term that we encountered in Chapter 7, and $S_{\hat{Y}}^2$ is the variance in Y that is predictable from knowing a person's score on X using the least-squares regression equation. $S_{\hat{Y}}^2$ is an index of the amount that our error in predicting Y is reduced by knowing X and applying the least-squares regression equation.

▶ Now You Do It

Let's take another look at the partitioning of a score using the 12 test scores from Now You Do It on page 204. Use test 1 to predict scores on test 2. You should get a regression equation of $\hat{Y} = -1.264X + 37.18$. The mean of test 2 is 20.33. Next compute the deviations of the raw scores from the mean; then compute the predicted scores. When you have these values for each of the 12 students, calculate the predicted deviation from the mean for each person ($\hat{Y} - \overline{Y}$) and the error of prediction ($Y - \hat{Y}$). Next, find SS_T, SS_P, and SS_E. Your results should match the table shown below. You should confirm that $SS_T = SS_P + SS_E$.

	Test 1	Test 2	$Y - \overline{Y}$	$\hat{Y}$	$\hat{Y} - \overline{Y}$	$Y - \hat{Y}$
1	11	22	1.67	23.28	2.95	−1.28
2	14	20	−.33	19.49	−.84	.51
3	15	17	−3.33	18.23	−2.10	−1.23
4	15	14	−6.33	18.23	−2.10	−4.23
5	12	23	2.67	22.02	1.69	.98
6	13	23	2.67	20.75	.42	2.25
7	10	25	4.67	24.55	4.22	.45
8	12	22	1.67	22.02	1.69	−.02
9	14	18	−2.33	19.49	−.84	−1.49
10	16	17	−3.33	16.96	−3.37	.04
11	15	24	3.67	18.23	−2.10	5.77
12	13	19	−1.33	20.75	.42	−1.75
Sums of Squares			124.66		58.55	66.12

Equation 8.7 shows two important things. First, the total variation in a set of scores can be broken down into a portion that is predictable from another variable and a portion that is independent of that other variable. The total individual differences in scores on quiz 2 (the variance) is 1.56 units, and this can be broken down into .60 units of differences that can be predicted by knowing the scores people earned on quiz 1, and .96 units of differences that remains unexplained by quiz 1. A portion of the individual differences between students in their scores on quiz 2 is predictable from their individual differences on quiz 1. This is the portion of quiz 2 that is accounted for by the regression line. Second, both $S_{\hat{Y}}^2$ and $S_{Y \cdot X}^2$ must be less than or equal to $S_{\hat{Y}}^2$ because a variance can never be negative. Clearly, it is not possible for more than all of the variance in Y to be predictable from X. Neither is it possible for more than all of the variance to be error.

The last thing we must do to get an index of how well one variable can be predicted from another is to convert equation 8.7 into a proportion so that we have an index that does not depend directly on the size of S_Y^2. Dividing both sides of equation 8.7 by S_Y^2 yields

$$\frac{S_Y^2}{S_Y^2} = 1 = \frac{S_{\hat{Y}}^2}{S_Y^2} + \frac{S_{Y \cdot X}^2}{S_Y^2} \tag{8.8}$$

The first term on the right, $S_{\hat{Y}}^2/S_Y^2$, is the proportion of total variance predictable from X, and the second term is the proportion that remains error. These are individual differences that cannot be predicted from X. Solving for the first term, we obtain

$$\frac{S_{\hat{Y}}^2}{S_Y^2} = 1 - \frac{S_{Y \cdot X}^2}{S_Y^2}$$

The left-hand term in this equation is given the label r^2 (r-squared) because, as we shall see shortly, it is the square of the correlation coefficient from equation 8.1. Therefore,

$$\left.\begin{array}{c} r_{YX}^2 = \dfrac{S_{\hat{Y}}^2}{S_Y^2} \\[2ex] r_{YX}^2 = 1 - \dfrac{S_{Y \cdot X}^2}{S_Y^2} \end{array}\right\}$$

(8.9)

The **square of the correlation coefficient (r^2)** is the proportion of variance in Y-scores that is linearly related to X or is linearly predictable from X.

These equations assert that the **square of the correlation coefficient *is the proportion of variance in Y-scores that is linearly related to X*** or is linearly predictable from X. The squared correlation coefficient is often called the *coefficient of determination* to reflect the fact that it reveals the extent to which scores on one variable can be viewed as being determined by the other variable. Because $S_{\hat{Y}}^2$ must be no greater than S_Y^2 (you can't predict more than all of the individual differences), and $S_{Y \cdot X}^2$ also must be no greater than S_Y^2, r^2 can take on values between zero (when $S_{Y \cdot X}^2 = S_Y^2$ and $S_{\hat{Y}}^2 = 0$) and 1.00 (when $S_{\hat{Y}}^2 = S_Y^2$ and $S_{Y \cdot X}^2 = 0$).

The fact that r^2 is limited to values between zero and 1.00 means that the correlation coefficient itself can only have values between -1.00 and $+1.00$. We know that a correlation can be positive or negative from our discussion surrounding equation 8.1, but because r^2 cannot exceed 1.00, the absolute value of r cannot exceed 1.00 either.

Let's take a look at how these equations work using the 20 pairs of scores in Table 8-1. The regression equation for predicting quiz 2 from quiz 1 was $\hat{Y} = .644X + .753$. The pairs of raw scores, the predicted scores, and the errors of prediction are shown in Table 8-4.

		Quiz 1	Quiz 2	$\hat{Y}$	$Y - \hat{Y}$
1		3	4	2.69	1.31
2		4	4	3.33	.67
3		6	5	4.62	.38
4		4	2	3.33	−1.33
5		5	4	3.97	.03
6		5	3	3.97	−.97
7		5	4	3.97	.03
8		4	4	3.33	.67
9		2	3	2.04	.96
10		6	4	4.62	−.62
11		3	2	2.69	−.69
12		3	1	2.69	−1.69
13		4	2	3.33	−1.33
14		3	3	2.69	.31
15		2	2	2.04	−.04
16		2	1	2.04	−1.04
17		3	3	2.69	.31
18		4	3	3.33	−.33
19		3	4	2.69	1.31
20		5	6	3.97	2.03
Sums		76	64	64.0	0.0
Sums of Squares		318	236	217.08	19.10
Variance					.96
Standard Deviation					.98

TABLE 8-4   Raw Scores, Predicted Scores, and Errors of Prediction for 20 Pairs of Scores from Table 8-1

There are some interesting features in Table 8-4. First, notice that the sum of scores for quiz 2 is 64, and so is the sum of predicted scores. This will always be true, and this means that the mean predicted score will always equal the mean observed score. Second, the sum of squared predicted scores plus the sum of squared errors is 236.18, which differs from the total sum of squared scores for quiz 2 only by rounding error. This too will always be true. It is a consequence of equation 8.5. Now, if you compute the variances of the quiz 2 scores, the predicted scores, and the errors, you will get

$$S_Y^2 = 1.56$$

$$S_{\hat{Y}}^2 = .60$$

$$S_{Y \cdot X}^2 = .96$$

which is what we found in Table 8-3. When we compute the ratios in equation 8.9, we get

$$r^2 = \frac{S_{\hat{Y}}^2}{S_Y^2} = \frac{.60}{1.56} = .385$$

$$r^2 = 1 - \frac{S_{Y \cdot X}^2}{S_Y^2} = 1 - .615 = .385$$

Both equations tell us that r^2 is .385, and r is $\sqrt{.385} = .620$. This, of course, is the value we got using equation 8.3, the raw score equation for r, within a rounding error of .003. If you compute the correlation using equation 8.3, you need only to square it to find the proportion of variance that X accounts for in Y. For example, in Now You Do It on page 206, we found the correlation between college GPA and score on the statistics final to be .83. By simply squaring this value, we learn that 69 percent of the variance in final exam score is predictable from college grade point average.

▶ *Now You Do It*

Let's use the sums of squares that we found in the previous Now You Do It to confirm that $r_{YX}^2 = S_{\hat{Y}}^2/S_Y^2$ and $r_{YX}^2 = 1 - (S_{Y \cdot X}^2/S_Y^2)$. You should have found that $SS_T = 124.66$, $SS_P = 58.55$, and $SS_E = 66.12$. Dividing each of these terms by $N(=12)$ gives us variances of $S_Y^2 = 10.39$, $S_{\hat{Y}}^2 = 4.88$, and $S_{Y \cdot X}^2 = 5.51$. For the first equation you should get

$$r_{YX}^2 = \frac{4.88}{10.39} = .4697$$

and for the second equation you should get

$$r_{YX}^2 = 1 - \frac{5.51}{10.39} = .4697$$

Taking the square root of either result gives us .685, which is (within rounding error) the correlation we found between these two sets of quiz scores in Now You Do It on page 204. Note, however, that we do not know from the ratios we computed here whether the correlation is positive or negative. We need to do the computations using equation 8.3 to find the sign, which in this case is negative.

The Correlation Coefficient Can Be Interpreted as Degree of Shared Variance

Another useful way to view the correlation coefficient is as an index of the degree of shared variance, or overlap, between two variables. If you think of each variable as a circle, with the circle representing the variance, two variables are correlated if their circles overlap. This way of viewing correlation is illustrated in Figure 8-4.

In Figure 8-4 the circles are the same size, indicating that the variances of the two variables are equal. This feature reflects the fact that correlation always refers to variables in standard score form. Therefore, the area of each circle is 1.00. The area where the circles overlap is labeled r^2, and the areas that do not overlap are labeled $1 - r^2$.

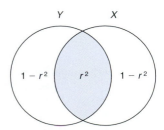

Figure 8-4 The square of the correlation coefficient can be viewed as the degree of overlap between the two variables.

This way of viewing the correlation between two variables is particularly useful when we are not thinking about predicting one variable from another, but only about the degree of relationship. If two variables are unrelated, their circles do not overlap, and they share no variance. On the other hand, two variables that correlate perfectly would overlap completely, indicating that they are the same variable. Degrees of overlap between these two extremes indicate varying levels of less than perfect correlation, but some degree of association.

The Correlation Coefficient Can Be Used to Make Predictions

The regression line that we developed in Chapter 7 for making least-squares predictions is appropriate for raw scores. The value of B is the slope of the regression line, and the intercept is A. Now, let us consider the regression line for standard (Z) scores.

The definitional formula for B (equation 7.3) was

$$B_{Y \cdot X} = \frac{\Sigma (Y_i - \overline{Y})(X_i - \overline{X})}{\Sigma (X_i - \overline{X})^2}$$

which we can rewrite as deviation scores ($y_i = Y_i - \overline{Y}$ and $x_i = X_i - \overline{X}$), giving us

$$B_{Y \cdot X} = \frac{\Sigma y_i x_i}{\Sigma x_i^2}$$

Standard scores are a special kind of deviation score because each Z-score is a deviation from the mean of its distribution. All deviation-score distributions have a mean of zero, and standard scores fulfill that condition; however, Z-scores have the additional property that $S_Z^2 = 1$ for all standard-score distributions. If we insert standard scores (Z_X and Z_Y) into the above equation for x_i and y_i, we obtain

$$B_{Z_Y Z_X} = \frac{\sum_{i=1}^{N} Z_{X_i} Z_{Y_i}}{\sum_{i=1}^{N} Z_{X_i}^2}$$

This equation gives us the slope of the regression line for standard scores. Notice that we have changed the subscripts on B to reflect this fact.

It can be shown that

$$\sum_{i=1}^{N} Z_{X_i}^2 = N$$

Therefore, we can rewrite the equation for the *slope of the least-squares regression line for standard scores* as

$$B_{Z_Y \cdot Z_X} = \frac{\sum_{i=1}^{N} Z_{Y_i} Z_{X_i}}{N}$$

This is the definition we gave for the correlation coefficient in equation 8.1. Therefore, the correlation coefficient, r, is the slope of the least-squares regression line for predicting standard scores on Y from standard scores on X.

The **correlation coefficient** has all the properties of a slope coefficient and, because the variables are in standard score form, some additional ones as well. For example, recall that there are two different slopes, $B_{Y \cdot X}$ and $B_{X \cdot Y}$, for raw or deviation scores. The slope for Z-scores is

$$r_{YX} = \frac{\Sigma Z_{Y_i} Z_{X_i}}{N}$$

The numerator of the right-hand term is the same regardless of which variable is doing the predicting, and so is the denominator. This means that for a given set of data, there is only one correlation coefficient, whereas there are two regression slopes for raw scores.

*The **correlation coefficient** (**r**) is the slope of the least-squares regression line for standard scores.*

There are two regression slopes for raw scores, but there is only one correlation coefficient.

The reason for this difference is shown in Figure 8-5. When raw scores are put in a scatter plot, the shape of the scatter plot is affected by the variances of the two variables. Part (a) illustrates the situation when the predictor has the larger variance, for example, predicting grade point averages (smaller variance) from final exam scores at Big Time U (larger variance) or attitudes from income for Dr. Cross's data on the Searchers. A typical scatter plot when the criterion variable has a larger variance than the predictor is shown in part (b). This is what we would get if we were predicting final exam scores from grade point average or income from attitudes. The case for standard scores is shown in part (c); here, because the variables are in the form of Z-scores, they have equal variances, and the scatter plot is the same regardless of which variable is the predictor.

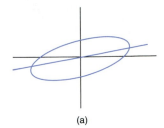

(a)

The correlation coefficient can be used just like B in a raw score regression equation to make predictions. The standard score regression equation is

$$\hat{Z}_{Y_i} = r_{YX} Z_{X_i} \qquad (8.10)$$

As we noted before, the sign of r has the same interpretation as the sign of B, and the magnitude of r is also the number of units of increase in $Z_{\hat{Y}}$ for each unit of increase in Z_X.

We can illustrate the equivalence of predictions using the raw score regression equation and the correlation coefficient with the data from Table 8-1. There we had scores for 20 students on two quizzes. Our last student, George, had a score of 5 on quiz 1. The raw score regression equation was

$$\hat{Y} = .644X + .753$$

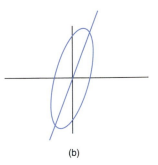

(b)

This yields a predicted score for George of 3.97. Using the mean (3.8) and standard deviation (1.21) of quiz 1, we find a Z-score for George of $+.99$. The correlation of quiz 1 with quiz 2 is .623. Inserting these two values into equation 8.10, we obtain

$$\hat{Z}_Y = .623(+.99) = +.617$$

That is, George's predicted Z-score on quiz 2 ($\hat{Z}_Y$) is $+.617$. If we now convert this Z-score back into the scale of the original quiz 2 scores using the mean (3.2) and standard deviation (1.25) of quiz 2, we get

$$\hat{Y} = 1.25(+.617) + 3.2 = 3.97$$

As you can see, if we convert the predictor to the standard score scale, apply equation 8.10, and then convert the resulting predicted standard score back into the scale of the criterion variable, we get the same result that we would get using the raw score equation. However, because it is seldom convenient to convert observed scores to Z-scores, equation 8.10 is not ordinarily used to make predictions.

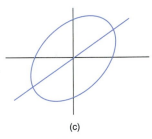

(c)

Figure 8-5 Which variable is the predictor makes a difference in the slope of the regression line for raw or deviation scores (a) and (b), but not for standard scores (c).

▶ *Now You Do It*

Let's use the set of 12 scores from Now You Do It on page 204 to try equation 8.10 out another time. The fifth person, call her Pamela, had a score on test 1 of 12. We have found the following statistics for these data.

$$\bar{X} = 13.33 \qquad \bar{Y} = 20.33$$
$$S_X = 1.75 \qquad S_Y = 3.22 \qquad r_{XY} = -.685 \qquad \hat{Y} = -1.264X + 37.18$$

What is Pamela's predicted Y-score? What is her $\hat{Z}_Y$? Show that the two results are equivalent. Be careful with the signs of Z and r.

$$\hat{Y} = -1.264(12) + 37.18$$
$$= 22.01$$
$$Z_X = -.76$$
$$\hat{Z}_Y = (-.685)(-.76)$$
$$= .52$$
$$\hat{Y} = 3.22(.52) + 20.33$$
$$= 22.00$$

The important feature of this example is that the negative slope of the regression line means that someone who is below the mean on the predictor variable is above the mean in predicted score.

The Relationship Between r and B

We have seen that both the correlation coefficient and B are slopes of regression lines. Both can be computed from the same basic data, but they are not, in general, equal. For example, the correlation between college grade point average and score on the statistics final at Big Time U is $+.83$, but the slope of the regression line for predicting final exam score from grade point average is $+14.15$. The slope of the regression line for predicting grade point average from final exam score is $+.049$. There is a fairly simple principle governing the relationship between the correlation coefficient and the regression slope for raw scores that we can demonstrate by substituting deviation score equivalents for the Z-scores in equation 8.10. Remembering that $x_i = X_i - \overline{X}$ and $\hat{y}_i = \hat{Y}_i - \overline{Y}$, one of the forms of a Z-score is

$$Z_{X_i} = \frac{x_i}{S_X} \quad \text{and} \quad \hat{Z}_{Y_i} = \frac{\hat{y}_i}{S_Y}$$

We substitute these equivalents into equation 8.10 to obtain

$$\frac{\hat{y}_i}{S_Y} = r_{YX}\left(\frac{x_i}{S_x}\right)$$

Multiplying both sides by S_Y yields

$$\hat{y}_i = \left(r_{YX}\frac{S_Y}{S_X}\right)x_i$$

Substituting $B_{Y \cdot X}$ for the quantity in parentheses gives us the deviation-score form of the regression equation. Therefore,

$$B_{Y \cdot X} = r_{YX}\frac{S_Y}{S_X} \tag{8.11}$$

which clearly reveals that the relationship between r and B is determined by the ratio of the standard deviation of the *predicted* (criterion) variable to the standard deviation of the *predictor* variable. For example, the standard deviations of the grade point average and final exam variables in the BTU data are $.79$ and 13.48, respectively. Inserting these values into equation 8.11, we get

$$B_{gpa \cdot final} = .83\left(\frac{.79}{13.48}\right) = .83(.0586) = .0486$$

Checking the relationship in the other direction, we get

$$B_{final \cdot gpa} = .83\left(\frac{13.48}{.79}\right) = .83(17.06) = 14.16.$$

The **relationship between r and B** is determined by the ratio of the standard deviation of the predicted (criterion) variable to the standard deviation of the predictor variable.

Both of these values are within rounding error of the regression coefficients we would obtain by direct computation from the raw scores. Also, equation 8.11 shows that $B_{Y \cdot X}$ can only equal $B_{X \cdot Y}$ if the two variables have equal standard deviations because only then will they both equal the correlation between the variables. Otherwise, they will differ by an amount that depends on the relative magnitudes of their standard deviations.

◆ *Properties of the Correlation Coefficient*

At this point we have developed three properties of the correlation coefficient.

1. The correlation coefficient is the slope of the best-fit regression line for a scatter plot of standard scores. As such, it reveals the general direction of the relationship, positive or negative.
2. The square of the correlation coefficient is the proportion of variance in Y that is predictable from X. r^2 reveals the strength of the relationship.
3. Because r^2 can take on values between zero and 1.0, r can only assume values in the range -1.00 to $+1.00$.

We have been using two examples to illustrate the properties of the correlation coefficient, and we have seen that use of the correlation coefficient results in the same predicted score that we get from the raw score regression equation. Now, use the data from Table 8-1 to confirm equation 8.11.

We start with the following statistics.

$$S_X = 1.21 \qquad r_{YX} = .623$$
$$S_Y = 1.25 \qquad B_{YX} = .644$$

Putting these values into equation 8.11 yields

$$B_{Y \cdot X} = \frac{1.25}{1.21}(.623) = (1.03)(.623) = .644$$

What do you think the regression coefficient for predicting X from Y would be? Can you think of a reason why the regression coefficients are so similar to the correlation coefficient?

To find the regression coefficient for predicting X from Y we simply reverse the roles of the two variables, producing

$$B_{X \cdot Y} = \frac{S_X}{S_Y} r_{XY}$$

$$= \frac{1.21}{1.25}(.623) = .968(.623) = .603$$

If you test this out by computing the regression coefficient from the raw data, you will get exactly this result, with a value for A of 1.872. The reason why the two regression coefficients are so similar to the correlation coefficient is that the two variables have almost equal standard deviations.

The Relationship Between r and $S_{Y \cdot X}$

We have now completed two separate definitions, or conceptualizations, of the correlation coefficient. On one hand, the slope of the best-fit regression line for a scatter plot of standard scores is the mean cross product of standard scores, and this we labeled r. This is the line that minimized the errors of prediction of Z_Y from Z_X. If we wish to predict Z-scores on the statistics final at BTU from college grade point average, the slope of the best-fit regression line is $+.83$. On the other hand, we have seen that it is possible to partition the variance of a criterion variable into two parts, one which is related to the predictor and one which is unrelated to the predictor. We gave the label r^2 to the ratio of the predictable variance to the overall variance. Sixty-nine percent of the individual differences in final exam scores can be predicted from college grades. This implies a relationship between the correlation as the slope of a regression line and the magnitude of the errors of prediction. Let's examine this relationship.

First, remember that in equation 8.9 one definition of r^2 was

$$r_{YX}^2 = 1 - \frac{S_{Y \cdot X}^2}{S_Y^2}$$

Multiplying both sides by S_Y^2 and rearranging terms produces

$$S_{Y \cdot X}^2 = S_Y^2 - r_{YX}^2 S_Y^2$$

which, by factoring out S_Y^2, may be rewritten as

$$S_{Y \cdot X}^2 = S_Y^2(1 - r_{YX}^2) \tag{8.12}$$

This equation is *the usual expression for the variance error of estimate.* We can obtain an expression for the standard error of estimate by taking the square root of both sides. This produces

$$S_{Y \cdot X} = S_Y \sqrt{(1 - r_{YX}^2)} \tag{8.13}$$

We would usually calculate the standard error of estimate using equation 8.13. That is, rather than computing a predicted score for each person and finding the standard deviation of the errors of prediction, as we did in Chapter 7 and Table 8-3, we would calculate a correlation between the variables and use equation 8.13.

We calculated the correlation for the data in Table 8-3 to be .623, so r^2 is .39. S_Y was 1.25, so equation 8.13 gives us

$$S_{Y \cdot X} = 1.25 \sqrt{(1 - .39)} = .976$$

This is what we found in Table 8-4, and it is obviously much easier to use equation 8.13.

In an actual research or prediction problem, we would seldom compute the standard error of estimate ($S_{Y \cdot X}$) from the errors of prediction. The only time it would be necessary to use such a method is when the assumption of equal variances in Y for all values of X is not met. The more usual course is to compute the correlation and then use equation 8.13. We could then use $S_{Y \cdot X}$ to determine prediction intervals as described in Chapter 7.

HOW THE COMPUTER DOES IT

Although the most common method of computing the standard error of estimate uses the correlation between the predictor and the criterion, prediction intervals usually are used in regression analyses. For this reason, most programs produce the standard error of estimate as part of the regression procedure rather than in the correlation routine. To find the standard error of estimate for predicting scores on the statistics final at BTU from college GPA, select the Regression procedure from the STATISTICS menu, insert Final as the dependent variable and CGPA as the independent variable, and click OK. You should get results that look like this.

Model Summary

Model	R	R Square	Adjusted R Square	Std. Error of the Estimate
1	.830	.688	.684	7.64

[a]Predictors: (Constant), College Grade Point Average

There are several things to keep in mind when using this procedure. First, it does many things that you will not need and that you may have trouble understanding at this stage in your statistics education. If you explore the STATISTICS options within the program, for example, you will find a box for "standard errors." These standard errors are different from what we have been discussing. At this stage, the only standard error you need is the one provided with the correlation output from the Regression procedure.

As we noted before, the Regression procedure performs an analysis called multiple correlation/regression in which more than one predictor is used simultaneously. This type of analysis is beyond our scope. To make sure you get results that agree with what we are describing, you should include only one independent variable in a given analysis. Otherwise, the program will put all of the variables you specify into the equation and give you something you don't want.

Finally, remember that SPSS and most other statistics programs use a standard deviation computed with $(N - 1)$ in the denominator. Therefore, the standard error of estimate produced by the program will be somewhat larger than you would obtain from the procedures we have described. If you want to check your results against ours, the best thing to do is to compute the standard deviation the way we have shown you and use the correlation coefficient from the computer output. An easy way to get the correct descriptive standard deviation is to divide the Sum-of-Squares Total from the table labeled ANOVA in the Regression procedure (see the output you got from How the Computer Does It on page 188) by N to obtain the sample variance; then take the square root. Dividing the Sum-of-Squares Residual from that table by N will give you the variance error of estimate directly.

What is the standard error of estimate for predicting score on the statistics final exam from college grade point average at Big Time U? What is the standard error of estimate for predicting grade point average from score on the final? Why are the two values so different?

We have the following statistics:

$$r_{YX} = .83 \qquad S_Y = 13.48 \qquad S_X = .79$$

Therefore,

$$S_{Y \cdot X} = 13.48 \sqrt{(1 - .83^2)} = 13.48 \sqrt{.3111} = 7.52$$

$$S_{X \cdot Y} = .79 \sqrt{(1 - .83^2)} = .44$$

The two standard errors of estimate are so different because the two variables have very different standard deviations. Note that the value in the $\sqrt{}$ is the same in both cases.

Correlation and the Shape of the Scatter Plot

In Chapter 7 we introduced the bivariate scatter plot as the frequency distribution for the scores of people on two variables taken together. Just as the normal distribution is really a family of distributions that share some general properties but differ depending on the mean and standard deviation of the variable, so also the bivariate distribution displayed in the scatter plot is a family of distributions. In this case, there is an additional factor that affects the shape of the distribution, and that factor is the magnitude of the correlation between the variables.

We saw in Chapter 7 that the sign of the regression coefficient, B, described the orientation of the scatter plot. When the sign is positive, the long axis of the ellipse goes from the lower left of the figure to the upper right, and when the sign is negative the ellipse goes from the upper left to the lower right. The same rule applies to the correlation coefficient as the slope of the regression line for standard scores. That is, the sign of the correlation coefficient describes the orientation of the scatter plot.

The magnitude of the correlation coefficient (in its role as an index of the degree of association between the variables) reflects the shape of the scatter plot rather than its orientation. Remember that the relative magnitudes of the standard deviations of the variables affected the slope of the regression line. When the two standard deviations are equal (for example, Z-scores) the scatter plot always has its longest axis at a 45 degree angle to the X- and Y-axes, the orientation depending on the sign of the correlation. The magnitude of the correlation coefficient reveals the degree of elongation of the ellipse formed by the scatter plot. The higher the correlation, the longer and skinnier the scatter plot. If the correlation is zero, there is no relationship between the variables, the slope of the regression line is zero, and the scatter plot has the shape of a circle. When the correlation is perfect, $S_{\hat{Y}}^2$ is equal to S_Y^2, $S_{Y \cdot X}^2$ is zero, and there is no scatter of observations around the regression line. All observations fall on the line, and the scatter plot (for standard scores) is a single line at a 45 degree angle to the X- and Y-axes. Degrees of correlation between these two extremes result in scatter plots that are elliptical in shape, with the degree of elongation reflecting the degree of correlation. Some examples are given in Figure 8-6. As you can see, the higher the correlation, the more elongated the ellipse, regardless of the sign of the correlation.

Because of its versatility, the correlation coefficient is one of the most widely used statistics in educational and psychological research. Not only does the correlation coefficient reveal the direction of a relationship through its sign, but the magnitude of the correlation coefficient tells us how strong the relationship is, and the standard error of estimate tells us how much uncertainty remains. In addition, special forms of the correlation coefficient, one of which we discuss in Chapter 14, can be used with nominal and ordinal data. The correlation coefficient also bears a close relationship to the techniques of analysis of variance that we discuss in Chapter 16.

Figure 8-6 Effect of the magnitude of the correlation coefficient on the shape of the scatter plot. The scatter plot is a circle for a correlation of zero. As the correlation gets larger, the scatter plot becomes an elongated ellipse. The long axis of the ellipse shows the sign of the correlation.

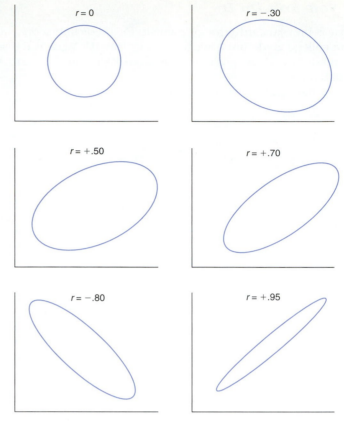

SUMMARY

An observed Y-score can be partitioned into a portion that is predictable from X and a portion that is error. Likewise, the variance of a set of Y-scores can be partitioned into variance that is related to X or is predictable from X and variance that is independent of X. The variance in Y that is unrelated to X is called the **variance error of estimate,** and its square root is the **standard error of estimate.**

An index of the strength of relationship or degree of association between two variables may be obtained from the ratio of the **predictable variance of Y** to the **total variance of Y.** The nature and degree of their association is expressed by the **correlation coefficient r.** The correlation coefficient is the slope of the regression line for predicting Z_Y from Z_X. It is related to $B_{Y \cdot X}$ by the ratio $S_{\hat{Y}}^2/S_Y$. The square of the correlation coefficient is the proportion of variance in Y that is linearly related to X. r can be computed directly from raw scores and used to obtain the standard error of estimate without computing predicted scores. The correlation coefficient affects the shape of the scatter plot and the bivariate normal distribution.

DEGREE OF ASSOCIATION IN THE SEARCHERS STUDY

Professor Cross and his students collected data on several variables that can yield measures of association. Which ones might reasonably be viewed as dependent or outcome variables, and which as independent or predictor variables? Clearly, the form of questionnaire a man received is an independent variable, but what about the others? A man's age or income might serve as a predictor of his attitudes and degree of religious involvement, as might his religious denomination and whether he is a member of the Searchers. One of the problems in most correlational studies is that the directions of any influences are often difficult to determine. We will return to this issue when we discuss the kinds of inferences we can draw from our data, but for now we focus on describing the degree of relationship.

The continuous variables are age, income, degree of religious involvement, and attitude toward equality. The discrete variables are denomination, membership status in the Searchers, and question-

TABLE 8-5	Correlations Among the Continuous Variables for the Searchers Study				
Variable		1	2	3	4
Attitude Toward Equality (1)		1.00	.84	.64	.40
Age (2)		.84	1.00	.74	.38
Income (3)		.64	.74	1.00	.35
Degree of Relgious Involvement (4)		.40	.38	.35	1.00

naire form. We should use the correlation coefficient to describe the relations between the continuous variables. You should get the results shown in Table 8-5.

It is quite clear from the results in this table that all of the continuous variables show modest to strong positive correlations. The smallest, between income and degree of religious involvement, is .347. The square of this correlation shows that about 12 percent of the individual differences in income are related to involvement, and the positive sign means men with higher incomes are reporting higher levels of involvement. At the other extreme, age shows a strong positive correlation with attitudes and income. Older men report higher incomes, as we would expect, and they also report more liberal attitudes toward equality for women, with 55 and 71 percent of the variance accounted for, respectively. (Remember, these data are hypothetical. Don't draw conclusions about human nature from them!)

REFERENCES

Darlington, R. B. (1990). *Regression and linear models.* New York: McGraw-Hill.

Edwards, A. L. (1984). *An introduction to linear regression and correlation* (2nd ed.). New York: Freeman.

Fox, J. (1997). *Applied regression analysis, linear models, and related methods.* Thousand Oaks, CA: Sage.

Galton, F. (1886). Family likeness in stature. *Proceedings of the Royal Society of London, 40,* 42–73.

Hamilton, L. C. (1992). *Regression with graphics: A second course in applied statistics.* Pacific Grove, CA: Brooks/Cole.

Ott, R. L. (1993). *An introduction to statistical methods and data analysis* (4th ed.). Belmont, CA: Duxbury.

Pedhazur, E. J. (1982). *Multiple regression in behavioral research: Explanation and prediction* (2nd ed.). New York: Holt, Rinehart and Winston.

Younger, M. S. (1985). *A first course in linear regression* (2nd ed.). Boston: Duxbury.

EXERCISES

1. Educational researchers have hypothesized a relationship between academic self-esteem and school achievement. The data for these two variables are listed below.

Participant	Self-Esteem	Achievement
1	25	52
2	34	66
3	43	78
4	21	48
5	48	85
6	29	62
7	33	72
8	41	70
9	36	66
10	18	40
11	27	50
12	24	43

a. For the data above, calculate the correlation between the two variables using formula 8.1.

b. For the data above, calculate the correlation between the two variables using formula 8.2.

c. For the data above, calculate the correlation between the two variables using formula 8.3.

2. The following information represents students' average weekly quiz scores and their scores on their final exam in history.

Student	Weekly Quizzes Average	History Final Exam Score	Student	Weekly Quizzes Average	History Final Exam Score
1	8	42	6	5	33
2	7	39	7	3	23
3	7	40	8	6	36
4	4	23	9	6	36
5	7	43	10	7	40

a. Calculate the correlation between the average score on the weekly quizzes and the score on the final exam using formula 8.1.

b. Calculate the correlation between the average score on the weekly quizzes and the score on the final exam using formula 8.3.

3. Given the following data:

Student	Attitude About Science	Science Final Exam Score	Student	Attitude About Science	Science Final Exam Score
1	50	70	11	55	80
2	40	60	12	65	90
3	40	65	13	70	95
4	60	85	14	50	68
5	65	85	15	55	74
6	58	78	16	62	88
7	48	68	17	55	78
8	50	73	18	68	94
9	60	84	19	45	64
10	60	82	20	55	70

a. Calculate the correlation between attitude toward science and the score on the final exam in science using formula 8.3. Given the correlation you calculated, explain what this correlation means. What is the value of r_{YX}^2? What is the interpretation of this value?

b. Calculate $S_{attitude}$ and S_{exam}. Use these values of the standard deviations of attitude toward science and final exam score in science and the correlation from part (a) to calculate the slope of the regression line when science attitude is the predictor variable and final exam score in science is the criterion variable. In addition, calculate the Y-intercept of the regression line.

c. Calculate the standard error of estimate for the final exam score in science given the science attitude score.

d. Suppose Pedro had a score of 52 on the measure of attitude toward science. What would his predicted standard score and predicted raw score be for the final exam in science?

e. Use the data from parts (a) and (b) to calculate the slope of the regression line when final exam score in science is the predictor variable and science attitude is the criterion variable. In addition, calculate the Y-intercept of the regression line.

f. Calculate the standard error of estimate for the final exam score in science given the science attitude score.

4. Given the following information:

Student	Attitude Toward Mathematics (X)	Final Exam Score (Y)
1	59	114
2	58	114
3	45	100
4	52	100
5	46	72
6	45	102
7	60	120
8	36	70
9	44	64
10	49	76
11	43	84
12	41	68
13	60	120
14	46	76
15	49	94
16	42	80
17	43	52
18	33	62
19	23	76
20	42	80
21	46	84
22	60	108
23	43	74
24	51	84
25	51	66
26	51	102
27	32	68
28	37	74
29	42	56
30	44	93

a. Calculate the correlation between attitude toward math and the score on the final exam in math using formula 8.3. Given the correlation you calculated, explain what this correlation means. What is the value of r_{YX}^2? What is the interpretation of this value?

b. Calculate $S_{attitude}$ and S_{exam}. Use these values of the standard deviations of attitude toward math and final exam score in math and the correlation from part (a) to calculate the slope of the regression line when math attitude is the predictor variable and final exam score in math is the criterion variable. In addition, calculate the Y-intercept of the regression line.

c. Calculate the standard error of estimate for the final exam score in math given the math attitude score.

d. Suppose Camille had a math attitude score of 66. What would her predicted standard score and predicted raw score be for the final exam in math?

e. Use the data from parts (a) and (b) to calculate the slope of the regression line when final exam score in math is the predictor variable and math attitude is the criterion variable. In addition, calculate the Y-intercept of the regression line.

f. Calculate the standard error of estimate for the final exam score in math given the math attitude score.

5. Given the following data:

Participant	Neuroticism (X)	Social Phobia (Y)
1	6	11
2	4	12
3	19	22
4	8	10
5	5	12
6	9	17
7	9	15
8	6	13
9	15	15
10	14	23
11	10	15
12	11	17
13	20	26
14	19	18
15	20	26
16	16	22
17	9	14
18	6	14
19	17	20
20	14	23

a. Calculate the correlation between neuroticism and social phobia using formula 8.3. Given the correlation you calculated, explain what this correlation means. What is the value of r_{YX}^2? What is the interpretation of this value?

b. Calculate $S_{neuroticism}$ and $S_{social\ phobia}$. Use these values of the standard deviations of neuroticism and social phobia and the correlation from part (a) to calculate the slope of the regression line when neuroticism is the predictor variable and social phobia is the criterion variable. In addition, calculate the Y-intercept of the regression line.

c. Calculate the standard error of estimate for the social phobia score given the neuroticism score.

d. Suppose Naomi's neuroticism score is 18. Find her standard neuroticism score, predicted standard social phobia score, and predicted raw social phobia score.

e. Calculate the total sum of squares, SS_T, the sum of squares due to prediction, SS_P, and the sum of squares due to error, SS_E, for the data above. What are the values of the total variance, the predictable variance, and the variance error of estimate?

f. Calculate r_{YX}^2 using equation 8.9 and the values in part (e). How does this compare to the value of r_{YX}^2 you found in part (a)?

6. The following data were collected to determine the relationship between empathy and aggression:

Participant	Empathy	Aggression
1	27	30
2	25	31
3	26	29
4	24	22
5	22	20
6	23	22
7	22	18
8	21	17
9	23	21
10	23	24
11	22	26
12	15	30
13	22	14
14	19	39
15	28	12

a. Calculate the correlation between empathy and aggression using equation 8.3. Given the correlation you calculated, explain what this correlation means. What is the value of r_{YX}^2? What is the interpretation of this value?

b. Calculate the standard error of estimate for a predictor variable of empathy and a criterion variable of aggression.

c. Calculate the slope of the regression line and the Y-intercept if the predictor variable is empathy and the criterion variable is aggression.

d. Suppose Jim's empathy score is 16. Find Jim's standard empathy score, predicted standard aggression score, and predicted raw aggression score.

e. Calculate the total sum of squares, SS_T, the sum of squares due to prediction, SS_P, and the sum of squares due to error, SS_E, for the data above. What are the values of the total variance, the predictable variance, and the variance error of estimate?

f. Calculate r_{YX}^2 using equation 8.9 and the values in part (e). How does this compare to the value of r_{YX}^2 you found in part (a)?

7. Use the SPSS program and SPSS data bank to find the Pearson product moment correlation for each of the following.

a. The correlation between post-instruction attitude (postatt), and the mathematics score on the Iowa Test of Educational Development for 9th Graders (ited9) from the data file math.sav.

b. The correlation between confidence in mathematics (confmath) and the final grade in mathematics (fingrd) from the data file career.sav.

c. The correlation between empathic concern (empathy) and altruism (altruism) from the data file donate.sav.

8. Sketch a scatter plot for each of the following correlations.

a. $r_{YX} = 1.00$
b. $r_{YX} = -1.00$
c. $r_{YX} = .00$
d. $r_{YX} = .25$
e. $r_{YX} = -.75$
f. $r_{YX} = -.33$
g. $r_{YX} = .67$

9. Estimate the correlations represented by each of the following scatter plots.

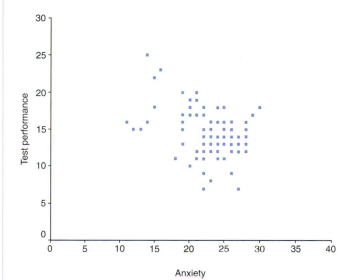

(a)

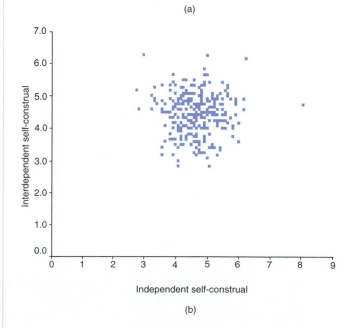

(b)

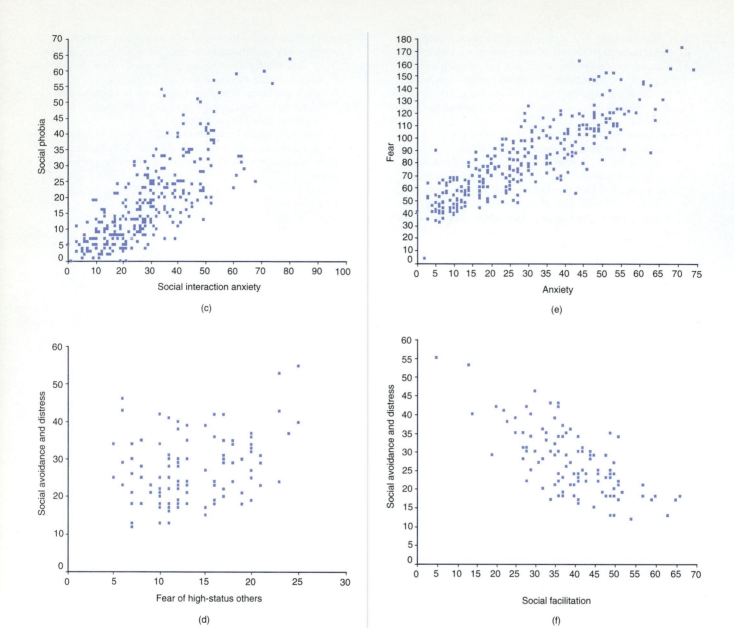

(c)

(e)

(d)

(f)

Part III

Logical Considerations in the Design and Analysis of Research

Chapter 9

Samples and Populations

CHAPTER OBJECTIVES

After studying this chapter, you should be able to:

❏ Differentiate between samples and populations.

❏ Identify sampling and treatment populations from a description of a study.

❏ State the reasons why it may be necessary or desirable to use a sample to represent a population.

❏ Identify the characteristics of a good sample.

❏ Identify the various kinds of samples from descriptions of research studies.

❏ Determine the most appropriate sampling method given a research problem.

❏ Identify situations where effects are confounded.

❏ Determine the number of participants needed for each cell of a stratified random sample.

The Nobel Prize–winning scientist Linus Pauling claimed that very large doses of vitamin C promote good health and, in fact, can prevent and cure colds. Pauling based his claims on his own experience and that of others who followed his advice. Suppose you were asked to design a study to test these claims scientifically. How should you proceed? Whom should you include in your study? To whom would you want your results to apply? How could you reach a conclusion without subjecting everyone to large doses of vitamin C? In order to provide a scientific foundation for research in psychology and education, we must be able to provide answers to questions like these.

Introduction

In Part III of this book we will move away from the mathematics of statistics to examine (1) how we obtain the data to which statistical analyses are applied, (2) how statistical questions are phrased, and (3) why statistical questions are phrased the way they are. Our discussion will not enable you to design and conduct a study of your own. However, a basic introduction to these research issues should help you to understand the relationship between finding answers to substantive questions in psychology and education and the methods of data analysis that make up statistics.

This chapter presents issues involved in deciding who will serve as participants in a study. For example, to test Pauling's claims about the value of vitamin C scientifically, we would have to begin by identifying populations of interest, samples from these populations, and their associated parameters and statistics, concepts mentioned briefly in Chapter 1. Because sampling methods are critical to sound statistical inference, the conclusions we could draw from our study would depend on how we drew samples from the populations of interest. The terms *sample* and *population* have precise meanings in the fields of educational and psychological research and statistics. It is necessary to keep the meanings of these terms clearly in mind in order to follow the process of generalizing from the observations on a sample to reach conclusions about a population.

Populations

The primary unit for research in psychology, education, and science generally is the **population.** *The population is composed of all of the individuals about whom we want to make statements or draw conclusions.* We are free to define the population in any way we choose; it can include the members of a statistics class, all freshmen at college X, all students at university Y, the residents of city P, state Q, or nation Z. The definition of the population actually consists of two parts, both determined by the investigator. The first part is the definition of the entities being studied, whether these are people, animals, plants, social institutions, or anything else of interest. For example, in a study of the value of vitamin C, we would probably want to be able to apply our results to all people. Therefore, humankind would be our population.

The population in one study may be the single entity, often called a participant, or subject, in another. For example, a psychologist might conduct a questionnaire study of the attitudes of individuals toward birth control to learn about factors influencing changing birth rates within a country. For the psychologist, the individual people in that country make up the population. A demographer, on the other hand, might use the same country (and perhaps even the same data) as one entity in a study of worldwide birth rates. In the latter case, individual attitudes and births to individual mothers still constitute the fundamental data, but individual responses are aggregated across people to provide a "score" for the nation. In turn, the nation as a whole functions as the subject, and the population is all nations.

The second part of the definition of a population is the limit for inclusion in the population. Once we have defined the general type of entity that we want to study, we must specify in more detail who qualifies for membership in the population. Does the population of interest include all people, as in our study of vitamin C, or does it include only registered voters, or male college freshmen, or some other group such as members of fraternities and sororities? This step is very important because it defines the individuals to whom we can apply the conclusions of our study. In general, we can generalize our findings only to individuals who are like those whom we have studied. For example, if the proper definition of the population is college students (all participants in the study are college students), then it would not be appropriate to say that the results of our study are applicable to middle-aged business executives. The population that we specify prior to drawing a sample is known as the **sampling population.**

A **population** consists of all of the individuals about whom we want to make statements or draw conclusions.

The **sampling population** is the population that is specified before the sample is drawn.

SPECIAL ISSUES IN RESEARCH

Defining Samples and Populations

Researchers in the natural sciences such as chemistry and physics seldom worry much about defining their populations because, as far as we know, the entities and properties they study do not show individual differences. All molecules of water, for example, "behave" in the same way under a given set of conditions, so it does not matter whether the sample of water molecules comes from a lake in Wisconsin or a well in Arizona. As long as the impurities have been removed so that we are dealing with pure water, the results will be the same and will generalize to all other water molecules under the same conditions.

Researchers studying the behavior of living organisms, particularly humans, have a much more difficult problem. Research in cross-cultural psychology over the last 25 years has made it increasingly clear that what is found in one group may not generalize to another. At the very least, findings from one group, such as middle-class Americans of European ancestry, need to be replicated in groups from other traditions before any claim to have discovered something about "human nature" can be sustained. This means that we must be very careful to define our populations clearly and to draw samples from those defined populations.

Samples

A **sample** is a subgroup of a population. The sample is taken from the population. This implies that the population has been defined, that entities in the sample have been chosen from the population by some means, that these entities have been measured in some way, and that we can draw conclusions about the sample and, hopefully, about the other members of the population from the measurements. The usual purpose of drawing a sample is to reach conclusions about the population. For example, if we wanted to test Dr. Pauling's claims about vitamin C for a population that includes all of humankind, the only practical way to do so would be to administer vitamin C to a small subgroup of this population.

▶ *Now You Do It*

Proper interpretation of research results requires that we be able to identify the population from which a sample was drawn. For each of the following situations, what do you think is the most appropriate definition of the population?

1. Statistics America has used census records to draw a sample of American households to determine the education levels of American adults.
2. The United States Golf Association (USGA) has sent a questionnaire to a sample of its members to assess opinion about a possible rule change.
3. Professor MacBeth has computed the correlation between mathematics anxiety scores and scores on the statistics final in his course.

Answers

1. The intended population is all American households, and since the sample was drawn from the population defined by the census, the generalization will be appropriate.
2. The intended population probably is all golfers, but since the questionnaire was sent only to members of the USGA, it is appropriate to generalize only to this population.
3. Professor MacBeth probably would like to be able to say something about the effects of mathematics anxiety on statistics students generally, but since his sample came only from his own class, he can properly generalize only to his own students, and maybe only to those from the current year.

Why Sample?

We have said that one of the main purposes of statistics is to enable us to draw conclusions about the characteristics of a population from measurements made on a sample. But why not avoid the whole problem and measure the entire population?

There are several reasons why using a sample rather than the entire population might be preferable, or even necessary. The most obvious is **economy.** Every observation requires time or money (or both) to obtain. Suppose an interview takes an hour of a research assistant's time and $20 in materials. Then every additional participant may cost an extra $40. If we can reach reasonably accurate conclusions on the basis of a small number of observations, why go to the effort and expense of collecting every possible observation?

A second reason to use a sample is **timeliness.** It is often important to be able to reach conclusions or make forecasts quickly (or within a reasonable period of time). Voter polls are an obvious example. Of what use is it to a candidate to know how every member of the population plans to vote if the results of the poll cannot be known until after the election? Similarly, while the census collects information on the entire population every 10 years, it takes so long to organize and analyze the data that they may be obsolete before the analyses can be completed.

Third, the study may **alter or destroy the entities** being observed. The physiological psychologist studying the effects of increased exposure to novel stimuli during infancy on brain tissue in mice must sacrifice the subjects to make the necessary measurements. An automobile manufacturer testing the collision resistance of a car model must generalize the results from a sample to a population of intact cars because the testing procedure destroys the product. In either case, a sample is needed.

The final reason why we sample is that **the entire population may just not be available.** There are cases where the population is functionally infinite, where there are so many potential measurements that it would not be possible, even theoretically, to collect them all. This is particularly true of infinitely repeatable processes such as the coin tosses we used to develop probability theory in Chapter 6. The same basic issue arises, however, with processes that change over time. No matter how many observations we collect on a person's psychological development, we can never measure those events that are still in the future. If we wish to make forecasts of any sort, such as predicting the academic performance of next year's freshman class, we must make them on the basis of incomplete information. This is why we use statistics to test hypotheses about experimental manipulations and to draw conclusions about treatments.

▶ *Now You Do It*

We have discussed four reasons why one might need to take a sample rather than measure the entire population. For each of the following situations, see if you can identify the primary reason for drawing a sample.

1. Freda just completed her degree in psychology and has taken a job as a quality control specialist for Tres Padres Mexican Foods. Her first task is to test the quality of the canned refried beans by having volunteers taste samples of the beans.
2. Micah works as a data analyst in the Chicago Public Schools system. He has been assigned the task of measuring parental support for a proposed bond issue.
3. Saul is working on a research project with his psychology professor. They are studying the effects of weather on people's moods. Each evening, a group of participants records the weather they have experienced that day, and they fill out a mood questionnaire.

Answers
1. Although economy is no doubt an issue for Tres Padres, the testing destroys the object as well.
2. Micah's primary problems will be timeliness and economy.
3. Saul has an infinite population of possible observations.

Characteristics of a "Good" Sample

There are two properties that samples should have in order to allow proper generalizations about the population. The first, and most important from the perspective of the sampling population, is that the sample should be **representative of the population.** This means that the sample should have the same general characteristics with the same relative frequencies as the population. The match need not be exact, but if there are substantial differences between the sample and the population, the results from the sample will not reflect the characteristics of the population, and generalizations will be in error. Any systematic failure of the sample to accurately represent the population is an **error in sampling.** An error in sampling was what caused the editor in the Truman-Dewey election mentioned in Chapter 1 to be so far wrong. When errors in sampling result in consistent differences between a statistic and its parameter, the sample is said to be **biased.** A good sample is an unbiased sample.

Any time sampling occurs, there will be variations from sample to sample. This is a random process and the result is called **sampling error.**

There are two factors that contribute to sampling error. The first is individual differences. Each sample contains a different group of people, and these people bring their particular characteristics to the sample. The samples differ simply because the people in them differ.

The second cause of sampling error is the occurrence of random events in the data collection process. For example, someone may misread an instrument, or there may be an error in scoring a test. Random events like these are called measurement errors or experimental errors, and they contribute to variability between samples that is not related to the purpose of the study.

A good sample is one that is **representative of the population** *in the sense that it has the same general characteristics with the same relative frequencies as the population does.*

Errors in sampling occur *when the sample differs in systematic ways from the population.*

Random deviations of samples from their population are the result of **sampling error.**

Sampling error cannot be avoided, but one of the purposes of statistical inference is to estimate its effect and qualify our conclusions in light of its probable influence. This is what news reports of political polls are referring to when they say that a poll has a margin of error of plus or minus three percent. Errors in sampling, as contrasted with sampling error, are systematic and result in statistics computed on the sample that are different from their respective population parameters in a nonrandom way.

In addition to being unbiased, it is desirable that samples include an element of **randomness** in the selection of their members. That is, there should be some uncertainty as to which members of the population will be included in the sample. One of the problems with some of Dr. Pauling's claims regarding vitamin C is that the results were based on the experiences of people who chose to try the vitamin and had success. They were not selected randomly to participate in a study. As we saw in Chapter 6, the laws of probability are based on an assumption that the events are independent. If the process of choosing the sample does not include an element of randomness, the assumption of independence is violated and the laws of probability cannot properly be employed. A study of vitamin C based on people who have used it and gotten positive results violates this requirement of independence because those who were included in the study had already gotten positive results. Inclusion in the study was not independent of outcome.

▶ *Now You Do It*

Proper interpretation of research findings requires that you be able to identify potential sources of bias. Consider each of the following situations. Is there reason to suspect that the sample is biased? If so, in what way?

1. A clinical psychologist is developing norms for her new test of adjustment. She bases her norms for the general public on the responses of the relatives of her severely disturbed clients.
2. An educational psychologist is studying the effect of peer tutoring on student achievement. The school where he is conducting the research is in a wealthy suburb of Boston.
3. A public opinion research company has been hired to conduct a poll of opinions of landowners toward water quality in a rural county. They conduct the poll by telephone and call every tenth number in the county telephone directory.

Answers
1. Relatives of people with severe psychological disturbances probably differ in systematic ways from the general population of Americans, so the sample is likely to give a distorted impression of how "normal" people would respond. However, several popular instruments in fact have been normed this way.
2. There is likely to be a relationship between income or social status and emphasis on education. This might distort the impact of peer tutoring and lead to an erroneous conclusion about its effectiveness.
3. This procedure is likely to produce a reasonably good sample if the population of the county has telephones, which we can expect to be the case for almost all landowners.

Types of Samples

In the last 70 years, great strides have been made in using samples to represent populations. The considerable success of public opinion polls, sometimes using samples of about 1,000 to represent as many as 100 million people, testifies to the accuracy with which population parameters can be estimated from sample statistics when the samples are well chosen. While a detailed treatment of sampling techniques is beyond our scope, a brief discussion of the major categories of samples and how they achieve or fail to achieve the qualities of good samples will help you to be an intelligent critic of the research literature. As you read research literature, be particularly attentive to how the researchers obtained their samples. Remember, accurate conclusions cannot be reached from poor samples.

Samples of Convenience

One of the most common types of samples found in the research literature of psychology and education is what is called the **sample of convenience,** or the **available cases sample.** This is the type of sample that occurs when there is no prior definition of the population. The issue of population is not mentioned in drawing a sample of convenience, and the implicit definition of the population is "whatever group might have given us a sample like this one." A student of psychology would obtain an available cases sample by standing outside the dining hall, handing out questionnaires to whoever will take them, and tabulating the results of those that are returned. The educator who tests only those students whose teachers are willing to donate class time for testing is similarly working with an available cases sample.

> A **sample of convenience** is obtained whenever the people studied are those who volunteer or just happen to be available.

Selected Cases Sampling

Worse than the sample of convenience is the **selected cases sample.** A selected cases sample occurs when there is some aspect of the data collection procedure that through either intent or ignorance produces a clear bias in the results. If we asked a group of supporters of a particular issue how they feel toward that or a related issue and then treated the results as representative of the general population, we would be using a selected cases sample. For example, we might ask members of the American Association of Retired Persons whether they support a reduction in Medicare benefits and then argue on the basis of the results that the general American population believes Congress should not reduce Medicare benefits. This tactic is relatively common in politics and persuasive advertising, but it has no place in scientific inquiry. The preferred alternatives to the available cases sample and the selected cases sample all require that the population be defined before the sample is obtained so that it is possible to state whether any given entity does or does not belong to the population.

> A **selected cases sample** occurs when there is some aspect of the data collection procedure that produces a clear bias in the results.

Random Sampling

Random sampling is the simplest method of drawing a sample from a defined population. In a random sample, every member of the population has an equal and independent chance of being included in the sample. Every member of the population starts with the same chance of being selected for the sample, and the selection of any one individual

> In **random sampling** each individual in the population has an equal and independent chance of being included in the sample.

SPECIAL ISSUES IN RESEARCH

Problems in Sampling from Human Populations

The problems of defining the population and sampling are particularly acute for research on human beings: Researchers are usually restricted to using volunteers in their studies, and this has an immediate effect on the population definition. For example, we can very carefully define the population and select people to take part, but then find that, when we ask them to participate, many refuse. White rats and other nonhuman subjects seldom do this, so generalizations are more straightforward. With humans, we never know what the effect of having to use more or less willing volunteers will be on our results.

When refusal is random, our ability to generalize is not impaired, but if willingness to participate is related to the topic under investigation, then differential participation rates can have a profound effect on the outcome by producing a biased sample. For example, if we wish to study the relationship between religious convictions and attitudes toward birth control, and people who have strong religious convictions tend to refuse to participate in our study, the results from our sample will not represent the entire population.

does not affect the probability of any other individual's being included. Samples taken at random from the population will, in the long run, be representative of the population. That is, a single random sample may differ substantially from the population, but if we take a large number of random samples, they will, on average, be representative of the population.

Suppose, for example, that we want to draw a random sample of 10 people from a statistics class with 100 members. Before anyone is selected for the sample, each member of the class has an equal chance of 10/100, or .10, of being included in the sample. Next, each member of the class is given a number between 00 and 99. We then consult a table of random numbers. Since each class member is represented by a two-digit number, we read pairs of digits from the table, and include each individual whose number comes up until we have 10 people. If a person's number comes up twice before we have completed our sample, we skip the number the second time. When we are finished, we have a random sample of 10 from our defined population of 100.

The advantage of using a table of random numbers for selecting our sample is that the probability that an individual's number will come up remains constant at 1/100 throughout the sampling process; each member retains a 10/100 chance of being included in the sample. On the other hand, if we had drawn names out of a hat, the probability of inclusion on the first draw would have been 1/100, on the second, 1/99, on the third, 1/98, and so on. The chance of inclusion would have changed as the sample was drawn. It would have been dependent on the order of drawing. For most research purposes, this difference is not very important compared with other effects, such as errors of measurement. Also, it can be overcome by what is called sampling with replacement: putting the person who was drawn back into the population after each draw. In addition, as the size of the population available for sampling gets large, the change in probability from one draw to the next becomes so small that it can be safely ignored. However, since random number tables are easy to use, assure random sampling, and can be applied to almost any situation for which true random sampling is appropriate, it is a good idea to use a table of random numbers when the research plan calls for drawing a random sample to assure that the requirements of equal and independent probability of inclusion in the sample are met as well as conditions allow.

Virtually all statistical procedures for generalizing from a sample to a population involve an assumption that the sample is a random sample from the population. Unfortunately, the conditions necessary for the simple type of random sampling we have just described seldom exist outside statistics books. In the real world we are seldom confronted with a clearly defined population that can be sampled randomly. Moreover, even when we can sample randomly, a random sample of manageable size may not be a good sample.

A good sample is one that is representative of the population; that is, the individuals in the sample possess many of the same characteristics in the same relative degree as do members of the population. A random sample of white rats or pigeons or fruit flies may be quite satisfactory because there is little variability among rats, pigeons, or flies. Frequently members of these species have been bred with the purpose of creating populations that are homogeneous with respect to specific characteristics. The random sample, in this case, resembles the population closely because there are few dimensions where the individuals differ; however, when people are the entities being sampled, the problem becomes more complex.

The difficulty with people as research subjects is that they vary on so many dimensions, and the status of individuals on those dimensions affects their behavior. Age is often an important factor; so are social status, level of education, and number of parents living at home, for example. Some of these dimensions may be important for one study, others for another.

Generalization to a population requires a sample drawn from the population *with respect to all relevant variables in the population*. When the population shows variation in some characteristic but the sample does not, the sample is not drawn from the population *with respect to that characteristic*. If we use the sample, we must redefine the population as a new population (a subpopulation of the original one) where the characteristic is constant. In other words, we cannot generalize across levels of a variable trait in the population, such as age or gender or socioeconomic status, unless that trait is represented as a variable in the sample.

An example may help clarify how characteristics that are variables in the population can become constants in a sample from that population and thereby restrict our ability to generalize to the population of interest. Suppose we want to do research on teaching methods: We want to compare the effectiveness of phonics and the whole-language method for teaching reading. There are two teachers available to teach two first-grade classes. Students are randomly assigned to one class or the other. Teacher A uses phonics for a year, while Teacher B uses the whole-language method. At the end of first grade, we measure reading achievement and find that the children in the class with Teacher B obtained higher scores. What conclusions can we draw?

We must be careful how we talk about the treatments in a study such as this one. In this study, the definitions of the treatments included the personalities and abilities of the teachers themselves, as well as the teaching methods they used. We cannot generalize the results to other teachers because each teacher used only one method and each method was used by only one teacher. Therefore, we cannot tell whether the results were due to the teacher or the method or a combination of the two. A statistician would say that the effect of the teachers is **confounded** with the effect of the treatments. We cannot generalize to other grades either because only one grade was represented. Also, even if the students were randomly sampled from all the first graders in the district, there is a question whether we can generalize to other districts because this one may be unique. Such factors as the ethnic mixture of the classes and parents' socioeconomic status, for example, have been ignored.

*Two effects are **confounded** when we cannot tell which one, or what combination, produced the results we observe.*

The point, which is crucial for good research studies in the behavioral and social sciences, is that generalizations can be made only to those segments of the population that are represented in the sample. Research with humans is particularly difficult in this regard because the population is diverse in many important ways. It is seldom necessary to consider variables like hair color when defining the population because hair color, although a variable, does not seem to be related to most of the behavioral variables of interest to psychologists and educators. However, many other variables such as age, gender, height, weight, number of siblings, and area of residence, for example, may be related in important ways to the variables under study. Unless the sample varies on these traits, the results of the study cannot be generalized across them.

Where studies of humans are involved, the characteristics of the sample often place constraints on the definition of the population. We can never generalize except to a population that might reasonably have yielded the sample we have measured. A sample of White Anglo-Saxon Protestant males between the ages of 18 and 22 is unlikely to have resulted by random selection from the general American population, but there have been attempts to draw conclusions about the nature of human beings from such groups. This is not to say that using such nonrepresentative groups for research is illegal, immoral, unethical, or in any way improper. What is improper is claiming that the results apply to other groups that differ in systematic ways from the group that was measured. Stratified random sampling, which is discussed in a later section, is one way to overcome this difficulty.

We must be clear about one other aspect of generalizing from sample results. The fact that we have obtained a sample of convenience and are using a limited selection of treatment conditions does not mean that the effect we have found in our study does not apply to individuals who are not members of the population and to treatment conditions we have not included. Such individuals may very well be members of the population; if they or people like them had been exposed to treatment condition X, they might have responded as the individuals in our study did. However, if such people were not included in the definition of the population, we would be on shaky ground if we included them in our generalizations. Thus, while all adults might respond to treatment X as college students do, we should conduct a study of treatment X with other age and education level groups before concluding that the effect we have found with college students is true for all people. Alternatively, there are some general strategies for obtaining samples that do permit us to generalize to the population of interest.

▶ Now You Do It

Identifying potential confounding conditions is very important for the correct interpretation of research results. Consider each of the following research designs and see if you can spot the potential confounding conditions.

1. Moshe is studying the effect of a prenatal educational pamphlet on the birth weight of babies. At one hospital, prospective mothers are given a pamphlet describing proper prenatal nutrition, and at another they are not. The average weight of the babies born in the two hospitals will be compared.
2. Pricilla is a sports psychologist interested in the effectiveness of imaging on athletes' speed in developing skills. At her university she trains the members of the tennis team in imaging skills. The members of the golf team do not receive training. She compares the national rankings of the two teams.

Answers
1. Birth weight is confounded with hospital, and with all variables on which the hospitals differ. For example, if one hospital is in a poor inner-city neighborhood and the other is in a wealthy suburban town, differences in social status could be the important variable.
2. Type of sport is confounded with the imaging training. Also, the outcome variable can be affected by school recruiting policies in the two sports, and many other factors.

Systematic Random Sampling

Systematic random sampling involves listing the members of the population in some order, determining a random starting place in the order, and taking individuals at a systematic interval from that point on.

One procedure that is popular when all members of the population can be listed in some order is the **systematic random sample.** Suppose the population of interest is all students at your university. We could start by listing all students in order by their student identification number (or listing them alphabetically, or in any other order). Next we use a table of random numbers to generate a starting place, to select the first member of our sample. If we want to take a sample that includes one percent of the population (or any other percentage, called *p*-percent), we would start with our randomly selected student and then choose every 100th student in order from the list. Assuming that there is nothing about a student's ID number that is related to the topic of the study, the result should be a representative and random sample of the population, and one that is easy to obtain. A variant on this procedure is to use listings in the telephone directory, with the one caveat that people who do not have telephones or have unlisted numbers are excluded from the population.

Stratified Random Sampling

*In **stratified random sampling** the population is divided into subpopulations, and random samples are drawn from the subpopulations in proportion to their frequency in the entire population.*

This method refers to the class of procedures that often are used to obtain representative samples from diverse human populations. There are a variety of special procedures that can be used in one application or another, but there are two basic elements that are common to them all: The first is stratification of the population into subpopulations, and the second is a systematic (but at least partially random) sampling of entities within subpopulations.

The first part of the procedure, stratification, requires identifying important status variables in the population as a whole. For example, if we want to do research on voting behavior (or perhaps take a political poll), it is necessary to begin by identifying important variables, or dimensions, that are related to voting; for example, age, gender, and social status. There are, of course, many others that would be included by a political research firm, but these will suffice to describe the process. Such dimensions are called **stratification variables.**

Stratification variables are variables used to describe the population before drawing a stratified random sample.

Once a set of stratification variables has been identified, the next step is to determine the relative frequency of each category of each variable in the population. In our example, let us assume that age has been divided into four categories, social status into three, and gender into two. The three stratification variables divide the population into 24 unique categories, or combinations. There are, for example, some females of middle social status who are in age range 2, and there are 23 other combinations of our three stratification variables. Our problem now is to determine the relative frequency of each combination in the population. This is often done with census records or information from other appropriate sources.

A sample that is representative of the population will contain the same relative frequency of each type of individual that the population does. Thus, if the relative frequency of middle-status females in age range 2 is .06 in the population and we want to draw a sample of 300 individuals, 18 of our 300 participants should be middle-status women in this age range. Similarly, each unique combination of the stratification variables is represented in the sample by an appropriate number of people.

Stratification ensures that the sample is representative of the population on the stratification variables. If information about the frequency of each combination of stratification variables is not obtainable, it is necessary to assure that the relative frequency of each of the categories of each stratification variable is properly represented in the sample. For example, if we cannot determine the relative frequency of middle-status females in age range 2, we should make sure that the sample contains the right proportion of females, the right proportion of people of middle social status, and the right proportion of people who are in age range 2. If women make up 52 percent of the population, there should be about 156 women in the sample, and if people who are of middle social status constitute about 45 percent of the population, there should be about 135 such individuals in the sample. Similarly, if people in age range 2 make up 28 percent of the population, there should be about 84 such people in the sample.

Once the relative frequency of each category in the population and the number of individuals of each type needed for our sample have been determined, we can proceed to draw our sample. There are many different ways of doing this. One simple way that can be used when the population is small is to sort all members of the population into their respective categories and draw a random sample of the needed size from each category. We take small random samples from defined subpopulations that are homogeneous with respect to our stratification variables. When these small random samples are combined, the result is a stratified random sample that is representative of the population in terms of the stratification variables and, we hope, also in terms of the variables of interest in our study. It is not a true random sample, but stratification has assured representativeness with a smaller sample size. Thus, using stratification variables is a more efficient sampling procedure than sampling at random from the entire population, and that is its primary virtue.

An example may help to clarify the role of stratification variables in a study. When the authors of the Wechsler Intelligence Scale for Children—Third Edition (WISC—III) set out to provide norms for the test, they defined several stratification variables. The primary variable was age. Two hundred children would be sampled at each age covered by the test. The next stratification variable was gender. Equal numbers of boys and girls would be included. Then, the authors consulted the census and found the percentage of different ethnic/racial groups in the United States population, and included appropriate numbers of children from each group. Other stratification variables included geographic region and socioeconomic status. The variable of interest was intelligence as measured by the WISC—III, but their stratification procedure assured that the sample would be representative of the population in terms of age, gender, and the other stratification variables, each of which is related to intelligence. Therefore, the sample should also be representative with respect to intelligence.

▶ *Now You Do It*

The Office of Institutional Research at Big Time University has been asked by the president to obtain a sample of opinions from 10 percent of graduating seniors on a proposed change in the university's core liberal arts requirements. How might the director of the office proceed? What sampling plans might she use, and what variables should she take into account?

Systematic random sampling is one possible method she could use. A questionnaire could be sent to every 10th student in the list of graduating seniors. However, it would probably be better to draw a sample that represents the different major areas of the school. If there are 4,000 graduating seniors, and they are distributed among major areas as follows, what should the frequencies in the sample be?

Arts	10%	Natural Sciences	15%
Social Sciences	15%	Life Sciences	20%
Humanities	15%	Engineering	10%
Business	15%		

Because the 10 percent sample will contain 400 students, the major areas should be represented as follows: Arts-40, Natural Sciences-60, Social Sciences-60, Life Sciences-80, Humanities-60, Engineering-40, Business-60.

There are many other ways of drawing stratified samples, some random and some not so random. Television networks report voting results from carefully selected precincts that have particularly representative voting histories. In another sampling procedure, called *cluster sampling,* researchers use school districts or city blocks or other naturally occurring aggregations as sampling units, and measure all individuals in the unit. Cluster sampling, where the clusters (schools, city blocks, and so forth) have been selected to represent particular segments of the general population, is the method most often used to provide norms for standardized achievement tests. Whatever the method, the objective is a sample that is representative of the population, so the results from the sample can be generalized to the population.

Careful selection of the stratification variables is crucial to the success of a sampling plan. It is these variables that assure the representativeness of the sample in heterogeneous human populations. It is often necessary to use the published reports of other studies to identify variables that are related to the variable we want to study. These reports can also help determine the relative frequencies of the categories of the stratification variables in the population we want to study. After identifying and defining the population, we can proceed with the study, confident that we shall be able to generalize the results to the population we have defined.

▶ *Now You Do It*

We have now studied several types of sampling. For each of the following situations, state which type of sampling is involved: (1) a sample of convenience, (2) a selected cases sample, (3) a simple random sample, (4) a systematic random sample, or (5) a stratified random sample.

1. Senator Slade says that, based on letters his office has received, 90 percent of the American public supports gun control.
2. Jesse draws a sample of three students from his first-grade class by putting every child's name in a hat and drawing out three names.
3. The city library board draws a sample of voters by taking every 20th voter on the roster of registered voters.
4. Gerhardt recruits 20 students from the introductory psychology class to participate in his study of obedience.

2, 3, 4, 1

Two Kinds of Populations

At the beginning of this chapter we defined *sample* as "a subgroup of a population." A sample is always a sample from some population. The question is, what population?

As you may recall, Linus Pauling based his claim regarding the efficacy of vitamin C on a sample that included those who tried vitamin C and reported whether it helped them. Who were the members of his population? They were people willing to try vitamin C and report on the results. To whom could he properly generalize the findings? He could generalize his findings to people willing to try vitamin C and report on the results, that is, people like those in his sample. In effect, the sample is defining the population, and, in turn, the population can only be defined as the group of people who tried vitamin C and reported whether it helped them. This is a characteristic of the available cases sample, and it has the problem that it is circular. We can avoid such circularity for some applications of scientific reasoning by identifying two separate populations.

The interpretation and application of research findings in the social and behavioral sciences involves (1) the **sampling population,** and (2) the **outcome** or **treatment population.** The sampling population is the population whose definition depends upon

*The **sampling population** is one defined by the investigator; it is the population from which the sample is drawn.*

the investigator. The definition of this population should be determined prior to drawing a sample, but often, as in the available cases sample, it is not. The sampling population is the population from which the sample is drawn and which is most frequently referred to in statistical discussions of populations. This is the population we were referring to in our discussion of sampling methods.

The second kind of population, the **outcome** or **treatment population,** is particularly useful in interpreting research results when there is a manipulated independent variable. This population is hypothetical; it does not actually exist as a population, but it is *the population that would exist if all members of the sampling population were exposed to the same experiences to which the sample was exposed.* The treatment population is particularly useful in interpreting the results of a study involving manipulated variables because it allows us to talk about what other people would be like *if they had received the treatment.*

> The **treatment population** is the hypothetical population that would exist if all members of the sampling population received the treatment being studied.

To clarify the differences between these two ways of viewing populations, imagine that we want to study the effects of vitamin A on spelling achievement in school children. Our research design might involve giving one group, the experimental group, the experimental treatment of interest (a pill that contains vitamin A), while a second group, the control group, receives a placebo (a sugar pill). We could define our sampling population as all elementary school children in the local school district whose parents are willing to have them participate in our study. This is a very restricted population, similar to the population in Pauling's study of vitamin C. We proceed by drawing a sample from this population and assigning half of the sample (which we shall now call the vitamin A group) to the vitamin condition of our study, while the other half (the control group) receives the sugar pill.

Note that we started with the sampling population, all elementary school children in the local school district whose parents would permit them to participate. We drew our sample from the sampling population, and we randomly divided these children into two groups. Now, we expose the two groups to the vitamin pill and sugar pill conditions, respectively, along with normal school work for a period of six months. At the end of this time we measure their spelling achievement, and, based on an appropriate analysis of the data, we conclude that the vitamin A group members are better spellers than the control group members. Each of the two groups derived from the single original sample is now *a sample from a separate treatment population.* A treatment population is a hypothetical population that describes what the members of the sampling population would be like if all of them had received the treatment. The vitamin A group is a sample from the hypothetical population that *would exist* if all members of the sampling population were given the vitamin A treatment. The control group is a sample from a hypothetical population of students that would exist if all members received the sugar pill. Each condition in the study, whether it is a treatment of interest, such as the vitamin treatment, or a control condition such as the sugar pill, creates a treatment population.

Two Subdivisions of Research Studies

Now that we have defined treatment populations, it is important to identify research situations and purposes that do and do not involve these hypothetical entities. As we stated in Chapter 1, there are two broad categories into which we can divide research studies: those that attempt to estimate what the present parameters of the sampling population are; and those whose objective is to estimate the consequences of *doing something* to the members of the sampling population. The former involves only the sampling population and is typical of what we called a survey or questionnaire or correlational study in

---

SPECIAL ISSUES IN RESEARCH

Control Groups

A control group is a group of participants who are treated just like the experimental group in every way except that they do not receive the treatment whose effect we are studying. In our example of the effect of vitamin A on spelling achievement, unless we have a group who receives the noneffective pill, we cannot tell whether it was the vitamin or just the act of receiving a pill that produced any effect we find. One of the most common research designs in psychology and education compares one or more experimental treatments with a control group.

Chapter 1. For example, the food services manager at your school may want to assess student opinion of the selection of lunches that are available by taking a sample of students who are eating in the dining hall. In studies such as these, the purpose is to estimate the population mean, standard deviation, correlation coefficient, and so forth. We then generalize to other members of the sampling population. Clear definition of the sampling population is particularly important for studies of this kind.

Generalization to treatment populations occurs when one or more of the independent variables in the study is a manipulated variable. Our study of the effects of vitamin A is an example of this type of research. What would happen to people like those in our study if we give them the treatment condition instead of the control condition? We want to conclude from this type of study that what we have done to the participants, our treatment, changed them so they are now a sample from a new population, and it is to this hypothetical new population, the treatment population, that we want to generalize the effects of our manipulation.

The relationship among sampling populations, samples, and treatment populations is illustrated in Figure 9-1. Part (a) illustrates the situation where a sample, such as students who fill out the food service manager's questionnaire, is used to draw inferences about the characteristics of the sampling population. The sample is drawn and measured on some characteristics, and we infer that the population has the characteristics of the sample. Part (b) of Figure 9-1 shows the division of the sample into two treatment conditions and illustrates that inferences are not to the sampling population but to the treatment populations that exist only after the treatments have been applied.

It is important for the first type of study to define a sampling population carefully and then draw a sample from that population because the relationship between the two is the foundation of much scientific inference as it is practiced in the behavioral and social sciences. We will go into the details of the logical system for generalizing the sample results to the population in the next chapter.

Figure 9-1 Inferences from sample data can be of two kinds. In a correlational study shown in part (a), inferences are drawn about the sampling population. In experimental studies where the participants have been exposed to a treatment, as in part (b), inferences are to the treatment population.

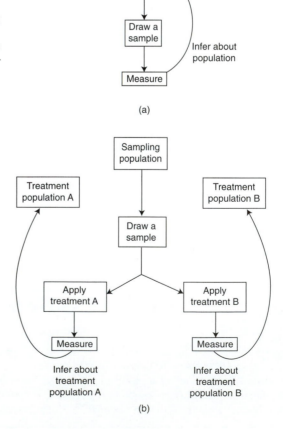

There are many types of studies that educational and psychological researchers have used, but most of them involve some aspects of the two mentioned here. For example, there are many studies where only levels of a single treatment condition, such as the magnitude of a reinforcer or delay of reinforcement, have been investigated. There are also many studies in the areas of personality and social psychology where only status variables such as personality traits and demographic characteristics have been the object of investigation. At a more complex level, there are examples where combinations of manipulated and status variables have been used as independent variables. For example, an investigator might use gender and a personality variable, such

as anxiety, in combination with a manipulated variable such as the type of visual stimulus presented in a study of reactions to fear stimuli. An increasingly popular type of study involves multiple measures of the participants over time. Books on research methods provide details about these experimental designs. Two classic evaluations of research designs by Campbell and Stanley (1963) and Cook and Campbell (1979) give the strengths and weaknesses of many design options. A thorough review of the most recent thinking on methods for measuring the change caused by treatments over time is provided by Collins and Horn (1991).

▶ *Now You Do It*

Properly generalizing the results of a research study requires that we be able to identify the appropriate populations. For each of the following situations from Now You Do It on page 236, identify the sampling population and, if appropriate, the treatment population. How valid would each generalization be?

1. Senator Slade says that, based on letters his office has received, 90 percent of the American public supports gun control.
2. Jesse draws a sample of three students from his first-grade class by putting every child's name in a hat and drawing out three names.
3. The city library board draws a sample of voters for a questionnaire study by taking every 20th voter on the roster of registered voters.
4. Gerhardt recruits 20 students from the introductory psychology class to participate in his study of obedience. Students' attitudes toward civil disobedience will be measured before and after seeing a film on police brutality.

Answers
1. Senator Slade's sampling population is those who write to him about gun control, but he generalizes to all Americans. The generalization is not appropriate.
2. Jesse's sampling population is the members of his class. Generalization is appropriate.
3. The sampling population is all eligible voters. Generalization is appropriate if the return rate from questionnaires is reasonably high.
4. The sampling population is introductory psychology students, and the treatment population is intended to be people who are exposed to the film. To the extent that college students might react to the film differently than other people, the generalization is weakened.

Parameters and Statistics

As we saw in Chapter 1, there is a distinction in numerical indices that parallels the difference between populations and samples. Numerical indices that apply to a population are known as **parameters.** They are usually represented by Greek letters, and *their values are true and unvarying for their populations*. For example, the mean of a population is generally indicated by the Greek letter μ (lower case mu). The parameter is the value of the index—for example, the mean or the correlation coefficient—that we would obtain by appropriate computation from measurements taken on every member of the population.

The numerical index that applies to a sample is called a **statistic.** The mean computed on measurements from a sample is a statistic, as is the standard deviation. Statistics are usually symbolized with English letters (such as *S* for the standard deviation). *Their values*

A **parameter** is a characteristic of a population and is usually represented by a Greek letter.

Table 9-1 Summary of Labels for Most Commonly Used Parameters and Statistics		
Index	**Parameter Label**	**Statistic Label**
Mean	μ (mu)	$\overline{X}$ (X-bar)
Standard deviation	σ (sigma)	S
Correlation	ρ (rho)	r

Statistical inference is the logical and mathematical system for relating observed sample statistics to unobserved population parameters.

are true and correct for the sample from which they were computed, but may not be the same as their corresponding population parameters. We rely on **statistical inference,** the logical and mathematical system for relating observed sample statistics to unobserved population parameters, to generalize our sample results to the population. For example, a sex researcher who wishes to estimate the frequency of affectionate relationships among older adults in the American population could obtain questionnaire results from a sample of couples. The procedures of statistical inference would enable her to get an accurate estimate of how the population would have responded to her questionnaire. The symbols for the most commonly used parameters and statistics are given in Table 9-1.

It is important to take note of the close relationship between the statistic-parameter distinction and the sample-population issue. When the population has a proper prior definition and the sample is drawn from it, the logical processes and mathematical procedures of statistical inference described and developed in subsequent chapters can be applied to generalize from the observed statistic to the unknown parameter. In the absence of a properly defined population and a properly drawn sample, such generalizations from statistics to parameters cannot appropriately be made. If the individuals we have measured are the only ones in whom we are interested, statistical inference and the procedures discussed in this and the following chapters, of course, are not needed. In such cases, the sample and the population are the same thing, and the conclusions reached are, within the limits of measurement errors, known to be accurate.

SUMMARY

The primary unit for research is the *population,* which includes all individuals to whom we want to apply the results of a study. There are two kinds of populations: the *sampling population* from which the sample is drawn and to which generalizations about status variables are made, and the *treatment population* which is hypothetical and to which generalizations about manipulated variables are made. A *sample* is a subgroup of a population. The sample is the group on whom the study is conducted with the intention of generalizing the findings in the sample to the appropriate population. A summary index such as the mean, when computed for a sample, is called a *statistic.* The same index computed for the population is called a *parameter.*

Samples may be obtained in several ways. A poor, but frequently used, approach is the *sample of convenience.* This method uses those subjects who are most readily available and is poor because there is no defined population to which the results can be generalized other than the treatment population. One preferred alternative is the *random sample,* where the population is defined and all members have an equal chance of appearing in the sample. Several *stratified random sample* methods have been developed to ensure representative samples, and some nonrandom representative procedures have also proven useful. The most important feature of the sample is that it must be representative of the sampling population from which it was drawn.

REFERENCES

Campbell, D. T., & Stanley, J. C. (1963). Experimental and quasi-experimental designs for research. In N. L. Gage (Ed.), *Handbook of research on teaching.* Chicago: Rand McNally.

Collins, L. M., & Horn, J. (1991). *Best methods for the analysis of change.* Washington, DC: American Psychological Association.

Cook, T. D., & Campbell, D. T. (1979). *Quasi-experimentation: Design and analysis issues for field settings.* Chicago: Rand McNally.

EXERCISES

1. Differentiate between a sample and a population.
2. Discuss why it might be preferable, or even necessary, to use data from a sample rather than data from a population.
3. Discuss the characteristics of a good sample.
4. Describe or define the various kinds of samples.
5. For each of the following, indicate the type of sampling used.
   a. The Hemingford School District wants to conduct a survey of the adult citizens (defined as people who are 21 years of age or older) within a school district that focuses on their perceptions of the quality of education that the young people of the community are receiving. Since Hemingford is a metropolitan area, not every adult citizen can be surveyed. The researchers decide that the most critical variable to be considered is socioeconomic status. Since certain neighborhoods within Hemingford can generally be associated with a particular socioeconomic status, those neighborhoods are identified and placed into groups based on socioeconomic status. As a result of the 2000 census, the proportion of each socioeconomic status group within Hemingford is known. Researchers then randomly select neighborhoods within each socioeconomic status group and randomly select blocks within each neighborhood. Each adult person on each selected block is then interviewed about the quality of education that the youth of Hemingford are receiving.
   b. Members of the legislative branch of a state government want to determine what people's attitudes are regarding capital punishment. The first question they want to address is whether or not the voters of the state support capital punishment. A subgroup within the legislature decides that it will determine the level of support for capital punishment by surveying those who are most affected by crimes of violence. In this regard, they survey the members of the organization Victims of Violent Crimes.
   c. A joint committee of the university community and the civic leaders of a small city where Smalltown University is located want to determine how the students at the university perceive the working relationship of the university and the city. To this end, the computer bank that lists all of the currently registered students is used to generate a sample that contains 15% of the current student body such that each student has an equally likely chance of being selected.
   d. Professor Chance is interested in college students' tendencies to procrastinate. Since he is teaching an introductory psychology course with 450 students, Professor Chance administers a procrastination questionnaire to his students.
   e. The Junction City School District is interested in determining how many of its high school students who are of a legal driving age (defined as age 16 in the state where Junction City is located) have tried to convince another student from the school not to drive a vehicle while under the influence of alcohol. Since the school computer contains all of the names and ages of students, students who are 16 years of age or older are selected and listed alphabetically. The computer is instructed to randomly select a student on this list and then select every 10th student after that who is on the list until 150 students are selected.
   f. Students enrolled in one of five major universities in a state are surveyed on their spending patterns to determine issues of setting tuition and fee rates. Since the number of students enrolled in each school is different, the sample is selected in such a way that it reflects the relative proportions of the five universities. Furthermore, two of the schools contain 60% females, two of the schools contain 58% males, and the fifth school has an equal proportion of males and females. Care is taken such that the proportions of males and females in the samples from each university are consistent with the proportion in the population of each university.
6. Describe the two different types of bias associated with sampling.
7. What does it mean to have results of a study confounded by another variable? Provide an example to illustrate this concept.
8. Define a sampling population and a treatment population in such a way that the differences are highlighted. Provide an example of each type of population that will illustrate the difference.
9. An educational psychologist was interested in how well sixth graders in the United States made the social, emotional, and cognitive adjustments from elementary school to middle school. The educational psychologist decided to initiate a Transitions pilot program to help sixth grade students in her school district with these adjustments. She randomly selected 5% of the incoming students to participate in this pilot project. The educational psychologist also formed a sample that matched the pilot group on important characteristics such as gender, socioeconomic status, grades in elementary school, and so forth. She labeled this group the control group. At the end of six weeks of school, the students in the Transitions program were compared to the control group on measures of social, emotional, and cognitive adjustment.
   a. What is the sampling population for this research study?
   b. What are the treatment populations, if any, for this study?
   c. What is the accessible population for this research study?
   d. What is the target population for this study?
   e. What type of research study is the study described above (i.e., what type of inference can be made?)?
10. In order to determine if the color of a room had an effect on the level of physical and verbal aggression of preschool-aged children (2–5) in the United States, a researcher randomly assigned preschool-aged children to either a room painted red or a room painted blue. Children were videotaped during 20-minute play sessions. Measures of physical and verbal aggression were then recorded for each child from the videotapes.
   a. What is the sampling population for this research study?
   b. What are the treatment populations, if any, for this study?
   c. What is the accessible population for this research study?

d. What is the target population for this study?

e. What type of research study is the study described above (i.e., what type of inference can be made?)?

11. A researcher is interested in malnutrition from birth to age two and intellectual development at age five. A sample of five-year-olds who were known to have suffered from malnutrition during the period from birth to age two and a sample of five-year-olds who did not suffer from malnutrition during the period from birth to age two were all administered IQ tests at age five. Children from three major metropolitan areas in the United States were selected randomly from each category.

a. What is the sampling population for this research study?

b. What are the treatment populations, if any, for this study?

c. What is the accessible population for this research study?

d. What is the target population for this study?

e. What type of research study is the study described above (i.e., what type of inference can be made?)?

12. For which of the following research methodologies is the risk high for making errors in our conclusions about the population?

a. The National Rifle Association conducted a survey of its members to determine the level of support for gun control legislation of citizens in the United States.

b. A political pollster randomly sampled the registered voters in the United States regarding their preference for President of the United States.

c. The Newton School District has seven high schools in its district. School administrators used the standardized achievement scores from an upper-middle-class high school to convince state officials and parents that their students were achieving at higher levels than students in other school districts in the state.

d. A random sample of members of the American Association of Retired Persons was surveyed in order to determine if there was national support for increasing Social Security payments in the United States.

e. Public school teachers in the state of Nebraska were randomly sampled to determine the level of support among public school teachers for legislation in support of year-round school.

f. Automobile manufacturers around the world conducted tests of fuel efficiency on a randomly selected number of all makes and models of new cars that were equipped with a special carburetor.

13. Indicate what type of sample is involved in each description below.

a. In order to determine if there is a relationship between empathic concern and physical aggression, a researcher utilized undergraduates enrolled in an introductory psychology course.

b. A political analyst surveyed a sample of United States citizens regarding their support of raising the mandatory retirement age to 70. Individuals were randomly selected from ethnic classification, age group, gender, and occupation based on the most current United States Census Bureau information.

c. The School Board of Washington High School is interested in determining the level of student support for requiring that a school uniform be worn to school. Student identification numbers are utilized in the sampling process since the 2,000 students at Washington High School are assigned student numbers. A table of random numbers is utilized to select the students until 50 students have been selected. These students are then asked to state their level of support for a proposal to initiate a policy that requires students to wear a uniform to school.

d. Members of the National Rifle Association are polled to determine if there is support for gun control legislation in the United States.

e. The leaders of a local union want to determine if there is majority support for an employee strike to improve working conditions at a factory that employs 5,000 workers. Because they are in the early phase of planning, the leaders want to sample 1% of the employees to see if they favor a strike before they put a lot of time and effort into planning the strike and negotiating. The employees are given a union number upon joining the union. Since these numbers are entered into a computer, it is easy to arrange them from smallest to largest. A membership number is then randomly drawn and every 100th member listed after that number is included in the sample until 50 names have been selected.

14. For each of the following, determine the number of participants needed for each cell of a stratified random sample.

a. Suppose that a high school district within a metropolitan area has 20,000 graduating seniors. They want to survey 5% of these students one year after graduating about the quality of their educational experience. They have decided to survey students based on the following criteria: attended a four-year college or university (20%), attended a two-year community college (15%), attended a vocational/technical school (40%), or worked or enlisted in the military (25%).

b. Suppose a researcher is interested in determining if there is a gender (female vs. male) and age category [adolescents (age 12–18) vs. young adults (age 19–25)] difference in the United States on a measure of responsibility. Assume that the proportion of males and females and the proportion of adolescents and young adults is the same in the United States. Find the number of participants in each category if the researcher wants a sample size of 1,500.

c. A social psychologist is interested in conducting a survey of people who are of legal voting age in the United States. However, the researcher wants a gender breakdown (female vs. male) and an educational breakdown (no high school diploma, high school diploma only, some post-secondary education credits, college/university degree) to be reflected in the final sample. Suppose the gender breakdown for this group based on the latest available data is 52% female and 48% male. In addition, suppose the educational breakdown is such that 10% do not have a high school diploma, 25% have a high school diploma only, 35% have some post-secondary education credits, and 30% have a college/university degree. Suppose the desired sample size is 1,000. Given only the information above, how many participants should be in each group?

Chapter 10

Hypotheses and Data Collection

CHAPTER OBJECTIVES

After studying this chapter, you should be able to:

- ❏ Given a description of a study, identify the independent and dependent variables.

- ❏ Explain the difference between status variables and manipulated variables.

- ❏ Describe the differences between research studies and evaluation studies.

- ❏ Given the description of a study, classify it as primarily a research study or an evaluation study.

- ❏ Explain why scientific research cannot prove something to be true.

- ❏ Explain the differences between inductive and deductive reasoning.

- ❏ Identify the assumed premises and facts of a claimed association.

- ❏ Given a theory, produce a hypothesis to test the theory.

- ❏ Given the description of a study, identify plausible rival hypotheses and state how their influences might be controlled or eliminated.

Do you remember the challenge confronting Dr. Numbercruncher and the faculty of the SAD Department at Big Time U? They were faced with the task of designing a study to determine whether students' mathematics anxiety might be affecting their performance in the basic statistics class. What specific questions do you think the faculty will be trying to answer with their study? Which of the variables will serve as independent variables, and which will be dependent variables? Score on the statistics final will clearly be a dependent variable, but can you see any others? Lab section will be one of the independent variables, but what about mathematics anxiety itself? Which of the variables are status variables, and which are manipulated variables? How will each of the variables that the researchers have selected to include in their study help to answer their questions? How can they phrase their questions so that the data they collect will provide answers to those questions?

Introduction

We noted in Chapter 1 that the difference between research studies and evaluation studies depends primarily upon the use that is to be made of the results. For example, the question of whether the lab section a student is enrolled in affects her performance on the statistics final is more an evaluation question than a research question because the SAD faculty want to know whether one particular way they are now teaching the material works better than other ways they are using. On the other hand, the question of whether there is a relationship between mathematics anxiety and performance in statistics classes is more a research question than an evaluation question because the goal is to obtain knowledge about the effects of math anxiety. Such differences in purpose have important effects on the logical structure that underlies what we do in a study. In this chapter, we will define some relevant terms and discuss how evaluation studies differ from research studies in their interpretation and execution. Then we will examine the logical structure that underlies the testing of scientific theories and of research hypotheses.

Independent and Dependent Variables

Both evaluation studies and research studies share an interest in the relationship between variables. For example, both the question of whether lab section makes a difference in performance on the statistics final at BTU and the question regarding the correlation between anxiety and performance are asking whether there is a relationship between two variables. Determining the presence (and perhaps magnitude) of a relationship between such variables is the primary reason for performing a study.

It is common practice in psychological and educational research to select the values for one variable and to leave the other variable free to vary. For example, we can select certain shock levels in an avoidance study and record the number of trials at each shock level a rat needs in order to learn the avoidance response. Or we can select particular teaching methods and measure educational achievement after instruction by these methods. The variable whose values are selected by the investigator is called the **independent variable.** Obviously, the independent variable must be a variable; it must have more than one value. For example, gender can be an independent variable when both males and females are included in the study; if only females are included, gender is a constant, or controlled, condition, not an independent variable. In the BTU study, lab section can be a variable if there are two or more sections, but it would be a constant if all students in the study were in the same section. The defining feature of the independent variable is that the investigator exercises direct control over the values of it that occur in the study.

The variable that is left free to vary is known as the **dependent variable;** its values are seen as depending upon the independent variable. The dependent variable is the measured outcome of the study. In the two examples given above, we can say that speed of learning the avoidance response (number of trials) depends upon level of shock, and educational achievement depends upon method of teaching; at least, this is probably what the investigator would like to be able to say. Both the number of trials and educational achievement are measured outcomes in their studies. Thus, they are dependent variables. In the BTU study, score on the statistics final is also a measured outcome, and, therefore, a dependent variable.

The distinction between independent and dependent variables has proven very useful in closely controlled studies in the natural sciences. One common form of this kind of study exposes a sample to one level or condition of the independent variable and measures the dependent variable, then changes the level or amount of the independent variable and observes whether the dependent variable changes. For example, a chemist studying the effects of changing temperature on the pressure of a gas might measure the temperature and pressure of the gas in a container, then increase the temperature, and observe whether the pressure changes. In another version sample A is exposed to one level or category of the independent variable, sample B is given a different level or category, and the dependent variable is measured in both samples. For example, a medical researcher studying the effects of a new drug on the AIDS virus might expose one sample of the virus to one drug and another sample of the virus to a second drug. In this type of study, a difference in the dependent variable between the two samples is attributed to the difference between the levels of the independent variable.

Manipulated and Status Variables

In younger and less well developed areas of investigation, such as psychology and education, the distinction between independent and dependent variables is often less clear because many characteristics that are studied as independent variables are not really under

Independent variables are those whose values are chosen or assigned by the investigator.

Dependent variables are the measured outcomes of a study.

the investigator's control. The investigator may not be able to obtain a representative sample from the population and to *assign* each sample to a level of the independent variable. For example, the BTU faculty may wish to study the influence of mathematics anxiety on statistics performance, but they cannot assign their students to levels of anxiety; this is a characteristic each student brings to the study. This problem has given rise to a distinction between variables which investigators can *manipulate* and those which they cannot. Experimental treatments such as shock level or instructional method can be manipulated. Because the investigator can change (manipulate) an individual's level on these variables, they are called **manipulated variables.**

Gender, age, height, weight, ability level, political preference, and many other characteristics often studied as independent variables in educational and psychological research are actually *status characteristics* of the study participants and are not under the investigator's control in the sense that individuals cannot be assigned to levels of these variables. If they function as variables in the study, they are called **status variables.** The investigator may choose to select individuals on the basis of their level on one or more status variables, thereby making them function like independent variables, but the participants nevertheless come to the study with predetermined membership in one group or another. For example, a developmental psychologist may include a group of six-year-olds and a group of eight-year-olds in a study of language development, but the children are *selected* for their age, not *assigned* to their age. This makes it difficult to draw certain kinds of conclusions.

Manipulated variables are independent variables where the investigator controls the level or category of the variable to which the participant is assigned.

Status variables are characteristics that participants bring to the study by virtue of who they are. They cannot be assigned by the investigator.

▶ *Now You Do It*

The variables collected as part of the mathematics anxiety study at BTU were listed in Table 2-1. They included age, sex, major, high school GPA, college GPA, number of math courses taken, math aptitude test score, level of mathematics anxiety, number of homework assignments completed, homework grade, course section, and final exam score. Which of these variables might serve as independent variables, and which would more likely be dependent variables? Which of the independent variables can be manipulated, and which are status variables? Are there any variables that could serve as dependent variables in one analysis and independent variables in another?

Answers

The variables that are clearly independent variables are age, sex, major, high school and college GPAs, number of mathematics courses taken, mathematics aptitude test score, and course section. Of these, only course section can be manipulated because it is the only variable to which faculty can assign students. All other variables are pre-existing characteristics of the students, and therefore status variables. Mathematics anxiety, homework completed, and homework grade might serve as dependent variables in some analyses (for example, is there a relationship between age or mathematics aptitude and mathematics anxiety). Score on the final exam is clearly a dependent variable.

Evaluation and Research

Over the last several years, there has been an increasing awareness that many studies are conducted to determine whether some social policy or government program has had the desired effect and that these studies may be different in important ways from more traditional scientific studies. For example, your university may have instituted a writing requirement in courses for science majors in an effort to improve writing skills. A study of the writing skills of science majors might be conducted to see whether the curricular change has had the desired effect. A study of this kind is quite different in purpose from one in which an investigator wishes to determine whether there is a difference in perceptual acuity between Inuit people from Alaska and urban residents from the lower 48 states. The terms *evaluation* and *policy research* have been used to refer to studies of the former type, while *basic research* and *experiment* are terms suggested for the latter. The difference in purpose between the two types of studies (did it work versus what is the nature of perceptual acuity) means that some rules and procedures may be required for one and not the other.

There is actually a continuous gradation of studies from the most pure, or basic, research to the most socially oriented evaluation. At one extreme is the pure research study whose ultimate objective is to discover scientific truth. This is dispassionate, amoral science, unaffected by anything outside the laboratory and conducted without regard to its potential uses or consequences. Examples of this type of research include investigations of the role of brain chemicals in learning and studies of classical conditioning. This type of study, which provides the model for scientific logic in psychology and education, is becoming rare in a world that is increasingly technological and sensitive to the interconnectedness of all knowledge.

At the other extreme is what we might call a pure evaluation study where the goal is to find out whether we have done what we set out to do. Questions are not phrased in terms of the nature of the universe but as whether our manipulation of the physical or social environment had the desired effect. For example, a school district might increase class size for some of its classes and not others and then conduct a study to compare the level of achievement of students in the larger classes with that of students in the smaller classes. If students in the larger classes perform just as well as students in the smaller classes, the school district may increase the size of all classes in order to save money. The focus is on the fact that the change did not have a detrimental effect.

This last example points out one of the most common differences between evaluation and research. Evaluation studies usually lead to direct consequences in terms of action: A program is started, expanded, reduced, or canceled as a result of the outcome of the evaluation. Pure research seldom has such effects. Another difference relates to the investigator's ability and desire to generalize results to a population. Sometimes in evaluation studies every member of the population has been included, so that there are no additional cases to which to generalize. Thus, data analysis may stop at the descriptive level because this form of data summarization provides answers to the investigator's questions. The descriptive indices are population parameters in this case. For example, a company might decide to institute a flex-time program, allowing employees to come to work at different times of the day that fit their personal schedules and needs. Company managers would monitor productivity, and if productivity did not drop, the program would be retained. The company's entire workforce is the population of interest, and the productivity of the entire workforce is the outcome of interest.

An investigator engaged in a research study, on the other hand, usually will wish to generalize the findings of the study to the widest possible population in order to state general conclusions that can be applied in any similar set of conditions. The results of the study are considered to be statistics, which are imperfect indicators of the desired parameters. For example, studies of classical or operant conditioning are intended to expose the workings of the normal human mind, usually without regard to cultural barriers. Most studies conducted by educators and psychologists, such as those that examine different types of therapy or different educational treatments, fall between these two extremes: They are motivated both by a desire to understand some phenomenon and by the prospect of having social or policy consequences.

Because of their different contexts and objectives, evaluation studies and research studies are conducted under somewhat different sets of rules. Although *careful planning and the best possible measurements are essential for useful results in either case,* studies done for research purposes are, in general, more carefully controlled and more restricted in the types of conclusions and generalizations that are drawn. A loose, but not entirely unfair, characterization of the way an evaluation study might be interpreted is that it *does not contradict* what its designers set out to show. This could mean that the variable under study is cost-effective, that it achieves the same results in less time than its competitors, for example. A study of the effect of class size on achievement falls into this category. Interpretation of an evaluation study may be quite casual and informal, in contrast to the rigidly prescribed logical structure of a research conclusion.

Another distinction that is often drawn between evaluation and research concerns values. An evaluation study often is conducted to show that something is true. The investigator has a clear stake in which way the results turn out. Research, on the other hand, is often described as value-free in the sense that the investigator is seen as being interested only in the truth and is indifferent to the direction of the outcome. The school district studying the effects of class size wants to reduce expenses, but the research psychologist studying the effects of delay of reinforcement in operant conditioning is characterized as not caring whether the experimental treatment enhances or retards performance. As recent

RESEARCH AND EVALUATION STUDIES

Research Studies	Evaluation Studies
1. Investigator assumes an indifferent attitude to the study outcome.	1. Investigator has a clear preference for one outcome over another.
2. Interpretation of results follows strict rules of inference.	2. Results are often interpreted as not inconsistent with the policy position that motivated the study.
3. Study results are generalized as widely as possible.	3. Results applied to policy or program that motivated study.
4. Study seldom has direct consequences in action.	4. Study usually has direct implications for action.

philosophers of science have pointed out, this is not really the case. The scientific researcher always approaches the study from a particular theoretical position, and the study is designed to support one belief or refute another. Thus, while the scientist may play the research game under a tighter set of rules than those that apply to policy analysis and evaluation, the study is still intended to show that someone's interpretation of reality is correct.

▶ *Now You Do It*

Consider the two studies that we have been looking at throughout the first nine chapters, the BTU study and the study of the Searchers. Is either of these an evaluation study? Is either of them basic research? Could you justify drawing a distinction between them?

The SAD faculty at BTU have designed a study that is more toward the evaluation end of the spectrum. If you review the description of the problem in Chapter 1, the study was designed to find out whether mathematics anxiety was related to students' test performance and whether changing the way lab sections were taught might improve the situation. What is the faculty likely to do if they find that one way of teaching the labs is clearly superior to the others?

Dr. Cross and his students are conducting a study that is more toward the basic science end of the continuum. The outcome of their study will not have particular policy implications. They may find that certain variables are associated with more liberal or conservative attitudes toward the role of women in society, but even if their results strongly support or strongly refute the accusations of POW (the Political Organization of Women, whose public statements about the Searchers initiated the study), there are no direct actions that are suggested.

The Logic of Research

For the student who is just beginning to study the way research is done in psychology and education the traditional process often seems backward: The investigator makes a statement of belief about the phenomenon under study, called a hypothesis, and then sets out to prove that the hypothesis is wrong! Why can't we simply say what we believe will happen and, if our expectation is confirmed, conclude that we were right? The source of the apparently illogical logic of research in all scientific disciplines lies in philosophy. For centuries philosophers have struggled with ways of discovering truth and ways of knowing about causal effects. The general trend of their thinking, as it applies to scientific inquiry, is that we can never conclude that something, a theory or hypothesis, for instance, is true from empirical evidence. However, we can conclude that it is false. We can never know what something is, but we can reach conclusions of reasonable certainty about what it is not.

It is sometimes said that science is the process of ruling out wrong explanations. There are two reasons philosophers offer for the statement that we can only rule out wrong explanations, not conclude that any explanation is true. The first, which we have already encountered, is the approximate nature of measurement operations: All measurements are only approximations of reality. The second, which extends beyond the bounds of measurement and science and goes all the way back to Aristotle, is the nature of the reasoning process itself.

Approximations in Measurement

A simple example can illustrate what the philosophers are saying about approximations in measurement. Suppose that we have a table and we want to measure its length. (We could equally well be measuring intelligence, weight, creativity, or any other continuous variable as an example.) We can use the best measurement procedures yet devised and still never be certain of the exact length of the table. Our measurements can rule out most of the possible values; we may, in fact, be able to rule out values below 59.9995 inches and above 60.0005 inches, but so long as the trait (not the measurement scale but the trait itself) is continuous, we can never say that any one value *is* the length of the table. We only have a range of uncertainty, albeit in this case a very narrow one.

A related issue in measurement is the dependence of the observed value on the conditions under which it was made. It is very important to make observations under standardized conditions, but it is impossible to assure absolute equivalence. Variations in the conditions of measurement result in measurement error and less than perfect reliability, and hence in some uncertainty as to the exact value of an observation. Of course, beyond some point, our ignorance loses its importance for practical applications of measurement; however, this principle of approximation governs all scientific applications of measurement.

The Reasoning Process

It is useful to distinguish two quite different logical processes when trying to understand the nature of scientific inquiry because the nature of the reasoning is not the same. The first of these processes is called **deduction.** Deductive reasoning proceeds from a set of facts that are known to be true to a conclusion about a necessary consequence. In its simplest form, the deductive reasoning chain is as follows:

General fact	All members of this class understand the mean.
Specific fact	Mary is a member of this class.
Specific conclusion	Mary understands the mean.

> *Deduction* involves reasoning from a general principle which is known to be true to a specific conclusion.

Given that the facts are true, the conclusion must be true. Of course, one or more of the facts may actually be false, but that is not a part of the reasoning process itself.

Contrast the example of deduction with the following set of statements that form an **inductive** chain of reasoning:

Specific fact	Mary understands the mean.
Specific fact	Mary is a member of this class.
General conclusion	All members of this class understand the mean.

> *Inductive reasoning* involves reasoning from specific facts to general conclusions.

The three sentences are identical in the two examples but appear in different order. Even given the truth of both facts in the second case, the conclusion does not necessarily follow; it could be either true or false even if both facts are true. We are using inductive reasoning when we argue from specific facts to general principles.

The difference between these two examples lies in the nature of the facts and conclusions. Deductive reasoning involves going from a general fact or principle (all members of group X have property Y) and the inclusion of a specific case within the group (Mary is a member of X) to a particular conclusion (Mary has property Y). The property of the particular case is *deduced* from the known general principle, and the conclusion can be correctly identified as either true or false. Technological applications of confirmed scientific findings—for example, the force of gravity and the effect of air resistance in the ballistics of rocket flight—involve this kind of process. However, since the objective of scientific inquiry is to find the general principles themselves, *science cannot use the deductive reasoning process as a final arbiter of truth*.

When, as in the case of scientific investigation in search of a general principle, we apply inductive reasoning, we reason from the particular fact (Mary has property Y) and fact of class membership (Mary is a member of X) to the general principle (all members of X have property Y). *No number of confirming particular observations can prove a general principle to be true*. We are no better off even if we have a substantial collection of specific facts, such as

> Peter understands the mean.
> Paul understands the mean.
> Mary understands the mean.
> Peter, Paul, and Mary are members of this class.

The only valid conclusion from inductive reasoning is a negative one. For example, given the facts,

> Edith *does not* understand the mean.
> Edith *is* a member of this class.

the conclusion, "*Not all* members of the class understand the mean," is a proper logical conclusion. That is, the fact that Mary understands the mean does not allow us to conclude that all students understand the mean, but the fact that Edith does not understand the mean does allow us to conclude that not all students understand the mean. Note that the location of *not* is crucial: The statement, "All members of the class do not understand the mean," which contains the same words (plus "do") in slightly different order, is an improper conclusion.

At this point, we can state two general principles about inductive reasoning in scientific research. First, *no number of confirming particular facts can assure that the general principle is true,* except for the trivial case where every member of the population is included as a fact. Second, *a single disconfirming fact, or exception to the general principle, proves the principle false.* Therefore, because science only has access to specific instances, or observations, the only way science can make progress is by ruling out incorrect alternatives. There are many possible explanations for any phenomenon, and a scientific study is designed and conducted to rule out some of the possible explanations. All scientific research, whether in physics, biology, sociology, education, or psychology, is part of an ongoing process of excluding incorrect explanations for phenomena.

▶ *Now You Do It*

Logical arguments do not ordinarily come packaged neatly as premises and conclusions. You have to dissect statements to find what premises are being assumed and what conclusions are claimed. Try your hand with the following statements that might appear as items on the evening news. State whether each involves deductive or inductive reasoning, and identify the principles claimed, the facts involved, and the conclusions asserted.

1. Joe Camel died prematurely because he was a smoker.
2. Professional golfers make lots of money. Just look at Tiger Woods and Annika Sorenstam.
3. The movies *Jurassic Park* and *Godzilla* were tremendously popular, showing that Americans love movies about dinosaurs.

Answers

1. The facts asserted are that Joe smoked and that he died prematurely. The principle claimed is that smoking causes premature death. The conclusion is that smoking caused Joe's death. The reasoning is deductive, but the tobacco industry would challenge the principle claimed.
2. Two specific facts are asserted, that Tiger Woods and Annika Sorenstam make a lot of money, and that they are professional golfers. The conclusion claimed as a general principle is that all professional golfers make a lot of money, and the reasoning is inductive, but not valid.
3. Again, two specific facts are claimed, that *Jurassic Park* and *Godzilla* are movies about dinosaurs and that they were very popular. The conclusion claimed and the general principle asserted is that these facts show Americans love this general class of movies. It is invalid inductive reasoning.

COMPARISON OF INDUCTIVE AND DEDUCTIVE REASONING

Deduction	Induction
1. Proceeds from a general principle to a specific conclusion.	1. Proceeds from specific instances to a general conclusion.
2. Requires that general principle be known.	2. Leads to the discovery of general principles.

Theories and Hypotheses

The philosophical school which developed the argument that science progresses by ruling out wrong explanations is known as falsificationism. Its most influential spokesman, Karl Popper, argued that the purpose of science is to rule out incorrect explanations. According to Popper, science proceeds by showing one explanation after another to be incorrect. Someone proposes an explanation for a phenomenon, which is known as a **theory** about the phenomenon. The task of science, in turn, is to show how the theory is false. Under this logical system, all theories ultimately are false, and currently held theories have the status "Not yet shown to be false." (There are other logical systems that have been proposed to govern scientific inquiry, but falsificationism underlies the logic of statistical inference as it is practiced in educational and psychological research. An introduction to the alternatives is given by Phillips [1987]. Cook and Campbell [1979] also discuss issues relating study design to the philosophy of science.)

> A **theory** is a proposed explanation for some phenomenon.

In order to rule out, or eliminate, an explanation as true, we must phrase the explanation in terms of its possible consequences. That is, if the explanation is untrue or false, we expect to observe that certain events occur. This means that the explanation must be stated in such a way that it is possible to observe and measure the events related to it. The proposed explanation of the phenomenon is the theory, and statements that specify observable consequences of the theory, what we would expect to find if the theory is true, are called **hypotheses.**

> **Hypotheses** are predictions derived from theories. A hypothesis predicts what we should expect to observe if the theory is true.

A hypothesis usually takes the form of an if-then statement. If theory X (children learn to read more quickly using phonics) is true, then under conditions A, B, and C (children instructed by the phonics method in normal classroom settings for a year by their regular teacher), using measurement procedure Y (a specific reading test), we should observe Z (children in the phonics group earn higher scores than children in other groups). Failure to observe Z implies that theory X is untrue, at least under conditions A, B, and C. If the theory that the world is flat is true, then a boat sailing west from San Francisco should fall off the edge and never be heard from again. If what we observe is that, after dodging various land masses that may be in the way and avoiding assorted sea serpents, the vessel arrives back at its starting point from the opposite direction, this implies that the theory is false because the observation is inconsistent with the theory's predictions. In considering this example, the important point is that no amount of research using sailing vessels (or those powered by any other means) could ever prove that the world is flat. Although the theory may state that you will fall off the edge of the flat world if you get too close, failure of your ship to return does not permit the conclusion that the theory is true, only that you were unable to prove it false. There are a variety of other reasons why you might fail to return, including shipwreck, being eaten by sea serpents, and deciding to settle on a delightful island in the South Pacific. The test of the hypothesis merely failed to disconfirm the theory.

Suppose, on the other hand, that I observe you sailing west from San Francisco, and some months later I observe you approaching the same point from the southeast. Can I conclude from these observations that the world is not flat? Not necessarily, because there are other explanations—other theories—that would cause us to expect to see you approaching from that direction. One possibility is that you cheated and, circling back out of sight, are now approaching as though having sailed around the world. Another is that the flat world rests on the back of a giant turtle and that when you reached the edge, the turtle caught your falling ship and returned it to the surface of the ocean on the other side of the world. There are countless other alternatives.

This far-fetched example contains several types of hypotheses that are important in scientific investigations. First, there is the hypothesis that comes from the theory of interest; this is often known as the **research hypothesis.** *The research hypothesis states what you expect to happen if your theory is true.* The theory that the world is flat leads to the prediction, the research hypothesis, that you will not return.

Next, there is the opposite of the first hypothesis. This second hypothesis, often termed the **alternative hypothesis,** arises from an alternative theory. In the scientific world an alternative hypothesis often is the negation of the research hypothesis. The alternative theory to our flat world research theory, that the world is *not* flat, leads to the prediction, the alternative hypothesis, that you will return. The alternative hypothesis and the research hypothesis make contradictory and mutually exclusive predictions.

Finally, there is a large class of other explanations for what we find that are consistent with either or both of the possible outcomes predicted by the research and alternative hypotheses. These other explanations, that you are a cheat, that you were shipwrecked or devoured by sea serpents, and so on, lead to **plausible rival hypotheses.** Plausible rival hypotheses provide other reasonable explanations for why the study turned out as it did, whichever way the results come out. (See Huck and Sandler [1979] for a list of cases where rival hypotheses provided alternative explanation for research findings.)

> The **research hypothesis** is the outcome you would expect if your theory is true.
>
> An **alternative hypothesis** is a hypothesis derived from a theory other than the one being tested in a study.
>
> **Plausible rival hypotheses** provide other reasonable explanations for why the study turned out as it did, whichever way the results come out.

▶ *Now You Do It*

Consider again the questions posed by the BTU faculty. Dr. Hardy said she felt that some students knew the course material better than was indicated by their test performance. What are some of the hypotheses that she might be entertaining about the relationship of final exam performance with the other variables in the study? What would the alternative hypothesis be in each case? Can you identify any plausible rival hypotheses that would account for the findings?

Answers

Clearly, Dr. Hardy thinks math anxiety is related to exam performance in a way that is different from its influence on homework assignments and class behavior. She probably would predict that there would be a negative relationship between math anxiety and test scores and that math anxiety would not be related to other variables. She might also expect that students who had taken more math courses would have lower anxiety (notice that she predicts relationship but does not infer causation) and that students who hand in more homework assignments might get higher test scores (perhaps causal?). Teaching each section using a different instructional strategy would test a hypothesis that some ways of teaching statistics result in greater learning of the material. The general alternative hypothesis would state that there is no relationship between any of these variables and that all teaching methods are equally effective. Plausible rival hypotheses include the influence of variables not considered in the study, special effects of having one lab instructor being particularly proficient with one teaching method so section differences are due to instructors, not methods, and an almost infinite variety of other factors.

Designing a Study

In any area of scientific inquiry there are theories. Often there are competing theories that offer different explanations for the phenomena. There are always plausible alternative theories that no one has yet considered. The goal of anyone who is conducting research is to plan the study so that the information collected will lead to a clear and unambiguous conclusion about the nature of truth. No study ever reaches this ideal stage of clarity, but there are various things the investigator can do to rule out some of the wrong alternatives. Carefully planning the study to rule out as many of the competing explanations as possible is essential to scientific progress. No amount of elaborate analysis can recover information from data when the planning of the study was inadequate. Remember that in Chapter 1 we said numbers become data when they have additional meaning beyond numberness and that it is the measurement process and the research design, the process by which the numbers were collected, that gives them their meaning. Statistical analysis cannot eliminate plausible rival hypotheses; only good research design can do that.

Ruling Out Plausible Rival Hypotheses

Control. One approach to ruling out various types of plausible rival hypotheses, which has had a long and honored tradition in many scientific disciplines, entails using *control variables* or *experimental controls*. In either sense, control means holding some condition constant to remove the possibility that it will have an effect on the dependent variable. This principle is employed, for example, by the physicist who takes all experimental measurements at the same temperature. Because the temperature is controlled and not allowed to vary, this rules out differences in temperature as an explanation for differences in the dependent variable. Experimental control is also employed when a psychologist uses only female subjects in a learning experiment or an educator studies teaching methods only in third-grade classes. Any characteristic that represents a constant condition across all observations can be ruled out as an explanation for any differences that are found and thus is eliminated as a plausible rival hypothesis. However, as we discussed in Chapter 9, there is a cost to control. It limits the range of conditions over which we can generalize our results.

Measurement Issues. As previously suggested, quality measurement procedures are equal partners with good research design in evaluating the research hypothesis. One of the most serious sources of plausible rival hypotheses is inadequate measurements. It is essential that the measures be reliable, thus reducing random errors of measurement that would make testing the hypothesis more difficult. The measures must also be valid, or the subjects' scores will not mean what we think they mean; thus, our conclusion would almost certainly be wrong because it would be drawn about something we were not intending to measure. Quality measurements of the trait or traits of interest coupled with good research design and careful data collection are necessary prerequisites if statistical analysis is to lead to proper conclusions. A brief review of measurement concerns is provided in Appendix B.

Randomization. Given adequate controls and good measurement procedures, a powerful way to eliminate alternate explanations is called **randomization,** or **random assignment of individuals to conditions.** This procedure is used to eliminate systematic differences between groups of study participants in order to equate the samples at the start of a study. For example, suppose we want to compare the effects of two teaching methods, a traditional rote-memory approach and an insight approach, on third graders learning the multiplication tables. If we let each child's parents determine which method their child will use, there is the danger that systematic differences among the parents will result in two groups of students who differ in systematic ways. These systematic but unmeasured differences in the samples provide plausible rival explanations for any differences we observe between the groups. By randomly assigning each child to one teaching method or the other, we rule out the possibility that systematic, or selective, differences will exist between the groups. Random assignment of participants is the most powerful method that social and behavioral scientists have of eliminating plausible rival hypotheses.

There are many other procedures that have been devised to rule out various undesirable factors in a research study—so many, in fact, that we cannot begin to cover them here. A number of books have been written on the topic, some of which are listed at the end of this chapter. Anyone who will be designing an original research study should consult such a source *before beginning the study* because there is no way to use statistical analysis to salvage meaning from a badly designed study. It is the design of the study and the quality of the measurements that permit us to isolate the research hypothesis from plausible rival hypotheses and to test the research hypothesis against its alternative.

> *Random assignment* is a way to eliminate any systematic differences between groups and to assure that samples differ only in random ways at the beginning of a study.

▶ *Now You Do It*

Dr. Grossman is studying individual differences in cognitive abilities. He believes cognitive ability is largely inherited. His research design calls for studying pairs of identical twins (twins from a single fertilized egg who are therefore genetically the same) who have been reared in separate homes. He recruits twin pairs and measures their cognitive abilities using a test he has developed himself. He compares the results with measurements taken on same-sex ordinary siblings reared in the same home. What is his research hypothesis? What is the alternative hypothesis? What are some plausible rival hypotheses if his results are in favor of the research hypothesis? What are some reasons, other than the alternative hypothesis, that his results might not support his theory?

Answers

If genes are major determiners of cognitive ability, twins should be more similar than ordinary siblings, and their similarity should not be affected by being reared in different homes. Therefore, Dr. Grossman's research hypothesis is that his twin pairs will show a higher correlation than will pairs of ordinary siblings.

The alternative theory is that genes are not important for cognitive ability, so that individual differences are largely due to environmental effects. Therefore, the alternative hypothesis is that the correlation between ordinary siblings will be at least as high as that for identical twins reared apart.

What are some other reasons why the results might be consistent with Dr. Grossman's theory? One plausible rival hypothesis concerns the ages of the participants. Identical twins are necessarily the same age, while siblings are usually at least a year apart in age. Greater age similarity might make the twins more similar. It is also possible that the person administering the test, who knows Dr. Grossman's theory, administers the test in a more similar way to the twin pairs than to the siblings. What other explanations can you think of?

What are some reasons why the results might fail to confirm Dr. Grossman's theory? One concerns the test. There are well-established tests with proven characteristics available. The fact that Dr. Grossman constructed his own test may mean that the measure of cognitive ability is not reliable. A second possibility is that the people raising the twins know that they are twins and have intentionally made their environments different. What other reasons can you come up with?

Drawing Conclusions

We noted earlier that according to the thinking of most philosophers, inductive reasoning in science can only reach the conclusion that something is false, not that it is true. As we shall see in the next chapters, statistical analysis never permits us to prove anything to be either true or false, only to reach a conclusion that a parameter falls in a particular range or that a hypothesis is unlikely to be true, with a specifiable probability that our conclusion is in error. We may, for example, conclude that there is a 95 percent probability that the population mean gain in reading ability using the phonics method of instruction is at least 3.7 points on the reading test we have used. We must always remember that statistical procedures have this limiting feature and view with considerable caution some of the more grandiose claims that are based on statistical analyses.

The conclusions that we can reach as a result of our study are of two basic kinds. Remembering that even in the most simple study there are at least two variables, we can conclude that there is a relationship between the variables under study or that there is not. In the natural sciences and in a few of the more highly developed areas of behavioral and social science, the hypothesis may specify the degree of relationship; in that case, our conclusion would be that the relationship is of that specified degree or it is not. In either case, we have (we hope) designed the study so that the data permit us to choose between only two alternatives, for example, that rehearsal leads to improved memory, or that it does not.

Correlation and Causation

Sometimes it is possible to draw stronger conclusions than that the variables are related; some research designs permit a statement about the nature of the relationship. The most popular of these statements is that the relationship is causal; that is, that changes in one variable cause changes in the other. This is the kind of statement that the chemist is able to make about the effects of heat on the pressure of a gas; increasing the temperature of the gas causes an increase in pressure.

Regardless of whether we can ever truly infer causation or not, many investigators reach the conclusion that there is a causal relationship between the variables in their study. Some even go so far as to say that the type of statistical analysis performed determines whether we can conclude that a relationship is causal. There is a timeworn saying in statistics that we cannot infer causation from a correlation. Unfortunately, this saying has been misinterpreted by some students, and even some teachers of statistics, as referring to the statistical analyses rather than to the data collection procedure. It is the design of the study, the choice of variables, the way they are used and measured, that determines whether causal inferences are warranted. If a treatment, such as using rehearsal in a memory task, causes improved memory, the causal effect occurs when the treatment takes place, not at the time of data analysis. Careful design and execution of a study *may* permit the inference that a relationship *revealed* by statistical analysis is causal, but the statistical analysis itself can only reveal meaning and information that are already in the data. The confusion has arisen because research designs that are likely to permit inferences about causal effects have traditionally been analyzed by one set of procedures (those of Chapters 13 and 16) while studies with weaker designs typically have used the descriptive correlational procedures of Chapters 7 and 8.

There are two primary factors in determining whether we can infer that a relationship is causal or not. One that we have already discussed at some length is a design that rules out other noncausal alternatives. The second is the nature of the variables under study. Some of the variables that are of great interest to social and behavioral scientists cannot be separated from a complex of other variables. Such a situation makes causal inferences impossible.

The major useful distinction for whether causal inferences can be drawn is between variables that can be manipulated, or assigned, by the investigator and variables that are status characteristics of the individual participants. *Manipulated variables* are those variables whose values are chosen by the investigator, and *their values can be changed by the investigator.* If, under appropriately controlled conditions, a change in variable X from value X_1 to value X_2 is consistently followed by a change in variable Y from Y_1 to Y_2, we can infer that there is a causal relationship between the change in X and the change in Y. The inference is particularly strong if reversing the change in X also reverses the change in Y. For example, if a given amount of a gas at constant volume is exposed to a heat source, we observe a rise in both temperature and pressure. The manipulated variable is the presence or absence of a source of heat. We can infer that the addition of that heat caused both the increase in temperature and the increase in pressure. (Had we allowed the volume to change, we would have observed something quite different, but that variable was controlled.)

Status variables are nonmanipulable characteristics possessed by the entities under study. The investigator has no control over these variables; they include such obvious properties as the age, height, weight, and gender of study participants. But as mentioned earlier, variables such as intellectual ability, psychomotor ability, motivation, cultural background, and personality, for example, should also be considered status variables. While their values may change systematically (we are all getting older at a constant rate) or erratically (the social interactions we experience may have quite unpredictable effects), the investigator can do little about them other than attempt to hold them constant by participant selection, equate their effects by random assignment of individuals to groups when the independent variable is manipulable, or observe the presence or absence of relationships among them. It is seldom, if ever, possible to conclude that a relationship between status variables is causal because the variables cannot be isolated from their context.

Consider again our package of gas. When we first observe it, it is at a given temperature, pressure, and volume. These are status characteristics of this particular package of gas. If we observe other packages of the same type of gas, or of different gases, each package will have its own temperature, pressure, and volume. We can, after extensive observation of packages of gas, conclude that there is a highly consistent relationship among these three variables. For example, for packages of gas that are all one liter in size, we would observe a correlation of about +.99 between temperature and pressure. However, we cannot, without manipulation, conclude that the relationship is causal.

Inability to manipulate variables does not necessarily mean that a science lacks power or importance. For example, astronomers have never been able to manipulate the variables they study, but they have determined the quantitative properties of the relationships they investigate to a very high degree of accuracy, and they borrow freely from other physical sciences and mathematics. Consequently, they are able to predict very accurately the occurrence and magnitude of future extraterrestrial events. Well-documented and stable relationships can be very useful and important even when causation cannot be inferred.

Psychologists and educational researchers have also discovered many very useful relationships between variables that cannot be manipulated. For example, discoveries about relationships between personality characteristics have led to useful theories about how personalities develop and change. Similar research has led to important theories about ability development and how learning occurs. Research in these areas may never involve manipulated variables, but it has already led to stable findings that can be generalized to many human populations.

▶ *Now You Do It*

At the end of Chapter 1 you were introduced to a study of the relationship between religious beliefs and attitudes toward equality for women (the Searchers study). Suppose Dr. Cross and his students found the following relationships between variables in the study:

1. Members of the Searchers have higher mean scores on the attitudes toward equality scale than do nonmembers.
2. The conservative form of the scale yielded lower mean scores than the other two forms.
3. The correlation between form of scale and scale score was .22.
4. Roman Catholics had lower scores on the attitudes toward equality scale than Protestants.

For which of these findings might a conclusion of causal relationship be justified?

Answers
1. Membership is a status variable, so cause is not justified.
2. Form of scale was randomly assigned, so we may entertain the possibility of a causal relationship.
3. This is another way of expressing the relationship in #2, so cause is a possible interpretation.
4. Denomination is a status variable.

SUMMARY

In this chapter, we drew a distinction between *independent* and *dependent variables*. Independent variables are those that the investigator manipulates during the course of a study or those for which he or she selects the values. Variables of the latter type are often status characteristics of the subjects. Dependent variables are those that are measured as an outcome of the study. The same variable may be an independent variable in one study and a dependent variable in another.

Evaluation studies usually attempt to determine whether some treatment has a desired effect. The investigator often has an interest in one of the possible outcomes, and decisions based on study outcomes may include cost-effectiveness and other social values. Formal testing of hypotheses derived from theories is often of secondary interest. *Research studies* are generally independent of social or monetary values. Their primary goal in testing hypotheses derived from theories is advancement of knowledge.

Science advances using *inductive reasoning,* reasoning that proceeds from specific observations to general principles. Inductive reasoning can show some general principle only to be false, never true. Therefore, science proceeds by ruling out wrong explanations for phenomena.

A research study is conducted to test a hypothesis. The hypothesis is a statement of the expected outcome of the study derived from a set of beliefs (called a theory) about the phenomenon of interest. There are three types of hypotheses: the *research hypothesis* derived from the theory; the negation of the research hypothesis, called the *alternative hypothesis;* and a potentially very large set of *plausible rival hypotheses,* which are alternate ways of explaining the study's results. Scientific progress is made through a continuing process of disproving research hypotheses, thus eliminating incorrect alternatives and theories.

Careful research design is necessary to rule out plausible rival hypotheses and provide a clear test of the research hypothesis. *Experimental controls* and *random assignment* of participants are two ways to focus the study on the research hypothesis. Careful selection of reliable and valid measures of the variables is also an essential feature of good research design. The conclusions that we draw from a research study usually involve the presence or absence of a relationship between the independent and dependent variables. Manipulated independent variables may allow us to infer a causal relationship with the dependent variable, while status independent variables do not. Highly accurate descriptions of the relationship between variables have proven very useful even when causal inferences cannot be drawn.

REFERENCES

Allen, M. J. (1995). *Introduction to psychological research*. Itasca, IL: Peacock.

Cook, T. D., & Campbell, D. T. (1979). *Quasi-experimentation: Design and analysis issues for field settings*. Chicago: Rand McNally.

Cozby, P. C. (1993). *Methods in behavioral research* (5th ed.). Mountain View, CA: Mayfield.

Huck, S. W., & Sandler, H. M. (1979). *Rival hypotheses: Alternative interpretations of data based conclusions*. New York: Harper & Row.

Myers, A., & Hansen, C. (1993). *Experimental psychology* (3rd ed.). Pacific Grove CA: Brooks Cole.

Phillips, D. C. (1987). *Philosophy, science and social inquiry*. Oxford, UK: Pergamon.

Ray, W. J. (1993). *Methods: Toward a science of behavior and experience* (4th ed.). Pacific Grove, CA: Brooks Cole.

Rosenthal, R., & Rosnow, R. L. (1991). *Essentials of behavioral research: Methods and data analysis* (2nd ed.). New York: McGraw-Hill.

Singleton, R. A., Straits, B. C., & Straits, M. M. (1993). *Approaches to social research* (2nd ed.). New York: Oxford.

Solso, R. L., Johnson, H. H., & Beal, M. K. (1998). *Experimental psychology: A case approach* (6th ed.). New York: Longman.

EXERCISES

1. In your own words, explain the difference between a status variable and a manipulated variable. Provide an example of each type of variable.

2. A researcher is interested in determining the differences in participants' perceptions of the expertise of a counselor based on the counseling approach that was utilized and the gender of the participant. In particular, participants were asked to rate the counselor's expertise on a 7-point scale. Participants were randomly assigned to watch a role-play video session of a counselor using either a rational-emotive approach to counseling or a client-centered approach to counseling. In addition, participants' genders (female/male) were recorded. The counselor and the client were portrayed by the same people in both approaches, and the presenting problem was the same in both approaches. Each video session lasted for 40 minutes.
   a. For the research study above, what is the dependent variable?
   b. For the research study above, what is (are) the manipulated independent variable(s), if any?
   c. For the research study above, what is (are) the status variable(s), if any?

3. An equal number of male children and female children aged five and eight were tested to determine their level of perspective-taking as measured by their answers to an orally administered test. The researcher was interested in whether there was an age difference as well as a gender difference.
   a. For the research study above, what is the dependent variable?

   b. For the research study above, what is (are) the manipulated independent variable(s), if any?
   c. For the research study above, what is (are) the status variable(s), if any?

4. A researcher was interested in the type of recruitment strategies that could be utilized on a college campus to increase the number of students who donated blood. One group of students was given a factual presentation that outlined the need for an increase in the number of blood donors whereas a second group of students received this factual presentation in addition to a video presentation of the stories of young children who were in need of donated blood. The number of students from each group who donated blood at the next campus blood drive was recorded.
   a. For the research study above, what is the dependent variable?
   b. For the research study above, what is (are) the manipulated independent variable(s), if any?
   c. For the research study above, what is (are) the status variable(s), if any?

5. Individuals who were mandated by the court to complete an anger management course were randomly assigned to one of two anger control classes: one that involved a cognitive-behavioral approach only and one that involved a cognitive-behavioral approach with role-playing. At the end of the six-week course, the participants were asked to keep a journal of the number of times that they were unable to control their anger in the next three months. At the end of the three-month recording period, the number of uncontrollable anger epi-

sodes was computed to determine if one anger manage-ment training program was more effective than the other.

   a. For the research study above, what is the dependent variable?

   b. For the research study above, what is (are) the manip-ulated independent variable(s), if any?

   c. For the research study above, what is (are) the status variable(s), if any?

6. Participants from Japan, a collectivistic culture, and the United States, an individualistic culture, completed a measure of achievement motivation that included two subscales: socially oriented achievement motivation and individually oriented achievement motivation. In addi-tion, there were an equal number of men and women from each country who completed the questionnaires. The social psychologist who conducted the research was interested in determining whether the motivation to achieve differed according to the values of the culture and according to the gender of the participant.

   a. For the research study above, what is (are) the depen-dent variable(s)?

   b. For the research study above, what is (are) the manip-ulated independent variable(s), if any?

   c. For the research study above, what is (are) the status variable(s), if any?

7. In your own words, explain the difference between an evaluation study and a research study.

8. For the following descriptions, make a determination whether the study could be most appropriately labeled an evaluation study or a research study. Provide a justifi-cation for your answer.

   a. An aggressive advertisement campaign was released by a company to improve the sale of a particular brand of athletic shoe. The company tracked the sales of the shoe over a six-week period of time in order to determine whether the aggressive ad campaign should be continued.

   b. An important component of reading comprehension is the encoding strategy that readers utilize. According to the Levels of Processing perspective, readers should recall more information if it is encoded at a deeper level. Two encoding strategies, paraphrasing what was read and recording the key words for what was read, were identified as involving two different levels of processing. Participants were given instruc-tions to read a passage one and only one time. Upon completion of the reading, they were instructed to ei-ther write down the central ideas in their own words (paraphrase) or list the key words in the passage (key-word). One week later the participants were asked to write down as many ideas from the passage as they could recall in a 40-minute time period.

   c. Participants were randomly assigned to view video-tape of a counseling session that involved marital problems of a couple. In introducing the study, the participants were either informed that the counselor was a masters level clinical student in training or that the counselor held a doctoral degree in clinical psy-chology. After viewing the same 30-minute session, the participants were asked to rate their perceptions of the effectiveness of the counselor on a 7-point scale.

   d. An environmental protection group wanted to deter-mine the effectiveness of signs strategically placed in

a mountain meadow describing the consequences to the plant life of failing to walk on the designated trail to determine if the state and federal government should fund a program to place this type of sign in all public park areas where the natural vegetation was at risk of dying as a result of human traffic.

9. Provide at least two original examples that illustrate eval-uation studies and research studies.

10. For Exercise 8(b) and (c), describe the control variables that were used to eliminate plausible rival hypotheses. For each study, what other variables should have been controlled and how would you have controlled them?

11. A researcher is interested in comparing attitude and achievement differences utilizing a metacognitive ap-proach to teaching algebra versus a traditional approach to teaching algebra. Prior to instruction, students are given a measure of their attitude toward mathematics as well as a measure of algebra achievement. Students are then randomly assigned to one of two 2nd-period alge-bra classes. Both teachers are in their eighth year of teaching algebra and both have been cited for their out-standing teaching. At the end of the school year, the measures of attitude toward mathematics and achieve-ment in algebra are once again administered. The re-searcher predicted that students who were taught alge-bra using the metacognitive approach would have a more positive attitude toward mathematics and score higher on the algebra achievement test when pre-instruction attitudes and achievement were considered than students who were taught algebra using a tradi-tional approach.

   a. For the research study above, what is (are) the depen-dent variable(s)?

   b. For the research study above, what is (are) the manip-ulated independent variable(s), if any?

   c. For the research study above, what is (are) the status variable(s), if any?

   d. For the research study above, what are the processes that were utilized to rule out plausible rival hypotheses?

   e. Name at least one major plausible rival hypothesis that might account for any differences that were found between the two groups. If you were the re-searcher, how would you design this study differently to eliminate this rival hypothesis?

12. From the research descriptions provided in Exercises 2–6, 8(b), and 8(c), indicate which research questions can be determined to have a causal answer, and justify your answer. Also keep in mind that some of the ques-tions address more than one research question.

13. In your own words, explain why scientific research can-not "prove" something to be true.

14. In your own words, explain the difference between in-ductive and deductive reasoning.

15. The following statements were taken from a local news-paper. For each of the following, identify the principle(s) claimed, the facts involved, and the conclusions that are asserted. In addition, state whether the argument repre-sents deductive or inductive reasoning.

   a. It is evident that the higher salaries paid to profes-sional athletes are responsible for the decline in inter-est in professional sports. Within the last decade, the salaries of professional baseball players have in-creased by 248% while attendance at the baseball parks is down 15% and television viewing of baseball games is down 18%.

b. Jolene Carr was 68 years of age when she died. Doctors noted that she was on a diet and lost weight despite having a normal body mass index for her age group. Doctors cited a study that indicated that women between the ages of 65 and 99 who have a normal body mass index and lose weight are four times more likely to die than women in this age category who maintained their weight or gained some weight.

c. Dentists caution against smoking if you have dental implants because they will fail. Jose Garcia and Julie Chan are both smokers who had dental implants that failed.

16. a. Within the area of cognitive psychology the levels of processing approach emphasizes that we can analyze information in many different ways: from the shallow, sensory kind of processing that may focus on judgments about the presence of a particular letter in a set of words that comprise a prose passage to a deeper, more complex kind of processing involved in judgments about the meaning of a passage. The theory can be summarized more succinctly as the more deeply a person processes prose information, the more likely she or he will be to recall that information in the future. Provide a hypothesis to test this theory and describe how you might design a research study to test the hypothesis.

b. From research on counselor effectiveness, the theory that is advanced is that a client will rate a counselor as more effective if the client perceives the counselor to have a high level of expertise. Provide a hypothesis to test this theory and describe how you might design a research study to test the hypothesis

c. In working with rival groups, social psychologists have noted that if you involve the rival groups in a cooperative task with a clearly defined goal, intergroup hostility will be reduced. Provide a hypothesis to test this theory and describe how you might design a research study to test the hypothesis.

17. For the following descriptions of research studies, list, if present, the control variables, the manipulated independent variables, and the status independent variables. Justify why you have placed each variable in the category that you selected. What are the research hypothesis and alternate hypothesis of each study? Provide a rival hypothesis that might account for the predicted effect. What steps were taken in the research design to eliminate plausible rival hypotheses? What other steps might have been taken?

a. Clinical psychologists and psychiatrists often prescribe Ritalin for controlling the behavior of hyperactive children. Several problems are associated with the use of Ritalin for this purpose. Among the problems discussed are the cost factor, the tendency to have a child who is so medicated that he or she cannot perform even simple tasks or appropriate behaviors, and variability in the effectiveness of Ritalin in controlling inappropriate behavior. Some clinical and counseling psychologists have argued that cognitive-behavioral approaches to teaching a child to control her or his inappropriate behavior are at least as effective as Ritalin but do not have many of the side-effects and prob-

lems associated with the use of medication. In an effort to determine the relative effectiveness of Ritalin and cognitive-behavioral approaches in reducing inappropriate behavior, a researcher selected children who had recently been diagnosed as hyperactive. All children were observed for 10 school days by observers trained to record the number of inappropriate behaviors exhibited by each child. All observations took place at the same time of day, with two observers recording the behaviors of each child for a one-hour time frame.

At the end of this observation period, baseline scores were established by averaging the activity scores of all observers for the 10-day observation period. The resulting score was the average number of inappropriate behaviors in a one-hour time period. The two raters' scores were compared and an interrater reliability was calculated to be .92. With parental approval, children were randomly assigned to one of two treatment conditions: Ritalin or cognitive-behavioral therapy. Children in the Ritalin condition were given appropriate dosages of Ritalin, which was monitored over a one-month time frame by the appropriate medical personnel. Children in the cognitive-behavioral therapy condition were taught to control their inappropriate behavior by using mediating responses and receiving rewards for behaviors that competed with inappropriate behaviors. The training period took place over a one-month period of time. At the end of one month, the children's behaviors were observed and recorded as outlined for the pretreatment condition. At the end of 10 days of observation an average number of inappropriate behaviors for a one-hour period was calculated. The pretreatment score was subtracted from the post-treatment score to indicate the change in the number of inappropriate behaviors. A negative difference indicated a decline in inappropriate behaviors while a positive difference indicated an increase in inappropriate behaviors.

b. Cigarette smoking among adolescents in the United States has been a focus of concern for a number of years. As smoking became more ingrained in the adolescent population, people looked to the educational system for help in decreasing, if not eliminating, cigarette smoking in the adolescent population. A common approach has been to provide factual information about the health risks involved in cigarette smoking. However, adolescents increasingly viewed this approach as another tactic that adults used to try to manipulate them. In many instances adolescents dismissed the information as "scare tactics." Recently, educators have been more successful in decreasing adolescent smoking behavior by using a social skills and reinforcement approach that emphasizes the social skills involved in refusing to smoke and social rewards for not smoking. Both of these approaches have as their goal changing attitudes toward smoking and as a result changing smoking behavior. An educational psychologist was interested in comparing the effects of these two educational approaches on attitudes toward cigarette smoking in adolescents. The educational psychologist predicted that students who were exposed to the social skills and reinforcement approach to teaching about cigarette smoking would express more negative attitudes (i.e., lower attitude scores) toward cigarette smoking than adolescents who received factual information about smoking. A unit on cigarette smoking was incorporated into an eighth grade health class. Students in the first- and third-period classes were given factual information associated with cigarette smoking while students in the second- and fourth-period classes received the social skills and reinforcement approach to smoking behavior. At the end of two days of instruction a measure of attitude toward cigarette smoking was obtained from each student in the study.

c. A psychologist interested in studying perception wanted to determine if lighting conditions (natural vs. fluorescent) differentially affected the reading speed of dyslexic and nondyslexic adult readers. Participants were given material to read under natural lighting conditions. The researcher predicted that dyslexic readers would read more slowly under fluorescent lights than nondyslexic readers while they would read at an equal speed under natural lighting conditions. Participants were asked to read a three-page passage one and only one time. The reading level of the passage was rated at the sixth grade level which is defined as the typical level of writing of most government publications. The amount of time it took the participant to read the passage was recorded. Participants were then exposed to conditions of fluorescent lighting and were asked to read a second three-page passage one and only one time. This second passage was rated at the same reading level as the first passage. Reading time for each participant was recorded.

Chapter 11

Sampling Distributions and Statistical Estimation of the Mean

LIST OF KEY TERMS AND CONCEPTS

Central Limit Theorem
Confidence Interval
Degrees of Freedom (*df*)
Expected Value
Interval Estimate
μ (mu)
Point Estimate
Sampling Distribution
σ (sigma)
Standard Error
Standard Error of the Mean
t Distribution
Unbiased Estimate of σ_X^2

CHAPTER OBJECTIVES

After studying this chapter, you should be able to:

❑ Determine the mean and standard deviation of a sampling distribution for the binomial distribution.

❑ For a research study where the sample size and standard deviation are given, determine the standard error of the mean.

❑ State the consequences of the central limit theorem for the sampling distribution of the mean.

❑ When σ^2 is known, determine the limits of the confidence interval for the mean for any specified probability.

❑ Compute the unbiased estimate of σ_X^2 from raw data or from sample statistics.

❑ Use the correct degrees of freedom to determine the required critical value of *t* for a given probability.

❑ Determine the confidence interval for the mean when σ_X is not known.

For several weeks in 1998 Congress debated a controversial proposal to do away with the Department of Education. In general, Republicans favored the move, while Democrats opposed it. The Democratic party commissioned a research polling organization to conduct a poll of voter opinion. The poll showed that 53% of eligible voters opposed the measure, with a "margin of error" of 4%. Not to be outdone, the Republican party sponsored its own poll, which showed that the public favored the legislation by a 52% to 48% margin. Do the two polls in fact differ, as the figures seem to suggest? What are some reasons why the two polls might come to differing conclusions? What can we say about the real beliefs of the 200 million eligible voters? Would you want to take an action based on a poll like one of these?

Introduction

In the last chapter, we discussed the general problem of testing scientific hypotheses and drawing conclusions. This chapter is devoted to the particular problems encountered when statistical analysis is used to estimate a population parameter from sample results. In the case of the two political polls above, for example, there is a population of voters who hold opinions about an issue. Each polling organization draws a sample from that population and attempts to estimate the proportion of the population (a parameter) that favors the proposal from the proportion of the sample (a statistic) that favors it. Making estimates of parameters from statistics requires that we develop a new kind of distribution, called a *sampling distribution*.

Sampling Distributions

To begin our discussion of sampling distributions, let's consider a population such as the outcomes from attempting to guess the color of each of 10 playing cards in an ESP experiment. This population is infinite. There are 11 possible kinds of outcomes from each trial, ranging from none correct to 10 correct, but there is no limit to the number of trials that could occur. If we think of each trial as a potential observation yielding a score in the range from 0 to 10, we can draw a random sample from the population. The random sample will be made up of *N* individual observations. For example, suppose that you and three friends each take turns at guessing the color, red or black, of 10 playing cards. The number of correct guesses on your 10 cards is your "score." You and your friends have drawn a random sample of four observations from the infinite population of guesses about 10 cards.

We can compute the mean and standard deviation for your four observations and record their values. Our sample has been drawn randomly from a population with a known mean ($\mu = 5.0$) and standard deviation ($\sigma = 1.58$) (see Chapter 6). [The lowercase Greek letter μ (pronounced mu) is the symbol usually used to denote the mean of a population. The symbol σ (lowercase Greek letter sigma) stands for the population standard deviation.] We know these values ($\mu = 5.0$ and $\sigma = 1.58$) are true for the population because this is a binomial distribution (the theoretical distribution of binary events) and the mean and standard deviation of any binomial distribution are given by equations 6.5 and 6.6. However, if you actually perform this little experiment, you will almost certainly find that your sample does not have a mean ($\overline{X}$) of exactly 5.0 nor a standard deviation (S) of 1.58. In fact, it is unlikely that the statistics from any sample will exactly equal the parameters of the population; instead, statistics from many samples will show a distribution of values around the parametric value.

It's a Saturday afternoon and there is no football game, so you and your friends decide to repeat your little experiment several more times; in effect, you take several more samples from the population. If you then compute some summary statistics ($\overline{X}$, *Mdn*, and *S*, for example) for each sample, you can prepare a frequency distribution for each type of statistic: a frequency distribution of means, one of medians, and another of standard deviations. Each of these frequency distributions will have a mean, median, and standard deviation of its own. Such a frequency distribution of sample statistics is called a **sampling distribution of the statistic.** The *frequency distribution of sample means is called the sampling distribution of the mean*. The frequency distribution of standard deviations is the sampling distribution of the standard deviation, and the frequency distribution of medians is the sampling distribution of the median. Every statistic has a sampling distribution which, in effect, treats the statistic from each sample as if it were a single observation. The notion of sampling distributions is fundamental to the entire logical structure of statistical estimation and inference because we reach conclusions about statistical hypotheses on the strength of probabilities based on sampling distributions.

> A **sampling distribution** is the frequency distribution of values we would get for a statistic if we drew a very large number of random samples of a given size, N, from a population.

▶ *Now You Do It*

As an exercise, and to give you a more intuitive feel for the idea of sampling distributions, try this little experiment. Get a group of three friends together over lunch or some other convenient occasion. Take a large glass or paper cup (make sure it's empty) and put 10 coins in it. Think of each coin as representing your response to a true-false test item on a subject about which you know nothing, perhaps a vocabulary test in the classical Hawaiian language or a test on the care and feeding of yaks. Heads, you get the answer right; tails, you get the answer wrong. Shake up the coins, dump them on a table, and count the number of heads. What was your score? Now pass the cup to a friend and see how he or she does on the "test." Repeat the process, each time recording each person's score until everyone has taken the test once. You now have a sample of $N = 4$ from a population, and unless someone cheated or didn't shake the cup, it is a random sample.

Next, have each person repeat the process, giving you a second sample. Do the experiment three more times, so you have five samples, each composed of four observations. The results might look something like this.

	Sample 1	Sample 2	Sample 3	Sample 4	Sample 5
	6	4	5	5	4
	2	5	5	5	2
	4	6	6	5	2
	8	6	3	9	4
Mean	5.0	5.25	4.75	6.0	3.0

If you look at the 20 individual scores, they have a mean of 4.8 and a standard deviation of 1.78, both fairly close to the population values that we know should be 5.0 and 1.58, because this is a binomial distribution of 10 events. However, when we consider the distribution of sample means; that is, the sampling distribution of the mean, we get a mean of 4.8 and a standard deviation *among the means* of .9925. There is noticeably less variability among the means than there is among the individual observations, even with samples as small as 4. Why do you think this might happen?

If you don't have three friends available, try the same exercise with the BTU data for any continuous variable such as GPA, anxiety score, or final exam score. In this case, assume the 68 students constitute a population, so the mean and standard deviation for $N = 68$ are population values. Next, pick a random ID number and sample every 17th person until you have a sample of size 4 (a systematic random sample). Repeat this process, each time picking a random starting place until you have results for five samples. Then find the mean and standard deviation for the five samples and compare these values to the population mean and standard deviation. How do the results compare?

Biased and Unbiased Estimates

We noted that the means of several samples will not, in general, be equal and will not equal the population mean μ; they form a sampling distribution around the population mean. If you did the exercise in the preceding Now You Do It, you have demonstrated this for yourself. It can be shown (this phrase is statisticians' jargon for "the proof is too complex to be included in this book"[1]) that *the mean of the distribution of means* of random samples from a population is equal to the mean of the population. That is, the population mean μ is the mean of the sampling distribution of $\overline{X}$. For example, if you continued the sampling experiment in the Now You Do It and collected a very large number of samples, the mean of the sample means would gradually converge on the population mean. This is a very important fact because it allows us to specify the location of the sampling distribution on the measurement scale. When we know the population mean, as we do in the case of our card-guessing and coin-tossing examples, we know that the distribution of sample means will center on this value. The mean of the sampling distribution of a statistic is known as the **expected value** of the statistic. When the expected value of a statistic is equal to the population parameter, we say that the statistic computed on a random sample from that population is an **unbiased estimate** of the parameter. The mean is an unbiased statistic. The mean of any random sample from a population is an unbiased estimate of μ. Therefore, each mean you collected from a sample of 10 card guesses or 10 coin tosses was an unbiased estimate of the population mean of 5.0.

Not all statistics are unbiased estimators of their parameters. Even when a sample is randomly drawn from the population, some of its statistics are biased; that is, the mean of the sampling distribution is either larger or smaller than the parameter. This is a feature of several statistics, most prominent of which is the sample standard deviation S. The mean of the sampling distribution of S ($\overline{S}$) is not equal to the population standard deviation, σ. $\overline{S}$ is always smaller than σ, and the amount of bias is determined by the size of the sample.

*The **expected value** of a statistic is the mean of the sampling distribution of the statistic.*

*A statistic is an **unbiased estimate** of a parameter when the expected value of the statistic is equal to the parameter.*

[1] This proof, along with many others related to topics we cover, can be found in Hays, W. L. (1994). *Statistics* (5th ed.). Fort Worth, TX: Harcourt.

If you compute the sample standard deviations for your card guesses or coin tosses, you will find that they average somewhat smaller than the population value of 1.58. We will explain the reason for the bias in S later in the chapter.

Standard Errors

Every sampling distribution has a mean, and every sampling distribution also shows some degree of spread, or variability, around its mean, which can be expressed as a standard deviation. If you repeated the card-guessing experiment or the coin-toss experiment in the Now You Do It several times, you would find that the sample results you obtained would differ from trial to trial. We can express the variability of sample means around μ by computing a standard deviation for the sampling distribution.

The standard deviation of the sampling distribution of a statistic is known as the **standard error** of the statistic. (Remember that when we introduced the standard error of estimate in Chapter 7 we said that the term *standard error* was used to refer to a standard deviation of something other than raw scores. This usage includes the standard deviations of the distributions of sample statistics.) The standard deviation of sample means around μ is known as the **standard error of the mean** and is given the symbol $\sigma_{\bar{X}}$. The subscript $\bar{X}$ indicates that we are talking about the standard deviation of the means of samples. There is also a standard error of the median (σ_{Mdn}) that is the standard deviation of the sampling distribution of the median; a standard deviation of the sampling distribution of the correlation coefficient (σ_r), called the standard error of r, and so on. The worst tongue twister of the lot is the standard error of the standard deviation, σ_S. This, of course, is the standard deviation of the sampling distribution of standard deviations.

Each sample statistic we compute is an estimate, either biased or unbiased, of its population parameter. Values computed from individual samples vary around that parameter. As sample size increases, we tend to find less variability among the means (or other statistics) from the various samples. In fact, the standard error of a statistic is, in part, an inverse function of the size of the sample on which it is based. Assuming that we know σ_X, the standard deviation of individual observations in the population, it can be shown that the standard error of the mean is

$$\sigma_{\bar{X}} = \frac{\sigma_X}{\sqrt{N}} \tag{11.1}$$

In our card-guessing and coin-tossing experiments, we had 10 cards or coins and a known standard deviation of 1.58. Using $\sigma_X = 1.58$ and equation 11.1, the standard error of the mean for samples of size 4 from a population of responses to these 10-item experiments is

$$\sigma_{\bar{X}} = \frac{1.58}{\sqrt{4}} = \frac{1.58}{2} = .79$$

If we were to take a very large number of random samples of size 4 from either the coin-toss or the card-guess population, compute the sample means, and calculate the standard deviation of the distribution of these means, we would expect to get a value of about .79.

Equation 11.1 states that as sample size increases, the standard error of the mean decreases. If you and 99 friends tossed 10 coins or guessed at 10 cards, this would reduce the value of $\sigma_{\bar{X}}$ to

$$\sigma_{\bar{X}} = \frac{1.58}{\sqrt{100}} = .158$$

When samples are larger, we see less variability among their means.

Increasing sample size has the same general effect on the sampling distributions of all statistics. The equations for the standard errors of several commonly used statistics are listed in Table 11-1. Every one of them is an inverse function of sample size. As N increases, the standard error decreases. Note that there are limitations on σ_S and σ_r. Because under certain conditions the sampling distributions of these statistics do not follow the normal

The **standard error** of a statistic is the standard deviation of its sampling distribution.

The standard error of the mean is the standard deviation of the sampling distribution of the mean. It is the standard deviation of sample means around μ.

When samples are larger, there is less variability among their means. The value of $\sigma_{\bar{X}}$ is smaller.

Table 11-1   Standard Errors of Some Common Statistics

	Sample Statistic	Standard Error
Mean	$\overline{X}$	$\sigma_{\overline{X}} = \dfrac{\sigma_X}{\sqrt{N}}$
Median	Mdn	$\sigma_{Mdn} = \dfrac{1.253\sigma_X}{\sqrt{N}}$
Standard deviation	S	$\sigma_S = \dfrac{\sigma_X}{\sqrt{2N}}$
Semi-interquartile range	Q	$\sigma_Q = \dfrac{0.7867\sigma_X}{\sqrt{N}}$
Correlation coefficient*	r	$\sigma_r = \sqrt{\dfrac{1-\rho^2}{N-1}}$

*Generally, we use Greek letters for population parameters. The letter used for the population correlation coefficient is the Greek lowercase letter ρ (rho).

distribution, these standard errors are not always appropriate and it is necessary to use special procedures with them unless the sample size is quite large. These special procedures will be presented in Chapter 14.

▶ *Now You Do It*

Look back at the previous Now You Do It where we conducted our little sampling experiment. The results gave us a mean of the means of 4.8 and a standard deviation of the means of .9925. What does statistical theory say the standard error of the mean should be? Equation 11.1 says it should be $1.58/\sqrt{4}$ or .79. The means from our experiment show somewhat more variability than we would predict, but if you increase the number of samples by repeating the experiment you will see that the results approach those predicted by equation 11.1. Also, perhaps, if you did conduct the experiment yourself, you got results more consistent with theory.

Now, suppose you had conducted your experiment with eight friends instead of three. Since there are nine of you, what should the standard error of the mean be? Equation 11.1 tells us it should be

$$\sigma_{\overline{X}} = \frac{1.58}{\sqrt{9}} = .527$$

This confirms that as sample size increases, the means should show less variability around μ.

The Central Limit Theorem

We have seen that increasing sample size has the effect of decreasing the standard error of a statistic. Sample size has an important effect on another characteristic of the sampling distribution, its shape. This effect is specified by the **central limit theorem,** which states, in part, that as sample size increases, the sampling distribution of the mean approaches the normal distribution, regardless of the shape of the population distribution. This effect is not strictly true for all statistics (it is true for the mean), but it is close enough so that we can handle the exceptions fairly easily.

The central limit theorem applies strictly to the sampling distribution of the means of random samples, the focus of most research attention. As sample size increases, the sampling distribution of $\overline{X}$ approaches a normal distribution with a mean equal to μ and a standard deviation of $\sigma_{\overline{X}}$, *regardless of the shape of the population distribution*. It is this last fact that gives the central limit theorem its importance. Not all populations are normal in shape; some, such as distributions of reaction time or depression, are definitely nonnormal, being seriously skewed or bimodal. The central limit theorem states that the sampling

*The **central limit theorem** states that as sample size increases, the sampling distribution of the mean approaches the shape of the normal distribution, regardless of the shape of the underlying population distribution.*

distribution of $\overline{X}$ approaches the shape of the normal distribution as sample size increases even when the samples are drawn from populations that are known to be nonnormal.

This principle, which is illustrated in Figure 11-1, also holds approximately for the sampling distributions of several other statistics as well, particularly when N is large. In the figure, the distribution of depression scores for a population of college students is shown. If we draw samples from this distribution, most of our scores will be very low, indicating little depression, because that is where most of the people are. For the smallest possible sample size, $N = 1$, the sampling distribution of the mean is identical with the population distribution. For small samples, the sampling distribution will center on μ, but will be skewed due to the intense skewing of the population. As sample size increases, the distribution becomes more nearly normal.

The central limit theorem is significant because it allows us to apply the properties of the normal distribution in making statements about sampling distributions. For example, can we take the results of a particular ESP experiment in which four students guessed correctly the color of 6 of 10 cards to be evidence for the existence of ESP? Could this result reasonably have occurred by chance? Because areas under the normal curve can be equated to probabilities, we can specify the probability that a sample with a mean in a particular range will occur. Although we are focusing our attention on the mean, the most widespread application of probability and the sampling distribution, the same logic can be applied to other statistics with only minor modifications.

Probability and the Sampling Distribution

What is the probability that a sample of $N = 4$ students who have taken our card-guessing test will achieve a mean greater than 6? As you will recall, our sample of four card-guessing test scores was drawn from a population with $\mu = 5.0$ and $\sigma_X = 1.58$. Since the mean of any sample drawn from this population is an unbiased estimate of μ, the mean of the sampling distribution of $\overline{X}$ is also 5.0, and the standard error of $\overline{X}$ for samples of size 4 is $1.58/\sqrt{4}$, or .79. The fact that the population is known to have a nearly normal distribution implies that we do not have to worry about the normality of the sampling distribution. (If we had reason to suspect the possibility of nonnormality, we might want to collect a larger sample to be reasonably confident of a sampling distribution with normal-curve properties.)

We know the sampling distribution is normal, the mean of the distribution is $\mu = 5.0$, and the standard deviation is .79, so we can determine the proportion of the sampling distribution that lies beyond the point $\overline{X} = 6.0$. We do this by computing the **Z-score for the given $\overline{X}$, 6.0, in the sampling distribution of $\overline{X}$.** (Note that at this point, because we have means rather than the discrete observations which occupied us in Chapter 6, we have a nearly continuous scale. For this reason, it is no longer necessary to consider the real limits of score intervals.)

$$Z_{\overline{X}} = \frac{\overline{X} - \mu}{\sigma_{\overline{X}}} \qquad (11.2)$$

With the values given, we obtain

$$Z_{6.0} = \frac{6.0 - 5.0}{.79} = +1.27$$

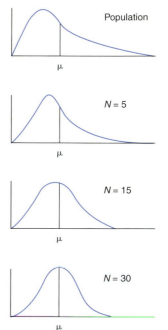

Figure 11-1 As sample size increases, the sampling distribution of the mean approaches the normal distribution.

Looking this value up in Appendix Table A-1, we find that the proportion of the normal curve beyond $Z = +1.27$ is .1020. We can therefore say that the probability of getting a mean test score of 6.0 *or more* in a sample of four students is about .102 or about 10 percent.

The same series of logical steps can be used to determine the probability that a sample mean will fall between two values. For the same set of conditions $\mu = 5.0$, $\sigma_{\bar{X}} = .79$, the probability that we would get a sample with a mean between 4.5 and 5.2 can be determined as follows:

$$Z_{4.5} = \frac{4.5 - 5.0}{.79} = -.63$$

$$Z_{5.2} = \frac{5.2 - 5.0}{.79} = +.25$$

The area between $Z_{4.5}$ and the mean is .2357, and the area from the mean to $Z_{5.2}$ is .0987. Therefore, the area between $Z_{4.5}$ and $Z_{5.2}$ is .2357 + .0987 = .3344, and the probability of obtaining a sample whose mean is in this range is about one-third.

▶ *Now You Do It*

Let's take a slightly different problem to check your expertise with probabilities and sampling distributions. Suppose we know (don't ask how!) that the distribution of statistics ability in the population has a mean of 500 and a standard deviation of 100. What is the probability that a randomly selected sample of 25 students from this population will have a mean ability level above 540? Below 490? Between 485 and 520?

First, we calculate the standard error of the mean to be $\sigma_{\bar{X}} = 100/\sqrt{25} = 20$. Next, we find the Z-scores for each sample mean in the sampling distribution of the mean for samples of size 25 by subtracting the population mean from the sample value and dividing by the standard error of the mean.

$$Z_{540} = (540 - 500)/20 = +2.00$$
$$Z_{490} = (490 - 500)/20 = -0.50$$
$$Z_{485} = (485 - 500)/20 = -0.75$$
$$Z_{520} = (520 - 500)/20 = +1.00$$

Using Appendix Table A-1 we find the area above +2.00 to be .0228, the area below −0.50 to be .3085, and the area between −0.75 and +1.00 to be .2734 + .3413 or .6147. This leads us to conclude that the probability of drawing a sample with a mean above 540 is .0228, the probability of drawing a sample with a mean below 490 is .3085, and the probability that our sample will be one with a mean between 485 and 520 is .6147. Once again, remember that the issue is not what the mean of a given sample will be, but whether the sample we actually draw will be one of the ones which has the property in question.

Now repeat the foregoing steps with $N = 100$. What do you get for $\sigma_{\bar{X}}$? What happens to the spread of the sampling distribution? What values do you get for the Z-scores? What happens to the probabilities?

$$\sigma_{\bar{X}} = \frac{100}{\sqrt{100}} = 10$$

$Z_{540} = +4.00$ $p(Z > +4.00) = .00003$

$Z_{490} = -1.00$ $p(Z < -1.00) = .1587$

$Z_{485} = -1.50$ $p(Z \text{ between } -1.50 \text{ and } +2.00) = .9104$

$Z_{520} = +2.00$

Z-Scores or the Original Scale?

At this point we must draw a distinction that is of substantial importance. We have been talking about the sampling distribution of means, and we have found, through application of the central limit theorem, that this sampling distribution is normal in shape. In order to make probability statements about the sampling distribution, we have converted it to the metric of Z-scores. Transforming a set of scores into Z-scores is a linear transformation involving subtraction of one constant (μ) and division by another constant ($\sigma_{\overline{X}}$). Because the original sampling distribution was normal, the resulting Z-score distribution is also normal (a linear transformation does not change the shape of the distribution) and we can use the table of the standard normal distribution to derive probability statements. Our probability statements are phrased in the metric of the Z-score distribution. We can convert the Z-score values back to the scale of the original scores by reversing the transformation (multiply by the standard deviation, then add the mean), thereby applying our Z-score probability statements to values in the scale of the original data, but the probabilities are determined from the Z distribution. Thus, in the example we considered above, we can say that the probability is .33 that we will obtain a sample with a mean between 4.5 and 5.2, but we had to find the Z-scores in order to reach this conclusion.

The Importance of How We Phrase Probability Statements

The way that we phrase our statements about the probability that a mean falls in a particular range is very important because we are about to apply the probability to statistical estimation. We must remember that *the population has a mean that does not vary or change;* the same thing is true of *each particular sample once that sample has been drawn.* The population mean and the mean of each particular sample are fixed; each has a probability of 1.00, or 100 percent, of occurring **in that population** or **in that *particular* sample.** What is uncertain, and therefore open to probabilistic statements, is which sample we will get.

The importance of this point for the logic of statistical inference cannot be overstated. In a greatly simplified example, consider a population with four possible samples, each of which is equally probable ($P = .25$) and each of which has a different mean. Suppose $\overline{X}_1 = 1$, $\overline{X}_2 = 2$, $\overline{X}_3 = 3$, and $\overline{X}_4 = 4$. If we reach into the population to obtain a sample and get sample 2, the mean of this sample is certain to be $\overline{X}_2 (=2)$, whether we ever collect the data and compute the mean or not. There is a 25 percent chance that we will draw this particular sample, but once it is drawn, only measurement and computational errors can affect the outcome. Therefore, when applying normal curve procedures and the principles of probability to problems involving sampling distributions, statements or conclusions about probability must be restricted to the sampling phase. *The question is not what value the sample will yield but which sample we will get.* When we ask what the probability is that our sample will have a mean of 5.13 or greater, what we are really asking is "What is the probability that we will draw one of the samples that has a mean of 5.13 or greater?"

The population mean and each particular sample mean are fixed. What is open to probabilistic statements is which sample we will get.

Point and Interval Estimation

Up to now we have been discussing fairly artificial situations—ones where we know the population parameters—and our question has been about the probability of obtaining a sample result in a particular range. In the real world of scientific inquiry we hardly ever know the parameters; we conduct our studies to make estimates of the parameters from sample data. It is this problem to which we now turn.

A parameter has a specific value for a population. If, for example, we were able to measure the trait of introversion or of sociability for every person in the country, the mean of that set of data would give us a parametric value. But for reasons of timeliness, economy, and so forth (see Chapter 9), such sets of data are hardly ever collected. Rather, we use the results from a sample—a statistic—to estimate the unknown parameter. The problem is that the statistic, like the parameter, is a single value, and it is therefore unlikely to be exactly equal to the parameter.

The central limit theorem tells us that the sample mean is our best estimate of μ, but hitting a parameter with a statistic is a little like trying to hit a baseball in flight with a rifle shot at 200 yards; it is next to impossible. If we had a shotgun, we would have a better chance. A special kind of probability interval known as a **confidence interval,** or **CI,** provides a good statistical approximation to a shotgun. A single statistic like $\overline{X}$, because it is a single value, is known as a **point estimate** of the parameter. If the sample mean is 22.5, this is a point estimate of μ. A point estimate, although it may be the best single

A single statistic is a **point estimate** of a parameter.

value we can obtain from a sample, can be in error by a substantial amount. A confidence interval, on the other hand, is a band of probability placed around the statistic. A confidence interval, because it covers all values within the range of the interval, provides what is known as an **interval estimate** of the parameter. The confidence interval is also called an *interval of uncertainty* because we do not know where in the interval μ is, but we do know, with a specified probability of being wrong, that the interval includes the parameter. A confidence interval running from 20 to 25 provides an interval estimate 5 units wide. This interval, because it covers a wider range of possible values, is much more likely to include μ than a point estimate is to hit the value of μ exactly.

*A **confidence interval** is a band of uncertainty around a statistic that provides an **interval estimate** of a parameter.*

We can apply confidence intervals to estimate population parameters in exactly the same way that we applied prediction intervals in Chapter 7 to put an interval of uncertainty around a person's predicted score. There, we knew the person's predicted score, $\hat{Y}_i$, and the prediction interval based on $\hat{Y}_i$ and the standard error of estimate provided a range that had a specified probability of including the person's unknown actual criterion score, Y_i. $\hat{Y}_i$ was a point estimate of Y_i, and the prediction interval gave a band of uncertainty, or a range of possible values. To estimate μ from $\overline{X}$ we proceed in an identical fashion.

Confidence Intervals When σ Is Known

We have seen earlier in this chapter that the standard error of the sampling distribution of $\overline{X}$ is given by

$$\sigma_{\overline{X}} = \frac{\sigma_X}{\sqrt{N}}$$

If we know the value of σ_X, then we can calculate $\sigma_{\overline{X}}$ for our sample size. Once this value is known, we simply decide what probability we wish to use, look up the appropriate value of Z, and compute our critical values. For example, if we wish to estimate the mathematics ability of eighth graders in our state and we know that in previous years a statewide testing program produced a population standard deviation of 15 (forgetting for the moment that this value may no longer be correct), we could draw a sample of 100 eighth graders from across the state and test them. Suppose we obtained a mean score on the mathematics test of 67.7. What is the 95% confidence interval?

Looking in Appendix Table A-1 for the value of Z that leaves 2.5% of the distribution in each end (thus including 95% between the two values), we would find $Z = 1.96$. This means that the interval from 1.96 standard errors of the mean below μ to 1.96 standard errors above μ will include 95% of the sampling distribution. That is, this interval will include the means of 95% of the possible samples from this population.

This is where things get a little tricky. If we place a 95% probability interval around μ and place intervals of the same width around each $\overline{X}$ in the sampling distribution, then for any sample whose mean falls within the 95% interval around μ, the 95% interval around that sample mean will include μ. This principle is illustrated in Figure 11-2. First, we show the 95% interval around μ. $\overline{X}_1$ and $\overline{X}_2$ fall within this interval. Therefore, 95% confidence intervals around $\overline{X}_1$ and $\overline{X}_2$ include the value of μ, as is shown in the next two lines of the figure. $\overline{X}_3$ falls outside the 95% interval around μ, and its 95% CI does not include μ. If we draw only one sample from this population, the probability is 95% that the sample we draw will be one of the samples falling in the 95% interval around μ, and if we place a 95% confidence interval around that sample mean, we may say that there is a 95% probability that this interval is one of the ones whose 95% CI includes μ.

In the example above, we drew a sample of 100 students from our state's population of eighth graders and found their mean to be 67.7. We know that the standard error of the mean for samples of this size from this population is $15/\sqrt{100} = 1.5$ and that the Z we need is 1.96. There is a 95% probability that our sample is one of the ones falling within 1.96 standard errors of the mean of μ. We can calculate the limits of this interval in the scale of raw scores as follows, where CI_U is the upper limit of the confidence interval and CI_L is the lower limit:

$$CI_L = [-1.96(1.5) + 67.7] = 64.76$$
$$CI_U = [+1.96(1.5) + 67.7] = 70.64$$

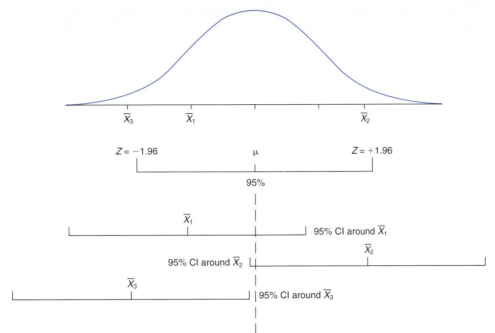

Figure 11-2 The sampling distribution of the mean is normal and centers on μ. A 95% probability interval around μ will enclose 95% of the sample means. If an interval of the same size is placed around each sample mean, 95% of the intervals around the sample means will include μ.

Therefore, there is a 95% probability that the 95% confidence interval around our sample mean, the interval running from 64.76 to 70.64, includes μ. The interval 64.76 to 70.64 is the 95% confidence interval for μ based on this sample.

One nice feature of the confidence interval is that we can choose the level of risk we wish to run of being wrong. If only a loose estimate of μ is needed, we could use a narrow interval such as the 50% CI. The appropriate value of Z is then .67 and the confidence interval in our example would run from 66.7 to 68.7. There is a 50% probability that this interval includes μ. To be very sure that our interval includes μ, we could use a 99.9% CI, thus leaving only one chance in 1,000 that our interval does not include μ. The required value of Z from Appendix Table A-1 in this case is 3.30. This very conservative confidence interval runs from 62.75 to 72.65 for our example. We don't know where μ is within this interval, but we can be very confident that it is somewhere in the interval.

We can see the effect of sample size on the CI by taking a sample of 25 rather than 100. In that case the value of $\sigma_{\overline{X}}$ is

$$\sigma_{\overline{X}} = \frac{15}{\sqrt{25}} = 3.0$$

Clearly, $\sigma_{\overline{X}}$ is exactly twice as large. Samples of 25 will show twice as much variability around μ as samples of 100 will, so the confidence interval at a given probability level will be twice as wide. The important implication of this fact is that we can always improve our estimate of μ by taking a larger sample. For a given value of σ_X, the larger the sample, the smaller $\sigma_{\overline{X}}$, and the narrower the interval that has the chosen probability of including the parameter.

▶ *Now You Do It*

The Stanford-Binet Intelligence Scale is defined as having a standard deviation of 16. You are a school counselor and have been charged by the district superintendent to estimate the level of intelligence of students in the district as part of a program to develop educational excellence. Your budget provides enough money to test 85 students. You draw a random sample of students from the district's schools and, after obtaining the parents' permission, test all 85 pupils. You find that the sample mean is 103.7. What is the 99% confidence interval?

You know that σ_X is 16, so your first task is to find $\sigma_{\overline{X}}$. N is 85, so $\sigma_{\overline{X}} = 16/\sqrt{85} = 16/9.2 = 1.74$.

You have been asked to determine the 99% CI. What value of Z do you need? Looking in Appendix Table A-1, you find that .5% of the distribution falls beyond a Z of 2.58, so the interval from -2.58 to $+2.58$ *in the standard normal distribution* provides the 99% CI.

To find the critical values in the metric of the original data, you must transform the Z-scores back into the scale of test scores. You do this by finding $Z(\sigma_{\overline{X}})$. The lower value will be $1.74(-2.58) = 4.49$ units below $\overline{X}$ and the upper value is $1.74(+2.58) = 4.49$ above $\overline{X}$. Therefore, the 99% CI is 99.2 to 108.2.

Interpreting the Confidence Interval

The technically correct interpretation of a confidence interval must be phrased in terms of our discussion earlier in this chapter of statistics and parameters. Suppose, for example, that the mean mathematics achievement of students in our state, the parameter, is 70.1. We said that the value of the parameter is a single true and correct value for the population and that the value of the statistic is true and correct for the sample. The same statement applies to confidence intervals. Our confidence interval running from 64.76 to 70.64 is the 95% CI for this sample. A different sample would provide a different confidence interval. This means that a given confidence interval either contains the parameter within its range or it doesn't; there is no probability about it. In our example, the parameter of 70.1 falls within the interval from 64.76 to 70.64, but there are other samples from this population whose CI's do not include the value 70.1. When we speak of a 95% confidence interval, we are really talking about confidence intervals for all possible samples from this population. A 95% confidence interval around the sample mean includes the value of the parameter for *95% of the possible samples from this population.* In other words, there is a 95% chance that our sample is one of the samples whose 95% confidence interval covers a range that includes μ. This statement is usually, although somewhat incorrectly, interpreted to mean that "there is a 95% chance that μ falls in the range covered by the confidence interval." For most research applications, the difference between these two ways of thinking about confidence intervals is not critical, but the former is the technically correct interpretation.

A **95% confidence interval** around the sample mean includes the value of the parameter for 95% of the possible samples from this population.

Estimating σ

The procedures described so far assume that we know the population standard deviation. This assumption is rarely met in practice, so it is usually necessary to estimate σ_X from the sample. This has some important consequences for working with the sampling distribution of the mean.

In Chapter 4, we learned that the variance of a set of observations is

$$S_{\overline{X}}^2 = \frac{\Sigma(X_i - \overline{X})^2}{N} = \frac{\Sigma x_i^2}{N}$$

This value is correct for the particular group, but it will tend to be smaller than the population variance. Therefore, S_X^2 is *biased* as an estimate of σ_X^2; it tends to provide a consistent underestimate of σ_X^2.

A small example may help to illustrate this point. Suppose we have a population of 20 scores as shown in Table 11-2. The mean and variance of this population are 27.9 and 90.7. Now, we draw a random sample of 5 scores from the population. It is shown as Sample 1 in the table. Next, we replace the five cases and draw a new random sample, Sample 2. Finally, we repeat the process a third time. For each sample, we have computed $\overline{X}$ as an estimate of μ. Each sample mean is somewhat smaller than μ, but this can happen just by chance. We have also calculated the sample variance (S_X^2) for each sample. You can see that each sample variance is substantially smaller than the population variance, revealing the bias in S_X^2. We will now explore why S_X^2 is a biased estimate of σ_X^2.

There are several explanations for why S_X^2 underestimates σ_X^2. The one most useful to us involves the fact that we are also using $\overline{X}$ to estimate μ_X. The definition of σ_X^2 is

$$\sigma_X^2 = \frac{\Sigma(X_i - \mu)^2}{N} \tag{11.3}$$

| Table 11-2 | A Population of 20 Scores and 3 Samples of $N = 5$ from the Population | | |

Population		Sample 1	Sample 2	Sample 3
25		25	25	25
29		40	34	40
34		19	16	33
40		24	28	16
16		16	27	21
18				
37	$\Sigma(X - \mu)$	−15.5	−9.5	−4.5
33	$\Sigma(X - \overline{X})$	0.0	0.0	0.0
19	$\Sigma(X - \mu)^2$	391	188	370
16	$\Sigma(X - \overline{X})^2$	342.8	170	366
24				
16				
45	$\overline{X}$	24.8	26.0	27.0
33	S_X^2	68.6	34.0	73.2
44	$\hat{S}_X^2$	85.7	42.5	91.5
28	$\hat{\sigma}_X^2$	78.2	37.6	74.0
21				
15		$\mu = 27.9$		
27		$\sigma^2 = 90.69$		
38				

where the summation is *across all members of the population,* all 20 scores in Table 11-2, for example. If we knew μ_X we could use it to estimate σ_X^2 from the sample data using the formula:

$$\hat{\sigma}_X^2 = \frac{\Sigma(X_i - \mu)^2}{N}$$

Here the summation is *across the individuals in the sample,* that is, across the five scores in each of our samples. The resulting value in each sample would be an unbiased estimate of σ_X^2. However, we seldom know μ, so we must use $\overline{X}$ in its place.

As you can see in Table 11-2, the sample mean is seldom equal to the population mean. $\overline{X}$ is an unbiased estimate of μ_X, but it is also at the least squares center of the sample. The principle of least squares, which we discussed in Chapter 4, tells us that in any given *sample* $\Sigma(X_i - \overline{X})^2$ will be less than $\Sigma(X_i - \mu)^2$ unless $\overline{X} = \mu$. In the three cases in Table 11-2, each sum of squares around $\overline{X}$ is smaller than the corresponding sum of squares around μ. For the first sample, the values for the sums of squares are 342.8 and 391, respectively. The differences between the two sums of squares are smaller in the other two samples because sample means are closer to μ, but the sum of squares around $\overline{X}$ is always smaller than around μ. S_X^2 will be smaller than $\hat{\sigma}_X^2$ because we must use $\Sigma(X_i - \overline{X})^2$ instead of $\Sigma(X_i - \mu)^2$.

$\Sigma(X_i - \overline{X})^2$ will always be less than $\Sigma(X_i - \mu)^2$ unless $\overline{X} = \mu$.

Degrees of Freedom

The amount of the bias in S_X^2 is related to the size of the sample. The larger the value of N, the better the estimate. When we use $\overline{X}$ to estimate μ_X, we are placing a numerical restriction on the data that is greater for small samples. This numerical restriction comes from the fact that, although in our sample the sum of deviations from the mean must equal zero, $\Sigma(X_i - \mu_X)$ does not have to equal any particular value. You can see this from an examination of the deviations of the scores in Sample 1 from Table 11-2. The deviations of the five scores from the sample mean sum to zero, but the deviations of the same scores from μ sum to −15.5. The other two samples also show $\Sigma(X - \overline{X}) = 0$, but the sums of deviations from μ are −9.5 and −4.5, respectively.

The consequence of this numerical restriction that the sum of deviations from the mean equals zero is that once $(N - 1)$ of the deviation values have occurred, there is only one possible number for the last one. The first four deviations in Sample 1 are +.2, +15.2, −5.8, and −.8. The sum of these scores is +8.8. Therefore, the last deviation must be

−8.8. ($N − 1$) of the values are free to vary without restriction, but the last is completely determined by the others. (Note that this restriction is placed on the entire set of values, not on any particular one of them. There is no specific score that must take a particular value. Rather, the mean is restricted in its location so that the deviations from it sum to zero.) The general term used to refer to the number of values that are not restricted is **degrees of freedom** (usually symbolized df). In the samples we are considering, each sample has four degrees of freedom. When we consider a set of N deviation scores, ($N − 1$) of them are not restricted, but the last one must cause the sum of the N deviations to be zero, so there are ($N − 1$) degrees of freedom.

▶ *Now You Do It*

Let's try out the idea of degrees of freedom with a few small examples. First, consider the numbers 1, 2, and 3. The mean of these numbers is 2. What are the three deviations from the mean? Obviously, they are −1, 0, and +1, and they sum to zero.

Next, take the three numbers 4, 8, and 9. What are the deviations from the mean? You should get −3, +1, and +2, because the mean is 7.

Now, suppose that the sum of 5 scores is 50 and that the first four scores are 8, 9, 12, and 15. What is the last score? Because the mean of the scores must be 10 (50/5), the four deviations that we know are −2, −1, 2, and 5. These deviations sum to +4. Because the sum of all deviations must be zero, the last one must be −4, and the missing score is 6. The last number was not free to vary once the first four and the mean had been selected.

The Unbiased Estimate of σ^2

We are now in a position to develop an unbiased estimate of σ^2. We will illustrate the process using the data in Table 11-2, where we had a small population of 20 scores and we took three samples from the population. Each sample provides an estimate of the population.

We have encountered the notion of a *sum of squares* or, more specifically, a sum of squared deviations from a mean, several times. In Chapter 4 we defined the sample variance as SS/N or $\Sigma(X_i − \overline{X})^2/N$, and we have now seen that this statistic is biased because the deviations have fewer degrees of freedom than there are scores being summed (that is, there are N deviations and, because they must sum to zero, there are $N − 1$ degrees of freedom). It can be shown that *a sum of squares for a sample, divided by its degrees of freedom* **(SS/df)**, *provides an* **unbiased estimate** *of the variance in the population from which the sample was drawn*.

If we use the symbol $\hat{S}_X^2$ to represent an unbiased estimate of the population variance σ_X^2 (once again we use the caret symbol ^ to indicate an estimate), our best estimate of σ_X^2 from the data in our sample is

$$\hat{S}_X^2 = \frac{\Sigma(X_i − \overline{X})^2}{N − 1} = \frac{SS}{N − 1} \tag{11.4}$$

and our best estimate of σ_X is provided by

$$\hat{S}_X = \sqrt{\frac{\Sigma(X_i − \overline{X})^2}{N − 1}} \tag{11.5}$$

A computing formula for $\hat{S}_X$ equivalent to equation 4.12 is

$$\hat{S}_X = \sqrt{\frac{N\Sigma X_i^2 − (\Sigma X_i)^2}{N(N − 1)}} \tag{11.6}$$

Applying equation 11.4 to the samples in Table 11-2, we find the unbiased estimates of σ^2 to be 85.7, 42.5, and 91.5, respectively. All are much closer to σ^2 than the sample values, and the first and third are quite good.

In Chapter 4 we defined S_X^2 as SS/N. It should be clear that the difference between S_X^2 and $\hat{S}_X^2$ in equation 11.4 is simply the difference between N and ($N − 1$) in the denominator; in fact, a little algebra shows that

$$(N - 1)\hat{S}_X^2 = NS_X^2$$

This produces the relationships

$$\hat{S}_X^2 = \frac{N}{(N - 1)} S_X^2$$

$$S_X^2 = \frac{(N - 1)}{N} \hat{S}_X^2$$

(11.7)

Clearly, as sample size becomes large, the difference of 1 unit between N and $(N - 1)$ becomes small, the ratio of N to $(N - 1)$ approaches 1, and S_X^2 approaches $\hat{S}_X^2$. By analogy to equation 11.1, our estimate of the standard error of the mean becomes

$$\hat{S}_{\bar{X}} = \frac{\hat{S}_X}{\sqrt{N}}$$

(11.8)

and we can use this value in place of $\sigma_{\bar{X}}^2$ to find confidence intervals when σ^2 is not known.

The t Distribution

Suppose Dr. Numbercruncher wants to estimate the mathematics anxiety for all students in the statistics course at BTU. These students constitute her population, and the 68 students we have been studying are a sample from that population. Unfortunately, Dr. Numbercruncher does not know either μ or σ. How should she proceed?

We have just seen that $\hat{S}_X^2$ provides an unbiased estimate of σ_X^2. It would seem reasonable to use $\hat{S}_X$ in place of σ_X to describe the sampling distribution of the mean when σ_X is unknown. We can do this, and it is the proper thing to do; however, the addition of an estimated value for the standard deviation complicates the problem a little.

Remember that the sampling distribution of the mean is normal in shape. When we convert a sample mean to the scale of Z-scores, we are performing a linear transformation of the original distribution, which does not change the shape of the distribution. The new distribution is the standard normal, or Z, distribution. We make this transformation because we know the probability of any Z-score in the normal distribution. By reversing the process, we are able to make probability statements in the scale of the original data.

The linear Z transformation involves subtracting a constant (μ) from every value in the distribution and dividing the result by a constant (σ_X). That is,

$$Z = \frac{Sample\ Mean - Constant}{Constant}$$

The reason we can use the standard normal distribution is because the same transformation is applied to every possible sample from the population. There is a one-to-one correspondence between the original sample means and their Z-scores. However, when we have only the data from our sample, we must estimate σ_X from the sample data. Each sample from the population will yield a different estimate of σ_X, so we no longer have a simple linear transformation. $\hat{S}_X$ varies from one sample to another. Therefore, our transformation becomes

$$\frac{Sample\ Mean - Constant}{Variable}$$

The statistic that results from this operation is called t and is defined as

$$t = \frac{\overline{X} - \mu}{\hat{S}_{\overline{X}}}$$

(11.9)

The values that result from this transformation no longer have a normal distribution because using a variable quantity in the denominator results in a nonlinear transformation. This fact was first discovered by the English statistician William Gossett, who, writing under the pen name Student, published a description of the actual distribution under these conditions. He showed that there is a unique unimodal, symmetric probability distribution corresponding to each possible number of degrees of freedom (df) and that as ($N - 1$) becomes large (over about 100 for practical purposes), the distribution approaches the normal distribution. He called this family of distributions the **t distributions.** Since in a given study such as Dr. Numbercruncher's, we have a single sample size (68) and single value for df (67), we use only one of the distributions at a time. Therefore, we will refer to the one distribution we are using as the t distribution.

The t distribution is an important concept in the practical application of statistics because it fits a wide variety of research problems. As we said, it is really a family of distributions, one for every possible number of degrees of freedom from 1 to infinity. Each distribution can be used in exactly the same way that we use the normal distribution to represent probabilities; that is, *the area under the curve beyond a certain point is the probability that a value of the t statistic at least that large will occur.* Likewise, the area between two values is the relative frequency (probability) with which we would expect events in that range to occur. The major difference between a t distribution (any one of them) and the normal distribution is that the t distribution will have a greater proportion of its area beyond any given distance from its mean. That is, a t distribution has a wider spread than the normal distribution.

This feature is illustrated in Figure 11-3. With two degrees of freedom the curve of the t distribution is flat with high tails. As df increases, the curve becomes more narrow and peaked until with infinite df it becomes the standard normal curve.

The existence of all these different distributions would present quite a problem for users of statistics, were it not for the fact that interest has centered on the values appropriate for the most commonly used confidence intervals and for testing hypotheses (described in the next chapter). This narrowing of interest has made it feasible to put the most popular critical values for a range of degrees of freedom into a single table, such as Appendix Table A-2, rather than to have a table like Appendix Table A-1 for each possible value of df. Part of Appendix Table A-2 is reproduced here (Table 11-3). This table is known as the t table. The entries across the top are the proportions of area in one or two tails of the distribution (that is, the probability levels which include one tail or both tails of the distribution); entries in the left-hand column are degrees of freedom (df); and the elements in the body of the table are values of the t statistic for that probability with the given degrees of freedom. Each row of Appendix Table A-2 contains the values of t for the given df that correspond to the values that appear in boldface in Appendix Table·A-1. The

Figure 11-3 t distributions for different degrees of freedom are unimodal symmetric but have different degrees of spread.

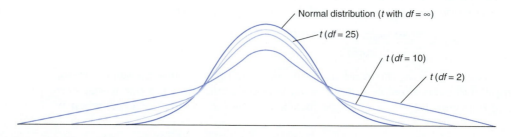

Normal distribution (t with $df = \infty$)

t ($df = 25$)

t ($df = 10$)

t ($df = 2$)

Table 11-3 Partial Table of Critical Values for the *t* Distributions

df	One tail Two tails	Proportion in Tail of Distribution					
		.10 .20	.05 .10	.025 .05	.01 .02	.005 .01	.0005 .001
1		3.078	6.314	12.706	31.821	63.657	636.619
2		1.886	2.920	4.303	6.965	9.925	31.598
.							
.							
.							
10		1.372	1.812	2.228	2.764	3.169	4.587
.							
.							
25		1.316	1.708	2.060	2.485	2.787	3.725
.							
.							
∞		1.282	1.645	1.960	2.326	2.576	3.291

statistic itself, covered briefly here and in detail in Chapter 13, is virtually identical to *Z* in its computation and interpretation. The difference is that it produces a different probability distribution.

Inspecting Table 11-3 shows clearly the relationship between degrees of freedom and the shape of the related *t* distribution. If we choose the .05 level of probability using both tails (the column headed .05 for two tails), we see a value of 1.960 in the bottom row with infinite (∞) degrees of freedom. This is exactly the value of *Z* that we used to form the 95% confidence interval in the normal distribution. That is, *t with infinite degrees of freedom is equal to Z.* As we move up the column to 25 *df,* the critical value for *t* increases to 2.060. This means that we must be farther from the center of this *t* distribution to contain 95 percent of the area under the curve between our two critical values (or have $2\frac{1}{2}$ percent of the area in each tail). As *df* drops to 10, the critical value increases to 2.228, and in the extreme case of *df* = 1, we must go all the way to a *t* of ±12.706 to encompass 95 percent of the distribution and leave only $2\frac{1}{2}$ percent in each tail. This principle is illustrated in Figure 11-4 for *df* values of 2, 10, 25, and infinity.

t with infinite degrees of freedom is equal to Z.

Confidence Intervals for the Mean Using *t*

The most common situation in which we would want to form a confidence interval for the mean occurs when we have only the data from our sample to guide us. Suppose, for example, that we have taken a sample of *N* = 26 students randomly selected from our university and have measured their self-esteem on a scale that has a score range from 0 (a feeling of total worthlessness) to 100 (unbounded self-esteem). Our sample data provide us with a mean of 60 and a *sample* standard deviation, S_X, of 15. We wish to find the 95% confidence interval for the mean self-esteem score at the institution. From what we have just seen about using $\hat{S}_X$ to estimate σ_X and the consequences of using this estimated value, the most logical course would be to use the sample mean ($\overline{X}$), our best estimate of σ_X ($\hat{S}_X$), and the *t* distribution to form our confidence interval for μ.

To form a confidence interval for μ from sample data only, use the sample mean ($\overline{X}$), the best estimate of σ_X ($\hat{S}_X$), and the t distribution.

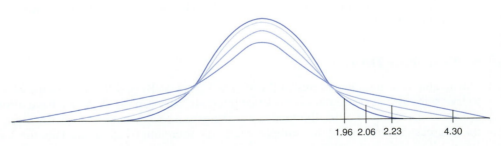

1.96 2.06 2.23   4.30

Figure 11-4 .05 critical values for several *t* distributions. With fewer degrees of freedom you must go farther from the mean to include 95% of the distribution.

We proceed to find the confidence interval exactly as we have in the past, but with the appropriate substitutions. The first thing we must check is which form of the standard deviation we have. We see that we have the sample standard deviation, not the unbiased estimate of σ_X^2. To convert this value into the $\hat{S}_X$ that we need we must first find S_X^2. Using the relationship in equation 11.7, we find

$$S_X^2 = 15^2 = 225$$

Then

$$\hat{S}_X^2 = \frac{N}{N-1} S_X^2 = \frac{26}{25}(225) = 234$$

and

$$\hat{S}_X = 15.3.$$

Now that we have the appropriate estimate of σ_X, we can proceed to find $\hat{S}_{\overline{X}}$, our best estimate of the standard error of the mean. (If we had obtained $\hat{S}_X$ directly from our calculator or computer, we could have ignored the above steps and proceeded directly to estimate σ_X. Most calculators and computer programs do produce $\hat{S}_X$ directly, and the better ones give you a choice, but you need to know which standard deviation you have in order to get the correct answer.) We find the necessary value by using equation 11.8.

$$\hat{S}_{\overline{X}} = \frac{\hat{S}_X}{\sqrt{N}}$$

$$\hat{S}_{\overline{X}} = \frac{15.3}{\sqrt{26}} = 3.0$$

That is, our best estimate of the standard error of the mean is 3.0.

We can also use the relationship in equation 11.7 to produce an alternate and more direct formula for $\hat{S}_{\overline{X}}$. If we have the sample standard deviation, the equation

$$\hat{S}_{\overline{X}} = \frac{S_X}{\sqrt{N-1}} \tag{11.10}$$

$$\hat{S}_{\overline{X}} = \frac{15.0}{\sqrt{25}} = 3.0$$

produces an identical result. The advantages of equation 11.8 are that it uses the quantity most likely to be produced by your software and it has exactly the same structure as equation 11.1 (rather than having $N - 1$ in the denominator).

Whichever method we employ to calculate $\hat{S}_{\overline{X}}$, we can now use it with $\overline{X}$ and t to determine a confidence interval for the mean level of self-esteem at our institution. With $N = 26$ we have $df = 25$, so the critical value of t for the 95% confidence interval is 2.06. The upper and lower limits of the confidence interval are calculated in the usual way, but with t substituting for Z.

$$CI_{\text{upper}} = +t(\hat{S}_{\overline{X}}) + \overline{X} \tag{11.11}$$
$$= +2.06(3) + 60 = 66.18$$
$$CI_{\text{lower}} = -t(\hat{S}_{\overline{X}}) + \overline{X} \tag{11.12}$$
$$= -2.06(3) + 60 = 53.82$$

The 95% confidence interval for the mean self-esteem score in the population from which this sample was drawn runs from 53.82 to 66.18. The probability is 95 percent that this interval includes μ.

▶ Now You Do It

Our example of the 68 students in the statistics class at BTU will give you the chance to try your hand at computing confidence intervals from sample data. We have already seen that the value of $\hat{S}_X$ for the final exam scores is 13.58. What value shall we use for the degrees of freedom? With a sample size of 68, there are 67 df, so we take this value and look in Appendix Table A-2. Unfortunately, the only values we find there are a row for $df = 60$ and one for $df = 120$. What to do?

Inspecting the table, you will see that there are entries for every df value up to 30, but beyond that point there are entries only for 40, 60, 120, and ∞. The reason is that the critical values are not changing very quickly, so more widely separated values of df can be used to save space. Whenever the exact value of df that you need is not provided, the appropriate response is to take the values associated with the next *smaller df*. The result will be a confidence interval slightly wider than necessary, but it is generally better to err on the side of caution and not claim more than we know. With this in mind, if we want a 95% confidence interval, the value of t we should use is 2.00 (95% critical value for $df = 60$) and we find

$$\hat{S}_{\bar{X}} = \frac{13.58}{\sqrt{68}} = 1.65$$

Alternatively, we could find

$$\hat{S}_{\bar{X}} = \frac{13.48}{\sqrt{67}} = 1.65$$

In either case, the 95% confidence interval extends from $(2.0)(1.65) = 3.3$ points below the mean of 34.24 to 3.3 points above. That is, the confidence interval is from 30.94 to 37.54. We may claim, with a 95 percent chance of being correct (or a 5 percent chance of being in error) that these students are a sample from a population with a mean between 30.94 and 37.54.

HOW THE COMPUTER DOES IT

Most software packages will compute confidence intervals, but it may be hard to find the place where they do it. The programs are also likely to use exact values for t rather than the approximate values we have to look up in a t table. For example, SPSS computes confidence intervals for the mean as part of its EXPLORE procedure within the SUMMARIZE section of the STATISTICS menu. To access the confidence interval commands select "Statistics" EXPLORE. The screen then looks like this.

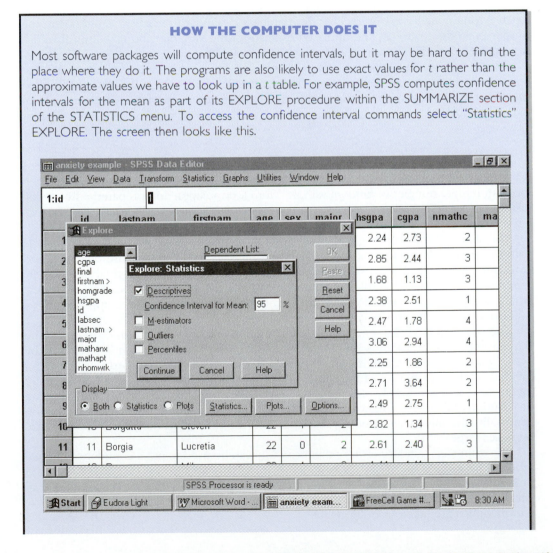

The program uses the 95% confidence interval as its default, so if you wish some other CI, such as 90% or 99%, you must change the value in the confidence interval box. Because it uses the exact *t* value for 67 degrees of freedom rather than the tabled value for 60 *df*, the confidence interval runs from 30.95 to 37.52 rather than from 30.94 to 37.54 as we found in Now You Do It. Confidence intervals for the mean are also produced as by-products of some other programs, but EXPLORE is the most direct place to find them.

The default options for this routine also produce measures of skewness and kurtosis, a stem-and-leaf plot, and a box plot. If you run this procedure you will notice that there are standard errors reported for the skewness and kurtosis indices. Like the mean, these statistics also have sampling distributions, and these standard errors have exactly the same interpretation that the standard error of the mean does. We return to this topic in Chapter 14.

Suppose we had treated our sample standard deviation as if it were a parameter. What would the consequences have been? Using our S_X as if it were σ_X and a sample size of 26 (that is, using equation 11.1), we would have found a standard error of the mean of 2.94 ($15/\sqrt{26}$). The critical value for Z is 1.96, so our confidence interval would run from $(2.94)(1.96) = 5.76$ units above the mean to $(2.94)(-1.96) = 5.76$ units below. Using the incorrect values would produce an interval about 8/10th of a unit too narrow. This may not seem like much, but when we come to use procedures similar to the ones we have just discussed to test statistical hypotheses, using the wrong procedures can lead us to reach wrong conclusions. Whenever the population standard deviation is unknown, the proper procedure is to use the unbiased estimate of the variance and take critical values from the *t* table with the appropriate degrees of freedom.

As we will see in Chapter 13, the *t* distribution fits a wide variety of situations that researchers encounter in data analysis. The reason it is so useful is that in most applied situations we do not know the population standard deviation, so we must estimate it from the sample data.

SUMMARY

A *sampling distribution* is the frequency distribution that we would expect to get if we drew an infinite number of random samples from a population and prepared a frequency distribution of values of the statistics computed on these samples. All descriptive statistics have sampling distributions.

The shape of the sampling distribution of a statistic may be approximately normal, as with the mean and median, or it may be nonnormal, as with the standard deviation and correlation coefficient. Most sampling distributions approach being normal in shape as sample size increases, regardless of the shape of the parent distribution. This tendency, which is precisely true for the sampling distribution of the mean and is approached by other statistics, is due to the *central limit theorem*.

The standard deviation of a sampling distribution is known as its *standard error*. The standard error of the mean is equal to the population standard deviation (σ_X) divided by $\sqrt{N}$ if σ_X is known. Otherwise, the estimate of the population standard deviation ($\hat{S}_X$) divided by the square root of the sample size is used. Standard errors of all statistics are inversely related to sample size. Larger samples show smaller variability. The sampling distribution of the mean follows the *t* distribution when σ_X is not known. The *t* distribution is actually a family of distributions that approach the normal distribution as sample size becomes large.

The estimated standard error of the mean ($\hat{S}_X$) can be used to put a confidence interval around the sample mean ($\overline{X}$). A confidence interval is an interval that has a specified probability of including the value of μ_X, the population mean. When σ_X is known, critical values are taken from the table of the normal distribution. When $\hat{S}_X$ is used to estimate σ_X, critical values are taken from the *t* table.

Table 11-4   Computing Confidence Intervals for the Continuous Variables in the Searchers Data

Variable	$\overline{X}$	$\hat{S}_X$	$\hat{S}_{\overline{X}}$	$\hat{S}_{\overline{X}}(t_{.05})$	$\hat{S}_{\overline{X}}(t_{.01})$	95% CI	99% CI
Age	39.72	11.59	1.50	3.00	3.99	36.72–42.72	35.73–43.71
Income	45,254.21	7270.77	938.65	1877.30	2496.81	43,376.91–47,131.51	42,757.40–47,751.02
Attitudes toward equality	40.40	8.29	1.07	2.14	2.85	38.26–42.54	37.55–43.25
Religious involvement	18.10	5.25	0.68	1.36	1.81	16.74–19.46	16.29–19.91

STANDARD ERRORS AND CONFIDENCE INTERVALS FOR THE SEARCHERS DATA

We can use the information computed in Chapters 3 and 4 to find confidence intervals for the means of our continuous variables in the Searchers study that Dr. Cross and his students conducted. We have already computed the values for $\hat{S}_X$ (S-estimate in Table 4-16), so most of the work is done. The values of $\overline{X}$ and $\hat{S}_X$ are given in the first two columns of Table 11-4. Dividing $\hat{S}_X$ by $\sqrt{60}$ for each variable produces the values of $\hat{S}_{\overline{X}}$ in column 3. The t table does not contain an entry for 59 df, so we will use 60. As we stated earlier, ordinarily when the df falls between two tabled values we would either interpolate or take the one with the fewer df, but in this case we are so close to the tabled value that the difference is negligible. The 5% value of t for df = 60 is 2.00, and the 1% value is 2.66. Multiplying the standard errors in column 3 by these t values produces the confidence interval values in columns 4 (5% interval) and 5 (1% interval). Finally, the intervals themselves ($\overline{X} \pm \hat{S}_{\overline{X}}[t]$) are shown in the last two columns. These values are the ones you would obtain from running SPSS on these data with the exception that for large numbers such as the income variable SPSS may give slightly different results because it keeps more decimal places than you would, both in the standard deviation and in the value of t it uses. Also, of course, the program uses the exact value of t for 59 degrees of freedom. The difference between the .05 critical value of t for df = 60 and the one for df = 59 is .0007.

REFERENCES

Hays, W. L. (1994). *Statistics* (5th ed.). Fort Worth, TX: Harcourt Brace.

Kirk, R. E. (1995). *Experimental design: Procedures for the behavioral sciences* (3rd ed.). Pacific Grove, CA: Brooks/Cole.

EXERCISES

1. In your own words, define each of the following terms:
   a. sampling distribution of the mean
   b. sampling distribution of the median
   c. sampling distribution of the standard deviation
   d. sampling distribution of the correlation

2. Suppose you and 14 friends are given a true/false exam with 20 questions. In addition, suppose this is an exam for which you and your friends know none of the answers and are merely guessing. If a score represents the number of times answered correctly, what would be the mean and standard deviation of the population? Determine the mean and the standard deviation of the sampling distribution of mean scores from this binomial distribution.

3. In your own words, define each of the following terms:
   a. standard error of the mean
   b. standard error of the variance
   c. standard error of the correlation

4. Suppose you administered an aggression questionnaire to the prison population and found that the distribution of scores was negatively skewed with a mean of 78 and a standard deviation of 10. Suppose you conducted a study on aggression in which you randomly sampled the prison population obtaining a sample of size 64. Sup-

pose that you continued this process for a large number of samples of size 64 being careful to calculate the mean score each time. Furthermore, suppose you constructed a frequency distribution of the means from each sample of size 64. Describe the characteristics of this distribution. How would this frequency distribution be affected if you had randomly selected samples of sizes of 144 instead of 64?

5. The SAT Quantitative Test has a population mean of 500 and a population standard deviation of 100. Suppose you randomly select a sample of 25 participants and administer the SAT Quantitative Test to each participant.
   a. What is the expected mean of this sample?
   b. What is the standard error of the mean for samples of size 25 from this population?
   c. What is the probability that this sample will be one of the samples that will have a mean SAT Quantitative score greater than 530?
   d. What is the probability that this sample will be one of the samples that will have a mean SAT Quantitative score less than 450?
   e. What is the probability that this sample will be one of the samples that will have a mean SAT Quantitative score between 475 and 525?

f. Suppose we obtain a mean SAT Quantitative score of 515. What is the 95% confidence interval for these data?

g. Suppose we obtain a mean SAT Quantitative score of 487. What is the 80% confidence interval for these data?

6. Suppose you intend to administer an IQ test that has a published population mean of 100 and a population standard deviation of 15 to a sample of 81 participants.

a. What is the expected mean of this sample?

b. What is the standard error of the mean for samples of size 81 from this population?

c. What is the probability that this sample will be one of the samples that will have a mean IQ score above 95?

d. What is the probability that this sample will be one of the samples that will have a mean IQ score above 102?

e. What is the probability that this sample will be one of the samples that will have a mean IQ score below 97?

f. What is the probability that this sample will be one of the samples that will have a mean IQ score between 96 and 104?

g. Suppose that when you administer the IQ test to the sample of 81 participants you get a mean IQ score of 96. What is the 99% confidence interval for these data?

h. Suppose you administer the IQ test to the sample of 81 participants and obtain a mean IQ score of 102. What is the 75% confidence interval for these data?

7. Suppose a sample of 50 participants is administered the Creativity Test with a population mean of 80 and a population standard deviation of 12.

a. What is the expected value of the mean of this sample?

b. What is the standard error of the mean for samples of size 50 from this population?

c. What is the probability that this sample will be one of the samples that will have a mean creativity score greater than 84?

d. What is the probability that this sample will be one of the samples that will have a mean creativity score less than 75?

e. What is the probability that this sample will be one of the samples that will have a mean creativity score between 77 and 82?

f. What is the 95% confidence interval for this Creativity Test if the mean score from a sample of 50 participants is 83?

g. Suppose that we have 150 participants instead of 50 as in part (f). Suppose also that the mean creativity score of this new sample is also 83. What is the 95% confidence interval for this Creativity Test?

8. Suppose a sample of 75 participants is administered the Patience Test with a population mean of 50 and a population standard deviation of 9.

a. What is the expected value of the mean of this sample?

b. What is the standard error of the mean for samples of size 75 from this population?

c. What is the probability that this sample will be one of the samples that will have a mean patience score greater than 52?

d. What is the probability that this sample will be one of the samples that will have a mean patience score less than 49?

e. What is the probability that this sample will be one of the samples that will have a mean patience score between 47 and 53?

f. What is the 95% confidence interval for this Patience Test if the mean score from a sample of 75 participants is 51.5?

g. Suppose that we have 150 participants instead of 75 as in part (f). Suppose also that the mean patience score of this new sample is also 51.5. What is the 95% confidence interval for this Patience Test?

9. Suppose the following scores represent a measure of empathic concern.

47   57   44   58   49   49   51   54   50   51

a. Calculate the variance (S^2) of the set of scores.

b. If the set of scores listed above represents a sample of scores from the population, what is the unbiased estimate ($\hat{S}^2$) of the population variance (σ^2)?

c. What is the degrees of freedom for the sample of empathic concern scores?

d. What are the critical values of t from Appendix Table A-2 for a 95% confidence interval? 99% confidence interval? 99.9% confidence interval?

10. Suppose the following scores represent a measure of verbal aggression.

17   26   24   18   15   21   12   14   22   17   28   26

a. What is the variance (S^2) of the set of scores?

b. If the set of scores listed above represents a sample of scores from the population, what is the unbiased estimate ($\hat{S}^2$) of the population variance (σ^2)?

c. What is the degrees of freedom for the sample of verbal aggression scores?

d. What are the critical values of t from Appendix Table A-2 for a 90% confidence interval? 95% confidence interval? 99% confidence interval?

11. Suppose you are told that the variance of a set of 28 scores is 169. What is the unbiased estimate of the population variance?

12. Suppose you are told that the unbiased estimate of the population variance is 67.36 for a sample of 42 scores. What is the variance of the set of scores?

13. For the set of scores listed in Exercise 9, what is the
a. 95% confidence interval?
b. 99% confidence interval?
c. 99.9% confidence interval?

14. For the set of scores listed in Exercise 10, what is the
a. 90% confidence interval?
b. 95% confidence interval?
c. 99% confidence interval?

15. Find the 95% confidence interval for the scores on the Anger Scale listed below:

28   21   19   33   10   15   18   21   32   28   22   25
38   24   20   18   19   29   31   30   13   18   21   26

16. Find the 90% confidence interval for the scores on the Anxiety Scale listed below:

52   54   62   66   48   45   58   54   55   61   63   56
58   60   66   49   44   42   40   46   48   52   55   67
62   60   58   54   47   53   59   61   69   55   56   51

17. Suppose you administer a self-esteem measure to 100 participants. The mean self-esteem score for these participants is 52.45 and the unbiased estimate of the popula-

tion standard deviation is 6.28. What is the 95% confidence interval?

18. Suppose you administer a hostility measure to 48 participants. The mean hostility score for these participants is 37.79 and the unbiased estimate of the population standard deviation is 5.37. What is the 95% confidence interval?

19. Use the SPSS program and SPSS data bank to complete this exercise. From the data set donate.sav, find the 90% and 99.9% confidence intervals for the following variables: perspective-taking (perspect) and empathic concern (empathy).

20. Use the SPSS program and SPSS data bank to complete this exercise. For the data set career.sav, find the 95% and 99% confidence intervals for the following variables: confidence in doing mathematics (confmath) and anxiety about doing mathematics (anxiety).

Chapter 12

Statistical Hypotheses, Decisions, and Error

CHAPTER OBJECTIVES

After studying this chapter, you should be able to:

❑ Given the description of a research study, differentiate between null hypotheses and alternate hypotheses.

❑ Identify the null hypothesis that a given research study is designed to test.

❑ Describe the types of decision errors and correct decisions that result from applying statistical reasoning to a research problem and identify examples of each.

❑ Differentiate between "statistically significant" and "scientifically important."

❑ Identify the assumptions being made in defining a null hypothesis.

❑ Identify directional and nondirectional null hypotheses and state the advantages and disadvantages of each.

❑ Determine the critical values and rejection regions needed to test a given null hypothesis.

❑ Determine what happens to power, Type I error, and Type II error as a result of changing N or α.

Think back to the study that the faculty of the SAD Department at Big Time U are conducting. They want to test the hypothesis that math anxiety negatively influences student performance in statistics courses. However, as we have seen, they cannot test their research hypothesis simply by finding examples that fit it; instead, they must find examples that are inconsistent with it. This situation seems paradoxical: How can the faculty support their hypothesis when the only avenue open to them is to try to prove that the hypothesis they are testing is NOT true? What hypothesis could the SAD faculty test where rejecting the hypothesis would support their belief that math anxiety adversely affects test performance?

Introduction

In Chapter 10 we discussed the general problem of testing hypotheses and drawing conclusions from a theoretical perspective. In this chapter we will consider the particular problems encountered when statistical analysis is used to test hypotheses in educational and psychological research. Our initial challenge is to find a way to state our research hypothesis—the outcome we would expect if our theory is true—in a form that will enable us to test it. Since no number of confirming observations can prove our theory to be true, how can we proceed, as the SAD faculty—or other researchers in psychology or education—must, by ruling out incorrect alternatives? In turn, how we can apply sampling distributions and probability through statistical analysis to reach a conclusion?

Statistical Hypotheses: Null and Alternate

The research hypothesis that the SAD faculty wish to test is that students with higher levels of math anxiety will earn lower scores on the statistics final. To test a research hypothesis like this one using statistics, it is first necessary to phrase it as a **statistical hypothesis.** That is, to test the research hypothesis using statistical methods, the SAD faculty must restate their hypothesis as a hypothesis about the distribution of test performance that they would expect to observe in samples from the population of BTU statistics students. A *statistical hypothesis is stated in terms of the distribution implied by the research question under investigation.* There are two kinds of statistical hypotheses, one called the *null hypothesis,* and the other called the *alternate hypothesis.* When we run a study to test a research hypothesis, we construct both a null hypothesis and an alternate hypothesis from the research hypothesis.

> A **statistical hypothesis** is stated in terms of the distribution implied by a research question.

Remember that according to inductive reasoning we can only prove a hypothesis to be false, we can never prove it to be true. Therefore, testing hypotheses using statistical methods must involve finding the hypothesis under test to be false: To support a research hypothesis like the one of interest to the SAD faculty, we must test a hypothesis that we would like to disprove. Statisticians designate the hypothesis that is tested and shown to be false the **null hypothesis.** The null hypothesis makes a prediction that is contrary to the research hypothesis; finding the null hypothesis to be false supports the research hypothesis. The null hypothesis includes the specification of the population and the characteristics of the sampling distribution for the statistic being tested *if the research hypothesis is not true.* For example, the SAD faculty could test their research hypothesis by converting it into a null hypothesis stating that the correlation between test scores and math anxiety is either zero or positive. If they can reject this null hypothesis, they have found support for their research hypothesis.

> The **null hypothesis** specifies the properties of a particular sampling distribution if the research hypothesis is not true.

Statisticians use the term **alternate hypothesis** to refer to *negation of the null hypothesis.* Note carefully that this alternate hypothesis is different from the alternative hypotheses discussed in Chapter 10. There we discussed hypotheses from alternative, or competing, theories and plausible rival hypotheses that provided alternatives to any theory. Here, the alternate hypothesis is a purely statistical hypothesis that contains no substance or basis in theory; it is merely the negation of the null hypothesis. It is not a hypothesis in the same sense that the null hypothesis is because no population is specified, but it does state a general belief in the untruth of the null hypothesis. The alternate hypothesis says that we believe the null hypothesis is not true, or, in other words, that our sample did not come from the population specified in the null hypothesis. The SAD faculty's alternate hypothesis is that their null hypothesis is false; that is, their sample did not come from a population in which the correlation between math anxiety and final exam performance is zero or positive.

> The **alternate hypothesis** states that the null hypothesis is false, or that our sample did not come from the population specified in the null hypothesis.

Let's take a slightly simpler example for our first look at how the null hypothesis works. We know that the population distribution for intelligence has a mean of 100 and a standard deviation of 15 because the tests are standardized to have these characteristics. We have been told that large doses of carrot juice in childhood will increase intelligence, and we propose as our *research hypothesis* that people who have consumed large quantities of carrot juice will have a mean intelligence test score greater than 100. The *null hypothesis* in this case would be that the mean test score for people who drank carrot juice is not greater than 100.

A null hypothesis can occur in many forms, but all can be reduced to one essential statement. The null hypothesis states that *sample S of size N was drawn from population P, which has the parameters* μ, σ, ρ (Greek letter ρ [rho] stands for the population correlation coefficient), and any other characteristics the research question may require. That is, the null hypothesis specifies that we believe our sample of N observations came from a population with a particular mean, μ, a particular standard deviation, σ, and any other characteristics, such as a correlation coefficient, ρ, that we might specify that bear on our research question. In our carrot juice example, the only parameter of interest is μ. The null hypothesis would therefore state that our sample of carrot juice drinkers came from a population with a mean intelligence test score no greater than 100. If we can reject this hypothesis, we can conclude that our carrot juice drinkers came from a population in which the mean test score is greater than 100, a conclusion that is consistent with our research hypothesis.

The null hypothesis is the hypothesis under direct test in any analysis involving statistical hypothesis testing. Since inductive reasoning is the only avenue open to us, we must proceed by negation. The only strong conclusion that we can reach is that the hypothesis under test is false; that is, we can reach the conclusion that sample *S* was *not* drawn from population *P*, which has the characteristics specified. Statistical inference *never* permits the conclusion that sample *S was* drawn from population *P*. Even if $\overline{X}$ computed from sample *S* is exactly equal to the μ specified in the null hypothesis, we cannot say that *S* was drawn from *P* because there are many other populations that could also yield sample *S*. Even if our sample of carrot juice drinkers produced a mean of exactly 100, we could not conclude that they came from a population with a mean of 100, only that the data are not inconsistent with that conclusion.

It is never possible to conclude that the null hypothesis is true, only that we cannot reject it.

If we find that our carrot juice drinkers have a mean intelligence test score of 110, we might conclude that the null hypothesis is false, which is the same as accepting the alternate hypothesis. Accepting the alternate hypothesis leads us to the conclusion that our sample did not come from a population with a mean of 100. We do not know what the population mean is, but we are confident that it is not 100.

Formal Statement of the Null and Alternate Hypotheses. As we have seen, statistical hypothesis testing requires selecting one of two mutually exclusive and exhaustive statistical hypotheses, the null hypothesis and the alternate hypothesis. The null is either true, or it is not true. Therefore, the null hypothesis and the alternate hypothesis must be stated in such a way that they cover all possible outcomes. This means that the null hypothesis must state one expected value or range of values for the parameter, and the alternate hypothesis must cover the rest of the possibilities. For example, if we believe that carrot juice improves intelligence, the null hypothesis must state the opposite of this belief. We would formally say that the mean intelligence test score of carrot juice drinkers is less than or equal to 100 ($\mu \leq 100$). The alternate hypothesis is the negation of the null, and therefore covers all other bases. The formal statement of the alternate hypothesis is that the mean is not less than or equal to 100. Since this statement is equivalent to our research hypothesis, rejecting the null hypothesis supports the research hypothesis, leading us to conclude in this example that $\mu > 100$.

▶ *Now You Do It*

Look back to the end of Chapter 1, where we met the faculty of the Department of Statistics for Analyzing Data at Big Time U. What were their two major research questions? If they wanted to apply the principles of statistical inference, what would their null hypotheses be?

The department members wondered whether math anxiety might be affecting students' performance on exams. What is the research hypothesis, and what would be an appropriate null hypothesis?

The department also wondered whether the way lab sections were taught affected students' performance. What would the research and null hypotheses be? Does the fact that each lab section is taught by a different instructor affect the clarity of the conclusions one might draw from this study?

The department collected scores on a math anxiety measure to provide information about students' anxiety. The faculty expect that students with higher anxiety will earn lower scores on statistics exams. Therefore, a loose statement of the null hypothesis would be that there is no relationship between math anxiety scores and final exam scores. We can form a proper testable null hypothesis in either of two ways. First, we could divide the students into two groups, those high in anxiety and those low in anxiety. Those with math anxiety scores above the median would be in the high-anxiety group (*H*); those below the median in the low-anxiety group (*L*). The null hypothesis would then be that the mean final exam score of the high-anxiety group would be greater than or equal to the mean of the low group; $\mu_H \geq \mu_L$. If the results are inconsistent with this hypothesis, the faculty can reject it and conclude that higher-anxiety students get lower final exam scores. Methods for testing this hypothesis will be presented in Chapter 13.

The faculty could address the same question with a correlation coefficient. Their null hypothesis would then be that there is a zero or positive correlation between anxiety scores and final exam scores: $\rho \geq 0$. If the data yield a large negative correlation, this would cause them to reject the null and conclude that their research hypothesis is correct. How to test this hypothesis will be described in Chapter 14.

The question about teaching methods in the different lab sections is similar to the first null hypothesis, but because there is no basis on which to expect one method to be better than another, the null hypothesis would be that all sections' means would be equal. Because there are three means involved, we must wait until Chapter 16 to answer this question.

The primary threat to the hypothesis about teaching methods is a confound: teaching method is confounded with teacher. If we find the means are unequal, we do not know whether it is the teaching method or the skill of the instructor that is responsible.

Testing Null Hypotheses

At the most fundamental level, a null hypothesis is a hypothesis about sampling. The null hypothesis is that our sample is a random sample drawn from a population with certain stated parameters. For example, our null hypothesis about the effects of carrot juice on intelligence says that our sample of test scores of carrot juice drinkers is a sample from a population of test scores with a mean of 100.

For a given sample size ($N = 25$), we know the shape of the sampling distribution of our chosen statistic (in this case, the mean) around the parameter ($\mu = 100$), and we know its spread ($\sigma_{\overline{X}} = \sigma_X/\sqrt{N} = 3$, so we can determine the probability that we would draw a sample with the observed value of the statistic from a population with that sampling distribution. If the probability is small that we would get such a sample from that population, we reach the conclusion that the null hypothesis is not true. We conclude that the sample was not drawn from a population with the parameters stated in the null hypothesis. This, in a nutshell, is how testing hypotheses with statistical methods works.

Suppose, for example, that we have found 25 people who have consumed the required amount of carrot juice. We measure their intelligence, and we find the sample mean to be 106. Our null hypothesis specifies a sampling distribution with a mean of 100 and a standard deviation of 15. How likely is it that we would draw a sample of 25 observations with $\overline{X} = 106$, if the hypothesis that $\mu = 100$ is true?

We can use the sampling distribution of the mean and areas under the normal curve to determine the probability that we would obtain a sample of 25 with a mean as great as 106 from a population with $\mu = 100$ and $\sigma = 15$ by computing a standard score for the sample mean in the sampling distribution implied by the null hypothesis. With $N = 25$, this is a sampling distribution with $\mu = 100$ and $\sigma_{\overline{X}} = 3$. Therefore, we obtain

$$Z_{106} = \frac{106 - 100}{3} = 2.00$$

The proportion of the normal distribution that lies *at or beyond* a Z of $+2.00$ is .0228. If we were to draw 100 independent random samples of $N = 25$ from the population specified in the null hypothesis ($\mu = 100$, $\sigma = 15$), we would expect to obtain about 2 samples with a mean of 106 or greater. Another way of saying this is that the odds are about 50 to 1 against getting a sample like this by chance from the specified population. The question then becomes: Do we still believe that our sample came from the population specified in the null hypothesis?

Reaching a Decision About the Null Hypothesis

The data collected in an investigation permit us to reach a decision regarding the null hypothesis. We can come to one of two decisions: We may decide that the null hypothesis is false; this is called *rejecting the null*. Rejecting the null hypothesis means we conclude that our sample of carrot juice drinkers did not come from a population with $\mu \leq 100$ and $\sigma = 15$.

If we reject the null hypothesis, we accept the alternate hypothesis that our sample of carrot juice drinkers came from a population with a mean greater than 100. On the other hand, we may conclude that the evidence does not justify rejecting the null hypothesis. In this case, we *fail to reject* the null hypothesis. That is, *we may reject the null hypothesis, or we may fail to reject the null. We cannot conclude that the null is true because to do so would violate the rules of inductive logic.* It is customary to use the term *null*, or the symbol H_0, to refer to the null hypothesis. The term *alternate*, or the symbol H_1, is used to indicate the alternate hypothesis: the conclusion that the null is false. If we reject H_0, we accept H_1.

*We have two choices with regard to the null hypothesis; we may reject it or we may fail to reject it. We may **not** conclude that the null is true.*

Our decision that the null is false is based on the probability that a sample such as ours would be drawn by random sampling from the population specified in the null hypothesis. If the probability of obtaining such a sample is small, we say that the difference between the sample statistic and the hypothesized population parameter is **statistically significant,** and we reach a decision to reject the null hypothesis. If we find the results from our carrot juice study to be statistically significant, this means that it is unlikely that we would have drawn a sample with a mean as large as the one we obtained by chance from a population in which the true mean was no greater than 100.

*A decision that a result is **statistically significant** means that the result is unlikely to have occurred by chance under the conditions specified in the null hypothesis.*

It is important to understand exactly what these terms mean and exactly what conclusions we can draw from a statistical analysis because the terms are often misinterpreted and the conclusions are frequently overdrawn. The term *statistically significant* does *not* mean *important;* it does *not* mean *meaningful;* it does *not* even mean *true.* All that the term *statistically significant* means is that *the results are unlikely to have occurred if all of our assumptions are true.*

The assumptions that we make in conducting a study include the following: The variables are good measures of the underlying traits we wish to study; the variables were accurately measured; appropriate controls have been applied; sampling has occurred with a random element to it; and the conditions about population parameters specified in the null hypothesis are true. If we have conducted the study correctly, only one of our assumptions, the null hypothesis, can be untrue. (Failure of any of the other assumptions, such as that the test we used to measure the intelligence of carrot juice drinkers does not measure intelligence, gives rise to a plausible rival hypothesis.) Statistically significant results are, by definition, unlikely to occur when our assumptions are true, and we conclude, therefore, that one of our assumptions, the null hypothesis, is false.

When do we say that the difference between the sample statistic and the population parameter specified in the null hypothesis is significant? Mathematical statistics does not have an answer, but editors of education and psychology journals have adopted some minimum values as conventions. When the probability of the sample occurring under the conditions specified by the null hypothesis is less than 5 percent, the result is said to be significant at the .05 level. A result that would occur less than one time in 100 is said to be significant at the .01 level. A result is usually considered statistically significant only if

---

ASSUMPTIONS IN TESTING STATISTICAL HYPOTHESES

1. The variables are valid measures of the constructs specified in our theory.
2. The variables have been accurately (reliably) measured.
3. All necessary controls have been included to rule out plausible rival hypotheses.
4. The sample is representative of the population and there has been a random element to its selection.
5. The conditions about the population specified in the null hypothesis are true.

In testing a hypothesis we assume all these conditions are true. If the result obtained is unlikely to have occurred by chance given these conditions, we assume that one of them, assumption 5, is not true.

Table 12-1	Possible Outcomes from Testing a Null Hypothesis	
	Decision	
True State	*Reject Null*	*Do Not Reject Null*
Null false	Correct decision (power)	Type II error (β)
Null true	Type I error (α)	Correct decision

the probability that it would occur by chance from the null hypothesis distribution is less than one or the other of these conventional values; that is, that it would occur by chance in fewer than 5 (or 1) samples out of 100. We saw in our calculations above that the probability that we would obtain a sample of 25 carrot juice drinkers with a mean intelligence test score of 106 from a population in which $\mu \leq 100$ and $\sigma = 15$ was about .02. Therefore, we would probably reject the null hypothesis.

Any time we reach a decision on the basis of statistical analysis, there is some chance that our decision is incorrect. There *is* a true state of affairs. From the point of view of testing statistical hypotheses, the true state may be that the null hypothesis is true or that the null is false. Our decision must be to reject the null or not to reject the null; this yields the four possible outcomes listed in Table 12-1. When the null is true and our decision is not to reject it, we have reached a correct decision. Likewise, when the null is false and we decide to reject it, we are correct; otherwise, our decision is in error.

Errors of Inference

There are two basic types of errors that we can make. We can *reject the null hypothesis when it is true;* this is called an **error of the first kind,** or a **Type I error.** If we conclude that carrot juice does have an effect on intelligence and, in reality, it does not, we have made a Type I error. Conversely, we can decide not to reject the null when it is false; this is a *failure to detect a false null hypothesis* and is called a **Type II error,** or an **error of the second kind.** (There is yet another type of error that we will consider shortly.)

Statistical significance relates to the probability of making a Type I error. In reaching a statistical decision, we can specify in advance the risk that we are willing to run of making a Type I error; this probability is given the symbol α (Greek letter alpha). We can specify any values we choose for α, but convention tends to dictate the use of the values .05 and .01 mentioned earlier.

A **Type I error** occurs if we reject the null hypothesis when it is true.

A **Type II error** occurs if we fail to reject the null hypothesis when it is false.

α specifies the probability that we will make a Type I error.

▶ *Now You Do It*

Consider each of the following situations. Is the decision reached a correct one? If not, what type of error has been made?

1. Jennifer has conducted a study of the effects of early stimulation on the cognitive development of mice. Her results lead her to conclude that early stimulation promotes cognitive development. In reality, early stimulation promotes cognitive development.
2. Clark has tested the relationship between anxiety and the development of eating disorders. His results lead him to conclude that there is no relationship. In reality, anxiety causes eating disorders.
3. Juan has collected some data on whether people who follow a vegetarian diet are more introverted. His results cause him to conclude that vegetarians are more introverted than non-vegetarians. In reality, vegetarians are not more introverted than non-vegetarians.

Answers
1. correct decision
2. Type II error
3. Type I error

Directional Hypotheses

Suppose that we are government gambling inspectors investigating allegations that slot machines have a bias in favor of the house. The slot machines simulate tosses of 10 coins. That is, each of 10 windows can show either heads or tails. Some disgruntled players have complained that casino operators have rigged the machines to favor the casino. (The same problem might arise with respect to roulette wheels, blackjack decks, craps dice, or other gambling apparatus, but the probability distributions are much more complex.) Our job is to collect a sample of plays and decide whether the machines are fair. The rules of the game as practiced at the casino in question say that if a play comes up five heads nobody wins; six or more heads is a win for the house, and four or fewer heads is a win for the player. What is the null hypothesis?

The need to state a null hypothesis brings up an interesting and important issue: It is possible for the machines to be unfair in favor of the house *or* the player. A loose statement of the null is that the game is fair, that for each window the probability of a head is .5. Formally, for machines with 10 windows, this statement implies a population distribution with parameters $\mu_X = 5.0$ and $\sigma_X = 1.58$ because this situation is identical to the theoretical distribution for a toss of ten coins in which $P = .5$ for each coin (see Chapter 6). For samples of $N = 10$ plays of the machine, this population yields a sampling distribution of the mean with $\mu_{\bar{X}} = 5.0$ and $\sigma_{\bar{X}} = 1.58/\sqrt{10} = .500$ as shown in Figure 12-1. However, there are different ways that we can consider this null to be wrong. As gambling investigators, we are probably only concerned that the public is not being cheated. Since the null is that the game is fair, the alternate hypothesis is that the game is not fair, but we are only interested in one type of unfairness. We don't care if the game is unfair in favor of the public, only if it favors the house. Therefore, we are concerned to reject the null only if our sample result favors the house by more than we would expect by chance. This is called a **one-tailed test,** or a **one-tailed null hypothesis,** because we would reject the null only if we get an extreme result in one particular tail of the null hypothesis distribution.

*In a **one-tailed test** (one-tailed null hypothesis) we would reject the null hypothesis only if the result comes out in the direction we specify.*

This situation gives the following statement of the null hypothesis and its alternate:

H_0: $\mu \leq 5.0$ (our sample is from a population with a mean less than or equal to 5.0)

H_1: $\mu > 5.0$ (our sample is from a population with a mean greater than 5.0)

We will reject the null hypothesis in favor of its alternate H_1 if the results from our sample are unlikely to have occurred by chance from the population specified in the null hypothesis, but only if the discrepancy is in the direction specified in H_1. That is, we would reject H_0 in favor of H_1 only if $\bar{X}$ is enough *larger* than μ to qualify for statistical significance. This is the situation that we found in our study of carrot juice. The null specified that carrot juice improved intelligence. Therefore, we would reject H_0 only if $\bar{X}$ was enough greater than μ that the difference was statistically significant.

Figure 12-1 Sampling distribution of the mean when $\mu = 5$.

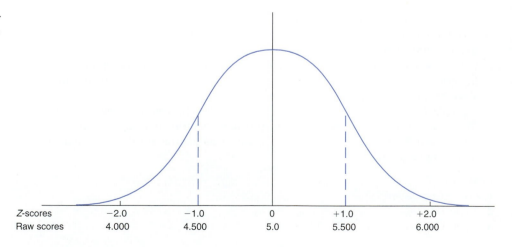

Z-scores	−2.0	−1.0	0	+1.0	+2.0
Raw scores	4.000	4.500	5.0	5.500	6.000

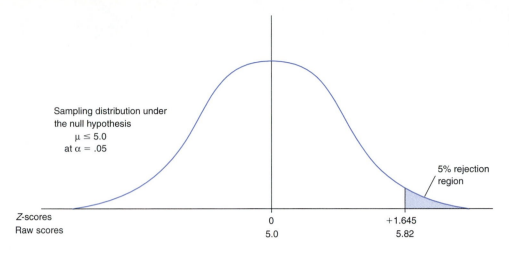

Sampling distribution under
the null hypothesis
$\mu \leq 5.0$
at $\alpha = .05$

5% rejection
region

Z-scores
Raw scores

0
5.0

+1.645
5.82

Rejection Regions. Here we must reintroduce α, the probability of a Type I error, or the probability that we will reject the null hypothesis when it is true. Consider Figure 12-2. In this figure, we again have the sampling distribution under the null hypothesis. We can select any level of α that we wish, but in this case, we have chosen the .05 level; that is, we have decided to reject the null if the sample results are such as would occur in only 5 samples out of 100 from this population. The shaded part of the figure is that portion of the sampling distribution that meets the condition specified in H_1, that $\overline{X}$ is so much larger than the hypothesized μ that it would occur in only 5 percent of samples. This area, where the conditions specified in the alternate hypothesis are met, is called the **rejection region.** A sample result that falls in this region will cause us to reject the null hypothesis.

Critical Values. The value of Z for defining the rejection region is found in the same way that we find the Z-value marking off any other specified percent of the distribution. After we decide on the value of α, we go to the table of the normal distribution and look up the **critical value** of Z beyond which α percent of the distribution falls. A Z of $+1.645$ cuts off 5 percent in the upper tail of the normal distribution, so this is the 5% critical value for Z. Values beyond this point are so large that they would occur in less than 5 percent of samples if H_0 is true.

Critical values can be expressed either in the scale of Z or in the scale of the raw data. Converting the critical value of Z back into a value in the scale of the raw data gives us the value we need in the sampling distribution. With a standard error of the mean of .500 and a μ of 5.00, this procedure gives the critical value of 5.82 shown in Figure 12-2. A sample of 10 plays yielding a mean of 5.82 or greater would occur only 5 percent of the time if the null hypothesis that $\mu = 5.0$ is true. Therefore, as gambling inspectors concerned for the public welfare, we would select a value of 5.82 as our critical value.

Now, consider the point of view of the casino operator. The boss wants to be certain that the machines are not biased in favor of the players, but he would not object to machines that favor the house. The operator's concern is that μ is not less than 5.0 (which would favor the players), so the null and alternate hypotheses of concern to the casino operator would be

H_0: $\mu \geq 5.0$

H_1: $\mu < 5.0$

These hypotheses imply the same sampling distribution as before, but the rejection region is located at the other end, as shown in Figure 12-3. As casino operator, we select $\alpha = .05$ and find the rejection region to be as shown in the figure. Because the rejection region is in the lower tail of the distribution, we use a Z of -1.645 with the same mean and standard deviation as before. If the sample mean is less than 4.18, we will decide to reject the null (and probably to get some new machines too!).

*The **rejection region** is the portion of the null hypothesis sampling distribution that is so extreme that samples this deviant would occur only α percent of the time if the null hypothesis is true.*

*A **critical value** is the value of Z (or any other test statistic such as t) that leaves α percent of the distribution in the rejection region.*

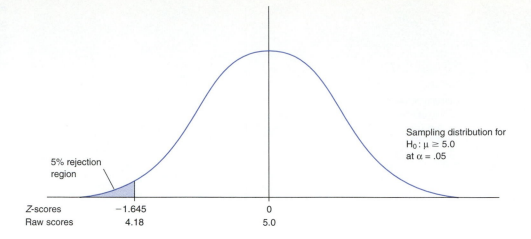

Figure 12-3 5% rejection region for a one-tailed null hypothesis that $\mu \geq 5.0$.

Sampling distribution for $H_0: \mu \geq 5.0$ at $\alpha = .05$

5% rejection region

Z-scores   −1.645   0
Raw scores   4.18   5.0

▶ *Now You Do It*

For each of the research questions given below, state the null and alternate hypotheses.

1. Jennifer believes early stimulation promotes cognitive development. She plans to test her hypothesis using two groups of mice, one of which receives stimulation (group S) and one which is reared under normal cage conditions (group N).
2. Clark plans to test his hypothesis that anxious people are likely to suffer eating disorders by comparing the mean anxiety scores of people who suffer from an eating disorder (group D) with people who do not have this disorder (group N).
3. Juan's hypothesis is that vegetarians are less introverted than non-vegetarians. He plans to test this hypothesis by comparing the mean introversion scores of a group of vegetarians (group V) with a group of non-vegetarians (group N).

Answers

1. $H_0:\ \mu_S \leq \mu_N$ $H_1:\ \mu_S > \mu_N$
2. $H_0:\ \mu_D \leq \mu_N$ $H_1:\ \mu_D > \mu_N$
3. $H_0:\ \mu_V \leq \mu_N$ $H_1:\ \mu_V > \mu_N$

Nondirectional Hypotheses

Suppose now that our only concern is to be sure that the game is fair to both parties; that is, we want to be certain that neither the casino operator nor the player has an advantage. In this case, we will reject the null hypothesis if the sample results in too many heads *or* too few. That is, we will have rejection regions at both ends of the sampling distribution, as shown in Figure 12-4. Using the same α-level (.05), we put half of the rejection region

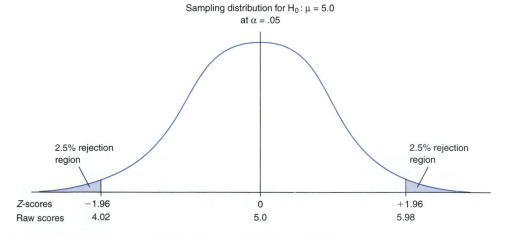

Figure 12-4 Rejection region for a two-tailed null hypothesis that $\mu = 5.0$.

Sampling distribution for $H_0: \mu = 5.0$ at $\alpha = .05$

2.5% rejection region

2.5% rejection region

Z-scores   −1.96   0   +1.96
Raw scores   4.02   5.0   5.98

in each tail of the sampling distribution. This null hypothesis is called a **two-tailed null hypothesis** because there is a rejection region in each tail of the distribution, and the test of the null is a two-tailed test. The two-tailed null and its alternate are symbolically stated as

$$H_0: \mu = 5.0$$
$$H_1: \mu \neq 5.0$$

where $\neq$ means "not equal to." This condition can be satisfied by a larger than expected difference in either direction, so another name for this type of null hypothesis is **nondirectional.**

We can summarize our discussion by saying that null hypotheses may be either one-tailed or two-tailed. In a one-tailed test, the null is rejected only if the value of the sample mean (or other statistic) is in the specified direction from the mean of the sampling distribution, while in a two-tailed test a difference in either direction can lead to rejection of the null. If we use the symbol K to stand for the value at the center of the null distribution ($K = 5.0$ in the casino example, $K = 100$ in our juice drinkers study), the two forms of the one-tailed null hypothesis are

(a) $H_0: \mu \geq K$

(b) $H_0: \mu \leq K$

while the two-tailed null hypothesis has the general form

(c) $H_0: \mu = K$

The three possible forms that the null hypothesis distribution can take are shown in Figure 12-5. In part (a), we would reject the null only if $\overline{X}$ is sufficiently less than K; no value of $\overline{X}$ that is greater than K would lead to rejection of the null. Conversely, in (b), the only values of $\overline{X}$ that would lead to rejection are those sufficiently greater than K. The two-tailed null is rejected whenever $\overline{X}$ is different enough from K in either direction to fall in the rejection region.

When the sampling distribution of the statistic is normal, as it is in our example, we can use the table of areas under the normal curve to compute critical values. Suppose that we set α at .05. Looking in Appendix Table A-1 under "Area Beyond Z," we find that 5 percent of the distribution falls beyond a Z of 1.645. This means that the critical value for a one-tailed hypothesis is *1.645 standard errors of the mean* ($\sigma_{\overline{X}}$s) away from the mean of the sampling distribution. Whether it is above or below the mean depends on the direction specified in the null hypothesis. For null hypotheses of the form $\mu \geq K$, the 5% critical value corresponds to a Z of -1.645, while for the null $\mu \leq K$ the value is $+1.645$.

It is important to note that the values given in Figure 12-5 are standard scores; that is, the distribution is expressed with a mean of zero and a standard deviation of 1. We can convert these values to the sampling distribution of our slot machine example by the usual linear transformation from Z-scores to raw scores (equation 5.4). This yields

$$CV_U = (+1.645)(.500) + 5 = (+.82) + 5 = 5.82$$
$$CV_L = (-1.645)(.500) + 5 = (-.82) + 5 = 4.18$$

where the standard deviation is the standard error of the mean (because we have a distribution of means), and the mean is as specified by the null hypothesis ($K = 5.0$).

Directionality and Critical Values

The first difference between one-tailed and two-tailed hypotheses, as we have seen, is a logical one: Are you interested in only one type of outcome? The second difference is a quantitative one: How large is the critical value? In a two-tailed hypothesis, we have rejection regions in both tails of the distribution. If we use one-tailed critical values ($+1.645$ and -1.645), we will have 5 percent of the distribution in each tail and a total of 10 percent of the distribution in the rejection region. Thus, if our hypothesis is two-tailed, we must adjust our critical values to keep 5 percent of the distribution in the total rejection region and maintain our risk of Type I error at 5 percent. We must put $2\frac{1}{2}$ percent of the distribution in each rejection region. (It is not necessary to put $2\frac{1}{2}$ percent in each tail. We could just as well put 4 percent in one tail and 1 percent in the other, but it would be difficult to

Figure 12-5 Critical values of Z and 5% rejection regions for two possible one-tailed null hypotheses and a two-tailed null hypothesis.

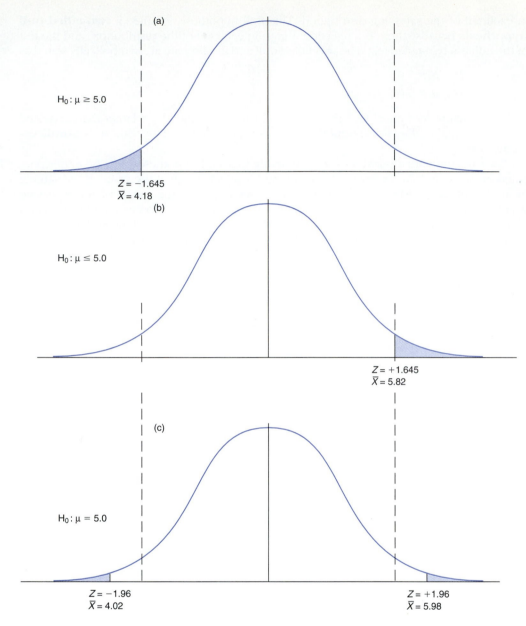

(a)

$H_0: \mu \geq 5.0$

$Z = -1.645$
$\bar{X} = 4.18$

(b)

$H_0: \mu \leq 5.0$

$Z = +1.645$
$\bar{X} = 5.82$

(c)

$H_0: \mu = 5.0$

$Z = -1.96$
$\bar{X} = 4.02$

$Z = +1.96$
$\bar{X} = 5.98$

justify this action in most cases. It is conventional to put half of the rejection region in each tail.) Going to the normal curve table, we find the necessary Z-values to be $+1.96$ and -1.96; these Z's give us critical values in the raw score scale of

$$CV_U = (+1.96)(.500) + 5 = (+.98) + 5 = 5.98$$
$$CV_L = (-1.96)(.500) + 5 = (-.98) + 5 = 4.02$$

which are the values that appear in Figure 12-4.

The consequences of stating our null as a two-tailed hypothesis should now be clear: The critical values are larger, and it is therefore more difficult to reject the null, but we can reject it with an outcome in either direction. These facts have led some people to be prejudiced against one-tailed null hypotheses. They reason, in part, that if an effect is big enough to be meaningful, it should be big enough to exceed the larger critical value required by the two-tailed test. They also argue that we have no sure way to know that the investigator actually did predict the direction of the outcome ahead of time. If the investigator did not predict the direction of the outcome, but used a one-tailed test anyway, then the correct value for α would be twice the value claimed.

For each of the research questions given below, find the critical value(s) of Z.

1. Jennifer believes that early stimulation promotes cognitive development. She plans to test her hypothesis at the $\alpha = .05$ level.
2. Clark plans to test his hypothesis that anxious people are likely to suffer eating disorders at the $\alpha = .01$ level.
3. Juan's hypothesis is that vegetarians are different from non-vegetarians in their level of introversion. He plans to test this hypothesis at the $\alpha = .05$ level.

Answers
1. This is a one-tailed test. $Z = +1.645$.
2. This is also a one-tailed test. $Z = +2.326$.
3. This is a two-tailed test. $Z = \pm 1.96$.

One Tail or Two: Converting Research Hypotheses Into Statistical Hypotheses

Rita, an aspiring undergraduate psychology major, has a theory that larger rewards lead to faster learning. Before running a study to test this theory, she frames her hypotheses. Since she is using rats for subjects (they are cheaper than children), a maze as her learning task, and food pellets as a reward, her hypotheses will be stated in terms of pellet size and running time in the maze. Her research hypothesis, derived from her theory, is that the rats receiving larger pellets will run faster (take less time to run the maze) than those receiving smaller pellets.

After consulting the research literature and her instructor, Rita realizes that her theory makes a directional prediction that μ_L (the mean running time for rats in the large pellet treatment population) will be less than μ_S (the mean running time for rats in the small pellet treatment population). The appropriate prediction, derived from the theory, is

$$\mu_L < \mu_S$$

which leads to the one-tailed null hypothesis

$$\mu_L \geq \mu_S$$

This is the appropriate null hypothesis. A two-tailed null hypothesis cannot properly be used to test a theory that makes a directional prediction because neither the null hypothesis nor its negation (the alternate) reflects directly on the truth or falsehood of the research hypothesis. In rejecting a null hypothesis, Rita can accept only the alternate hypothesis. Unless H_1 corresponds to her theory's prediction, she cannot claim that her results support her theory. *In order to draw proper logical conclusions from research, we must frame our statistical hypotheses to be logically consistent with the predictions from our theories.* There are two possible outcomes; either we reject the null in favor of the alternate hypothesis or we do not reject the null. Formally, the alternate hypothesis is the negation of the null; unless either the null or its negation is exactly equivalent to our research hypothesis, our study does not test the predictions made from our theory. Clearly, Rita should state as her

◆ *Steps in Testing the Null Hypothesis*

The formal testing of a null hypothesis involves the following steps.

1. Specify the sampling distribution implied by the null hypothesis.
2. Select a level for α (usually .05 or .01).
3. Determine whether the null hypothesis is one-tailed or two-tailed.
4. Determine the value of Z (or other appropriate statistic as described in later chapters) that cuts off the required portion of the sampling distribution.
5. Calculate the critical value(s) for the statistic of interest (usually the mean).
6. Compare the sample value with the critical value.
7. If the sample value falls in the rejection region, reject the null hypothesis.

null hypothesis that $\mu_L \geq \mu_S$ so that in rejecting it she can conclude what she wants to conclude, that $\mu_L < \mu_S$.

Rita's situation exemplifies a more general issue relating to statistical hypothesis testing. When we first begin to study an area it is reasonable to merely look for relationships without predicting their direction or magnitude. If Rita had had no reason to expect that one group would perform better than the other, a two-tailed null would have been appropriate. But once we have collected enough observations to begin to formulate theories, we have reached the point where we can no longer claim ignorance about at least the direction of the relationship we expect to find. As in Rita's case, most statistical tests should test directional hypotheses that have been justified on the basis of theory and prior research.

▶ *Now You Do It*

Formulating hypotheses is a crucial part of the research process. Try your hand at identifying the null hypothesis implied by each of the following research problems. State each hypothesis both verbally and as an equation involving parameters.

1. Titania believes that children learn mathematics more quickly when taught in same-sex groups than in mixed groups.
2. Oberon believes that women have more accurate memories than men for social interactions.
3. Hal believes that social introversion is related to interpersonal competence, but he is not sure how.

Answers

1. Titania's null hypothesis would be that the mean for same-sex groups will be less than or equal to the mean for mixed groups.

$$H_0: \; \mu_{SS} \leq \mu_{\text{mixed}} \qquad H_1: \; \mu_{SS} > \mu_{\text{mixed}}$$

2. Oberon's null hypothesis is that men's memories will be at least as accurate as women's.

$$H_0: \; \mu_M \geq \mu_W \qquad H_1: \; \mu_M < \mu_W$$

3. Hal's null hypothesis can be stated with regard to either group means or a correlation coefficient. The null hypothesis states that (1) the means of two groups that differ in social introversion (high and low) will not differ in interpersonal competence or that (2) the correlation between scores on a social introversion scale and those on an interpersonal competence scale will be zero.

$$H_0: \; \mu_H = \mu_L \; \text{ or } \; \rho = 0 \qquad H_1: \; \mu_H \neq \mu_L \; \text{ or } \; \rho \neq 0$$

SPECIAL ISSUES IN RESEARCH

Directional Versus Nondirectional Hypotheses

Some investigators feel that specifying a directional null hypothesis might inhibit them from exploring the implications of a statistically significant result that was in the opposite direction. This might be true at the beginning of a program of research, and might justify a two-tailed test at that stage. But it is a very weak theory that cannot predict the direction of an outcome. If a result is large enough to be statistically significant but is in the direction opposite from that suggested by the theory, we believe the result should be confirmed in a repeated experiment before it is published.

Other researchers argue that insisting on two-tailed tests guards against researchers publishing marginal results because of the larger difference required. They suspect that unethical researchers would claim to have made a directional prediction just to be able to use the smaller critical value. However, science tends to be self-correcting. It is never a single study that determines "truth", but the accumulation of consistent findings from several studies. Therefore, subsequent research should disconfirm incorrect marginal results.

As knowledge progresses, it seems appropriate that research should be held to a higher standard, and although the critical values may be less extreme for one-tailed hypotheses, the need to specify the direction of outcome more than offsets this advantage *so long as the directional prediction follows from theory*. A result that goes against a prediction reasonably derived from theory should be checked in a new study before publication. As research progresses in the social and behavioral sciences we should be able to predict not only the direction, but also the size of the relationship.

Decision Errors Revisited

We have seen that a Type I error—a decision to reject the null when it is true—is directly related to the concept of statistical significance. The investigator can select and control the probability of a Type I error by her/his choice of a critical value. In fact, as we saw, the investigator actually chooses the probability of a Type I error first and then computes the critical value.

Type I Errors

Considering again the population of outcomes from our slot machines will help us to see exactly what is involved in a Type I error. Since this is a hypothetical example, we have the advantage of being able to specify what truth is. To start with, we set truth equal to the null hypothesis; that is, our population has the parameters $\mu = 5.0$ and $\sigma_X = 1.58$, yielding a sampling distribution of the mean for samples of $N = 10$ of $\mu_{\bar{X}} = 5.0$ and $\sigma_{\bar{X}} = .500$.

Suppose that we specify a one-tailed null hypothesis that $\mu \leq 5.0$ and an α of .05. The critical value is 5.82, and our rejection region is the portion of the scale above the critical value. We would conclude that the null is false whenever we obtain a sample result where $\bar{X} > 5.82$. *Under the condition that the null is true,* we would expect five samples in 100 to have means in this region. The chance that our sample is one of these five is 5/100, or 5 percent, the selected value for α. Each of these five samples would lead us to decide—incorrectly—that the null is false. We would reach the incorrect decision 5 times and the correct decision 95 times. Since the decision to reject the null is a Type I error (remember, we know truth), we say that the Type I error rate is 5 percent.

In the real world, we do not draw 100 independent random samples; rather, we draw a single sample from the population and reach our conclusion on the basis of the results from that sample. The distribution specified by the null hypothesis yields a rejection region that would contain 5 percent (or any other value we might choose for α) of the samples. The chance that our single sample is one of those that falls in the rejection region when the null hypothesis is true is 5 percent, the selected value for α. Therefore, the probability that we will make a Type I error by rejecting the null hypothesis when it is true is also α (.05 in this case).

Type II Errors

Now consider what happens if the truth is something else. Suppose that each window in the slot machine has a probability of .60 of coming up heads. This means that the real characteristics of the population are $P = .60$, $Q = .40$, and that if we spun each wheel a large number of times, each window would come up heads on 60 percent of the trials. For a slot machine such as this, the mean of the population (NP) is 6.0 and the standard deviation ($\sqrt{NPQ}$) is 1.5492. Samples of size 10 (10 trials on the slot machine) from this population yield a sampling distribution of the mean with $\mu_{\bar{X}} = 6.0$ and $\sigma_{\bar{X}} = .4899$ ($1.5492/\sqrt{10}$). If we assume that the population distribution is normal (it isn't quite normal because binomial distributions are symmetric only when $P = Q$, but the sampling distribution will be very close to normal due to the central limit theorem), we can use Appendix Table A-1 to represent areas under the curve. Our null hypothesis was that $\mu \leq 5.0$. If we were omniscient, we would know that H_0 is false and that the alternative (H_1: $\mu > 5.0$) is correct. However, we do not have such knowledge, so we must run an experiment and attempt to infer the true state of affairs from the results.

The two sampling distributions we have described are shown in Figure 12-6. The one on the left is the sampling distribution that would exist if H_0 were true ($\mu_{\bar{X}} = 5.0$). The one on the right is the sampling distribution for $\mu_{\bar{X}} = 6.0$. This is one of the many sampling distributions that satisfies the negation of the null hypothesis and, therefore, satisfies the alternate hypothesis. A similar distribution could be drawn for $\mu_{\bar{X}} = 5.1$, $\mu_{\bar{X}} = 5.2$, and so on.

The 5% critical value for the null, 5.82, is shown in Figure 12-6; this value marks the decision point. A sample result to the left of (less than) this value will lead to the decision not to reject the null hypothesis. Since we now know (as abstract observers, not as practicing scientists) that $\mu_{\bar{X}} = 6.0$, a decision not to reject H_0 is an error; more specifically, it is a Type II error, a failure to detect a false null hypothesis.

Just as α is the probability of making a Type I error when the null hypothesis is true, β is the probability of making a Type II error, of failing to reject H_0 *when a given alternative*

β is the probability of making a Type II error, of failing to reject H_0 when a given alternative to the null is true.

Figure 12-6 Power and the probabilities of Type I (α) and Type II (β) errors for a true μ of 6 for a one-tailed hypothesis.

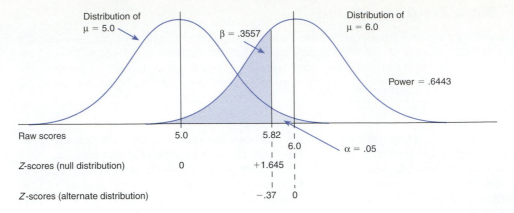

to the null is true. We have seen that the investigator can specify α, so that the probability of a Type I error is under her/his control. This is possible because there is only one sampling distribution specified by the null hypothesis. Since the alternate hypothesis does not specify a single distribution but a class of distributions and there is no way of knowing which one of them is correct, it is not possible to specify or control β. It will have one value if $\mu_{\bar{X}} = 5.1$, another if $\mu_{\bar{X}} = 6.0$; there are as many different values for β as there are different values for $\mu_{\bar{X}} \neq 5.0$.

We have seen that α is the proportion of the area of the null hypothesis sampling distribution that falls in the rejection region. β *is the proportion of the area of a particular alternate sampling distribution* (for example, $\mu_{\bar{X}} = 6.0$) *that **does not** fall in the rejection region* (that falls in the region of nonrejection). In Figure 12-6, this is the proportion of the upper distribution (centered on 6.0) that falls below 5.82.

We can calculate the value of β for any given alternative to the null. The procedure is to calculate the Z-score of the null hypothesis critical value *in the alternate distribution* and look up the appropriate area associated with that Z-score in Appendix Table A-1. The value of β for the alternate $\mu_{\bar{X}} = 6.0$ is calculated as follows:

$$CV = 5.82 \text{ (from the null)}$$
$$\mu_{\bar{X}} = 6.0 \qquad \sigma_{\bar{X}} = .4899$$
$$Z_{5.82} = \frac{5.82 - 6.0}{.4899} = \frac{-.18}{.4899} = -.37$$
$$\beta = area\ below\ Z = .3557.$$

This value can be interpreted to mean that *if the true value of $\mu_{\bar{X}}$ is 6.0,* there is about a 36 percent chance that our experiment with a sample of 10 will *fail* to lead to the conclusion that the null hypothesis is false. The probability of a Type II error, that we will fail to detect that the slot machines are biased in favor of the house, is .36.

Effect of a Two-Tailed H_0

A slightly different situation arises when the null is two-tailed. This would be the case, for example, if we wanted to make sure that the slot machines were fair, both to the players and to management. Our null hypothesis would be that $\mu = 5.0$, and the alternate would be that $\mu \neq 5.0$. If we keep the null distribution the same but put a $2\frac{1}{2}$ percent rejection region in each tail, we get Figure 12-7. Here the true state of affairs is still $\mu_{\bar{X}} = 6.0$, but the critical values are more extreme. We know from Figure 12-4 that the critical values for this null hypothesis are 4.02 and 5.98. The probability of a Type I error remains at 5 percent because we have set it there by our choice of α. What is the probability of a Type II error under these conditions? Again, we find the answer by calculating the Z-score of the critical values in the alternate distribution. The Z for 5.98 is

$$Z_{5.98} = \frac{5.98 - 6.0}{.4899} = -.04$$

which yields $\beta = .484$. The effect of going to a two-tailed null is to increase the probability of a Type II error for a constant value of α.

*β is the proportion of the area of a particular alternate sampling distribution that **does not** fall in the rejection region.*

The effect of going from a one-tailed to a two-tailed null is to increase the probability of a Type II error for a constant value of α.

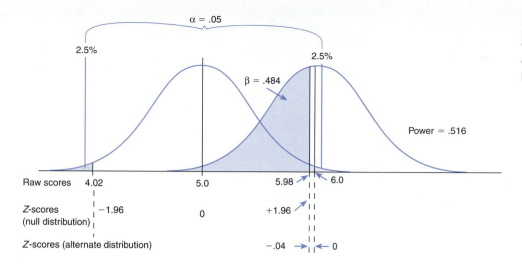

◆ *Now You Do It*

As a check on your understanding of how to determine β, let's consider a situation where the null hypothesis says that $\mu = 20$ and the standard deviation is 4. Determine the value of β if the true state of affairs is that $\mu = 22$, $\alpha = .05$, and the sample size is 9. What are the steps we must use?

The first thing to do is determine the value of the standard error of the mean for the null hypothesis distribution with $N = 9$.

The next step is to compute the critical value in the null distribution. You should have gotten 1.33 for $\sigma_{\bar{x}}$. With a two-tailed null, what is the critical value for Z? Looking in the table of the normal distribution, you should find $Z = 1.96$.

Next, you should calculate the critical value in the raw score scale. The formula is $CV = (\pm 1.96)(\sigma_{\bar{x}}) + \mu$. You should get 22.61.

We must now determine the Z-score of 22.61 in the alternate distribution. This distribution has a mean of 22 and a standard error of 1.33. Therefore, 22.61 has a Z-score of $+.46$. What part of the alternate distribution falls below the critical value? Looking up $+.46$ in the normal curve table yields an area between Z and the mean of .1772. Therefore, $\beta = .6772$; there is about a 68 percent chance that, under these conditions, we would fail to detect that $\mu \neq 20$ when the true value of μ is 22.

Type III Errors

There is another interesting effect of going from a one-tailed to a two-tailed null hypothesis. It introduces a new kind of error, which McNemar (1969) has called Type III error. A **Type III error,** which has no particular symbol, occurs when we reject the null hypothesis, but in the wrong direction. We have been dealing with a case where the basic null hypothesis is that $P = .50$ (which for 10 windows in our slot machine yields a null hypothesis mean of 5.0). The situation leading to Type III errors is illustrated in Figure 12-8, which shows a two-tailed null distribution and an alternate "true" distribution that centers on $\mu_{\bar{x}} = 5.1$. When the true population parameter is not very different from the value specified in the null hypothesis (but the null is still false), part of the alternate distribution falls in each rejection region. There is then a possibility that chance will produce a result in the rejection region in the wrong tail of the distribution, leading us to conclude (in this case) that $\mu_{\bar{x}} < 5.0$ when the truth is that $\mu_{\bar{x}} > 5.0$.

*A **Type III error** occurs when we reject the null hypothesis in the wrong direction.*

The probability of a Type III error is calculated in the same way as β, that is, by calculating the Z-score of the null hypothesis critical value in the alternate distribution. We can calculate the standard deviation of the alternate distribution in the same way that we have in the past.

$$\sigma_X = \sqrt{10(.51)(.49)} = 1.58$$
$$\sigma_{\bar{x}} = 1.58/\sqrt{10} = .500$$

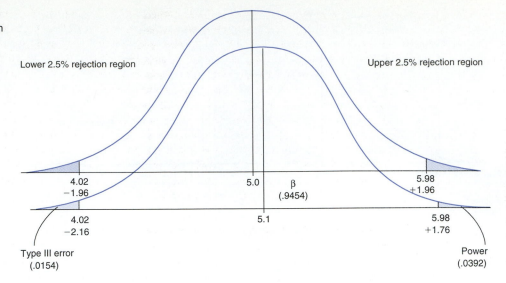

Figure 12-8 Probability of a Type III error is the proportion of the alternate distribution falling in the wrong rejection region of the null hypothesis distribution.

Lower 2.5% rejection region

Upper 2.5% rejection region

4.02
−1.96

5.0
β
(.9454)

5.98
+1.96

4.02
−2.16

5.1

5.98
+1.76

Type III error
(.0154)

Power
(.0392)

For the situation shown in Figure 12-8, the Z for the *lower critical value* in the alternate distribution is

$$Z_{4.02} = \frac{4.02 - 5.1}{.500} = -2.16$$

so the probability of a Type III error is the area beyond this Z or .0154. That is, there is a better than one percent chance that we will find a statistically significant difference in the wrong direction.

Type III errors can also occur with one-tailed hypotheses, but since the rejection region is in only one tail they can only occur when we have made the larger error of misspecifying the direction of the effect we expect. For example, if Rita had predicted that larger rewards would result in faster learning and reality is that they retard learning, finding a significant result in favor of large rewards would be a Type III error. However, if prior research and theory are sufficient to cause us to make a directional prediction, the size of the relationship will ordinarily be large enough that the probability of an error in direction is unlikely.

▶ *Now You Do It*

In the previous Now You Do It you determined the value for β. Let's use the same basic principles to find the probability of a Type III error for these data. What are the steps?

First, we must find the lower .025 critical value in a distribution with $\mu = 20$ and $\sigma_{\bar{X}} = 1.33$. This value is 17.39.

Next, calculate the Z-score for this critical value in the alternate distribution. You should get −3.47. The probability of a Z this extreme is .0003, so there is little chance that we would commit a Type III error under these conditions.

Power

Power *is the probability that a statistical test will lead to a correct rejection of the null hypothesis.*

The value of β is the probability of making a Type II error; its complement $(1 - \beta)$ is called **power.** *Power is the probability that a statistical test will lead to a correct rejection of the null hypothesis* (see Table 12-1). In the first example above, β was .3557, so the power of the test is $(1 - .3557)$ or .6443. That is, given the conditions of α and the alternate described, the probability that we would obtain a result that would lead us to reject the null is .6443. There is a 64 percent chance that we would correctly conclude from a study such as the one we have described that the slot machines are biased in favor of the casino.

The issue of Type III error requires that we be careful in how we express the notion of power. If we reject the null hypothesis in the wrong direction, that is not a correct decision. Therefore, in situations where there is a nonzero probability of a Type III error, power must be defined as $1 - (\beta + \text{probability Type III})$. As an alternative, we could define power as the proportion of the alternate distribution falling in *the correct rejection region* of the null hypothesis distribution. This is illustrated as the unshaded part of the alternate distribution in Figures 12-6 and 12-7. This definition leads us to calculate the Z-score of the critical value in the correct tail of the null distribution and determine the area of the alternate distribution falling in that region directly.

Power *is the proportion of the alternate distribution that falls in the rejection region of the null distribution.*

Keep in mind that the power of a test varies inversely with β and that their relative magnitudes depend on the particular alternate distribution that is being considered and on the value of α chosen by the investigator. We can use variants of our slot machine example to illustrate these points.

▶ *Now You Do It*

As a check on your understanding of how to determine β and power, let's confirm that for H_0: $\mu \leq 5.0$ with $N = 100$ and $\alpha = .05$, the β's for the two alternatives in our slot machine example are .0000 for $\mu = 6.0$ and .0630 for $\mu = 5.5$, respectively. What are the steps we must use?

The first thing to do is determine the values of the standard errors of the mean for the null hypothesis distribution and these alternative distributions with $N = 100$. For the null distribution

$$\sigma_{\bar{X}} = 1.58/\sqrt{100} = .158$$

We have two alternative situations, $\mu_{\bar{X}} = 6.0$ (therefore, $P = .60$) and $\mu_{\bar{X}} = 5.5$ ($P = .55$). The standard deviations are 1.55 and 1.57 respectively, so the standard errors are .155 and .157.

The next step is to compute the critical value in the null distribution. This is 5.26 and will be the same for both alternatives. Then we compute $Z_{5.26}$ in the two alternate distributions. For the alternative that $\mu_{\bar{X}} = 5.5$ we find that $Z_{5.26}$ is -1.53, which has an area beyond it (β) of .0630 and produces a power of .9370. The alternative that $\mu_{\bar{X}} = 6.0$ produces a Z of -4.77, which is beyond the limits of our table.

Relationships Among α, β, and Power

Changing α. The effect of changing α (for a constant alternate true state of affairs such as $\mu_{\bar{X}} = 6.0$) is to change the critical value. Lowering α to .01 changes the needed Z from 1.645 to 2.326 and raises the critical value for our slot-machine example from 5.82 to 6.16 [$5.0 + (2.326 \cdot .500)$]. The situation is illustrated in Figure 12-9. The value of $Z_{6.16}$ on the alternate distribution (where $\mu_{\bar{X}} = 6.0$) is

$$Z_{6.16} = \frac{6.16 - 6.0}{.4899} = +.33$$

which produces $\beta = .629$ and power $= .371$. Thus, reducing the chance of making a Type I error by 4 percent means that we increase the chance of a Type II error by about 27 percent and reduce the power of the test by that amount. The price to the investigator of reducing the probability of one kind of error is always an increase in the risk of making another kind of error. It is up to the investigator to decide which type of error it is more important to avoid.

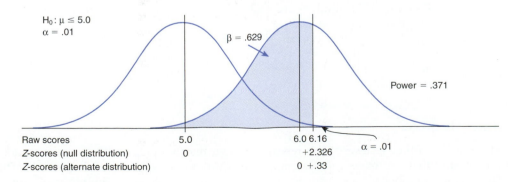

H_0: $\mu \leq 5.0$
$\alpha = .01$

$\beta = .629$

Power $= .371$

Raw scores	5.0	6.0 6.16	$\alpha = .01$
Z-scores (null distribution)	0	+2.326	
Z-scores (alternate distribution)		0 +.33	

Figure 12-9 Effect of changing α from .05 to .01 is to increase β and reduce power.

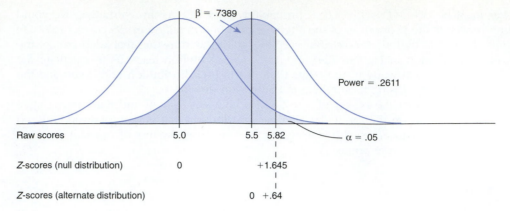

Figure 12-10 Power and the probabilities of Type I (α) and Type II (β) errors for a true μ of 5.5. Note the increase in β.

In many practical situations, the choice of which error to minimize is dictated by the consequences of a wrong decision. Most investigators are more concerned with Type I errors than with Type II (few ever consider Type III errors because their probability is always low). However, in areas such as medical research, where there are ethical considerations surrounding withholding a potentially beneficial treatment from one group, it may be appropriate to focus on minimizing β.

Different Alternate True State. Consider now what happens when the alternate distribution (truth) is assumed to have $\mu_{\bar{x}} = 5.5$ and $\sigma_{\bar{x}} = .4975$ (which we calculate from $P = .55$, $Q = .45$). This does not in any way affect α or the critical values because these are defined on the null hypothesis distribution, but it has a profound effect on β. If $\alpha = .05$, the one-tailed critical value returns to 5.82. The situation is diagrammed in Figure 12-10. Calculating $Z_{5.82}$ on the new alternate distribution, we obtain

$$Z_{5.82} = \frac{5.82 - 5.5}{.4975} = \frac{+.32}{.4975} = +.64$$

$\beta = .7389$ and power is .2611. We are much less likely to detect the fact that the null hypothesis is false if the difference between truth and the null distribution is smaller.

▶ *Now You Do It*

Let's use the example from Now You Do It on page 297 to examine further the effects of changing α and the alternative. In the previous example, we had a null hypothesis distribution with a mean of 20, $\alpha = .05$ and $\sigma_{\bar{x}} = 1.33$. The mean of the alternate distribution was 22. Under those conditions, we found that β was .68 and power was .32.

1. If we change α to .01, what happens to β and power?
2. If we change the alternative mean to 24, what happens to β and power?

Answers
1. Since our null hypothesis was two-tailed, the critical value for Z becomes 2.576 and the critical values in the raw score scale are 23.43 and 16.57. If the true μ is 22, Z's for the critical values are $+1.08$ and -4.08. Power is .14 and the probability of a Type III error is nil. β is .86.
2. Changing the alternative mean to 24 (but keeping α at .05), we must determine the Z's for critical values of 17.39 and 22.61 in a distribution with $\mu = 24$. These Z's are -4.97 and -1.05. Therefore, $\beta = .15$ and *power* $= .85$. Clearly, it is easier to detect a larger true difference.

Sample Size

There is one other feature of the research design that is often under the investigator's control and that can have a considerable effect on the test of the null hypothesis. This factor is the number of observations collected, or the sample size. The effect of sample size is readily apparent from equation 11.1, which reads

$$\sigma_{\bar{X}} = \frac{\sigma_X}{\sqrt{N}}$$

For a given standard deviation, the standard error of the mean is reduced when sample size is increased. In the slot-machine study, for example, an increase in sample size from $N = 10$ to $N = 25$ reduces the value of $\sigma_{\bar{X}}$ in the null hypothesis sampling distribution from .500 $[(1.58)/\sqrt{10} = .500]$ to .316 $[(1.58)/\sqrt{25} = .316]$. This in turn reduces the critical value for the one-tailed null hypothesis with $\alpha = .05$ from 5.82 $[.500(1.645) + 5.0 = 5.82]$ to 5.52 $[.316(1.645) + 5.0 = 5.52]$. A further increase in sample size to $N = 100$ reduces $\sigma_{\bar{X}}$ to .158 $[(1.58)/\sqrt{100} = .158]$ and the critical value to 5.26 $[.158(1.645) + 5.0 = 5.26]$.

Obviously, the effect of increasing sample size is to reduce variability among samples. The sampling distribution is still centered in the same location, but it shows reduced spread. The same effect occurs with the sampling distributions under the various alternate hypotheses. The result of increasing N, as computations from our slot-machine alternatives (still 10 windows, but with 25 or 100 plays of the machine) show, is that if we keep α and true alternative constant, β is reduced and power is increased. For $N = 25$, when the true $\mu = 6.0$ and $\sigma_X = 1.55$, then

$$\sigma_{\bar{X}} = 1.55/\sqrt{25} = .31$$

With a critical value of 5.52, we calculate the Z for 5.52 in the alternate distribution to be

$$Z_{5.52} = (5.52 - 6.0)/.31 = -.48/.31 = -1.55$$

The proportion of the area of the alternate distribution that falls below the critical value is .0606; therefore,

$$\beta = .0606$$

and

$$power = .9394$$

This is quite an improvement over the values of β (.3557) and power (.6443) that we obtained for $N = 10$. Raising N reduced β and increased power without any change in the risk of a Type I error.

The same calculations for our other alternative ($\mu_{\bar{X}} = 5.5$, $\sigma_X = 1.57$) give the following results:

$$\sigma_{\bar{X}} = 1.57/\sqrt{25} = .31$$

The critical value is still 5.52, so $Z_{5.52}$ in the alternate distribution is

$$Z_{5.52} = (5.52 - 5.5)/.31 = +.0645$$

The associated value of β is .52 with power = .48. For $N = 100$, the reduction in β and the increase in power would be even more substantial.

The investigator's ability to control sample size provides indirect control over β. As we noted before, α is under your direct control simply by changing the critical value; on the other hand, β depends in part on the true value of $\mu_{\bar{X}}$ in the population. If H_0 is that $\mu_{\bar{X}} = K$ and truth is that $\mu_{\bar{X}} \neq K$, then β is inversely related to N and also to the difference between the real $\mu_{\bar{X}}$ and K. By letting N get very large, we can reduce β to a small amount for any value of $\mu_{\bar{X}} \neq K$. Thus, while we can never know the actual value of β, we can calculate it for any alternative true state and we can reduce it by taking a larger sample.

Let us now review again briefly the process involved in testing a statistical hypothesis.

1. First, we state a research hypothesis that is derived from a theory or belief. This is what we expect to find in our study. In the example of the government investigator's study of the casino's slot machines, the research hypothesis was that the mean number of heads that would show up in the 10 dials is greater than 5.0.
2. We then determine a null hypothesis that is the obverse of the research hypothesis. The null hypothesis implies a distribution with a particular set of parameters for the population (for example, $\mu_X = K$ for the mean, or $\rho = K$ for the correlation coefficient) if the null is true. In the slot-machine study, the parameters implied by the

null hypothesis were $\mu \leq 5.0$ and $\sigma_X = 1.58$. If our research hypothesis is true, data that we collect in a study should be inconsistent with the null hypothesis. This will lead us to reject the null hypothesis in favor of the alternate hypothesis (that the null hypothesis is false), thereby supporting our research hypothesis.

3. Next, we select values for N and α, which are used along with the null hypothesis value of the parameter and the standard error of the sampling distribution to calculate a critical value for the statistic. This critical value is the value in the null hypothesis sampling distribution of the statistic (usually the mean) that cuts off the extreme α percent of that distribution (called the rejection region) consistent with the research hypothesis. The rejection region may be in the upper tail, the lower tail, or both tails of the sampling distribution, depending on the null hypothesis itself. In the slot-machine study we selected $N = 10$, $\alpha = .05$, and our rejection region was in the upper tail. These decisions gave us a critical value of 5.82.

4. Once we have determined the rejection region, we collect our data and compute the value of the statistic of interest. If the value of the statistic falls in the rejection region, we conclude that the null hypothesis is false—that the sample did not come from a population with the parameters stated in the null hypothesis. A value of the statistic in the rejection region leads to the conclusion that the difference between the sample value and the hypothesized value for the parameter is statistically significant or is unlikely to have occurred by chance in sampling from the population specified in the null hypothesis. If the mean for our sample of 10 trials on the slot machine exceeds 5.82, we will conclude that the null hypothesis is false. If the null is false, the machines are biased in favor of the casino.

Sample size occupies a very important place in this sequence of events because of its effect on the standard error of the sampling distribution. As sample size increases, the standard error decreases, bringing the critical value closer to the parameter specified in the null. (Remember that in our preceding example the effect of increasing N from 10 to 25 to 100 reduced the critical value from 5.82 to 5.52 to 5.26.) This means that smaller and smaller differences will be needed to lead to rejecting the null hypothesis.

▶ *Now You Do It*

Let's bring all of the various aspects of error and power together in one last problem. Shawna is designing a study to investigate ESP. She believes that psychokinesis (thoughts influencing the physical world) exists, although its influence may be very small. To test her theory, she designs a study using dice. Each of her trials will involve rolling two dice. On each even-numbered trial she will project a wish for an even number for the sum of the dice, and on each odd-numbered trial she will project a wish for an odd number as the sum. If the result of a roll agrees with her wish, she will score it as a hit. Otherwise, she will score the trial as a miss. She recruits 16 classmates to serve as participants in her study, and each participant will roll the dice 50 times. Each person's score will be the number of hits. She decides to test her hypothesis using $\alpha = .05$.

1. What is the null hypothesis and what are the mean and standard deviation of the null hypothesis sampling distribution? If the effect of psychokinesis is 5 percent, what are the mean and standard deviation of the alternate distribution? What is the critical value, and what are the values of power and β?

2. Now suppose that Shawna decides to run her test with $\alpha = .01$. What happens to the critical value, and what are the values of power and β?

3. If the effect of psychokinesis is really 10 percent instead of 5 percent, what are the mean and standard deviation of the alternate distribution? Using $\alpha = .05$, what are power and β? If Shawna sets $\alpha = .01$, what are power and β?

4. Suppose Shawna is able to recruit 32 friends to serve in her study. What happens to the null distribution? What happens to the alternate distribution? (Assume that the effect of psychokinesis is 5 percent.) Using $\alpha = .05$, what are power and β?

5. After getting a federal grant, Shawna is able to expand her study to include 64 participants. Keeping $\alpha = .05$, what is the standard error of the mean? (Again, assume that the effect of psychokinesis is 5 percent.) What are the values of power and β?

Answers

The various intermediate results you should get and the values of power and β for each of these scenarios are given below. The null hypothesis states that $\mu \leq 25$. Notice that at a given level of α, as sample size increases, our power to detect an effect of a particular size also increases. We can always make a study more powerful by using a larger sample.

	N	μ	σ	$\sigma_{\bar{x}}$	μ	σ	$\sigma_{\bar{x}}$	.05	.01	Power	β
		H_0			**Alternate**			**CV**			
1	16	25	3.54	.88	26.25	3.53	.88	26.45		.41	.59
2	16	25	3.54	.88	26.25	3.53	.88		27.05	.18	.82
3	16	25	3.54	.88	27.50	3.52	.88	26.45		.88	.12
	16	25	3.54	.88	27.50	3.52	.88		27.05	.70	.30
4	32	25	3.54	.63	26.25	3.53	.62	26.03		.64	.36
5	64	25	3.54	.44	26.25	3.53	.44	25.73		.88	.12

Interpreting Research Results

We mentioned before that statistical significance refers to the probability that the difference between the value of the parameter stated in the null hypothesis and the value of our

---

SPECIAL ISSUES IN RESEARCH

Power Curves

The man who has done the most to bring issues of power analysis to researchers in education and psychology is Jacob Cohen. Beginning in the early 1960s, Cohen pointed out that many studies lacked sufficient power to detect even moderately large differences between groups. Gradually, statistics books began to include graphical ways to determine power for a given sample size and mean difference. Because power varies continuously with both N and the size of the true difference between the treatment populations, the way it has most often been displayed is in the form of power curves. There is a different power curve for each possible combination of sample size and level of α. The abscissa of the graph is marked off in units of difference between the means, expressed as Z-scores, and the ordinate gives the power of the test. A power curve might look like the graph below. Note that the curve begins at a difference of zero and a level equal to α, and rises to 1.00. We are virtually certain to detect very large differences.

More recently, computer programs have become available that will compute power for a wide variety of situations. Programs such as Power and Precision (Borenstein, Cohen, and Rothstein, 1997) can be very useful for computing the sample size needed to achieve a desired level of power for a given mean difference.

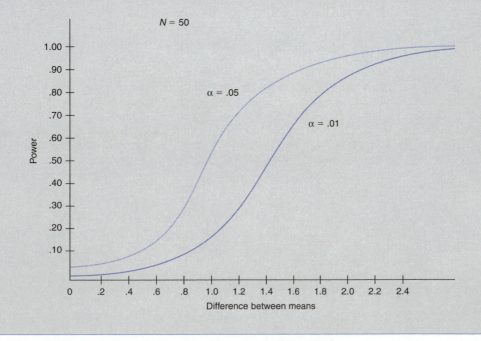

---

sample statistic would occur by chance in random samples from that population. Statistical significance, or the probability of a Type I error, is related to the probability of our study's results if H_0 is true, not to their importance or meaningfulness. If we couple this understanding with the knowledge that increasing N will lead us to find smaller differences to be statistically significant, we can see that using large samples can result in meaningless and unimportant differences being found statistically significant; thus, there is a disadvantage (aside from the logistic problems of dealing with large sets of data) to using large samples.

The moral of this story is that we must use our heads as well as statistical analysis both in the design and in the interpretation of a study. If we are going to perform a study to test a hypothesis, the study should be designed with the statistical analyses in mind. We

SPECIAL ISSUES IN RESEARCH
The Hypothesis-Testing Controversy

The presentation we have given regarding hypothesis testing follows the traditional approach found in psychological and educational research. At this time there is a major controversy raging in both disciplines over the value of testing null hypotheses in the way we have described. Two leaders on the side of change are Frank Schmidt (1996, Schmidt & Hunter, 1997) and Jacob Cohen (1990, 1994). Both of these authors have argued that statistical hypothesis testing as it has been practiced in psychology and education has been more an impediment to progress than a help. Schmidt has gone so far as to state that nothing can ever be learned by testing null hypotheses and that the practice of conducting such tests should be banned from the research literature. The American Psychological Association Publications Board has appointed a committee to review the arguments. Statements on both sides of the debate can be found in Harlow, Mulaik, and Steiger (1997).

The primary alternatives that have been offered to the significance test are the use of confidence intervals, the reporting of effect sizes, and the use of an approach called Bayesian statistics. We have already examined confidence intervals in some detail, and we will take a closer look at effect size measures in the next chapter. The methods of Bayesian analysis require procedures we will not be covering, but an introduction is given by Pruzek (1997) in the volume edited by Harlow et al. (1997).

At this point it may be useful to examine some relationships. One fact that has been pointed out by several authors and may have occurred to you in your study of null hypothesis testing and confidence intervals so far is that these two methods of analysis are, in some senses, opposite sides of the same coin. That is, the limits of a confidence interval around a sample statistic are the reverse of null hypothesis critical values. If the population mean under the null hypothesis, which we have called K, falls within the confidence interval around the sample mean, then the sample mean necessarily falls within the area of nonrejection around K. If K does not fall within the confidence interval, the sample mean would fall in the rejection region and we would reject H_0. Thus, some authors ask, what's the argument about?

The issue, as Rozeboom (1960) and Meehl (1967) pointed out many years ago, is that as sample size increases the confidence interval becomes more precise; we know with greater confidence where the population parameter is. It is somewhere within the confidence interval (with a probability of α that we are wrong). With the null hypothesis, we experience no such increase in information. As sample size increases, the area of nonrejection becomes narrower. Therefore, if we reject a null hypothesis with a large sample, we reduce our uncertainty about the location of μ less than we do with a small sample. As Meehl (1967) noted, in the limiting case a two-tailed null hypothesis cannot be correct (it specifies a point, and a point has zero probability), so a perfectly precise experiment in which the null hypothesis is rejected tells us nothing! Meehl refers to this as the quasi-always false null hypothesis.

Effect sizes provide a different way of looking at research questions. We have already met the idea of an effect size in our discussion of the correlation coefficient. Effect sizes are measures of the amount of variance (hence uncertainty) that one variable accounts for in another. The argument in favor of effect size measures is that they not only tell us that there is a relationship between the variables in our study, but how large, in a standardized sense, that relationship is. Schmidt (1996) argues that they are the most meaningful statistics.

Many authors, for example, Mulaik, Raju, and Harshman (1997) and Abelson (1997), take a more moderate position. Mulaik et al. note that most objections to null hypothesis testing object to the mindless use of zero as a value for K when testing the difference between groups. If a rational value for K that is derived from theory or from past research is tested, they argue, then many of the objections to significance testing vanish. In this case, however, the role of the null hypothesis has been shifted to correspond to the research hypothesis, so the theory is supported by nonrejection. Abelson points out some of the shortcomings of the alternatives to significance testing and argues that if we abandon tests of significance, we will eventually have to reinvent them.

From the perspective of learning about research practices in psychology and education, the fact of the matter is that there is half a century of research that has relied heavily on significance tests. It is important for you to know about significance testing in order to be able to read that literature, even if practices change in the near future. In this book we present three alternative ways to appraise the outcome of a study in the hope that you will be prepared to handle whatever the future holds.

should know what major analyses will be performed before the data are collected, and sample size should be selected with a view to the size of difference that it is useful to interpret. After the statistical analyses have been completed, a logical analysis should be used to determine what the outcome of the study means for the hypotheses under test. Do the results make sense? Are they consistent with what is known about the phenomenon under investigation? Are they what we expected? A negative answer to any one of these questions should be cause for alarm. Far too often research reports are submitted for publication because the results reached statistical significance rather than because the findings make sense or advance our understanding of an area. Remember, *statistically significant means nothing more than that the result is unlikely to have occurred by chance under the conditions stated in the null hypothesis.* Statistical analysis is a tool to be used to provide information for decisions, not the decisions themselves.

SUMMARY

Statistical inference involves using sampling distributions to test *statistical hypotheses.* A sampling distribution is the frequency distribution of values of a statistic that would be obtained from a large number of random samples of a given size drawn from a population. All statistics have sampling distributions.

There are two kinds of statistical hypotheses. The *null hypothesis* (H_0) states the characteristics of the population that are to be tested and usually includes a statement of the mean and standard deviation. The *alternate hypothesis* (H_1) is the negation of the null. The null hypothesis may be *one-tailed* or *two-tailed.*

Statistical inference proceeds by calculating a *critical value* for the statistic in its assumed sampling distribution. If the statistic computed from the sample data falls in the *rejection region,* the decision is made to reject the null hypothesis.

An experimental result is said to be *statistically significant* when it is unlikely that the sample result could have been obtained by chance from a population where the conditions specified in the null hypothesis are true; such a finding leads to a decision to *reject the null.* A *Type I error* occurs when the decision is reached to reject the null, but it is actually true. The probability that we will make a Type I error is controlled by specifying α, usually as .05 or .01. When the true situation is that the null is false, but the decision is reached not to reject it, a *Type II error* has been made. β is the probability of a Type II error. The *power* of an analysis refers to its ability to detect a false null hypothesis. For a constant level of α, increasing sample size (N) reduces β and increases power. Likewise, if the difference between the true state of affairs and the null hypothesis conditions is increased, power increases and β drops. Reducing the probability of a Type I error will, other things equal, increase β and reduce power.

REFERENCES

Abelson, R. P. (1997). A retrospective on the significance test ban of 1999 (If there were no significance tests, they would be invented). In Harlow et al. (Eds.), *What if there were no significance tests?* (pp. 117–141). Mahwah, NJ: Erlbaum.

Borenstein, M., Cohen, J., & Rothstein, H. (1997). *Power and precision.* Mahwah, NJ: Erlbaum.

Cohen, J. (1988). *Statistical power analysis for the behavioral sciences* (2nd ed.). Mahwah, NJ: Erlbaum.

Cohen, J. (1990). Things I have learned (so far). *American Psychologist, 45,* 1304–1312.

Cohen, J. (1994). The earth is round ($p < .05$). *American Psychologist, 49,* 997–1003.

Harlow, L. L., Mulaik, S. A., & Steiger, J. H. (1997). *What if there were no significance tests?.* Mahwah, NJ: Erlbaum.

McNemar, Q. (1969). *Psychological statistics* (4th ed.). New York: Wiley.

Meehl, P. E. (1967). Theory-testing in psychology and physics: A methodological paradox. *Philosophy of Science, 34,* 103–115.

Mulaik, S. A., Raju, N. S., & Harshman, R. A. (1997). There is a time and place for significance testing. In Harlow et al. (Eds.), *What if there were no significance tests?* (pp. 65–115). Mahwah, NJ: Erlbaum.

Pruzek, R. M. (1997). An introduction to Bayesian inference and its applications. In Harlow et al. (Eds.), *What if there were no significance tests?* (pp. 287–318). Mahwah, NJ: Erlbaum.

Rozeboom, W. W. (1960). The fallacy of the null hypothesis significance test. *Psychological Bulletin, 57,* 416–428.

Schmidt, F. L. (1996). Statistical significance testing and cumulative knowledge in psychology: Implications for the training of researchers. *Psychological Methods, 1,* 115–129.

Schmidt, F. L., & Hunter, J. E. (1997). Eight common but false objections to the discontinuation of significance testing in the analysis of research data. In Harlow et al. (Eds.), *What if there were no significance tests?* (pp. 37–64). Mahwah, NJ: Erlbaum.

EXERCISES

1. State the null and alternate hypotheses for each of the following research predictions.
   a. Rats who are reinforced for pressing a lever using a variable schedule of reinforcement will require more sessions to extinguish their lever-pressing behavior than rats who are reinforced on a constant schedule.
   b. Adolescents from upper income families will have higher self-esteem scores than adolescents from the lowest income families.
   c. Eighth grade students from rural schools and eighth grade students from metropolitan schools will differ on moral reasoning scores.
   d. Job satisfaction ratings of entry level employees will be different than the ratings of middle management.
   e. Children raised in conditions of emotional deprivation will have lower than normal intelligence. (NOTE: Intelligence will be measured by an IQ test with a population mean of 100.)
   f. Children who are taught to read using a phonics approach will have better reading comprehension scores than children who are taught to read using a whole language approach.
   g. Older adults who exercise regularly will have lower scores on measures of depression than older adults who do not exercise regularly.
   h. Men will score higher than women on a spatial rotation task.
   i. The resting heart rates of marathon runners are lower than the resting heart rates of sprinters.
   j. Children who are malnourished have lower body temperatures than normal (the normal body temperature is 98.6°).

2. In testing the null hypothesis, a researcher must make a decision about the null hypothesis based on a statistical test of the data. With respect to how things are in reality versus the researcher's decision about the null hypothesis, there are four possible outcomes. List these outcomes and describe what each outcome means. Provide an example to illustrate each outcome.

3. In your own words describe the difference between the null hypothesis and the alternate hypothesis.

4. Suppose that a researcher conducts a study, analyzes the data, and reports that the results were "statistically significant using an alpha of .01." What does the researcher mean when he or she reports this? What is the relationship, if any, between a statistically significant research finding and a scientifically important research finding?

5. In testing the null hypothesis several assumptions are made. Describe each assumption.

6. For each of the following descriptions, determine whether the decision reached by the researcher in the first sentence is correct given the information in the subsequent sentences. If the decision was incorrect, indicate the type of error that was committed.
   a. On the basis of the data that were collected, a researcher made the claim that men and women do not differ on empathic behaviors. In reality, women behave in ways that demonstrate empathy more often than men.
   b. After a series of tests, a medical researcher claimed that serum A did not cure a particular medical problem and the results were published in a reputable journal of the time. However, twenty-five years later a number of medical researchers had found that serum A did, in fact, cure the particular medical problem and it has been used since that time.
   c. A researcher who studied literacy, concluded that children who were raised in homes where the parents read to their children on a consistent basis learned to read faster than children whose parents did not read to them. This result has been consistently demonstrated.
   d. Researchers originally claimed that home schooling was detrimental to academic achievement for the children involved. Subsequent research has conclusively demonstrated that home schooling, in general, enhances achievement scores of children.
   e. A researcher found that stress is not related to disordered eating patterns. However, in reality, stressful situations lead to either overeating behaviors or anorectic behaviors.
   f. A researcher found that individuals who adhered to a strict vegan diet did not have any lower levels of activity than individuals who had other diets as long as the diets were all nutritionally sound. In reality, the activity levels of individuals did not differ as long as they have nutritionally sound diets.

7. Considering the information contained in the table below, provide the critical value, find the probability of committing a Type II error given the true population mean, and find the power of the test given the true population mean for each of the following.

Null Hypothesis	Null Hypothesis Population Mean	σ	Sample Size	True Population Mean	α
a. H_0: $\mu \leq 100$	100	15	25	110	.05
b. H_0: $\mu \geq 500$	500	100	50	450	.01
c. H_0: $\mu = 25$	25	3	15	25.8	.05

8. What is a Type III error? For the population mean in Exercise 7(c), calculate the probability of making a Type III error.

9. For each of the following examples, find the standard error of the mean, critical value(s), and the rejection region(s) for the given null hypothesis tested at the given α level.

Population Mean	Population Standard Deviation	Sample Size	Null Hypothesis	α
a. 100	15	9	H_0: $\mu = 100$	.05
b. 500	100	25	H_0: $\mu \leq 500$	.01
c. 16	1.33	30	H_0: $\mu = 16$	.01
d. 18.8	1.75	20	H_0: $\mu \geq 18.8$	.05
e. 63.2	9.73	80	H_0: $\mu = 63.2$	.001
f. 100	15	36	H_0: $\mu \leq 100$	.05

10. Suppose the sample means are given below for the corresponding examples in Exercise 9. What decision should you make about the null hypothesis in each case?
    a. $\overline{X} = 92$
    b. $\overline{X} = 550$
    c. $\overline{X} = 18$
    d. $\overline{X} = 20$
    e. $\overline{X} = 60$
    f. $\overline{X} = 104$

11. a. Suppose you want to test a sample of 36 people who are exposed to a class situation that will alter their level of creativity. You have decided to measure their creativity by using the Creativity Test in which the reported population mean is 80 and the population standard deviation is 9. Thus, suppose that you predict that the treatment population from which this sample was drawn will differ from the given population mean on the Creativity Test (i.e., H_0: $\mu = 80$). What are the consequences for power and Type II error of testing the hypothesis at $\alpha = .01$ instead of $\alpha = .05$ if the true treatment population mean is 84? Please justify your answers.

b. Suppose that you want to conduct a one-tailed test of the hypothesis about the Creativity Test scores of people from a particular treatment population. The Creativity Test has a reported population mean of 80 and a population standard deviation of 9. Furthermore, suppose you have a theoretical basis for making a research prediction that the treatment population mean will be higher than the published population mean on the Creativity Test (i.e., H_0: $\mu \leq 80$). What are the consequences for power and Type II error of testing the hypothesis with a sample of 64 people instead of 49 people if the true treatment population mean is 84 and $\alpha = .01$? Please justify your answers.

c. Suppose that you want to test a hypothesis about the Creativity Test scores (with a published population mean of 80 and a population standard deviation of 9) of people from a particular treatment population. Furthermore, suppose that you randomly select a sample of 25 people from this treatment population to test the null hypothesis at $\alpha = .05$. Although you have strong evidence to test the research hypothesis that the mean Creativity Test scores from this treatment population will be higher than the published population mean (i.e., H_0: $\mu \leq 80$), you decide to test the research hypothesis that the mean of the treatment population will differ from the published mean (i.e., H_0: $\mu = 80$). What are the consequences for power and Type II error of testing a nondirectional null hypothesis compared to testing a directional null hypothesis if the true treatment population mean is 84? Please justify your answers.

d. Given the results of parts (a), (b), and (c), indicate how a researcher might alter the power and the probability of making a Type II error.

12. Discuss the relative advantages and disadvantages of testing a directional vs. a nondirectional null hypothesis.

Part IV

Inferential Statistics in Action

Chapter 13

Testing Hypotheses About Means

CHAPTER OBJECTIVES

After studying this chapter, you should be able to:

❏ Describe how test statistics are used to test hypotheses.

❏ Given the description of a research study involving one sample, select the appropriate test statistic and conduct the analysis.

❏ Given a description of a two-sample research study, select the appropriate test statistic and conduct the analysis.

❏ State the advantages of a matched pairs or repeated measures design over an independent samples design.

❏ Given a research study, explain the interpretation of both r_{pb}^2 and d as measures of effect size.

❏ Use the relationships between power, α, N, and effect size to determine the sample size needed for given values of power, α, and effect size.

❏ Compute confidence intervals for one- and two-sample Z- and t-tests.

An ESP researcher reports that hypnotized participants have a much higher rate of accurate telepathic transmission than do nonhypnotized individuals. The result is statistically significant at the .05 level. A social psychologist reports that children who view a film about the dangers of smoking report significantly more negative attitudes toward tobacco products than children who have not seen such a film. The difference between the mean of the group seeing the film and the mean of the control group is significant at the .01 level. In both of these studies, the report of the research claims that the results are unlikely to have occurred by chance. The studies are similar to many reported in the research literature of psychology and education. How do researchers determine whether a difference between means is unlikely to have occurred by chance? How do they apply the principles of probability to reach conclusions about the differences between means? What do claims that results are statistically significant at a specified level tell you about the size or importance of the measured effects?

Introduction

In the final part of this book, Chapters 13–16, we will be applying the principles we have learned to answer substantive research questions in psychology and education. A researcher poses a question which he/she then restates as a statistical null hypothesis. The null hypothesis postulates an expected pattern in the descriptive statistics. As in the case of the ESP researcher and the social psychologist testing the effect of the film about smoking, the hypothesis most often involves means, but it may be phrased in terms of variances, correlation coefficients, or other descriptive statistics. This chapter presents methods for drawing conclusions from one or two means using the Z and t distributions. Chapter 14 covers ways to test hypotheses and form confidence intervals for variances and correlation coefficients. Chapter 15 will show you how to test hypotheses about frequencies and ordinal measurements. Finally, in Chapter 16 we introduce the basic principles of the analysis of variance. Odd though it may seem from the title, this last chapter is once again about means, but now more than two of them.

The majority of the hypotheses tested in the behavioral and social sciences concern the mean or means of one or more groups of research participants. In the last chapter we described in some detail the logical structure underlying the testing of statistical hypotheses. The general procedure described there was to select a statistic of interest, specify the sampling distribution of that statistic under the null hypothesis, calculate the value of the test statistic in the sample data, and reject the null if the probability of obtaining that extreme a sample value of the test statistic was less than a chosen α-level. We'll begin this chapter by considering more closely what a test statistic really is, then we'll apply the concept to a set of successively more complex studies.

Test Statistics

Suppose that we are studying a person who says he can read people's minds. We pose the following task for him. There is a deck of 50 cards, half with a diamond printed on them, half with a star. Twenty students in our statistics class volunteer to serve as subjects. Each is to shuffle the card deck, draw a card, and try to "transmit" the pattern on the card to the mind reader. Each participant is to repeat this process 6 times, then pass the deck to the next student. Therefore, with each of 20 participants, the mind reader can have a score from 0 correct to 6 correct guesses. The scores of the 20 students are shown in Table 13-1. How can we analyze the data to test whether this person can read minds?

First, we need to determine what the null and alternate hypotheses are. If the person cannot read minds, how many times should he correctly identify the pattern the subject is "sending"? What is the standard deviation of the expected distribution? The answers, from the binomial distribution formulas in Chapter 6, are $\mu = 3.0$ and $\sigma = 1.22$. With a sample size of 20, $\sigma_{\bar{x}}$ is .27. Because only a success rate higher than 3 would indicate the ability to read minds, we will reject H_0 only if our sample result is greater than 3. That is, this is a one-tailed test with H_0: $\mu \leq 3.0$, H_1: $\mu > 3.0$.

In this example the null hypothesis is a hypothesis about means, but, as we saw in Chapter 12, the statistic upon which we base our decision to reject or not to reject the null hypothesis is Z, not $\bar{X}$. In this case, Z is the **test statistic.** The term *test statistic* refers to the numerical index used to test a hypothesis. The test statistic is computed from the descriptive statistics of the sample *using the conditions assumed by the null hypothesis,* and may involve sample means, sample variances, or both.

*A **test statistic** is a numerical index used to test statistical hypotheses.*

Three Ways to Test a Null Hypothesis

In our mind-reading example, the null hypothesis is $\mu \leq 3.0$. There are three slightly different ways to approach testing this null hypothesis. The first approach, which we used extensively in the last chapter, is to

1. select a value for α (let's use .05),
2. look up the associated critical value of the test statistic ($Z = +1.645$), and
3. use the linear transformation procedure to calculate critical values *in the scale of the raw data.*

Table 13-1	Scores of the Mind Reader for 20 Students								
3	5	4	2	5	4	2	3	5	5
1	5	3	4	3	2	4	3	2	1

For the example we are considering, this produces a critical value of 3.44 (+1.645(.27) + 3.0 = 3.44). If the sample mean exceeds this critical value, we will reject the null hypothesis. The actual mean for the sample in Table 13-1 is 3.30. Clearly, we will not reject the null hypothesis on the basis of these data.

The second procedure, which we also encountered (but less explicitly) in the last chapter, is to

1. select α,
2. look up the associated value of the test statistic, and
3. transform the descriptive sample statistic into a value *in the scale of the test statistic*.

If the value of the test statistic computed from the sample data exceeds the critical value from the table, we reject the null hypothesis. Our sample had produced a mean of 3.30. Using μ and $\sigma_{\bar{x}}$, we find Z to be +1.11. This value does not exceed the critical value of Z for a one-tailed null hypothesis at $\alpha = .05$, so we would not reject the null hypothesis. This procedure is logically equivalent to the first, differing only in the metric in which the decision is made. It is somewhat more commonly used in reporting research results, but its ties to the sampling distribution are less obvious, a slight disadvantage.

The third alternative, which is logically distinct from the first two, was introduced to psychological and educational research in the 1930s. Using this approach, the researcher

1. calculates the value of the test statistic from the sample data,
2. goes to a table of the test statistic (such as appendix Table A-1 or A-2), and
3. looks up the *probability that a result of the size obtained would occur*.

In our example, a Z of 1.11 has a probability of occurring of .1335. The information presented in the research report would be the sample statistic ($\mu = 3.30$), the value of the test statistic ($Z = 1.11$), and the probability of this value occurring under the conditions specified in the null hypothesis ($P = .13$). The critical logical difference between this and the previous two procedures is that here *no decision is made before the fact concerning the risk of a Type I error*. That is, we do not commit ourselves to a particular level of risk of a Type I error before we look at the data. We are free to claim whatever significance level we find in the data.

Some Test Statistics

The primary test statistic (Z) we encountered in Chapter 12 is called the **critical ratio.** This was one of the earliest test statistics to be developed and was widely used prior to 1915. Since then, its use has dropped off markedly because it can only be used properly with means when we also know the population standard deviation. It has been replaced by the *t*-statistic, which has the advantage of being more appropriate for many research problems because it does not require that we know σ. Other test statistics that we will encounter in later chapters are the F and χ^2 (chi-square) statistics.

The choice of which test statistic to use to test a particular hypothesis depends on several factors. First, what descriptive property of the sample is being examined? Hypotheses about means and correlation coefficients may involve Z, t, or F, depending on the number of means specified in the null hypothesis or the particular kind of correlation coefficient.

THREE WAYS TO TEST A NULL HYPOTHESIS

1. Decide on the risk of Type I error. Look up the associated critical value of the test statistic. Calculate a critical value for the sample statistic in the metric of the measurement scale of the dependent variable. If the statistic exceeds the critical value, reject H_0.
2. Decide on the risk of Type I error. Look up the associated critical value of the test statistic. Calculate the value of the sample statistic in the metric of the test statistic. If the value calculated from the sample exceeds the critical value, reject H_0.
3. Calculate the value of the sample statistic in the metric of the test statistic. Look up the probability of the test statistic in the appropriate table. Report the value of the test statistic and its probability.

Hypotheses about variances, such as that $\sigma^2 = 50$ or that the variance for males on some trait is the same as the variance for females, generally use F or χ^2 as the test statistic.

Second, how much is known about the population affects which test statistic we will choose. The more we know about the population, the more powerful our test will be. Test statistics differ in their power (for example, Z is more powerful than t), and when we know more about the population (for example, we know the population variance) we can use statistics with greater power.

Finally, the level of scale used to make measurements must be considered in selecting a test statistic. Most of the popular procedures involving Z, t, and F require that the dependent variable be measured on an interval scale and assume that the population distribution is normal. Test statistics that do not assume these conditions are generally called **nonparametric,** or *distribution-free statistics*. These statistics are appropriate for nominal or ordinal data.

> **Nonparametric** *statistics are test statistics that do not assume a normal distribution of the dependent variable in the population.*

When σ Is Known—One-Sample Z

Olga works as an assistant in a laboratory doing research on the reactions of monkeys to various drugs. The lab has been getting some unusual results recently, and the researchers suspect that the drugs they have been getting from their supplier may not have been measured correctly. The manufacturer has assured the lab that their 10 milligram tablets have a mean dosage of 10.0 mg, with a standard deviation of .01 mg. Olga has been given a sample of 10 tablets and has been told to test whether the manufacturer's claims are valid. How should she go about answering this question?

As we learned in Chapter 11, the use of Z as a test statistic is appropriate in situations where the standard deviation of X in the population (σ_X) is known; this was the case in our coin-toss and ESP experiments, and it is true for Olga's problem as well. We know that $N = 10$ and $\sigma = .01$. The null hypothesis is $\mu = 10.0$. Olga decides to test H_0 at $\alpha = .01$. The standard error of the mean is

$$\sigma_{\bar{X}} = \frac{\sigma_X}{\sqrt{N}} = .0032$$

and the test statistic is

$$Z_{\bar{X}} = \frac{\bar{X} - \mu}{\sigma_{\bar{X}}} \qquad \text{(eq. 11.2 repeated)}$$

She has obtained a sample mean of 10.0067. What is her decision?

The table of areas of the normal distribution is used to determine whether the probability that the Z from our sample was drawn from the null hypothesis distribution is less than α. With $\alpha = .01$ and a two-tailed null (H_0: $\mu = 10.0$), the critical value of Z is 2.576. The Z from Olga's study is

$$Z_{\bar{X}} = \frac{10.0067 - 10.000}{.0032} = 2.09$$

This Z does not exceed the critical value, so Olga's data do not contradict the manufacturer's claims. She would not reject H_0.

Note that this procedure corresponds to the second alternative described in the box on ways to test a null hypothesis (see page 312). If we take our sample Z, 2.09, and go to the table to find the probability of the sample Z ($P = .0366$, which is not less than .01), we are using the third. We could, of course, also use the first alternative by computing a critical value for $\bar{X}$ using the linear transformation equation

$$\bar{X}_{CV} = Z_\alpha(\sigma_{\bar{X}}) + \mu$$
$$= 2.576(.0032) + 10.000 = 10.0082$$

and comparing our sample mean to this value directly. Obviously, Olga's sample mean does not exceed this critical value.

Had Olga chosen to test her hypothesis at the .05 level using either of the first two procedures from the box on page 312, she would have rejected it, but she cannot do so at her chosen level. However, using the third way to test the null, she would report that $Z = 2.09$, $P < .04$. The readers of her report could then decide for themselves if the evidence is strong enough to support the manufacturer's claim.

An essential point to keep in mind throughout this and future discussions of hypothesis testing is the difference between the population parameter and its sample estimate (in this case μ and $\overline{X}$) and the test statistic (here Z). Our hypotheses are about the population parameters, but our decisions about those hypotheses are reached on the basis of test statistics. As we saw, Olga's hypothesis was about μ, but her test statistic was Z. Just as there is a sampling distribution of the sample statistic around its parameter, so, also, is there a sampling distribution of the test statistic around the value it would have if the null hypothesis were true. The sampling distribution of $\overline{X}$ around μ is precisely paralleled by the sampling distribution of $Z_{\overline{X}}$ around the value it would have (0) if the null hypothesis were true. This distinction was blurred in Chapter 12 to illustrate the relationship between sample statistics, probability, and decisions about hypotheses. In cases where only one or two sample means are involved, it is easy to go back and forth between the two concepts: we can calculate $\overline{X}$ from Z and $\sigma_{\overline{x}}$, or we can calculate Z from the sample data. However, when we attack more complex problems, such as hypotheses involving three or more means, it is not possible to regenerate sample statistics from the test statistic. In these cases, *it is the sampling distribution of the test statistic that is used to test the hypothesis,* not the sampling distribution of the sample statistic around the parameter. Therefore, our decision whether to reject H_0 or not is always based on the test statistic.

▶ *Now You Do It*

Let's review the one-sample critical ratio with an example from the BTU data. The registrar at BTU has been collecting admissions data for 50 years on the incoming freshman class; the mean high school GPA in this population is 2.50 with a standard deviation of .773. Dr. Numbercruncher wants to test the hypothesis that this year's students are different from those in previous years. Using an α of .05 and a two-tailed null hypothesis, what conclusion will she reach?

The appropriate null hypothesis is $\mu = 2.50$, which is a two-tailed null. What is the standard error of the mean? You should get

$$\sigma_{\overline{X}} = \frac{.773}{\sqrt{68}} = .0937$$

The only piece of information you need from the data on this year's students is the mean high school GPA, which we found in Chapter 3 was 2.43. You can then enter that value into equation 11.2 to obtain a Z of

$$Z = \frac{2.43 - 2.50}{.0937} = -.747$$

Our obtained Z does not exceed the critical value of ± 1.96, so Dr. Numbercruncher will conclude that the data do not support the hypothesis that this year's class is different from previous ones.

When σ Is Known—Two-Sample Z

There are relatively few research problems where we would select the one-sample Z-test just described. The most frequently encountered application is with tests involving the correlation coefficient (see Chapter 14) and in studies of ESP and similar phenomena where the null hypothesis distribution is known from probability theory. Somewhat more frequently, we are interested in hypotheses involving two samples that differ in some way. This difference may involve status characteristics of the individuals—gender, for example—or it may involve some sort of manipulation; in either case, it is necessary to modify the critical ratio.

Sampling Distribution of the Difference Between Means

First, consider two different populations, males and females, for example, and a dependent or measured variable, perhaps score on a verbal ability test. We know that the male population has a frequency distribution on this test with mean, μ_M, of 99.9 and standard deviation of 15.1. Likewise, the female population has a frequency distribution with mean, μ_F, of 100.1 and standard deviation of 14.9. If we were to draw a sample of $N = 25$ from each population, what should we expect to get in terms of the sample means?

We know that there is a sampling distribution of $\overline{X}_F$ around μ_F that is formed by taking samples of size N_F. In our example, this sampling distribution will have mean $\mu_{\overline{X}_F} = 100.1$ and standard error

$$\sigma_{\overline{X}_F} = \frac{\sigma_F}{\sqrt{N_F}} = \frac{14.9}{\sqrt{25}} = 2.98$$

Likewise, there is a sampling distribution of $\overline{X}_M$ for samples of size N_M, which has $\mu_{\overline{X}_M} = 99.9$ as its mean and $\sigma_{\overline{X}_M} = 3.02$ as its standard error.

Now, suppose we draw a sample of males and a sample of females, compute $\overline{X}_M = 100.3$ and $\overline{X}_F = 99.8$ and find the difference between them ($\overline{X}_M - \overline{X}_F = .5$), and record this difference between the means. Then we go back to the populations and draw another sample from each, compute the means (say they are $\overline{X}_M = 99.6$ and $\overline{X}_F = 101.0$) and find and record their difference ($99.6 - 101.0 = -1.4$). We repeat this process many, many times. Each pair of samples produces a difference between their means (for example, the values of .5 and -1.4 between $\overline{X}_M$ and $\overline{X}_F$ above). We now have the materials to make a frequency distribution of the *difference between means*. This frequency distribution is called the **sampling distribution of the difference between means.** The elements in the distribution are the observed sample differences ($\overline{X}_M - \overline{X}_F$).

> The **sampling distribution of the difference between means** is the frequency distribution of differences between pairs of sample means.

We express the mean of this sampling distribution as $\mu_{\overline{X}_M - \overline{X}_F}$. It is *a mean of the distribution of differences between means* and it can be shown (See Hays, 1994, pp. 321–323) that

$$\mu_{\overline{X}_M - \overline{X}_F} = \mu_M - \mu_F$$

That is, *the mean of the sampling distribution of the differences between sample means is equal to the difference between the population means.* In the case we have been considering,

$$\mu_{\overline{X}_M - \overline{X}_F} = 99.9 - 100.1 = -.2$$

> The mean of the sampling distribution of the differences between sample means is equal to the difference between the population means.

This sampling distribution of differences between means has a standard deviation around its mean, which is known as the **standard error of the difference between means** and is symbolized $\sigma_{\overline{X}_M - \overline{X}_F}$.

$$\sigma_{\overline{X}_1 - \overline{X}_2} = \sqrt{\frac{\sigma_M^2}{N_M} + \frac{\sigma_F^2}{N_F}}$$

> The **standard error of the difference between means** is the standard deviation of the sampling distribution of the differences between means.

For our example, we would find the standard error of the difference between means to be

$$\sigma_{\overline{X}_1 - \overline{X}_2} = \sqrt{\frac{228.01}{25} + \frac{222.01}{25}} = \sqrt{18} = 4.24$$

The standard deviation of the distribution of differences between means (the standard error of the difference) is 4.24. Note carefully that the formula requires that we use variances, not standard deviations. In general, if we use the numerals 1 and 2 to designate the groups, the above equations become

$$\mu_{\overline{X}_1 - \overline{X}_2} = \mu_1 - \mu_2 \tag{13.1}$$

$$\sigma_{\overline{X}_1 - \overline{X}_2} = \sqrt{\frac{\sigma_1^2}{N_1} + \frac{\sigma_2^2}{N_2}} \tag{13.2}$$

The process of forming the sampling distribution of differences between means is shown graphically in Figure 13-1. In the figure, we see four pairs of samples drawn from the two populations ($\overline{X}_{11}$, $\overline{X}_{12}$, $\overline{X}_{13}$, and $\overline{X}_{14}$ from population 1; and $\overline{X}_{21}$, $\overline{X}_{22}$, $\overline{X}_{23}$, and $\overline{X}_{24}$ from population 2). Next, the figure shows that the difference between each pair of means

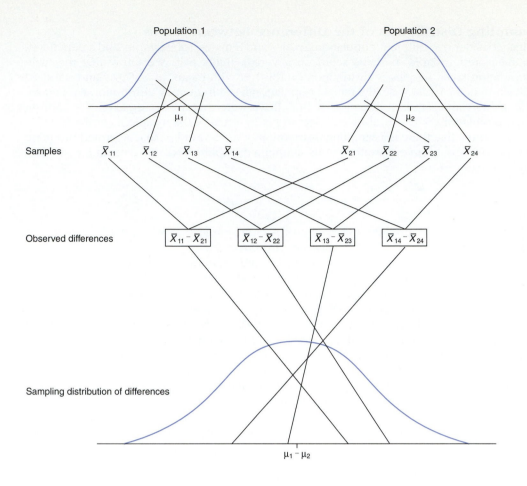

Figure 13-1 Construction of the sampling distribution of the differences between means.

Population 1

Population 2

Samples $\overline{X}_{11}$ $\overline{X}_{12}$ $\overline{X}_{13}$ $\overline{X}_{14}$ $\overline{X}_{21}$ $\overline{X}_{22}$ $\overline{X}_{23}$ $\overline{X}_{24}$

Observed differences $\boxed{\overline{X}_{11} - \overline{X}_{21}}$ $\boxed{\overline{X}_{12} - \overline{X}_{22}}$ $\boxed{\overline{X}_{13} - \overline{X}_{23}}$ $\boxed{\overline{X}_{14} - \overline{X}_{24}}$

Sampling distribution of differences

$\mu_1 - \mu_2$

(for example, $\overline{X}_{12} - \overline{X}_{22}$) is computed. Finally, the differences themselves are plotted in a new distribution. *It is the resulting sampling distribution of differences between means that is the basis for the test statistic.*

Suppose, for example, we know that the distribution of lengths of male babies at birth has a standard deviation of 1.3 inches, and that the standard deviation of the distribution of lengths of female newborns is 1.2 inches. If we were to take samples of 25 male and 20 female babies, what is the standard error of the difference between means? From equation 13.2,

$$\sigma_{\overline{X}_M - \overline{X}_F} = \sqrt{\frac{1.69}{25} + \frac{1.44}{20}} = \sqrt{.1396} = .374$$

That is, if we took a large number of pairs of samples of 25 males and 20 females, computed the sample means, found the differences between the means, and computed the standard deviation of this distribution of differences, we would get a value of .374. Again, note that this calculation requires variances, not standard deviations.

An important feature of the sampling distribution of the difference between means is that it will be normal in shape, regardless of the shapes of the distributions in the two populations, if the samples are reasonably large. This is a result of the central limit theorem and makes it possible to use Z as a test statistic when σ_1 and σ_2 are known.

General Form of the Test Statistic

We have already seen that the general form of the Z-test is that of a Z-score of the sample statistic in the sampling distribution of that statistic. The Z for a two-group comparison has exactly the same form, provided we remember that *the sampling distribution is a distribution of differences.* The equation is

$$Z_{\overline{X}_1 - \overline{X}_2} = \frac{(\overline{X}_1 - \overline{X}_2) - (\mu_1 - \mu_2)}{\sigma_{(\overline{X}_1 - \overline{X}_2)}} \tag{13.3}$$

where the quantities in parentheses are single quantities in the sampling distribution of differences. For the case of our gender differences in verbal ability, where we found $\overline{X}_M = 100.3$, $\overline{X}_F = 99.8$, and the standard error of the difference to be 4.24, equation 13.3 yields

$$Z_{\overline{X}_M - \overline{X}_F} = \frac{(100.3 - 99.8) - (99.9 - 100.1)}{4.24} = \frac{(.5) - (-.2)}{4.24} = +.165$$

Clearly, a result like this one is not at all unlikely, given the population values.

If you compare equation 13.3 with the equation for Z-scores in Chapter 11, you should notice a similarity. In fact, we can state a general form of the critical ratio Z. The same structure also applies to the t-test that we will encounter shortly. Both test statistics have the same elements arranged in the same way. The elements are (1) a result observed in the data from the sample(s), (2) a value specified as the center of the sampling distribution under the null hypothesis, and (3) the standard deviation of the sampling distribution. The organization is

$$Test\ statistic = \frac{(sample\ result) - (null\ hypothesis\ value)}{standard\ deviation\ of\ sampling\ distribution}$$

The sample result can be a single mean ($\overline{X}$) or a difference between two means ($\overline{X}_1 - \overline{X}_2$). The null hypothesis value can be a single population mean (μ) or a difference between two population means ($\mu_1 - \mu_2$). The value in the denominator will be the appropriate estimated standard error (we present five alternatives in this chapter).

You can see the three basic elements in the example of verbal ability test scores for men and women. The sample result is the observed difference between the sample means ($+.50$). The null hypothesis value is the known difference between the populations ($-.20$), and the standard deviation of the sampling distribution is the standard error of the difference (4.24). Each test statistic in this chapter shares this structure.

The Z computed in equation 13.3 is treated in the same way as the Z in equation 11.2. We can use the normal curve table to find the probability that a Z this large or larger would occur, and if this probability is less than our chosen α, we reject the null hypothesis. Of course, it is also possible to have determined the critical value of Z or of ($\overline{X}_1 - \overline{X}_2$) beforehand; then the decision can be reached without further reference to tables. That is, any of the three approaches to hypothesis testing can be used.

Hypotheses Revisited

Before considering an extended example of computing a two-sample critical ratio, we must discuss the hypothesis we are testing. Naturally, the hypothesis under test is a null hypothesis; it has the general form

$$H_0: \begin{array}{l} \mu_1 - \mu_2 \leq K \\ \mu_1 - \mu_2 \geq K \end{array} \right\} \quad \text{One-tailed}$$

$$\mu_1 - \mu_2 = K \quad \text{Two-tailed}$$

That is, the null may state that the difference between population means is less than or equal to some specified value, K, greater than or equal to K, or equal to K; and the value

---

GENERAL FORM OF THE Z- AND t-TEST STATISTICS

All Z- and t-tests have the general form

$$Z \text{ or } t = \frac{(sample\ value) - (null\ hypothesis\ value)}{standard\ deviation\ of\ the\ sampling\ distribution}$$

If one sample is involved, the sample value is a single sample statistic, the null hypothesis value is an assumed parameter value, and the standard deviation is the standard error of the statistic. If two samples are involved, the sample result is a difference between statistic values in the two samples, the null hypothesis value is the assumed difference between the parameters (often 0), and the standard deviation is the standard error of the difference between the statistics.

of K is chosen with reference to the theory that gave rise to the hypothesis. In our verbal-ability score example, a two-tailed null hypothesis would be that $\mu_M - \mu_F = -.2$. Note that these null hypotheses are structurally the same as the one-sample null, but that the hypothesis is not about a mean but a difference between means.

An Example. Suppose that we believe in the power of thought processes to alter events in the physical world (a phenomenon known as psychokinesis), and we wish to conduct an empirical test of this belief. We have two sets of 10 coins, and we are going to project a wish onto these coins; for one set (A), we shall "think heads," and for the other set (B), we shall "think tails." We know from the binomial distribution (see Chapter 6) that $\mu_A = 5.0$ and $\mu_B = 5.0$ if there is nothing to our theory; also, from our knowledge of the binomial distribution, $\sigma_A = \sigma_B = 1.58$. At this stage in our research, we have no idea about the magnitude of effect to expect, but our theory clearly predicts that μ_A, the mean number of heads for the set of coins where we think heads, should exceed μ_B, the mean number of heads for the other set, if thinking has the power to affect events in the physical world.

The appropriate null hypothesis in this example is the directional hypothesis that $\mu_A - \mu_B \le 0$. The null value for both μ_A and μ_B is 5.0, but our theory makes a directional prediction. A result in the opposite direction, no matter how big, does not support our theory of psychokinetic effects. We should not reject the null in favor of our theory unless $\overline{X}_A$ is larger than $\overline{X}_B$ by an amount that exceeds the critical value for our chosen α-level.

We now have the null hypothesis and its alternate:

$$H_0: \mu_A - \mu_B \le 0 \qquad (\text{or } \mu_A \le \mu_B)$$
$$H_1: \mu_A - \mu_B > 0 \qquad (\text{or } \mu_A > \mu_B).$$

We decide to flip each set of coins 10 times. With $\sigma_A = \sigma_B = 1.58$, we can calculate the standard error of the sampling distribution of differences between means. This value, which is also called simply the *standard error of the difference*, is, by equation 13.2,

$$\sigma_{\overline{X}_A - \overline{X}_B} = \sqrt{\frac{1.58^2}{10} + \frac{1.58^2}{10}}$$
$$= \sqrt{\frac{2.5}{10} + \frac{2.5}{10}}$$
$$= \sqrt{.5}$$
$$= .7071$$

Knowing that there is a general unwillingness in the scientific community to accept the proposition that thoughts affect the physical world, we decide to guard quite carefully against a Type I error. We decide that there should be a risk of only one chance in 100 that we are in error if we decide to reject the null hypothesis. Therefore, choosing $\alpha = .01$ and going to Appendix Table A-1, we find that the smallest value of Z for which the area beyond is no more than .01 is 2.326. (Had our hypothesis been two-tailed, we would have used $\alpha/2 = .005$ in each tail and arrived at $Z = 2.576$.) The critical value for our

Table 13-2 Outcomes from the Psychokinesis Study of Tosses of 10 Coins			
Group A "Think Heads"		Group B "Think Tails"	
5	7	4	3
6	5	3	4
6	6	6	4
4	8	5	6
5	7	5	4

test statistic is $Z = +2.326$ because our rejection region is only in the upper tail of the sampling distribution.

We are now ready to conduct our experiment. The 10 coins of group A are placed in a shaker, and while they are shaken and dumped on the table, we earnestly "think heads." We keep on "thinking heads" for nine more trials, each time recording the number of heads. This is followed by 10 observations of the 10 coins in group B, only now we project a wish for tails. Our study is now complete. We obtained the data shown in Table 13-2. All that remains is to calculate the value of the test statistic and reach a decision about our null hypothesis.

SPECIAL ISSUES IN RESEARCH

How to Express Statistical Significance

This value of Z for our psychokinesis study (next page) puts us in a bind because if we had used an α of .05, we would reject the null hypothesis and conclude that our wishing had affected the coins; however, Z does not reach the value we specified for an α of .01. The probability that a difference this large or larger would occur by chance is somewhere between .05 and .01. One solution to this problem that has become popular in recent years (the third alternative we described earlier) is to give the value of the test statistic and the probability that such a value would occur. If we were to adopt that approach, we would report that $Z = 2.12$ and $P < .02$. (The area beyond a Z of 2.12 is given in Appendix Table A-1 as .0170. However, it is customary to specify P as less than some quantity and to give only one nonzero numeral.)

There is one advantage to this alternative, but there are two disadvantages. The advantage is that readers of the research report have more information about the probability of a Type I error and can reach their own decisions to reject or not to reject the null hypothesis. The first, less serious disadvantage is that the formal logical structure of the hypothesis test has been violated by changing the rules after the fact. We stated a value of .01 at the outset and have dropped our standards to $\alpha = .05$ because the data did not give us sufficiently extreme results. This is ethically questionable. The second and more serious problem is that the level of statistical significance has become a metric for judging the importance of a research finding. Results having a lower likelihood of Type I error are viewed by the unwary as more important. Relevance to the theory being tested, freedom from misinterpretation, and predictive power are more valuable criteria for judging research results, but they are often overlooked in favor of the simple number provided by the α-level. Meehl (1978) has called this practice asterisk-

counting because it is common for research reports to indicate different levels of statistical significance by different numbers of asterisks following the test statistics. A further discussion of this topic is beyond our scope, but if you are interested, you should consult the paper by Meehl and other sources listed at the end of this chapter and Chapter 12.

In addition to the level of significance problem, the psychokinesis study has a weakness in design that you may have identified. Remember that one of our goals in designing an experiment is to eliminate alternate explanations and provide a clear test of the research hypothesis. In the design of our psychokinesis study, we overlooked one obvious plausible rival hypothesis: a pre-existing bias in the two sets of coins. It may be, and we have no way of knowing, that the coins in group A were biased to come up heads. A better design that would have eliminated this alternative would have been to use the same 10 coins for both conditions and to alternate our treatment from trial to trial. (That is, think heads on one trial, and tails on the next. This alternative assumes, of course, that the coins do not have a memory for what happened on the last trial and are not permanently altered by previous events.) Another alternative would be to assign coins to treatment groups by a random process (which we probably did). Failure to eliminate this plausible rival hypothesis of a pre-existing difference by properly designing the study would make the issue of α-level moot and the results of the study uninterpretable. This is often a problem with research in psychology and education, where it is common to use pre-existing groups. Some of the dangers of this practice and some alternative designs were discussed by Campbell and Stanley (1966). As we have emphasized repeatedly, no amount of sophisticated analysis or subtle interpretation can save a poor research design.

The first thing we do is calculate the means for the two groups. We find that $\Sigma X_A = 59$ and $\Sigma X_B = 44$, which gives us 5.9 and 4.4 for the values of $\overline{X}_A$ and $\overline{X}_B$, respectively. Since we know that $\sigma_{\overline{X}_A - \overline{X}_B} = .7071$, we calculate Z as follows:

$$Z = \frac{5.9 - 4.4}{.7071}$$

$$Z = +2.12$$

▶ *Now You Do It*

The Metropolitan Achievement Tests (MAT) are tests of academic achievement that are standardized to have a standard deviation of 15 points. You have been conducting an intensive mathematics enrichment program for a group of minority children, and you wish to test whether there is a difference between boys and girls in their math achievement after exposure to your curriculum. There are 15 boys and 12 girls in your class, and their mean scores on the MAT are 113 and 119 respectively. What is the null hypothesis, and what conclusion do you reach? Run the appropriate test at the $\alpha = .05$ level.

You have no reason to predict a difference between boys and girls, so the null is two-tailed. Therefore, the critical value for Z is ± 1.96 and the difference you expect is 0.

Questions
1. What is the standard error of the difference between means?
2. What is the critical value for the difference between the means?
3. What conclusion do you reach?

Answers
1. You should get 5.81.
2. You should get ± 11.39.
3. $Z = 1.03$, and the difference between the means is 6.

Neither value falls in the critical region, so you do not reject the null.

When σ Is Unknown—One-Sample t

The two procedures described so far both assume that the population standard deviations are known by the investigator. This assumption is rarely met in practice, so it is usually necessary to estimate the population values from those found in the sample. For example, although the faculty at Big Time U may know what the test scores and grades of past classes have been, it would be questionable to call these population values. In most cases, the investigator does not even have the benefit of data from previous identical or highly similar analyses. For this reason, studies like those of the SAD faculty and of Dr. Cross and his students must rely on the data from their own study to estimate σ.

We learned in Chapter 11 that this has some important consequences for the sampling distribution of the mean. There we saw that an unbiased estimate of σ^2 could be computed from the sample data by dividing the sum of squares by its degrees of freedom (equation 11.4). We also found that we had to use the t distribution for making probability statements and forming confidence intervals on the sampling distribution of the mean. Here we shall see how to use $\hat{S}_{\bar{X}}^2$ and the t distribution to test hypotheses about means when σ is not known.

The t statistic provides exactly the same hypothesis testing procedure that we encountered with Z, except that we do not assume that we know σ_X. Suppose, for example, that we have a theory about learning that includes the proposition that plants can learn. To test this proposition we must design a situation where plants might learn, expose a sample of plants to the appropriate stimuli, and measure their behavior.

There are plants, such as the Venus flytrap, that emit measurable behaviors, so we decide to use a sample of five of them. The flytrap closes on anything that stimulates its ingestive area, so we shall test our proposition by seeing whether we can train the plant to close to the sound of a bell. First, we pair the bell with a fly's being placed on the plant's ingestive area. This is repeated 100 times. Then, we omit the fly and just ring the bell, at the same time measuring the amount of closure by the plant in millimeters.

Our research hypothesis is that learning will take place or, stated in proper quantitative form, that the amount of closure at the sound of the bell is greater than zero. The null hypothesis, then, is

$$H_0: \mu \leq 0$$

This must be a one-tailed test because it is not possible for there to be less than no closure. The alternate hypothesis is, of course, consistent with our research hypothesis.

$$H_1: \mu > 0$$

Since we do not know the population standard deviation, we must use a t-test to reach a decision about our hypothesis. Our experiment provided the data shown in Table 13-3. The scores are millimeters of maximum closure within 10 seconds after ringing the bell.

The test statistic for a hypothesis involving one sample and one unknown population standard deviation is called Student's t, in honor of the pen name used by Gossett, who discovered the t distribution. It is identical in structure to the one-sample critical ratio and tests the same null hypothesis, namely that $\mu = K$, $\mu \leq K$ or $\mu \geq K$. The formula is

$$t_{df} = \frac{\bar{X} - K}{\hat{S}_{\bar{X}}} \tag{13.4}$$

The numerator in t is the same as in Z, but there is a difference in the denominator. Because we do not know the population standard error of the mean, we must estimate it from the

---

Table 13-3 Computing a One-Sample t for the Learning in Plants Study

Plant	Score (X)	X^2
1	0	0
2	1	1
3	3	9
4	2	4
5	0	0
	$\Sigma X = 6$	$\Sigma X^2 = 14$

Calculations

$\bar{X} = 1.2$ $\quad S_X = 1.1662$ $\qquad \hat{S}_X = 1.3038$

$\qquad\qquad S_X^2 = 1.36$ $\qquad\qquad \hat{S}_X^2 = 1.70$

$$\hat{S}_{\bar{X}} = \frac{\hat{S}_X}{\sqrt{N}} = \frac{1.3038}{\sqrt{5}} = .583$$

$$\hat{S}_{\bar{X}} = \frac{S_X}{\sqrt{N-1}} = \frac{1.1662}{\sqrt{4}} = .583$$

$$t_4 = \frac{\bar{X} - \mu}{\hat{S}_{\bar{X}}} = \frac{1.2 - 0}{.583} = 2.058$$

sample data. As we saw in Chapter 11, this estimated standard error of the mean, $\hat{S}_{\bar{X}}$, is found by substituting $\hat{S}_X$ for σ_X in equation 11.1.

$$\hat{S}_{\bar{X}} = \frac{\hat{S}_X}{\sqrt{N}}$$

(13.5)

The value of t obtained from equation 13.4 is just like a Z; it is a deviation score divided by a standard deviation. In this case, the deviation score is the deviation of a sample mean from the mean of the sampling distribution, and the standard deviation is the estimated standard error of the mean. The difference between t and Z lies in the shape of the sampling distribution of the test statistic. The critical ratio Z has a sampling distribution that is always normal in shape. The t statistic with $(N - 1)$ degrees of freedom has a sampling distribution that is the t distribution for that number of degrees of freedom, hence the df subscript in the formula. (The actual numerical value of df, for our example, 4, is usually given here.)

The steps necessary to compute the quantities to test our hypothesis are shown in Table 13-3. First we compute $\bar{X}$ and find it to be 1.2, the mean number of millimeters of closure within 10 seconds after ringing the bell. Next we must compute our estimate of σ. The computational formula

$$\hat{S}_X = \sqrt{\frac{N \sum X_i^2 - (\sum X_i)^2}{N(N - 1)}}$$

(11.6 repeated)

yields a value of 1.3038. Next, we compute $\hat{S}_{\bar{X}}$ using equation 13.5 and find a value of .583. Finally, we compute $t_{(4)}$ (t with $df = 4$) and find it to be 2.058. We look up the critical value for $\alpha = .05$ in Appendix Table A-2 for four degrees of freedom ($N - 1$) and find 2.132 for a one-tailed test. We must conclude that we cannot reject the null hypothesis at the .05 level.

▶ *Now You Do It*

Let's suppose that over the last 20 years the faculty in the SAD Department have found that their students come into the basic statistics course with an average college GPA of 2.50. Can we reasonably conclude that this year's class is a sample from a population with a mean of 2.50? Since we have no basis in theory to predict whether this class is better or worse than others, we should use a two-tailed test, and because we do not know σ_X, t provides the appropriate sampling distribution and test statistic. We have the following information from the data

$$\bar{X} = 2.27 \qquad \hat{S}_X = .796 \qquad N = 68$$

From this, we calculate

$$\hat{S}_{\bar{X}} = \frac{.796}{\sqrt{68}} = .0965$$

$$t_{67} = \frac{2.27 - 2.50}{.0965} = -2.38$$

Looking in Appendix Table A-2 under $df = 60$ we find the two-tailed .05 critical value for t to be 2.000 and the .02 critical value to be 2.390. Clearly, we can reject the null hypothesis that this sample was drawn from a population with a mean of 2.50 with a risk of Type I error of less than 5 percent. Because the second critical value is so close to our obtained t (2.39 versus 2.38) and our df is larger than the tabled value, we would probably be safe in concluding that our risk of Type I error is less than .02. Indeed, this is what the computer tells us. The 95% confidence interval for the mean of the population from which this sample was drawn runs from 2.08 to 2.46. Notice, however, that a 99% confidence interval would include the value 2.50.

HOW THE COMPUTER DOES IT

Most computer programs will compute a one-sample t-test. In SPSS the procedure is located in the Compare Means submenu of the STATISTICS menu under the heading One-Sample T Test. The program asks you to identify the variable you wish to analyze and the hypothesized value of the parameter, called the "Test Value." The program automatically produces a 95% confidence interval *on the difference between the sample mean and the test value* and uses the exact critical value of t for the degrees of freedom. This confidence interval is not one for the population parameter but is more closely associated with the hypothesis test. If the interval includes the value zero, you would not reject H_0 at the two-tailed .05 level. That is, the confidence interval focuses not on estimating μ but on estimating the true difference between the sample mean and the test value. You have to do some extra work to get the confidence interval for μ.

You can alter the confidence interval using the Options subcommand and simply typing in the probability value you want. The screen looks like this:

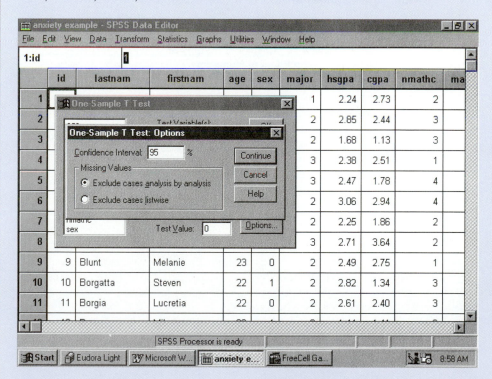

Output includes the values of $\overline{X}$, $\hat{S}_{\overline{X}}$, t, and the *two-tailed probability* of t as well as the confidence interval around $\overline{X}$. There is no option to run the test as a one-tailed test, and there is no way to specify a level for α. Thus, this program—and all other widely used ones—allows only the third form of the hypothesis test, which may exacerbate the problem mentioned earlier of P-values being taken as a metric for the importance of a result.

Two Independent Samples

Situations where a design using a single sample provides a test of the hypothesis of interest are relatively rare. A much more normal situation is a research design involving two samples, where one sample gets a treatment of some kind and the other acts as a control, or where the members of one group differ from those in the other on some status variable such as gender, age, or ability level. It is then necessary to estimate the standard deviations in both samples and use these estimates to find the standard error of the difference between the means.

The data in Table 13-4 could have come from a typical two-group study. Suppose, for example, that as consumer psychologists we believe that the presence of soft background music will induce people to feel more relaxed when shopping, resulting in increased sales. Since audio systems are fairly expensive, we decide to run a study to see whether our hunch is likely to be true before recommending that our clients install such systems.

The first thing we do is to set up our hypotheses. We decide to associate the null hypothesis with the negation of our expectation because we want to guard against recommending the expense of installing a system if it will not result in more sales. We have direct control over the probability of a Type I error, so we control the probability of mistakenly installing the system by our choice of the null hypothesis and α. Loosely stated, our null is that the presence of music will not result in increased sales. We shall reject this null in favor of its alternative and recommend installing audio systems only if the probability that the null is true is sufficiently small. Note that there are several other null hypotheses that we could have chosen; for example, that the music would not reduce buying or that music would increase (or decrease) buying by a certain amount. We could even have had as our research hypothesis simply that the presence of music would change buying behavior without specifying the direction of change, thus making our null hypothesis two-tailed. The choice of null hypothesis is up to you as the investigator and you should make it carefully to provide the most direct answer to the research question of interest.

Table 13-4 Results of a Study of the Effect of Music on Purchasing Behavior		
Group 1	Group 2	Summary Statistics
3	6	
9	5	$\Sigma X_1 = 232$
7	2	$\Sigma X_2 = 181$
17	11	$\Sigma X_1^2 = 2584$
8	14	$\Sigma X_2^2 = 1773$
3	5	$\Sigma X_1 X_2 = 1884$
11	9	
10	15	$\overline{X}_1 = 9.28$
8	2	$\overline{X}_2 = 7.24$
4	3	$\hat{S}_{X_1} = 4.24$
9	9	$\hat{S}_{X_2} = 4.39$
14	9	$S_{X_1} = 4.15$
13	14	$S_{X_2} = 4.30$
13	13	
11	0	$\hat{S}_p = 4.31$
6	6	$\hat{S}_p^2 = 18.62$
9	9	
9	7	$\hat{S}_{\overline{X}_1 - \overline{X}_2}^2 = 1.49$
17	5	$\hat{S}_{\overline{X}_1 - \overline{X}_2} = 1.22$
14	10	$r_{X_1 X_2} = .458$
13	8	
5	1	
12	10	
4	0	
3	8	

The formal statement of our hypotheses is

$$H_0: \mu_{music} - \mu_{nonmusic} \leq 0$$
$$H_1: \mu_{music} - \mu_{nonmusic} > 0$$

The population mean dollar sales to people in the music condition (μ_{music}) is contrasted with the population mean dollar sales to people who are not exposed to music ($\mu_{nonmusic}$) through sample estimates ($\overline{X}_{music}$ and $\overline{X}_{nonmusic}$) of the parameters. The one-tailed hypothesis was chosen because we are not interested in the alternative that music reduces sales. If it does not increase sales we will not recommend the change.

Next, we must consider some aspects of the design of our study. Since we have no population parameters available, we cannot use a critical ratio. Fluctuations in the economy make it unwise to compare our new condition (music) with average sales for the last six months. We decide to run a study using two groups of people with 25 shoppers in each group in order to have samples large enough to be representative of their respective populations. This means that we are going to analyze our data and test our hypothesis using a two-sample t-test.

This problem is very much like the one we encountered with the two-sample Z-test. Again, we are dealing with a sampling distribution that involves differences between means, and again, our decision whether to reject the null hypothesis will be based on the deviation of the difference between our samples ($\overline{X}_1 - \overline{X}_2$) from the mean of the sampling distribution of differences under the assumptions of the null hypothesis. The change from the two-sample Z is that the sampling distribution of the test statistic is a t distribution rather than the normal distribution.

A two-sample t-test with observations on 50 different people has 48 degrees of freedom. We decide to use an α of .05 and since our hypothesis is one-tailed, we look in column 2 of Appendix Table A-2 to find the critical value. This reminds us that beyond $df = 30$, most tables of the t distribution are incomplete because the change in critical values from one df to the next is relatively small. The next lower tabled entry, for $df = 40$, is 1.684, so we use this as our critical value. If the t obtained from the comparison $\overline{X}_{music} - \overline{X}_{nonmusic}$ exceeds $+1.684$, we shall reject the null hypothesis, conclude that music increases sales, and recommend to our clients that they install audio systems. However, to run the test correctly, there is one additional change that we must make.

Pooled Variance. As we saw in Chapter 11, the larger the degrees of freedom on which an estimate of the population variance is based, the better the estimate. This principle becomes important in the case of two samples because an assumption underlying the null hypothesis is that the two samples both come from the same population. We assume that they are two random samples from the same sampling population or two random samples from two sampling populations with equal variances, so they have the same population variance. While it would not be appropriate to combine the two samples and use the deviations of the ($N_1 + N_2 = 50$) observations from a single mean to estimate the value of σ_X, we can pool the two separate sample estimates, pool their degrees of freedom, and obtain a better single estimate of the population standard deviation.

Remember that a sum of squares divided by its degrees of freedom provides an unbiased estimate of the population variance. If we let $\hat{S}_1^2$ (based on N_1 observations) be the unbiased estimate of σ^2 in sample 1 (our 25 shoppers in the music condition) and $\hat{S}_2^2$ (based on N_2 observations) be the unbiased estimate of σ^2 from sample 2 (our 25 nonmusic shoppers), then the improved **pooled estimate of σ^2** is given by the pooled sums of squares divided by the pooled df.

An improved **pooled estimate of the population variance** is given by the pooled sums of squares divided by the pooled degrees of freedom.

$$\hat{S}_p^2 = \frac{SS_1 + SS_2}{df_1 + df_2} = \frac{(N_1 - 1)\hat{S}_1^2 + (N_2 - 1)\hat{S}_2^2}{(N_1 - 1) + (N_2 - 1)}$$

or

$$\hat{S}_p^2 = \frac{(N_1 - 1)\hat{S}_1^2 + (N_2 - 1)\hat{S}_2^2}{N_1 + N_2 - 2} \tag{13.6a}$$

The pooled variance estimate $\hat{S}_p^2$ is based on ($N_1 - 1$) + ($N_2 - 1$) or ($N_1 + N_2 - 2$) degrees of freedom.

Another form of $\hat{S}_p^2$ is useful in certain situations. When only the *sample* standard deviations are available, the relationships in equation 11.7 can be used and equation 13.6a can be written as

$$\hat{S}_p^2 = \frac{N_1 S_1^2 + N_2 S_2^2}{N_1 + N_2 - 2} \qquad (13.6b)$$

In conducting our study we randomly divided 50 people into two groups and admitted them to a store. Group 1 had background music, group 2 did not; the dollar sales volume for each customer is recorded in Table 13-4 along with the sums and sums of squares needed for computation. The means are found to be

$$\overline{X}_{music} = 232/25 = 9.28$$

and

$$\overline{X}_{nonmusic} = 181/25 = 7.24$$

The *sample standard deviations* are

$$S_{music} = \sqrt{\frac{25(2584) - (232)^2}{25^2}} = 4.15$$

and

$$S_{nonmusic} = \sqrt{\frac{25(1773) - (181)^2}{25^2}} = 4.30$$

while the *estimates of the population standard deviation* are

$$\hat{S}_{music} = \sqrt{\frac{25(2584) - (232)^2}{25(24)}} = 4.24$$

and

$$\hat{S}_{nonmusic} = \sqrt{\frac{25(1773) - (181)^2}{25(24)}} = 4.39$$

Either of these pairs of standard deviations can be used in one of the variations of equation 13.6 to compute the pooled variance and standard deviation (the difference in the results below is due to rounding).

$$\hat{S}_p^2 = \frac{24(4.24)^2 + 24(4.39)^2}{48} = 18.62 \qquad (13.6a)$$

$$\hat{S}_p^2 = \frac{25(4.15)^2 + 25(4.30)^2}{48} = 18.60 \qquad (13.6b)$$

$$\hat{S}_p = 4.31$$

Computing the Estimated Standard Error of the Difference

Once we have the pooled variance estimate, it is used in exactly the same way as σ^2 to compute the estimated standard error of the difference between means. By substituting $\hat{S}_p^2$ for σ_1^2 and σ_2^2 in equation 13.2, we obtain

$$\hat{S}_{\overline{X}_1 - \overline{X}_2} = \sqrt{\frac{\hat{S}_p^2}{N_1} + \frac{\hat{S}_p^2}{N_2}} \qquad (13.7)$$

The value of $\hat{S}_p^2$ is then used to compute the standard error of the difference between means

$$\hat{S}_{(\overline{X}_{music} - \overline{X}_{nonmusic})} = \sqrt{\frac{18.62}{25} + \frac{18.62}{25}} = 1.22$$

This standard error is then used to compute t, the test statistic

$$t_{df} = \frac{(\overline{X}_1 - \overline{X}_2) - (\mu_1 - \mu_2)}{\hat{S}_{\overline{X}_1 - \overline{X}_2}} \qquad (13.8)$$

which produces a t of

$$t_{48} = \frac{(9.28 - 7.24) - 0}{1.22} = \frac{+2.04}{1.22} = +1.672$$

(Note the 0, which is the difference between the populations assumed under the null hypothesis. This value is an integral part of the conceptual base of the formula, but because in most cases it is assumed to be zero, it is often omitted.) This t has ($N_1 + N_2 - 2$) degrees of freedom, which means that we shall find the necessary critical values in the ($N_1 + N_2 - 2$)th row of Appendix Table A-2.

A two-sample t-test with observations on 50 different people has 48 degrees of freedom. We decide to use an α of .05, and since our hypothesis is one-tailed, we look in column 2 of Appendix Table A-2 to find the critical value. This reminds us that beyond $df = 30$, most tables of the t distribution are incomplete because the change in critical values from one df to the next is relatively small. Therefore, only the critical values for selected distributions are included. In general, the proper thing to do is to use the critical value for the next lower tabled df. The tabled entry for $df = 40$ is 1.684, so we use this as our critical value. If the t obtained from the comparison $\overline{X}_{music} - \overline{X}_{nonmusic}$ exceeds +1.684, we shall reject the null hypothesis, conclude that music increases sales, and recommend to our clients that they install audio systems.

The observed value of t, 1.672, is not quite as large as the critical value we specified, so we would not reject the null hypothesis; however, the outcome was in the predicted direction. Depending on the cost of audio systems and the perceived long-term sales return, we may choose to let the matter drop and do research on some other theory, or we may repeat the study with a larger sample, thus increasing our degrees of freedom and our power by reducing the critical value for t and (probably) the estimated standard error of the difference. Note that if our df had been 60 or more, we would have rejected the null hypothesis.

A study such as the one just described brings up an additional consideration in research: How large does a difference have to be before it is useful? Where monetary concerns are at issue, the answer is often quite clear. If an audio system costs $1,000 per month to operate, the effect it produces must exceed that amount for the expense to pay off. But what about scientific theory? Hypothesis testing can tell us whether it is reasonable to view the effect of some treatment as greater than zero, but that is not enough. We can show almost any effect to be nonzero if we make the sample size large enough. This fact has led Meehl (1978) to refer to the "quasi-always false null hypothesis." The answer to the question is that there is no pat answer. We return to this issue at the end of the chapter.

▶ *Now You Do It*

Let's test the null hypothesis that men and women earn the same grades at BTU to review the use of the independent samples t-test. First, what kind of hypothesis do we have? Because the null assumption is one of equality of the means, it is a two-tailed null hypothesis.

The basic data we need are the means and standard deviations. If you haven't already computed these somewhere else, the values you should get are as follows:

$$\overline{X}_F = 2.25 \qquad \hat{S}_F = .764 \qquad S_F = .751$$
$$\overline{X}_M = 2.28 \qquad \hat{S}_M = .828 \qquad S_M = .817$$

We can use either form of the standard deviation to calculate $\hat{S}_p^2 = .642$ and $\hat{S}_p = .801$. The standard error of the difference between means is .1965, which yields a nonsignificant t of $-.153$.

If you ran the analysis using SPSS, you should have gotten slightly different values. Because it keeps more decimal places than we reported, the difference between means is .0343 and the standard error is .1967, yielding a t of $-.174$. Note that the confidence interval for the mean difference includes 0.

HOW THE COMPUTER DOES IT

Commercially available computer programs such as SPSS or SYSTAT are written to accommodate the needs of both new users and advanced statistical analysts. For this reason, the output often includes information that is not directly meaningful to the novice. We encountered this feature in the correlation and regression programs, and it is present here as well.

The test we have just been describing is called the independent samples *t*-test. It is included in the Compare Means section of the STATISTICS menu under that name. You are asked to insert a "test variable" which is the dependent variable (dollars of sales in the above example) and a "grouping variable" which is the independent variable (music condition). When you insert a variable name in the grouping variable box, the Define Groups button is illuminated and you must specify the numerical or alphabetical identifiers of the groups you wish to analyze. You will get a screen that looks like this:

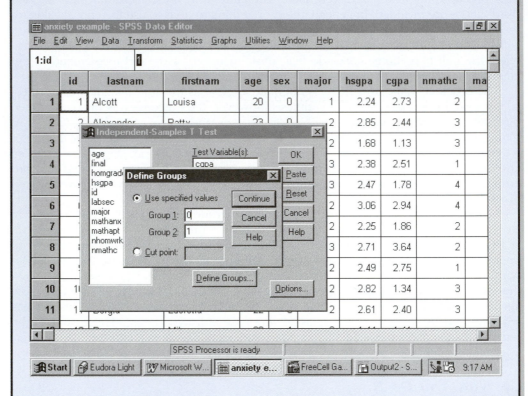

Suppose we wish to test the hypothesis that there is no difference between male and female students in statistics performance at Big Time U. Females are coded 0 and males are coded 1, so we would enter these values as shown on the screen. If we wanted to compare education majors with "other" majors, we would use 2 and 3 as the group codes after specifying "major" as the grouping variable. The options button controls the same type of confidence interval that we encountered with the one-sample *t*-test.

The output includes two kinds of *t*-tests. The first one is the one we have been using. It assumes the variances of the groups are equal and uses the pooled estimate of σ^2. The output in the row for this test also includes a test of the null hypothesis that the variances are equal called Levene's test. It is a statistic similar to the *F*-test we introduce in Chapter 14. The second *t*-test, which has quite different degrees of freedom, does not assume the groups have equal variances. It is beyond our scope, but it tests the same basic null hypothesis. As was true of the one-sample *t*-test, a two-tailed null hypothesis is assumed in the calculation of the probability for *t*. If your hypothesis is one-tailed, the correct probability value is one-half the value given in the output.

Matched Pairs t

There is one other alternative that we could use to raise power instead of increasing our sample size. If it is reasonable under the research conditions to consider observations as occurring in pairs, it is possible to use a test statistic called the **matched pairs t**, or the **t test for repeated observations.** Suppose, for example, that instead of using two distinct groups of 25 shoppers to test the effectiveness of music, we had tested the same people twice, once with music, and once without music. This procedure has the advantage of reducing $\hat{S}_{(\bar{X}_1 - \bar{X}_2)}$ in most situations, thereby increasing power and making it possible to reject the null hypothesis at a given level of α with a smaller difference between the means.

One necessary condition for using a matched pairs t is that the observations can be logically paired. The pairing can be accomplished in any of three general ways. The first is to make a pair of observations on each person, one under each condition; in our effect of music on shopping example, this would involve observing each shopper once under the music condition and once under the no-music condition. This is called a *repeated-observations* or *repeated-measures design*.

The second procedure is to take naturally occurring pairs of individuals—identical twins, siblings, married couples, and so on—and assign one member of each pair to each condition; this technique is commonly used in genetic and social research. If we had 25 pairs of siblings, we could have used this technique to test our music-shopping hypothesis. The third procedure is to form pairs by matching otherwise unrelated individuals on the basis of one or more status characteristics. It is common in educational research, for example, to form pairs matched by age, ability, or socioeconomic status. The last two designs are called *matched-pairs designs*.

Covariance Method

Once the pairs have been formed, the study is conducted in the usual way to test the same general hypothesis that would be tested using the t-test for independent samples. The difference is that instead of using equation 13.7 to compute $\hat{S}_{(\bar{X}_1 - \bar{X}_2)}$, we use the formula

$$\hat{S}_{\bar{X}_1 - \bar{X}_2} = \sqrt{\frac{\hat{S}_p^2}{N_1} + \frac{\hat{S}_p^2}{N_2} - 2r_{X_1 X_2}\left(\frac{\hat{S}_p}{\sqrt{N_1}}\right)\left(\frac{\hat{S}_p}{\sqrt{N_2}}\right)} \tag{13.9}$$

The difference between equations 13.7 and 13.9 is the term that is subtracted,

$$2r_{X_1 X_2}\left(\frac{\hat{S}_p}{\sqrt{N_1}}\right)\left(\frac{\hat{S}_p}{\sqrt{N_2}}\right)$$

Any time the correlation between paired observations is positive, this quantity will be positive, thereby reducing $\hat{S}_{(\bar{X}_1 - \bar{X}_2)}$. X_1 is the measurement of a person under condition 1 (music) and X_2 is the measurement of the same individual (or the pairmate) under condition 2 (nonmusic). Notice that if the correlation between the two variables is negative, equation 13.9 will yield a larger value than equation 13.7. Therefore, the matched pairs t should be used only when there is reason to believe that $r_{X_1 X_2}$ is positive. (Actually, equation 13.7 for the standard error of the difference based on independent samples is a special case of 13.9 where $r_{X_1 X_2}$ is zero because there is no basis for pairing observations.)

The consequences of matching can be illustrated using the data from Table 13-4. Suppose that we had actually observed each of 25 individuals twice instead of observing 50 different people, 25 under each condition. We compute the sum of cross products $\Sigma X_1 X_2$ to be 1884 and find

$$r_{X_1 X_2} = \frac{25(1884) - 232(181)}{\sqrt{25(2584) - (232)^2} \sqrt{25(1773) - (181)^2}} = +.458$$

This value is combined with the necessary standard deviations in equation 13-9 to yield

$$\hat{S}_{\bar{X}_1 - \bar{X}_2} = \sqrt{\frac{(4.31)^2}{25} + \frac{(4.31)^2}{25} - 2(+.458)\left(\frac{4.31}{\sqrt{25}}\right)\left(\frac{4.31}{\sqrt{25}}\right)} = .90$$

This standard error of the difference between means is substantially smaller than the value of 1.22 we found for independent samples. Substituting this value into our equation for t,

The **matched pairs,** or **repeated observations t-test,** is used when testing the difference between the means of two variables on the same group of people or the scores of two people who have been matched in some way.

In a **repeated-measures** study each person is observed under both levels of the independent variable.

In **matched-pairs studies** each person is paired with one other person, either as a result of a natural pairing (e.g., siblings) or on the basis of being similar on some other variable such as age or ability.

The important thing to notice in this example is that the same data, the same set of numbers, can lead to different decisions, depending on the assumptions we make and the way we design our study. It is often possible to make a study more sensitive to differences between means by matching individuals or making multiple observations of people. It is usually desirable to design the study to be as sensitive as possible. Note that had we collected our data from a repeated-measures/matched-pairs design and analyzed them with an independent samples

t we would have reached an incorrect decision because we applied the wrong test. We would have made a Type IV error. Type IV errors can be made in either direction; we can err in concluding the null hypothesis is false because we used an inappropriately sensitive test (Z instead of t), or we can err in failing to reject the null hypothesis because we used a test that was too conservative. It is important to know what assumptions can be made about the data when selecting a statistical test.

we obtain

$$t_{(24)} = \frac{(9.28 - 7.24) - (0)}{.90} = 2.27$$

This t is based on *25 pairs of observations;* therefore, there are only 24 degrees of freedom associated with this test. (We shall use a different approach shortly to illustrate this principle.) Going to Appendix Table A-2 with $df = 24$, we find the .05 one-tailed critical value for t to be 1.711, leading us to conclude that we should reject the null hypothesis and conclude that music does increase sales. In fact, our obtained t exceeds the one-tailed critical value for $\alpha = .025$.

Direct Difference Method

In the **direct difference method,** the differences between paired observations are found, then statistics are computed and statistical tests conducted directly on the difference scores.

Paired observations offer us another way of looking at differences. Instead of grouping the observations and then finding the difference between groups, we can find the differences between the paired observations first and then compute statistics on the differences. With paired shoppers in our example, we could find the difference between what shopper 1 spent when music was playing and what he/she spent without music. We would repeat this computation for each individual, thereby obtaining 25 differences in the amount spent under the two conditions. This is sometimes called the *direct difference method.* It involves computing a simple difference score D_i for each individual

$$D_i = X_{i1} - X_{i2}$$

and performing our analyses on these difference scores. For the shopping scores in Table 13-4, the difference score for the first shopper is $3 - 6 = -3$. The difference score for each of the other 24 shoppers is shown in Table 13-5.

Using difference scores has the effect of changing our formal hypotheses. For our music example,

$$H_0: \mu_{music} - \mu_{nonmusic} \leq 0 \text{ becomes } \mu_D \leq 0$$
$$H_1: \mu_1 - \mu_2 > 0 \text{ becomes } \mu_D > 0$$

There are similar changes in the one-tailed hypotheses. These changes are slight and in most cases do not require a great change in our thinking.

We now have a single sample of D's, as shown in the fourth column of Table 13-5, rather than two samples of X's, as we had in Table 13-4, so we are back to a one-sample problem. The mean difference in the sample, $\overline{D}_X$, is our sample estimate of μ_D and the sampling distribution of $\overline{D}_X$ will have an estimated standard error,

$$\hat{S}_{\overline{D}} = \frac{\hat{S}_D}{\sqrt{N}} \qquad (13.10)$$

where $\hat{S}_D$ is computed just like any other estimate of a population standard deviation

$$\hat{S}_D = \sqrt{\frac{N \sum D_i^2 - (\sum D_i)^2}{N(N-1)}} \qquad (13.11)$$

Table 13-5 Computing a Matched Pairs *t* for the Data from Table 13-4

Individual	X_1	X_2	D	D^2	Summary Statistics
1	3	6	−3	9	
2	9	5	4	16	$\Sigma D_i = 51$
3	7	2	5	25	$\Sigma D_i^2 = 589$
4	17	11	6	36	
5	8	14	−6	36	$\overline{D} = 2.04$
6	3	5	−2	4	$\hat{S}_D = 4.495$
7	11	9	2	4	$\hat{S}_{\overline{D}} = .90$
8	10	15	−5	25	$t = \dfrac{2.04}{.90} = 2.27$
9	8	2	6	36	
10	4	3	1	1	
11	9	9	0	0	
12	14	9	5	25	
13	13	14	−1	1	
14	13	13	0	0	
15	11	0	11	121	
16	6	6	0	0	
17	9	9	0	0	
18	9	7	2	4	
19	17	5	12	144	
20	14	10	4	16	
21	13	8	5	25	
22	5	1	4	16	
23	12	10	2	4	
24	4	0	4	16	
25	3	8	−5	25	

Of course, in both equations N refers to the number of D's or the number of pairs of original observations. The test of the null hypothesis has the general form of a one-sample test (H_0: $\mu_D = K$). It is a *t*-test of the form

$$t_{df} = \frac{\overline{D} - K}{\hat{S}_{\overline{D}}} \tag{13.12}$$

and it is easy to see that since $\overline{D}$ is based on N differences, the *t*-test will have $(N - 1)$ degrees of freedom.

Table 13-5 shows the computations for a direct difference analysis. Summing the values in the D column to get $\Sigma D_i = 51$ and dividing by $N = 25$ yields

$$\overline{D} = 51/25 = 2.04$$

which is exactly the value we obtained for $\overline{X}_1 - \overline{X}_2$. In fact, it will always be true that

$$\overline{D}_X = \overline{X}_1 - \overline{X}_2$$

Next, we use $\Sigma D_i^2 = 589$, $N = 25$, and $\Sigma D_i = 51$ to compute

$$\hat{S}_D = \sqrt{\frac{25(589) - (51)^2}{25(24)}} = 4.495$$

Applying this result in equation 13.10 produces the standard error of the difference between means,

$$\hat{S}_{\overline{D}} = \frac{4.495}{\sqrt{25}} = .90$$

which is the same value that we obtained for $\hat{S}_{\overline{X}_1 - \overline{X}_2}$ when we applied equation 13.9 to the data in Table 13-4. The final step is to compute *t* with $df = 24$.

$$t_{(24)} = 2.04/.90 = 2.27$$

Obviously, this also is exactly the same value that we obtained before.

The routine for computing a matched-pairs or repeated measures *t*-test in SPSS is called a Paired Samples T-Test and is found in the Compare Means procedures of the STATISTICS menu. The unique feature of this program is that it assumes the data are already included in a paired form. That is, each data record is assumed to represent a pair of observations, included as two different variables. In the case of multiple measures on each person, either repeated measurements of the same variable at different times or for comparison of different variables with common measurement scales (as in the GPA example we include below) this is quite straightforward. However, if the data are on paired individuals, it may be necessary to do some work on the data file in order to make the data ready for analysis unless you anticipated the needs of this specific program when the data were entered into the file. This is one reason why you should know what analyses you plan to conduct before running the study and preparing the data.

To use the paired samples program you must specify two different variables for the analysis. Click on the two variable names, then click the arrow to select them. The paired variables will appear on one line in the window. Click OK and the program will perform the analysis with the second variable subtracted from the first. Output includes the correlation between the variables as well as the means, standard deviations, mean difference, standard error of the difference, *t*-test with two-tailed probability, and 95% confidence interval for the difference between means.

The two different procedures for computing *t* for matched pairs or repeated measures data are exactly equivalent; which one we use is a matter of convenience. The direct difference method has some appeal because the computations are less involved. However, there is the added step in data preparation of calculating the *D*'s. Many modern calculators and programs for personal computers include built-in statistical routines that will compute $\overline{X}_1$, $\overline{X}_2$, $\hat{S}_1$, $\hat{S}_2$, and $r_{X_1 X_2}$ automatically from the raw paired observations. If this feature is available, it may be simpler to compute *t* by equations 13.7 and 13.8 than by 13.10–13.12. Which computation you use is less important than how well you design and conduct your study.

▶ *Now You Do It*

It is sometimes claimed that college grades are lower than high school grades. The SAD faculty at BTU decided to test this assertion with the data they collected on their students. What do you think they would find? You have high school GPA (HSGPA) and college GPA (CGPA) in your data file for the 68 students in their class. What answer do you get?

First, let's consider what the null hypothesis should be. Should it be one-tailed or two-tailed? Folklore claims CGPA will be lower than HSGPA. This is a one-tailed prediction, but we may also suspect that grade inflation has been more severe in college than in high school. We must decide whether we want to allow for this possibility or not. Since we are testing the folklore prediction, we probably should use a one-tailed test:

$$H_0: \mu_{bs} \leq \mu_c \text{ or}$$
$$\mu_{bs} - \mu_c \leq 0$$

Our preliminary computations reveal

$$\overline{X}_{bs} = 2.43 \qquad \hat{S}_{bs} = .72 \qquad r_{bs,c} = +.50$$
$$\overline{X}_c = 2.27 \qquad \hat{S}_c = .80$$
$$\overline{X}_{bs} - \overline{X}_c = .16 \qquad \hat{S}_{\overline{X}_{bs} - \overline{X}_c} = .09$$
$$t_{67} = 1.78$$

Looking in Appendix Table A-2 for $\alpha = .05$, $df = 60$, and a one-tailed test we find a critical value of 1.671, which would cause us to reject H_0. However, the two-tailed critical value is 2.00. Therefore, had we not used the directional hypothesis implied by folklore, we would not have rejected the null. Remember, failure to reject the null does not mean that we can accept $\mu_{bs} = \mu_c$, only that we do not have evidence that they are different.

If you ran the analysis on your computer you got very slightly different values. Specifically, t was 1.77 and its probability was reported as .08. The numerical difference is due to rounding, and the probability is for a two-tailed hypothesis. Therefore, we can get the appropriate value for the one-tailed test to be .08/2 or .04, which is less than our chosen α.

Effect Size and Power

Recently researchers have begun to ask questions that go beyond those involved in testing null hypotheses. They ask not only "Is this difference unlikely to have occurred by chance under the conditions specified in the null hypothesis?", but also "Is this difference large enough to be meaningful and worthwhile?" We asked this question when we considered the design of our study of the effect of music on shopping. How large does the difference in amount spent have to be before it justifies the increased cost of the audio system? We have seen that sample size affects power, which is the probability of rejecting the null hypothesis when it is false. The null hypothesis is false only when the independent variable has an effect on (or is related to) the dependent variable. The size of this relationship is known as the **effect size.** The larger the change in the dependent variable that is attributable to the independent variable, the greater the effect size. For example, music that increases buying behavior by $10 per shopper has a larger effect than music that increases the amount spent by only $2 per shopper. Likewise, an educational program that increases student achievement by 10 points has a larger effect than one that increases achievement by 5 points. The problem is how to express effect size in meaningful units.

> **Effect size** is a measure of the amount that the dependent variable changes as a result of a change in the independent variable.

There are two ways to think of effect size, both of which involve the proportion of variance in the dependent variable that is related to change in the independent variable. Their metrics are very different. On one hand, a measure of effect size can be viewed as essentially the same as the square of a correlation coefficient. Remember that r^2 can be interpreted as the proportion of variance in a dependent variable, Y, that can be predicted from (or is related to) an independent variable, X. Thus, if we have a continuous independent variable, such as score on the SAT, r^2 gives us an index of the size of the effect that one variable has on another one. An equivalent index for use in an analysis where a t-test is called for is

$$r^2_{pb} = \frac{t^2}{t^2 + df} \qquad\qquad (13.13)$$

For the set of data in Table 13-5, where we found $t = 2.27$, equation 13.13 gives us a value of

$$r^2_{pb} = \frac{2.27^2}{2.27^2 + 24} = \frac{5.15}{5.15 + 24} = .18$$

That is, whether one is shopping in the music condition or not accounts for 18 percent of the variance in the amount shoppers spent.

This index, known as the *squared point biserial correlation coefficient,* is actually a squared product-moment correlation between a continuous dependent variable and a dichotomous independent variable. That is, if we represent shopping without music as a score of zero on the X-variable "shopping condition," and shopping with music as a score of 1, then the r^2 between shopping condition and amount spent is .18. Both r^2 and r^2_{pb} are descriptive of the relationship in the sample. Both can be interpreted as the proportion of variance in the dependent variable that is related to the independent variable.

▶ *Now You Do It*

In the previous Now You Do It we encountered an example of a repeated measures t-test that showed a significant difference between high school GPA and college GPA. What proportion of variance does this significant result explain?

The t that we found previously was 1.78. Putting this value into equation 13.13 produces

$$r_{pb}^2 = \frac{1.78^2}{1.78^2 + 68 - 1} = \frac{3.168}{70.168} = .045$$

That is, this statistically significant result accounted for less than 5 percent of the variance in grade point averages. It is not unusual for results that look very promising when viewed from the perspective of statistical significance to appear much more modest when cast in the cold light of variance explained.

An alternative measure of effect size that has become popular in efforts to summarize research across several studies (known as meta analysis) is to express the differences between means in relation to the standard deviation in the study. We may, for example, express the difference between means in our two shopping conditions as a function of the pooled estimate of the standard deviation, $\hat{S}_p$. In the two-group case which we consider here, the procedure is quite direct and the relationship of the measure of effect size to power is clear. As we proceed to more complex experimental designs in Chapter 16 the principle remains the same, but the procedures are less obvious.

The usual two-group measure of effect size has been given the label d and is defined simply as the observed difference between the means divided by the pooled estimate of the population standard deviation,

$$d = \frac{(\overline{X}_1 - \overline{X}_2) - (\mu_1 - \mu_2)}{\hat{S}_p} \tag{13.14a}$$

where each of the terms has the usual definition. For our data from the shopping example in Table 13-4, we find

$$d = \frac{(9.28 - 7.24) - 0}{4.31} = \frac{2.04}{4.31} = .47$$

That is, being in the music condition increases spending by about half a standard deviation. Note carefully that the term in the denominator is the pooled estimate of the population standard deviation computed from equation 13.6, *NOT* the estimate of the standard error of the difference between the means (equation 13.7 for independent samples, equation 13.9 for matched pairs or repeated measures). We use the former term because our purpose is to describe the magnitude of the difference between means relative to the variation among individuals in the population, not the variation among samples from the population.

▶ *Now You Do It*

In the previous Now You Do It you calculated r_{PB}^2 for the data from Now You Do It on page 332. We can also calculate d for these data. We found the difference between means was .16, and the pooled estimate of the standard deviation is

$$\hat{S}_p = \sqrt{\frac{(67)(.8)^2 + (67)(.72)^2}{67 + 67}} = .76$$

Therefore,

$$d = \frac{.16}{.76} = .21$$

We use the pooled estimate of σ^2 even though this is a repeated-measures research design because the two sets of scores will not ordinarily have the same standard deviation, and our estimate of σ^2 is better if it is based on more observations.

As a measure of effect size, d does not have the same interpretation as r_{pb}^2; r_{pb}^2 is limited to the range between 0 and 1.0 and can be interpreted as the proportion of variance in one variable that is accounted for by the other. In contrast, d is not limited to any particular range, and it cannot be interpreted in terms of variance accounted for. Although what constitutes a large value of d therefore has not been clearly determined, the index is seeing widespread use in summaries of research.

Equation 13.14a can be modified for situations where σ_X is known and for one-sample tests. For the single sample situation you have an estimate of σ^2 based on only one group, so the appropriate formula is

$$d = \frac{\overline{X} - K}{\hat{S}_X} \tag{13.14b}$$

When σ_X is known, you can simply substitute it for $\hat{S}_X$ or $\hat{S}_p$ in the appropriate formula. For example, we can calculate the effect size for our mind reader who provided the data for the one-sample critical ratio. In that case, K was 3.0, the observed mean was 3.30, and σ_X was 1.22. Substituting these values in equation 13.14b yields

$$d = \frac{3.30 - 3.0}{1.22} = .25$$

That is, the mind reader scored about .25 standard deviations above chance. In the two-sample case where the population standard deviations differ, it is appropriate to use the weighted average of the standard deviations in the denominator of equation 13.14a. The weighted average of the standard deviations is

$$\sigma_{AVE} = \sqrt{\frac{N_1 \sigma_1^2 + N_2 \sigma_2^2}{N_1 + N_2}}$$

where all terms have their usual meanings.

Relationship to Power. Effect size is a measure of the actual difference between treatment population means, and it is estimated from sample results by equation 13.14a or 13.14b. In our discussion of power we said that the power of a test is its ability to detect the fact that a null hypothesis is false, which we called the alternate state of affairs. *A null hypothesis is false whenever the effect size is not zero,* and effect size is another name for a nonzero alternate state. The larger the effect size, the greater the difference between the population means, and, other things being equal, the greater the power. Since power is determined by (1) effect size, (2) sample size (larger N produces greater power), and (3) choice of α level (power will be greater for a test at $\alpha = .05$ than $\alpha = .01$), if we know any three of the values (power, d, α, and N) we can determine the fourth.

Finding Needed Sample Size

One of the major uses of effect size and power is to determine the sample size needed to achieve a desired level of power for an effect size that is considered large enough to be meaningful. Meehl (1978) and Rozeboom (1960), among others, have pointed out that if we use sufficiently large samples we can find trivial differences to be statistically significant. Such large-scale studies are not uncommon in government and military personnel research. They may also be present in large educational studies. The problem is to select a sample size that is large enough to detect meaningful and important effects, but small enough so that meaningless differences will not lead to rejection of the null hypothesis. If, for example, we had observed 10,000 shoppers in our study of the effect of music on spending, a difference in spending of less than 10 cents per shopper would have been statistically significant. The solution to the problem requires trial and error (a process called *iteration* in statistical analysis).

When σ Is Known

The first step in determining appropriate sample size is to select values for effect size, power, and α. Deciding what is a meaningfully large effect size is a problem that is answered by conducting pilot studies or consulting previous research. You might, for example, say that an increase in spending is not meaningful unless it is at least 8/10th of a standard deviation. A common suggestion for a reasonable level of power is .50, a 50 percent chance that the null hypothesis will be rejected if the selected effect size exists.

Some authors argue for a power of as much as .80. One of the usual values for α (.05 or .01) is most commonly employed.

After these basic decisions are made, a trial value for N is used with σ_X to calculate power for the chosen effect size. If power is too low, a larger N is chosen and the process is repeated. Different values of N are tried until one is found that produces approximately the desired power. Once sample size has been selected, the study can be conducted with reasonable assurance that results of a meaningful magnitude will be detected, and trivial results will be rejected.

Let us assume that we want to achieve a power of .50 for an effect size of $d = .80$ when σ is 6.0 and we choose an α of .05 for a two-tailed null hypothesis, $\mu_1 - \mu_2 \leq 0$.

$$Power = .50 \qquad \alpha = .05 \qquad d = .80 \qquad \sigma = 6.0$$

What is the power if we have $N = 10$ subjects in each of our two samples?

Remember that power is the proportion of the alternate distribution that falls in the rejection region of the null distribution. The first thing we must do is convert our effect size into a difference between the means so we know where the alternate distribution we are interested in is located. Solving equation 13.14a for the difference between the means $(\overline{X}_1 - \overline{X}_2)$, when $d = .80$ and $\sigma = 6.0$ we get

$$\overline{X}_1 - \overline{X}_2 = d\sigma = 4.80$$

An effect size of .80 corresponds to a difference between means of 4.80. Therefore, the alternate distribution is centered at +4.80 on the scale specified in the null hypothesis. Next, we must find the standard error of the difference between means. This gives us the standard deviations of the null and alternate distributions. From σ^2 and N we compute

$$\sigma_{\overline{X}_1 - \overline{X}_2} = \sqrt{\frac{36.0}{10} + \frac{36.0}{10}} = 2.68$$

The critical value of Z for a one-tailed hypothesis at $\alpha = .05$ is +1.645 in the null distribution, Remember, the null distribution centers on a value of zero. With $\sigma_{\overline{X}_1 - \overline{X}_2} = 2.68$, the critical value in the scale of the raw data is $2.68(1.645) = 4.41$. If the location of the alternate distribution, (expressed as a difference between means rather than as d) is +4.80, then the Z-score of a score of 4.41 in this *alternate sampling distribution* with a mean of 4.80 and a standard deviation of 2.68 is $-.15$ $[(4.41 - 4.80)/2.68 = -.15]$. The probability of obtaining a Z-value of $-.15$ *or greater* is about .56 or 56 percent (from Appendix Table A-1). (Note that we are interested in the entire part of the alternate distribution beyond the critical value.) Our power just about equals the value we set, so we can proceed with the current design with the knowledge that our study has the desired level of power under the assumptions we have made. These calculations show us what we can expect if our assumptions about effect size and variance are approximately correct.

▶ Now You Do It

Suppose you have been assigned the task of identifying the world's most effective statistics textbook. There are two contenders for this prestigious honor and you decide to conduct a study to select the winner. To make the contest fair, you decide to set your α-level at .01. You have determined that an effect size of 1.5 would be suitable. A review of the literature reveals that σ is 17.3 and you feel power must be at least .95. How large a sample will you need?

The first thing to do is determine the values you can look up or compute directly. You know d is 1.5 and $\sigma = 17.3$. From these facts $\overline{X}_1 - \overline{X}_2$ must be at least 25.95. You can now try some values for N, because all other calculations depend on N.

If you start with samples of size 10, you should find that the standard error of the difference is 7.74 and the critical value of Z for an α of .01 (two-tailed because we do not have a directional hypothesis) is 2.576. These pieces of information yield a critical value in the raw score scale of 19.94, which has a Z-score in the alternate distribution of $-.78$. Looking in Appendix Table A-1, we find the probability of a Z this large or larger to be .78. Samples of 10 yield a power of .78 under the conditions we have specified. We must try a larger value for N.

If you repeat the process outlined here for $N = 15$, you should get a Z of -1.53, which corresponds to a power of .94. To achieve a power of at least .95, we will need to have 16 observations in each sample, yielding a Z of -1.67.

When σ Is Not Known

Estimating needed sample size when σ is not known is somewhat more difficult than the procedure described above. There is a process that exactly parallels the steps we described, but it requires a special set of distributions known as the noncentral t distributions. Unlike the normal distribution, a t distribution changes shape for a situation other than the null hypothesis. The degree of change depends on the effect size. However, two much simpler approaches than using the noncentral t are available.

First, as we noted in Chapter 12, there are computer programs available to compute power. These programs can also be used to compute the sample size needed to achieve a specified level of power for a specified effect size. There is even a Web site that is devoted to power at http://www.surveysystem.com/sscalc.htm.

Second, many researchers use graphs, such as the ones we described in Chapter 12, that have been prepared to show the relationships among power, effect size, α, and N. These graphs are known as *power curves*. Power curves, and instructions for their use, are provided in more advanced statistics books such as Kirk (1995) and Keppel (1991). Each curve represents a single combination of N and α. By selecting a desired effect size, you can read out the power your test will have for that effect size. N is determined by using the curve that produces the desired power for the chosen effect size.

Testing Hypotheses in the BTU Data

There are two variables that we might wish to use as independent variables to test differences between means in the data from the study of math anxiety and performance in statistics at BTU: (1) gender and (2) math anxiety as a dichotomized variable. (It is a common, although perhaps questionable, practice in research to take a continuous variable such as anxiety and create two groups, a high-anxiety group and a low-anxiety group, by dichotomizing at the median. Those people above the median are considered high anxiety; those below are low anxiety.)

The results for all possible comparisons based on gender are shown in Table 13-6. As you can see, none of the differences is statistically significant and all of the 95% confidence intervals for differences between means contain 0. These data suggest that any differences between men and women on these variables in the population from which this sample was drawn are small (note that we cannot conclude that any of them is zero).

The situation is quite different for the dichotomized math anxiety groups. Thirty-four students scored at or below 18, so anyone with a score of 18 or below would be placed in the low-anxiety group and anyone with a score above 18 is considered to have a high level of math anxiety. The results are shown in Table 13-7. The correlational relationships that we found between math anxiety, math aptitude, and final exam scores in Chapter 8

Table 13-6	t-Tests on the Differences Between Male and Female Statistics Students at Big Time U						
Variable/Group		$\overline{X}$	$\hat{S}_X$	$\hat{S}_p^2$	$\hat{S}_{\overline{X}_1 - \overline{X}_2}$	t	95% CI
HSGPA	Males	2.42	.59	.48	.17	$-.12$	$-.36$ to .32
	Females	2.44	.81				
Final Exam	Males	34.28	14.13	190.45	3.38	.02	-6.69 to 6.83
	Females	34.21	13.34				
Homework	Males	10.70	2.49	9.47	.75	1.09	$-.68$ to 2.32
	Females	9.88	3.73				
Math Anxiety	Males	18.48	5.17	30.71	1.36	$-.25$	-3.06 to 2.38
	Females	18.82	6.01				
Math Aptitude	Males	53.45	11.15	132.06	2.82	$-.05$	-5.78 to 5.50
	Females	53.59	11.94				
Number of Math	Males	2.76	1.27	1.53	.30	.83	$-.35$ to .85
Courses	Females	2.51	1.19				

We have seen in previous chapters that we can place confidence intervals around a sample statistic to get an interval of uncertainty—a confidence interval—for the population parameter. It should therefore not be surprising that we can apply the same strategy with differences between means. For example, we found a difference between the means of the two groups in our shopping example of 2.04 dollars. What is the 95% confidence interval for the difference between the means of the two populations?

From Table 13-5, we found $\hat{S}_{\overline{D}} = .90$. Because we are estimating σ^2, we must use the t distribution. The two-tailed critical value of t for $df = 24$ (we had 25 pairs of scores) and $\alpha = .05$ is 2.064. Therefore, the 95 percent CI runs from $+.18$ to $+3.90$. We may conclude, with a 95 percent chance of being right, that the mean difference between buying behavior with music present and buying behavior without music is at least as great as 18 cents and no greater than $3.90.

The confidence interval for the difference between means is what SPSS will give us if we request a confidence interval for a two-sample t-test. However, *we can actually use confidence intervals in place of conventional hypothesis tests without loss of information.* Even better, confidence intervals provide us with a bonus, a specific interval estimate for the difference between population parameters. Some current theorists (for example, Cohen [1994] and Schmidt [1996]) have suggested that confidence intervals should replace significance testing entirely. Let us examine how confidence intervals can be applied to hypothesis testing problems.

The first piece of the puzzle is to realize that the α-level critical values we calculate for a hypothesis test are really the limits of a $1 - \alpha$ confidence interval around the parameter value specified in the null hypothesis (assuming a two-tailed test; otherwise we have one side of a $1 - 2\alpha$ confidence interval). That is, null hypothesis testing has been using confidence intervals all along, but confidence intervals on the null value rather than on our best estimate of what the population parameter actually is. Suppose that instead of placing our confidence interval around the null value, we place it around the sample estimate of the parameter (this will be either $\overline{X}$ or $\overline{X}_1 - \overline{X}_2$, depending on whether our study involves one or two samples). For a given α-level and standard error, the confidence interval is the same width regardless of whether we place it

around the null hypothesis value or around the sample estimate of the parameter. Therefore, if the sample value falls in the *rejection region* for the α-level two-tailed null hypothesis test, the null hypothesis value will fall *outside* the $1 - \alpha$ confidence interval around the sample statistic. This principle is shown in Figure 13-2.

To illustrate the principle, consider the following problem. We have two groups with 10 participants in each group and have selected an α of .05 for a two-tailed null hypothesis that $\mu_1 = \mu_2$. The standard error of the difference between means is 1.0. That means that our critical values for t would be ± 2.10. If we observe a difference of $\overline{X}_1 - \overline{X}_2 = +2.5$, we would get a value for t of $+2.5$ for our test statistic and would reject H_0. However, if we place a 95% confidence interval around our obtained difference of $+2.5$, the limits of that confidence interval would be $+.40$ and $+4.60$. This interval *does not include the null value* (0), so, given our sample result, *the null value is unlikely to be the population parameter.* Computing the value of t for the null value in the sampling distribution around the obtained value produces the same number, 2.5, as our null hypothesis test, but with a minus sign. Therefore, both procedures lead to the same conclusion regarding the null hypothesis with the same risk of error.

The confidence interval approach has two additional advantages over testing the hypothesis in the traditional way. First, it is intuitively more direct. We conduct a study to estimate what the population parameter is, and the confidence interval tells us how large an error we are likely to make. It just seems to make more sense to test the prediction from your theory rather than its negative. Second, we focus on **what is** rather than on **what is not.** If we increase sample size, we improve the accuracy of our estimate (reduce our band of uncertainty) rather than make the area around the null hypothesis value smaller. The relative benefits of hypothesis testing versus the use of confidence intervals are being hotly debated in the professional literature of both psychology and education (see the references at the end of Chapter 12, Thompson [1993], and various papers in issues of *The American Psychologist* and *Psychological Methods*). Both approaches are currently being used, and a reader of the literature needs to be able to appraise research results using either framework.

Figure 13-2 The confidence interval provides all of the information in the test statistic. If the sample mean falls in the α% rejection region, the null hypothesis value falls outside the $(1 - \alpha)$% confidence interval.

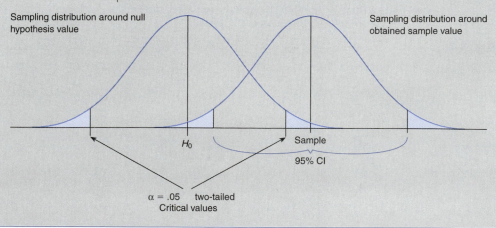

Sampling distribution around null hypothesis value

Sampling distribution around obtained sample value

H_0

Sample

95% CI

$\alpha = .05$ two-tailed
Critical values

Table 13-7   *t*-Tests on the Differences Between High- and Low-Anxiety Statistics Students at Big Time U

Variable/Group		$\bar{X}$	$\hat{S}_X$	$\hat{S}_p^2$	$\hat{S}_{\bar{X}_1-\bar{X}_2}$	t	95% CI
Final Exam	Low	40.71	12.67	144.11	2.91	4.45	7.13 to 18.77
	High	27.76	11.30				
Homework	Low	10.37	3.60	10.80	.80	.36	−1.31 to 1.89
	High	10.08	2.94				
Math Aptitude	Low	58.35	11.00	110.90	2.55	3.78	4.54 to 14.74
	High	48.71	10.04				

are revealed in significant differences between the high- and low-anxiety groups in this analysis. However, according to these (fictional) data, homework grades are not significantly different for the groups.

We have two ways to examine the magnitude of the relationship between math anxiety and the aptitude and final exam scores. Using equation 13.13, we find the squared point biserial correlations of anxiety with final exam score and with aptitude to be .23 and .18 respectively (the correlation with homework grade is .002). About 23% of final exam score and about 18% of aptitude score are accounted for by whether you are in the high- or low-anxiety group. In the metric of *d* from equation 13.14 we get an effect size of 1.08 for final exam scores and one of .92 for math aptitude. Both of these are large effect sizes, but the metric does not give the clear interpretation that correlations do.

SUMMARY

Hypotheses about group means are tested by comparing the value of a test statistic with a critical value for that statistic. The *critical value* of the statistic is the value that cuts off a chosen percent (α) of the sampling distribution that the statistic would have under the assumption that the null hypothesis is true. When the value of the test statistic exceeds the critical value, we can reject the null hypothesis with a risk no greater than α of being wrong.

In research situations where the population variance is known, the test statistic is a *critical ratio Z.* The sampling distribution of the critical ratio is the normal distribution. There is a critical ratio for single sample studies that involves the *standard error of the mean,* and there is a two-sample critical ratio where the *standard error of the difference between means* is used. In either case, the numerator is the difference between the sample result and the value expected under the null hypothesis.

There are few research situations where the population variance is known. When the population variance is not known, it is necessary to use an unbiased estimate of the population variance. *The unbiased estimate is obtained by dividing the sum of squares by its degrees of freedom* and is used to estimate the standard error of the mean.

The sampling distribution that results when the difference between the sample mean and the hypothesized population value is divided by the estimated standard error of the mean is known as a *t distribution* and the test statistic is called *t.* There is a different *t*

SPECIAL ISSUES IN RESEARCH

Why Not Run All Possible *t*-Tests?

In this example we have computed all possible *t*-tests. This is not a good practice in real research situations because we are running the risk of capitalizing on chance relationships. We should expect that one out of every 20 *t*-tests would be statistically significant at the .05 level by chance in random data. If we run two tests, both at the .05 level, the chance that **either** the first **or** the second will be significant is about .10, not .05. The more tests we run, the greater the chance of making a

Type I error. Therefore, only tests that examine predictions from our theory should be conducted. When there are many possible comparisons between means and we have no clear idea which to run, the primary alternative is to run a single test for the equality of all the means, a procedure known as analysis of variance which we cover in Chapter 16, and then analyze selected comparisons using a correction for the number of means.

Table 13-8 Statistics Needed to Compute Independent Samples *t*-Tests for Attitude Data in the Searchers Study

	N	$\bar{X}$	$\hat{S}_x$	$\hat{S}_{\bar{X}_1-\bar{X}_2}$	$\bar{X}_1 - \bar{X}_2$	t
Denomination						
Roman Catholic	30	39.83	7.47	2.15	−1.13	−.527
Protestant	30	40.97	9.12			
Member of Searchers						
Yes	30	41.50	8.56	2.14	2.20	1.029
No	30	39.30	8.00			

distribution for each possible number of degrees of freedom. As *df* gets large, the *t* distribution approaches the normal distribution.

A one-sample *t* is known as *Student's t*. When the study involves two independent samples, the two sample variances are pooled to obtain a better estimate of the population variance. The *pooled variance estimate* is used to compute an estimated standard error of the difference between means, and this is used to compute a *t* that has $df = N_1 + N_2 - 2$.

It is often possible to obtain a more sensitive test of our hypothesis by using *related samples*. This may be done by using naturally occurring pairs of individuals, by making observations of the same people under different conditions, or by matching individuals on status variables. The major benefit obtained from matching is a reduction in the standard error of the difference between means.

Effect size is the magnitude of relationship between the independent and dependent variables, or the size of the causal effect of one on the other. It is expressed as a proportion of variance or as a difference between means divided by the standard deviation of the sample and can be used in conjunction with power to determine the sample size needed to detect an effect of a given size. Confidence intervals can be used in place of formal null hypothesis tests or to supplement the information from the hypothesis testing procedure.

TESTS OF HYPOTHESES IN THE SEARCHERS DATA

The primary dependent variable in Dr. Cross's study of the Searchers is scores on the attitudes toward equality for women scale. There are two variables that we might use as independent variables: denomination and whether the man is a member of the Searchers. Both of these variables require that we use an independent samples *t*-test. The values you should get are shown in Table 13-8. As you can see, neither of these variables shows significant differences in attitude.

REFERENCES

Campbell, D. T., & Stanley, J. C. (1966). *Experimental and quasi-experimental designs for research*. Chicago: Rand McNally.

Cohen, J. (1994). The earth is round ($p < .05$). *American Psychologist, 49,* 997–1003.

Harlow, L. L., Mulaik, S. A., & Steiger, J. H. (1997). *What if there were no significance tests?* Hillsdale, NJ: Erlbaum.

Hays, W. L. (1994). *Statistics* (5th ed.). Fort Worth, TX: Harcourt Brace.

Keppel, G. (1991). *Design and analysis: A researcher's handbook* (3rd ed.). Upper Saddle River, NJ: Prentice Hall.

Kirk, R. E. (1995). *Experimental design: Procedures for the behavioral sciences* (3rd ed.). Pacific Grove, CA: Brooks/Cole.

Meehl, P. E. (1978). Theoretical risks and tabular asterisks: Sir Karl, Sir Ronald, and the slow progress of soft psychol-

ogy. *Journal of Consulting and Clinical Psychology, 46,* 806–834.

Mulaik, S. A., Raju, N. S., & Harshman, R. A. (1997). There is a time and place for significance testing. In Harlow et al., *What if there were no significance tests?* (pp. 65–116). Mahwah, NJ: Erlbaum.

Rozeboom, W. W. (1960). The fallacy of the null hypothesis significance test. *Psychological Bulletin, 57,* 416–428.

Schmidt, F. L. (1996). Statistical significance testing and cumulative knowledge in psychology: Implications for the training of researchers. *Psychological Methods, 1,* 115–129.

Thompson, B. (Ed.). (1993). Statistical significance testing in contemporary practice [Special issue]. *Journal of Experimental Education 61* (4).

EXERCISES

1. Explain the logic involved in using test statistics to test hypotheses.

2. Social, developmental, and educational psychologists were interested in studying the effect of social isolation

during childhood on the intellectual development of children. Psychologists studying this issue believe that social interaction (an environmental component of intelligence) is necessary for normal intellectual development and thus predicted that the IQ scores of children raised in social isolation would be lower than the population norm. They administered a standard IQ test with a population mean of 100 and a population standard deviation of 15 to 20 children who were randomly selected from among a group of children who were determined to have been raised in conditions that conformed to an operational definition of isolated.

a. Provide a symbolic representation of the null and alternate hypotheses.

b. Test the null hypothesis at an α of .05 for the IQ scores of the children presented below:

85, 90, 95, 110, 85, 90, 95, 75, 120, 85, 100, 105, 85, 90, 95, 80, 90, 100, 95, 85

c. What can you conclude from the statistical test in part (b)?

3. An educational psychologist is interested in creating a course that will increase students' creativity. The course will emphasize teaching skills that have been determined to be involved in creative thinking. The educational psychologist will use the Test of Creative Thinking, which has a published mean of 80 and a standard deviation of 10, to determine if the course has been successful in enhancing students' creativity. Twenty-five students are randomly selected to be a part of a pilot program to test this course before it is adopted into the curriculum. At the end of the pilot program, students are administered the Test of Creative Thinking.

a. Provide a symbolic representation of the null and alternate hypotheses.

b. Test the null hypothesis at an α of .01 for the creativity scores listed below:

78, 84, 88, 83, 89, 79, 80, 81, 86, 87, 82, 80, 76, 79, 80, 79, 83, 86, 82, 81, 84, 80, 83, 81, 79

c. What can you conclude about the creativity course based on part (b)?

4. A school psychologist is interested in determining if the current senior class differs significantly at an α of .05 from the norm for scores on the verbal test of the Scholastic Aptitude Test which has a mean of 500 and a standard deviation of 100. Students from this small school generally do not attend college in large numbers and thus there are only 15 students who had taken the test at the time of the study.

a. Provide a symbolic representation of the null and alternate hypotheses.

b. Test the null hypothesis at an α of .05 if the scores for each student are as follows:

620, 550, 480, 510, 490, 450, 750, 590, 610, 440, 570, 660, 520, 500, 490

c. Given the results of part (b), what can you conclude about the students at this school?

5. Psychologists are interested in determining whether males and females differ on measures of verbal ability. On the Verbal Ability Test, 14-year-old females have a population standard deviation of 90 whereas 14-year-old males have a population standard deviation of 110. The psychologists believe that verbal ability is driven primar-

ily by biological considerations such as dual hemispheric development and biochemical processes such that females will always have a verbal advantage in general. Thus, the psychologists randomly select 20 males and 20 females of age 14 and test their verbal ability.

a. Provide a symbolic representation of the null and alternate hypotheses.

b. Test the null hypothesis at an α of .05 for the scores on the Verbal Ability Test given below:

Females: 510, 530, 520, 540, 450, 600, 620, 480, 510, 530, 540, 550, 520, 560, 500, 540, 570, 530, 520, 540

Males: 490, 560, 510, 500, 580, 440, 510, 470, 520, 540, 510, 490, 480, 520, 540, 510, 500, 520, 510, 500

c. What can you conclude about verbal ability of the genders based on part (b)?

6. A measure of interdependent self-construal has population standard deviations of 8 for people raised in metropolitan areas and 9 for people raised in rural areas. Social psychologists believe that people who are raised in rural areas will have higher levels of interdependent self-construal because they must depend upon each other more for economic survival and because they tend to socialize more with a larger proportion of the community. Suppose we randomly select 15 people who were raised in a metropolitan community and 15 people who were raised in a rural community to test this hypothesis.

a. Provide a symbolic representation of the null and alternate hypotheses.

b. Test the null hypothesis at an α of .01 for the following:

Metropolitan: 45, 42, 36, 49, 32, 52, 28, 37, 43, 48, 54, 49, 39, 44, 47

Rural: 48, 52, 43, 45, 39, 35, 55, 49, 56, 60, 45, 46, 42, 39, 51

c. What can you conclude about interdependent self-construal and the environment in which a person is raised from part (b)?

7. Suppose that the spatial reasoning population standard deviations for males and females are 4.5 and 6.9, respectively. The research literature supports a hypothesis that males will score higher on spatial reasoning than females.

a. Provide a symbolic representation of the null and alternate hypotheses.

b. Test the null hypothesis at an α of .05 given the data below:

Males: 42, 36, 46, 52, 44, 40, 36, 45, 43, 43, 47, 40, 45, 40

Females: 43, 40, 39, 40, 40, 28, 47, 30, 32, 27, 43, 34, 50, 36

c. Given the results of part (b), what can you conclude about gender and spatial reasoning?

8. A cognitive psychologist believes that if algebra were taught using the principles of cognitive psychology as opposed to using modeling combined with extensive drill and practice, algebra students would be better at solving novel algebra problems. However, it is also possible that students who are taught using the principles of cognitive psychology are more confused because the approach is novel and thus would score worse on a novel

algebra problem-solving test. Suppose we know that the population standard deviations on the novel algebra problem-solving test are 9 and 12 for the cognitive approach and the modeling with drill and practice approach, respectively.

a. Provide a symbolic representation of the null and alternate hypotheses.

b. Test the null hypothesis at an α of .05 for the scores on the novel algebra problem-solving test given below:

Cognitive: 18, 22, 16, 20, 17, 23, 25, 10, 15, 23, 24, 21, 20, 19, 22, 23, 20, 22

Modeling: 15, 18, 19, 21, 11, 8, 21, 23, 18, 12, 15, 16, 19, 20, 21, 18, 17, 14

c. What can you conclude about the teaching approaches based on part (b)?

d. Compute the confidence intervals for these data.

9. You are hired by a government agency to investigate a company for truth in advertising. You will be called as a witness in a class action suit against a company that has been accused of making false claims about their product. In particular, this company has claimed that on average each of its chocolate chip cookies contains 16 chips. The people who have initiated the suit have claimed that the advertisements are false because the average number of chocolate chips in each cookie is less than 16. You decide to randomly sample 16 stores that carry this cookie. You randomly select a bag of cookies from the shelf and after paying the clerk for the cookies you randomly select a cookie from the bag. If you select a cookie that is less than a whole cookie, you continue to randomly select a cookie until you have selected a whole cookie. You then count the number of chocolate chips in this cookie and record the number. You continue this process until you have recorded the number of chocolate chips in one cookie from each of 16 different bags of cookies from 16 different supermarkets.

a. Provide a symbolic representation of the null and alternate hypotheses.

b. To be especially careful, you have decided to test the null hypothesis at an α of .01. Given the values of the sampling provided below, test the null hypothesis.

18, 14, 13, 17, 16, 15, 17, 18, 13, 12, 15, 16, 15, 16, 14, 15

c. What can you conclude about the average number of chips in a cookie from part (b)?

10. Recently people have become increasingly concerned about the effect of competitive attitudes on young, impressionable athletes. In particular, young female gymnasts who train in the prestigious gymnastics clubs in the United States have been the focus of concern with respect to eating disorders. Because of the pressure to maintain a thin body shape, people fear that these young female gymnasts are at a higher risk for eating disorders. An eating disorder inventory is administered to a group of 20 randomly selected young gymnasts who are enrolled in these prestigious clubs. The cutoff score for being at risk for an eating disorder has been determined to be 80. Thus, it is believed that these young female gymnasts will score higher than 80 on this instrument.

a. Provide a symbolic representation of the null and alternate hypotheses.

b. Test the null hypothesis at an α of .05 given the scores below:

85, 79, 78, 87, 88, 90, 75, 83, 85, 88, 88, 77, 76, 89, 88, 75, 76, 84, 85, 84

c. Given the results in part (b), what can you conclude about the risk factor for an eating disorder for young female gymnasts enrolled in prestigious clubs?

11. At the urging of the school counselor, a school has recently included a section on the effects of smoking cigarettes on the health of adolescents in a health education class. The counselor made this suggestion based on a study during the previous five years of the number of adolescents who smoked cigarettes. The counselor found the average number of cigarettes smoked per day by students who smoke is 14.3. The counselor is interested in determining whether the new material helped to reduce the number of cigarettes smoked per day by adolescents in his school compared to the previous years when no information about smoking was available in the curriculum. At the end of the new section, the counselor randomly selected 20 student smokers and asked them to record the number of cigarettes they smoked in the ensuing week. An average number of cigarettes smoked per day was then computed.

a. Provide a symbolic representation of the null and alternate hypotheses.

b. Test the null hypothesis at an α of .05 using the data below:

12, 15, 11, 10, 9, 15, 8, 11, 10, 13, 14, 11, 12, 16, 12, 11, 6, 19, 21, 14

c. What can you conclude about the new section of the health education course based on the results from part (b)?

12. A social psychologist is interested in testing whether children who were the only child in a family expressed a different need for solitude than the population value of 20. She did not have a solid theoretical reason for suspecting a directional difference. In fact, she was able to construct arguments that would support findings either greater than the norm or less than the norm. She randomly selected 15 children from a list of "only children" and administered a privacy scale.

a. Provide a symbolic representation of the null and alternate hypotheses.

b. Test the null hypothesis at an α of .05 for the scores below:

12, 16, 22, 25, 13, 15, 18, 19, 24, 20, 21, 18, 17, 19, 21

c. What can you conclude about "only children" from the results of part (b)?

13. A counseling psychologist was interested in determining whether the perception of a counselor's expertise was affected by the approach the counselor utilized. In particular, the counseling psychologist was interested in contrasting a rational-emotive approach with a client-centered approach using an instrument that asked participants to rate their perceptions of the counselor's expertise in working with a client utilizing a number of questions on a 7-point Likert-like rating scale. Responses to each item were added together to form an expert score.

A high score indicated a high level of perceived expertise. Twenty participants were randomly assigned to watch one of two videotaped counseling sessions. The client in the tapes presented the same problem for each counseling approach. Participants were asked to try to project themselves into the place of the client while viewing the tape. Upon completion of the tape, participants were asked to complete the counselor expertise rating scale. The researchers predicted that there would be a difference in the expertise ratings although it was not clear which approach would be better.

a. Provide a symbolic representation of the null and alternate hypotheses.

b. Given the following scores, test the null hypothesis at an α of .05:

   Rational-emotive: 56, 42, 38, 41, 42, 25, 66, 38, 55, 61

   Client-centered: 51, 45, 43, 40, 48, 22, 61, 51, 64, 56

c. What can you conclude about the counseling approaches given the results of the statistical test in part (b)?

d. Calculate the effect size using both r_{pb}^2 and d. What are the interpretations of each of these measures of effect size?

e. What is the confidence interval for this study? What would you conclude about the results of this study using the confidence interval?

14. A clinical psychologist is interested in determining if relaxation training through deep breathing can control panic attacks. Thirty participants are asked to first record the number of panic attacks experienced during a one-month period to establish a baseline. Participants are then randomly assigned to one of two conditions: control (where no treatment is initiated), and relaxation training using deep breathing (where participants are taught how to relax using deep breathing techniques). Upon the completion of relaxation training for the treatment group, all of the participants were asked to report the number of panic attacks they experienced in the subsequent month. The baseline number of panic attacks was then subtracted from the number of panic attacks. A negative difference score would then indicate a decrease in the number of panic attacks while a positive difference score would indicate an increase in the number of panic attacks. According to the theory, relaxation training should lower the levels of anxiety which are thought to serve as a trigger for a panic attack of the subject. Thus, participants who were taught to relax should have a larger decline in panic attacks than control participants. The difference scores for each subject are provided according to the treatment condition to which they are assigned.

a. Provide a symbolic representation of the null and alternate hypotheses.

b. Test the null hypothesis at an α of .05 for the scores listed below. Remember that a negative score means that there has been a decrease in the number of panic attacks from baseline levels.

   Control: 0, −1, 1, −1, 3, −2, 2, 0, 2, 0, 0, 2, −3, −2, 0

   Relaxation: −2, −1, 0, −1, 1, 2, −3, −2, −1, −2, −2, −3, −1, −2, −2

c. Given the results of part (b), what can you conclude about the effectiveness of relaxation training in controlling panic attacks?

d. Calculate the effect size using both r_{pb}^2 and d. What are the interpretations of each of these measures of effect size?

e. What is the confidence interval for this study? What would you conclude about the results of this study using the confidence interval?

15. At times a horrifying event or a tragedy immediately follows a joyous occasion. Clinical psychologists have noted that their clients have seemed to express more depression when this occurred than clients who did not experience a joyous event just prior to a horrifying event or tragedy. Many clinical psychologists believe that the different level of depression is related to the client's rating of horror associated with the event. A researcher decided to investigate this idea. He predicted that participants who viewed horrifying events preceded by humorous events would report higher levels of experienced horror than participants who viewed horrifying events only. Twenty participants were randomly assigned to one of two conditions: horror preceded by humor and horror only. Participants were presented videotapes of these two conditions and were asked to rate their feelings of horror the moment the videotape ended. Higher scores indicated greater feelings of horror.

a. Provide a symbolic representation of the null and alternate hypotheses.

b. Use the scores below to test the null hypothesis at an α of .05:

   Horror Preceded by Humor: 5, 6, 7, 6, 5, 3, 6, 7, 4, 6

   Horror only: 4, 3, 6, 4, 5, 2, 4, 5, 7, 4

c. Given the results of part (b), what can you conclude regarding feelings of horror?

d. Calculate the effect size using both r_{pb}^2 and d. What are the interpretations of each of these measures of effect size?

e. What is the confidence interval for this study? What would you conclude about the results of this study using the confidence interval?

16. Adolescents in the United States are increasingly isolated from significant elderly people in their lives. In addition, U.S. society has become increasingly oriented toward the young. These factors, in combination with other things, may account for the increasingly negative view of adolescents toward the elderly. Social psychologists have theorized that these attitudes can be changed if adolescents are involved in cooperative projects with the elderly. Thus, adolescents were randomly assigned to one of two groups: an intergenerational project group and a control group. Measures of the adolescents' attitudes toward the elderly were taken prior to the beginning of the intergenerational project and at the end of the intergenerational project. These scores were subtracted to yield a change in attitude toward the elderly score. Thus, a positive difference score means an improvement in the attitude toward the elderly.

a. Provide a symbolic representation of the null and alternate hypotheses.

b. Given the change in attitude toward the elderly scores below, test the hypothesis at an α of .05.

Control: $-1, 2, 3, -2, -1, 2, -3, 0, 2, 4, -2, -1, 1, -1, 0$

Intergenerational project: $4, 5, 3, 4, 5, 1, 0, -1, 2, -2, 4, 2, 2, 3, -1, -1, 6, 2$

c. Given the results of part (b), what can you conclude about the intergenerational project?.

d. Calculate the effect size using both r_{pb}^2 and d. What are the interpretations of each of these measures of effect size?

e. What is the confidence interval for this study? What would you conclude about the results of this study using the confidence interval?

17. What are the advantages of using a matched-pairs or repeated-measures design over an independent-samples design?

18. A cognitive psychologist is interested in enhancing the recall of related textbook materials. One way of enhancing recall of text materials is to include some type of signal to increase the salience of related information. According to information processing theories, information that is signaled in a chapter will increase recall of related information in subsequent chapters. For the present study, 10 participants were exposed to a two-chapter booklet of reading material. The information provided in each chapter was novel to the participants and contained 12 related topic paragraphs with eight sentences that expressed a single idea. Six of the paragraphs in Chapter 1 contained the signal, "As you will see in Chapter 2," while the remaining paragraphs contained no signals. The paragraphs in each chapter were arranged in random order. Participants read Chapter 1 one and only one time, completed an interpolated task unrelated to the prose material for 10 minutes, read Chapter 2 one and only one time, completed some demographic information for 5 minutes, and were given 10 minutes to recall as many of the ideas from the two chapters as they could remember. Two individuals scored each passage with an interrater reliability of .95. Only recall of Chapter 2 information was recorded for analysis with separate scoring for the unsignaled and signaled passages for each subject.

a. Provide a symbolic representation of the null and alternate hypotheses.

b. Test the null hypothesis at an α of .05 using the following scores:

Participant	Unsignaled	Signaled
1	8	9
2	5	10
3	5	8
4	6	8
5	5	6
6	7	9
7	6	11
8	5	9
9	4	12
10	7	13

c. What can you conclude from this study given the result of part (b)?

19. A cognitive psychologist was interested in whether participants could recall more text information if they paraphrase what they read or if they read the information

again. Since the research literature and theories provided no basis from which a directional prediction could be derived, the cognitive psychologist predicted only that there would be a difference. The psychologist constructed two reading passages that were determined to be equal in reading level and novelty of information. Half of the participants received Passage A first while the other half received Passage B. Upon reading their passage one and only one time and completing a 5-minute interpolated task, half of the participants for a particular passage were asked to reread the passage one and only one more time while the other participants were asked to paraphrase the content of the passage. Students were then asked to complete a second interpolated task for 10 minutes and then write down the ideas they could remember from the passage they read. The number of correct responses was recorded for each subject. On the second day the procedure was the same except participants received the alternate passage and were instructed to complete the alternate procedure after the first interpolated task.

a. Provide a symbolic representation of the null and alternate hypotheses.

b. Test the null hypothesis at an α of .01 for the recall scores below:

Participant	Reread	Paraphrase
1	11	15
2	12	10
3	9	13
4	14	14
5	12	14
6	14	15
7	10	14
8	11	15
9	8	9
10	11	14
11	16	17
12	22	26
13	18	15
14	15	11
15	10	11
16	8	8
17	12	14
18	15	17
19	6	8
20	11	11

c. What can you conclude from this study given the result of part (b)?

20. For some individuals, the contrast of black print on white paper is so great that it slows down their reading speed. Several researchers have suggested that masking the page with colored overlays will reduce the contrast and thus increase the reading speed of individuals with this contrast sensitivity problem. In particular, light gray overlays have been suggested as being beneficial in this regard. The researcher created two equivalent reading passages and printed half on gray paper and half on white paper. Half of the participants received the passage on gray paper first and half received the passage on white paper first. Participants were instructed that both reading speed and reading comprehension were equally important in this study. In addition, participants were instructed to read the passage only one time. When they completed the reading of each passage, participants' reading times were recorded in seconds.

a. Provide a symbolic representation of the null and alternate hypotheses.
b. Test the null hypothesis at an α of .05 given the data below.

Participant	White	Gray
1	220	190
2	210	170
3	210	160
4	200	185
5	250	220
6	300	375
7	210	200
8	235	240
9	245	225
10	275	225
11	300	275
12	240	220

c. What can you conclude from this study given the result of part (b)?

21. Suppose that we have two independent samples from which we have collected data and want to conduct a two-tailed test. Compute the sample size that would be needed given the following information:
   a. For $\alpha = .05$, $d = .80$, and $\sigma^2 = 8$, what size would the sample have to be to achieve a power of .60?
   b. For $\alpha = .01$, $d = 1.80$, and $\sigma^2 = 15$, what size would the sample have to be to achieve a power of .75?

22. The data in the data set named math.sav are a pilot study in which two approaches to teaching algebra (method; 1 = lecture, 2 = metacognitive method) were compared for their effect on the students' attitudes toward mathematics (postatt) and the students' achievement in mathematics (ITED9). To make certain that certain rival hypotheses were eliminated, preinstruction measures of attitude toward mathematics (preatt) and achievement in mathematics (ITED8) were obtained. Using your SPSS data bank, perform an independent samples t-test using a two-tailed test with an α of .05 to answer the following questions. Use the computer printout to support your answers.
   a. Did the students in the two groups differ on their attitude toward mathematics before instruction in algebra was provided?
   b. Did the students in the two groups differ in their level of mathematics achievement before instruction in algebra was provided?
   c. Did the students in the two groups differ on their attitude toward mathematics after instruction in algebra was provided?
   d. Did the students in the two groups differ in their level of mathematics achievement after instruction in algebra was provided?

23. The data in the data set named self.sav were used to test cultural differences (citizen; 1 = Japanese, 2 = American). Using your SPSS data bank, perform an independent samples t-test using a two-tailed test with an α of .05 to answer the following questions. Use the computer printout to support your answers.
   a. Do Japanese and Americans differ on independent self-construal (indsc)?
   b. Do Japanese and Americans differ on interdependent self-construal (intsc)?
   c. Do Japanese and Americans differ on individualistic self-esteem (indse)?
   d. Do Japanese and Americans differ on collectivistic self-esteem (cses)?
   e. Do Japanese and Americans differ on individually-oriented achievement motivation (ioam)?
   f. Do Japanese and Americans differ on socially-oriented achievement motivation (soam)?

24. The data from the data set named career.sav from your SPSS data bank contain information that is related to career decisions of middle school and high school students. In particular, students' perceptions of attitudes toward mathematics for those important people in their environment may be important components in selecting or deselecting particular careers. Because attitudes are often a function of the gender of the person, participants were coded by gender (1 = male, 2 = female) in the data set. Using your SPSS data bank, perform an independent samples t-test using a two-tailed test with an α of .05 to answer the following questions. Use the computer printout to support your answers.
   a. Do females and males differ in their confidence in doing mathematics (confmath)?
   b. Do females and males differ in their perception of mother's attitude toward mathematics (mother)?
   c. Do females and males differ in their perception of father's attitude toward mathematics (father)?
   d. Do females and males differ in their perception of mathematics as a male domain (maledom)?
   e. Do females and males differ in their perception of the usefulness of mathematics (useful)?
   f. Do females and males differ in their anxiety about doing mathematics (anxiety)?

Chapter 14

Inferences About Variances and Correlations

CHAPTER OBJECTIVES

After studying this chapter, you should be able to:

❏ Find the critical value of χ^2 that cuts off a specified proportion of the χ^2 distribution for a given *df*.

❏ Test the null hypothesis that a sample was drawn from a population with a specified variance.

❏ Determine a confidence interval for σ^2 given N and $\hat{S}^2$.

❏ Find the critical value of F that cuts off a specified proportion of the F distribution.

❏ Conduct a test of the hypothesis that two samples are from populations with equal variances.

❏ Transform a correlation into Z_F and back again.

❏ Conduct a test of the hypothesis that $\rho = 0$.

❏ Use Z_F to (1) test hypotheses about r for values other than 0, (2) compute confidence intervals for correlations, (3) test the equality of a correlation in two samples, and (4) average correlations.

❏ Use t to test whether two variables correlate equally with a third variable.

Mary Littlejohn is the director of testing services for a large urban school district. For the last several years she has been collecting data on the performance of the district's students on required standardized tests of mathematics achievement and reading comprehension for eleventh graders. She has noticed that the girls, on average, score slightly higher than the boys on the reading comprehension test, but that the top scores almost always go to boys. The tests are standardized to have means of 100 and standard deviations of 15. She would like to test whether the district shows the same standard deviation in reading comprehension as the national standardization group, and she would also like to test whether the boys show greater variability than the girls. In addition, she would like to know whether the correlation between test scores and grades is the same for boys as for girls. What null hypotheses are implied by her questions? How might she go about testing these hypotheses?

Introduction

Our discussion in Chapter 13 concerned testing hypotheses about the mean. Depending on the nature of the study and whether we can assume we know the population variance, we can use the one- or two-sample form of the Z- or t-test to draw conclusions about means. In this chapter, we turn our attention to testing hypotheses about two other statistics, variances and correlation coefficients. For example, Dr. Littlejohn would like to know whether her district shows greater variation than the national standardization sample. She also wants to test a hypothesis concerning the variability of her male students compared to her female students. To answer either of these questions, we must first discuss two additional sampling distributions. Then we will see how they can be applied in drawing inferences about variances. Finally, we turn our attention to how Dr. Littlejohn might draw an inference about the correlation between test scores and grades.

The Chi-Square Distribution

The first question Dr. Littlejohn asked was whether her sample showed greater variability than the standardization sample. To answer this question and understand how probability relates to the sampling distribution of the standard deviation, we must first develop a new probability distribution called the chi-square distribution. We can then use this distribution to answer some questions about standard deviations in the same way that we have used the normal and t distributions to answer questions about means.

The statistic we need, χ^2 (the Greek lowercase letter chi, squared), can be defined by the equation

$$\chi^2_{df} = \frac{(N-1)\hat{S}^2_X}{\sigma^2_X} \tag{14.1}$$

The term in the numerator is $(N-1)$ times the unbiased estimate of the population variance from sample data (equation 11.4). The denominator is the population variance. Since Dr. Littlejohn knows the population variance, we can use χ^2 to answer her question of whether her students show greater variability than the norm sample, if we know the shape of the sampling distribution of χ^2.

We know from our work with the mean in Chapter 12 that the shape and spread of sampling distributions depend on sample size. χ^2 is rather like t in this regard; not only the spread of the distribution, but also its shape and location, depend on the sample size. There will be a different χ^2 distribution centering on a different point for each possible sample size. As was the case with t, we can use the concept of degrees of freedom to indicate which of the many χ^2 distributions applies to a given problem.

There is another important feature to keep in mind concerning the χ^2 distribution; the numerator is a sum of squares $[(N-1)\hat{S}^2_X = SS]$. Since sums of squares cannot be negative, the χ^2 distribution has zero as its minimum possible value. Figure 14-1 illustrates the nonnegative property of several χ^2 distributions. Note that with small df's, the curves show marked positive skew. As sample size increases, the curves shift to the right and become more nearly symmetrical.

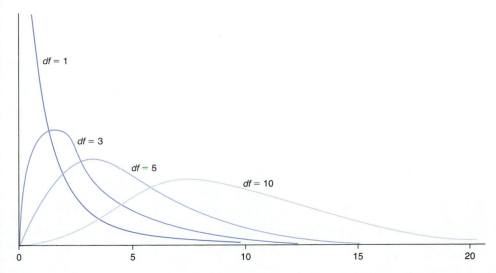

Figure 14-1 The distribution of χ^2 has extreme positive skew for small df. As the degrees of freedom increases, the curve moves to the right and becomes more nearly symmetric.

Many quantities that describe sample results have the same or nearly the same distribution as χ^2. We cover one of these, sample variances (and, therefore, standard deviations because the standard deviation is the square root of the variance), in this chapter, while others will be described in Chapter 15. The χ^2 family of distributions probably has more different applications in research and statistics than any other sampling distribution.

Critical values of the χ^2 distribution are included in Appendix Table A-3, a portion of which is reproduced here as Table 14-1. As in the case of the t distributions, there is a family of curves, so only a few selected values are given for each member of the family (each df value). The value at the head of each column is the proportion of the curve that falls *to the right* of the given value of χ^2. The first column, labeled df, indicates the degrees of freedom, or which member of the family of distributions the critical values in that row refer to. For example, the value .06, which falls in the first row ($df = 1$) under the heading .80, is the value of χ^2 with $df = 1$ for which 80 percent of the distribution falls to the right; another way of looking at this value is that it cuts off the bottom 20 percent of the curve. Likewise, the value 11.1 in the .05 column of row 5, is the value of χ^2 that cuts off the top 5 percent of the distribution when $df = 5$. We can also say that 90 percent of this $df = 5$ distribution is included between the values $\chi^2 = 1.1$ and $\chi^2 = 11.1$ (1.1 cuts off the bottom 5 percent or the top 95 percent, so there is 90 percent between the two values). Note that we are treating these values of χ^2 in exactly the same way that we treated the values in the t distribution. The difference is that t is symmetric around zero, while χ^2 is always positive and only approaches symmetry for large degrees of freedom.

The χ^2 table stops at $df = 30$ because the distribution is very nearly normal above this value. For $df > 30$ an approximate value for a critical ratio Z can be computed by using the expression

$$Z = \sqrt{2\chi^2} - \sqrt{2\,df - 1} \qquad\qquad (14.2)$$

Table 14-1 Percentile Values of the Chi-square Distribution

df	$\chi^2_{.990}$	$\chi^2_{.980}$	$\chi^2_{.975}$	$\chi^2_{.95}$	$\chi^2_{.90}$	$\chi^2_{.80}$	$\cdots$	$\chi^2_{.10}$	$\chi^2_{.05}$	$\chi^2_{.025}$
1	—	—	—	—	.02	.06	....	2.7	3.8	5.0
2	.02	.04	.05	.10	.21	.45	....	4.6	6.0	7.4
3	.11	.18	.22	.35	.58	1.00	....	6.3	7.8	9.4
4	.30	.43	.48	.71	1.1	1.6	....	7.8	9.5	11.1
5	.55	.75	.83	1.1	1.6	2.3	....	9.2	11.1	12.8
6	.87	1.1	1.2	1.6	2.2	3.1	....	10.6	12.6	14.4
7	1.2	1.6	1.7	2.2	2.8	3.8	....	12.0	14.1	16.0
8	1.7	2.0	2.2	2.7	3.5	4.6	....	13.4	15.5	17.5
9	2.1	2.5	2.7	3.3	4.2	5.4	....	14.7	16.9	19.0
10	2.6	3.1	3.2	3.9	4.9	6.2	....	16.0	18.3	20.5
11	3.1	3.6	3.8	4.6	5.6	7.0	....	17.3	19.7	21.9
12	3.6	4.2	4.4	5.2	6.3	7.8	....	18.5	21.0	23.3
13	4.1	4.8	5.0	5.9	7.0	8.6	....	19.8	22.4	24.7
14	4.7	5.4	5.6	6.6	7.8	9.5	....	21.1	23.7	26.1
15	5.2	6.0	6.3	7.3	8.5	10.3	....	22.3	25.0	27.5
16	5.8	6.6	6.9	8.0	9.3	11.2	....	23.5	26.3	28.8
17	6.4	7.3	7.6	8.7	10.1	12.0	....	24.8	27.6	30.2
18	7.0	7.9	8.2	9.4	10.9	12.9	....	26.0	28.9	31.5
19	7.6	8.6	8.9	10.1	11.7	13.7	....	27.2	30.1	32.9
20	8.3	9.2	9.6	10.9	12.4	14.6	....	28.4	31.4	34.2
21	8.9	9.9	10.3	11.6	13.2	15.4	....	29.6	32.7	35.5
22	9.5	10.6	11.0	12.3	14.0	16.3	....	30.8	33.9	36.8
23	10.2	11.3	11.7	13.1	14.8	17.2	....	32.0	35.2	38.1
24	10.9	12.0	12.4	13.8	15.7	18.1	....	33.2	36.4	39.4
25	11.5	12.7	13.1	14.6	16.5	18.9	....	34.4	37.7	40.6
26	12.2	13.4	13.8	15.4	17.3	19.8	....	35.6	38.9	41.9
27	12.9	14.1	14.6	16.2	18.1	20.7	....	36.7	40.1	43.2
28	13.6	14.8	15.3	16.9	18.9	21.6	....	37.9	41.3	44.5
29	14.3	15.6	16.0	17.7	19.8	22.5	....	39.1	42.6	45.7
30	15.0	16.3	16.8	18.5	20.6	23.4	....	40.3	43.8	47.0

Source: Adapted from R.A. Fisher and F. Yates, *Statistical Tables for Biological, Agricultural, and Medical Research,* published by Oliver & Boyd Ltd., Edinburgh, 1963. Reprinted by permission of Addison Wesley Longman Ltd. Values less than .01 are left blank.

The table of the normal distribution is then used to find the probability associated with Z. Because there is an equation for the χ^2 distribution like the equation for the normal distribution, computer programs often calculate exact probability of χ^2 for any value of df.

χ^2 and Variance

Chi-square is the test statistic we would need to answer Dr. Littlejohn's question about whether the students in her district show the same variability in test scores as the national standardization group. Let us see how χ^2 relates to the variance of a sample.

One-Sample Hypothesis Test

The chi-square distribution gives us an equivalent of the one-sample t-test that we can use to test hypotheses about variances such as Dr. Littlejohn's. Remember from Chapter 13 that for a single sample, $t_{df} = (\overline{X} - K)/\hat{S}_{\overline{X}}$, where we were testing the hypothesis that our sample came from a population with $\mu = K$. Assuming that we know the population variance, we can use χ^2 to test the hypothesis that our sample was drawn at random from that population. The hypothesis that Dr. Littlejohn would test has a very similar form. It would be

$$H_0: \sigma^2 = K$$

And the alternate hypothesis would be

$$H_1: \sigma^2 \neq K$$

In this case, we know $K = 225$ because the standard deviation in the national sample was defined to be 15.

Dr. Littlejohn has taken a sample of 25 students and she computed the value of $\hat{S}_X^2$ to be 133. Her null hypothesis is that this is a random sample from a population with a variance of 225. The appropriate number of *degrees of freedom* for this χ^2 test is 24 ($N - 1$). Selecting $\alpha = .05$ (2.5 percent of the distribution in each tail), we find the critical values for χ^2 with $df = 24$ to be 12.4 and 39.4. From the data, we calculate χ^2 to be

$$\chi^2 = \frac{24(133)}{225} = 14.2$$

Since the sample result falls between the critical values, we decide not to reject the null hypothesis and conclude that our sample *might be* a random sample from a population with a variance of 225. Remember, failure to reject the null hypothesis does not mean we can accept it. Note carefully that χ^2 uses variances, not standard deviations, but because the standard deviation is the square root of the variance, the conclusion we draw applies to both.

THE COMPUTER DOESN'T DO IT

One of the things we have mentioned several times is that the statistical methods used by psychological and educational researchers tend to emphasize the mean as the statistic of interest. Nowhere is this more obvious than in the way computer programs ignore inferences involving the variance and standard deviation. The only forms of the χ^2 statistic that SPSS implements are the ones related to nonparametric statistics such as those we discuss in Chapter 15. Therefore, if your research question involves equation 14.1 and inferences about variances, you will have to do the computations yourself. The same is true for most of the other procedures covered in this chapter. This is surprising because the SPSS Frequencies and Descriptives procedures provide standard errors and confidence intervals for the skewness and kurtosis indices described in Chapter 4, but not the variance. The reason for including skewness and kurtosis CI's is that they are used to test whether a distribution is normal in shape.

▶ *Now You Do It*

The Federal Government has set standards regarding the accuracy with which milk containers must be filled. Some parents in your school district claim their children report receiving containers with widely varying amounts of milk. The government standard allows a variance of .20 ounces. As the district statistician, you draw a sample of 10 milk containers, carefully measure the contents in each, and obtain a value for $\hat{S}_X^2$ of .39 ounces. Should you conclude that the company is violating federal packaging standards? What is the null hypothesis?

Since the question you seek to answer involves a claimed population value for a variance, you are going to use a χ^2. The null hypothesis is that the producer's packaging does not differ in variability from government standards, but is the test one- or two-tailed? Are you concerned if the producer is more consistent than the standards require? Probably not, so the appropriate test is one-tailed. Why does it matter? Greater power. As we have seen in several contexts, a one-tailed test will detect a false null hypothesis more readily than a two-tailed test at equal risk of a Type I error. With a sample of 10 containers and setting α at .05 the critical value for χ^2 is 16.9. Your χ^2 should be

$$\chi_9^2 = \frac{(9).39}{.2} = \frac{3.51}{.2} = 17.55$$

The value of χ^2 for your sample result exceeds the critical value. Therefore, you should reject the null hypothesis that your sample came from a population with a variance of .2. The manufacturer's packaging shows greater variability than you would expect in a random sample from a population that meets the federal guidelines.

Confidence Intervals for σ^2

We encounter one-sample hypotheses involving known parameters infrequently with variances, so this use of χ^2 to test hypotheses seldom occurs in the real world of research. However, the relationship between χ^2 and variance does allow us to use χ^2 and the value of the variance estimate from our sample to calculate a confidence interval for the unknown parameter, σ^2. The logic here is the same as it was using Z or t to establish confidence intervals for the mean. Just as we would expect 95 percent of the values of Z to fall between the upper and lower 2.5 percent critical values, we would also expect 95 percent of the χ^2's to fall between the upper and lower 2.5 percent critical values. Thus, we can set up the inequalities

$$\chi_{.975}^2 \leq \frac{(N-1)\hat{S}_X^2}{\sigma_X^2} \leq \chi_{.025}^2$$

That is, 95 percent of observed χ^2's should fall at or above the lower critical value and at or below the upper critical value. By solving these inequalities for σ_X^2, we obtain

$$\frac{(N-1)\hat{S}_X^2}{\chi_{.025}^2} \leq \sigma_X^2 \leq \frac{(N-1)\hat{S}_X^2}{\chi_{.975}^2} \qquad (14.3)$$

Phrased in the format of limits for the confidence interval, we can say that our lower and upper CI limits for σ^2 are

$$CI_L = \frac{(N-1)\hat{S}_X^2}{\chi_{.025}^2}$$

and

$$CI_U = \frac{(N-1)\hat{S}_X^2}{\chi_{.975}^2} \qquad (14.4)$$

Of course, other confidence intervals can be computed by simply substituting other critical values for χ^2. For Dr. Littlejohn's data ($\hat{S}_X^2 = 133$), the 95% confidence interval for σ_X^2 is

$$\frac{24(133)}{39.4} \leq \sigma_X^2 \leq \frac{24(133)}{12.4}$$

$$81.0 \leq \sigma_X^2 \leq 257.4$$

In other words, this confidence interval permits us to be confident (with a 5 percent chance of error) that Dr. Littlejohn's sample was drawn from a population with a variance somewhere between 81.0 and 257.4. Note that, just as we saw in the last chapter, the information provided by the confidence interval includes the results of the hypothesis test. That is, if the confidence interval includes the value specified in the null hypothesis (225), we would decide not to reject H_0. To express the confidence interval in the metric of the standard deviation, we can simply take the square roots of the values we obtained for the variance; the confidence interval for the population standard deviation runs from 9.0 to 16.0.

▶ *Now You Do It*

Let's complete our analysis of the great milk scandal from the previous Now You Do It. The sample estimate of the population variance was .39. What are the 90% and 98% confidence intervals? (We use the 90% CI because we tested our hypothesis at the $\alpha =$ .05 level, one-tailed.) Looking in the χ^2 table (Appendix Table A-3) we find the following critical values in the $df = 9$ row:

$$.99 = 2.1 \qquad .95 = 3.3 \qquad .05 = 16.9 \qquad .01 = 21.7$$

Therefore, the 90% CI is

$$\frac{(9).39}{16.9} \le \sigma_X^2 \le \frac{(9).39}{3.3}$$

$$.21 \le \sigma_X^2 \le 1.06$$

and the 98% CI is

$$\frac{(9).39}{21.7} \le \sigma_X^2 \le \frac{(9).39}{2.1}$$

$$.16 \le \sigma_X^2 \le 1.68$$

That is, we may be 90% certain that our sample came from a population with a variance between .21 and 1.06, and 98% certain that the sample came from a population with a variance between .16 and 1.68. Note that the first CI does not include the null hypothesis value (.20) and we rejected H_0 at the $\alpha = .05$ level, one-tailed. The second CI does include .20, and the χ^2 from our sample did not exceed the .01 χ^2 critical value. If we had chosen the more restrictive α-level, we would have concluded that the company's product fell within federal guidelines.

The confidence intervals we have computed illustrate the point made in Chapter 4 that the sampling distribution of the variance is not symmetric. If it were, the CI's would be symmetric around the sample value, but we can see that the CI extends much farther above $\hat{S}_X^2$ than it does below it. As the degrees of freedom increase, you will find this tendency to decrease; the CI's will be more symmetric because the sampling distribution is more symmetric.

The *F* Distribution

We are now ready to answer questions such as Dr. Littlejohn's second question, do the scores of boys and girls on the standardized achievement test show equal variability? To do this, we will need to develop another very useful distribution. Suppose we have two independent estimates of the same population variance, perhaps from two independent samples such as boys and girls; this gives rise to two χ^2's,

$$\chi_1^2 = \frac{(N_1 - 1)\hat{S}_1^2}{\sigma_X^2} \qquad \chi_2^2 = \frac{(N_2 - 1)\hat{S}_2^2}{\sigma_X^2}$$

Since these χ^2's can be based on different degrees of freedom (and therefore refer to different χ^2 distributions), we must reduce them to the same scale by dividing each by its degrees of freedom (the appropriate $N - 1$). Doing this produces

$$\frac{\chi_1^2}{df_1} = \frac{\hat{S}_1^2}{\sigma_X^2} \qquad \frac{\chi_2^2}{df_2} = \frac{\hat{S}_2^2}{\sigma_X^2}$$

We can now define a function, called F (after the great English statistician R. A. Fisher), as

$$F = \frac{\dfrac{\chi_1^2}{df_1}}{\dfrac{\chi_2^2}{df_2}} = \frac{\dfrac{\hat{S}_1^2}{\sigma_X^2}}{\dfrac{\hat{S}_2^2}{\sigma_X^2}} \qquad\qquad (14.5)$$

Under the assumption that both samples have been drawn from the same population (that is, the same σ_X^2 applies to both samples), we can simplify the fraction in the right-hand term by inverting the denominator and multiplying

$$F = \frac{\hat{S}_1^2}{\sigma_X^2} \cdot \frac{\sigma_X^2}{\hat{S}_2^2} = \frac{\hat{S}_1^2}{\hat{S}_2^2} \qquad\qquad (14.6)$$

The **F distribution** is the sampling distribution of the ratio of two unbiased estimates of a population variance.

It should be clear from the fact that we have degrees of freedom terms associated with each of the χ^2's in equation 14.5 that the sampling distribution of F is affected by two degrees-of-freedom terms. The numerator variance estimate is based on df_1 and the denominator variance estimate is based on df_2. Because these two df's are independent (neither sample size depends on the other), there is a unique sampling distribution of F for every possible *combination of numerator and denominator degrees of freedom*. Also, F, which is the ratio of two positive numbers (variances must be positive), must itself be positive.

There is a unique sampling distribution of F for every possible combination of numerator and denominator degrees of freedom.

Critical values for many of the sampling distributions of F can be found in Appendix Table A-4, part of which is reproduced here as Table 14-2. Because there are so many distributions (900 possibilities for the combinations of df of 30 or less), only critical values for the conventional levels of α (.05, .01, and .001) are included in the table. Also, the F ratio is most frequently used in situations where the value of the numerator df is small, so only selected df values are included.

The F table is entered using the two degrees-of-freedom values. The numerator degrees of freedom (df_1) is used to find the proper column and the denominator degrees of freedom (df_2) to find the proper row of the table. The three values at the intersection of the df_1 column with the df_2 row are the .05, .01, and .001 *upper critical values* for the F distribution that has df_1 and df_2 as its degrees of freedom. For example, with df_1 of 10 and df_2 of 15, the critical values are 2.54, 3.80, and 6.08, respectively.

SPECIAL ISSUES IN RESEARCH

Why We Only Use the Upper Tail of the F Distribution

Although each of the F distributions is a complete sampling distribution with extreme values at both the upper and lower ends, interest centers on the upper, or right side, of the distribution. This is possible because the decision as to which variance estimate to call $\hat{S}_1^2$ in equation 14.6 is arbitrary. For example, if males and females actually have equal population variances, half of the time the variance estimate for females should be greater than that for males, and half the time the reverse would be true. Therefore, if we always put the variance estimate for females in the numerator when the null hypothesis is true, half of the time we would expect to get a value for F less than 1, and half of the time we would get a value greater than 1.

To save space, it has become conventional to place the larger of the two sample variance estimates in the numerator, which has the effect of allowing us to deal with only the upper half of each F distribution. This practice also has the effect of *making all hypotheses tested with F nondirectional hypotheses*

unless specific and careful logical steps are taken in constructing the hypotheses prior to the analysis. Procedures for one-tailed hypotheses and confidence intervals can be found in advanced texts such as Hays (1994, pp 361-362).

The values in the F table are for two-tailed hypotheses, which means that the .05 critical value 2.54 that we looked up earlier actually cuts off the upper 2.5 percent of the area of the F distribution for 10 and 15 degrees of freedom, but since we arbitrarily place the larger variance estimate in the numerator, we must double the probability. This is so because by arbitrarily placing the larger variance estimate in the numerator of the F-ratio we are considering only the top half of the distribution (a value less than 1.0 is not possible), and 2.5 percent of the whole distribution is 5 percent of the top half of the same distribution. This doubling is taken care of in the labeling of Appendix Table A-4.

Table 14-2 Selected Critical Values of the F Distributions

df_2	df_1	1	2	3	4	5	6	7	8	9	10	12	15	20	30	∞
1	.05	161.4	199.5	215.7	224.6	230.2	234.0	236.8	238.9	240.5	241.9	243.9	245.9	248.0	250.1	254.3
	.01[1]	40.52	50.00	54.03	56.25	57.64	58.59	59.28	59.81	60.22	60.56	61.06	61.57	62.09	62.61	63.66
	.001[2]	40.53	50.00	54.04	56.25	57.64	58.59	59.29	59.81	60.23	60.56	61.07	61.58	62.09	62.61	63.66
2	.05	18.51	19.00	19.16	19.25	19.30	19.33	19.35	19.37	19.38	19.40	19.41	19.43	19.45	19.46	19.50
	.01	98.50	99.00	99.17	99.25	99.30	99.33	99.36	99.37	99.39	99.40	99.42	99.43	99.45	99.47	99.50
	.001	998.50	999.00	999.17	999.25	999.30	999.33	999.36	999.37	999.39	999.40	999.42	999.43	999.45	999.47	999.50
3	.05	10.13	9.55	9.28	9.12	9.01	8.94	8.89	8.85	8.81	8.79	8.74	8.70	8.66	8.62	8.53
	.01	34.12	30.82	29.46	28.71	28.24	27.91	27.67	27.49	27.34	27.23	27.05	26.87	26.69	26.50	26.12
	.001	167.03	148.50	141.11	137.10	134.58	132.85	131.58	130.62	129.86	129.25	128.32	127.37	126.42	125.45	123.47
⋮																
14	.05	4.60	3.74	3.34	3.11	2.96	2.85	2.76	2.70	2.65	2.60	2.53	2.46	2.39	2.31	2.13
	.01	8.86	6.51	5.56	5.04	4.70	4.46	4.28	4.14	4.03	3.94	3.80	3.66	3.51	3.35	3.00
	.001	17.14	11.78	9.73	8.62	7.92	7.44	7.08	6.80	6.58	6.40	6.13	5.85	5.56	5.25	4.60
15	.05	4.54	3.68	3.29	3.06	2.90	2.79	2.71	2.64	2.59	2.54	2.48	2.40	2.33	2.25	2.07
	.01	8.68	6.36	5.42	4.89	4.56	4.32	4.14	4.00	3.89	3.80	3.67	3.52	3.37	3.21	2.87
	.001	16.59	11.34	9.34	8.25	7.57	7.09	6.74	6.47	6.26	6.08	5.81	5.54	5.25	4.95	4.31
16	.05	4.49	3.63	3.24	3.01	2.85	2.74	2.66	2.59	2.54	2.49	2.42	2.35	2.28	2.19	2.01
	.01	8.53	6.23	5.29	4.77	4.44	4.20	4.03	3.89	3.78	3.69	3.55	3.41	3.26	3.10	2.75
	.001	16.12	10.97	9.01	7.94	7.27	6.80	6.46	6.20	5.98	5.81	5.55	5.27	4.99	4.70	4.06
⋮																
25	.05	4.24	3.39	2.99	2.76	2.60	2.49	2.40	2.34	2.28	2.24	2.16	2.09	2.01	1.92	1.71
	.01	7.77	5.57	4.68	4.18	3.86	3.63	3.46	3.32	3.22	3.13	2.99	2.85	2.70	2.54	2.17
	.001	13.88	9.22	7.45	6.49	5.89	5.46	5.15	4.91	4.71	4.56	4.31	4.06	3.79	3.52	2.89
26	.05	4.23	3.37	2.98	2.74	2.59	2.47	2.39	2.32	2.27	2.22	2.15	2.07	1.99	1.90	1.69
	.01	7.72	5.53	4.64	4.14	3.82	3.59	3.42	3.29	3.18	3.09	2.96	2.82	2.66	2.50	2.13
	.001	13.74	9.12	7.36	6.41	5.80	5.38	5.07	4.83	4.64	4.48	4.24	3.99	3.72	3.44	2.82
27	.05	4.21	3.35	2.96	2.73	2.57	2.46	2.38	2.31	2.25	2.20	2.13	2.06	1.97	1.88	1.67
	.01	7.68	5.49	4.60	4.11	3.78	3.56	3.39	3.26	3.15	3.06	2.93	2.78	2.63	2.47	2.10
	.001	13.61	9.02	7.27	6.33	5.73	5.31	5.00	4.76	4.57	4.41	4.17	3.92	3.66	3.38	2.75
28	.05	4.20	3.34	2.95	2.71	2.56	2.45	2.36	2.29	2.24	2.19	2.12	2.04	1.96	1.87	1.65
	.01	7.64	5.45	4.57	4.07	3.75	3.53	3.36	3.23	3.12	3.03	2.90	2.75	2.60	2.44	2.06
	.001	13.50	8.93	7.19	6.25	5.66	5.24	4.93	4.69	4.50	4.35	4.11	3.86	3.60	3.32	2.69
29	.05	4.18	3.33	2.93	2.70	2.55	2.43	2.35	2.28	2.22	2.18	2.10	2.03	1.94	1.85	1.64
	.01	7.60	5.42	4.54	4.04	3.73	3.50	3.33	3.20	3.09	3.00	2.87	2.73	2.57	2.41	2.03
	.001	13.39	8.85	7.12	6.19	5.59	5.18	4.87	4.64	4.45	4.29	4.05	3.80	3.54	3.27	2.64
30	.05	4.13	3.32	2.92	2.69	2.54	2.42	2.33	2.27	2.21	2.16	2.09	2.01	1.93	1.84	1.62
	.01	7.56	5.39	4.51	4.02	3.70	3.47	3.30	3.17	3.07	2.98	2.84	2.70	2.55	2.39	2.01
	.001	13.29	8.77	7.05	6.12	5.53	5.12	4.82	4.58	4.39	4.24	4.00	3.75	3.49	3.22	2.59
40	.05	4.08	3.23	2.84	2.61	2.45	2.34	2.25	2.18	2.12	2.08	2.00	1.92	1.84	1.74	1.51
	.01	7.31	5.18	4.31	3.83	3.51	3.29	3.12	2.99	2.89	2.80	2.66	2.52	2.37	2.20	1.80
	.001	12.61	8.25	6.59	5.70	5.13	4.73	4.44	4.21	4.02	3.87	3.64	3.40	3.14	2.87	2.23

[1]Entries in this row must be multiplied by 100.
[2]Entries in this row must be multiplied by 10,000.

Source: R.A. Fisher and F. Yates, Statistical Tables for Biological, Agricultural, and Medical Research, published by Oliver & Boyd Ltd., Edinburgh, 1963. Permission of Addison Wesley Longman Ltd.

F and Variance

We are now ready to test Dr. Littlejohn's second hypothesis. Chi-square provided a test of hypotheses about a single sample variance relative to an assumed σ^2, and F gives us a way to deal with two samples. That is, the F statistic gives us a way to test hypotheses about variances that is equivalent to the two-sample t-test for means. Suppose Dr. Littlejohn has drawn a sample of 16 male students and 16 female students from the school records. If we assume that her sample of males and her sample of females come either from the same population or from two populations with equal variances, equation 14.6 tells us that the ratio of the sample estimates of those variances should come from an F distribution with $df_1 = (N_1 - 1)$ and $df_2 = (N_2 - 1)$. Thus, the condition that we assume to be true, our null hypothesis for Dr. Littlejohn, can be stated as

$$H_0: \sigma_M^2 = \sigma_F^2$$

The more general form of this null hypothesis is

$$H_0: \sigma_1^2 = \sigma_2^2$$

If this condition is true, the ratio $\hat{S}_1^2 / \hat{S}_2^2$ will have a sampling distribution that is the same as the F distribution with df_1 and df_2.

Dr. Littlejohn computed the variance estimates in her two samples and found the variance estimate for the female students' scores to be 99.7. The male students produced a variance estimate of 166.3. She wishes to use an α of .05. To test her null hypothesis, we determine the critical value of F at the $\alpha = .05$ level of protection from Type I error by reference to Appendix Table A-4 with $df_1 = 15$ and $df_2 = 15$. The .05 critical value is 2.40. Then we compute the ratio of the two sample variance estimates, *placing the larger variance estimate in the numerator and being careful that we have the df terms associated with their correct variance estimates*. This produces

$$F = \frac{166.3}{99.7} = 1.67$$

When using F, place the larger variance estimate in the numerator and be careful that the df terms are associated with the correct variance estimates.

If the value of F computed from the sample data had exceeded the critical value from the table, we would decide to reject the null hypothesis and conclude either that the two samples did not come from the same population (first assumption) or that the two populations from which these samples were drawn do not have equal variances (second assumption). In Dr. Littlejohn's case, however, the evidence does not contradict the null hypothesis, so we would not reject it.

As a second example of using F, suppose we wish to test a hypothesis that the variances in scores on the statistics final exam are the same in the spring and fall semesters, which amounts to a null hypothesis that $\sigma_S^2 = \sigma_F^2$ (or that the students in the two semesters are random samples from a single population). We collect a sample of spring scores from 16 students and a sample of fall scores from 23 students. The two classes give us $\hat{S}_S^2 = 71.7$ and $\hat{S}_F^2 = 30.4$. Since we must put the larger variance estimate ($\hat{S}_S^2$) in the numerator, $df_1 = 15$ and $df_2 = 22$ (as with t and χ^2, we go to the next *lower df* when our exact value is not in the table). Assume that we set α at .01. Looking in Appendix Table A-4, we find the .01 critical value for 15 and 22 degrees of freedom to be 2.98. We compute the value of F from our study and find

$$F = \frac{71.7}{30.4} = 2.36$$

This F does not exceed the critical value, so we conclude that we cannot reject the null hypothesis. (Had we used an α of .05 and a critical value of 2.15, our decision would have been different.) As with all tests of hypotheses, we cannot conclude that σ_S^2 and σ_F^2 are equal, only that we cannot reject the possibility that they are. (Note that if we had gotten our degrees of freedom reversed and looked up the .01 critical value for F with 22 and 15 df [the closest value] we would have found 3.37. This error would in some cases lead us to draw the wrong conclusion, thus making a Type IV error. For a given pair of df values the critical values for F are different depending on which one is in the numerator.)

The F statistic is widely used in psychological and educational research. We have seen in this section that it allows us to test a hypothesis about the equality of two variances.

This hypothesis is important both for cases where the equality of variances is of interest in its own right and when we wish to check whether the assumption of equal variances is met for a t-test. In Chapter 16 we will use the F statistic again, there to test hypotheses about means using the procedures of analysis of variance.

▶ *Now You Do It*

The faculty of the SAD Department at BTU want to test the hypothesis that men and women do not differ in variability in their math anxiety scores. Can you help them?

Obviously, this is a null hypothesis that the two variances are equal ($\sigma_M^2 = \sigma_F^2$), calling for an F-test. The two sample variance estimates are $\hat{S}_M^2 = 26.69$ based on 39 cases and $\hat{S}_F^2 = 36.15$ with $N = 29$. Remembering that we must put the larger variance estimate in the numerator and that the df terms must be kept with their variance estimates, we will have 28 numerator df and 38 denominator df. Going to the table of the F distribution (Appendix Table A-4) for 20 and 30 degrees of freedom, we find the three entries to be 1.93 (.05), 2.55 (.01), and 3.49 (.001). The ratio 36.15/26.69 yields an F from the data of 1.35, which does not exceed any critical value, so we conclude that we cannot reject the null hypothesis. Again, this does not mean the variances are equal, just that we cannot say they are unequal.

Inferences About Correlation Coefficients

Like all statistics, the correlation coefficient is subject to variation from sample to sample. For a given population, the sample correlation coefficient (r) has a sampling distribution around the population parameter ρ (Greek lowercase letter rho), and this distribution has a standard error, the standard error of the correlation coefficient, σ_r. However, the sampling distribution of r is somewhat more complex than the other sampling distributions we have encountered. We will first consider the situation where ρ is assumed to be zero. Then, we will discuss a transformation of the correlation that allows us to test other hypotheses and to obtain other useful results.

Testing the Hypothesis $\rho = 0$

There are several general situations that we encounter in testing hypotheses concerning correlations; most frequently, we are faced with a test of the null hypothesis that $\rho = 0$. For example, the SAD faculty at Big Time U might want to test whether the correlation between math anxiety and scores on the statistics final can be considered a random deviation from zero. In this case, their null and alternate hypotheses are

$$H_0: \rho = 0$$
$$H_1: \rho \neq 0$$

With this null hypothesis, the sampling distribution of r *under the null hypothesis* will be symmetrical. For large samples (N at least 100), an appropriate standard error for the correlation coefficient is

$$\sigma_r = \frac{1}{\sqrt{N}} \qquad (14.7)$$

which can then be used in the critical ratio formula

$$Z = \frac{r - \rho}{\sigma_r} \qquad (14.8)$$

where the value of ρ is 0 and is included only to remind you of the general structure of Z-tests. This is a *normal deviate*, or *critical ratio Z*, of exactly the same kind we encountered in Chapter 13 and has the same three parts: the sample statistic, the assumed population value under the null hypothesis, and the standard deviation of the sampling distribution. If the SAD faculty had collected data on all 340 statistics students in their class, rather than the 68 who volunteered for their study, this would be the appropriate test. For a sample this large, the standard error of the correlation would be

$$\sigma_r = \frac{1}{\sqrt{340}} = .054$$

The two-tailed .05 critical value for Z is ± 1.96. Multiplying these critical values by σ_r gives $\pm .106$. Therefore, if they had data on the whole class and the correlation exceeded .106 in either direction, the SAD faculty would be reasonably safe in concluding that the correlation in the population of statistics students is not zero.

In reality, the SAD faculty have data from only 68 students, making it inappropriate to use equation 14.8. Testing the hypothesis that $\rho = 0$ using smaller samples ($N < 100$) requires a test involving the t distribution. The test statistic

$$t_{N-2} = \frac{r - \rho}{\sqrt{(1 - r^2)/(N - 2)}} \qquad (14.9)$$

follows the t distribution with $(N - 2)$ degrees of freedom. That is, the quantity

$$\frac{r - \rho}{\sqrt{(1 - r^2)/(N - 2)}}$$

has a sampling distribution with the same shape as the t distribution with $(N - 2)$ degrees of freedom.

The null hypothesis tested by either equation 14.8 or 14.9 is that $\rho = 0$ (or possibly some other value near zero, but this is quite unlikely). If the value of the test statistic exceeds the critical value at the chosen α-level (.05 or .01), we reject the null at that level of confidence. Of course, a theory or prior experience may call for a one-tailed null hypothesis, in which case the critical value and rejection region are determined accordingly.

The correlation between math anxiety and final exam scores for the 68 students participating in the SAD statistics study was $-.55$. Inserting the appropriate values into equation 14.9 we get

$$t_{(66)} = \frac{-.55 - 0}{\sqrt{\frac{(1 - .55^2)}{(68 - 2)}}} = \frac{-.55}{.103} = -5.34$$

As a standard part of their output, most programs that compute correlations provide a test of the null hypothesis that $\rho = 0$ using equation 14.9. SPSS allows you to specify whether the test is to be one- or two-tailed. As usual, you select two (or more) variables. Then you can choose optional output that includes sums of squares and sums of cross products and covariances. The missing data option allows you to include in the analysis only individuals who have scores on all variables (listwise exclusion) or to include people whenever they have scores on a given pair of variables in the analysis (pairwise exclusion). The screen looks like this:

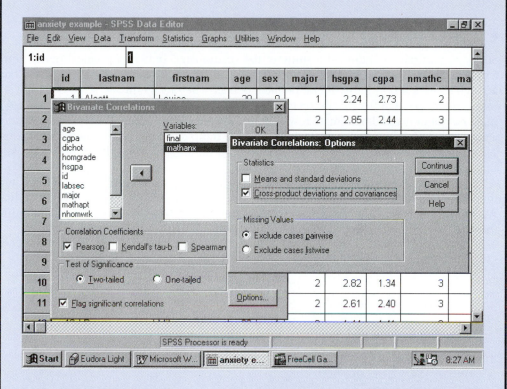

Output includes the correlations, the number of cases on which each correlation was computed, and the probability of a Type I error, but not, unfortunately, the value of the test statistic. The "Flag significant correlations" box causes the program to put asterisks on statistically significant correlations.

The .001 critical value of t with $df = 60$ is 3.46, so we can confidently reject the null hypothesis that $\rho = 0$. The data suggest that the correlation is moderately large and negative. Note that the sign of the t-test is the same as the sign of the correlation coefficient. If we had elected to run a one-tailed test to give ourselves greater power, we would have to pay attention to the sign because the rejection region would then be in only one tail of the distribution, and if the sign of the t-test had placed us in the wrong tail, we could not reject H_0, regardless of how large t was.

▶ *Now You Do It*

One question that might be of considerable interest to the faculty of the SAD Department at BTU is whether there is reason to believe that the number of homework assignments completed is related to course performance. What do you think?

Once you have computed the correlation coefficient and found it to be $+.447$ you start to think that doing your homework just might lead to better grades, but is the correlation significant? Only a t-test can tell us. Plugging the known values into equation

14.9 yields

$$t_{66} = \frac{.447}{\sqrt{(1 - .447^2)/(68 - 2)}} = \frac{.447}{.110} = 4.06$$

A t of 3.46 with $df = 60$ is statistically significant at the $P < .001$ level, so we can tell the SAD faculty that there is a positive correlation between the number of homework assignments completed and score on the final. If you ran this test on the computer, your correlation came out with two asterisks on it, indicating significance at the .01 level.

> ### ◆ Now You Do It

Suppose you have conducted a study of the relationship between political beliefs and attitudes toward racial equality. The political beliefs scale is a measure of liberalism; that is, high scorers report more liberal political beliefs. The racial equality scale is scored so that higher scores reflect more egalitarian attitudes. You have obtained data from 45 male respondents and 65 female respondents. For the entire set of data you have found a correlation of +.30, and for the male and female samples you have obtained correlations of +.27 and +.40 respectively. How should you go about testing the null hypotheses that all correlations are zero?

In testing the overall correlation, it is appropriate to use the Z-test because N is greater than 100. For each of the separate correlations, you need to use the t-test because the N's are less than 100. The degrees of freedom are 40 and 60 respectively. When you run the tests, you should obtain the following results:

$$Z = \frac{.30 - 0}{1/\sqrt{110}} = \frac{.30}{.095} = 3.15$$

$$t_{40} = \frac{.27 - 0}{\sqrt{\dfrac{1 - .27^2}{43}}} = \frac{.27}{.147} = 1.84$$

$$t_{60} = \frac{.40}{\sqrt{\dfrac{1 - .40^2}{63}}} = \frac{.40}{.115} = 3.46$$

Now, here's a question for you to think about. The results for the overall group and for the female respondents are statistically significant, but the correlation for the men is not significantly different from zero. How can you reconcile these findings? What interpretations are appropriate?

Testing Hypotheses Where $\rho \neq 0$

The most difficult problem that we face when dealing with the correlation coefficient is that the sampling distribution has a different shape, depending on the value of ρ. Remember that r is limited to the range of values between +1.0 and −1.0. If the value of ρ is about zero (that is, in the middle of this range), sample deviations can occur equally in either direction. However, as ρ departs from zero, this symmetry is lost because one end of the sampling distribution is more restricted than the other due to the limiting values of ±1.0. There is more space on one side of the parameter than on the other. For extreme values of ρ, the sampling distribution of r becomes markedly skewed due to the restriction that all values must fall between ±1.0. The basic problem is illustrated in Figure 14-2.

Fisher's Z Transformation. A solution to this problem was developed by R. A. Fisher, the person for whom the F statistic is named. Fisher proposed a transformation for r that would normalize the distribution. The resulting index is called Fisher's Z or Z_F. It is unfortunate that the same symbol has been applied to both this transformation and the critical ratio statistic, but the usage is so widespread and firmly entrenched that we must

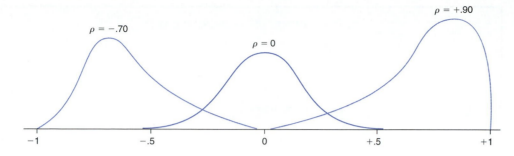

Figure 14-2 The sampling distribution of the correlation coefficient is symmetric when $\rho = 0$. As ρ departs from zero in either direction, the curve becomes increasingly skewed due to the limits of ± 1.0 for r.

learn to live with it. The statistic Fisher developed,

$$Z_F = (1.1513) \log_{10} \left(\frac{1 + r}{1 - r} \right) \tag{14.10}$$

has the advantage that its sampling distribution is almost exactly normal for any value of ρ and has a standard error of

$$\sigma_{Z_F} = \frac{1}{\sqrt{N - 3}} \tag{14.11}$$

It is not necessary to use equation 14.10 to compute Z_F. Because there are a limited number of values that r can have (to two decimal places) and the distribution is symmetric around zero, it is possible to prepare tables of the necessary values. Two tables are provided in Appendix A for the Fisher's Z transformation; one, Table A-5, gives the value of Z_F for any value of r to two decimal places and is used to transform r to Z_F. The other, Table A-6, is used to reverse the transformation after any necessary computations have been completed. That is, you enter Table A-5 with r and extract Z_F; you enter Table A-6 with Z_F and extract r. Both tables are entered with the first digit as the row and the second digit as the column. The desired value is then read at the intersection. For example, to find the value of Z_F for a correlation of .63 we would enter Table A-5 at the row labeled .6 and go across to the column headed .03. The value at the intersection, .741, is the value of Z_F for a correlation of .63. Z_F retains the sign of r. That is, if r is negative, Z_F is negative.

▶ ***Now You Do It***

What are the values of Z_F associated with the following correlations?

$$r = .45 \qquad r = -.67 \qquad r = .98$$

What are the correlations associated with the following values of Z_F?

$$Z_F = -1.08 \qquad Z_F = -.72 \qquad Z_F = .45$$

Answers
Z_F .485, −.811, 2.298
r −.79, −.62, .42

Applying Z_F to Hypothesis Testing. Sometimes the situation can arise where we wish to test a hypothesis that ρ is some other value than 0. The SAD faculty might, for example, wish to test the null hypothesis that the correlation between anxiety and final exam score is −.75. Because the sampling distribution of r around a population parameter as high as this is definitely not symmetric, we must use Fisher's Z transformation: we transform the r and the ρ to Z_F's and carry out the test using the formula

$$Z = \frac{Z_{F_r} - Z_{F_\rho}}{\frac{1}{\sqrt{N - 3}}} \tag{14.12}$$

where Z_{F_r} is the transformed value of the observed correlation and Z_{F_ρ} is the transformed value of the hypothesized population correlation. The test is performed using $Z_{\bar{F}}$

transformed values for the sample and population correlations and the standard error of Z_F. If Z from equation 14.12 exceeds the usual critical value (1.96, 2.576, or another value we might choose) we would reject the null hypothesis that $\rho = -.75$.

The SAD faculty would proceed to test the null hypothesis that the correlation between math anxiety and final exam scores is $-.75$ at $\alpha = .05$ as follows. Their null and alternate hypotheses are

$$H_0: \rho = -.75$$
$$H_1: \rho \neq -.75$$

They have data from a sample of 68 students and the observed correlation is $-.55$. The transformed values are $Z_{F_r} = -.618$ and $Z_{F_\rho} = -.973$, so with this sample size we obtain

$$Z = \frac{-.618 - (-.973)}{\frac{1}{\sqrt{65}}} = \frac{+.355}{.124} = +2.86$$

Since a Z of 2.86 exceeds the critical value of 1.96, we would reject the hypothesis that $\rho = -.75$ in the population that produced this sample at the $\alpha = .05$ level. That is, it is unlikely that this sample was drawn from a population in which $\rho = -.75$. The fact that the Z is positive suggests that the sample was probably drawn from a population in which ρ is closer to zero than $-.75$ (that is, not as large a negative correlation), but we do not at this stage know what that value might be.

We can see how important Fisher's Z transformation is if we compare the result above with what we would get if we tested the same hypothesis without the transformation. Using Equation 14.9 produces

$$t_{66} = \frac{-.55 - (-.75)}{\sqrt{\frac{(1 - .55^2)}{(68 - 2)}}} = \frac{+.20}{.103} = +1.95$$

A t of 1.95 with $df = 60$ is not significant at the two-tailed $\alpha = .05$ level, so we would not reject the hypothesis that $\rho = -.75$. Again we see that using the wrong test can lead us to the wrong conclusion, a Type IV error. Had we been testing the hypothesis that $\rho = 0$, either test would have been appropriate and either would have led to the same conclusion. The difference is caused by the asymmetry of the sampling distribution of r for large values of ρ.

▶ *Now You Do It*

The publisher of the B-Projective Psychokinesis Test (BPPT) claims that the test predicts performance in graduate study in education and psychology with a correlation of $+.75$. (The authors claim that high scorers can intuit what the test questions will be.) You doubt that the correlation is that high and conduct a study to test your hypothesis. In a sample of 67 students you find a correlation of .48. Can you reject the null hypothesis?

The first question we must answer is what is the null hypothesis? You believe that the correlation is lower than .75. Therefore, you must test a directional hypothesis. Your correct null and alternate hypotheses are

$$H_0: \rho \geq +.75$$
$$H_1: \rho < +.75$$

Since the hypothesized population value is quite different from zero, we will have to use Fisher's Z transformation and the associated test. The two values of Z_F are .523 and .973. The test of significance produces the following:

$$Z = \frac{.523 - .973}{\frac{1}{\sqrt{64}}} = \frac{-.45}{.125} = -3.60$$

Clearly, you can reject the null hypothesis and conclude that your results are not consistent with the publisher's claims. The correlation between the BPPT and performance in graduate school is not as high as they claimed.

Confidence Intervals for Correlations

A second situation that we occasionally encounter and which calls for Z_F is the need to place a confidence interval for ρ on an observed r. Since this usually involves a nonzero r, it is generally necessary to use Z_F. For example, in the section above we were able to conclude that it is unlikely that the population from which the SAD Department drew their students is one in which the correlation between math anxiety and statistics achievement is $-.75$, but what is a reasonable range for this correlation? A confidence interval provides an answer.

The procedure for finding confidence intervals is a direct generalization of principles we have already discussed. The first step is to determine Z_F for the obtained r from Appendix Table A-5, then calculate σ_{Z_F} (equation 14.11). Using the appropriate normal deviate Z's for our chosen α (Z_α) from Appendix Table A-1 (for example, ± 1.96 for the 95% confidence interval), we compute the upper and lower CI limits for Z_F using the usual linear transformation

$$CI_L = (-Z_\alpha)\sigma_{Z_F} + Z_{F_r}$$

and

$$CI_U = (+Z_\alpha)\sigma_{Z_F} + Z_{F_r}$$

These limits, which are in the metric of Z_F, are then transformed by Appendix Table A-6 back into correlation values.

To find the 95% confidence interval for the relationship between math anxiety and performance on the statistics final, the BTU faculty can therefore proceed as follows: The obtained r in their sample was $-.55$ computed on 68 students. Appendix Table A-5 shows Z_F for a correlation of $-.55$ to be $-.618$. The standard error of Z_F is

$$\frac{1}{\sqrt{65}} = \frac{1}{8.06} = .124$$

The upper- and lower-limit values for the confidence interval in the metric of Z_F are

$$(-1.96)(.124) + (-.618) = -.85$$

and

$$(+1.96)(.124) + (-.618) = -.37$$

Appendix Table A-6 shows the corresponding r's to be $-.70$ and $-.35$. This confidence interval is symmetric in terms of probability and in terms of Z_F, but not in terms of r. The confidence interval in the scale of r goes from $-.70$ to $-.35$, with its center at $-.55$, reflecting the fact that the sampling distribution of r in this region is positively skewed. (The sampling distribution would be negatively skewed for positive correlations.) Had we used σ_r from equation 14.7 and not transformed to Z_F, we would have obtained a different, *and erroneous* pair of limits.

▶ *Now You Do It*

We can note here once again that confidence intervals include within them the information from a null hypothesis test. Suppose, for example, we look at the correlation between number of homework assignments completed and math anxiety score in the BTU data.

The correlation is $-.30$. First, let's test the null hypothesis that $r = 0$, then that $r = -.60$. Finally, we will find the 95% and 99% confidence intervals.

We can test the first null hypothesis using either Z_F or a t-test. For a correlation of $-.30$, $Z_F = .31$, which shows that for correlations of this size the sampling distribution is still nearly symmetric. The t-test produces a value of -2.55 with $df = 66$, which is significant at the .05 level two-tailed (the exact probability of a t with a magnitude this large or larger is .013). The Z-test yields a value of -2.50, which has a two-tailed probability of .012. That is, for testing the null hypothesis that $\rho = 0$ the two procedures agree almost exactly, although, because $N < 100$, the t-test is the appropriate one to use.

Now, run the test $H_0 : \rho = -.60$. You should get the same result (with sign reversed) from the t-test ($+2.55$), but the proper test, the Z-test using Z_F, yields a value of

$$Z = \frac{-.310 - (-.693)}{.124} = \frac{+.383}{.124} = +3.09$$

which is significant at the .002 level. We can conclude with confidence that this sample did not come from a population in which the correlation is $-.60$. The reason for the difference is the increasing asymmetry of the sampling distribution as ρ departs from zero.

The 95% and 99% confidence intervals for ρ from these data in the metric of Z_F are $-.55$ to $-.07$ and $-.63$ to .01 respectively. In the metric of r, the CI's are $-.50$ to $-.07$ and $-.56$ to .01. The 95% confidence interval does not include zero, and the significance test rejected the null hypothesis at the .05 level. The 99% CI does include zero, and we were unable to reject the null that $\rho = 0$ at the .01 level. Note that the 99% CI in the metric of Z_F does not include the value $-.693$ (the Z_F of a correlation of $-.60$). Also, the 99% CI in the metric of r does not include the value $-.60$. Both results are consistent with the correct test of that null hypothesis (the one using Z_F).

Hypotheses About Two Correlations

Independent Samples. We occasionally encounter a situation where we want to test a null hypothesis that two correlations are equal, or that they differ by a specified amount. If the two correlations involve the same variables in two different samples—for example, testing whether the correlation is the same for men as it is for women—Z_F can be used. First, transform both r's to Z_F's, then calculate the standard error of the difference between two Z_F's:

$$\sigma_{(Z_{F_1} - Z_{F_2})} = \sqrt{\frac{1}{N_1 - 3} + \frac{1}{N_2 - 3}} \tag{14.13}$$

A critical ratio Z of the usual form then provides the test statistic for the null hypothesis that $\rho_1 = \rho_2$ ($\rho_1 - \rho_2 = 0$) or any other hypothesized difference between the ρ's.

$$Z = \frac{Z_{F_{r1}} - Z_{F_{r2}} - (Z_{F_{\rho1}} - Z_{F_{\rho2}})}{\sigma_{(Z_{F_1} - Z_{F_2})}} \tag{14.14}$$

Suppose we found the correlations between midterm and final examination scores for students in a statistics class. For the 17 men in the class, the correlation is .78, and for the 24 women, r is .55. The null hypothesis that $\rho_M = \rho_F$ is tested by finding

$$\sigma_{(Z_{F_1} - Z_{F_2})} = \sqrt{\frac{1}{17 - 3} + \frac{1}{24 - 3}} = \sqrt{.119} = .345$$

and

$$Z = \frac{(1.045 - .618) - 0}{.345} = \frac{+.427}{.345} = 1.24$$

Clearly, we cannot reject the null hypothesis of equal correlations on the basis of these data, but we also cannot reach the conclusion that they are equal.

One Sample. Another problem that arises in some research contexts is to test whether the correlation of variable A with variable B is the same as the correlation of variable A with variable C. For example, we might have data for 50 adolescents on a measure of (A)

extroversion, a measure of (B) thrill-seeking, and a measure of (C) aggressiveness, and we might want to test whether the relationship between extroversion and thrill-seeking is equal to the relationship between extroversion and aggressiveness. A test of the hypothesis that $\rho_{AB} = \rho_{AC}$ is provided by the following Z-test.[1]

$$Z = \frac{\sqrt{N}(r_{AB} - r_{AC})}{\sqrt{(1 - r_{AB}^2)^2 + (1 - r_{AC}^2)^2 - 2r_{BC}^3 - (2r_{BC} - r_{AB}r_{AC})(1 - r_{AB}^2 - r_{AC}^2 - r_{BC}^2)}}$$

$$(14.15)$$

If the obtained Z exceeds the tabled value, we may conclude that the two variables (B and C) do not correlate equally with the third variable (A) (that is, $r_{AB} \neq r_{AC}$).

Assume we have tested our sample of 50 high school students on our three variables. Each person took all three measures, and we obtained the following correlations:

$$r_{AB} = +.60 \qquad r_{AC} = +.30 \qquad r_{BC} = +.40$$

Putting these quantities in the proper arrangement in equation 14.15, we get

$$Z = \frac{\sqrt{50}(.60 - .30)}{\sqrt{(1 - .60^2)^2 + (1 - .30^2)^2 - 2(.40)^3 - [2(.40) - (.60)(.30)](1 - .60^2 - .30^2 - .40^2)}}$$

$$= \frac{2.12}{\sqrt{.41 + .83 - .13 - (.8 - .18)(1 - .36 - .09 - .16)}}$$

$$= \frac{2.12}{\sqrt{1.24 - .13 - .24}} = 2.27$$

This value of Z exceeds the two-tailed .05 critical value for Z (1.96 from Appendix Table A-1), so we may conclude that the correlation of extroversion with thrill-seeking is different from its correlation with aggressiveness.

▶ Now You Do It

Let's use the BTU data to review the last two significance tests. One interesting question we might want to ask is whether the correlation between aptitude and achievement is the same for highly anxious students as for those reporting low levels of anxiety. In Chapter 13, we created two groups by dichotomizing the anxiety variable at the median. We can use the same strategy here to create the same two groups and find the correlation between math aptitude and final exam score in each group. You should find correlations of .80 in the low-anxiety group and .91 in the high-anxiety group, which have Z_F's of 1.099 and 1.528 respectively. Equation 14.13 produces a standard error of .254, and equation 14.14 yields a Z of +1.69. If, on the basis of theory, we had predicted a higher correlation in the high-anxiety group, we could take these results as modest support, because the test statistic is significant at the $P < .05$ level for a one-tailed test (the critical value is 1.645); however, from the perspective of a two-tailed test, where the critical value is ±1.96, the test statistic fails to reach the critical value, so without predicting direction we cannot conclude that the correlation is different depending on anxiety level.

Another potentially interesting question might be to ask whether math aptitude test score (MA) correlates more highly with score on the statistics final (SF) or with overall college GPA. It should correlate more highly with statistics performance. That is,

$$H_0: \ = \rho_{MA,SF} \leq \rho_{MA,GPA}$$

$$H_1: \rho_{MA,SF} > \rho_{MA,GPA}$$

The relevant correlations are

$$r_{MA,SF} = .88 \qquad r_{MA,GPA} = .71 \qquad r_{SF,GPA} = .83$$

[1]Several formulas have been proposed to test the difference between correlations obtained from a single sample. None of them has been found completely satisfactory. The one we have provided was developed by Olkin (1967). A study by May and Hittner (1997) found that this formula generally gave the most satisfactory results. Formulas by Hotelling (1940) and Williams (1959) based on the t distribution, yielded significant results too often. A formula by Meng, Rosenthal, and Rubin (1992) gave results similar to Olkin's formula.

Inserting these values into equation 14.15 you should get

$$Z = \frac{\sqrt{68}\,(.88 - .71)}{\sqrt{(1 - .88^2)^2 + (1 - .71^2)^2 - 2(.83)^3 - [2(.83) - (.88)(.71)](1 - .88^2 - .71^2 - .83^2)}}$$

$$Z = \frac{(8.246)(.17)}{\sqrt{.051 + .246 - 1.144 - (1.035)(-.967)}} = 3.57$$

This Z reaches the $P < .001$ significance level, so we can reject the null hypothesis. The relationship is in the expected direction, so it lends strong support to the claim that the math aptitude test measures something more specific than general academic ability.

Averaging Correlations

The final situation that we cover is the problem of averaging the correlation between two variables across several groups of people. For example, research on the Wechsler Intelligence Scale for Children sometimes involves several groups of children who differ in age. For some research questions it is desirable to collapse the results across ages to obtain larger samples. It is not appropriate simply to consider all children to be a single sample, because cognitive ability is related to age. Therefore, correlations are computed within each age group and then averaged across ages.

Because the sampling distribution of the correlation coefficient is highly skewed for large values of ρ, the meaning of differences between correlations changes depending on where we are in the range of possible values (again because of the limits of ± 1.0). For this reason, we cannot use simple averaging techniques with r's. We can, however, use Z_F because it does represent a variable on an interval scale. Whenever it is necessary to find an average of several correlations, particularly when they differ by more than about .10, the appropriate procedure is

1. transform all r's to Z_F's,
2. find $\overline{Z}_F$ (the mean of the Z_F's), and then
3. reverse the transformation using $\overline{Z}_F$ to find the mean r, $\overline{r}$.

When the correlations being averaged are based on samples of different sizes, it is necessary to compute a weighted average. Under these conditions, the proper weighting is given by the formula

$$\overline{Z}_F = \frac{(N_1 - 3) Z_{F_1} + (N_2 - 3) Z_{F_2} + \cdots + (N_K - 3) Z_{F_K}}{(N_1 - 3) + (N_2 - 3) + \cdots + (N_K - 3)} \tag{14.16}$$

The mean correlation, $\overline{r}$, is found by transforming $\overline{Z}_F$ back into the metric of r using Appendix Table A-6. Hypotheses concerning this mean correlation should be tested using $\overline{Z}_F$, which has a standard error of

$$\sigma_{\overline{Z}_F} = \frac{1}{\sqrt{(N_1 - 3) + (N_2 - 3) + \cdots + (N_K - 3)}}. \tag{14.17}$$

The critical ratio test statistic for testing null hypotheses concerning $\overline{r}$ is

$$Z = \frac{\overline{Z}_F - Z_{F_\rho}}{\sigma_{\overline{Z}_F}} \tag{14.18}$$

As usual, Z_{F_ρ} is the Z_F of ρ, the correlation under the null hypothesis. The null hypothesis ρ is usually zero, but may take on any appropriate value.

To illustrate this procedure and the appropriate use of equations 14.8 and 14.9, imagine that we have the correlation between the midterm examination scores and final examination scores for each of three classes in statistics. The data are

$$
\begin{array}{lll}
r_1 = .61 & N_1 = 10 & Z_{F_1} = .709 \\
r_2 = .37 & N_2 = 26 & Z_{F_2} = .389 \\
r_3 = .15 & N_3 = 107 & Z_{F_3} = .151
\end{array}
$$

Because $N_3 > 100$, we can test the hypothesis that $\rho = 0$ using equation 14.8 on the sample 3 data.

$$\sigma_r = \frac{1}{\sqrt{107}} = \frac{1}{10.34} = .097$$

$$Z = \frac{.15}{.097} = 1.55$$

We cannot reject the null hypothesis that $\rho = 0$ based on the data from sample 3 alone.

The correct test for the hypothesis $\rho = 0$ based on the data from group 1 or group 2 employs equation 14.9, because the sample sizes are less than 100. Using the group 1 data

$$t_{(8)} = \frac{+.61}{\sqrt{\frac{(1.0 - .61^2)}{(10 - 2)}}} = \frac{+.61}{\sqrt{.078}} = 2.18$$

with $df = 8$, while for group 2, with $df = 24$

$$t_{(24)} = \frac{+.37}{\sqrt{\frac{(1.0 - .37^2)}{(26 - 2)}}} = \frac{+.37}{\sqrt{.036}} = 1.95$$

Neither of these test statistics exceeds the $P < .05$ two-tailed critical value for its sampling distribution either. Still, because these can be viewed as three random samples from one population, the mean correlation should be a better estimate of ρ than the correlation from any single sample. We may therefore find $\bar{r}$ and test the hypothesis that the mean correlation is zero. This is done by using equations 14.16–14.18 to find

$$\bar{Z}_F = \frac{7(.709) + 23(.389) + 104(.151)}{7 + 23 + 104} = \frac{29.614}{134} = .221$$

$$\bar{r} = .22$$

$$\sigma_{\bar{Z}_F} = \frac{1}{\sqrt{134}} = .0864$$

$$Z = \frac{.221}{.0864} = 2.56$$

from which we can conclude that the mean correlation of .22 *does* differ significantly from zero. By combining all the data, we increase the power of our test.

We can place a confidence interval on this mean correlation in the usual way. For the 95% CI we start with

$$\bar{Z}_F = .221 \qquad \sigma_{\bar{Z}_F} = .0864 \qquad Z_{0.95} = \pm 1.96$$

Finding the lower and upper limits of the CI for $\bar{Z}_F$ yields

$$CI_{lower} = -1.96(.0864) + .221 = .052$$

$$CI_{upper} = +1.96(.0864) + .221 = .390$$

Converting these Z_F's back into correlations produces a confidence interval centering on $+.22$ and running from $+.05$ to $+.37$. Once again, this confidence interval is symmetric in terms of probability, but has a slight negative skew due to the asymmetry of the sampling distribution of r.

SPECIAL ISSUES IN RESEARCH

Mean or Median Correlation?

A popular alternative way to find the central tendency of a set of correlations is to determine the median correlation. This method has the advantage of simplicity, but it does not allow sample size to affect the result and does not use all the available information, nor does it permit easily interpretable tests of the collective correlation. We recommend that when the samples differ in size or when the correlations spread out over a range of more than about .10 units the methods described here be used.

◆ Now You Do It

Let's use the data from the Big Time U study of statistics performance to review the procedures for averaging correlations and finding confidence intervals. There are three distinct groups in the sample, psychology majors ($N = 20$), education majors ($N = 26$), and other majors ($N = 22$). The correlation between mathematics aptitude test score and overall college grade point average is .81 for the psychology majors, .74 for the education majors, and .58 for the others. You can verify these correlations either by hand computation using the formulas in Chapter 8, or you can use the SELECT CASES option on the DATA Menu in SPSS to analyze each major separately. What are the next steps?

The first thing to do is to convert each correlation into a Z_F. Looking in Appendix Table A-5, we find the necessary values to be 1.127, .950, and .663. Placing these values into equation 14.16 yields

$$\overline{Z}_F = \frac{17(1.127) + 23(.950) + 19(.663)}{17 + 23 + 19} = \frac{19.159 + 21.85 + 12.597}{59} = .909$$

Converting $\overline{Z}_F$ back into the metric of r yields an average correlation of .72. We then find the standard error of $\overline{Z}_F$ from equation 14.17 to be .13. This value, combined with the critical values from the normal distribution, produces the following confidence interval in the metric of $\overline{Z}_F$.

$$CI_L = (-1.96)(.13) + .909 = .65$$
$$CI_U = (+1.96)(.13) + .909 = 1.16$$

Using Appendix Table A-6, we can convert these Z_F's back into correlations. We find the limits of the 95% CI to be .57 and .82. From these data, we would conclude that our best estimate of the correlation between mathematics aptitude test score and college grade point average is .72, and there is a 95 percent probability that ρ is between .57 and .82.

SUMMARY

It is possible to draw inferences about other statistics and parameters than the mean; the *chi-square* (χ^2) family of distributions is particularly useful for drawing inferences about variances. There is a unique χ^2 distribution for each number of degrees of freedom. χ^2 can also be used to place a confidence interval for σ^2 around $\hat{S}_X^2$.

A second family of distributions that is useful in several hypothesis testing situations is called the *F distribution*. An F distribution with df_1 and df_2 degrees of freedom is the sampling distribution of the ratio of two χ^2's each divided by its *df*. It can also be viewed as the sampling distribution of the ratio of two independent, unbiased estimates of a population variance. An F distribution is used to test hypotheses about the equality of variances from two samples.

A critical ratio or normal deviate Z can be computed directly from r to test the hypothesis that $\rho = 0$ when sample size is larger than 100, while a *t*-test is used to test this hypothesis for smaller samples. Hypotheses about correlation coefficients use the *Fisher Z transformation* to provide a sampling distribution that is normal in shape when $\rho \neq 0$. Confidence intervals and tests of hypotheses involving several correlations also require using Fisher's Z.

ANALYSES OF THE SEARCHERS STUDY DATA

Let's review the various tests we have learned in this chapter using the data from Dr. Cross's study of the Searchers. There are a number of analyses we might be interested in running on these data. First, let us address the variances. We might want to test whether there was reason to believe that men from different religious denominations differed in variability on any of the continuous variables, and we might ask the same question regarding members of the Searchers versus nonmembers.

As we have seen in several other contexts involving significance tests, we can approach this problem in either of two ways. First, we can compute confidence intervals around the sample variances. If the variance from one sample falls within the confidence interval for the other, we

Table 14-3 Confidence Intervals and *F*-Tests for the Variances from the Searchers Data

Variable (Group)	$\hat{S}^2_x$	$(N-1)\,\hat{S}^2_x$	98% Confidence Interval	F
Age				
Catholic	124.8	3619.2	73.0–253.1	1.09
Protestant	136.5	3958.5	79.8–276.8	
Attitude				
Catholic	55.8	1618.2	32.6–113.2	1.49
Protestant	83.2	2412.8	48.6–168.7	
Religious Involvement				
Catholic	18.9	548.1	11.1–38.3	1.83
Protestant	34.6	1003.4	20.2–70.2	
Age				
Member	142.8	4141.2	83.5–289.6	1.18
Nonmember	120.7	3500.3	70.6–244.8	
Attitude				
Member	73.2	2122.8	42.8–148.4	1.15
Nonmember	63.9	1853.1	37.4–129.6	
Religious Involvement				
Member	23.2	672.8	13.6–47.0	1.41
Nonmember	32.8	951.2	19.2–66.5	

should not reject the null hypothesis. As an alternative, we can compute an *F* statistic from the ratio of the two variances and test the null hypothesis that $\sigma^2_1 = \sigma^2_2$. The results of the 98% confidence interval and the *F*-test are presented in Table 14-3. Because all confidence intervals are at the same probability level and involve the same degrees of freedom (29), we only need to look up two χ^2 values: the one for the lower tail which cuts off the bottom 1 percent (99 percent above) and the one for the upper tail cutting off the top 1 percent. These values are 14.3 and 49.6 respectively. Both the confidence intervals and the *F*-tests lead to the same conclusion: in no case can we reject the hypothesis that the variances are equal.

The correlations between the four continuous variables based on all 60 men in the sample are given in Table 14-4. Below the correlations, a test of the null hypothesis that $\rho = 0$ using the *t*-test and the smallest correlation is given. Note that from this test we can tell that all correlations are statistically significant at $\alpha = .01$ for the two-tailed hypothesis. Significance is unrelated to the sign of the correlation. In effect, we are testing r^2.

We might also wish to determine a confidence interval for the correlation between income and attitudes toward equality. Noting that the observed correlation falls exactly between two tabled values of Z_F in Appendix Table A-5, we might interpolate to obtain a Z_F of .766 for the correlation of .645. Finding a value of $1/\sqrt{57} = .132$ for σ_{Z_F} and choosing a 95% confidence interval, we would multiply .132 by ± 1.96 to get $\pm.259$. Adding .766 to both the positive and negative values produces limits in the metric of Z_F of .507 and 1.025. Using Appendix Table A-6 to convert these back to *r*'s we get a confidence interval running from .47 to .77. There is a probability of .95 that our sample was drawn from a population in which the correlation between these two variables is in the range .47 to .77.

Next, suppose we wanted to test whether the level of correlation between variables is the same for Roman Catholic men and Protestant men. We will take the correlation between attitudes

Table 14-4 Correlations Between Four Continuous Variables and Test of Significance

Variable	Attitudes	Age	Income	Religious Involvement
Attitudes	1.000			
Age	.839	1.000		
Income	.645	.739	1.000	
Religious Involvement	.399	.376	.347	1.000

$$t_{(58)} = \frac{.347 - 0}{\sqrt{\dfrac{1 - .347^2}{58}}} = \frac{.347}{.123} = 2.82$$

toward equality and degree of religious involvement as our example. The correlation for the 30 Catholic men is .54, and the correlation for the 30 Protestant men is .16. Is it reasonable to conclude that these are random samples from a single population? Here are the results.

Catholic Men

$r = .54$

$Z_F = .604$

Protestant Men

$r = .16$

$Z_F = .161$

$$\sigma_{(F_{RC} - Z_{F_P})} = \sqrt{\frac{1}{27} + \frac{1}{27}} = .272$$

$$Z = \frac{.604 - .161}{.272} = 1.63$$

On the basis of these data we cannot conclude that these correlations are significantly different. The two samples might have come from a common population. However, as always, the fact that the difference is not statistically significant does not mean that we can conclude that it is zero.

If we wished to find the average of the Catholic and Protestant correlations above, we would get

$$\bar{Z}_F = \frac{27(.161) + 27(.604)}{27 + 27} = .38 \qquad \bar{r} = .36$$

Note that in the entire sample of 60 men the correlation was .40. Although the weighted average of subgroup means will always equal the mean of the entire group, this is not necessarily true of the correlation coefficient. Sometimes, particularly when the group means differ, the correlation computed on the total group can be much larger than the average of the within group correlations, or even higher than either of them.

Finally, suppose we wish to test whether the correlation between income (I) and attitudes toward equality (A) is significantly different from the correlation between income and degree of religious involvement (R). Here we have two correlations on data from the same 60 individuals, so we must use equation 14.15. The three correlations are $r_{IA} = .645$, $r_{IR} = .347$, $r_{AR} = .399$. The computations go as follows.

$$Z = \frac{\sqrt{60}\,(.64 - .35)}{\sqrt{(1 - .64^2)^2 + (1 - .35^2)^2 - 2(.40)^3 - [2(.40) - (.64)(.35)](1 - .64^2 - .35^2 - .40^2)}}$$

$$= \frac{2.246}{.902} = 2.49$$

This Z is statistically significant, so we can conclude that income correlates more highly with attitudes toward equality than it does with religious involvement, but we do not know what the difference is. From these data our best guess would be a difference of about .30, but all we know with reasonable certainty is that there is a difference.

REFERENCES

Hays, W. L. (1994). *Statistics* (5th ed.). Fort Worth, TX: Harcourt Brace.

Hotelling, H. (1940). The selection of variates for use in prediction, with some comments on the general problem of nuisance parameters. *Annals of Mathematical Statistics, 11,* 271–283.

May, K., & Hittner, J. B. (1997). Tests for comparing dependent correlations revisited: A Monte Carlo study. *Journal of Experimental Education, 65,* 257–269.

Meng, X. L., Rosenthal, R., & Rubin, D. B. (1992). Comparing correlated correlation coefficients. *Psychological Bulletin, 111,* 172–175.

Olkin, I. (1967). Correlations revisited. In J. C. Stanley (Ed.), *Improving experimental design and statistical analysis* (pp. 102–128). Chicago: Rand McNally.

Williams, E. J. (1959). Significance of difference between two nonindependent correlation coefficients. *Biometrics, 15,* 135–136.

EXERCISES

1. For each of the following, use Appendix Table A-3 to find the critical value of χ^2 that cuts off the specified portion of the χ^2 distribution for the given degrees of freedom.

   a. Find the value of χ^2 for which 90 percent of the distribution falls to the right if $df = 1$.

   b. Find the value of χ^2 for which 5 percent of the distribution falls to the left if $df = 2$.

c. Find the value of χ^2 for which 5 percent of the distribution falls to the right if $df = 3$.

d. Find the value of χ^2 that cuts off the bottom 30 percent of the distribution if $df = 10$.

e. Find the values of χ^2 between which 80 percent of the distribution falls if $df = 6$.

f. Find the values of χ^2 between which 98 percent of the distribution falls if $df = 5$.

2. Given the following information, test the hypothesis that the sample was drawn from the population with the given variance.

a. $\sigma^2 = 128$, $N = 25$, $\hat{S}^2 = 108$ if the research hypothesis is that the variance in the treatment population should be less than the given population variance for $\alpha = .05$.

b. $\sigma^2 = 180$, $N = 20$, $\hat{S}^2 = 200$ if the research hypothesis is that the variance in the treatment population should be greater than the given population variance for $\alpha = .05$.

c. $\sigma^2 = 250$, $N = 30$, $\hat{S}^2 = 300$ if the research hypothesis is that the variance in the treatment population should be different from the given population variance for $\alpha = .10$.

3. Determine the 90% and 98% confidence intervals for Exercise 2(c).

4. Given the following information, test the hypothesis that the sample was drawn from the population with the given variance. Test the hypothesis at $\alpha = .05$.

a. An educational psychologist believes that dyslexic children will have greater variability in their scores on the Nelson-Denny Reading Comprehension Test ($\mu = 80$, $\sigma = 9$). The following scores were obtained from the sample of dyslexic children:
77, 82, 71, 88, 68, 60, 79, 89, 66, 62, 90, 78

b. The norms for a measure of depression ($\mu = 30$, $\sigma = 6$) were established for the population. A clinical psychologist believes that individuals who are clinically depressed will be less variable on this measure of depression since the norms were established from the general population. Use the following scores from a sample of clinically depressed individuals to test this hypothesis:
38, 40, 41, 35, 34, 33, 37, 39, 42, 36

5. A test of creativity ($\mu = 80$, $\sigma = 9$) is given to a sample of individuals who have been determined to be creative using another method. The researcher believes that creative individuals will differ from the general population on their variability of this measure. Test this hypothesis at $\alpha = .10$ for the data given below:
88, 95, 100, 98, 88, 80, 110, 82, 84, 99, 92, 90, 88, 86

6. Determine the confidence intervals for Exercises 4(a), 4(b), and 5.

7. Given the following information, use Appendix Table A-4 to find the critical values of F associated with the given values of α.

a. $df_1 = 3$, $df_2 = 40$, $\alpha = .05$

b. $df_1 = 10$, $df_2 = 60$, $\alpha = .01$

c. $df_1 = 20$, $df_2 = 120$, $\alpha = .001$

8. For the following, test the hypothesis that the two samples are from populations with equal variances.

a. $\hat{S}_a^2 = 40$, $N_a = 20$; $\hat{S}_b^2 = 90$, $N_b = 25$; $\alpha = .05$

b. $\hat{S}_a^2 = 37.4$, $N_a = 18$; $\hat{S}_b^2 = 82.1$, $N_b = 16$; $\alpha = .01$

c. $\hat{S}_a^2 = 383.3$, $N_a = 25$; $\hat{S}_b^2 = 98.5$, $N_b = 25$; $\alpha = .001$

9. For the following data, test the hypothesis that the two samples are from populations with equal variances at $\alpha = .05$.

a. Japanese participants' scores on independent self-construal: 66, 68, 67, 52, 57, 54, 53, 59, 63, 69; U.S. participants' scores on independent self-construal: 65, 78, 81, 64, 74, 77, 66, 59, 79, 66

b. Female participants' scores on verbal reasoning: 54, 58, 49, 56, 55, 62, 66, 54, 58, 60, 70, 50; Male participants' scores on verbal reasoning: 48, 42, 56, 44, 38, 59, 62, 66, 44, 48, 56, 75

10. Given the following information, test the hypothesis that the correlation in the population is zero for the given alpha using a two-tailed test.

a. $r = .36$ for $N = 60$ at $\alpha = .05$

b. $r = -.25$ for $N = 36$ at $\alpha = .01$

c. $r = .15$ for $N = 150$ at $\alpha = .05$

11. Given the following data, test the hypothesis that the correlation in the population is zero for $\alpha = .05$ using a two-tailed test.

a. The following pairs of scores represent an independent self-construal score in the first position and an interdependent self-construal score in the second position.
(42, 38) (55, 26) (44, 58) (49, 26) (59, 54) (26, 28) (32, 55) (30, 44) (44, 45) (58, 38)

b. The following pairs of scores represent empathy in the first position and aggression in the second position.
(32, 64) (48, 53) (55, 46) (65, 49) (54, 40) (36, 56) (66, 32) (44, 51) (48, 48)

c. The following pairs of scores represent shyness in the first position and depression in the second position.
(23, 31) (33, 37) (55, 49) (48, 42) (42, 40) (38, 39) (24, 29) (51, 50) (41, 43) (44, 45) (56, 52) (54, 54)

12. For the following correlations, transform them into the appropriate Z_F scores using Appendix Table A-5.

a. .53

b. .88

c. .23

d. .46

e. .14

13. For the following Z_F scores, transform them into the appropriate correlations using Appendix Table A-6.

a. .56

b. .15

c. .89

d. 1.83

e. .32

14. For each of the following, test the hypothesis that the correlation in the population, ρ, is a given value when we know (or can compute) the correlation of a sample, r, of size N.

a. At $\alpha = .05$, test the hypothesis that $\rho = .60$ when $r = .48$ for a sample of size $N = 36$.

b. At $\alpha = .05$, test the hypothesis that $\rho = -.85$ when $r = -.66$ for a sample of size $N = 100$.

c. At $\alpha = .01$, test the hypothesis that $\rho = .55$ when $r = .78$ for a sample of size $N = 75$.

d. At $\alpha = .01$, test the hypothesis that $\rho = -.38$ when $r = -.22$ for a sample of size $N = 50$.

15. Calculate the confidence intervals for each of the questions in Exercise 14. Explain what these confidence intervals mean.

16. For each of the following, test the null hypothesis that the two correlations are equal at $\alpha = .05$.
    a. Suppose the correlation between spatial ability and mathematics achievement is .58 for a sample of 52 men and .45 for a sample of 48 women.
    b. Suppose the correlation between empathy and aggression is $-.34$ for a sample of 46 men and $-.64$ for a sample of 60 women.
    c. Suppose the correlation between life satisfaction and depression is $-.23$ for 28 older adult Americans and $-.36$ for 23 young adult Americans.

17. For each of the following the *research* hypothesis is given such that there is a directional statement of differences in two correlations. Test the null hypothesis regarding these correlations at $\alpha = .05$.
    a. A researcher hypothesized that the correlation between verbal reasoning and writing skills would be higher for women than for men. The researcher administered tests of verbal reasoning and writing skills to a sample of 40 women and 35 men. In collecting the data, the researcher found a correlation of .68 for the women and .39 for the men. What should the researcher conclude?
    b. A researcher hypothesized that the correlation between grade point average and the score on a locus of control test would be higher for people who are more internal in their locus of control than for people who are more external. In collecting the data, the researcher found a correlation of .35 for the 28 internal locus of control participants and .18 for the 22 external locus of control participants. What should the researcher conclude?
    c. A researcher predicted that there would be a stronger negative correlation between computer anxiety and computer performance among the elderly than among young adults. In collecting the data the researcher found a correlation of $-.54$ for the 34 elderly participants and $-.28$ for the 27 young adult participants. What should the researcher conclude?

18. For each of the following, find the average correlation.
    a. There are four sixth grade math classes during the first period at Cottonwood Middle School. What is the average correlation between math anxiety and math achievement for the sixth grade students in these four classes given the following correlations and sample sizes: $r_1 = -.34$, $N_1 = 38$; $r_2 = -.24$, $N_2 = 22$; $r_3 = -.48$, $N_3 = 33$; $r_4 = -.41$, $N_4 = 29$?
    b. Three separate studies were conducted to determine the relationship between anger control scores and empathy scores. The correlations and sample sizes for each study were as follows: $r_1 = -.45$, $N_1 = 120$; $r_2 = -.26$, $N_2 = 222$; $r_3 = -.38$, $N_3 = 163$. What is the average correlation?
    c. There is a relationship between the number of hours of practice learning a motor skill and the score on a test of that motor skill. Suppose the following correlations were found in studies with the designated sample sizes as follows: $r_1 = .34$, $N_1 = 12$; $r_2 = .25$, $N_2 = 18$; $r_3 = .31$, $N_3 = 16$; $r_4 = .22$, $N_4 = 21$; $r_5 = .27$, $N_5 = 29$; $r_6 = .18$, $N_6 = 13$. What is the average correlation?

19. For each of the following, three measures of behavior were taken. Correlations were calculated for variables 1 and 2 and for variables 1 and 3. Test the null hypothesis that the correlations between variables 1 and 2 and between variables 1 and 3 are equal.
    a. Suppose a social psychologist is interested in measuring hostility, anger, and aggression. The psychologist assessed 100 adolescents on these measures and found that the correlation between hostility and aggression was .45 while the correlation between anger and aggression was .58 and the correlation between anger and hostility was .36. Test the null hypothesis that the correlations between anger and aggression and hostility and aggression are equivalent in the population.
    b. An educational psychologist is trying to determine if measures of behaviors associated with self are equally correlated with self-esteem. In particular, self-handicapping and procrastination are both correlated with self-esteem. The correlation between self-handicapping and self-esteem is $-.59$ whereas the correlation between procrastination and self-esteem is $-.38$. Furthermore, the correlation between self-handicapping and procrastination is .43 in a sample of 75 students. Test the null hypothesis that the correlations between procrastination and self-esteem and self-handicapping and self-esteem are equivalent in the population.

20. A social psychologist believes that aggression is more highly negatively correlated with empathy than it is with perspective-taking. To test this belief the social psychologist administered measures of aggression, empathy, and perspective-taking to 48 participants. The correlation between aggression and empathy was $-.63$ whereas the correlation between aggression and perspective-taking was $-.51$. Furthermore, the correlation between empathy and perspective-taking was .47. What should the social psychologist conclude in regard to his research hypothesis?

Chapter 15

Inferences with Variables Measured on Nominal and Ordinal Scales

CHAPTER OBJECTIVES

After studying this chapter, you should be able to:

- ❏ Given a set of data, prepare a contingency or crosstab table with cell frequencies and marginal frequencies.

- ❏ From a contingency table, compute the expected cell frequencies and the χ^2 test of independence.

- ❏ Given a set of frequencies or proportions, test the fit of a set of data to those expectations using the χ^2 goodness-of-fit test.

- ❏ Given a set of before and after measurements, use McNemar's χ^2 test of change to test whether there has been a greater change in one direction than the other.

- ❏ Compute the correlation between two variables using ranked data.

- ❏ Test a one- or two-tailed null hypothesis concerning central tendency for ranked data with the Wilcoxon-Mann-Whitney test using the exact probability tables for small samples and the Z-test for large samples.

- ❏ Test a one- or two-tailed null hypothesis concerning central tendency for ordinal data from a matched-pairs or repeated-measures design with the Wilcoxon signed ranks test using the exact probability tables for small samples and the Z-test for large samples.

Consider the data the Big Time U faculty gathered in their study of statistics anxiety. Test scores can be viewed as interval scale data (although some would question the practice), but the variables of gender, academic major, and lab section are clearly nominal, and academic standing probably is best considered an ordinal variable. The faculty might be interested in determining whether there are relationships between these variables. For example, is there a relationship between gender and choice of a college major? Similarly, Dr. Cross and his students collected data on religious affiliation and membership in the Searchers, clearly nominal variables. They also used different forms of their questionnaire. The forms might be considered to represent an ordered set of conditions on the variable "degree of positiveness toward feminism." For variables like these, the idea of a normal distribution makes no sense. How can we approach the analysis of nominal and ordinal data?

Introduction

Mary Stuart is a guidance counselor who helps students make career and college major decisions. As part of her job, she administers interest inventories that ask students to choose activities they prefer. Her supervisor has asked her to determine whether there are different patterns of major selection for men and women. She has data from 86 recent students with whom she has worked. She has coded the data using M for males, F for females, P for psychology majors, E for education majors, and S for sociology majors. The data are shown in Table 15-1. How can Mary answer her supervisor's question?

The inferential statistics we have examined in the last two chapters have been appropriate for dependent variables such as statistics final exam scores, or attitudes toward equality for women, that are measured on an interval or ratio scale. It is common, particularly in disciplines such as sociology and political science, but sometimes also in psychology and education, to encounter variables such as reaction time that are continuous but do not have a normal distribution, or variables such as preferences that lack interval scaling properties. In either case, using the mean of the observations may give a distorted picture of the data (in the first case because the mean is affected by extreme scores and in the second case because the intervals between scale values do not have a constant meaning). When, for either of these reasons, it is not appropriate to use the mean and other least-squares statistics, we may turn to one of a variety of special tests (see Siegal & Castellan, 1988, or Wilcox, 1987, for a fairly comprehensive list of alternatives and their applications). There are also cases where the variable or variables of interest, such as those confronting Mary Stuart, are measured on a nominal scale or an explicitly ordinal scale (for example, ranks). Special methods are required for these situations as well.

In this chapter we will first learn about three tests based on the χ^2 distribution, tests which generally involve nominal variables. Then we will present a measure of correlation appropriate for ordinal variables. Finally, we will examine tests for the difference in central tendency of two samples that are appropriate when the dependent variable is ordinal. The applications we will present are among the most common situations you are likely to encounter, although there are many alternative procedures that may be more appropriate in specific cases.

Tests Based on the χ^2 Distribution

Mary Stuart's research question can be answered using a test based on the χ^2 distribution. In Chapter 14 we introduced the χ^2 distribution as a way to test hypotheses about single variances and to form confidence intervals for σ^2 around an observed $\hat{S}^2$. Now we will

Table 15-1 Gender and College Major Data for Mary Stuart's 86 Career Counseling Clients

Gender	Major	Gender	Major	Gender	Major	Gender	Major	Gender	Major
M	P	F	E	M	S	M	E	F	P
F	P	M	S	M	E	M	E	M	P
F	E	F	P	F	E	M	P	F	S
M	P	F	S	M	P	M	P	F	P
M	E	M	P	M	P	F	E	F	E
F	E	F	P	M	E	F	E	M	S
M	P	F	E	F	E	F	E	M	E
F	S	F	P	F	E	M	E	M	S
M	P	F	E	M	E	F	E	F	P
F	S	F	E	M	P	M	S	M	E
F	E	F	P	F	S	F	E	F	E
M	S	M	P	F	E	F	E	F	P
M	P	F	E	F	E	M	S	F	E
M	E	F	S	M	P	F	E	F	E
M	S	M	P	F	P	F	S	M	E
M	S	M	E	M	P	M	P	F	E
F	E	M	P	M	S				
M	P								

examine three statistics whose distributions approximate the χ^2 distribution; one appropriate for Mary's question, one for testing whether a distribution such as a set of test scores has a particular shape, and one that can be used to test whether people have changed categories from one testing to another. Although all three traditionally go by the name χ^2, they are not strictly χ^2's; however, the sampling distributions of these test statistics approach the χ^2 distribution.

Frequency Comparisons

The most widely known use of the χ^2 distribution is for testing hypotheses about the relationship between nominal variables such as gender and college major. Recall from our discussion in Chapter 1 that nominal variables do not contain information about order or amount. The letters or numerals assigned to the categories are not quantitative and, therefore, cannot be treated as scores in the sense that we have used that term. Assigning the number 1 to psychology majors and 2 to education majors does not imply more or less of anything. The only quantitative information that we can obtain from a nominal scale is information about the frequency with which each category occurs. For this reason, tests of hypotheses concerning nominal variables such as Mary's generally involve comparing the observed frequencies in the categories with those that would be expected to occur if some null hypothesis were true. For example, Mary wishes to test a null hypothesis that there is no relationship between gender and selection of college major. Both of these are nominal variables, and the null hypothesis really says that the patterns of frequencies for selecting majors would be the same for men as for women.

Tests of hypotheses with nominal variables usually compare the observed frequencies in categories with those that would be expected to occur if the null hypothesis were true.

Probability Revisited

In Chapter 6 we introduced the idea of probability and the distribution of chance events. At that time our attention focused on a single quantitative variable such as the number of heads in the toss of some coins or the number of correct guesses in an ESP experiment. When we test hypotheses about frequencies, the question we want to ask usually takes a form similar to that of a correlation; for example, is there a relationship between two variables such as gender and college major? To answer this question we must broaden our understanding of probability to include two new concepts, **joint probability** and **conditional probability.**

Suppose you have the student numbers of Mary's 86 career counseling clients on slips of paper in a jar. Some members of the class are psychology majors, some are education majors, and the rest are sociology majors. Also, some are male and some are female. What is the likelihood that the student number we draw from the jar belongs to a female psychology major? This is a question in joint probability because two conditions, female *and* psychology major, must be met. **Joint probability** refers to the likelihood that a given observation will satisfy two conditions simultaneously. Remember from Chapter 6 that the probability that both A and B will occur is the product of their separate probabilities (the multiplication law). Since joint probability involves satisfying two conditions, we should expect the multiplication law to apply unless there is something else going on.

Joint probability refers to the likelihood that a given observation will satisfy two conditions simultaneously.

Contingency Tables. The entries in Table 15-2 show how we would arrange the data from Table 15-1 to examine Mary's question. A table like this is called a **contingency table** or a **cross-tabulation** (crosstab). The contingency, or crosstab, table is very similar to the scatter plot. It is prepared by setting up two axes, one for each variable. Each axis is divided into as many parts as there are categories of the variable. *Every* individual has

A **contingency table** contains the frequencies of all possible combinations of values for two (or more) categorical variables.

Table 15-2   Contingency (Cross-Tabulation) Table for Gender and College Major

College Major	Gender		Row Marginal
	Male	*Female*	
Psychology	19	10	29
Sociology	10	7	17
Education	12	28	40
Column Marginal	41	45	86

a score on *both* variables, and each person is tabulated as a frequency in the appropriate cell of the table. Each cell contains the number of observations that satisfy that particular *combination* of conditions. For example, the students we are considering include 19 male psychology majors and 28 female education majors. The values in the cells (19, 28, and so on) are called **cell frequencies** while the values along the bottom and right sides (29, 17, 40, and so on) are called **marginal frequencies** (because they are in the margins of the table). The entry in the lower right corner, 86, is the total number of clients on whom Mary has data. You can see that the cell frequencies in any row of the table sum to that row marginal, and the cell frequencies in a column sum to that column marginal. The row and column marginal frequencies both sum to the total number of observations.

▶ Now You Do It

Let's use some of the data from the BTU study that is similar to Mary's data to practice preparing a contingency table. We have discussed the issue of whether men and women tend to choose majors with the same frequency. What do the data say? What are the steps we must take?

First, we must lay out the axes for our table. The categories are male and female for one axis and psychology, education, and other for the second axis. After laying out the axes, we must tally each person in the cell where he or she belongs. Your results should look like this.

	Psychology	Education	Other	Row Marginal
Female	9	11	9	29
Male	11	15	13	39
Column Marginal	20	26	22	68

The data in Table 15-2 allow us to compute the empirical joint probabilities of the various possible outcomes. These are shown in Table 15-3. The empirical probabilities are computed by simply dividing each cell entry by the total number of students. Thus, the empirical joint probability that an individual observation will be both a male and a psychology major is 19/86, or .22. The other empirical joint probabilities have been calculated in the same way. As with the frequencies, the cell probabilities sum to their respective marginal probabilities (within rounding error). This is simply an application of our familiar addition rule of probability; the probability that an observation will be in *either* cell R_1C_1 (row 1, column 1) or cell R_1C_2 (row 1, column 2) is the sum of their separate probabilities.

Conditional Probability. Looking again at the data in Table 15-2, we can see that if we know nothing about a person, the probability that the individual is an education major is simply

$$\frac{N_{Ed}}{N_{Tot}} = \frac{40}{86} = .47$$

the number of education majors in the class divided by the total number of students in the class. However, if we impose the condition (that is, we know) that the person is a woman, then the probability that the person is an education major becomes 28/45 = .62.

Table 15-3   Cell and Marginal Probabilities (Relative Frequencies)

College Major	Gender		Row Marginal
	Male	*Female*	
Psychology	.22	.12	.34
Sociology	.12	.08	.20
Education	.14	.33	.47
Column Marginal	.48	.52	1.00

The value .62 is the probability that a person is an education major, *given the condition* of being female. Knowing the person's gender has changed the probability pattern for major selection. What can this tell us about the relationship between college major and gender?

As you will recall from Chapter 6, the addition and multiplication laws of probability apply when three conditions are met. First, the events must be mutually exclusive; second, they must be exhaustive; third, they must be independent in the sense that one event does not influence another. These same three conditions apply to the two variables—gender and college major—individually in Table 15-2, and the conditions of exhaustiveness and exclusiveness apply to the cells as well. The condition of independence applies to the joint probabilities in the sense that the occurrence of one joint event (such as male psychology major) does not affect the probability of another joint event. However, in testing hypotheses about frequencies, the question of interest generally involves independence in a slightly different sense. We are interested in the question of whether there is a difference in pattern, whether males tend to select different majors than females. Stated another way, we want to know whether gender and the majors people select are independent, or whether they are related. Our null hypothesis is that they are independent.

We can state this null hypothesis in a way that will help us develop the concept of **conditional probability.** If gender and major are independent, the probability that a person is an education major is the same (.47) regardless of his or her gender. However, given that we know the individual selected is a female, we have seen that the probability that she will be an education major is .62 rather than .47; .62 is the probability of an education major, given the condition that the person is female. *Conditional probability refers to the probabilities that hold **within a single row or column** of the contingency table.* The condition is being in a particular row or column of the table. For example, gender and college major would be independent if the probability that a person in Mary Stuart's sample was an education major remained .47 regardless of whether the person is male or female.

There are ways to calculate conditional probabilities directly from the contingency table, but our interest in them is restricted to what they can tell us about the association between nominal variables. *Two nominal variables are independent if the probabilities for the categories of the first variable do not change when you move from one category to another of the second variable.* Two nominal variables are related or associated if the probabilities for the categories of the first variable do change when you move from one category to another of the second variable. Another way to say this is that gender and college major are related if the pattern of relative frequencies of majors in Table 15-3 is different for males than for females.

Expected Frequency. As you will recall from Chapter 6, the addition rule of probability dealt with questions of the *either A or B* form, and the multiplication rule of probability was used for questions about the probability of *both A and B*. Each cell in a contingency table satisfies two conditions, the row condition and the column condition. Therefore, if the variables of major and gender are independent, the multiplication rule tells us that each of the joint probabilities in Table 15-3 should be equal to the row probability multiplied by the column probability. The probability of a male psychology major should be $.48 \times .34 = .16$, and the probability of a female education major should be $.52 \times .47 = .24$. The multiplication rule *will apply* if the variables are independent, and it is this property that gives us the pattern of frequencies we should expect if the null hypothesis of no relationship between the variables is true. Our null hypothesis—that the variables are independent—predicts that each observed cell probability, such as the probability of a female sociology major, will equal the product of its marginal probabilities ($.52 \times .20$). If our data are not consistent with this expectation, then we reject the null hypothesis and conclude that the variables are not independent.

As we have just seen, when two variables are independent, the multiplication rule tells us what cell probabilities we should expect. If two variables are independent, the pattern of relative frequencies we observe in our data should closely resemble the expected frequencies. It can be shown that the distribution of the *discrepancies between observed and expected frequencies* very closely approximates the χ^2 distribution. When one or both of the variables has at least three categories and there is a fairly large number of observations, the approximation is very good. This will allow us to use the χ^2 distribution to test the null hypothesis that two nominal variables are independent of one another.

Conditional probability is the probability of being in category X (for example, education major) of one variable, **given** that the person is in category Y (for example, female) of the other variable.

Two nominal variables are independent if the probabilities for the categories of the first variable do not change when you move from one category to another of the second variable.

Two nominal variables are associated if the probabilities for the categories of one variable change depending on the category of the other.

Table 15-4	Cross-Tabulation of Marital Status and Life Satisfaction			
	Life Satisfaction			
Marital Status	*High*	*Moderate*	*Low*	**Marginal Frequencies**
Never married	31	53	61	145
Now married	102	112	104	318
Divorced or Widowed	10	47	7	64
Marginal Frequencies	143	212	172	527

Let's take a slightly larger example to illustrate the use of χ^2 for testing hypotheses about the independence of two variables. Suppose that we are conducting an investigation of marital status and life satisfaction. The categories of our two variables are shown in Table 15-4. For the purposes of this analysis, each observation yields two pieces of information about an individual: her/his marital status and stated life satisfaction. If Ahab states that he has never been married and is very well satisfied with his life, a tally mark representing him is placed in the upper-left cell; Jezebel, who is married and claims moderate life satisfaction, would be tabulated in the middle-middle cell, and all other people in the sample would be handled in a similar way, each contributing one observation to the frequency in the appropriate cell. Data we might collect in such a study are included in Table 15-4. The values in the cells are called the **observed cell frequencies.**

Observed cell frequencies *are the actual numbers of cases occurring in each cell of the contingency table.*

The first thing that we must do to analyze the results of our study is to compute the **expected cell frequencies** under the null hypothesis that the two variables are unrelated. The marginal frequencies for each variable give the number of individuals in each category *of each variable considered singly.* If there is no relationship between the variables, we should expect the cell frequencies in each column to be proportional to the marginal frequencies on the right. Likewise, the cell frequencies in each row should be proportional to the marginals on the bottom. That is, the multiplication rule should apply.

The **expected cell frequencies** *are the cell frequencies calculated from the marginal frequencies. They are the frequencies we would expect if the null hypothesis is true.*

The marginal frequencies give us the probability distribution for each variable *without consideration of the other variable.* For example, the chance that a randomly selected individual from Table 15-4 will express moderate life satisfaction is 212/527 or .40228; likewise, the probability that a person will never have been married is 145/527 or .27514.

If (and only if) the two variables are statistically independent, the cell frequencies within each row (or column) of the table will be proportional to the row (or column) marginal frequencies *within random variations due to sampling.* For example, if the two variables of life satisfaction and marital status are statistically independent, the cell frequencies in the never-married row should occur in the same proportions as the marginal frequencies at the bottom of the table. Likewise, the cell frequencies for the high satisfaction column should be proportional to the marginals on the right of the table, as should those for the moderate and low satisfaction columns.

Under the assumption (our null hypothesis) that the two variables are unrelated (that the relative distributions are the same in each column), we use the multiplication rule to calculate the probability that a person will be both moderately satisfied and never married (that is, fall in the top-middle cell of the table). The probability under the null hypothesis is

$$(.40228)(.27514) = .11068$$

This is the probability, or relative frequency, with which we would expect to find never-married, moderately satisfied people in our sample *if the two variables are independent.* This relative frequency can be converted into an expected cell frequency by multiplying it by the total number of observations; in our example, the result is

$$.11068 \cdot 527 = 58.33$$

Using the subscripts r and c to designate any particular row or column, expected cell frequencies can be computed directly from the marginal frequencies by the general equation

$$E_{rc} = \frac{N_r N_c}{N_T}$$

(15.1)

Table 15-5 Computing Chi-Square for Data in Table 15-4

Marital Status	Life Satisfaction			Marginal Frequencies
	High	Moderate	Low	
Never Married	$O = 31$ $E = 39.35$ $(O - E)/E = 1.77$	$O = 53$ $E = 58.33$ $(O - E)/E = 0.49$	$O = 61$ $E = 47.32$ $(O - E)/E = 3.96$	145
Now Married	$O = 102$ $E = 86.29$ $(O - E)/E = 2.86$	$O = 112$ $E = 127.92$ $(O - E)/E = 1.98$	$O = 104$ $E = 103.79$ $(O - E)/E = 0.00$	318
Divorced or Widowed	$O = 10$ $E = 17.37$ $(O - E)/E = 3.13$	$O = 47$ $E = 25.75$ $(O - E)/E = 17.54$	$O = 7$ $E = 20.89$ $(O - E)/E = 9.24$	64
Marginal Frequencies	143	212	172	527

Calculations

$\chi^2 = 1.77 + 0.49 + 3.96 + 2.86 + 1.98 + 0.00 + 3.13 + 17.54 + 9.24$
$= 40.97$

where

E_{rc} is the expected frequency in cell rc, the cell at row R and column C,
N_r is the marginal frequency in row R,
N_c is the marginal frequency in column C, and
N_T is the total number of observations.

For our example, equation 15.1 yields

$$E_{12} = \frac{145(212)}{527} = 58.33$$

as the expected frequency in cell$_{12}$, which is the same value we obtained using the product of probabilities. We can use equation 15.1 to compute an expected frequency for each cell of the table.

The expected frequencies computed with equation 15.1 are then used with the observed frequencies to compute a value of χ^2 for the entire table. The formula for this application of χ^2 is

$$\chi^2 = \sum_{r=1}^{R} \sum_{c=1}^{C} \frac{(O_{rc} - E_{rc})^2}{E_{rc}} \tag{15.2}$$

For each cell rc we (1) find the *squared difference* between the observed cell frequency O_{rc} and the expected cell frequency E_{rc} and (2) divide the squared difference by the expected cell frequency. These cell values are then (3) summed across the R rows and C columns to give a value for the entire table.

The steps for computing χ^2 in our example are illustrated in Table 15-5. The three numbers in each cell are (1) the observed frequency (from Table 15-4), (2) the expected frequency (calculated by equation 15.1), and (3) the value of $(O - E)^2/E$ for that cell. Thus, for the upper-left cell ($r = 1$, $c = 1$), we obtain

$$\frac{(31 - 39.35)^2}{39.35} = \frac{(-8.35)^2}{39.35} = \frac{69.7225}{39.35} = 1.77$$

The same procedure is repeated for each cell in turn. Then we add across all cells to obtain a total χ^2 for the entire table of 40.97.

▶ Now You Do It

In the previous Now You Do It we prepared the contingency table for gender and college major for the BTU study. What are the expected frequencies and what is the value of χ^2? The marginal frequencies for female and psychology major were 29 and 20 respectively. If you apply equation 15.1 to these marginals, you should get an expected frquency of 8.53. The other expected frequencies that you should get are given in their appropriate cells below, after the observed frequencies. The χ^2 for this table is .073.

	Psychology	Education	Other	Row Marginal
Female	9/8.53	11/11.09	9/9.38	29
Male	11/11.47	15/14.91	13/12.62	39
Column Marginal	20	26	22	68

Degrees of Freedom. We must have a value for degrees of freedom in order to look up the critical values for χ^2. Remembering that in each row of the table, the cell entries must sum to the marginal frequency, we can see that if there are C (the number of columns in the table) cells in each row, $(C - 1)$ of the cells are free to vary. Once the first $(C - 1)$ cell frequencies have occurred, the value of the last is fixed. Given values of 31 and 53 for the first two cells in row 1 of Table 15-4, the last must be 61 in order for the sum to be 145.

The same principle holds in each column. All but one of the cells are free to vary. There are $(R - 1)$ degrees of freedom in each column and $(C - 1)$ degrees of freedom in each row. Since the total number of cells is given by the product $R \cdot C$, the degrees of freedom for the whole table is given by

$$df = (R - 1)(C - 1) \tag{15.3}$$

In our example, this value is

$$df = (3 - 1)(3 - 1) = 4$$

Looking in Appendix Table A-3, we find the .05 critical value for χ^2 with $df = 4$ to be 9.5. The critical value for an α of .001 is 18.5. Clearly, our obtained result exceeds these values.

At this point we should consider again what hypothesis is being tested by this χ^2. Our reasoning proceeds as follows:

1. We assume that the marginal frequencies reflect the population with respect to the relative frequency of each category of each variable.
2. Our null hypothesis is that the two variables are independent. Under this condition, the expected cell frequencies are given by equation 15.1.
3. If the null hypothesis is true, the empirical cell frequencies in the population will be equal, within sampling error, to the expected frequencies.
4. If the null hypothesis is true and we draw a very large (in theory, infinite) number of samples from the population and compute a value of χ^2 as defined by equation 15.2 for each sample, the frequency distribution of these sample statistics will have approximately the shape of a χ^2 distribution with $(R - 1)(C - 1)$ degrees of freedom.
5. If the null hypothesis is true, the probability that we will draw a sample with a value of χ^2 greater than the critical value in the χ^2 distribution for our selected value of α (for example, $\alpha = .01$) is no greater than α.
6. If the sample value of χ^2 exceeds the critical value, we conclude that the null hypothesis is false. Because we are not predicting a direction of the differences, the null hypothesis is, by its nature, a two-tailed hypothesis.

Note that once again our null hypothesis has specified a sampling distribution (in this case for the test statistic, χ^2) that we would expect if the null is true, and we ask the question, "How likely is it that we would get results like those in our sample from a distribution like the one specified in the null hypothesis?" If it is sufficiently unlikely, the result is statistically significant and we reject H_0.

Interpreting the Results. The substantive conclusion that we reach upon rejecting the null hypothesis depends, of course, on the nature of our original research question. In the example we have been discussing, our research interest was in the relationship between life satisfaction and marital status. The null hypothesis was that the two variables are unrelated, which means that we expect to find no systematic tendency among people of a particular marital status to state a particular level of life satisfaction, and vice versa. Since our obtained χ^2 of 40.97 exceeds the .001 critical value, we must reject the null hypothesis and conclude that there is a relationship between the two variables. The size of the χ^2 for each cell indicates the deviation from the expected value. Inspecting the cells leads us to conclude that people who fall in the divorced/widowed category tend to claim moderate satisfaction much more frequently and high or low satisfaction much less frequently than we would expect. Also, there is a tendency for never-married people to report their level of satisfaction as low. (Uncovering the deeper meaning of these make-believe findings is left to your imagination.) However, we must remember that examining the cell frequencies after the fact should lead us to new and more precise hypotheses for future research rather than to a claim that we have explained something. We should be able to predict discrepancies from independence and test for goodness of fit to our predictions before we claim much.

It should be clear that χ^2 is a statistic that characterizes a sample, just as t and Z are. Like t, χ^2 has a sampling distribution, based on its degrees of freedom, that would occur if the null conditions were true. A sample value which is unlikely to have come from that sampling distribution leads us to reject the null hypothesis.

HOW THE COMPUTER DOES IT

All major statistics packages have routines for computing χ^2 tests of independence, although they may not be where you would expect them. For example, SPSS has a Nonparametric option on the STATISTICS menu, but this is not where you find the χ^2 test of independence. It is located as a procedure within the Summarize option under the heading Crosstabs. If you select this option, you will be prompted to enter a row variable and a column variable (the program will actually let you do more than two variables at a time, but let's keep it simple). You must also use the Statistics option button to select the χ^2 statistic or you will only get the table. The screen looks like this:

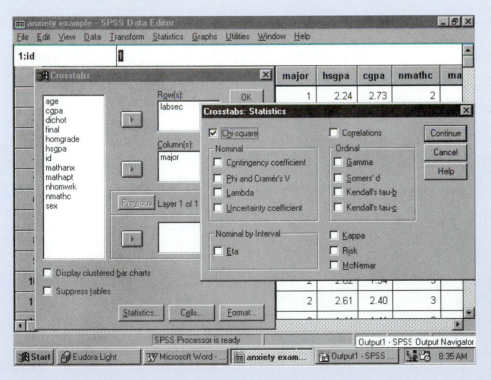

The options under Nominal, Ordinal, and Nominal by Interval provide other analysis alternatives that you will find described in texts such as those listed at the end of this chapter. In addition, you can use the "Cells" button to have the program provide expected cell frequencies and a variety of percentages and residuals. The output from the program includes the χ^2 we have been discussing, called Pearson's χ^2, and two other alternative indices as well as the proportion of the χ^2 distribution falling to the right of the obtained value.

Dr. T. Leary is interested in patterns of drug use on campus. Specifically, he would like to know if there is an association between class and the number of times students report using drugs. To test his research hypothesis that there is a relationship, he sets up a null hypothesis of independence and collects some data using a questionnaire. One item on the questionnaire asks students how often during the past term they have used illegal drugs. Students responded on a scale of not at all, once or twice, or more than twice. He collected responses from 239 students with the following results.

| Class | Frequency of Drug Use | | | |
	Never	1 or 2	>2	Totals
Freshman	16	20	17	53
Sophomore	8	25	40	73
Junior	30	22	15	67
Senior	28	10	8	46
Totals	82	77	80	239

This analysis clearly calls for a χ^2. What steps must we take?

The values in the table are the *observed* frequencies. What else do we need? First, we need the *expected* frequencies. We can compute these using equation 15.1. For the first cell, we obtain 53(82)/239 = 18.2. We must do the same for each of the remaining cells. We now have O for E for each cell.

The next step is to find $O-E$ for each cell. SPSS calls these values the residuals. The expected cell frequencies and residuals you should get are as follows. Notice that in each row and column the residuals sum to zero (with two small rounding errors). Do you think this might have something to do with degrees of freedom?

	Never	1 or 2	>2	Totals
Freshman	18.2/−2.2	17.1/2.9	17.7/−.7	53/0
Sophomore	25.0/−17.0	23.5/1.5	24.4/15.6	71/.1
Junior	23.0/7.0	21.6/.4	22.4/−7.4	67/0
Senior	15.8/12.2	14.8/−4.8	15.4/−7.4	46/0
Totals	82/0	77/0	80/.1	239

The only remaining steps are to (1) square the residuals, (2) divide each squared residual by the expected value in its cell, and (3) sum the results across cells. The result you get should be 41.53. Looking in the χ^2 table we find that the .01 critical value for $df = 6$ (remember, $df = (R − 1)(C − 1)$) is 16.8, so Dr. Leary can reject the null hypothesis and conclude that class standing and reported drug use are related. Inspecting the residuals, we find that the largest discrepancies are for the sophomores (more use reported than expected) and the seniors (less use reported than expected).

The χ^2 Goodness-of-Fit Test

Let's take a study of attitudes toward alcohol use on college campuses to examine another way that χ^2 can be used to answer research questions. A survey research firm has been hired to assess attitudes toward the use of alcohol. One of the statistical procedures they plan to use requires that the data have an approximately normal distribution. Therefore, prior to conducting the test, the researchers must determine whether the distribution of responses satisfies this condition. If the distribution is normal, approximately 16 percent of students should report attitudes at least one standard deviation below the mean (−1 SD), about 34 percent should fall between the mean and −1 SD, another 34 percent should fall from the mean to +1 SD, and about 16 percent should be above +1 SD. That is, the null hypothesis states that the responses will have a pattern of frequencies consistent with a normal distribution. The researchers have interviewed 586 students and obtained the results shown in Table 15-6.

Table 15-6	χ^2 Goodness-of-Fit Test for Attitudes Toward Alcohol				
	Below −1 SD	−1 SD to the Mean	The Mean to +1 SD	Above +1 SD	Totals
Expected	93.0	200.0	200.0	93.0	586
Observed	59	183	262	82	586
O–E	−34.0	−17.0	62.0	−11.0	0
χ^2	12.43	1.44	19.22	1.30	34.39

In this example the research question differs in a fundamental way from the questions we have asked in the past. Here, a specific distribution has been proposed, and the research question is whether the data are consistent with that distribution. The researchers think they know what the distribution looks like, and they are interested in supporting this belief. What they want to know is whether the data are consistent with the distribution they have proposed. Tests which ask whether the data are consistent with a particular proposed set of conditions, such as having a normal distribution, are known as goodness-of-fit tests because they ask whether there is a good fit between the proposal and the data.

The procedure for testing goodness-of-fit is a very simple modification of the χ^2 test we have just discussed. In the previous section we calculated our expected cell frequencies from the marginal frequencies. A larger-than-expected discrepancy between the E_{rc}'s and the O_{rc}'s caused us to conclude that the variables were not independent. In the goodness-of-fit application we assume that the expected frequencies are known from some other source (our model), so they do not have to be estimated. For example, if the null hypothesis that attitudes toward alcohol use are normally distributed is true, there should be about 93 students whose attitudes fall over one standard deviation below the mean, about 200 students in the next interval (−1 SD to 0), about 200 students within one SD above the mean, and another 93 students over one SD above the mean. The χ^2 test then answers the question, "Could the frequencies observed in my data reasonably have occurred in a random sample from a population with the characteristics assumed in the model?" The question is asked most clearly in terms of proportions or probabilities. If we know that in the general population 51 percent of live births are male babies, we can test whether some particular group has an abnormal balance of gender in their neonates. The fact that the proportions must sum to 100 percent tells us that if there are two categories, once we have specified the frequency in one, the frequency in the other is fixed, so there is one degree of freedom. Similarly, in a classification with six categories, five of the proportions are free to vary, but the last is fixed because the sum must be 1.0. Therefore, for the special case of a goodness-of-fit test, the χ^2 has $C - 1$ degrees of freedom, where C is the number of classification categories.

The expected frequencies in Table 15-6 showing student attitudes toward alcohol are obtained by multiplying the total sample size by the proportions expected in a normal distribution from Appendix Table A-1. For example, 15.87 percent of 586 is 93.0. This is

In the χ^2 goodness-of-fit test the researcher asks whether the observed frequencies could reasonably have come from a distribution specified as the model.

SPECIAL ISSUES IN RESEARCH

Goodness-of-Fit and Null Hypothesis Testing

In a way, tests of goodness-of-fit turn ordinary null hypothesis testing around and apply the logic backward (some would argue that goodness-of-fit testing is logical and null hypothesis testing is not). That is, in the normal application of goodness-of-fit procedures we believe we actually do know what "truth" is and we are asking whether our description provides adequate fit to the data at hand. In contemporary research the most common application of this logic is in testing what are called structural equation models. The researcher postulates a specific set of relationships, known as a model, among the variables he or she has observed in a research study. The object of the

analysis is to test whether the observed data might reasonably be considered a random sample from a population that is described by the model. A statistically significant result means that the model does not adequately account for the relationships observed in the data. The investigator's objective is to find a model that *does* fit; that is, one that *does not* lead to rejection of the null hypothesis. In an important way this is directly contrary to the approach used in traditional statistical inference. Most applications of this logic are well beyond our scope, but the χ^2 goodness-of-fit test is one place where we can use it and see how it works.

the frequency we would expect to find falling more than one standard deviation below the mean in a normal distribution. The observed frequencies, of course, come from the data. The values in the $O-E$ row, the differences between what we expect from our model and what we have found in our data, are the residuals which are useful for pinpointing the categories in which the discrepancies are largest. The final row, labeled χ^2, contains the results of equation 15.2 for that category. The total χ^2 is 34.39 with three degrees of freedom. The fact that this χ^2 is significant means that the data are not consistent with our null hypothesis. We cannot assume the data came from a population in which the distribution is normal.

▶ *Now You Do It*

Suppose the governing board of your university has decreed that the student body shall be composed of 25% freshmen, 20% sophomores, 25% juniors, 20% seniors, and 10% graduate students. One member of the board is an engineer who is a real stickler for compliance and detail, and she asks for a statistical analysis to prove that the institution is in compliance with the board's directive. You are selected to perform the analysis. How should you proceed?

There are 10,000 students at the university. The "population model" specified by the board calls for the proportions given above. Multiplying these proportions by 10,000 provides the expected frequencies listed in the table below. A count of the student body produced the observed frequencies listed in the table. Is there cause for concern?

Class	Observed Frequency	Expected Frequency	Discrepancy $(O-E)$	χ^2
Freshmen	2650	2500	+150	9.0
Sophomores	2050	2000	+50	1.25
Juniors	2375	2500	−125	6.25
Seniors	2075	2000	+75	2.81
Grad Students	850	1000	−150	22.5

The discrepancies between the observed and expected frequencies must sum to zero, so there are 4 df for this analysis. The χ^2 is computed using equation 15.2, producing the results at the right side of the table. The χ^2 for the entire table is the sum of the values for the five cells, 41.8. This value clearly exceeds all critical values for $df = 4$, so the composition of the student body differs by more than would be expected by chance from the distribution dictated by the board. Inspection of the table shows clearly that the lack of fit is due to an excess of freshmen and too few graduate students.

HOW THE COMPUTER DOES IT

SPSS computes the χ^2 goodness-of-fit test under the heading of "chi-square" within the Nonparametric procedures from the Statistics menu. First you specify the variable whose distribution is to be tested. Then you must decide whether all values of the variable are to be included (the default) or whether you wish to specify a smaller range of values. Finally, you must decide whether your model is that all categories have equal frequencies (the default) or you wish to specify other values. The program does not have an option for you to insert a cross-tabulation table, so you must be prepared to work from the raw data file. Suppose, for example, we had a model that said 50 percent of students in the BTU statistics class were education majors, 25 percent were psychology majors, and 25 percent were in the "other" category. Because psychology is coded 1, education is 2, and other is 3, you would set up the program like what is shown at top of the next page.

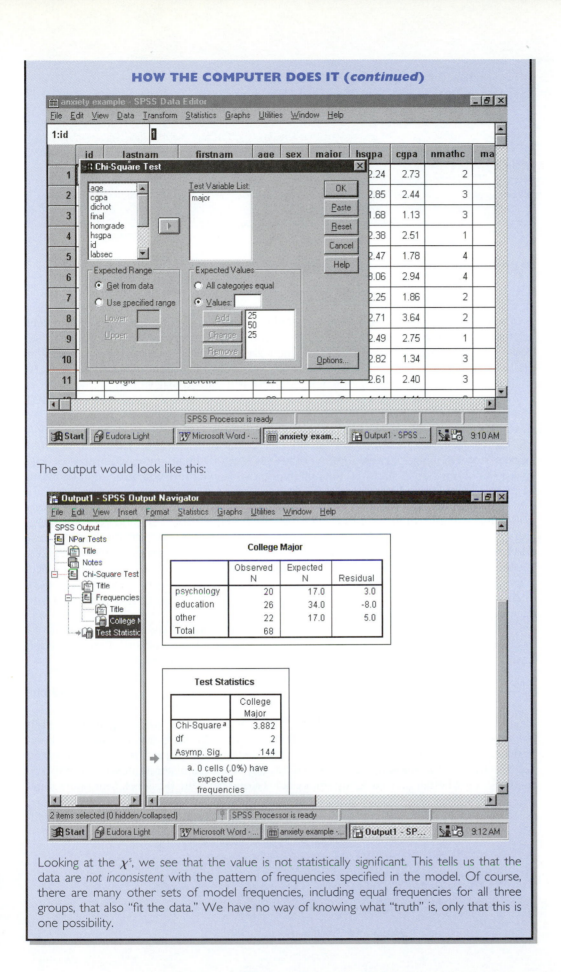

The output would look like this:

Looking at the χ^s, we see that the value is not statistically significant. This tells us that the data are *not inconsistent* with the pattern of frequencies specified in the model. Of course, there are many other sets of model frequencies, including equal frequencies for all three groups, that also "fit the data." We have no way of knowing what "truth" is, only that this is one possibility.

McNemar's χ^2 Test of Change

There are a number of areas of research in psychology and education where the variable of interest has two categories (it is a dichotomous variable) and the objective of the study is to see whether some treatment causes people to change categories. One popular current area of study for psychologists and health care professionals is smoking cessation. Likewise, contemporary educators are often concerned about whether an educational treatment changes students from nonmasters to masters of some set of educational objectives. In both cases, the dependent, or outcome, variable is a categorical variable (whether it is nominal or ordinal depends on whether you consider one category superior to the other) and the objective of the analysis is to determine whether the treatment has caused a change in category membership. The null hypothesis in a study like this is that *no more people change category in one direction than the other*. That is, if our concern is with whether people stop smoking, our research hypothesis is that our treatment causes more people to stop smoking than to start, and the null hypothesis is that there will be an equal number of people stopping and starting.

*Data organization for **McNemar's χ^2 test of change** involves a fourfold table in which one axis denotes status before intervention and the other axis denotes status after intervention.*

The basic structure of the data for a research question involving change in a dichotomous dependent variable is a *four-fold table* like the one shown in Table 15-7, and the appropriate test of the null hypothesis uses a variation of the χ^2 statistic known as **McNemar's χ^2 test of change.** The two axes of the table represent status before the treatment and status after the treatment. The generic representation of the categories is $(+, -)$, indicating that subjects may change from an unfavorable category such as smoker $(-)$ to a favorable category such as nonsmoker $(+)$ or vice versa. People who do not change ($[+,+]$ and $[-,-]$) are included in the table but omitted from the analysis.

The cells of the table in which change occurred are labeled A (change from $-$ to $+$) and D (change from $+$ to $-$), while the cells where no change occurred are labeled B and C. The χ^2 test of change is computed using the following formula:

$$\chi^2 = \frac{(A - D)^2}{A + D} \tag{15.4}$$

This χ^2, which has $df = 1$, is identical to what we would obtain using equation 15.2 but is easier to calculate and relates more directly to changes in status. Assume we have been conducting a study to determine whether exposure to an educational film about smoking has an effect on the smoking behavior of teenagers. We administer a questionnaire to 100 teens asking whether they currently smoke at least one cigarette per week (our operational

		Status Before Treatment	
Table 15-7 Contingency Table for McNemar's χ^2 Test of Change		$-$	$+$
Status After Treatment	$+$	A	B
	$-$	C	D

		Status Before Treatment	
		S	*NS*
Status After Treatment	*NS*	16	53
	S	25	6

Table 15-8 Contingency Table for Test of Change in Smoking

definition of being a smoker). Then we show them the film. A month later we again administer the questionnaire about smoking to the same students and obtain the results shown in Table 15-8. Can we conclude that the smoking behavior of this group has changed?

The table shows that 53 students reported being nonsmokers (NS) on both occasions, 25 were smokers (S) on both occasions, 6 students became smokers during our study, and

HOW THE COMPUTER DOES IT

You can access the McNemar test within the Crosstabs section of the Summarize procedure in the STATISTICS menu or in the section of the Nonparametric Statistics menu for two related samples. To use this test to assess change, your data must include two variables, one indicating status before "treatment" and the other status after treatment. Select one variable as the row variable, the other as the column variable, and then go to the Statistics optional menu. The screen should look like this:

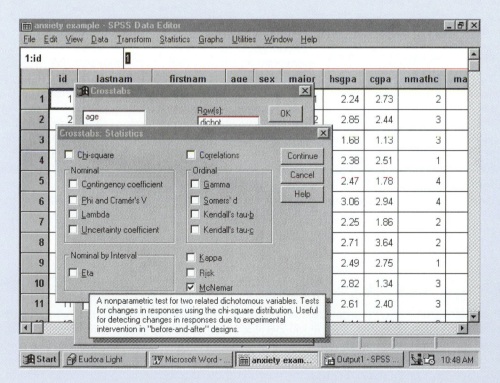

Click on the McNemar box, then click Continue and you will get your results. For the test to work properly, both variables must be coded in the same way. If, for example, "smoker" is coded 1 on the pretest variable, it must also be coded 1 on the posttest variable. The program uses the binomial distribution (see Chapter 6) to compute an exact probability when the number of people changing category is small. (This application of the binomial distribution is beyond our scope.)

16 students gave up smoking under our definition of smokers and nonsmokers (that is, 22 students changed category). If the null hypothesis, that change in either direction was equally likely, is true, we would expect 11 students to have started smoking and 11 to have given it up. The χ^2 computed by equation 15.4 tests this null hypothesis.

$$\chi^2 = \frac{(A - D)^2}{A + D} = \frac{(16 - 6)^2}{16 + 6} = 4.55$$

With one degree of freedom, this χ^2 is significant at $\alpha = .05$, but not at $\alpha = .01$. Therefore, there is some indication that a greater than random change occurred, and inspection of the cells of the table suggests that more students gave up smoking than started.

▶ Now You Do It

Dr. Numbercruncher, the chair of the SAD Department at BTU, has decided to conduct a study to see whether it is possible to change students' levels of math anxiety using a program of behavior modification. The 68 students who are participating in the department's larger study of anxiety and performance take part. Each student is hooked up to an instrument that measures anxiety and can administer an electric shock. Students are then shown math problems of varying complexity and, if they respond with an anxiety response, they are shocked. Dr. Numbercruncher believes the shocks will suppress the anxiety responses. She believes more students will change from being anxious to being nonanxious. What do you expect will happen?

The data Dr. N. obtained are shown here. What is her null hypothesis, and what decision should she reach?

		Status Before Treatment	
		Anxious (−)	Nonanxious (+)
Status After Treatment	Nonanxious (+)	5	14
	Anxious (−)	29	20

Answer

Dr. Numbercruncher's research hypothesis is a one-tailed hypothesis that her treatment will reduce anxiety, or that there will be more changes from anxious to nonanxious than vice versa. Therefore, her null hypothesis is that there will not be more changes from − to + than from + to −. Because the difference (A–D) is squared, the test is inherently two-tailed. The .01 tabled critical value for χ^2 is 6.6, but this is a two-tailed critical value for the same reason that F is inherently two-tailed. The χ^2 from the data is 9.0, so we can say with confidence that change is occurring on the basis of the statistical test, but examination of the cell frequencies shows that the change is in the direction opposite to that hypothesized by Dr. Numbercruncher. Therefore, she cannot reject her null hypothesis.

Spearman's Rank Order Correlation

There are sometimes situations in psychological and educational research where we wish to have a measure of the association between two variables when one or both of the variables are measured on an ordinal scale. For example, we might have a group of people who have been rated for physical attractiveness (call this Variable X) and sociability (call this Variable Y). We wish to determine whether there is a relationship between a person's perceived sociability and his or her rated physical attractiveness, but we are unwilling to assume that the ratings represent an interval scale.

The usual way to handle this problem is to treat the scores explicitly as ordinal information in the form of ranks. Ranks may be assigned so the person with the highest numerical score receives a rank of 1, the next person is ranked 2, and so on until everyone has been ranked, or the reverse order may be used. Be consistent.

A procedure for computing the similarity between two sets of ranks was proposed by Charles Spearman in 1902. Called Spearman's rho (ρ_S), the index uses the difference between ranks to compute a measure of association. (Note that this is *not* the population correlation, ρ, that we discussed in Chapter 14.) We proceed in very much the same way that we did for the matched-pairs t-test in Chapter 13. After converting the scores to ranks, we find the difference between the rank of each person on X and on Y. If we call this difference for person i d_i, then

$$d_i = (\text{rank on } X) - (\text{rank on } Y)$$

and the rank order correlation is given by the formula

$$\rho_S = 1 - \frac{6 \sum d_i^2}{N(N^2 - 1)} \tag{15.5}$$

where the second term on the right is a measure of lack of agreement. (If the ranks agreed perfectly, all of the d_i's would be zero, and the correlation would be 1.0.)

Consider the two sets of ratings for 10 people on our attractiveness and sociability variables listed in Table 15-9. The original data (ratings) are given in columns 2 and 3, the ranks based on the ratings in columns 4 and 5, the d_i's in column 6, and the d_i^2's in the last column. (Notice that the d_i's sum to zero.) The value for ρ_S is computed at the bottom of the table. This ρ_S of $+.30$ suggests that there is a moderate correlation between rated attractiveness and rated sociability.

A Spearman ρ_S is really a product moment correlation such as we computed in Chapter 8, but between two sets of ranks. If you apply the computational formula for the correlation (equation 8.3) to the ranks in Table 15-9, you will get the same value that we got in the table using equation 15.5. However, this is not the value you will get if you apply equation 8.3 to the original ratings. The result of that computation is $+.447$. If a set of scores reasonably can be assumed to have interval scale properties, the correlation should be computed on the raw scores because the ranking procedure throws away the interval information.

Test of Significance

The test for statistical significance of ρ_S (H_0: $\rho = 0$) is identical to the t-test for the significance of a correlation. That is, if you substitute ρ_S for r in equation 14.9, the resulting t with $df = N - 2$ is a test of the hypothesis that the correlation in the population is zero. For our example,

$$t_{(N-2)} = \frac{\rho_S - \rho}{\sqrt{\dfrac{(1 - \rho_S^2)}{N - 2}}} \tag{15.6}$$

$$= \frac{.30 - 0}{\sqrt{\dfrac{(1 - .30^2)}{10 - 2}}} = .889$$

Table 15-9 Computation of Spearman's Rank Order Correlation Between Physical Attractiveness and Sociability

	X Score	Y Score	X Rank	Y Rank	d_i	d_i^2
1	42	130	4	7	−3	9
2	36	133	8	5	3	9
3	44	140	2	2	0	0
4	45	119	1	9.5	−8.5	72.25
5	32	119	9	9.5	−.5	.25
6	37	135	7	3	4	16
7	27	123	10	8	2	4
8	41	134	5	4	1	1
9	40	132	6	6	0	0
10	43	142	3	1	2	4
						$\sum d_i^2 = 115.50$

$$\rho_S = 1 - \frac{6 \sum d_i^2}{N(N^2 - 1)} = 1 - \frac{6(115.5)}{10(100 - 1)} = 1 - .70 = .30$$

Obviously, this t with $df = 8$ is not statistically significant. Therefore, we cannot conclude that this sample came from a population in which ρ_S is other than zero.

There is one additional restriction that applies to ρ_S. Fisher's Z transformation is not appropriate with ρ_S, so we cannot test hypotheses that involve nonzero parameters for the correlation or find confidence intervals. Also, we cannot find a mean correlation, so if we wish to combine correlations from several groups we must use the median correlation.

Procedures for Handling Ties

Spearman's ρ_S ordinarily is used with small samples. One reason for this is that as the number of observations gets large we are likely to encounter an increasing number of tied ranks (we had one tie between subjects 4 and 5 in our data). A large number of tied ranks reduces the accuracy of the correlation. Also, with large samples the difference between the rank correlation and an ordinary correlation between the original scores diminishes, so investigators are likely to use an ordinary correlation coefficient with larger samples.

When two or more individuals receive the same score, and hence would be assigned the same rank, the usual practice is to average the ranks the people would have received if their scores had been slightly different and assign the average rank to all people who are tied. The next person in the list would receive the rank immediately below the lowest of the ranks that were averaged. For example, if we have the set of 10 scores

$$4, 5, 7, 7, 12, 16, 16, 16, 19, 22$$

rank 1 is assigned to the score of 22 and rank 2 to 19, but there are three people at 16. If these people were not tied, they would receive ranks 3, 4, and 5, so the correct procedure is to average these ranks and assign all three individuals a rank of 4 ($[3 + 4 + 5]/3$). The next person, with a score of 12, is assigned rank 6. The two people tied at 7 are both assigned the rank of 7.5 ($[7 + 8]/2$). Finally, ranks 9 and 10 are assigned to the last two scores. The pairing of the original scores with their ranks would be

Score	4	5	7	7	12	16	16	16	19	22
Rank	10	9	7.5	7.5	6	4	4	4	2	1

We use similar ranking methods and procedures for ties in relation to the next two statistical tests.

▶ *Now You Do It*

If you use a computer to carry out the analysis of the relationship between math anxiety and final exam score for the BTU data, you will need to transform the variables to ranks, then run the correlation program. You should get a correlation between these variables based on ranked data of $-.549$. (Be careful how you assign ranks. You could reverse the sign of the correlation if you rank lowest anxiety 1 and highest final score 1.)

For a smaller example, suppose we are conducting a study of the relationship between body weight and humor. We have a measure of responsiveness to humor and one of body weight in stones (an ancient Scottish unit of weight) for each of eight people. Because we are uncertain about the scale properties of the humor variable, we elect to use a Spearman correlation. The raw scores are as follows:

Weight	15	18	12	12	17	15	19	14
Humor	8	5	7	6	8	4	9	9

What are the ranks, and what is the correlation?

Answer

Weight	4.5	2	7.5	7.5	3	4.5	1	6
Humor	3.5	7	5	6	3.5	8	1.5	1.5
d_i	1	−5	2.5	1.5	−.5	−3.5	−.5	4.5

For the correlation you should get .196, which is not statistically significant ($t_{(6)} = .49$).

The Wilcoxon-Mann-Whitney Test

Bert Mancuso is a developmental psychologist who is interested in the development of aggressive behavior in children. As part of a larger study, he has been observing a group of eight children, five boys and three girls, in a period of free play, and he has rated their aggressiveness. He wants to test the null hypothesis that the two groups do not differ. Each child has received a rating on a scale from 1 (no aggressive behaviors) to 20 (frequent aggression), producing the following scores:

> Boys' Scores: 11, 4, 10, 8, 6
>
> Girls' Scores: 13, 5, 15

Bert is not willing to assume that his ratings represent an interval scale. How can he test the null hypothesis that the groups do not differ in aggressiveness?

It is not unusual in psychology and education to encounter situations like this one, where we have data from two groups, but the dependent variable does not meet the assumptions of an interval scale or it has a highly skewed distribution. When either of these cases occurs, we must have an alternative to the t-test as a test for the equality of location or central tendency for the two groups. In the case where the two groups are independent (different individuals compose each group), one popular test has been the Wilcoxon test and another has been the Mann-Whitney U-test. Recent work has shown that these are really equivalent tests, so we discuss them as a single procedure. They have the advantage of being among the most powerful of the nonparametric tests. In this section we present a version of the test modeled on the treatment by Siegel and Castellan (1988, pp. 128–136) which has the advantage over some other presentations that the associated tables for testing statistical significance offer both upper and lower critical values. This enables us to test either directional or nondirectional hypotheses concerning ranked data.

Bert's basic null hypothesis is that sets of ratings for boys and girls are random samples from a single population (or two equivalent populations). If the null hypothesis is true, when the two groups are combined and the scores are assigned ranks in the combined group, we should expect the average of the ranks for the boys to be equal to the average of the ranks for the girls, within a reasonable margin for sampling error.

The principle underlying the Wilcoxon-Mann-Whitney test is that for any given number of observations in the combined group, the *sum of the ranks* will be a constant equal to $N(N + 1)/2$. (In Bert Mancuso's case, the sum of the eight ranks will be 36.) If there is the same number of individuals in each of two groups, the sums of the ranks of the two groups should be approximately equal, and if the groups are of unequal size, the sums of the ranks should be approximately proportional to the sample sizes. We would reject the null when the sum of the ranks in one group is larger or smaller than would be expected by chance under the null hypothesis conditions.

We can rank the eight scores for the children in Bert Mancuso's study as a single group, keeping group identification. This produces

Score:	4	5	6	8	10	11	13	15
Rank:	1	2	3	4	5	6	7	8
Group:	B	G	B	B	B	B	G	G

The sum of these 8 ranks is $8(9)/2 = 36$ and the average rank is 4.5. If the null hypothesis is true, we should expect the average rank for the boys to be equal to the average rank for the girls. In terms of the test statistic, we should expect the sum of ranks for the boys to be about 5/8ths of the total, or 22.5, and the sum of ranks for the girls to be about 3/8ths, or 13.5. For these data, the average ranks are 5.67 for the girls and 3.80 for the boys. The sums are 19 for the boys and 17 for the girls. Is this difference larger than we would expect to obtain by chance?

The Wilcoxon-Mann-Whitney test statistic is the *sum of the ranks for the group with fewer members* (in this case, the girls). Call this sum W_S, the Wilcoxon sum for the smaller group. Summing the ranks for the three girls gives us a total of

$$W_S = 2 + 7 + 8 = 17$$

Tables for testing the significance of W_S for small groups are provided in Appendix Table A-7. There is a separate section of the table for each possible number of members of the *smaller* group (N_S) from $N_S = 3$ to $N_S = 10$. In each section of the table there is a column for each possible number of members of the larger group (N_L) from $N_L = N_S$ to $N_L = 10$ (more for small values of N_S). When the number of members of the larger group exceeds 10, a normal distribution approximation, described later, is used. Since there are 3 members in our smaller group, we go to the first section of the table, reproduced in part here as Table 15-10.

There are three basic kinds of information in the table. The first column gives a lower bound sum (S_{Low}) for W_S. After this, the columns occur in pairs. Let us look at the columns that apply to the example we are following. Because there are 5 children in the larger group, this is the pair of columns headed $N_L = 5$. The first column of the pair contains the probability of a value of W_S *as small as or smaller than* the sum in the S_{Low} column for $N_L = 5$. Then comes a column of upper bound sums (S_{Up}) for $N_L = 5$. The single lower bound column on the left is used in conjunction with each of the *pairs* of columns to the right.

The probabilities of various values of W_S for S_{Low} and S_{Up} are symmetric. Thus, for $N_S = 3$ and $N_L = 5$, the maximum possible sum of ranks for the any three of the eight individuals is $6 + 7 + 8 = 21$. Therefore, the probability that W_S will be 21 or less is 1.00, and the probability that it will be greater than 21 is 0.000. The probability that the sum

Table 15-10   Lower- and Upper-Tail Probabilities for W_S

			$N_S = 3$				
S_{Low}	$N_L = 3$	S_{Up}	$N_L = 4$	S_{Up}	$N_L = 5$	S_{Up}	
6	.0500	15	.0286	18	.0179	21	
7	.1000	14	.0571	17	.0357	20	
8	.2000	13	.1143	16	.0714	19	
9	.3500	12	.2000	15	.1250	18	
10	.5000	11	.3143	14	.1964	17	
11	.6500	10	.4286	13	.2857	16	
12	.8000	9	.5714	12	.3929	15	
13	.9000	8	.6857	11	.5000	14	
14	.9500	7	.8000	10	.6071	13	
15	1.0000	6	.8857	9	.7143	12	
16			.9429	8	.8036	11	
17			.9714	7	.8750	10	
18			1.0000	6	.9286	9	
19					.9643	8	
20					.9821	7	
21					1.0000	6	

will be 20 or less is .9821, and the probability that it will be greater than 20 is .0179 (the two probabilities sum to 1.00). If you look carefully at each column you will see that it is symmetric with the exception of the probability of 1.00 for the maximum possible that W_S can have for that number of cases.

The S_{Up} columns contain the value of W_S for the upper tail of the distribution. The probability paired with each of these sums is the probability of a W_S *as great as or greater than* the tabled entry. For $N_S = 3$ and $N_L = 5$, the probability of a sum of ranks of 21 (more is not possible) is .0179, and the probability of a sum of 20 or more (20 or 21) is .0357. The S_{Low} column looks at the probabilities from the bottom up, and the S_{Up} column looks at them from the top down (the probability of a sum of 6 or greater is 1.00 because a smaller value for three ranks is not possible, just as the probability of a sum of 21 or less is also 1.00). Because of this property of symmetry between the upper and lower sums, only nonrepeating portions of the table are given for larger values of N_S and N_L.

In our example, the sum for the smaller group is 17. We enter the table with $N_S = 3$, $N_L = 5$, and $W_S = 17$. Here we find in the first row the values 6 (from the S_{Low} column), .0179, and 21. The value .0179 is the probability that we would get a W_S of *6 or less* for the sum of ranks in the smaller group (that is, the probability that the three girls received ranks of 1, 2, and 3). It is also the probability of a sum of 21 (S_{Up}) *or more* for this group. These are the probabilities that are appropriate for a one-tailed hypothesis. (Of course, you must be careful to select the correct tail.) If, as was the case in our example, we did not specify a direction in our null hypothesis, then we must consider outcomes in either end of the distribution. The probability that we would find a sum for the smaller group that is either 6 or less or 21 or more would be .0179 + .0179 = .0358.

Now, moving down the S_{Low} column to the row containing 17 we find a value of .8750 in the probability column under $N_L = 5$. This is the probability that we would get a sample with a value of W_S of 17 or less in an experiment like this one (highly likely). Moving down the S_{Up} column to the value of 17, we find .1964. This is the probability of a sum of 17 or more for W_S under the conditions of this study. For our two-tailed null hypothesis, the probability of a Type I error if we decide to reject the null on this evidence would be 2(.1964) or .3928 because the tabled values are for only one tail of the distribution. Clearly, we would not reject the null.

When to Use S_{Low} and S_{Up}. We mentioned that, like the normal distribution, the probability distribution of sums for the Wilcoxon-Mann-Whitney test is symmetric. For this reason, only some of the probability values are listed for most combinations of N_S and N_L. Moreover, when you use Appendix Table A-7, you will use it somewhat differently, depending on whether the hypothesis is one- or two-tailed. If the hypothesis is one-tailed, use the S_{Low} column if the rejection region is in the lower tail and the S_{Up} column if the rejection region is in the upper tail. When you look up your sum, the probability in the table will be the probability of a sum that low or lower for S_{Low}. For S_{Up}, the probability in the table is for a sum that large or larger. In either case, if the probability associated with your sum is below your chosen value of α, you can reject the null hypothesis.

If your null hypothesis is two-tailed, it does not matter whether you consult the S_{Low} column or the S_{Up} column. Go to the appropriate probability column for your combination of N_S and N_L. If your sum appears in only one column, that is the probability to use, but you must double it to get the correct value for α. If your sum appears in both columns, as it may for smaller sample sizes and for values near the middle of the distribution, look in the rows for both entries. The probability to use is the smaller one associated with your sum. For example, if N_S is 3, N_L is 7, and the sum of ranks for the smaller sample is 24, this sum appears in both the S_{Low} and S_{Up} columns. The probability associated with an S_{Low} of 24 is .9667. This is the probability of a sum of *24 or less*. The probability associated with an S_{Up} of 24 is .0583. This is the probability of a sum for the smaller sample of *24 or more,* and this is the proper value to use. Because this is a two-tailed hypothesis, you must double the probability, so the chance that you would get a sum this extreme is .1166.

Large-Sample Methods

Suppose we are interested in the question of whether children who have seen others commit violent acts are themselves more prone to violence. We have shown 9 children from a class a film depicting one child aggressing against another. The other 13 children

◆ The Procedure in Words

The Wilcoxon-Mann-Whitney test is really much simpler than it seems. To test a null hypothesis of no difference between the groups, follow these steps:

1. Write down the scores in numerical order, keeping group identification.
2. Assign ranks to the combined group.
3. Find W_S, the sum of ranks in the group with fewer members.
4. Find the page in Appendix Table A-7 corresponding to the number of people in the smaller group.
5. Find the pair of columns corresponding to the number of observations in the larger group.
6. Find the row for $S_{Low} = W_S$. If the probability is less than .50, this is the value you need. For a one-tailed test, this is α if it is in the appropriate tail. For a two-tailed test, double the probability to find α.
7. If the probability found in step 6 is greater than .50, move to the S_{Up} column and find the appropriate sum. Proceed as in step 6.

in the class have not seen the film. We now observe the children for a period of a week and count the number of aggressive acts in which each child is involved. The results are shown in Table 15-11.

When both samples contain 10 or fewer observations, we must use the method described earlier to test the null hypothesis that the two groups do not differ. However, as sample size increases, the sampling distribution of the Wilcoxon-Mann-Whitney test quickly approaches the normal distribution. When the total number of observations is greater than about 15, as in the study of aggressive acts committed by 22 children, we can use a variation of the one-sample Z-test from Chapter 13 in place of Appendix Table A-7.

We proceed just as we did for the small-sample test. The children are ranked as a combined group, keeping track of group identity as shown in Table 15-11. Then W_S, the sum of the ranks for children in the smaller group, is computed. If we believe that seeing violent acts will result in the children committing more violent acts, our one-tailed null hypothesis is that the mean ranking of children seeing the violent film (μ_F) will not be higher than the mean ranking of the children not seeing the film (μ_{NF}). That is,

H_0: $\mu_F \leq \mu_{NF}$

Table 15-11 Number of Aggressive Acts by 9 Children Who Saw a Film Depicting Aggression and 13 Children Who Did Not See the Film

Saw Film		Did Not See Film	
Number of Aggressive Acts	Rank in Combined Group	Number of Aggressive Acts	Rank in Combined Group
3	13.5	0	2
6	19.5	4	16
5	17.5	1	5.5
1	5.5	0	2
5	17.5	7	21
2	9.5	2	9.5
2	9.5	3	13.5
6	19.5	1	5.5
8	22	0	2
		3	13.5
		3	13.5
		1	5.5
		2	9.5
Sum of Ranks	$W_S = 134$		

If we define $N_T = N_S + N_L$, then the mean of the distribution of W_S under the null hypothesis is

$$\mu_{W_S} = \frac{N_S(N_T + 1)}{2} \tag{15.7}$$

For our data from 22 children, equation 15.7 yields

$$\mu_{W_S} = \frac{9(23)}{2} = 103.5$$

The standard deviation of the sampling distribution of W_S is

$$\sigma_{W_S} = \sqrt{\frac{N_S(N_L)(N_T + 1)}{12}} \tag{15.8}$$

With $N_T = 22$, equation 15.8 produces a standard error of W_S of

$$\sigma_{W_S} = \sqrt{\frac{(9)(13)(23)}{12}} = 14.97$$

The statistical significance of W_S can then be tested with the formula

$$Z = \frac{W_S \pm .5 - \mu_{W_S}}{\sigma_{W_S}} \tag{15.9}$$

The correction of $\pm .5$ is applied because the sum statistic W_S is a discrete variable. By adding .5 for a lower-tail critical region or subtracting .5 for an upper-tail critical region, the rejection region includes all of the observed sum rather than going from the midpoint. Note that this is exactly the same procedure we used to correct the normal distribution approximation to the binomial distribution in Chapter 6.

Our hypothesis is one-tailed with the rejection region in the upper tail because we expect the children who have seen the film to behave more violently. With a rejection region in the upper tail, we must *subtract* .5 from our obtained W_S, so we calculate the Z for 133.5, the lower real limit of our obtained sum. Equation 15.9 produces

$$Z = \frac{(134 - .5) - 103.5}{14.97} = 2.004$$

which exceeds the .05 one-tailed critical value of 1.645. In fact, the probability of a Z of this size is about .023. Thus, we would conclude that the null hypothesis is false and our study does provide evidence that watching violence produces violent behavior.

HOW THE COMPUTER DOES IT

SPSS includes the Wilcoxon-Mann-Whitney test, listed as the Mann-Whitney test, as the default statistic under the Two Independent Samples option of the Nonparametric procedure on the STATISTICS menu. The screen prompts you to name a test variable (dependent) and a grouping variable (independent). You must also specify the coding of group membership for the grouping variable. This feature allows you to compare two groups from a multigroup independent variable, such as college major from the BTU data.

We mentioned earlier that the Wilcoxon test and the Mann-Whitney test are really the same test, but they approach the problem slightly differently and produce different but equivalent test statistics. The program provides both indices. For fewer than 30 cases, the program computes an exact probability for W_S such as the ones included in Appendix Table A-7. The printed value is two times the probability from our table (that is, a two-tailed test). The Z from equation 15.9 and its two-tailed probability are also provided, but this test does not employ the correction included in equation 15.9, so for small samples these results can be quite inaccurate and should be ignored. For more than 30 cases, only the normal distribution Z-test is used.

Six men and five women have been taking part in a student focus group on planning a new student union for Big Time U. At the end of the process, each was asked to rate the value of the focus group as a means of making decisions on a scale from 1 to 20. The results were as follows:

Men	4	7	10	11	13	18
Women	3	5	6	8	15	

Is there a tendency for one group to rate the focus group higher?

We could use a two-sample *t*-test for this analysis, but we might not want to assume that the ratings have a normal distribution or represent an interval scale. In that case, the most likely test to use is the Wilcoxon-Mann-Whitney. What steps must we follow?

First, we must determine the combined ranks for the two groups. There are 11 cases, so the ranks you should get are

Men	2	5	7	8	9	11
Women	1	3	4	6	10	

The sum of the ranks in the smaller group is 24. Looking in Appendix Table A-7 on the page for $N_S = 5$ in the column for $N_L = 6$ and the row for $S_{Low} = 24$ we find the entry .1645. This is the one-tailed probability, but since our hypothesis asked simply if there was a difference, we must double the value to obtain .329. We have no basis for concluding that the men and women differed in their perceptions of the focus group.

The Wilcoxon Signed Ranks Test

In our discussion of the *t*-test in Chapter 13 we saw that we could get a more sensitive test of our null hypothesis if the participants in the study were matched. The same principle can be applied to analyses that involve ranks. Suppose, for example, our study of the effect of viewing violence in films on the performance of aggressive acts was conducted on a sample of eight pairs of identical twins. One member of each pair was shown a film depicting children performing aggressive acts. The other member of each pair saw a similar film, but without any aggression between actors. Each group was then observed for two hours in a period of free play and the number of aggressive acts by each child was counted. Because we are not willing to consider the number of aggressive acts committed to be an interval variable, a *t*-test is not appropriate. However, the number of aggressive acts is clearly an ordinal variable, so the **Wilcoxon signed ranks test** is appropriate. The raw data from our study might look like the values in columns 2 and 3 of Table 15-12. The number of aggressive acts by the twin who viewed the violent film is given in the X_1 column, and the number of aggressive acts by the other twin is in the X_2 column.

The signed ranks test starts with the pairs of scores and finds the difference between them. That is,

$$d_i = X_1 - X_2$$

Table 15-12   Data From a Study of Aggressiveness for the Wilcoxon Signed Ranks Test

Pair	X_1	X_2	d_i	$\|d_i\|$	Rank	Signed Rank
1	5	3	2	2	3.5	3.5
2	8	4	4	4	6	6
3	5	4	1	1	1.5	1.5
4	3	5	−2	2	3.5	−3.5
5	7	2	5	5	7	7
6	7	1	6	6	8	8
7	4	5	−1	1	1.5	−1.5
8	6	3	3	3	5	5

d_i will be negative if X_2 is greater than X_1. The signed differences are shown in column 4 of the table. Next, the differences are rank ordered by their *absolute values* (the size of the difference without regard to sign). The absolute values of the differences ($|d_i|$) are entered in column 5, and their rank order, starting with the smallest difference, is shown in column 6. Finally, the sign for each difference is attached to the rank for that difference in the Signed Rank column (column 7). It is these signed ranks that provide the basis for the test statistic.

Null Hypothesis and Test Statistic. The null hypothesis that we test with the Wilcoxon signed ranks test is the same one we test with the matched-pairs *t*-test. The hypothesis may be one-tailed or two-tailed. In our example, we expect the children who have seen the violent film to be more aggressive. Therefore, the one-tailed null states that group 2 (the scores in column X_2) will be greater than or equal to group 1 (X_1). We will reject this hypothesis any time the members of group 1 systematically receive higher scores than their pairmates in group 2. The two-tailed null would simply state that the two groups will be equal. We would reject this hypothesis any time the differences are consistently in favor of either group.

The test statistic for the one-tailed test is the sum of the positive differences (S^+) or the sum of the negative differences (S^-), depending on which tail of the distribution contains the rejection region. If differences between the two groups are due to random events, the sum of the positive differences should be approximately equal to the sum of the negative differences. Whenever the sum we find in the predicted tail of the distribution is substantially larger than the other one, we conclude that the null hypothesis is false. Under our research hypothesis, we believe the twins in the X_1 column will perform more violent acts, which means we will reject the null hypothesis if S^+ is greater than we would expect. The test statistic for a two-tailed null hypothesis is the sum of the positive differences (S^+) or the sum of the negative differences (S^-), whichever is larger.

One-tailed probability values for the signed ranks test are given in Appendix Table A-8. In this table, N refers to the number of *pairs of observations*, eight in our example. There is a separate column for each value of N from 3 to 15. The values in the column labeled S are the possible values that the larger sum of the differences (S^+ or S^-) can assume. The values in the columns labeled 3 through 15 are the one-tailed probabilities of a sum that large or larger occurring for the given number of pairs. If your research hypothesis predicts that the values of the first variable will be larger than those of the second, as ours does, the positive sum should be greater than the negative sum, and this is the value you would use to enter the table. If your research hypothesis predicts that the values in the second column will be greater than those in the first column, the negative sum should be larger than the positive one, and you should enter the table with the negative sum. Remember that for a one-tailed test your hypothesis is supported only if the larger sum has the correct sign.

To use Appendix Table A-8, enter with the value of S and the number of pairs in your study. For example, the sum of the positive rank differences in our study is 31. Entering Appendix Table A-8 in the column labeled 8 and the row labeled 31, we find the value .0391. This is the probability of obtaining a sum of 31 or more from eight pairs of ranks if the null hypothesis is true. We can reject the null hypothesis at the $P < .05$ level. Because the values in Appendix Table A-8 are one-tailed probabilities, if your null hypothesis is two-tailed, you must double the probabilities in the table to determine your chances of making a Type I error.

Tied Ranks. Ties can occur in two ways in the Wilcoxon signed ranks test. The two members of a pair can each receive the same score, yielding a difference of zero, or two different pairs can have the same difference between their members, resulting in equal values of d_i. These two situations are handled differently.

The Wilcoxon signed ranks test assumes that the dependent variable is a continuous variable. This means that both members of a pair should not receive exactly the same score and that values of $d_i = 0$ should not occur. Some authors suggest that pairs for which $d_i = 0$ should be dropped from the analysis and that only the pairs for which d_i is not zero should be included in N. Other authors suggest that if there is an even number of pairs for which d_i is zero, the pairs should be divided between positive and negative signs. If there is an odd number of tied scores, one pair should be discarded and the others divided evenly between positive and negative. If you encounter this situation, you will have to decide how you wish to handle it.

Ties that involve equal differences in different pairs are handled just the way ties are handled in Spearman's ρ_S: the tied pairs are assigned the average of the ranks that they would have received if they were not tied. For example, pairs 3 and 7 in Table 15-12 both show a $|d_i|$ of 1. They would be ranked 1 and 2, but are both given a rank of 1.5. The same thing occurred for pairs 1 and 4, and they each received a rank of 3.5.

The Wilcoxon Signed Ranks Test With Large Samples. When the sample size exceeds 15, we can use the normal distribution as a good approximation to the exact probabilities for the Wilcoxon signed ranks test. It can be shown that the mean and standard deviation of S are

$$\mu_S = \frac{N(N+1)}{4} \tag{15.10}$$

$$\sigma_S = \sqrt{\frac{N(N+1)(2N+1)}{24}} \tag{15.11}$$

These values can then be used along with the value of S from the data to construct a critical ratio Z.

$$Z = \frac{S - \mu_S}{\sigma_S} = \frac{S - \left(\frac{N(N+1)}{4}\right)}{\sqrt{\frac{N(N+1)(2N+1)}{24}}} \tag{15.12}$$

For example, if we have conducted a study where 24 married couples have been asked to recall details of their relationship, we might ask whether men or women would recall more details. This would be a two-tailed null hypothesis. We form our differences by subtracting each husband's score from his wife's score. What are the mean and standard deviation under the null?

$$\mu_S = \frac{24(24+1)}{4} = 150$$

$$\sigma_S = \sqrt{\frac{24(24+1)(48+1)}{24}} = 35.0$$

If the value of S obtained from our sample is 223, then

$$Z = \frac{223 - 150}{35} = 2.09$$

This value exceeds the .05 critical value for a two-tailed critical ratio (1.96), so we would reject the null hypothesis that men and women recall equal numbers of details about their relationship.

HOW THE COMPUTER DOES IT

The Wilcoxon signed ranks test is available in SPSS in the Nonparametric Tests section of the Statistics menu. If you click on the 2 Related Samples option you will be prompted to enter the names of two variables. The Wilcoxon test will be performed by subtracting the second variable from the first. The Wilcoxon test is the default option in this procedure, but the McNemar test of change that we discussed earlier and a test called the Sign test are also available. Output from the Wilcoxon test includes the number of positive, negative, and tied ranks; the mean ranks, the positive and negative sums, and the significance test. The significance test is the Z statistic, regardless of sample size, and the P-value is always reported for a two-tailed hypothesis. For our eight-pair example, the program reports a Z of 1.825 and a probability of .068, a P-value which is close to the one we found in Appendix Table A-8.

▶ *Now You Do It*

George has been running a study of the effects of color contrasts on reaction time. Participants look at a computer monitor and color names are flashed on the screen in different colors. For example, the word RED might come on the screen in red letters or in letters of some other color. The task is to respond as quickly as possible by pressing one button if the name and color match and a different button if the name and letter color do not match. George believes that contrasting color and meaning will lead to slower reaction times (larger numbers). Participants serve in both conditions, so this is a repeated-measures study. Because reaction times show highly skewed distributions, the mean is not a good measure of central tendency. Therefore, he plans to analyze his data with a Wilcoxon signed ranks test. Scores for the two conditions are mean reaction times for 10 trials. Here are his data for 9 participants. See if you get the same results he did.

Name and Color Same	Name and Color Different
10.7	12.3
8.1	9.5
14.8	14.3
7.7	8.6
9.2	15.7
10.5	9.9
15.8	17.3
8.8	12.4
6.3	14.8

What steps must we take to analyze these data? The first task is to find the signed differences. Next, we must find the absolute values of the differences. Third, we rank the differences. Finally, we put the signs back on the differences and find the sums of the positive and negative differences. You should get what we show here.

X_1	X_2	d_i	$|d_i|$	Rank	Signed Rank
10.7	12.3	−1.6	1.6	6	−6
8.1	9.5	−1.4	1.4	4	−4
14.8	14.3	+.5	.5	1	+1
7.7	8.6	−.9	.9	3	−3
9.2	15.7	−6.5	6.5	8	−8
10.5	9.9	+.6	.6	2	+2
15.8	17.3	−1.5	1.5	5	−5
8.8	12.4	−3.6	3.6	7	−7
6.3	14.8	−8.5	8.5	9	−9

For these data, S^+ is 3 and S^- is 42. Because George has set up his data so that the differences are NO CONTRAST − CONTRAST, negative differences are consistent with his theory and the null hypothesis is $\mu_{S^+} \geq \mu_{S^-}$. He will reject the null hypothesis if S^- is sufficiently larger than S^+. Looking in Appendix Table A-8, we find that for an N of 9, the probability of getting a value of S^- of 42 or more is only .0098. George therefore has good reason to reject the null hypothesis and conclude that reaction times are slower when the color names are printed in contrasting colored letters.

SUMMARY

Special methods are required when the dependent variable in a study is measured on a nominal or ordinal scale. In this chapter we have covered three situations where the χ^2 distribution provides a test of our null hypothesis. The first and most common situation, the χ^2 *test of independence,* uses the addition and multiplication laws of probability to determine the set of frequencies that we would expect to find if the null hypothesis that the two variables are unrelated is true. This set of expected frequencies is compared with the observed frequencies in the data. The resulting χ^2 with $(R - 1)(C - 1)$ degrees of freedom is compared to the critical values of the χ^2 distribution. If the differences between the observed and expected frequencies are greater than would be likely by chance, the obtained χ^2 will exceed the critical value and we conclude that the variables are related; that is the pattern of frequencies for one variable is different at different levels of the other.

When a specific pattern of frequencies is expected on the basis of theory, the χ^2 *goodness-of-fit test* can be used to test whether the frequencies observed in our data differ from those predicted by our theory. In this special application there is only one row in the table, and the degrees of freedom are $(C - 1)$.

Another special χ^2 test is applied when the research question involves change in category membership. In this case, *McNemar's χ^2 test of change* is used. The columns of the data table refer to group before treatment and the rows to group after treatment. The data table includes four cells, those in the same category both before and after treatment, and those who have changed category. The null hypothesis is that people will have changed equally in both directions, and a significant value of χ^2 means that more people have changed in one direction than the other.

Spearman's rank order correlation (ρ_S) provides a measure of association for ordinal variables. Each person is ranked independently on each of two variables, and the correlation reflects the degree to which people tend to have similar ranks on both variables. The null hypothesis that $\rho_S = 0$ can be tested with a t-test that is identical to the one for use with the product moment correlation and interval variables.

The *Wilcoxon-Mann-Whitney test* can be used to test the null hypothesis that two groups do not differ on an ordinal variable. The two groups are ranked jointly, and the sum of the ranks for the smaller group is compared to critical values included in Appendix Table A-7. If there are more than 10 observations in the larger group, the one-group Z-test is used to test whether the sum in the smaller group differs from what would be expected, given the sizes of the two groups.

Finally, the *Wilcoxon signed ranks test* provides a nonparametric companion for the matched-pairs t-test. The absolute differences between paired observations are ranked. Then the signs are reassigned to the ranks and the positive and negative sums are computed. The appropriate sum is compared with the values in Appendix Table A-8 to find the probability of a sum that large or larger. The values in the table are one-tailed probabilities. For more than 15 pairs of observations there is a Z-test that compares the obtained sum with what would be expected if the null hypothesis is correct.

REFERENCES

Siegel, S., & Castellan, N. J. (1988). *Nonparametic statistics for the bebavioral sciences* (2nd ed.). New York: McGraw-Hill.

Wilcox, R. R. (1987). *New statistical procedures for the social sciences: Modern solutions to basic problems*. Hillsdale, NJ: Erlbaum.

EXERCISES

1. Theories of career development suggest that changes in career choices begin to emerge during the junior and senior year of high school when students become more realistic about their career choices. Freshmen (F) and seniors (S) at a particular high school were randomly selected to answer questions regarding their career choices. These responses were then coded into one of three categories: careers requiring no further education (N), careers requiring vocational/technical training (V), and careers requiring college training (C). The following data were collected:

Class	Educational Requirements of Career Choice	Class	Educational Requirements of Career Choice	Class	Educational Requirements of Career Choice
S	C	S	C	S	V
S	C	S	V	S	V
F	N	F	C	F	C
S	V	S	N	S	V
F	N	F	V	F	N
F	C	S	C	F	C
S	C	S	N	S	C
F	N	F	N	F	N
S	V	S	V	F	V
F	V	S	N	S	C
S	C	S	C	F	N
F	N	F	V	S	N
F	V	S	C	F	C
F	C	S	N	F	N

a. For the data above, prepare a contingency table (crosstab table) with cell frequencies and marginal frequencies.

b. Compute the expected cell frequencies for the contingency table.

c. Conduct a χ^2 test of independence.

d. What can you conclude from this test?

2. A researcher used Herzberg's motivation-hygiene theory to examine female police officers' job satisfaction. Measures of job satisfaction were taken for female police officers and female civilian personnel. Scores reflected a high, moderate, or low level of job satisfaction. The researcher hypothesized that more female police officers would express a low level of job satisfaction than female civilian personnel would. From this study, the researcher found that 118 female police officers and 98 female civilian personnel expressed low job satisfaction, 68 female police officers and 85 female civilian personnel expressed moderate job satisfaction, and 14 female police officers and 66 female civilian personnel expressed high job satisfaction.

a. For the data above, prepare a contingency table (crosstab table) with cell frequencies and marginal frequencies.

b. Compute the expected cell frequencies for the contingency table.

c. Conduct a χ^2 test of independence.

d. What can you conclude from this test?

3. A researcher examined the rates of antisocial risk-taking behavior of adolescents from four groups: Native Americans who live off the reservation, Native Americans who live on the reservation, Blacks, and Whites. From the risk-taking measure, three categories of risk-takers were identified: high antisocial risk-takers, moderate antisocial risk-takers, and low antisocial risk-takers. The researcher predicted that there would be a difference in the frequency of antisocial risk-taking behavior according to the four groups. The high antisocial risk-taker category contained 98 Native Americans who live off the reservation, 54 Native Americans who live on the reservation, 53 Blacks, and 58 Whites. The moderate antisocial risk-taker category contained 57 Native Americans who live off the reservation, 85 Native Americans who live on the reservation, 91 Blacks, and 96 Whites. The low antisocial risk-taker category contained 41 Native Americans who live off the reservation, 66 Native Americans who live on the reservation, 79 Blacks, and 71 Whites.

a. For the data above, prepare a contingency table (crosstab table) with cell frequencies and marginal frequencies.

b. Compute the expected cell frequencies for the contingency table.

c. Conduct a χ^2 test of independence.

d. What can you conclude from this test?

4. Students at a particular university had an outbreak of a skin rash. A medical doctor examined them before a researcher tested the effect of an experimental drug on the skin rash. College students either received or did not receive the experimental drug. Three days later these students were re-examined by a medical doctor for improvement of the rash. They were diagnosed to have either improved or not improved. The doctor was blind to the experimental condition to which each student was assigned. The researcher predicted that more students who received the experimental drug would show improvement. The data indicated that 24 students who re-ceived the experimental drug showed improvement while 11 did not improve. Furthermore, 3 students who did not receive the experimental drug showed improvement whereas 17 did not improve.

a. For the data above, prepare a contingency table (crosstab table) with cell frequencies and marginal frequencies.

b. Compute the expected cell frequencies for the contingency table.

c. Conduct a χ^2 test of independence.

d. What can you conclude from this test?

5. A university surveyed 485 senior students about their perception of counseling center use by gender. Participants were asked if they would expect to find considerably more men, considerably more women, or approximately equal numbers of men and women using the services of the counseling center. The researchers expected an equal number of people to endorse each category. The data for the study were that 189 students perceived the counseling center as used considerably more by women, 158 students perceived the counseling center as used almost equally by men and women, and 139 students perceived the counseling center as used considerably more by men. Test the expectancies of the researchers using the χ^2 goodness-of-fit test.

6. A school district was interested in enhancing the critical thinking skills of its students. They selected the Test of Critical Thinking Abilities in which the scores have been shown to be normally distributed. In particular, 34 percent of the scores should be between the mean of the test and one standard deviation above the mean of the test, 16 percent of the scores should be higher than one standard deviation above the mean, 34 percent of the scores should be between the mean of the test and one standard deviation below the mean of the test, and 16 percent of the scores should be lower than one standard deviation below the mean. The school district administered curriculum changes aimed at increasing critical thinking skills of its students. They then administered the Test of Critical Thinking Abilities at the end of the school year to determine if their students scored higher than expected on the test. They found that 28 students scored more than one standard deviation above the mean, 54 students scored between the mean and one standard deviation above the mean, 52 students scored between the mean and one standard deviation below the mean, and 14 students scored more than one standard deviation below the mean. Use the χ^2 goodness-of-fit test to determine if the students scored higher on the test than expected from a normal distribution.

7. Elderly drivers in the United States are often perceived as being involved in a disproportionate number of automobile accidents. This perception has led the public to call for more stringent policies regarding rights of the elderly to drive. This research study used 50 participants from each of four groups to test this perception: adolescent less (age 16–18), young adults (age 26–28), middle-aged adults (age 46–48), and elderly adults (age 76–78). Participants were asked to operate a driving simulation machine for a 20-minute drive in congested city traffic. An average number of driving errors had been established for this simulated driving program. The number of participants who exceeded the average number of errors was recorded for each group. Test the expectancy that an equal number of participants from each category would exceed the mean number of driving errors using

the χ^2 goodness-of-fit test. What can you conclude from this test if 24 adolescents, 21 young adults, 18 middle-aged adults, and 15 elderly adults exceeded the mean number of errors?

8. A pre- and posttest design was used to determine HIV risk-taking behaviors of injecting drug users. One hundred eighty-four injecting drug users reached by street outreach participated in this study. After an initial assessment of risk-taking injection behaviors, participants were given an educational program about HIV risk-taking behaviors and how to reduce these risks. At posttest, the participants were again assessed for risk-taking injection behaviors. Of the 184 participants reached by street outreach, 108 were assessed to be engaging in risk-taking injection behaviors at pretest. Eighty-two of these participants were still engaging in risk-taking injection behaviors at posttest. For participants who did not engage in risk-taking injection behaviors at pretest, only 6 engaged in risk-taking injection behaviors at posttest. Use McNemar's χ^2 test to determine if there was a greater change in risking-taking injection behaviors afte the educational program for those who initially engaged in risk-taking injection behaviors vs. those who did not initially engage in risk-taking injection behaviors.

9. High school juniors and seniors were asked to indicate whether they had driven a vehicle after consuming alcohol within the last 30 days. One hundred fifty-six of the 459 participants indicated that they had. They were then presented with an educational program that focused on the dangers of driving after drinking even a small amount of alcohol. Thirty days after the educational program, students were again asked whether they had driven a vehicle after consuming alcohol within the last 30 days. Of the students who indicated they had driven after consuming alcohol prior to the educational program, 114 indicated that they also had driven after consuming alcohol after the educational program. Of the students who indicated they had not driven after consuming alcohol prior to the educational program, 28 indicated that they had engaged in this behavior after the educational program. Use McNemar's χ^2 test to determine if there was a greater change in driving after consuming alcohol after the educational program for those who initially engaged in driving after consuming alcohol vs. those who did not initially engage in driving after consuming alcohol.

10. Violent offenders were assessed for anger control problems prior to incarceration. Of the 465 violent offenders, 293 were determined to have an anger control problem. Despite this fact, the policy of the prison was that every inmate convicted of a violent crime had to participate in an anger control program. Upon the completion of the program, 199 of the violent offenders who had an anger control problem were determined to still have an anger control problem. For the violent offenders who did not initially have an anger control problem, 15 were determined to have an anger control problem when they completed the program. Use McNemar's χ^2 test to determine if there was a greater change in anger control for those who initially had an anger control problem vs. those who initially did not have an anger control problem.

11. A social psychologist was interested in testing the relationship between empathy and aggression. However, in conducting the research, the psychologist determined that neither the empathy nor the aggression scores were normally distributed because of the small number of observations. Since these scores violate an assumption of the Pearson product moment correlation, use Spearman's rank order correlation to find the correlation between empathy and aggression for the scores listed below. Test this correlation to determine if it is significantly different from zero in the population.

Empathy	Aggression
24	47
48	42
33	51
62	43
24	35
27	40
44	36
38	38
55	25
52	26
47	31
34	42

12. A clinical psychologist is interested in testing the relationship between shyness and depression. However, in conducting the research, there is only a small number of observations. Thus, use Spearman's rank order correlation to find the correlation between shyness and depression for the scores listed below. Test this correlation to determine if it is significantly different from zero in the population.

Shyness	Depression
45	56
28	37
35	31
63	58
48	52
29	31
46	51
39	35
56	58
28	26

13. An educational psychologist is interested in determining the correlates of procrastination. She believes that procrastination should be positively related to fear of negative evaluation. To test this hypothesis, she conducted a pilot study that yielded the data below. Conduct a Spearman's rank order correlation to find the correlation between procrastination and fear of negative evaluation. Test this correlation to determine if it is significantly different from zero in the population.

Procrastination	Fear of Negative Evaluation
47	56
30	37
36	31
64	58
49	52
31	31
48	51
40	35
57	58
29	26

14. Young children are often said to be egocentric. As a result, some of their behaviors are viewed as selfish. In particular, during playtime young children are often unwilling to share their toys. A developmental psychologist created an animated film on the positive benefits of shar-

ing things with others. Some four-year-old children saw the film while others did not. The children were then observed for half an hour during playtime. The number of sharing behaviors they engaged in was recorded for each child. The child psychologist predicted that children who saw the animated film on sharing behavior would engage in more sharing behaviors that children who did not see the film. Test this hypothesis using the Wilcoxon-Mann-Whitney test with the data below.

Number of Sharing Behaviors

Saw the Film	Did Not See the Film
7	2
3	4
4	3
3	2
	1

15. Children were taught a dart-throwing skill using one of two approaches: physical practice of the technique without the dart or mental practice of the technique. After the practice opportunity, children played a dart game that required the dart-throwing skill they practiced. The researcher hypothesized that there would be a difference in scores as a function of the type of practice but he was uncertain of the direction of the effect. Test this hypothesis using the Wilcoxon-Mann-Whitney test.

Score on the Dart Game

Physical Practice	Mental Practice
23	17
18	15
25	11
16	14
18	
14	

16. A clinical psychologist was interested if patients in brief group therapy could benefit from this type of therapy to reduce their feelings of alienation. Thus, 11 participants who received brief group therapy were compared on a measure of alienation to 12 participants who were on a waiting list to receive brief group therapy. Test the hypothesis that the participants in brief group therapy would have lower alienation scores than the control participants using the Wilcoxon-Mann-Whitney test for the data below.

Alienation

Brief Group Therapy	Control
24	47
39	42
32	35
28	39
41	58
35	51
26	43
39	40
21	44
33	47
44	45
	47

17. Two groups of rats were trained to run a maze until they completed five consecutive errorless trials. One group of rats was then administered concentrated doses of caffeine while a control group did not receive caffeine. After a period of time, the rats were tested on the maze. The number of errors that the rats made were recorded.

The researcher predicted that the rats who were given concentrated doses of caffeine would make more errors in running the maze than the rats who were not given caffeine. Test this hypothesis using the Wilcoxon-Mann-Whitney test for the data below.

Errors

Caffeine	No Caffeine
4	3
5	4
9	6
6	5
8	0
4	2
11	4
8	1
3	7
6	5
9	2
10	

18. Researchers wanted to examine the effect of music therapy on the recall of negative memories as measured on the Negative Affect Schedule of eight elderly individuals in a residential care home who had been diagnosed as depressed. The measure of negative memories was taken prior to therapy and again after therapy. The researchers predicted that participants would recall fewer negative memories after music therapy than before music therapy. Test this hypothesis using the Wilcoxon signed ranks test for the data below.

Negative Memories

Pre-Therapy	Post-Therapy
12	6
6	5
5	7
13	6
9	5
4	7
7	4
8	3

19. Two groups of 5 adult male albino Wistar rats were trained to lever press in a Grason Stadler Operant Box to receive food pellets. Once lever-pressing behavior was learned it was maintained on a variable ratio reinforcement schedule until the behavior was stable for all of the rats. Rats were paired on their level of lever-pressing behavior per unit time. One rat in each pair was randomly selected to be injected with scopolamine hydrobromide using a .125 mg/kg dose level. After 3 days, rats were tested on the number of lever presses they engaged in a 20-minute period of time. The researcher predicted that the control rats would have more lever presses than the rats who were injected with scopolamine hydrobromide because it disrupts the attenuation mechanism by impairing the cholinergic reticular pathways. Test this hypothesis using the Wilcoxon signed ranks test for the data below.

Number of Lever Presses

Control Rats	Rats Injected with Scopolamine Hydrobromide
48	36
57	48
35	36
63	41
42	47

20. Educational psychologists matched 40 young developmentally delayed children on Stanford-Binet Intelligence Scale scores. One member of each pair was randomly selected to participate in an experimental language development program presented via a computer while the other member of the pair received the standard training in language development. At the end of nine months, participants were administered the Illinois Test of Psycholinguistic Abilities (ITPA). The researchers hypothesized that participants who received the experimental language development program would score higher on the ITPA than participants who were in the standard training program. Test this hypothesis using the Wilcoxon signed ranks test for the data below.

ITPA Score

Experimental Program	Standard Program
45	41
38	30
41	47
25	18
54	44
33	34
42	31
39	34
35	37
28	18
24	21
36	38
27	33
31	30
42	38
36	41
42	47
33	29
49	41
18	31

21. Researchers were interested in the relationship between acoustical stimulation and certin physiological changes. They formed two acoustical stimulation conditions, a recording of aircraft noise and white noise, to which participants were exposed during the eight-hour workday. Participants were initially matched on white cell counts. One member of each pair was randomly selected to be exposed to the aircraft noise while the other member received the white noise. At the end of the workday, a white cell count was taken. The researchers predicted that there would be a difference in the white cell counts for the two levels of acoustical stimulation. Test this hypothesis using the Wilcoxon signed ranks test for the data below.

White Cell Count

Aircraft Noise	White Noise
46	41
58	42
61	55
65	57
59	52
34	37
72	54
69	57
45	47
58	44
64	51
65	55
66	64

Aircraft Noise	White Noise
47	53
49	50
43	48
63	51

22. Use the SPSS data bank provided with this book to conduct a χ^2 test of independence analysis of the data contained in the data file donate.sav. Conduct an analysis to answer the question, "Is the recruitment strategy (recruit; 1 = control, 2 = information session, 3 = information session and empathic video) related to the number of people who donated blood (donate; 0 = did not donate blood, 1 = donated blood)?" Given the result of your analysis, what answer would you provide for this question?

23. Use the SPSS data bank provided with this book to conduct a χ^2 test of independence analysis of the data contained in the data file career.sav. Conduct an analysis to answer the question, "Are there gender differences (gender; 1 = male, 2 = female) in career interests (interest; 1 = fine arts, 2 = humanities, 3 = social sciences, 4 = biological sciences, 5 = physical sciences and math) among students in grades 7–10?" Given the result of your analysis, what answer would you provide for this question?

24. Suppose that students in grades 7–10 are expected to express career interests in the following areas for the given proportions: fine arts (.20), humanities (.15), social sciences (.25), biological sciences (.30), and physical sciences and mathematics (.10). Use the SPSS data bank provided with this book to conduct a χ^2 goodness-of-fit test of the data contained in the data file career.sav for career interests (interest; 1 = fine arts, 2 = humanities, 3 = social sciences, 4 = biological sciences, 5 = physical sciences and math) among students in grades 7–10. Given the result of your analysis, do the data fit the expected proportions?

25. Use the SPSS data bank provided with this book to conduct a one-tailed Spearman's rank order correlation of the data contained in the data file math.sav because the data may not be normally distributed. Find the correlation between post-instruction math attitude (postatt) and the final grade in algebra (algpa). Since a high score on the attitude measure implies a negative attitude toward math, the researcher has predicted that there would be a significant negative relationship between post-instruction math attitude and the final grade in algebra. Given the result of your analysis, what can you conclude about the researcher's prediction?

26. Use the SPSS data bank provided with this book to conduct a Mann-Whitney test of the data contained in the data file math.sav. In this study, two different methods of teaching algebra (method; 1 = lecture, 2 = metacognitive) are compared with respect to performance on the Iowa Test of Educational Development Math Subtest for 9th Graders (ITED9). The researcher predicted that students who were taught algebra using the metacognitive approach would demonstrate higher levels of achievement as measured by their score on the math subtest of the Iowa Test of Educational Development for 9th Graders. Based on your analysis, what can you conclude about the researcher's prediction?

Chapter 16

Analysis of Variance

CHAPTER OBJECTIVES

After studying this chapter, you should be able to:

❏ State the reasons it may be necessary or desirable to include more than two groups in a single study.

❏ Explain why it is not appropriate to test all possible pairs of means.

❏ Describe the partitioning of a score or sum of squares into its component parts.

❏ Given data from a research study, compute the sums of squares using either the definitional equations or the *df* components.

❏ Compute mean squares and the *F* statistic from the sums of squares.

❏ State the proper interpretation of an *F*, given its value and *df*'s.

❏ Define and test a planned contrast involving either pairs of means or weighted combinations of means.

❏ Use the Scheffé method to test a post hoc contrast.

The SAD faculty at Big Time U collected data on college major and lab section. There were three categories of each variable. How might they test whether there is a statistically significant difference in mathematics anxiety between the lab sections, or whether statistics proficiency, as measured by the final exam, was different for students with different majors or from different lab sections? Do you think there might be some problems with comparing each lab section with every other lab section or each major with every other major? What solutions might we find for these problems? How might Dr. Cross and his students solve similar problems with the data from their study?

Introduction

Like the SAD faculty at Big Time U, researchers in the behavioral and social sciences often find it desirable or necessary to compare more than two groups. For example, researchers studying the effect of drugs on behavior are likely to include a high-dose group, a low-dose group, and a group that is given a neutral substance that is not the drug at all, such as sugar or salt. Up to this point, we have discussed methods for comparing two groups; in this chapter, we will bring together ideas and principles you have already learned in order to develop a method for testing hypotheses that involve the means of more than two groups. The method of analysis that we will be using is one member of a family of analytic procedures that has the general name **analysis of variance,** ANOVA, for short.

Analysis of variance (AN-OVA) *is a family of methods for testing hypotheses that involve the means of more than two groups.*

The procedures of analysis of variance are among the most widely used by researchers in education and psychology.

Preliminary Concerns When Using More Than Two Groups

Why Use More Than Two Groups?

We may want to use more than two groups in a single study for several reasons. First, we may wish to study as the independent variable a variable that exists in several forms (different college majors, for example) or at many levels (such as different drug doses). Studying only two forms or two levels of such a variable would give us a distorted picture of how the independent variable and the dependent variable are related. A study of the relationship between gender and statistics performance or math anxiety is necessarily limited to the two conditions of the gender variable, but a study of whether statistics achievement is related to college major or whether drug dosage is related to reaction time is inherently more complex. There are many possible college majors, and drug dosage, of course, is a continuous variable. Selecting only two college majors would very much limit our ability to generalize our findings because so many values of the independent variable would not be included in the analysis. Selecting just two drug dosages not only has the problem of restricted generalizability, but might lead us to reach incorrect conclusions about the shape of the relationship.

The nature of this problem is illustrated in Figure 16-1, where a graph of the relationship between age and scores on a visual reasoning test is displayed. In the graph, scores on visual reasoning show an initial period of increase during the years of childhood and adolescence, followed by a period of stability during mature adulthood, and ending in decline after about the age of 60. If we sample the age variable (that is, choose to include levels of age for the independent variable) only during childhood and adolescence (points 1 and 2), we would conclude that ability constantly increases. Studying two adult age groups (points 3 and 4) would lead us to postulate a stable condition, while two samples, one age 60, the other age 75 (points 5 and 6), would reveal a decline in function. To get an accurate description of the phenomenon, we would need to draw samples at more than two ages.

A second reason for studying more than two groups at the same time is *economy*. Suppose the independent variable we are interested in exists in five different forms. If we conduct studies comparing only pairs of means, there are 10 possible pairings, requiring 10 studies. If we use 15 participants in each condition in each study, we must test 300 individuals (30 in each of 10 studies). By conducting a single study with five levels of the independent variable, we can complete the same study using only 75 participants without loss of information or accuracy. The saving in time and effort would make this an attractive alternative to the pairwise studies.

Figure 16-1 Graph of a curvilinear relationship between age and visual reasoning showing the need to have several levels of the independent variable to capture the shape of a relationship.

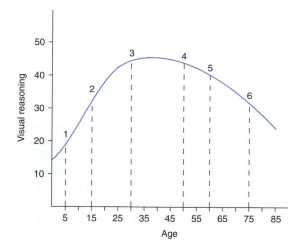

Selecting Levels of the Independent Variable

Once we decide to study an independent variable that has more than two levels, we need to choose what levels to select unless there are few enough that all may be included. If there are three kinds of drug—for example, aspirin, acetaminophen, and ibuprofin—available to treat headaches, and these are the only values of the independent variable that exist, it makes sense to include all of them in a study. But there are other cases, particularly where the independent variable is a continuous variable—such as age, or study time, or drug dosage—when it is not possible, even theoretically, to include all levels of the variable of interest. Then it becomes necessary to select particular levels to include in the study.

There are several factors to consider when selecting levels of the independent variable such as age or drug dosage. First, we want to be sure to include values that span the range of possibilities. Therefore, we should include one level near each extreme. If we are interested in the course of intellectual development between ages 2 and 25, then we should include groups at both of these levels. Next, we should include enough middle-level values that we can detect nonlinear relationships like the one described between age and visual reasoning in Figure 16-1. Finally, the practical realities of time and economics dictate that we cannot have a very large number of groups because of limits on the number of research participants available, the time needed to conduct the study, the cost of experimental materials, and so forth. All these factors are usually explored and the decision about study design is made based on one or more pilot studies in which small numbers of individuals are exposed to the proposed experimental conditions so the experimenter can get an idea what to expect in the complete study.

▶ *Now You Do It*

We have discussed some situations where you might want to include more than two groups in a study. Can you think of other situations where it would be desirable to include several groups? What advantages would having multiple groups provide in each case?

Studies of all sorts of developmental processes are good candidates for including several groups because this is the only way to track the full course of development. More generally, any time the independent variable is a quantitative variable, such as the amount of time spent in instruction, the magnitude or quality of a reinforcer, or the degree of meaningfulness of a set of stimuli, it is clearly possible to have multiple groups. In such situations, it is not possible to determine the overall shape of the relationship between the independent variable and the dependent variable without including a treatment group from each of several levels of the independent variable.

When there are several qualitatively different treatments, such as modes of instruction, or modes of sensory presentation, we might also want to include as many of the conditions in one study as possible simply because this is more economical. If there are four possible conditions, and we are comparing two groups, with 10 participants each, in each study, we would have to run six studies to examine all pairs of conditions. This would require 120 individuals. By conducting a single study with four groups, we could get the same information from 40 participants in one-third the time.

Why Not Test All Pairs of Means?

Prior to the 1920s, if someone conducted a study involving more than two groups, it was common practice to test hypotheses by what has been called the multiple t-test method: Each group was paired with every other group, and a t-test was performed to test the null hypothesis that the means of the two groups were not different. The drug example we described would require three comparisons: high drug (H) with low drug (L), high drug (H) with no drug (N), and low drug (L) with no drug (N). In this case, the problem does not look too bad; however, in a study with 10 groups, the number of comparisons required to test every pairing is 45.

Aside from the problems that one may encounter in attempting to interpret a large number of comparisons among means, there are two other factors that make the multiple t-test method inappropriate for testing hypotheses in the way that we have been doing it; both involve the risk of a Type I error. Because Type I error has occupied a central position in the thinking of many statisticians and research methodologists for 75 years, it is to be expected that much attention has been focused on the problem.

The first difficulty is purely a matter of probability. The value we choose for α is given in terms of a single study. It is the number of times we would expect to get a difference as large as the critical value in many repetitions of the study if the null hypothesis were true. For an α of .05, we would expect the sample results to exceed the critical value five times in 100 repetitions of the comparison. The chance that our study is one of those five is only .05; however, if we compute more than one t-test (or critical ratio, or χ^2, or any other test statistic), the probability that at least one of them will exceed the critical value is greater than 5 percent, and possibly much greater. For example, if the null hypothesis is *exactly* true (the population means are all equal) and we conduct 45 t-tests among 10 groups, we would expect two or three of the obtained t's to exceed the .05 critical value due to chance variations from sample to sample, and the chance that none of the t's would exceed the critical value is relatively small.

We can see the effect of using multiple comparisons by determining what is called the **familywise** error rate. This term (which in older texts is called experimentwise error) means the same thing as Type I error but refers to the situation when several tests are conducted within a single experiment (the ordinary Type I error rate applies to a single test). The familywise error rate, α_{FW}, when multiple comparisons are made is given by the equation

$$\alpha_{FW} \leq 1 - (1 - \alpha)^c \tag{16.1}$$

where c is the number of comparisons. If we conduct four comparisons at the $\alpha = .05$ level, the value of α_{FW} is

$$\alpha_{FW} \leq 1 - (.95)^4 = .185$$

That is, if we conduct four t-tests, each at the $\alpha = .05$ level, with data from populations where all means are equal, we run the risk of making at least one Type I error not 5 percent of the time, but possibly over 18 percent of the time. Clearly, as the number of comparisons gets large, the risk of erroneously concluding that a difference exists also

Familywise error is the rate of Type I errors made in all analyses run on the data from a given study.

gets large. For 10 comparisons, $\alpha_{FW} \le .40$, and for 45 comparisons $\alpha_{FW} \le .90$. A Type I error becomes highly likely with 45 comparisons. We discuss ways to control the familywise error rate later in the chapter.

A second problem compounds the first: The t-tests are not independent of each other, but involve overlapping pairs of means. For example, the three-group drug study yields three comparisons, H with L, H with N, and L with N, but these three comparisons are not independent; the values for any two of them determine the third. If $H = 15$, $L = 10$, and $N = 5$, then $(H - L) + (L - N) = (H - N)$. We can even say that there are two degrees of freedom among these means, so that once two of the tests have been run, the value of the third is no longer unknown or free to vary. As the number of groups increases, this second problem gets worse. For example, with four groups, there are six possible comparisons, but only three independent comparisons. The lack of independence of the comparisons combines with the familywise error problem to so alter the probability of a Type I error that the result of any single comparison among means may be uninterpretable.

Partitioning Variance

Let us consider some hypothetical data for a drug study such as the one described at the beginning of the chapter. We want to study the effects of drug X, claimed to be a memory enhancer, on learning a list of vocabulary words. We pretest a group of college students and determine that none of them knows any of the words in our list, words like fracid, oscitant, syzygy, and palindrome; we then draw three random samples of five students. Each student is to learn a list of 20 words under one of the three conditions: Group H receives a high dose of the drug by injection, group L a low dose (but equal volume) of the drug, and group N is given an equivalent volume injection of saline. The students study the vocabulary words and later are tested to determine how many of the words they learned. Scores are the total number of words correctly defined. The results of the study are shown in Table 16-1. How shall we decide whether the drug has had the claimed effect?

Partitioning a Score

We have already encountered the fundamental principle by which the problems surrounding multiple comparisons are solved. It is a variation on the separation of total variance into predictable and unpredictable portions that we encountered in our discussion of correlation in Chapter 8.

First, let us consider the idea that a score may be composed of several parts. This proposition is illustrated in Figure 16-2, where the data for our study of memory drugs from Table 16-1 are plotted. The scores of two individuals, X_{11} and X_{23}, (perhaps our friends Ahab and Jezebel) are equal; both are 11. But, if we bring in the information that they belong to different groups, we can say that their scores are composed of three pieces. Everyone in the study contributes information to the total mean, $\overline{X}_T$, so we may say that $\overline{X}_T$ is a part of each person's score; this part is labeled (**a**) in the figure. Those people in group 1 contribute information to $\overline{X}_1$, so their scores caused $\overline{X}_1$ to be different from $\overline{X}_T$. This difference $(\overline{X}_1 - \overline{X}_T)$, is the effect of being in group 1. It is part of everyone's score

N-dimension (Subject)	H	L	N
1	11	12	15
2	6	14	11
3	9	11	17
4	8	12	13
5	12	15	16
$\Sigma X =$	46	64	72
$\Sigma X^2 =$	446	830	1060
Mean $=$	9.2	12.8	14.4

Table 16-1 Data From a Three-Group Drug Study

J-dimension (Group)

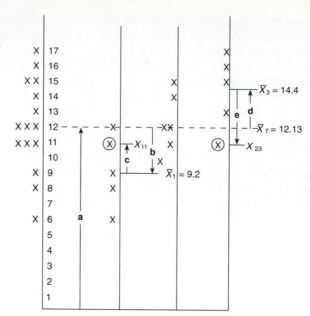

Figure 16-2 Frequency distribution of data from Table 16-1 showing that each score is composed of three parts.

who is in group 1 and is labeled (**b**) in the figure. Finally, each person contributes something unique to her or his own score, individual differences that make them more or less different from the group mean; for score X_{11} this part $(X_{11} - \overline{X}_1)$ is labeled (**c**). There is a comparable component for the score of each person in group 1.

A person in group 3 shares the (**a**) effect for the overall mean with everyone else; however, the members of group 3 all contribute to the difference between their group mean, $\overline{X}_3$, and the total mean $\overline{X}_T$. This effect of being in group 3 $(\overline{X}_3 - \overline{X}_T)$ is labeled (**d**) in the figure. Score X_{23} differs from the group mean by an amount $(X_{23} - \overline{X}_3)$, labeled (**e**); that is due to the unique properties of this person. Notice that (**b**) and (**d**) have the same meaning, the difference of a group mean (call it $\overline{X}_j$) from $\overline{X}_T$, but different numerical values. Similarly, (**c**) and (**e**) both represent the difference between a single score and a group mean, but are numerically different.

We now have scores that are composed of three additive parts; for each person, we have

$$X_{ij} = \overline{X}_T + (\overline{X}_j - \overline{X}_T) + (X_{ij} - \overline{X}_j) \tag{16.2}$$

where
$\overline{X}_T$ is the overall or total mean, the effect that is common to every person in the study.

$(\overline{X}_j - \overline{X}_T)$ is the difference between the mean of one treatment condition and the overall mean. This piece is common to everyone in a particular group and is the effect of being in that group.

$(X_{ij} - \overline{X}_j)$ reflects individual differences among the people in a group plus random effects from measurement errors that are unique to any single observation.

SPECIAL ISSUES IN RESEARCH

Double Subscript Notation

We collected data on 15 people in our drug study, five in each of three groups. For the purpose of keeping track of the data, we will use an expanded summation-notation scheme; the subscript i refers to a person within a group and j identifies the group. Thus, the general symbol for a score is X_{ij}, the score of the ith person in group j. The row subscript is given first. In general, there will be N_j individuals in group j, and the study will include J groups. A particular group is identified by substituting its number ($1 = H$, $2 = L$, $3 = N$) for the subscript j, while inserting numbers for both subscripts specifies a particular score; thus, the sum of scores for the high-drug group is symbolized

$$\sum_{i=1}^{5} X_{i1}$$

---

indicating summation across the 5 individual scores in group 1, and has the numerical value 46. Likewise, X_{41} in Table 16-1 has the value 8 and X_{22} is 14. The total number of participants (N_T) is the sum of the group N_j's ($N_T = \sum N_j$) or, when there are the same number of observations in each of the J groups, $N_T = JN$, the number of groups times the number of individuals in each group.

This double subscript notation allows us to specify our summation operations more easily. A single $\sum$ involves summation over only one dimension of the table, while the double summation ($\sum\sum$) means that we are using values from the entire table–summing across both the people dimension and the group dimension. We can then use our familiar formulas for means and variances, only changing the number of dimensions across which we sum. The mean and variance for the complete set of data in Table 16-1 are given by the formulas:

$$\overline{X}_T = \frac{\sum_j \sum_i X_{ij}}{\sum_j N_j}$$

$$S_T^2 = \frac{\sum_j \sum_i X_{ij}^2 - \frac{(\sum_j \sum_i X_{ij})^2}{\sum_j N_j}}{\sum_j N_j}$$

The subscripts j and i on the two summation signs indicate the order in which the summation occurs. We always start with the summation sign closest to the variable, so, for the data in Table 16-1, this symbolization indicates that we sum over the 5 levels of i for $j = 1$ (the 5 scores in column 1), then over

the 5 levels of i for $j = 2$ (column 2), then over i for $j = 3$. Finally, we sum the sums across j to get a total for the table. For example, the equation for S_T^2 tells us to:

1. square each individual score and sum down each column then across columns ($\sum_j \sum_i X_{ij}^2$),
2. sum the individual scores down each column then across columns and square this quantity [$(\sum_j \sum_i X_{ij})^2$],
3. divide the quantity in step 2 by the total number of individuals in the experiment ($N_T = JN$),
4. subtract the quantity in step 3 from the quantity in step 1 (at this stage, as you should recall from Chapter 4, we have a sum of squares),
5. divide the quantity in step 4 by the total number of individuals.

Both of these equations are written in a general form that allows for any number of groups. Applying these equations to the data in our example, we find $\overline{X}_T$ to be

$$\overline{X}_T = \frac{(46 + 64 + 72)}{(5 + 5 + 5)} = \frac{182}{15} = 12.13$$

The total variance S_T^2 is

$$S_T^2 = \frac{(1060 + 830 + 446) - \frac{182^2}{15}}{15}$$

$$= \frac{(2336) - 2208.27}{15} = 8.52$$

▶ *Now You Do It*

Let's take a small set of scores to make sure you understand how these summation signs work. In the table below there are scores for four people in each of five groups. Use these data to answer the following questions.

1. What is $\sum_i X_{i3}$?
2. What is $\sum_j X_{2j}$?
3. What is $\sum_j \sum_i X_{ij}$?
4. What is $\sum_j \sum_i X_{ij}^2$?
5. What is $\overline{X}_2$?
6. What is S_T^2?

10	8	7	12	6
7	11	9	10	8
9	10	12	8	12
12	14	9	6	9

Answers
1. This is the sum of column 3, = 37
2. This is the sum of row 2, = 45
3. This is the sum of all scores in the table, = 189
4. This is the sum of squared scores for the entire table, = 1879
5. This is the mean of column 2, = 10.75
6. This is the variance for the entire table, = 4.65

The two cases in our example can be decomposed to yield

$$X_{11} = 11 = 12.13 + (9.2 - 12.13) + (11 - 9.2)$$
$$= 12.13 + (-2.93) + (+1.8)$$
$$X_{23} = 11 = 12.13 + (14.4 - 12.13) + (11 - 14.4)$$
$$= 12.13 + (+2.27) + (-3.4)$$

Each individual's score is composed of the overall mean, plus the deviation of their group mean from the overall mean (due to being in group j), plus the deviation of the individual's score from the group mean (due to individual differences and random effects). The effect of being in group 1 is -2.93, while the effect of being in group 3 is $+2.27$.

By subtracting $\overline{X}_T$ from both sides of equation 16.2, we can convert the equation to deviations from the overall mean. The deviation of individual X_{ij} from $\overline{X}_T$ $(X_{ij} - \overline{X}_T)$ is composed of two parts, a deviation of the group mean from $\overline{X}_T$ and the individual's deviation from the group mean. As an equation, this is

$$(X_{ij} - \overline{X}_T) = (\overline{X}_j - \overline{X}_T) + (X_{ij} - \overline{X}_j) \tag{16.3}$$

For our cases of Ahab and Jezebel,

$$(11 - 12.13) = (9.2 - 12.13) + (11 - 9.2)$$
$$-1.13 = (-2.93) + (1.8)$$
$$(11 - 12.13) = (14.4 - 12.13) + (11 - 14.4)$$
$$-1.13 = (2.27) + (-3.4)$$

That is, being in group 1 (the high-drug group) reduces the score of Ahab and each other person in the group by 2.93 units, while being in group 3 (the no-drug group) produces an increase of 2.27 units for Jezebel and the other members of her group. We know from Chapter 3 that the sum of the deviations of a set of scores from their mean is always zero, so if we take the sum of each of these terms across all people in our study, each will have a value of zero.

However, we also recall from Chapter 4 that the sum of squared deviations is a sum of squares that can be used to calculate a variance. The sum of squares of all scores around $\overline{X}_T$ can be obtained by summing over both subscripts,

$$SS_T = \Sigma_j \Sigma_i (X_{ij} - \overline{X}_T)^2$$

Dividing this sum of squares by N_T produces the total variance of the set of scores in our study.

The partitioning of deviations in equation 16.3 shows that the total sum of squares can be broken down into two parts, one part due to the effect of differences in group means (differences in amount of the drug), and a second part due to the deviations of individuals from the means of their groups that result from individual differences and measurement errors. Writing these sums of squared deviation terms out fully yields

$$\Sigma_j \Sigma_i (X_{ij} - \overline{X}_T)^2 = \Sigma_j \Sigma_i (\overline{X}_j - \overline{X}_T)^2 + \Sigma_j \Sigma_i (X_{ij} - \overline{X}_j)^2$$

The partitioning of the 15 scores from our study of the memory-enhancing effects of drug X is shown in Table 16-2. Each person's score is shown in the first column of their

Table 16-2 Partitioning the Variation of Scores for Three Groups From the Drug Study in Table 16-1

	Group 1				Group 2				Group 3		
X_{i1}	$X_{i1} - \overline{X}_T$ $\overline{X}_T = 12.13$	$\overline{X}_1 - \overline{X}_T$ $\overline{X}_1 = 9.2$	$X_{i1} - \overline{X}_1$	X_{i2}	$X_{i2} - \overline{X}_T$ $\overline{X}_T = 12.13$	$\overline{X}_2 - \overline{X}_T$ $\overline{X}_2 = 12.8$	$X_{i2} - \overline{X}_2$	X_{i3}	$X_{i3} - \overline{X}_T$ $\overline{X}_T = 12.13$	$\overline{X}_3 - \overline{X}_T$ $\overline{X}_3 = 14.4$	$X_{i3} - \overline{X}_3$
11	−1.13	−2.93	1.8	12	−.13	.67	−.8	15	2.87	2.27	.6
6	−6.13	−2.93	−3.2	14	1.87	.67	1.2	11	−1.13	2.27	−3.4
9	−3.13	−2.93	−.2	11	−1.13	.67	−1.8	17	4.87	2.27	2.6
8	−4.13	−2.93	−1.2	12	−.13	.67	−.8	13	.87	2.27	−1.4
12	−.13	−2.93	2.8	15	2.87	.67	2.2	16	3.87	2.27	1.6
Column ΣX	−14.65	−14.65	0		3.35	3.35	0		11.35	11.35	0
Column ΣX^2	65.7	42.9	22.8		13.0	2.2	10.8		49.0	25.8	23.2

group, their deviation from the overall mean $(X_{ij} - \overline{X}_T)$ is shown in the second column, the deviation of their group's mean from the overall mean $(\overline{X}_j - \overline{X}_T$, the effect of being in group $j)$ is shown in the third column, and the individual's deviation from their group mean $(X_{ij} - \overline{X}_j$, the combination of individual differences and experimental error) is shown in the fourth column. For example, the first person in group 1 (Ahab) has a raw score of 11. This score is 1.13 units below $\overline{X}_T$, but the remaining two columns show that his deviation of −1.13 is due to a group effect of −2.93 and an individual differences/error effect of +1.8. The score of the second person in group 3 (Jezebel) is also 11, but her deviation of −1.13 units from $\overline{X}_T$ is due to a group effect of +2.27 and an individual differences/error component of −3.4.

The sums of the deviation scores and the sums of squares for each group are shown at the bottom of the table. There are two important points to notice here. First, the sum of the deviations of scores within each group around their group mean is zero. Second, the sum of squares due to the group effect plus the sum of squares due to individual differences/error is equal to the sum of squares around $\overline{X}_T$:

$$65.7 = 42.9 + 22.8$$
$$13.0 = 2.2 + 10.8$$
$$49.0 = 25.8 + 23.2$$

Summing these sums of squares for the entire experiment yields

$$127.7 = 70.9 + 56.8$$

That is, the 127.7 units of variability around $\overline{X}_T$ in the experiment can be broken up into 70.9 units that are due to effects of the drug (differences between treatments) and 56.8 units that are due to individual differences and error. We have analyzed the total variability into these two additive pieces.

There is one additional simplification that we can make. For every individual in group j, the quantity $(\overline{X}_j - \overline{X}_T)$ is the same; therefore, we can substitute, as we have in equation 16.4, $N(\overline{X}_j - \overline{X}_T)^2$ for the summation over the N individuals in the group, yielding

$$\sum_j \sum_i (X_{ij} - \overline{X}_T)^2 = \sum_j (N[\overline{X}_j - \overline{X}_T]^2) + \sum_j (\sum_i [X_{ij} - \overline{X}_j]^2) \qquad (16.4)$$

The two quantities to the right of the equals sign in equation 16.4 are the sums of squares in which we are interested. The first, which involves group means, includes only differences between groups; it is called the **between-groups sum of squares or SS$_B$.** The second term contains the deviations of individuals within a group from their own group mean; therefore, it is called the **within-groups sum of squares** or **SS$_W$.** We have placed brackets around the part of each term that involves summation across individuals within a group. The second term in parentheses on the right represents the sum of squared deviations of the people in group j from the mean for their group. This is the same quantity that we used in Chapter 13 to find the pooled sum of squares that was used to compute $\hat{S}_p^2$. SS$_W$ is therefore often called the *pooled within-groups sum of squares.* We can now write equation 16.4 in the form

$$SS_T = SS_B + SS_W \qquad (16.5)$$

The total *SS* is equal to the sum of its two independent parts, that which is due to differences between groups (and predictable if we know group membership) and that which is due to variation not associated with group membership (errors that remain after using group membership information).

▶ Now You Do It

Now it's your turn to practice the process of finding deviations and making sure they add up. This is not how we would compute the answer for most larger problems, but it will help you to conceptualize the logical processes.

Ahab and Jezebel have conducted a small ESP study in which participants are to guess the figures on cards under four conditions. Scores are the number of correct matches in 25 trials. In condition 1, the "sender" is in the same room as the subject and visible to the subject. In condition 2, the "sender" is in the same room, but hidden behind a screen. In condition 3, the "sender " is in an adjacent room, and in condition 4 the "sender " is in a building on the other side of campus. There were three participants in each condition. Here are their scores.

	Condition 1	Condition 2	Condition 3	Condition 4
1	15	10	6	5
2	11	6	5	4
3	13	5	4	6

The first thing you must do is calculate the means of the groups and the mean of the total study. For $\overline{X}_T$ you should get 7.5, and for the four groups you should get 13, 7, 5, and 5. Next, find the deviations, then square them and sum them. You should get the deviations shown in the table. The values in the last two rows are the sums and sums of squares. The three sums-of-squares terms for the entire experiment are $SS_T = 155$, $SS_B = 129$, $SS_W = 26$.

	Group 1			Group 2		
	$X_{ij} - \overline{X}_T$	$\overline{X}_j - \overline{X}_T$	$X_{ij} - \overline{X}_j$	$X_{ij} - \overline{X}_T$	$\overline{X}_j - \overline{X}_T$	$X_{ij} - \overline{X}_j$
	7.5	5.5	2	2.5	−.5	3
	3.5	5.5	−2	−1.5	−.5	−1
	5.5	5.5	0	−2.5	−.5	−2
ΣX	16.5	16.5	0	−1.5	−1.5	0
ΣX^2	98.75	90.75	8.0	14.75	.75	14.0

	Group 3			Group 4		
	$X_{ij} - \overline{X}_T$	$\overline{X}_j - \overline{X}_T$	$X_{ij} - \overline{X}_j$	$X_{ij} - \overline{X}_T$	$\overline{X}_j - \overline{X}_T$	$X_{ij} - \overline{X}_j$
	−1.5	−2.5	1	−2.5	−2.5	0
	−2.5	−2.5	0	−3.5	−2.5	−1
	−3.5	−2.5	−1	−1.5	−2.5	1
ΣX	−7.5	−7.5	0	−7.5	−7.5	0
ΣX^2	20.75	18.75	2.0	20.75	18.75	2.0

The Null Hypothesis

We should now consider what null hypothesis we are testing. The null hypothesis that is appropriate for our three-group drug study is

$$\mu_H = \mu_L = \mu_N$$

That is, if the drug has no effect, we would expect all of the population means to be equal. More generally, the null hypothesis is that all of the group means are equal:

$$\mu_1 = \mu_2 = \mu_3 = \cdot \cdot \cdot = \mu_J$$

Because ANOVA involves several groups, we cannot specify directional hypotheses in most cases. Special methods are needed if we want to test hypotheses about the order of differences among means, for example, that $\mu_H \geq \mu_L \geq \mu_N$. Such methods are available (see any of the books listed at the end of the chapter) but are beyond our scope. Here, we limit our concern to nondirectional, or two-tailed, null hypotheses. As we have seen, a two-tailed null hypothesis has the general form $\mu_1 = \mu_2$. When more than two groups are present in the study, the nondirectional null hypothesis specifies the equality of all means.

If the null hypothesis is true, the three groups in our drug study can be considered random samples from the same population (or from three populations with identical means and variances). A true null condition implies that differences among the groups in our study are no greater than would be expected to occur by chance under the conditions of random sampling, and this implies that the drug has no effect. The alternate hypothesis, as usual, is that the null hypothesis is false. In this case, if we reject the null, we conclude that *not all of the means are equal*. Some means may be equal to other means, but not all of them are equal. We would have to inspect the pattern of mean differences to see whether our research hypothesis, that the drug enhances memory, is supported by the data.

Another way to look at the question of whether the means of several groups are equal is to think in terms of variances. If we take several samples from the same population, their means will vary at random around the population mean because of individual differences and random errors. This is what we would expect if the null hypothesis is true and the drug has no effect. However, if the samples were actually drawn from different treatment populations, their population means would show variability. We can express this variation among the population means as σ_μ^2. The null hypothesis says that $\sigma_\mu^2 = 0$, and the alternate hypothesis is that $\sigma_\mu^2 \neq 0$. If our sample means show more variability than we would expect for samples that are all drawn from the same population, then we will reject the null hypothesis and conclude that at least some of the variability among the means is due to the effect of the drug. This is the path we will take to test the null hypothesis.

Mean Squares as Variance Estimates

In Chapter 13, we discussed two concepts that are central to understanding ANOVA: the first is the concept of degrees of freedom; the second is the principle that a sum of squares divided by its degrees of freedom is an unbiased estimate of the variance in the population from which the sample was drawn. From the second principle, we can see that each of the terms in equation 16.4, because it is a sum of squares, can be converted into an estimate of a population variance by dividing it by the appropriate *df*. The variance that is estimated and *df*'s appropriate to each sum of squares are

$$\frac{\sum_j \sum_i (X_{ij} - \overline{X}_T)^2}{JN - 1} = \frac{SS_T}{JN - 1} = MS_T$$

$$\frac{\sum_j (\sum_i [X_{ij} - \overline{X}_j]^2)}{J(N - 1)} = \frac{SS_W}{JN - J} = MS_W \qquad (16.6)$$

$$\frac{\sum_j (N[\overline{X}_j - \overline{X}_T]^2)}{J - 1} = \frac{SS_B}{J - 1} = MS_B$$

The three formulas in equations 16.6 are definitional formulas that show what the underlying processes of a statistic are. In each case, a mean is subtracted from each member of a set of scores. This difference is squared, then the squared differences are summed across all individuals in the study. These sums of squares are then divided by their respective degrees of freedom. Let's take the data from our three-group drug study in Table 16-1 to illustrate how these formulas work.

The three sum-of-squares components of each score from equation 16.4 are shown in Table 16-3. The column sums and sums of squares are shown in the last two rows of the table. These are the same deviations that we examined in Table 16-2, but rearranged to collect the sum-of-squares terms together by whether they are between groups, within groups, or total. You can see that for each effect (total, between groups, and within groups),

Table 16-3	Sums and Sums of Squares for Three Groups in the Study of Drug Effects on Learning								
	SS_T $(X_{ij} - \overline{X}_T)$			SS_B $(\overline{X}_j - \overline{X}_T)$			SS_W $(X_{ij} - \overline{X}_j)$		
	High	Low	None	High	Low	None	High	Low	None
1	−1.13	−.13	2.87	−2.93	.67	2.27	1.8	−.8	.6
2	−6.13	1.87	−1.13	−2.93	.67	2.27	−3.2	1.2	−3.4
3	−3.13	−1.13	4.87	−2.93	.67	2.27	−.2	−1.8	2.6
4	−4.13	−.13	.87	−2.93	.67	2.27	−1.2	−.8	−1.4
5	−.13	2.87	3.87	−2.93	.67	2.27	2.8	2.2	1.6
Sums	−14.65	3.35	11.35	−14.65	3.35	11.35	0	0	0
SS's	65.72	13.04	48.96	42.92	2.24	25.76	22.80	10.80	23.20

the sums are zero (within a small rounding error, $-14.65 + 3.35 + 11.35 \approx 0$). More importantly, we find the three sum-of-squares terms for the entire study to be

$$SS_T = 65.72 + 13.04 + 48.96 = 127.72$$

$$SS_B = 42.92 + 2.24 + 25.76 = 70.92$$

$$SS_W = 22.80 + 10.80 + 23.20 = 56.80$$

and we can see that the equality in equation 16.5 is verified because $127.72 = 70.92 + 56.80$.

The estimated variances in equations 16.6 are called **mean squares** or **MS**'s. If the null hypothesis that each of our three drug-treatment groups comes from the same population is true, each MS provides an unbiased estimate of the variance in the population from which the samples were drawn.

The MS_W term has an $N - 1$ term in the denominator for each of the J groups in the numerator. In our case, since there are 5 participants in each group, $N - 1 = 4$, and J is 3. This corresponds to the number of degrees of freedom *within* each group. We sum across groups in the numerator to obtain the pooled sum of squares within groups. This produces

$$SS_W = 22.80 + 10.80 + 23.20 = 56.80$$

as the pooled within-groups sum of squares. Next, we multiply the degrees of freedom associated with each group by the number of groups (which is the same as adding across

---

MS_W AS AN ESTIMATE OF σ_X^2

Under the assumptions that the null hypothesis is true and that our groups represent J random samples from a single population, we know that there is a population variance, σ_X^2. One estimate that we have of σ_X^2 is MS_T, the estimated variance based on JN individuals that loses one df for the overall mean; but, as we have seen, MS_T contains both the random variance among individuals and the systematic differences between groups, if any. However, MS_W is also an estimate of σ_X^2, but based on an average of several samples.

$$MS_W = \frac{\sum_j(\sum_i[X_{ij} - \overline{X}_j]^2)}{J(N-1)}$$

$$MS_W = \frac{\sum_j \frac{\sum_i(X_{ij} - \overline{X}_j)^2}{(N-1)}}{J}$$

$$MS_W = \frac{\sum_j \hat{s}_j^2}{J}$$

That is, assuming the null hypothesis is true, each of the J groups provides an independent and unbiased estimate of σ_X^2 from its own within-group variance,

$$\frac{\sum_i(X_{ij} - \overline{X}_j)^2}{N-1}$$

Because the sum of deviations (not squared deviations) must be zero *within each group*, there are $(N-1)$ degrees of freedom associated with the sum of squares within each group. But there are J independent groups, so the degrees of freedom associated with SS_W is the sum of the df within each group. If, as is usually the case, the groups are of equal size, df_W is $J(N-1)$ or $JN - J$, and MS_W is simply an average of these J unbiased estimates of σ_X^2. Therefore, if the null hypothesis is true, MS_W is an unbiased estimate of the same population variance, σ_X^2, that is estimated by MS_T. A null hypothesis that the samples are drawn from different populations but that the means and variances of these populations are the same gives us MS_W as an estimate of each of these variances. That is, if the null hypothesis is true,

$$MS_W = individual\ differences + experimental\ error$$
$$MS_W = \sigma_{ID}^2 + \sigma_{error}^2$$

all groups) to obtain a pooled df for the denominator. This gives us

$$df_W = 3(5 - 1) = 12$$

and

$$MS_W = 56.80/12 = 4.733$$

Equations 16.6 also say that we can calculate the sum of squares between groups by finding the difference between each group mean and the total mean $(\overline{X}_j - \overline{X}_T)$, squaring this difference, and multiplying the squared difference by the number of participants in the group. This process works because, as we discussed earlier and you can see in Table 16-3, a group effect is the same for every member of the group. For our data we have found the sum of squares between groups to be 70.92. Because the three between-group *sums* must themselves sum to zero, there are $J - 1 = 2$ degrees of freedom for the between-groups sum of squares. Therefore,

$$MS_B = 70.92/2 = 35.46$$

Although the SS's in equation 16.4 are additive, the MS's or estimated variances in equations 16.6 are not; that is, the SS's form an equation, but MS_T is not equal to $MS_W + MS_B$. The equality is destroyed because the terms in equation 16.4 are divided by different quantities to obtain the mean squares in equations 16.6. The degrees of freedom, however, are additive

$$JN - 1 = (JN - J) + (J - 1)$$
$$df_{total} = df_{within} + df_{between}$$

MS_B AS AN ESTIMATE OF $\sigma_{\overline{X}}^2$

MS_B, the between-groups mean square, involves the division of a sum of squared *deviations of means about their mean* by their df. If we think of $\overline{X}_T$ as our best estimate of μ, we can view the sample means as deviating around this value. This is the situation that gave rise to the concept of the standard error of the mean. It is as if we had taken several samples from a population and were estimating the standard error of the mean in the population from the variability among the means of our samples.

$$\hat{S}_{\overline{X}}^2 = \frac{\Sigma_j(\overline{X}_j - \overline{X}_T)^2}{J - 1}$$

But what can cause these sample means to vary? There are two possibilities. The first, which underlies the concept of the standard error of the mean, is random variation from sample to sample. The samples contain different people, so there are two factors that can produce variation: the random effect of individual differences between the people who happen to be in the different groups, and the (hopefully) random effect of experimental errors. That is, chance can cause the means to vary from each other. These two factors are the same ones that make up MS_W.

The other factor that can cause the means to be different is the way that the groups are systematically different, the independent variable. If we have chosen groups of different ages, the effect of age may produce group differences larger than the random variation we would expect in samples. Likewise, if we have exposed our samples to different treatments, the treatment conditions may have produced differences in the means that are greater than we would expect from random variation. These are the factors that would cause the population means to vary, σ_μ^2. That is, there are three possible factors that can affect MS_B: real differences between the groups, random individual differences, and experimental error.

$$MS_B = \text{group effects} + \text{individual differences} + \text{error}$$
$$MS_B = \sigma_\mu^2 + \sigma_{ID}^2 + \sigma_{error}^2$$

If H_0 is true, $\sigma_\mu^2 = 0$, so $MS_B = \text{individual differences} + \text{error}$. Therefore, MS_B and MS_W should differ only randomly if H_0 is true.

◆ *Now You Do It*

You calculated the deviations and sums of squared deviations for a small set of scores in Now You Do It on page 411. What are the three MS terms that come from those data?

Answer

The sums of squares were $SS_T = 155$, $SS_B = 129$, and $SS_W = 26$. What are the degrees of freedom?

There were three participants in each group, so there are $(N - 1) = 2$ degrees of freedom within each group. There were four groups, so $df_W = J(N - 1) = 8$. With four groups, $df_B = (J - 1) = 3$. Dividing each sum of squares by its df produces the following mean squares.

$$MS_B = 129/3 = 43$$

$$MS_W = 3.25$$

$$MS_T = 155/11 = 14.09$$

The *F* Statistic and Tests for the Equality of Means

In Chapter 14, we developed a test statistic, F, that is based on the ratio of two unbiased estimates of a population variance. We saw that the sampling distribution of F is complex, depending on the degrees of freedom for the numerator variance and the degrees of freedom for the denominator variance. The F statistic itself was used to test the null hypothesis that the two variances were estimates of the same population variance or represented two samples drawn from populations with equal variances. Let us take stock of where we are now.

1. We have converted our null hypothesis about equal means into one that involves the variance among the means. The null hypothesis states that $\mu_1 = \mu_2 = \mu_3 = \cdots = \mu_J$ or that $\sigma_\mu^2 = 0$.
2. We have two variances, MS_B with $(J - 1)$ degrees of freedom and MS_W with $(JN - J)$, or $J(N - 1)$, degrees of freedom, both of which are estimates of the same population variance *under the conditions specified in the null hypothesis.*
3. We have the F statistic as a way to test a null hypothesis of the equality of two variances.

Now, consider what MS_B and MS_W contain. MS_B, which is 35.46 in our example, contains group effects, individual differences, and error; while MS_W, which is 4.733 in the example, contains just individual differences and error. The individual differences and error components of these two mean squares should differ by only random amounts. Therefore, we would expect MS_B to be larger than MS_W if, and only if, the treatment effects are greater than zero, but MS_B should never be less than MS_W. The F statistic tests the hypothesis that two variances differ by only a random amount, so the statistic

$$F = \frac{MS_B}{MS_W} = \frac{SS_B/df_B}{SS_W/df_W} \tag{16.7}$$

tests the null hypothesis that the group effect is no greater than we would expect from random variations among samples. You can see this by substituting the contents of the mean squares in the F formula.

$$F = \frac{group\ effect + individual\ differences + error}{individual\ differences + error}$$

If group effects are zero, the two mean squares should differ by a random amount, but if group effects are present, MS_B should be larger than MS_W by an amount equal to σ_μ^2. For this reason, we always put MS_B in the numerator of the F statistic when we are testing a hypothesis related to ANOVA.

We can illustrate the computation of F with the drug study data that we have been considering from Table 16-3. Inserting the MS's in the equation for F yields

$$F = \frac{MS_B}{MS_W} = \frac{35.46}{4.733} = 7.49$$

which has $df_B = 2$ and $df_W = 12$. Looking in Appendix Table A-4, we find that the .01 critical value for an F with 2 and 12 degrees of freedom is 6.93. Since our obtained F exceeds this value, we would decide to reject the null hypothesis. This leads us to the conclusion that the means are not all equal, or, alternatively, that the variance among the means is greater than zero.

This F has $(J - 1)$ and $J(N_j - 1)$ for its numerator and denominator degrees of freedom, respectively, because these are the df's associated with MS_B and MS_W. A study that yields a value of F exceeding the appropriate critical value will cause us to reject the null hypothesis that gave rise to the study. Thus, a statistically significant value of F can be interpreted to mean that the value of the group effects variance (σ_μ^2) is not zero, which implies that we must reject the hypothesis that the means are equal.

▶ *Now You Do It*

What is the F for the data in the previous Now You Do It? You should get $43/3.25 = 13.23$. The $P = .01$ critical value for an F with 3 and 8 degrees of freedom is 7.59. We would therefore reject H_0.

Computing the Answer[1]

Computational Formulas

It should now be clear why the procedures described in this chapter are called analysis of variance even though the basic null hypothesis involves means. We have broken down the total variability among the study participants into its component parts. Those components are a between-groups effect, which is a combination of the differences between the population means, individual differences between the study participants, and errors of measurement; and a within-groups effect, which includes only individual differences and error. We tested our hypothesis with these components of variance. The null hypothesis states that there is no variance that is due to differences between population means. Therefore, if the between-groups component is significantly greater than the within-groups component (as it is in our example), we reject the null and conclude that the population means are not equal.

The general equations 16.6 are sometimes used to compute the quantities needed to perform an analysis of variance, but ordinarily we would not use them because they require working with decimals. There are several alternatives that allow us to work with sums and sums of squares computed directly from the data table. For all of them, the needed quantities can be obtained from the sums ($\sum_i X_{ij}$) and sums of squared scores ($\sum_i X_{ij}^2$) for each group.

The scheme presented here for computing the sums of squares needed to perform an analysis of variance is sometimes known as the **df-components method** because there is a particular total that is associated with each term in the degrees of freedom. These totals can be combined to form the needed sums of squares in exactly the pattern specified in the formula for degrees of freedom for that sum of squares.

*The **df-components method** uses the degrees of freedom for each MS to give rules for combining simple sums to obtain the sums of squares.*

To see how we can use the df-components method to compute the sums of squares for our memory-enhancing drug example, first consider the various df terms themselves; we have $df_T = JN - 1$, $df_B = J - 1$, and $df_W = JN - J$. There are three unique components here: JN, J, and 1; there is also a total which we can obtain from Table 16-1 that is associated with each of these terms. The expressions for df tell us how to compute the total associated with each term and how to combine these totals to obtain the sums of squares.

There are two dimensions to our data table in Table 16-1. There is a J (columns) dimension that refers to groups and has three levels, high, low, and none, and there is an N (rows) dimension with subscript i that refers to individuals and has five levels, one for

[1]To keep our notation as simple as possible, all equations are stated as though there were an equal number of participants in each group. When this is not the case, we must substitute the total number of participants for JN and the number of participants in each particular group for N in the calculation of the actual degrees of freedom, sums, and df components. However, the simple df terms can be used to combine df components to obtain the sums of squares.

each person in a group. Each of the *df* components involves these dimensions and tells us which quantities to square and add.

Step 1: To compute the total associated with a *df* component, we first **add across the dimensions *not* mentioned in the component.** For example, in the **1** component neither *J* nor *N* is mentioned. Therefore, the **1** component implies adding over both dimensions of the table as a first step to yield the single value

$$\mathbf{1} \rightarrow \Sigma_j \Sigma_i X_{ij}$$

The **1** component always refers to the simple total of all the scores in the data table. This value is 182 for the data in Table 16-1. Therefore, 182 is the quantity associated with the **1** component at this stage.

The **J** component mentions the letter *J*, so this term implies adding over the *N* (= 5 in our example) dimension as a first step, resulting in *J* separate totals. That is, we find each column total

$$\mathbf{J} \rightarrow \Sigma_i X_{ij}$$

For our example, the three column totals are 46, 64, and 72.

Finally, the **JN** component mentions both *J* and *N*, so this term implies no summations, resulting in *JN* (= 15 in our case) separate terms. The **JN** term always refers to the individual scores.

$$\mathbf{JN} \rightarrow X_{ij}$$

Because the **N** term never appears alone, we do not sum over just the *J* dimension.

Step 2: Next, we square each of the resulting sums. This produces the terms

$$\mathbf{1} \rightarrow (\Sigma_j \Sigma_i X_{ij})^2$$
$$\mathbf{J} \rightarrow (\Sigma_i X_{ij})^2$$
$$\mathbf{JN} \rightarrow X_{ij}^2$$

The **1** component produces the single term 182^2 (= 33124). The **J** component produces the three terms 46^2 (= 2116), 64^2 (= 4096), and 72^2 (= 5184). Finally, the **JN** term produces the 15 squared individual scores.

Step 3: Third, we divide each term that we computed in step 2 by the number of things we added in step 1 to obtain that term. As a set of formulas, this produces

$$\mathbf{1} \rightarrow \frac{(\Sigma_j \Sigma_i X_{ij})^2}{JN}$$

$$\mathbf{J} \rightarrow \frac{(\Sigma_i X_{ij})^2}{N}$$

$$\mathbf{JN} \rightarrow \frac{X_{ij}^2}{1}$$

Application of this step to our terms in step 2 produces the following values:

$$\mathbf{1} \rightarrow \frac{33124}{15} = 2208.27$$

$$\mathbf{J} \rightarrow \frac{2116}{5} = 423.2 \qquad \frac{4096}{5} = 819.2 \qquad \frac{5184}{5} = 1036.8$$

$$\mathbf{JN} \rightarrow \frac{121}{1} \qquad \frac{36}{1} \qquad \text{and 13 other similar terms}$$

Step 4: Finally, we add across the dimensions that *are* included in the component. Because neither the *J* dimension nor the *N* dimension is included in the **1** component, we don't need to add anything. For the other two components, we get the formulas

$$\mathbf{J} = \Sigma_j \frac{(\Sigma_i X_{ij})^2}{N_j}$$

$$\mathbf{JN} = \Sigma_j \Sigma_i X_{ij}^2$$

The total of the three **J** terms is $423.2 + 819.2 + 1036.8 = 2279.2$, and the sum of the squared single scores is 2336. Therefore, we have

$$\mathbf{1} = 2208.27$$
$$\mathbf{J} = 2279.20$$
$$\mathbf{JN} = 2336$$

As a general rule, we must square as many things as are specified in the component; the **1** component involves squaring a single quantity, the grand total of the raw scores $(\Sigma_j \Sigma_i X_{ij})$. The **J** component involves squaring J (in the case of our example, 3) quantities that have been summed over the i dimension $(\Sigma_i X_{ij})$. These are the column totals. Finally, the **JN** component requires squaring the raw scores themselves, 15 items $(JN = 3(5) = 15)$ in our example. More generally, the **JN** component involves squaring each individual score in the data table. Our rules for the df-components method are as follows:

1. Add across the dimensions that *are not* mentioned in the component.
2. Square each of the resulting totals.
3. Divide each of the squared totals by the number of things added.
4. Add across the dimensions that *are* mentioned in the component.

To apply these rules in a simple and straightforward manner, it is essential to lay the data out so that we can see what needs to be calculated. The following table shows where the terms are located in a data table such as Table 16-1.

	Group 1	Group 2	Group 3	
	$\mathbf{JN}\ (X_{ij}^2)$	$\mathbf{JN}\ (X_{ij}^2)$	$\mathbf{JN}\ (X_{ij}^2)$	
	$\mathbf{JN}\ (X_{ij}^2)$	$\mathbf{JN}\ (X_{ij}^2)$	$\mathbf{JN}\ (X_{ij}^2)$	
	$\mathbf{JN}\ (X_{ij}^2)$	$\mathbf{JN}\ (X_{ij}^2)$	$\mathbf{JN}\ (X_{ij}^2)$	
Column Sum	$\mathbf{J}\ (\Sigma_i X_{ij})^2$	$\mathbf{J}\ (\Sigma_i X_{ij})^2$	$\mathbf{J}\ (\Sigma_i X_{ij})^2$	$\mathbf{1}$ = Total Sum $(\Sigma_j \Sigma_i X_{ij})^2$

Now that we have the quantities associated with each term in the df expressions, we are ready to compute the sums of squares. We take the expression for the degrees of freedom for each mean square and substitute the values we computed for the df components to obtain the sums of squares. The expression for df_T is $(JN - 1)$, so we subtract the **1** component from the **JN** component

$$SS_T\ (\mathbf{JN} - \mathbf{1}) = 2336 - 2208.27 = 127.73$$

Treating the other two SS's in the same way, we get

$$SS_B\ (\mathbf{J} - \mathbf{1}) = 2279.20 - 2208.27 = 70.93$$
$$SS_W\ (\mathbf{JN} - \mathbf{J}) = 2336 - 2279.20 = 56.80$$

which are the same values, within rounding error, that we obtained using the definitional formulas.

HOW THE COMPUTER DOES IT

The simplest way to do an ANOVA like the ones in this chapter with SPSS is to use the One-Way ANOVA program in the Compare Means Subcommand of the STATISTICS menu. The program will prompt you for a variable and a factor. The variable is the dependent variable, and the factor is the independent variable. Each distinct numerical value of the independent variable will be considered a group. Output will be in the form of a summary table like the one in the next section. Other forms of ANOVA are included under the heading General Linear Model, but these cover more complicated designs and analysis options.

One advantage to the One-Way ANOVA program is that it includes several types of analyses that can be conducted after the initial ANOVA is run. Some of the options included under the Contrasts and Post Hoc buttons are described later in this chapter.

The value of this procedure is threefold: First, it is quick and easy. Second, it is general. Analysis of variance can involve very complex designs with multiple independent variables; nevertheless, the rules for computing and combining *df* components apply to most commonly used ANOVA designs. Third, the procedure provides a handy mnemonic for remembering what goes where. If you can figure out the degrees of freedom for a sum of squares, the *df* expression always tells you how to compute and combine the components. The complete computational formulas for the sums of squares for an analysis of variance like the one we have described here are as follows.

Computational Formulas for Analysis of Variance Using the *df*-Components Method		
Sum of Squares	***df***	**Computational Formula**
SS_T	$JN - 1$	$\sum_j \sum_i X_{ij}^2 - \dfrac{(\sum_j \sum_i X_{ij})^2}{JN}$
SS_B	$J - 1$	$\sum_j \dfrac{(\sum_i X_{ij})^2}{N} - \dfrac{(\sum_j \sum_i X_{ij})^2}{JN}$
SS_W	$JN - J$	$\sum_j \sum_i X_{ij}^2 - \sum_j \dfrac{(\sum_i X_{ij})^2}{N}$

▶ ## Now You Do It

Imagine that you are teaching a statistics class and that a student tells you that students in the class understand the mean, graphs, and standard deviations much better than they understand correlation. The student has asked 16 friends from the class to volunteer to take a 20-item quiz on one of the above topics. Four students took the quiz on the mean, four others took the quiz on graphs, four took the quiz on standard deviations, and four took the quiz on correlation. The student reports the following results. Do you agree that the conclusion is warranted? How should you reach a conclusion?

Student	Mean	Graphs	Standard Deviation	Correlation
1	13	11	13	10
2	14	12	16	10
3	14	17	15	7
4	12	11	11	4

Just looking at the numbers, it appears the student may be right, but is this a chance aberration? What is the null hypothesis, and what conclusion do the data suggest? It looks like a problem for ANOVA. The first thing to do is to calculate $\sum X_i$ and $\sum X_i^2$ for each group. You should get the following:

$\sum X_i$	53	51	55	31
$\sum X_i^2$	705	675	771	265

The null hypothesis is that the means are equal for all groups, but the data suggest that the students in the correlation condition got lower scores. From the cell totals you should compute

$$\sum_j \sum_i X_{ij} = 190 \quad \text{and} \quad \sum_j \sum_i X_{ij}^2 = 2416$$

The latter is the **JN** component. All that remains is to find the **J** component and the **1** component. Following our rules, we take the group sums, square them, divide by the number of observations in a group, and add the results across the groups. For example $53^2/4 = 702.25$. The total across all four cells is 2349. For the **1** component we square 190 and divide by 16 to obtain 2256.25. We thus have

$$\mathbf{JN} = 2416 \quad \mathbf{J} = 2349 \quad \mathbf{1} = 2256.25$$

Combining these quantities as dictated by the df's produces

$$SS_T = 2416 - 2256.25 = 159.75$$

$$SS_B = 2349 - 2256.25 = 92.75 \qquad df_B = 3 \qquad MS_B = 30.92$$

$$SS_W = 2416 - 2349 = 67 \qquad df_W = 12 \qquad MS_W = 5.58$$

The resulting F with 3 and 12 df is 5.54. The critical values for this F are 3.49, 5.95, and 10.80. The exact probability of this F from the computer output is .013. It looks like the student may have a point!

Presenting and Interpreting Results

The results of an analysis of variance often are presented in a table called a Summary Table. The Summary Table for our three-group drug study is given in Table 16-4 and includes five columns. The column labeled *Source* specifies which source of variance is being summarized in each row. The between-groups source is shown in the first row; the within-groups source in the second; and the total in the last row (some programs print the Total line first). Column 2, labeled *SS*, gives the values of the sums of squares; the df column gives the degrees of freedom associated with each source. The values in the *MS*, or mean square column, are, of course, the SS's divided by their degrees of freedom; since MS_T is not used in any computations, it is generally omitted. The F column gives the value of the F statistic, MS_B/MS_W; it is not uncommon for asterisks (*) to be used to indicate F's that exceed the standard critical values and are therefore considered statistically significant. An alternative you will see in many computer outputs is the probability of an F as large or larger than the one obtained in the study. For example, the SPSS output for the analysis in the previous Now You Do It includes a column labeled "Sig " and the value .013 for the F, indicating that 13 times in 1000 you would get an F this large or larger by chance if the null hypothesis were true.

Although the purpose of the analysis of variance is the same as that of the t-test, testing hypotheses about the equality of population means, its interpretation is considerably more complex. A statistically significant value for t is interpreted to mean that the difference between the two sample means is greater than would be expected by chance under the conditions specified by the null hypothesis. A statistically significant value for F implies that the differences among the several group means are greater than would be expected by chance under the null hypothesis conditions. Unfortunately, neither the F itself nor any of the mean squares give us information about which mean differences are responsible for the statistical significance. We cannot tell from Table 16-4 which means differ from which.

The solution to this problem involves going back to a point that is closer to the original data. First, we can go to the treatment-group means. Graphs of the means, such as the one in Figure 16-3, often help interpretation, particularly when the number of means to be considered is large. From this graph, it is easy to see that the low-drug group in our study of the memory-enhancing drug is more similar to the no-drug group than it is to the high-drug group. Also, from the graph and direct inspection of the group means, we might be tempted to conclude that the high-drug-group mean is statistically significantly different from the no-drug mean and that this difference is at least part of the cause of the significant F.

Table 16-4	Summary Table for an Analysis of Variance			
Source	SS	df	MS	F
Between	70.93	2	35.465	7.49**
Within	56.80	12	4.733	
Total	127.73			

$**P < .01$

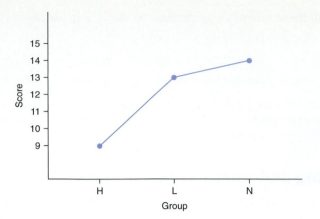

Figure 16-3 Plot of group means for drug study.

▶ *Now You Do It*

Prepare a plot of the group means to help you interpret what is happening for the statistics student data in the previous Now You Do It. You should get a graph that looks like this:

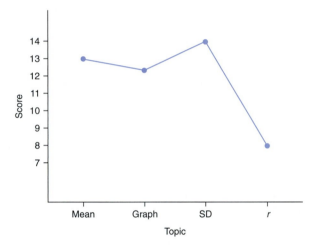

We begin to encounter real difficulties when we attempt to extend our interpretations. Can we conclude that the difference between any other pair of means is statistically significant? Unfortunately, we cannot draw such conclusions from the ANOVA itself; in fact, even our apparently safe decision about μ_H and μ_N is not entirely secure. In order to reach conclusions about specific pairs of means after a significant F, we must perform what are called *post hoc contrasts*. We address this issue after we present a larger example reviewing the basic procedures.

An Example with Five Groups

Suppose that you are a highly paid educational psychologist who has been asked to assist with a study of the effect that listening habits have on learning verbal material. The general design of the study calls for five treatment groups, each of which will study a reading passage for a specified time while experiencing one of five kinds of auditory stimulation. At the end of the allotted time, each participant is to complete a set of test questions that measure memory for and comprehension of what has been read. The participants are college freshmen and, in order to control possible familiarity with the text to be studied, the reading material has been taken from an obscure science fiction story.

The object of the study is to determine whether some types of auditory stimulation cause greater learning interference than others. You hypothesize that optimum performance should occur in a quiet environment, so you include one group with this as a base condition (Q group). All other conditions include some kind of stimulus occurring at an average

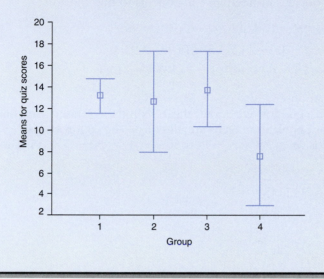
level of 50 decibels (moderately loud). One group hears random meaningless noise, sometimes called white noise (WN group). A second group hears relatively obscure classical instrumental music (CM group). Another group hears what is sometimes called elevator music (EM), a mixture of some vocal and some instrumental music of a generally nondescript character. The final group is exposed to contemporary rock music (RM). Fifty students are recruited and 10 are assigned on a random basis to each group. We decide to test the null hypothesis of equal group means at the $\alpha = .01$ level. The raw data, which are composed of group membership and score on the 25-item test of memory/comprehension, are shown in Table 16-5.

Table 16-5   Raw Data and Basic Sums for Noise/Studying Experiment

		Experimental Groups			
Subject	Quiet	White Noise	Classical Music	Elevator Music	Rock Music
1	21	19	14	18	13
2	17	15	17	15	19
3	22	18	19	21	15
4	20	21	16	20	11
5	17	14	21	13	14
6	20	16	17	16	15
7	18	18	18	16	14
8	17	17	16	18	12
9	20	15	17	17	18
10	19	17	18	15	17
$\overline{X}$	19.1	17.0	17.3	16.9	14.8
S_X	1.70	2.00	1.79	2.30	2.44
$\Sigma_i X_{ij}$	191	170	173	169	148
$\Sigma_i X_{ij}^2$	3677	2930	3025	2909	2250

$\Sigma_j \Sigma_i X_{ij} = 851$
$\Sigma_j \Sigma_i X_{ij}^2 = 14{,}791$

With this information we can proceed to compute the *df* component terms. The **JN** term is already available as $\sum_j\sum_i X_{ij}^2 = 14{,}791$. The two other terms are associated with the **J** and the **1** components, and are

$$\mathbf{J} = \left(\frac{191^2}{10} + \frac{170^2}{10} + \frac{173^2}{10} + \frac{169^2}{10} + \frac{148^2}{10}\right) = 14{,}577.5$$

$$\mathbf{1} = 851^2/50 = 14{,}484.0$$

We next use the degrees of freedom components to find the *SS*'s:

$$SS_T \rightarrow \mathbf{JN} - \mathbf{1} \rightarrow 14{,}791 - 14{,}484 = 307$$
$$SS_B \rightarrow \mathbf{J} - \mathbf{1} \rightarrow 14{,}577.5 - 14{,}484 = 93.5$$
$$SS_W \rightarrow \mathbf{JN} - \mathbf{J} \rightarrow 14{,}791 - 14{,}577.5 = 213.5$$

As a check we can observe that $SS_T = SS_B + SS_W$.

We are now in a position to compute the *MS*'s and the *F* ratio to test our null hypothesis that the treatment population means are all equal. The *df* for MS_B is $J - 1 = 5 - 1 = 4$, and that for MS_W is $JN - J = J(N - 1) = 5(10 - 1) = 50 - 5 = 45$. The results are

$$MS_B = 93.5/4 = 23.375$$
$$MS_W = 213.5/45 = 4.74$$

and

$$F = \frac{23.375}{4.74} = 4.93$$

with 4 and 45 degrees of freedom. Looking in Appendix Table A-4 we find that the critical value for $\alpha = .01$ with 4 and 40 degrees of freedom is 3.83 (as usual, we go to the next lower *df* when our exact value is not given). Under our decision rule, we reject the null hypothesis of equal treatment population means. The Summary Table would look like this:

Source	SS	df	MS	F	Sig
Between Groups	93.5	4	23.38	4.93	.002
Within Groups	213.5	45	4.74		
Total	307.0	49			

Our next step in analyzing the results of this or any experiment that results in a significant overall *F* will be to attempt to determine the source of the significant differences by comparing pairs of means between individual treatment conditions. Inspecting the graph of the group means in Figure 16-4, we see that the group that studied in quiet had the highest mean and the group that listened to rock music had the lowest mean. The other three groups appear to be about equal.

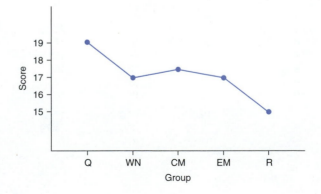

Figure 16-4 Plot of group means for five-group learning study.

▶ *Now You Do It*

Let's bring all the pieces of an analysis of variance together by using the Big Time U data to test the null hypothesis that the lab sections do not differ in their knowledge of statistics as shown by their final exam scores. This one is going to require that you be a little careful, because the labs do not all have the same number of students. How should we proceed?

Even if you do most of your analyses on a computer, try this one "by hand " to verify that you understand the pieces. What are the pieces of the df-components method?

There are three components we must find. The first is the **JN** component. This is simply the sum of the 68 individual X_{ij}^2's. It has a numerical value of 92,056. The second component is the **1** component. For this one, take the total sum of scores ($\Sigma_j \Sigma_i X_{ij}$), which is 2328, square it, and divide the result by 68. You should get a value of 79,699.77. Finally, we must find the **J** component. This one is a little tougher, and it is here that we must pay attention to sample size. The individual lab section sums are 895, 717, and 716 for sections 1, 2, and 3 respectively. Square each of these sums, and divide each squared sum by the number of scores on which it is based. This gives

$$\frac{895^2}{22} + \frac{717^2}{22} + \frac{716^2}{24} = 81{,}138.58$$

We now have our three components. What are the df terms? The degrees of freedom symbol for the total sum of squares is $JN - 1$. This gives us

$$SS_T = 92{,}056 - 79{,}699.77 = 12{,}356.23$$

The df for sum of squares between groups is $J - 1$, which yields

$$SS_B = 81{,}138.58 - 79{,}699.77 = 1438.81$$

Finally, the df for within-groups sum of squares is $JN - J$.

$$SS_W = 92{,}056 - 81{,}138.58 = 10{,}917.42$$

You can check these sums of squares using the One-Way ANOVA program in SPSS.

Finally, what are the mean squares and the F? Each MS is its SS divided by its df. Therefore,

$$MS_B = \frac{1438.81}{2} = 719.40$$

$$MS_W = \frac{10{,}917.42}{65} = 167.96$$

$$F = \frac{719.4}{167.96} = 4.28$$

Looking in Appendix Table A-4, we find the .05, .01, and .001 critical values for an F with 2 and 60 df to be 3.15, 4.98, and 7.77. Our obtained F is large enough to reject the hypothesis of equal means for the three lab sections at the $P < .05$ level. The computer printout tells us that the exact probability of an F this large is .018. Your summary table should look like this:

Source	SS	df	MS	F
Total	12,356.23	67		
Between Groups	1438.81	2	719.40	4.28
Within Groups	10,917.42	65	167.96	

Comparisons Among Pairs of Means

There are two basic ways to proceed when comparing pairs of means. The approaches differ on the issue of whether you started the study with the specific intention of comparing a particular small number of mean pairs or whether you will explore the data more or less without restriction to determine which differences are significant. Some authors state that if you have a small number of comparisons planned before you run the study, you should not even conduct the *F*-test for the overall experiment but should simply conduct the comparisons of interest, perhaps using a somewhat smaller value for α, for example .02 or .01 instead of .05. Everyone agrees, however, that if you are going to explore the data after the fact of a significant *F*, the methods for **post hoc comparisons** should be used.

The Bonferroni Method

The practice of protecting against a higher familywise rate of Type I errors when running more than one test by using a lower α level is called the **Bonferroni method.** The basic idea is that one can divide up the risk of a Type I error among the comparisons that are to be made. If we wish to keep our overall familywise error rate at no more than .05, and we wish to run two tests, the Bonferroni approach would call for conducting those tests with an α of .025 for each one. When a small number of comparisons are run, this procedure is simple and makes good intuitive sense. In the case of large studies with many comparisons it may not work too well. For example, if we wish to use the Bonferroni method for 10 comparisons and keep our familywise error at .05, we would have to use the .005 critical values for *t*. As the number of groups increases, the number of possible comparisons goes up very quickly.

Planned Comparisons

Before we begin our discussion of planned comparisons, we must define some terms. The mean square within groups (MS_W) is often called the error mean square or MS_E. MS_W is exactly like the pooled variance estimate that we used to perform our two group *t*-tests in Chapter 13 (they are identical for two groups); thus, we can use the MS_W just like an $\hat{S}_p^2$, where the pooling has involved more than two groups, for conducting comparisons between pairs of group means.

Next, we must define a **contrast.** A contrast (C) is the difference between two selected *means or averages of means.* As an example, consider our study of the memory-enhancing drug from Table 16-1. We found a significant overall *F*, indicating that some means were different from other means. To further explore the cause of this significant *F*, we might wish to form a contrast between the mean of the high-drug group and the average of the means of the low- and no-drug groups. This would produce a contrast of the form

$$C = \overline{X}_H - \frac{\overline{X}_L + \overline{X}_N}{2}$$

If the groups are of different sizes, each must be weighted by its sample size. Therefore, the general statement of a contrast between the mean of one group and the average of the means of two others would be

$$C = \frac{N_H \overline{X}_H}{N_H} - \frac{N_L \overline{X}_L + N_N \overline{X}_N}{N_L + N_N}$$

For the drug study contrast, *C* would be

$$C = \frac{5(9.2)}{5} - \frac{5(12.8) + 5(14.4)}{5 + 5} = 9.2 - \frac{12.8 + 14.4}{2}$$

SPECIAL ISSUES IN RESEARCH

Should You Use Planned Contrasts?

The use of planned contrasts is somewhat controversial because there is always the possibility that a researcher will, after looking at the data, select only the largest comparisons to run in hope of finding a statistically significant result. For this reason, it is important that any planned contrasts be ones that can be derived from the theory or research background that prompted the study. This is similar to the issue of one-tailed versus two-tailed tests of simple mean differences. In both cases, it is imperative (and the responsibility of the investigator) to provide a clear and convincing reason, based on theory or a review of the research literature, that a specific outcome is to be expected. Then the analysis of the data is used in a confirming way to show that the study results are consistent with what was predicted.

The difference between the mean of the high-drug group and the average of the means of the low- and no-drug groups is -4.4, and it is this difference that we would test by methods to be described shortly. As a second example, if we were interested in contrasting the mean of the quiet condition with the average of the means of the four noise conditions in our second study, we would have

$$C = \frac{10(19.1)}{10} - \frac{10(17.0) + 10(17.3) + 10(16.9) + 10(14.8)}{10 + 10 + 10 + 10}$$

$$= 19.1 - 16.5 = +2.6$$

HOW THE COMPUTER DOES IT

One of the advantages of using the One-Way ANOVA program in SPSS is that it can perform contrasts like the one we have just described. However, the program does it in a slightly different way. Each contrast is defined by a set of coefficients that are used to multiply the means. The sum of the coefficients should be zero. Thus, if we wish to contrast the high-drug group with the average of the low- and no-drug groups, the coefficients to use would be 2 for the high group and -1 for both of the others. (An alternative would be to use 1 for the high group and $-1/2$ for the other two.) To compare the quiet group with the average of the four noise groups in our study of the effects of noise on studying, we would use the coefficients 4, -1, -1, -1, -1. (Note that in all cases the coefficients sum to zero.) The value of t will be identical to what we would get using equations 16.8–16.10, but the values for the contrast and standard error may be different.

In order to run the analysis, select One-Way ANOVA from the Compare Means menu. Next, select the dependent variable and the factor. Then click on the Contrasts button. Enter the coefficient for the first group and click Add. Repeat this process until you have entered coefficients for all groups to be included in the contrast. You can exclude any group from a specific contrast by simply giving it a coefficient of 0, but there must be as many coefficients as groups. For our data on noise and studying, the screen might look like this:

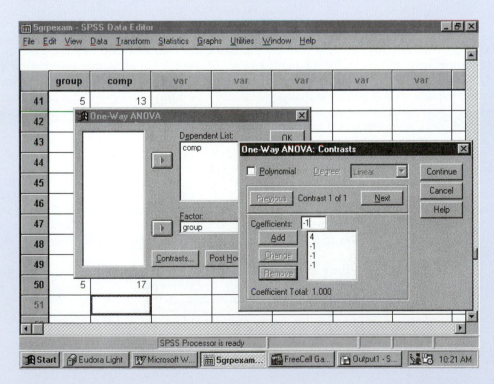

Note that the program keeps a running total of the coefficients you have entered. When you are done, this should equal 0.00.

The general form for such contrasts is

$$C = \frac{N_a \overline{X}_a + N_b \overline{X}_b + \cdots + N_k \overline{X}_k}{N_a + N_b + \cdots + N_k} - \frac{N_m \overline{X}_m + N_n \overline{X}_n + \cdots + N_z \overline{X}_z}{N_m + N_n + \cdots + N_z} \qquad (16.8)$$

That is, the value of the contrast C is the difference between the weighted average of the means on one side of the comparison and the weighted average of the means on the other side. Each mean is weighted by its sample size. C is treated just like the difference $\overline{X}_1 - \overline{X}_2$ in the ordinary t-test. It has a sampling distribution, and this distribution has a standard error.

As our next step, we must compute the standard error of the contrast. Recalling that MS_W is equivalent to $\hat{S}_p^2$, we can compute the standard error of C from the general equation.

$$\hat{S}_C = \sqrt{MS_W \left[\left(\frac{1}{N_a + N_b + \cdots + N_k} \right) + \left(\frac{1}{N_m + N_n + \cdots + N_z} \right) \right]} \qquad (16.9)$$

For our first problem this gives

$$\hat{S}_C = \sqrt{4.733 \left(\frac{1}{5} + \frac{1}{5 + 5} \right)} = \sqrt{4.733(.2 + .1)} = 1.19$$

The fact that we are treating this as a planned comparison (we intended to run this comparison before we looked at the data) means that we can compute an ordinary t-test as a test of significance. The value of t for this planned comparison is given by

$$t_{df_W} = \left(\frac{C - K}{\hat{S}_C} \right) \qquad (16.10)$$

where C is the value of the contrast from equation 16.8 and K is almost always zero, but is included here to remind you that this t-test has exactly the same form that we have been using in previous chapters. For our data, this yields

$$t = \frac{-4.4}{1.19} = -3.69$$

The degrees of freedom for MS_W provide the df for the t. In this case, we are using a t distribution with $df = 12$. The two-tailed $\alpha = .01$ critical value in this distribution is ± 3.055. Because the value of t from our study exceeds this critical value, we can reject the null hypothesis and conclude that performance under the high drug dosage condition is poorer than under low or no drug. This result implies that, contrary to the claims made for it, this drug seems to retard memory rather than enhance it.

▶ Now You Do It

Let's return to the data from our five-group study of noise and memory. The contrast that we derived from our theory was that noise of any kind would interfere with memory for what was studied. Therefore, we can justify comparing group Q with the average of the other groups. Equation 16.8 gave us a value for the contrast of 2.60. What else do we need?

The standard error of the contrast is obtained from equation 16.9. For these data we found MS_W was 4.74. Therefore,

$$\hat{S}_C = \sqrt{4.74 \left[\left(\frac{1}{10} \right) + \left(\frac{1}{10 + 10 + 10 + 10} \right) \right]} = \sqrt{4.74(.125)} = .7697$$

From equation 16.10 we find t to be

$$t_{45} = 2.60/.7697 = 3.38$$

Tabled values of t include $df = 40$ and $df = 60$. The .01 critical value for a one-tailed test (remember, our theory predicted higher performance for the quiet group) is 2.423, which would lead us to reject the null hypothesis and conclude that there is a difference between the quiet group and the average of the other four groups.

If you chose to enter these data into SPSS and run the analysis using One-Way ANOVA, the value for the contrast is 10.40 and the standard error is 3.08. These values still produce a t of 3.376, the same value we got above. Because the program can perform a much wider variety of contrasts than the simple ones we have considered, it uses a more complex and general set of equations, but it ends up at the same place.

Post Hoc Contrasts

Research in a well developed area of inquiry often involves the use of planned comparisons between group means, and the problem of familywise error when more than one comparison is conducted is often handled by the simple expedient of adjusting the α level by the Bonferroni method. When investigators enter a new area of study, however, they may not have clear expectations about what they will encounter. In this case, it may be desirable to explore the data for hints after running the overall F without the restriction of having to state ahead of time what you expect to find. Comparisons of this kind are called **post hoc,** or after the fact, contrasts because they are undertaken after we have looked at the data and found some differences we would like to explore.

*A **post hoc contrast** is one that is conducted after looking at the data from a study.*

Although it might seem logical to conduct post hoc contrasts by computing t-tests for any interesting pairs of means, this, unfortunately, is improper for the same reasons that caused us to develop the methods of ANOVA in the first place. Even if we only compared the two most extreme means, computing only one t, that t is suspect simply because we chose the largest of the several possible differences; we biased the comparison in our favor by selecting it after seeing that this was the largest difference.

A number of statisticians have developed procedures for dealing with this problem. Some are appropriate under certain conditions, others under different conditions; some analyze pairs of means, while others can be used in more complex situations. Many of these methods are described in the more advanced texts listed at the end of this chapter. Here, we present a method for post hoc comparisons developed by Scheffé (1953). It has the advantage that it is easy to understand and compute, can be applied in virtually any situation, uses the standard F table, and is conservative (that is, if a comparison is significant using this procedure, you can be confident that the relationship it represents would have been found to be significant if you had designed the original study to make only this comparison). The test maintains the familywise error rate at a level *no higher than* the chosen α regardless of how many comparisons are conducted. The price of this conservatism is considerably lower power.

As a first step, the Scheffé procedure uses the critical value of F to calculate what we can call a **corrected critical value** for t. We then compute a t in the ordinary way using C and $\hat{S}_C$ from equations 16.8–16.10. That is,

$$t = \frac{C - K}{\hat{S}_C}$$

If this t exceeds the corrected critical value for t, we conclude that the difference between the two portions of the contrast is statistically significant.

The corrected critical value for t is found by multiplying the appropriate critical value of F by the *between-groups degrees of freedom* and taking the square root of the resulting product. First, we look up the F at the chosen α for df_B and df_W. For our drug dosage example with $df_B = 2$ and $df_W = 12$ and $\alpha = .01$, the appropriate F is 6.93. This F is then multiplied by df_B. In our example, this gives 13.86, which has a square root of 3.723. This is the new, corrected, critical value of t for all comparisons conducted on these data.

*The **Scheffé procedure** uses a **corrected critical value** **for t** that is computed from df_B and the critical value of F_{df_B, df_W} for the overall study.*

$$t_{corrected\ critical} = \sqrt{(df_B)(F_{(df_B, df_W)})} \tag{16.11}$$

This corrected critical t is treated as having the same α level as the F.

The final step is to use equation 16.10 to compute a t for the contrast. For our drug dosage example we found

$$t = -4.4/1.1916 = -3.69$$

which is only slightly smaller than the corrected critical value. (The fact that the computed value of t is negative reflects the fact that the value of the contrast was negative. Post hoc tests are treated as nondirectional hypotheses, so the sign of the test is not of particular importance in deciding whether to reject the null hypothesis.) Our post hoc test does not lead us to reject the null hypothesis that the high-drug group is not different from the average of the other two. Of course, if we had selected the .05 level for α, we would have reached a different conclusion because the critical value would have been smaller. The decision we reach with the Scheffé procedure contrasts with the outcome of the same computations when the test was conducted as a planned comparison and illustrates the penalty that we pay in power for conducting exploratory analyses after looking at the data.

One of the advantages of the Scheffé procedure is that it can be used repeatedly with the same data, so that we can test other contrasts easily and without violating our chosen α level. For example, the contrast between the high- and no-drug groups would proceed as follows:

$$C = 9.2 - 14.4 = -5.2$$

$$\hat{S}_C = \sqrt{4.733\left(\frac{1}{5} + \frac{1}{5}\right)} = \sqrt{1.8932} = 1.376$$

$$t = -5.2/1.376 = -3.779$$

The value of $\hat{S}_C$ is different because the comparison involves the scores of only 10 people, but the corrected critical value for t remains the same for all contrasts from the study, so we would reject the null hypothesis that $\mu_H = \mu_N$ at the $\alpha = .01$ level on the basis of these data.

Planned and Post Hoc Contrasts for the Study of the Effect of Music on Studying and Recall.

Let us review the various methods for conducting comparisons among the groups of a study using the data from Table 16-5. We had five groups, each one studying the same material under differing noise conditions. The means for the groups were

Q	Quiet	19.1
WN	White Noise	17.0
CM	Classical Music	17.3
EM	Elevator Music	16.9
RM	Rock Music	14.8

First, let us examine two contrasts that we might have derived from theory and conduct them as planned comparisons. One interesting comparison would be between the groups who listened to sounds where speech was present and those where nonspeech sound was present. This comparison contrasts the average of the WN and CM group means with the average of the EM and RM group means. We compute the value of the contrast to be

$$C = \frac{10(17.0) + 10(17.3)}{10 + 10} - \frac{10(16.9) + 10(14.8)}{10 + 10}$$

$$= 17.15 - 15.85 = 1.3$$

The standard error for this contrast uses the MS_W of 4.74 from the complete study:

$$\hat{S}_C = \sqrt{4.74\left[\left(\frac{1}{10 + 10}\right) + \left(\frac{1}{10 + 10}\right)\right]} = .688$$

If we consider this to be the only planned comparison we will conduct, the result is

$$t = 1.30/.688 = 1.89$$

a value which does not reach the two-tailed .05 critical value for t. Clearly, this is not the cause of the significant overall F.

Another hypothesis we might have chosen to entertain for the cause of the significant F is that the quiet group differs from the four groups that had noise of some form present. This produces the contrast we examined in the last Now You Do It:

$$C = \frac{10(19.1)}{10} - \frac{10(17.0) + 10(17.3) + 10(16.9) + 10(14.8)}{10 + 10 + 10 + 10} = 2.6$$

The standard error for this contrast is

$$\hat{S}_c = \sqrt{4.74\left(\frac{1}{10} + \frac{1}{10 + 10 + 10 + 10}\right)} = .77$$

and the t for this contrast is

$$t = \frac{2.6}{.77} = 3.38$$

which exceeds the standard critical values. Therefore, had we planned, before conducting the study, to run this comparison as our only contrast, we could conclude that scores were significantly better under the quiet condition than in any condition where noise was present.

The fact that we have now conducted two tests on the same data presents us with a problem: an inflated risk of a Type I error. We have discussed two ways to control this familywise error: the Bonferroni method, which is appropriate for multiple planned comparisons, and the Scheffé method for post hoc comparisons. If we treat our two contrasts as planned comparisons, we should use the Bonferroni method. To apply the Bonferroni method properly, we must specify the number of comparisons we will conduct before any of them is performed because we will divide our risk of a Type I error among the contrasts we run. The results will differ depending on how much protection we choose to apply, and how many contrasts we choose to perform, but because we conducted our overall F-test at the $\alpha = .01$ level, the only value of t available to us from the limited selection of values in Appendix Table A-2 is 3.551 for an α of .001. This would be the appropriate value for conducting 10 comparisons. We have conducted only two comparisons, so we can expect these results to be quite conservative. Our obtained value of t was 3.38, leading us not to reject the null hypothesis.

Before we jump to conclusions, however, we must remember that without the support of a theory (which we claimed before) we would be engaged in blatant data exploration at this stage. If we had selected these particular contrasts because, after looking at the data, they appeared likely to be significant, we should treat them as post hoc tests. Under these circumstances, we should apply a correction to control familywise error and to adjust for the fact that these contrasts were selected because of their size. Treating them as post hoc tests, we will use the Scheffé procedure.

The contrast of the quiet condition with the four noise conditions using the Scheffé procedure would use the critical value of F from our overall test (3.83) and the degrees of freedom between groups. Applying equation 16.11 results in a corrected critical value for t of

$$t_{corrected\ critical} = \sqrt{(4)(3.83)} = 3.91$$

Our value of t obtained from the data was 3.38, so we would not reject the null hypothesis using this method, either.

Finally, (and perhaps in desperation!) we might decide to test the difference between the quiet group and the rock music group. The value for this contrast is

$$C = 19.1 - 14.8 = 4.3$$

SPECIAL ISSUES IN RESEARCH

Probabilities and the Bonferroni Method

Proper application of the Bonferroni method for two planned tests would divide α equally between the two tests, leading to a probability of a Type I error of .005 for each test when $\alpha = .01$. The needed critical value is not available in any standard table, which is a drawback of the Bonferroni method. However, if you conduct the test using a computer program that provides exact probabilities for t, you will find that the probability of a t of 3.38 with $df = 45$ is less than .003, so we should reject the null hypothesis using this procedure.

(Note that for pairs of means it is not necessary to weight them if the samples are of the same size.) The standard error for the contrast is

$$\hat{S}_C = \sqrt{4.74\left(\frac{1}{10} + \frac{1}{10}\right)} = .974$$

which produces a value of t of

$$t = 4.3/.974 = 4.41$$

This most extreme of the possible comparisons exceeds the corrected critical value for t, therefore reaching statistical significance. We can conclude that the difference between the quiet group and the rock music group is greater than zero. In fact, one feature of the Scheffé test is that if the overall F is significant, there is certain to be at least one post hoc comparison of some form that also is significant. It may not be a comparison that is of any substantive interest, but there will be at least one statistically significant result.

▶ *Now You Do It*

Let's take a look at the results of our study of students' knowledge of statistics described in Now You Do It on page 420. The four group means were 13.25, 12.75, 13.75, and 7.75. The student who brought these data to our attention claimed that knowledge was lower in correlation than in the other three areas. This suggests a planned contrast between the correlation mean and an average of the other three means. What coefficients should we use, and what is the outcome? Once you have found the answer to that question, let's do some additional exploring using Scheffé's method.

To compare three groups with one, we need the coefficients 1, 1, 1, −3 (or −1, −1, −1, 3). MS_W was 5.58, so

$$\hat{S}_C = \sqrt{5.58\left(\frac{1}{4} + \frac{1}{4 + 4 + 4}\right)} = 1.36$$

The value of the contrast is

$$C = 13.25 + 12.75 + 13.75 - (3)7.75 = 39.75 - 23.25 = 16.5$$

The t with $df = 12$ is 12.13, which exceeds all conventional critical values. The data support the student's assertion.

We might also wish to compare the correlation mean with each of the others individually. This implies a Scheffé procedure. The corrected critical values are

$$t_{.05} = \sqrt{(3)(3.49)} = 3.24$$
$$t_{.01} = \sqrt{(3)(5.95)} = 4.22$$
$$t_{.001} = \sqrt{(3)(10.80)} = 5.69$$

Clearly, our contrast would have been statistically significant even under the stringent rules of Scheffé. But how about the pairs of means? The standard error for each comparison is 1.67, and the mean differences are 5.0, 5.5, and 6.0. Therefore, the t's are 2.99, 3.29, and 3.59. Two are significant at the .05 level, but the third is not.

If you carried out the above analysis using a computer and chose the Tukey method instead of Scheffé's, you would have found all three comparisons statistically significant. This illustrates the conservative nature of the Scheffé procedure and its relative lack of power. However, only the Scheffé method allows you to test the four-group contrast, and there the added number of participants raised the power to the point where the effect of different materials was clearly found.

Effect Size and Proportion of Variance Explained

The F and t statistics and their associated probabilities tell us the risk of making a Type I error, but, as we have discussed, researchers have found them to be insufficient for conveying the full meaning of a set of results. There are two primary reasons that we need an index in addition to these test statistics. The first is that the values of t and F are highly dependent on sample size. We saw in our discussion of power that increasing N would increase t even when the difference between the means remained the same. As we approach the limiting case of infinite sample size, one of two things happens; either the experiment becomes so powerful that a difference approaching zero is judged to be statistically significant, or any real difference between means produces a t that is extremely large. Therefore, one reason to go beyond t and F is to have an index that is independent of sample size.

The second reason that researchers have sought some index other than the test statistics is that they tell us little or nothing about how much of the variance in the dependent variable is explained by the independent variable. The answer to this question requires an approach similar to that used in correlation. Recall that the squared correlation coefficient is defined as the proportion of variance in the predicted (dependent) variable that is explained by the predictor (independent) variable. A correlation-like statistic provides a very useful index of the strength of the effect that the independent variable has on the dependent variable.

Measures of Variance Accounted For

Eta-square (η^2). There are two commonly reported indices of the strength of an experimental effect in an ANOVA. The first is called η^2 (eta-square) and is also known as the correlation ratio. This statistic is simply the ratio of SS_B to SS_T, and it tells us the proportion of total variation in the dependent variable that is due to group differences.

$$\eta^2 = SS_B / SS_T \tag{16.12}$$

For the three-group drug study η^2 is

$$SS_B / SS_T = 70.92/127.72 = .555$$

and for our study of the effects of noise on learning we get

$$\eta^2 = 93.5/307 = .305$$

η^2 can be interpreted just like r^2. It is the proportion of variance in the dependent variable that is related to the independent variable.

Omega-square (ω²). Eta-square is a descriptive statistic for the sample data at hand. That is, it describes how much of the observed variation in the dependent variable is attributable to the treatment in this particular group of individuals. Like the sample variance, it tends to be biased as an estimate of the condition in the population. An alternative index that estimates the degree of association between the independent and dependent variables in the population is called ω^2 (omega-square). Omega-square uses the values for MS_W and MS_B in the population to form a ratio of variance explained to total variance.

Omega-square is defined as

$$\omega^2 = \frac{\sigma_\mu^2}{\sigma_\mu^2 + \sigma_W^2} \tag{16.13}$$

where σ_μ^2 is the variation among population means and σ_W^2 is the variation within random samples from the population (individual differences plus error). As given in equation 16.13, ω^2 involves unknown parameters, the σ^2's. We can estimate ω^2 from the sample data using the formula

$$\hat{\omega}^2 = \frac{SS_B - (J-1)MS_W}{SS_T + MS_W} \tag{16.14}$$

where J is the number of levels of the independent variable. Applying equation 16.14 to the data from our drug example produces the following results:

$$\hat{\omega}^2 = \frac{70.92 - (2)4.733}{127.72 + 4.733} = \frac{61.454}{132.453} = .464$$

The value for our second example, the effect of noise on studying, is

$$\hat{\omega}^2 = \frac{93.5 - (4)4.74}{307 + 4.74} = \frac{74.54}{311.74} = .239$$

Omega-square, like η^2, theoretically can take on values between zero and one. Because $\hat{\omega}^2$ estimates the situation in the population while η^2 reflects the sample, and hence includes sample-specific effects, $\hat{\omega}^2$ will always be less than η^2, sometimes much less. Therefore, we need some different guidelines to evaluate the magnitude of effects that $\hat{\omega}^2$ reveals. One guideline that has been offered is that values of $\hat{\omega}^2$ of less than about .03 should be considered to reflect small effects, those between .03 and .15 represent medium-sized effects, and any $\hat{\omega}^2$ over .15 can be considered to represent a fairly large effect. Reanalyses of results reported in various journals suggest that many studies produce $\hat{\omega}^2$'s in the range of about .05 to .10, that is, medium-sized effects.

Effect Size

In Chapter 13 we discussed a statistic known as d which expressed the magnitude of the difference between a pair of means in terms of the pooled standard deviation, $\hat{S}_p$. This statistic can also be used to characterize the differences between pairs of means in a multi-group study. All that is needed is to substitute $\sqrt{MS_W}$ for $\hat{S}_p$ in equation 13.14. The d for any pair of means, $\overline{X}_A$ and $\overline{X}_B$, is therefore

$$d = \frac{\overline{X}_A - \overline{X}_B}{\sqrt{MS_W}} \tag{16.15}$$

We can illustrate the use of equation 16.15 with the data from our learning study (Table 16-5). The five groups had the following means:

Quiet	19.1
White Noise	17.0
Classicial Music	17.3
Elevator Music	16.9
Rock Music	14.8

The mean square within groups was 4.74, which has 2.18 as its square root. Therefore, the d between quiet and rock music is

$$d_{Q,RM} = (19.1 - 14.8)/2.18 = 4.3/2.18 = 1.97$$

which clearly is a substantial effect, and the d between white noise and classical music is

$$d_{WN,CM} = .3/2.18 = .14$$

a very modest effect.

It is also possible to compute an average effect size for the entire study. Kirk (1995, p. 181) has shown that an average effect size value for a study, which he labels $\hat{f}$, can be computed from

$$\hat{f} = \sqrt{\frac{((J-1)/JN)(MS_B - MS_W)}{MS_W}} \qquad (16.16)$$

For the data from Table 16-5, equation 16.16 produces

$$\hat{f} = \sqrt{\frac{(4/50)(23.38 - 4.74)}{4.74}} = .56$$

This would ordinarily be interpreted as a moderately large effect, which is consistent with the fact that the F-test was significant at the $\alpha = .01$ level. Kirk also shows that $\hat{f}$ and $\hat{\omega}^2$ are intimately related by the formula

$$\hat{f} = \sqrt{\frac{\hat{\omega}^2}{1 - \hat{\omega}^2}}$$

which further illustrates the point that measures of effect size such as d and $\hat{f}$ are telling essentially the same story that is told by measures of association such as r^2, η^2, and $\hat{\omega}^2$; they just do so in different metrics. Which you use is a matter of preference and the way you wish to make your point.

Post hoc comparisons give us a way to tease out of a significant result from an analysis of variance those group differences that are large enough to merit further consideration. Measures of effect size give us a way to express how large these differences are in a metric that can be compared across studies. By combining these methods, ANOVA gives us a powerful way to explore the phenomena of interest to educators and psychologists.

▶ **Now You Do It**

As our final exercise, let's determine how much of the variability in final exam scores is accounted for by lab section in the Big Time U data. You have the necessary sums of squares and mean squares in Now You Do It on page 425. What is η^2? What is the value of $\hat{\omega}^2$? What are the effect sizes for comparing section 1 with section 2 and section 3? What is the effect size for the overall study?

Answers
You can find η^2 from the printout of the Compare Means procedure in SPSS, or you can compute it from SS_B/SS_T. If you use the program, be careful. It gives you both η and η^2. Eta does not have a useful interpretation in this context. The value you want is $\eta^2 = .116$. That is, about 11.6 percent of the individual differences in final exam scores are accounted for by lab section differences. The more conservative $\hat{\omega}^2$ produces

$$\hat{\omega}^2 = \frac{1438.81 - 2(167.96)}{12,356.23 + 167.96} = \frac{1102.89}{12,524.19} = .088$$

To compute effect sizes, we need a pooled standard error of the mean. When there are more than two groups, this statistic is estimated from the MS_W of the study. For our data,

$$\hat{S}_p = \sqrt{MS_W} = 12.96$$

Therefore, the d for comparing section 1 with section 2 is .62, and the d for comparing section 1 with section 3 is .84. The overall effect size from equation 16.16 is

$$\hat{f} = \sqrt{\frac{(2/68)(719.40 - 167.96)}{167.96}} = \sqrt{\frac{(.0294)(551.44)}{167.96}} = \sqrt{\frac{16.21}{167.96}} = .31$$

which we might interpret as a moderate effect.

SUMMARY

Many research problems in the behavioral and social sciences require a study design using more than two groups. Results of such studies should not ordinarily be analyzed by comparing pairs of means unless a theoretically justified and small number of such comparisons are *planned in advance* because *t*-tests so computed are not independent of each other. The results of one comparison affect the magnitude of others; also, multiple *t*-tests increase the risk of making a Type I error.

The alternative to multiple *t*-tests is the *analysis of variance,* or *ANOVA.* ANOVA involves partitioning the total variance among scores into two parts: a variance due to differences between groups and a variance within groups. These two variances are both independent estimates of the population variance under the conditions specified in the null hypothesis. The *F ratio* with degrees of freedom of $(J - 1)$ and $(J \cdot N_j - J)$ is used to test the equality of the two variances. A significant value for *F* leads to rejecting the null hypothesis and the conclusion that the variance among means is greater than would be expected by chance.

ANOVA requires computing SS_B and SS_W (which sum to SS_T). The *df-components method* can be used for these calculations. The sums of squares are then divided by their degrees of freedom to produce the mean squares, and $F = MS_B / MS_W$.

Interpreting results is aided by graphs of means and by performing *post hoc contrasts,* which involve combinations of means. MS_W is used as a pooled variance to compute a standard error for the contrast. In the *Scheffé procedure* a corrected critical value for *t,* computed from the *F* and df_B, is used for all post hoc contrasts in the study. The magnitude of the effect produced by an experimental treatment can be estimated using η^2 or $\hat{\omega}^2$. The effect size for any pair of means can be determined using the *d* statistic with $\sqrt{MS_W}$ substituted for $\hat{S}_p$. An effect size for the entire study is provided by $\hat{f}$.

ANALYSIS OF VARIANCE OF THE SEARCHERS DATA

When Dr. Cross and his students designed their study of attitudes toward equality for women in the men's religious group called the Searchers, one of the independent variables, form of the questionnaire, had three levels. Form 1 of the questionnaire was phrased in a neutral and factual way, form 2 was aggressively pro-feminist in tone, and form 3 was conservative. Dr. Cross included this variable to allow for a test of whether the language in which the questions were asked would affect the responses. The null hypothesis is that the form of questionnaire a man received would have no effect on his stated attitudes. In statistical form, the null is that the means of the three groups do not differ by more than a chance amount, or, $\mu_N = \mu_F = \mu_C$. Let us now test whether the data give us reason to believe that this null hypothesis is false.

The easiest way to perform the ANOVA by hand is to sort the data file by the form of questionnaire variable, then compute the sums and sums of squared scores. These are then used in computational equations on page 420 to obtain the necessary quantities. The sums and sums of squared scores you should get are

	Neutral	Pro-Feminist	Conservative	Totals
ΣX	839	828	757	2424
ΣX^2	35,777	35,392	30,811	101,980
$\dfrac{(\Sigma X)^2}{N}$	35,196.05	34,279.2	28,652.45	98,127.7

Putting these quantities in the formulas produces the following values for the *df* components:

$$\mathbf{JN} = 101,980$$

$$\mathbf{J} = 98,127.7$$

$$\mathbf{I} = (2424)^2/60 = 97,929.6$$

Combining these totals as directed by the expressions for the degrees of freedom gives us

$$SS_T = \mathbf{JN} - \mathbf{I} = 101,980 - 97,929.6 = 4050.4$$

$$SS_B = \mathbf{J} - \mathbf{I} = 98,127.7 - 97,929.6 = 198.1$$

$$SS_W = \mathbf{JN} - \mathbf{J} = 101,980 - 98,127.7 = 3852.3$$

Dividing each sum of squares by its *df* gives us the following mean squares:

$$MS_B = 198.1/2 = 99.05$$
$$MS_W = 3852.3/57 = 67.58$$

Finally, the ratio of these two mean squares produces an *F* of

$$F = \frac{99.05}{67.58} = 1.47$$

Looking in the *F* table for 2 and 40 degrees of freedom, we find the .05 critical value is 3.23. Dr. Cross cannot reject the null hypothesis. On the basis of these data there is no reason to believe that form of the questionnaire affected respondents' answers. Since the overall *F* is not significant, we would not conduct any post hoc tests.

REFERENCES

Hays, W. L. (1994). *Statistics* (5th ed.). Fort Worth, TX: Harcourt Brace.

Howell, D. C. (1992). *Statistical methods for psychology* (3rd ed.). Boston: PWS-Kent.

Keppel, G. (1991). *Design and analysis: A researcher's handbook*. (3rd ed.). Upper Saddle River, NJ: Prentice Hall.

Kirk, R. E. (1995). *Experimental design: Procedures for the behavioral sciences* (3rd ed.). Pacific Grove, CA: Brooks/Cole.

Scheffé, H. (1953). A method for judging all contrasts in the analysis of variance. *Biometrika 40*, 87–104.

EXERCISES

1. State the reasons why it may be necessary (or desirable) to include more than two groups in a single study.

2. Why isn't it appropriate to test all possible pairs of means in a research study?

3. In conducting some statistical test, we often partition the data into its component parts. Describe the components of this partitioning process.

4. What does it mean when we say "the results of the *F*-test yielded a statistically significant result"?

5. When is it appropriate to use planned comparisons in research?

6. When is it appropriate to use post hoc comparisons in research? Why are post hoc comparisons necessary?

7. To examine the relationship between an inferred hemispheric style and problem-solving performance, participants were classified as having preferences of left, integrated, or right hemispheric style on the basis of scores on Your Style of Learning and Thinking. Participants were then asked to solve the seven disk Tower of Hanoi problem. The researcher recorded the number of moves it took to solve the problem. The data are provided below.

Hemispheric Preference

Left	Integrated	Right
238	194	176
206	207	185
224	213	210
225	226	189
229	208	211
195	195	186
213	191	196
244	221	216
191	203	205
248	209	210

a. For the data above, compute the sums of squares using the definitional equations.

b. For the data above, compute the sums of squares using the *df* components.

c. Determine the degrees of freedom for each component of the sums of squares.

d. Compute the mean squares and the *F* statistic from the sums of squares.

e. From the calculations above, construct a summary table for this analysis.

f. Calculate the means and standard deviations for each condition.

g. Find the critical *F* value for this test from Appendix Table A-4 for the appropriate degrees of freedom and $\alpha = .05$.

h. What can you conclude given the value of *F* for the established *df* 's?

i. Test the following hypotheses using planned comparisons. What can you conclude as the result of each test?

  i. Participants who have a left hemispheric preference will require a different number of moves to solve the Tower of Hanoi problem than participants who have a right hemispheric preference.

  ii. Participants who have a left hemispheric preference will require a different number of moves to solve the Tower of Hanoi problem than participants who have an integrated hemispheric preference.

  iii. Participants who have an integrated hemispheric preference will require a different number of moves to solve the Tower of Hanoi problem than participants who have a right hemispheric preference.

  iv. Participants who have either an integrated or right hemispheric preference will require a different number of moves to solve the Tower of Hanoi problem than participants who have a left hemispheric preference.

  v. Participants who have an integrated hemispheric preference will not differ on the number of moves to solve the Tower of Hanoi problem from participants who have a right or left hemispheric preference.

j. Use the Scheffé method to test post hoc contrasts for the following hypotheses. What can you conclude as a result of each test?

   i. Participants who have a left hemispheric preference will require a different number of moves than participants who have a right hemispheric preference.

   ii. Participants who have a left hemispheric preference will require a different number of moves than participants who have an integrated hemispheric preference.

   iii. Participants who have an integrated hemispheric preference will require a different number of moves than participants who have a right hemispheric preference.

k. Are the results of 7i(i) and 7j(i), 7i(ii) and 7j(ii), and 7i(iii) and 7j(iii) consistent because we are testing the same hypotheses? If not, explain why not.

l. Calculate η^2, $\hat{\omega}^2$, and $\hat{f}$ for the F-test. What do these values indicate about the size of the treatment effect?

m. Calculate d for each pair of means. What does this value indicate for each pair of means?

8. There is increasing concern about the effect on memory of chemicals in the food and water supply that are introduced to the food and water supply either as additives or pollutants. To test this effect, thirty male albino mice were trained on a maze until they could complete it in criterion time. The mice were then randomly assigned to one of three drinking water conditions: purified drinking water (control), a 4% aluminum sulfate water solution, and an 8% aluminum sulfate water solution. At the end of six weeks, the running times for each mouse were recorded. The data are presented below.

Drinking Water

Purified (Control)	4% Aluminum Sulfate	8% Aluminum Sulfate
215	325	309
206	320	317
186	282	276
175	295	285
200	298	309
205	315	291
198	318	266
182	248	277
216	323	293
212	266	308

a. For the data above, compute the sums of squares using the definitional equations.

b. For the data above, compute the sums of squares using the df components.

c. Determine the degrees of freedom for each component of the sums of squares.

d. Compute the mean squares and the F statistic from the sums of squares.

e. From the calculations above, construct a summary table for this analysis.

f. Calculate the means and standard deviations for each condition.

g. Find the critical F value for this test from Appendix Table A-4 for the appropriate degrees of freedom and $\alpha = .05$.

h. What can you conclude given the value of F for the established df's?

i. Test the following hypotheses using planned comparisons. What can you conclude as the result of each test?

   i. Mice that drank purified water will run the maze faster than mice that drank a 4% aluminum sulfate water solution.

   ii. Mice that drank purified water will run the maze faster than mice that drank an 8% aluminum sulfate water solution.

   iii. Mice that drank a 4% aluminum sulfate water solution will run the maze faster than mice that drank an 8% aluminum sulfate water solution.

   iv. Mice that drank either a 4% aluminum sulfate water solution or 8% aluminum sulfate water solution will run the maze slower than mice that drank purified water.

j. Use the Scheffé method to test post hoc contrasts for the following hypotheses. What can you conclude as a result of each test?

   i. Mice that drank purified water will run the maze faster than mice that drank a 4% aluminum sulfate water solution.

   ii. Mice that drank purified water will run the maze faster than mice that drank an 8% aluminum sulfate water solution.

   iii. Mice that drank a 4% aluminum sulfate water solution will run the maze faster than mice that drank an 8% aluminum sulfate water solution.

k. Are the results of 8i(i) and 8j(i), 8i(ii) and 8j(ii), and 8i(iii) and 8j(iii) consistent because we are testing the same hypotheses? If not, explain why not.

l. Calculate η^2, $\hat{\omega}^2$, and $\hat{f}$ for the F-test. What do these values indicate about the size of the treatment effect?

m. Calculate d for each pair of means. What does this value indicate for each pair of means?

9. Suppose that you conducted a research study with 20 participants in each of four conditions. You decide to conduct post hoc tests on all possible pairwise comparisons. List all of the comparisons that you would make. What is the familywise error rate if you tested each comparison at $\alpha = .05$? Suppose you want to keep the familywise error rate at .05, using the Bonferroni method. What value should you use for α for each comparison?

10. A sports psychologist is interested in the effect of different types of practice conditions on performance of a motor skill. Participants for whom throwing darts was a novel task were randomly assigned to one of four conditions. In the control condition, participants did not engage in any form of practice for throwing darts. In the physical practice condition, participants were asked to practice the proper movements involved in throwing a dart without holding a dart. In the mental practice condition, participants were asked to imagine the movements involved in throwing a dart. In the combined practice condition, participants were asked to first imagine the movements involved in throwing the dart and then practice the proper movements involved in throwing a dart without holding the dart. After a defined amount of time,

participants were asked to play a dart game and their scores were recorded. The data are presented below. Conduct an *F*-test of the data at $\alpha = .05$. If the *F*-test is statistically significant, then use the Scheffé method to test all possible pairwise comparisons. What can you conclude as a result of these tests?

Type of Practice

Control	Physical	Mental	Combined
24	32	16	28
16	15	11	27
13	21	13	21
11	24	17	35
21	15	15	18
13	17	18	31
17	19	21	36
18	21	15	32
14	24	14	24
13	23	16	27
10	27	11	25
6	20	13	29

11. Conduct planned comparisons at $\alpha = .05$ for the data in Exercise 10 that will answer the following questions.
   a. Does practice (physical, mental, combined) lead to better performance on the dart-throwing task than no practice (control)?
   b. Does one form of single practice (physical, mental) lead to a better performance on the dart-throwing task than the other?
   c. Does combined practice lead to a better performance on the dart-throwing task than single practice?

12. Within cognitive psychology, the theory of levels of processing suggests that the more deeply information is processed, the more likely it is to be recalled at some future time. To test this theory, participants were randomly assigned to one of three conditions. In the first condition, participants were given a passage in which letters were randomly deleted. A blank space was provided in lieu of the letter. Participants in this condition were asked to write in the missing letter. This task was defined as a shallow level of processing. In the second condition, participants were given the passage but with each sentence on a separate slip of paper. They were instructed to place the strips of paper in an order that would yield a logically consistent passage. This task was defined as an intermediate level of processing. In the third condition, participants were asked to read the passage and to rewrite the passage in their own words. This task was defined as a deep level of processing. One week later the participants were given a recall test of the information in the passage. The data for this study are listed below. Construct a summary table for this data. What can you conclude if $\alpha = .01$?

Level of Processing

Shallow	Intermediate	Deep
6	11	10
4	12	13
9	14	15
10	11	14
11	10	16
7	9	15
8	14	16
13	15	9

13. Conduct the appropriate Scheffé post hoc comparisons for the data in Exercise 12 at $\alpha = .01$.

14. Cultural psychologists wanted to determine the factors that influence the problem-solving abilities of African-American college students. Participants were randomly assigned to one of the three presentation conditions: traditional analytic, social context, or culturally relevant. After the presentation of relevant material was given, participants in each condition were given a series of problem-solving tasks. The scores on this test are presented below. Test the hypothesis that there would be a difference in problem-solving scores as a function of the type of presentation that was given at $\alpha = .05$.

Type of Presentation

Traditional Analytic	Social Context	Culturally Relevant
18	21	23
21	19	22
24	22	27
16	14	17
13	18	21
13	11	16
13	21	28
19	17	23

15. Conduct the appropriate Scheffé post hoc comparisons for the data in Exercise 14 at $\alpha = .05$. Given the results from the test in Exercise 14 and the results of the Scheffé post hoc comparisons, explain the seemingly contradictory results.

16. According to the distinctiveness of encoding theory, individuals who process information in more distinctive ways should subsequently recall that information better. To test this theory, participants were randomly assigned to one of three reading strategies: read/reread, read/outline, or read/paraphrase. Since the read/reread strategy has only one encoding strategy, it has less distinctiveness than read/outline and read/paraphrase strategies which both have two distinctive encoding strategies. Thus, recall should be higher for read/outline and read/paraphrase strategies than for the read/reread strategy if the distinctiveness of encoding theory is valid. Participants were asked to read a passage the first time and either reread the passage, outline the passage, or paraphrase the passage the second time. Ten days later they were given a recall test on the passage. Scores for each participant are listed below. Test the hypothesis at $\alpha = .05$.

Encoding Strategy

Read/Reread	Read/Outline	Read/Paraphrase
10	17	19
18	11	15
16	22	15
22	16	24
15	19	18
17	20	21

17. Given the results of Exercise 16, conduct the appropriate Scheffé post hoc comparisons at $\alpha = .05$.

18. Intracerebroventricular (ICV) administration of either $La^{(3+)}$, $Gd^{(+3)}$, or saline (control) was conducted to determine the effect on the cocaine-induced activity level of

male Wiser rats. If either $La^{(3+)}$ or $Gd^{(3+)}$ reduces cocaine-induced motor activity then there may be involvement of either $La^{(3+)}$, or $Gd^{(3+)}$-sensitive calcium channels in the loco-motor stimulant effect of cocaine. The data are given below. Conduct a one-way analysis of variance on this data and the appropriate Scheffé post hoc comparisons at $\alpha = .05$.

IVC Infusion

Saline	$La^{(3+)}$	$Gd^{(3+)}$
48	36	36
39	32	54
41	41	39
45	45	47
42	37	41
51	32	42
38	43	46
44	35	49
47	29	37
41	36	41
45	44	40
48	31	42
40	37	43
50	38	47
35	43	39

19. Use the SPSS data bank provided with this book to conduct a one-way analysis of variance on data contained in the data file career.sav. There are developmental changes in adolescents' perceptions of certain academic subjects. For this analysis, determine if there is a grade level (gradelvl: 7th, 8th, 9th, 10th) difference in students' perceptions of mathematics as a fun activity (mathfun). Conduct a one-way analysis of variance on this data and Scheffé post hoc comparisons at $\alpha = .05$. In addition, obtain the descriptive statistics for perceptions of mathematics as a fun activity for each grade level. What can you conclude given this information?

20. Use the same data as in Exercise 19 but conduct Tukey post hoc comparisons instead of Scheffé. Are your conclusions different? If so, why does this difference exist?

21. Suppose that instead of doing the analysis in Exercise 19, we wanted to conduct some planned comparison analyses to test the following predictions: (1) Middle school students (7th and 8th graders) will perceive mathematics as less fun than high school students (9th and 10th graders), (2) Senior high school students (10th graders) will perceive mathematics as more fun than junior high school students (7th, 8th, and 9th graders), (3) 9th grade students will perceive mathematics as more fun than middle school students (7th and 8th graders), and (4) 8th grade students will perceive mathematics as more fun than 7th grade students. Use the SPSS data bank to conduct these planned comparisons. What is your familywise error rate? If you want to keep familywise error at .05, what should your adjusted α be? Given this adjusted alpha, what can you conclude based on the data analysis from SPSS?

Appendices

Appendix A

Table A-1   Areas Under Unit Normal Distribution $Z = \dfrac{X - \overline{X}}{S_x}$

Z	Area Between Z and $\overline{X}$	Area Beyond Z	Z	Area Between Z and $\overline{X}$	Area Beyond Z	Z	Area Between Z and $\overline{X}$	Area Beyond Z
0.00	.0000	.5000	0.59	.2224	.2776	1.18	.3810	.1190
0.01	.0040	.4960	0.60	.2257	.2743	1.19	.3830	.1170
0.02	.0080	.4920	0.61	.2291	.2709	1.20	.3849	.1151
0.03	.0120	.4880	0.62	.2324	.2676	1.21	.3869	.1131
0.04	.0160	.4840	0.63	.2357	.2643	1.22	.3888	.1112
0.05	.0199	.4801	0.64	.2389	.2611	1.23	.3907	.1093
0.06	.0239	.4761	0.65	.2422	.2578	1.24	.3925	.1075
0.07	.0279	.4721	0.66	.2454	.2546	1.25	.3944	.1056
0.08	.0319	.4681	0.67	.2486	.2514	1.26	.3962	.1038
0.09	.0359	.4641	0.68	.2517	.2483	1.27	.3980	.1020
0.10	.0398	.4602	0.69	.2549	.2451	1.28	.3997	.1003
0.11	.0438	.4562	0.70	.2580	.2420	1.29	.4015	.0985
0.12	.0478	.4522	0.71	.2611	.2389	1.30	.4030	.0968
0.13	.0517	.4483	0.72	.2642	.2358	1.31	.4049	.0951
0.14	.0557	.4443	0.73	.2673	.2327	1.32	.4066	.0934
0.15	.0596	.4404	0.74	.2704	.2296	1.33	.4082	.0918
0.16	.0636	.4364	0.75	.2734	.2266	1.34	.4099	.0901
0.17	.0675	.4325	0.76	.2764	.2236	1.35	.4115	.0885
0.18	.0714	.4286	0.77	.2794	.2206	1.36	.4131	.0869
0.19	.0753	.4247	0.78	.2823	.2177	1.37	.4147	.0853
0.20	.0793	.4207	0.79	.2852	.2148	1.38	.4162	.0838
0.21	.0832	.4168	0.80	.2881	.2119	1.39	.4177	.0823
0.22	.0871	.4129	0.81	.2910	.2090	1.40	.4192	.0808
0.23	.0910	.4090	0.82	.2939	.2061	1.41	.4207	.0793
0.24	.0948	.4052	0.83	.2967	.2033	1.42	.4222	.0778
0.25	.0987	.4013	0.84	.2995	.2005	1.43	.4236	.0764
0.26	.1026	.3974	0.85	.3023	.1977	1.44	.4251	.0749
0.27	.1064	.3936	0.86	.3051	.1949	1.45	.4265	.0735
0.28	.1103	.3897	0.87	.3078	.1922	1.46	.4279	.0721
0.29	.1141	.3859	0.88	.3106	.1894	1.47	.4292	.0708
0.30	.1179	.3821	0.89	.3133	.1867	1.48	.4306	.0694
0.31	.1217	.3783	0.90	.3159	.1841	1.49	.4319	.0681
0.32	.1255	.3745	0.91	.3186	.1814	1.50	.4332	.0668
0.33	.1293	.3707	0.92	.3212	.1788	1.51	.4345	.0655
0.34	.1331	.3669	0.93	.3238	.1762	1.52	.4357	.0643
0.35	.1368	.3632	0.94	.3264	.1736	1.53	.4370	.0630
0.36	.1406	.3594	0.95	.3289	.1711	1.54	.4382	.0618
0.37	.1443	.3557	0.96	.3315	.1685	1.55	.4394	.0606
0.38	.1480	.3520	0.97	.3340	.1660	1.56	.4406	.0594
0.39	.1517	.3483	0.98	.3365	.1635	1.57	.4418	.0582
0.40	.1554	.3446	0.99	.3389	.1611	1.58	.4429	.0571
0.41	.1591	.3409	1.00	.3413	.1587	1.59	.4441	.0559
0.42	.1628	.3372	1.01	.3438	.1562	1.60	.4452	.0548
0.43	.1664	.3336	1.02	.3461	.1539	1.61	.4463	.0537
0.44	.1700	.3300	1.03	.3485	.1515	1.62	.4474	.0526
0.45	.1736	.3264	1.04	.3508	.1492	1.63	.4484	.0516
0.46	.1772	.3228	1.05	.3531	.1469	1.64	.4495	.0505
0.47	.1808	.3192	1.06	.3554	.1446			
0.48	.1844	.3156	1.07	.3577	.1423	**1.645**	**.4500**	**.0500**
0.49	.1879	.3121	1.08	.3599	.1401			
0.50	.1915	.3085	1.09	.3621	.1379	1.65	.4505	.0495
0.51	.1950	.3050	1.10	.3643	.1357	1.66	.4515	.0485
0.52	.1985	.3015	1.11	.3665	.1335	1.67	.4525	.0475
0.53	.2019	.2981	1.12	.3686	.1314	1.68	.4535	.0465
0.54	.2054	.2946	1.13	.3708	.1292	1.69	.4545	.0455
0.55	.2088	.2912	1.14	.3729	.1271	1.70	.4554	.0446
0.56	.2123	.2877	1.15	.3749	.1251	1.71	.4564	.0436
0.57	.2157	.2843	1.16	.3770	.1230	1.72	.4573	.0427
0.58	.2190	.2810	1.17	.3790	.1210	1.73	.4582	.0418

Z	Area Between Z and $\overline{X}$	Area Beyond Z	Z	Area Between Z and $\overline{X}$	Area Beyond Z	Z	Area Between Z and $\overline{X}$	Area Beyond Z
1.74	.4591	.0409	2.16	.4846	.0154	2.57	.4949	.0051
1.75	.4599	.0401	2.17	.4850	.0150			
1.76	.4608	.0392	2.18	.4854	.0146	**2.576**	**.4950**	**.0050**
1.77	.4616	.0384	2.19	.4857	.0143			
1.78	.4625	.0375	2.20	.4861	.0139	2.58	.4951	.0049
1.79	.4633	.0367	2.21	.4864	.0136	2.59	.4952	.0048
1.80	.4641	.0359	2.22	.4868	.0132	2.60	.4953	.0047
1.81	.4649	.0351	2.23	.4871	.0129	2.61	.4955	.0045
1.82	.4656	.0344	2.24	.4875	.0125	2.62	.4956	.0044
1.83	.4664	.0336	2.25	.4878	.0122	2.63	.4957	.0043
1.84	.4671	.0329	2.26	.4881	.0119	2.64	.4959	.0041
1.85	.4678	.0322	2.27	.4884	.0116	2.65	.4960	.0040
1.86	.4686	.0314	2.28	.4887	.0113	2.66	.4961	.0039
1.87	.4693	.0307	2.29	.4890	.0110	2.67	.4962	.0038
1.88	.4699	.0301	2.30	.4893	.0107	2.68	.4963	.0037
1.89	.4706	.0294	2.31	.4896	.0104	2.69	.4964	.0036
1.90	.4713	.0287	2.32	.4898	.0102	2.70	.4965	.0035
1.91	.4719	.0281				2.75	.4970	.0030
1.92	.4726	.0274	**2.326**	**.4900**	**.0100**	2.80	.4974	.0026
1.93	.4732	.0268						
1.94	.4738	.0262	2.33	.4901	.0099	**2.81**	**.4975**	**.0025**
1.95	.4744	.0256	2.34	.4904	.0096			
			2.35	.4906	.0094	2.85	.4978	.0022
1.960	**.4750**	**.0250**	2.36	.4909	.0091	2.90	.4981	.0019
			2.37	.4911	.0089	2.95	.4984	.0016
1.97	.4756	.0244	2.38	.4913	.0087	3.00	.4987	.0013
1.98	.4761	.0239	2.39	.4916	.0084	3.05	.4989	.0011
1.99	.4767	.0233	2.40	.4918	.0082	3.10	.4990	.0010
2.00	.4772	.0228	2.41	.4920	.0080	3.15	.4992	.0008
2.01	.4778	.0222	2.42	.4922	.0078	3.20	.4993	.0007
2.02	.4783	.0217	2.43	.4925	.0075	3.25	.4994	.0006
2.03	.4788	.0212	2.44	.4927	.0073	3.30	.4995	.0005
2.04	.4793	.0207	2.45	.4929	.0071	3.35	.4996	.0004
2.05	.4798	.0202	2.46	.4931	.0069	3.40	.4997	.0003
2.06	.4803	.0197	2.47	.4932	.0068	3.45	.4997	.0003
2.07	.4808	.0192	2.48	.4934	.0066	3.50	.4998	.0002
2.08	.4812	.0188	2.49	.4936	.0064	3.60	.4998	.0002
2.09	.4817	.0183	2.50	.4938	.0062	3.70	.4999	.0001
2.10	.4821	.0179	2.51	.4940	.0060	3.80	.4999	.0001
2.11	.4826	.0174	2.52	.4941	.0059	3.90	.49995	.00005
2.12	.4830	.0170	2.53	.4943	.0057	4.00	.49997	.00003
2.13	.4834	.0166	2.54	.4945	.0055			
2.14	.4838	.0162	2.55	.4946	.0054	4.25	.49999	.00001
2.15	.4842	.0158	2.56	.4948	.0052			

Source: Values in table computed by using the normal distribution program in the standard library of the TI-58 calculator.

Degrees of Freedom	One Tail	.10	.05	.025	.01	.005	.0005
	Two Tails	.20	.10	.05	.02	.01	.001
1		3.078	6.314	12.706	31.821	63.657	636.619
2		1.886	2.920	4.303	6.965	9.925	31.598
3		1.638	2.353	3.182	4.541	5.841	12.941
4		1.533	2.132	2.776	3.747	4.604	8.610
5		1.476	2.015	2.571	3.365	4.032	6.859
6		1.440	1.943	2.447	3.143	3.707	5.959
7		1.415	1.895	2.365	2.998	3.499	5.405
8		1.397	1.860	2.306	2.896	3.355	5.041
9		1.383	1.833	2.262	2.821	3.250	4.781
10		1.372	1.812	2.228	2.764	3.169	4.587
11		1.363	1.796	2.201	2.718	3.106	4.437
12		1.356	1.782	2.179	2.681	3.055	4.318
13		1.350	1.771	2.160	2.650	3.012	4.221
14		1.345	1.761	2.145	2.624	2.977	4.140
15		1.341	1.753	2.131	2.602	2.947	4.073
16		1.337	1.746	2.120	2.583	2.921	4.015
17		1.333	1.740	2.110	2.567	2.898	3.965
18		1.330	1.734	2.101	2.552	2.878	3.922
19		1.328	1.729	2.093	2.539	2.861	3.883
20		1.325	1.725	2.086	2.528	2.845	3.850
21		1.323	1.721	2.080	2.518	2.831	3.819
22		1.321	1.717	2.074	2.508	2.819	3.792
23		1.319	1.714	2.069	2.500	2.807	3.767
24		1.318	1.711	2.064	2.492	2.797	3.745
25		1.316	1.708	2.060	2.485	2.787	3.725
26		1.315	1.706	2.056	2.479	2.779	3.707
27		1.314	1.703	2.052	2.473	2.771	3.690
28		1.313	1.701	2.048	2.467	2.763	3.674
29		1.311	1.699	2.045	2.462	2.756	3.659
30		1.310	1.697	2.042	2.457	2.750	3.646
40		1.303	1.684	2.021	2.423	2.704	3.551
60		1.296	1.671	2.000	2.390	2.660	3.460
120		1.289	1.658	1.980	2.358	2.617	3.373
∞		1.282	1.645	1.960	2.326	2.576	3.291

Table A-2 Critical Values of t Distributions

Source: R. A. Fisher and F. Yates, *Statistical Tables for Biological, Agricultural, and Medical Research,* published by Oliver & Boyd Ltd., Edinburgh, 1963, by permission.

Table A-3   Selected Percentile Values of the Chi-Square Distribution

df	$\chi^2_{.999}$	$\chi^2_{.995}$	$\chi^2_{.990}$	$\chi^2_{.980}$	$\chi^2_{.975}$	$\chi^2_{.95}$	$\chi^2_{.90}$	$\chi^2_{.80}$	$\chi^2_{.75}$	$\chi^2_{.70}$	$\chi^2_{.60}$	$\chi^2_{.50}$
1	—	—	—	—	—	—	.02	.06	.10	.15	.28	.46
2	—	.01	.02	.04	.05	.10	.21	.45	.58	.71	1.0	1.4
3	.02	.07	.11	.18	.22	.35	.58	1.0	1.2	1.4	1.9	2.4
4	.09	.21	.30	.43	.48	.71	1.1	1.6	1.9	2.2	2.8	3.4
5	.21	.41	.55	.75	.83	1.1	1.6	2.3	2.7	3.0	3.7	4.4
6	.38	.68	.87	1.1	1.2	1.6	2.2	3.1	3.5	3.8	4.6	5.4
7	.60	.99	1.2	1.6	1.7	2.2	2.8	3.8	4.3	4.7	5.5	6.4
8	.86	1.3	1.6	2.0	2.2	2.7	3.5	4.6	5.1	5.5	6.4	7.3
9	1.2	1.7	2.1	2.5	2.7	3.3	4.2	5.4	5.9	6.4	7.4	8.3
10	1.5	2.2	2.6	3.1	3.2	3.9	4.9	6.2	6.7	7.3	8.3	9.3
11	1.8	2.6	3.0	3.6	3.8	4.6	5.6	7.0	7.6	8.1	9.2	10.3
12	2.2	3.1	3.6	4.2	4.4	5.2	6.3	7.8	8.4	9.0	10.2	11.3
13	2.6	3.6	4.1	4.8	5.0	5.9	7.0	8.6	9.3	9.9	11.1	12.3
14	3.0	4.1	4.7	5.4	5.6	6.6	7.8	9.5	10.2	10.8	12.1	13.3
15	3.5	4.6	5.2	6.0	6.3	7.3	8.5	10.3	11.0	11.7	13.0	14.3
16	3.9	5.1	5.8	6.6	6.9	8.0	9.3	11.2	11.9	12.6	14.0	15.3
17	4.4	5.7	6.4	7.3	7.6	8.7	10.1	12.0	12.8	13.5	14.9	16.3
18	4.9	6.3	7.0	7.9	8.2	9.4	10.9	12.9	13.7	14.4	15.9	17.3
19	5.4	6.9	7.6	8.6	8.9	10.1	11.7	13.7	14.6	15.4	16.8	18.3
20	5.9	7.4	8.3	9.2	9.6	10.9	12.4	14.6	15.5	16.3	17.8	19.3
21	6.4	8.0	8.9	9.9	10.3	11.6	13.2	15.4	16.3	17.2	18.8	20.3
22	7.0	8.6	9.5	10.6	11.0	12.3	14.0	16.3	17.2	18.1	19.7	21.3
23	7.5	9.3	10.2	11.3	11.7	13.1	14.8	17.2	18.1	19.0	20.7	22.3
24	8.0	9.9	10.9	12.0	12.4	13.8	15.7	18.1	19.0	19.9	21.7	23.3
25	8.6	10.5	11.5	12.7	13.1	14.6	16.5	18.9	19.9	20.9	22.6	24.3
26	9.2	11.2	12.2	13.4	13.8	15.4	17.3	19.8	20.8	21.8	23.6	25.3
27	9.8	11.8	12.9	14.1	14.6	16.2	18.1	20.7	21.7	22.7	24.5	26.3
28	10.4	12.5	13.6	14.8	15.3	16.9	18.9	21.6	22.7	23.6	25.5	27.3
29	11.0	13.1	14.3	15.6	16.0	17.7	19.8	22.5	23.6	24.6	26.5	28.3
30	11.6	13.8	15.0	16.3	16.8	18.5	20.6	23.4	24.5	25.5	27.4	29.3

Source: Adapted from R.A. Fisher and F. Yates, *Statistical Tables for Biological, Agricultural, and Medical Research,* published by Oliver & Boyd Ltd., Edinburgh, 1963, by permission. Values less than .01 are left blank.

$\chi^2_{.40}$	$\chi^2_{.30}$	$\chi^2_{.25}$	$\chi^2_{.20}$	$\chi^2_{.10}$	$\chi^2_{.05}$	$\chi^2_{.025}$	$\chi^2_{.02}$	$\chi^2_{.01}$	$\chi^2_{.005}$	$\chi^2_{.001}$	df
.71	1.1	1.3	1.6	2.7	3.8	5.0	5.4	6.6	7.9	10.8	1
1.8	2.4	2.8	3.2	4.6	6.0	7.4	7.8	9.2	10.6	13.8	2
2.9	3.7	4.1	4.6	6.3	7.8	9.4	9.8	11.3	12.8	16.3	3
4.0	4.9	5.4	6.0	7.8	9.5	11.1	11.7	13.3	14.9	18.5	4
5.1	6.1	6.6	7.3	9.2	11.1	12.8	13.4	15.1	16.7	20.5	5
6.2	7.2	7.8	8.6	10.6	12.6	14.4	15.0	16.8	18.5	22.5	6
7.3	8.4	9.0	9.8	12.0	14.1	16.0	16.6	18.5	20.3	24.3	7
8.4	9.5	10.2	11.0	13.4	15.5	17.5	18.2	20.1	22.0	26.1	8
9.4	10.7	11.4	12.2	14.7	16.9	19.0	19.7	21.7	23.6	27.9	9
10.5	11.8	12.5	13.4	16.0	18.3	20.5	21.2	23.2	25.2	29.6	10
11.5	12.9	13.7	14.6	17.3	19.7	21.9	22.6	24.7	26.8	31.3	11
12.6	14.0	14.8	15.8	18.5	21.0	23.3	24.1	26.2	28.3	32.9	12
13.6	15.1	16.0	17.0	19.8	22.4	24.7	25.5	27.7	29.8	34.5	13
14.7	16.2	17.1	18.2	21.1	23.7	26.1	26.9	29.1	31.3	36.1	14
15.7	17.3	18.2	19.3	22.3	25.0	27.5	28.3	30.6	32.8	37.7	15
16.8	18.4	19.4	20.5	23.5	26.3	28.8	29.6	32.0	34.3	39.3	16
17.8	19.5	20.5	21.6	24.8	27.6	30.2	31.0	33.4	35.7	40.8	17
18.9	20.6	21.6	22.8	26.0	28.9	31.5	32.3	34.8	37.2	42.3	18
19.9	21.7	22.7	23.9	27.2	30.1	32.9	33.7	36.2	38.6	43.8	19
21.0	22.8	23.8	25.0	28.4	31.4	34.2	35.0	37.6	40.0	45.3	20
22.0	23.9	24.9	26.2	29.6	32.7	35.5	36.3	38.9	41.4	46.8	21
23.0	24.9	26.0	27.3	30.8	33.9	36.8	37.7	40.3	42.8	48.3	22
24.1	26.0	27.1	28.4	32.0	35.2	38.1	39.0	41.6	44.2	49.7	23
25.1	27.1	28.2	29.6	33.2	36.4	39.4	40.3	43.0	45.6	51.2	24
26.1	28.2	29.3	30.7	34.4	37.7	40.6	41.6	44.3	46.9	52.6	25
27.2	29.2	30.4	31.8	35.6	38.9	41.9	42.9	45.6	48.3	54.0	26
28.2	30.3	31.5	32.9	36.7	40.1	43.2	44.1	47.0	49.6	55.5	27
29.2	31.4	32.6	34.0	37.9	41.3	44.5	45.4	48.3	51.0	56.9	28
30.3	32.5	33.7	35.3	39.1	42.6	45.7	46.7	49.6	52.3	58.3	29
31.3	33.5	34.8	36.2	40.3	43.8	47.0	48.0	50.9	53.7	59.7	30

Table A-4 Critical Values of *F* Distribution

df_2	df_1	1	2	3	4	5	6	7	8	9	10	12	15	20	30	∞
1	.05	161.4	199.5	215.7	224.6	230.2	234.0	236.8	238.9	240.5	241.9	243.9	245.9	248.0	250.1	254.3
	.01[1]	40.52	50.00	54.03	56.25	57.64	58.59	59.28	59.81	60.22	60.56	61.06	61.57	62.09	62.61	63.66
	.001[2]	40.53	50.00	54.04	56.25	57.64	58.59	59.29	59.81	60.23	60.56	61.07	61.58	62.09	62.61	63.66
2	.05	18.51	19.00	19.16	19.25	19.30	19.33	19.35	19.37	19.38	19.40	19.41	19.43	19.45	19.46	19.50
	.01	98.50	99.00	99.17	99.25	99.30	99.33	99.36	99.37	99.39	99.40	99.42	99.43	99.45	99.47	99.50
	.001	998.50	999.00	999.17	999.25	999.30	999.33	999.36	999.37	999.39	999.40	999.42	999.43	999.45	999.47	999.50
3	.05	10.13	9.55	9.28	9.12	9.01	8.94	8.89	8.85	8.81	8.79	8.74	8.70	8.66	8.62	8.53
	.01	34.12	30.82	29.46	28.71	28.24	27.91	27.67	27.49	27.34	27.23	27.05	26.87	26.69	26.50	26.12
	.001	167.03	148.50	141.11	137.10	134.58	132.85	131.58	130.62	129.86	129.25	128.32	127.37	126.42	125.45	123.47
4	.05	7.71	6.94	6.59	6.39	6.26	6.16	6.09	6.04	6.00	5.96	5.91	5.86	5.80	5.75	5.63
	.01	21.20	18.00	16.69	15.98	15.52	15.21	14.98	14.80	14.66	14.55	14.37	14.20	14.02	13.84	13.46
	.001	74.14	61.25	56.18	53.44	51.71	50.52	49.66	49.00	48.48	48.05	47.41	46.76	46.10	45.43	44.05
5	.05	6.61	5.79	5.41	5.19	5.05	4.95	4.88	4.82	4.77	4.74	4.68	4.62	4.56	4.50	4.36
	.01	16.26	13.27	12.06	11.39	10.97	10.67	10.46	10.29	10.16	10.05	9.89	9.72	9.55	9.38	9.02
	.001	47.18	37.12	33.20	31.08	29.75	28.83	28.16	27.65	27.24	26.92	26.42	25.91	25.40	24.87	23.78
6	.05	5.99	5.14	4.76	4.53	4.39	4.28	4.21	4.15	4.10	4.06	4.00	3.94	3.87	3.81	3.67
	.01	13.74	10.92	9.78	9.15	8.75	8.47	8.26	8.10	7.98	7.87	7.72	7.56	7.40	7.23	6.88
	.001	35.51	27.00	23.70	21.92	20.80	20.03	19.46	19.03	18.69	18.41	17.99	17.56	17.12	16.67	15.74
7	.05	5.59	4.74	4.35	4.12	3.97	3.87	3.79	3.73	3.68	3.64	3.57	3.51	3.44	3.38	3.23
	.01	12.25	9.55	8.45	7.85	7.46	7.19	6.99	6.84	6.72	6.62	6.47	6.31	6.16	5.99	5.65
	.001	29.24	21.69	18.77	17.20	16.21	15.52	15.02	14.63	14.33	14.08	13.71	13.32	12.93	12.53	11.70
8	.05	5.32	4.46	4.07	3.84	3.69	3.58	3.50	3.44	3.39	3.35	3.28	3.22	3.15	3.08	2.93
	.01	11.26	8.65	7.59	7.01	6.63	6.37	6.18	6.03	5.91	5.81	5.67	5.52	5.36	5.20	4.86
	.001	25.42	18.49	15.83	14.39	13.48	12.86	12.40	12.05	11.77	11.54	11.19	10.84	10.48	10.11	9.33
9	.05	5.12	4.26	3.86	3.63	3.48	3.37	3.29	3.23	3.18	3.14	3.07	3.01	2.94	2.86	2.71
	.01	10.56	8.02	6.99	6.42	6.06	5.80	5.61	5.47	5.35	5.26	5.11	4.96	4.81	4.65	4.31
	.001	22.86	16.39	13.90	12.56	11.71	11.13	10.70	10.37	10.11	9.89	9.57	9.24	8.90	8.55	7.81
10	.05	4.96	4.10	3.71	3.48	3.33	3.22	3.14	3.07	3.02	2.98	2.91	2.84	2.77	2.70	2.54
	.01	10.04	7.56	6.55	5.99	5.64	5.39	5.20	5.06	4.94	4.85	4.71	4.56	4.41	4.25	3.91
	.001	21.04	14.90	12.55	11.28	10.48	9.93	9.52	9.20	8.96	8.75	8.44	8.13	7.80	7.47	6.76
11	.05	4.84	3.98	3.59	3.36	3.20	3.09	3.01	2.95	2.90	2.85	2.79	2.72	2.65	2.57	2.40
	.01	9.65	7.21	6.22	5.67	5.32	5.07	4.89	4.74	4.63	4.54	4.40	4.25	4.10	3.94	3.60
	.001	19.69	13.81	11.56	10.35	9.58	9.05	8.66	8.35	8.12	7.92	7.63	7.32	7.01	6.68	6.00
12	.05	4.75	3.89	3.49	3.26	3.11	3.00	2.91	2.85	2.80	2.75	2.69	2.62	2.54	2.47	2.30
	.01	9.33	6.93	5.95	5.41	5.06	4.82	4.64	4.50	4.39	4.30	4.16	4.01	3.86	3.70	3.36
	.001	18.64	12.97	10.80	9.63	8.89	8.38	8.00	7.71	7.48	7.29	7.00	6.71	6.40	6.09	5.42

df	α															
13	.05	4.67	3.81	3.41	3.18	3.03	2.92	2.83	2.77	2.71	2.67	2.60	2.53	2.46	2.38	2.21
	.01	9.07	6.70	5.74	5.21	4.86	4.62	4.44	4.30	4.19	4.10	3.96	3.82	3.66	3.51	3.17
	.001	17.82	12.31	10.21	9.07	8.35	7.86	7.49	7.21	6.98	6.80	6.52	6.23	5.93	5.63	4.97
14	.05	4.60	3.74	3.34	3.11	2.96	2.85	2.76	2.70	2.65	2.60	2.53	2.46	2.39	2.31	2.13
	.01	8.86	6.51	5.56	5.04	4.70	4.46	4.28	4.14	4.03	3.94	3.80	3.66	3.51	3.35	3.00
	.001	17.14	11.78	9.73	8.62	7.92	7.44	7.08	6.80	6.58	6.40	6.13	5.85	5.56	5.25	4.60
15	.05	4.54	3.68	3.29	3.06	2.90	2.79	2.71	2.64	2.59	2.54	2.48	2.40	2.33	2.25	2.07
	.01	8.68	6.36	5.42	4.89	4.56	4.32	4.14	4.00	3.89	3.80	3.67	3.52	3.37	3.21	2.87
	.001	16.59	11.34	9.34	8.25	7.57	7.09	6.74	6.47	6.26	6.08	5.81	5.54	5.25	4.95	4.31
16	.05	4.49	3.63	3.24	3.01	2.85	2.74	2.66	2.59	2.54	2.49	2.42	2.35	2.28	2.19	2.01
	.01	8.53	6.23	5.29	4.77	4.44	4.20	4.03	3.89	3.78	3.69	3.55	3.41	3.26	3.10	2.75
	.001	16.12	10.97	9.01	7.94	7.27	6.80	6.46	6.20	5.98	5.81	5.55	5.27	4.99	4.70	4.06
17	.05	4.45	3.59	3.20	2.96	2.81	2.70	2.61	2.55	2.49	2.45	2.38	2.31	2.23	2.15	1.96
	.01	8.40	6.11	5.18	4.67	4.34	4.10	3.93	3.79	3.68	3.59	3.46	3.31	3.16	3.00	2.65
	.001	15.72	10.66	8.73	7.68	7.02	6.56	6.22	5.96	5.75	5.58	5.32	5.05	4.78	4.48	3.85
18	.05	4.41	3.55	3.19	2.93	2.77	2.66	2.58	2.51	2.46	2.41	2.34	2.27	2.19	2.11	1.92
	.01	8.29	6.01	5.09	4.58	4.25	4.01	3.84	3.71	3.60	3.51	3.37	3.23	3.08	2.92	2.57
	.001	15.38	10.39	8.49	7.46	6.81	6.36	6.02	5.76	5.56	5.39	5.13	4.87	4.59	4.30	3.67
19	.05	4.38	3.52	3.13	2.90	2.74	2.63	2.54	2.48	2.42	2.38	2.31	2.23	2.16	2.07	1.88
	.01	8.18	5.93	5.01	4.50	4.17	3.94	3.77	3.63	3.52	3.43	3.30	3.15	3.00	2.84	2.49
	.001	15.08	10.16	8.28	7.27	6.62	6.18	5.85	5.59	5.39	5.22	4.97	4.70	4.43	4.14	3.51
20	.05	4.35	3.49	3.10	2.87	2.71	2.60	2.51	2.45	2.39	2.35	2.28	2.20	2.12	2.04	1.84
	.01	8.10	5.85	4.94	4.43	4.10	3.87	3.70	3.56	3.46	3.37	3.23	3.09	2.94	2.78	2.42
	.001	14.82	9.95	8.10	7.10	6.46	6.02	5.69	5.44	5.24	5.08	4.82	4.56	4.29	4.00	3.38
21	.05	4.32	3.47	3.07	2.84	2.68	2.57	2.49	2.42	2.37	2.32	2.25	2.18	2.10	2.01	1.81
	.01	8.02	5.78	4.87	4.37	4.04	3.81	3.64	3.51	3.40	3.31	3.17	3.03	2.88	2.72	2.36
	.001	14.59	9.77	7.94	6.95	6.32	5.88	5.56	5.31	5.11	4.95	4.70	4.44	4.17	3.88	3.26
22	.05	4.30	3.44	3.05	2.82	2.66	2.55	2.46	2.40	2.34	2.30	2.23	2.15	2.07	1.98	1.78
	.01	7.95	5.72	4.82	4.31	3.99	3.76	3.59	3.45	3.35	3.26	3.12	2.98	2.83	2.67	2.31
	.001	14.38	9.61	7.80	6.81	6.19	5.76	5.44	5.19	4.99	4.83	4.58	4.33	4.06	3.78	3.15
23	.05	4.28	3.42	3.03	2.80	2.64	2.53	2.44	2.37	2.32	2.27	2.20	2.13	2.05	1.96	1.76
	.01	7.88	5.66	4.76	4.26	3.94	3.71	3.54	3.41	3.30	3.21	3.07	2.93	2.78	2.62	2.26
	.001	14.20	9.47	7.67	6.70	6.08	5.65	5.33	5.09	4.89	4.73	4.48	4.23	3.96	3.68	3.05
24	.05	4.26	3.40	3.01	2.78	2.62	2.51	2.42	2.36	2.30	2.25	2.18	2.11	2.03	1.94	1.73
	.01	7.82	5.61	4.72	4.22	3.90	3.67	3.50	3.36	3.26	3.17	3.03	2.89	2.74	2.58	2.21
	.001	14.03	9.34	7.55	6.59	5.98	5.55	5.23	4.99	4.80	4.64	4.39	4.14	3.87	3.59	2.97
25	.05	4.24	3.39	2.99	2.76	2.60	2.49	2.40	2.34	2.28	2.24	2.16	2.09	2.01	1.92	1.71
	.01	7.77	5.57	4.68	4.18	3.86	3.63	3.46	3.32	3.22	3.13	2.99	2.85	2.70	2.54	2.17
	.001	13.88	9.22	7.45	6.49	5.89	5.46	5.15	4.91	4.71	4.56	4.31	4.06	3.79	3.52	2.89
26	.05	4.23	3.37	2.98	2.74	2.59	2.47	2.39	2.32	2.27	2.22	2.15	2.07	1.99	1.90	1.69
	.01	7.72	5.53	4.64	4.14	3.82	3.59	3.42	3.29	3.18	3.09	2.96	2.82	2.66	2.50	2.13
	.001	13.74	9.12	7.36	6.41	5.80	5.38	5.07	4.83	4.64	4.48	4.24	3.99	3.72	3.44	2.82

Table A-4 Continued

df_2	df_1	1	2	3	4	5	6	7	8	9	10	12	15	20	30	∞
27	.05	4.21	3.35	2.96	2.73	2.57	2.46	2.38	2.31	2.25	2.20	2.13	2.06	1.97	1.88	1.67
	.01	7.68	5.49	4.60	4.11	3.78	3.56	3.39	3.26	3.15	3.06	2.93	2.78	2.63	2.47	2.10
	.001	13.61	9.02	7.27	6.33	5.73	5.31	5.00	4.76	4.57	4.41	4.17	3.92	3.66	3.38	2.75
28	.05	4.20	3.34	2.95	2.71	2.56	2.45	2.36	2.29	2.24	2.19	2.12	2.04	1.96	1.87	1.65
	.01	7.64	5.45	4.57	4.07	3.75	3.53	3.36	3.23	3.12	3.03	2.90	2.75	2.60	2.44	2.06
	.001	13.50	8.93	7.19	6.25	5.66	5.24	4.93	4.69	4.50	4.35	4.11	3.86	3.60	3.32	2.69
29	.05	4.18	3.33	2.93	2.70	2.55	2.43	2.35	2.28	2.22	2.18	2.10	2.03	1.94	1.85	1.64
	.01	7.60	5.42	4.54	4.04	3.73	3.50	3.33	3.20	3.09	3.00	2.87	2.73	2.57	2.41	2.03
	.001	13.39	8.85	7.12	6.19	5.59	5.18	4.87	4.64	4.45	4.29	4.05	3.80	3.54	3.27	2.64
30	.05	4.13	3.32	2.92	2.69	2.54	2.42	2.33	2.27	2.21	2.16	2.09	2.01	1.93	1.84	1.62
	.01	7.56	5.39	4.51	4.02	3.70	3.47	3.30	3.17	3.07	2.98	2.84	2.70	2.55	2.39	2.01
	.001	13.29	8.77	7.05	6.12	5.53	5.12	4.82	4.58	4.39	4.24	4.00	3.75	3.49	3.22	2.59
40	.05	4.08	3.23	2.84	2.61	2.45	2.34	2.25	2.18	2.12	2.08	2.00	1.92	1.84	1.74	1.51
	.01	7.31	5.18	4.31	3.83	3.51	3.29	3.12	2.99	2.89	2.80	2.66	2.52	2.37	2.20	1.80
	.001	12.61	8.25	6.59	5.70	5.13	4.73	4.44	4.21	4.02	3.87	3.64	3.40	3.14	2.87	2.23
60	.05	4.00	3.15	2.76	2.53	2.37	2.25	2.17	2.10	2.04	1.99	1.92	1.84	1.75	1.65	1.39
	.01	7.08	4.98	4.13	3.65	3.34	3.12	2.95	2.82	2.72	2.63	2.50	2.35	2.20	2.03	1.60
	.001	11.97	7.77	6.17	5.31	4.76	4.37	4.09	3.86	3.69	3.54	3.32	3.08	2.83	2.55	1.89
120	.05	3.92	3.07	2.68	2.45	2.29	2.18	2.09	2.02	1.96	1.91	1.83	1.75	1.66	1.55	1.25
	.01	6.85	4.79	3.95	3.48	3.17	2.96	2.79	2.66	2.56	2.47	2.34	2.19	2.03	1.86	1.38
	.001	11.38	7.32	5.78	4.95	4.42	4.04	3.77	3.55	3.38	3.24	3.02	2.78	2.53	2.26	1.54
∞	.05	3.84	3.00	2.60	2.37	2.21	2.10	2.01	1.94	1.88	1.83	1.75	1.67	1.57	1.46	1.00
	.01	6.63	4.61	3.78	3.32	3.02	2.80	2.64	2.51	2.41	2.32	2.18	2.04	1.88	1.70	1.00
	.001	10.83	6.91	5.42	4.62	4.10	3.74	3.47	3.27	3.10	2.96	2.74	2.51	2.27	1.99	1.00

Source: R. A. Fisher and F. Yates, *Statistical Tables for Biological, Agricultural, and Medical Research*, published by Oliver & Boyd Ltd., Edinburgh, 1963, by permission.
[1]Entries in this row must be multiplied by 100.
[2]Entries in this row must be multiplied by 10,000.

Table A-5 Transformation of r to Z_F (Fisher's Z)

r	.00	.01	.02	.03	.04	.05	.06	.07	.08	.09
.0	0.000	0.010	0.020	0.030	0.040	0.050	0.060	0.070	0.080	0.090
.1	0.100	0.110	0.121	0.131	0.141	0.151	0.161	0.172	0.182	0.192
.2	0.203	0.213	0.224	0.234	0.245	0.255	0.266	0.277	0.288	0.299
.3	0.310	0.321	0.332	0.343	0.354	0.366	0.377	0.388	0.400	0.412
.4	0.424	0.436	0.448	0.460	0.472	0.485	0.497	0.510	0.523	0.536
.5	0.549	0.563	0.577	0.590	0.604	0.618	0.633	0.648	0.663	0.678
.6	0.693	0.709	0.725	0.741	0.758	0.775	0.793	0.811	0.829	0.848
.7	0.867	0.887	0.908	0.929	0.950	0.973	0.996	1.020	1.045	1.071
.8	1.099	1.127	1.157	1.188	1.221	1.256	1.293	1.333	1.376	1.422
.9	1.472	1.528	1.589	1.658	1.738	1.832	1.946	2.092	2.298	2.647
.99	2.647	2.700	2.759	2.826	2.903	2.995	3.106	3.250	3.453	3.800

Source: Values in table computed by using the equation

$$Z_F = 1.1513 \log_{10} \frac{1+r}{1-r}$$

Table A-6 Transformation of Z_F (Fisher's Z) to r

Z_F	.00	.01	.02	.03	.04	.05	.06	.07	.08	.09
0.0	.00	.01	.02	.03	.04	.05	.06	.07	.08	.09
0.1	.10	.11	.12	.13	.14	.15	.16	.17	.18	.19
0.2	.20	.21	.22	.23	.24	.24	.25	.26	.27	.28
0.3	.29	.30	.31	.32	.33	.34	.35	.35	.36	.37
0.4	.38	.39	.40	.41	.41	.42	.43	.44	.45	.45
0.5	.46	.47	.48	.49	.49	.50	.51	.52	.52	.53
0.6	.54	.54	.55	.56	.56	.57	.58	.58	.59	.60
0.7	.60	.61	.62	.62	.63	.64	.64	.65	.65	.66
0.8	.66	.67	.68	.68	.69	.69	.70	.70	.71	.71
0.9	.72	.72	.73	.73	.74	.74	.74	.75	.75	.76
1.0	.76	.77	.77	.77	.78	.78	.79	.79	.79	.80
1.1	.80	.80	.81	.81	.81	.82	.82	.82	.83	.83
1.2	.83	.84	.84	.84	.85	.85	.85	.85	.86	.86
1.3	.86	.86	.87	.87	.87	.87	.88	.88	.88	.88
1.4	.89	.89	.89	.89	.89	.90	.90	.90	.90	.90
1.5	.91	.91	.91	.91	.91	.91	.92	.92	.92	.92
1.6	.92	.92	.92	.93	.93	.93	.93	.93	.93	.93
1.7	.94	.94	.94	.94	.94	.94	.94	.94	.94	.95
1.8	.95	.95	.95	.95	.95	.95	.95	.95	.95	.96
1.9	.96	.96	.96	.96	.96	.96	.96	.96	.96	.96
2.0	.96	.96	.97	.97	.97	.97	.97	.97	.97	.97
2.1	.97	.97	.97	.97	.97	.97	.97	.97	.97	.98
2.2	.98	.98	.98	.98	.98	.98	.98	.98	.98	.98
2.3	.98	.98	.98	.98	.98	.98	.98	.98	.98	.98
2.4	.98	.98	.98	.98	.98	.99	.99	.99	.99	.99

Source: Values obtained by interpolation from Table A-5.

Table A-7 Lower- and Upper-Tail Probabilities for W_s, the Wilcoxon-Mann-Whitney Rank-Sum Statistic
(Entries are the probability that $W_s \leq S_{Low}$ and $W_s \geq S_{Up}$. W_s is the sum of ranks for the smaller group.)

$N_s = 3$

S_{Low}	$N_L = 3$	S_{Up}	$N_L = 4$	S_{Up}	$N_L = 5$	S_{Up}	$N_L = 6$	S_{Up}	$N_L = 7$	S_{Up}	$N_L = 8$	S_{Up}	$N_L = 9$	S_{Up}	$N_L = 10$	S_{Up}	$N_L = 11$	S_{Up}	$N_L = 12$	S_{Up}
6	.0500	15	.0286	18	.0179	21	.0119	24	.0083	27	.0061	30	.0045	33	.0035	36	.0027	39	.0022	42
7	.1000	14	.0571	17	.0357	20	.0238	23	.0167	26	.0121	29	.0091	32	.0070	35	.0055	38	.0044	41
8	.2000	13	.1143	16	.0714	19	.0476	22	.0333	25	.0242	28	.0182	31	.0140	34	.0110	37	.0088	40
9	.3500	12	.2000	15	.1250	18	.0833	21	.0583	24	.0424	27	.0318	30	.0245	33	.0192	36	.0154	39
10	.5000	11	.3143	14	.1964	17	.1310	20	.0917	23	.0667	26	.0500	29	.0385	32	.0302	35	.0242	38
11	.6500	10	.4286	13	.2857	16	.1905	19	.1333	22	.0970	25	.0727	28	.0559	31	.0440	34	.0352	37
12	.8000	9	.5714	12	.3929	15	.2738	18	.1917	21	.1394	24	.1045	27	.0804	30	.0632	33	.0505	36
13	.9000	8	.6857	11	.5000	14	.3571	17	.2583	20	.1879	23	.1409	26	.1084	29	.0852	32	.0681	35
14	.9500	7	.8000	10	.6071	13	.4524	16	.3333	19	.2485	22	.1864	25	.1434	28	.1126	31	.0901	34
15	1.0000	6	.8857	9	.7143	12	.5476	15	.4167	18	.3152	21	.2409	24	.1853	27	.1456	30	.1165	33
16			.9429	8	.8036	11	.6429	14	.5000	17	.3879	20	.3000	23	.2343	26	.1841	29	.1473	32
17			.9714	7	.8750	10	.7262	13	.5833	16	.4606	19	.3636	22	.2867	25	.2280	28	.1824	31
18			1.0000	6	.9286	9	.8095	12	.6667	15	.5394	18	.4318	21	.3462	24	.2775	27	.2242	30
19					.9643	8	.8690	11	.7417	14	.6121	17	.5000	20	.4056	23	.3297	26	.2681	29
20					.9821	7	.9167	10	.8083	13	.6848	16	.5682	19	.4685	22	.3846	25	.3165	28
21					1.0000	6	.9524	9	.8667	12	.7515	15	.6364	18	.5315	21	.4423	24	.3670	27
22							.9762	8	.9083	11	.8121	14	.7000	17	.5944	20	.5000	23	.4198	26
23							.9881	7	.9417	10	.8606	13	.7591	16	.6538	19	.5577	22	.4725	25
24							1.0000	6	.9667	9	.9030	12	.8136	15	.7133	18	.6154	21	.5275	24

$$N_s = 4$$

S_{Low}	$N_L=4$	S_{Up}	$N_L=5$	S_{Up}	$N_L=6$	S_{Up}	$N_L=7$	S_{Up}	$N_L=8$	S_{Up}	$N_L=9$	S_{Up}	$N_L=10$	S_{Up}	$N_L=11$	S_{Up}	$N_L=12$	S_{Up}
10	.0143	26	.0079	30	.0048	34	.0030	38	.0020	42	.0014	46	.0010	50	.0007	54	.0005	58
11	.0286	25	.0159	29	.0095	33	.0061	37	.0040	41	.0028	45	.0020	49	.0015	53	.0011	57
12	.0571	24	.0317	28	.0190	32	.0121	36	.0081	40	.0056	44	.0040	48	.0029	52	.0022	56
13	.1000	23	.0556	27	.0333	31	.0212	35	.0141	39	.0098	43	.0070	47	.0051	51	.0038	55
14	.1714	22	.0952	26	.0571	30	.0364	34	.0242	38	.0168	42	.0120	46	.0088	50	.0066	54
15	.2429	21	.1429	25	.0857	29	.0545	33	.0364	37	.0252	41	.0180	45	.0132	49	.0099	53
16	.3429	20	.2063	24	.1286	28	.0818	32	.0545	36	.0378	40	.0270	44	.0198	48	.0148	52
17	.4429	19	.2778	23	.1762	27	.1152	31	.0768	35	.0531	39	.0380	43	.0278	47	.0209	51
18	.5571	18	.3651	22	.2381	26	.1576	30	.1071	34	.0741	38	.0529	42	.0388	46	.0291	50
19	.6571	17	.4524	21	.3048	25	.2061	29	.1414	33	.0993	37	.0709	41	.0520	45	.0390	49
20	.7571	16	.5476	20	.3810	24	.2636	28	.1838	32	.1301	36	.0939	40	.0689	44	.0516	48
21	.8286	15	.6349	19	.4571	23	.3242	27	.2303	31	.1650	35	.1199	39	.0886	43	.0665	47
22	.9000	14	.7222	18	.5429	22	.3939	26	.2848	30	.2070	34	.1518	38	.1128	42	.0852	46
23	.9429	13	.7937	17	.6190	21	.4636	25	.3414	29	.2517	33	.1868	37	.1399	41	.1060	45
24	.9714	12	.8571	16	.6952	20	.5364	24	.4040	28	.3021	32	.2268	36	.1714	40	.1308	44
25	.9857	11	.9048	15	.7619	19	.6061	23	.4667	27	.3552	31	.2697	35	.2059	39	.1582	43
26	1.0000	10	.9444	14	.8238	18	.6758	22	.5333	26	.4126	30	.3177	34	.2447	38	.1896	42
27			.9683	13	.8714	17	.7364	21	.5960	25	.4699	29	.3666	33	.2857	37	.2231	41
28			.9841	12	.9143	16	.7939	20	.6586	24	.5301	28	.4196	32	.3304	36	.2604	40
29			.9921	11	.9429	15	.8424	19	.7152	23	.5874	27	.4725	31	.3766	35	.2995	39
30			1.0000	10	.9667	14	.8848	18	.7697	22	.6448	26	.5275	30	.4256	34	.3418	38
31					.9810	13	.9182	17	.8162	21	.6979	25	.5804	29	.4747	33	.3852	37
32					.9905	12	.9455	16	.8586	20	.7483	24	.6334	28	.5253	32	.4308	36
33					.9952	11	.9636	15	.8929	19	.7930	23	.6823	27	.5744	31	.4764	35
34					1.0000	10	.9788	14	.9232	18	.8340	22	.7303	26	.6234	30	.5236	34

$N_S = 5$

S_{Low}	$N_L = 5$	S_{Up}	$N_L = 6$	S_{Up}	$N_L = 7$	S_{Up}	$N_L = 8$	S_{Up}	$N_L = 9$	S_{Up}	$N_L = 10$	S_{Up}
15	.0040	40	.0022	45	.0013	50	.0008	55	.0005	60	.0003	65
16	.0079	39	.0043	44	.0025	49	.0016	54	.0010	59	.0007	64
17	.0159	38	.0087	43	.0051	48	.0031	53	.0020	58	.0013	63
18	.0278	37	.0152	42	.0088	47	.0054	52	.0035	57	.0023	62
19	.0476	36	.0260	41	.0152	46	.0093	51	.0060	56	.0040	61
20	.0754	35	.0411	40	.0240	45	.0148	50	.0095	55	.0063	60
21	.1111	34	.0628	39	.0366	44	.0225	49	.0145	54	.0097	59
22	.1548	33	.0887	38	.0530	43	.0326	48	.0210	53	.0140	58
23	.2103	32	.1234	37	.0745	42	.0466	47	.0300	52	.0200	57
24	.2738	31	.1645	36	.1010	41	.0637	46	.0415	51	.0276	56
25	.3452	30	.2143	35	.1338	40	.0855	45	.0559	50	.0376	55
26	.4206	29	.2684	34	.1717	39	.1111	44	.0734	49	.0496	54
27	.5000	28	.3312	33	.2159	38	.1422	43	.0949	48	.0646	53
28	.5794	27	.3961	32	.2652	37	.1772	42	.1199	47	.0823	52
29	.6548	26	.4654	31	.3194	36	.2176	41	.1489	46	.1032	51
30	.7262	25	.5346	30	.3775	35	.2618	40	.1818	45	.1272	50
31	.7897	24	.6039	29	.4381	34	.3108	39	.2188	44	.1548	49
32	.8452	23	.6688	28	.5000	33	.3621	38	.2592	43	.1855	48
33	.8889	22	.7316	27	.5619	32	.4165	37	.3032	42	.2198	47
34	.9246	21	.7857	26	.6225	31	.4716	36	.3497	41	.2567	46
35	.9524	20	.8355	25	.6806	30	.5284	35	.3986	40	.2970	45
36	.9722	19	.8766	24	.7348	29	.5835	34	.4491	39	.3393	44
37	.9841	18	.9113	23	.7841	28	.6379	33	.5000	38	.3839	43
38	.9921	17	.9372	22	.8283	27	.6892	32	.5509	37	.4296	42
39	.9960	16	.9589	21	.8662	26	.7382	31	.6014	36	.4765	41
40	1.0000	15	.9740	20	.8990	25	.7824	30	.6503	35	.5235	40

Table A-7 Continued

$N_S = 6$

S_{Low}	$N_L = 6$	S_{Up}	$N_L = 7$	S_{Up}	$N_L = 8$	S_{Up}	$N_L = 9$	S_{Up}	$N_L = 10$	S_{Up}
21	.0011	57	.0006	63	.0003	69	.0002	75	.0001	81
22	.0022	56	.0012	62	.0007	68	.0004	74	.0002	80
23	.0043	55	.0023	61	.0013	67	.0008	73	.0005	79
24	.0076	54	.0041	60	.0023	66	.0014	72	.0009	78
25	.0130	53	.0070	59	.0040	65	.0024	71	.0015	77
26	.0206	52	.0111	58	.0063	64	.0038	70	.0024	76
27	.0325	51	.0175	57	.0100	63	.0060	69	.0037	75
28	.0465	50	.0256	56	.0147	62	.0088	68	.0055	74
29	.0660	49	.0367	55	.0213	61	.0128	67	.0080	73
30	.0898	48	.0507	54	.0296	60	.0180	66	.0112	72
31	.1201	47	.0688	53	.0406	59	.0248	65	.0156	71
32	.1548	46	.0903	52	.0539	58	.0332	64	.0210	70
33	.1970	45	.1171	51	.0709	57	.0440	63	.0280	69
34	.2424	44	.1474	50	.0906	56	.0567	62	.0363	68
35	.2944	43	.1830	49	.1142	55	.0723	61	.0467	67
36	.3496	42	.2226	48	.1412	54	.0905	60	.0589	66
37	.4091	41	.2669	47	.1725	53	.1119	59	.0736	65
38	.4686	40	.3141	46	.2068	52	.1361	58	.0903	64
39	.5314	39	.3654	45	.2454	51	.1638	57	.1099	63
40	.5909	38	.4178	44	.2864	50	.1942	56	.1317	62
41	.6504	37	.4726	43	.3310	49	.2280	55	.1566	61
42	.7056	36	.5274	42	.3773	48	.2643	54	.1838	60
43	.7576	35	.5822	41	.4259	47	.3035	53	.2139	59
44	.8030	34	.6346	40	.4749	46	.3445	52	.2461	58
45	.8452	33	.6859	39	.6261	45	.3878	51	.2811	57
46	.8799	32	.7331	38	.5741	44	.4320	50	.3177	56
47	.9102	31	.7774	37	.6227	43	.4773	49	.3564	55
48	.9340	30	.8170	36	.6690	42	.5227	48	.3962	54
49	.9535	29	.8526	35	.7136	41	.5680	47	.4374	53
50	.9675	28	.8829	34	.7546	40	.6122	46	.4789	52
51	.9794	27	.9097	33	.7932	39	.6555	45	.5211	51

S_{Low}	$N_L = 7$	S_{Up}	$N_L = 8$	S_{Up}	$N_L = 9$	S_{Up}	$N_L = 10$	S_{Up}
28	.0003	77	.0002	84	.0001	91	.0001	98
29	.0006	76	.0003	83	.0002	90	.0001	97
30	.0012	75	.0006	82	.0003	89	.0002	96
31	.0020	74	.0011	81	.0006	88	.0004	95
32	.0035	73	.0019	80	.0010	87	.0006	94
33	.0055	72	.0030	79	.0017	86	.0010	93
34	.0087	71	.0047	78	.0026	85	.0015	92
35	.0131	70	.0070	77	.0039	84	.0023	91
36	.0189	69	.0103	76	.0058	83	.0034	90
37	.0265	68	.0145	75	.0082	82	.0048	89
38	.0364	67	.0200	74	.0115	81	.0068	88
39	.0487	66	.0270	73	.0156	80	.0093	87
40	.0641	65	.0361	72	.0209	79	.0125	86
41	.0825	64	.0469	71	.0274	78	.0165	85
42	.1043	63	.0603	70	.0356	77	.0215	84
43	.1297	62	.0760	69	.0454	76	.0277	83
44	.1588	61	.0946	68	.0571	75	.0351	82
45	.1914	60	.1159	67	.0708	74	.0439	81
46	.2279	59	.1405	66	.0869	73	.0544	80
47	.2675	58	.1678	65	.1052	72	.0665	79
48	.3100	57	.1984	64	.1261	71	.0806	78
49	.3552	56	.2317	63	.1496	70	.0966	77
50	.4024	55	.2679	62	.1755	69	.1148	76
51	.4508	54	.3063	61	.2039	68	.1349	75
52	.5000	53	.3472	60	.2349	67	.1574	74
53	.5492	52	.3894	59	.2680	66	.1819	73
54	.5976	51	.4333	58	.3032	65	.2087	72
55	.6448	50	.4775	57	.3403	64	.2374	71
56	.6900	49	.5225	56	.3788	63	.2681	70
57	.7325	48	.5667	55	.4185	62	.3004	69
58	.7721	47	.6106	54	.4591	61	.3345	68
59	.8086	46	.6528	53	.5000	60	.3698	67
60	.8412	45	.6937	52	.5409	59	.4063	66
61	.8703	44	.7321	51	.5815	58	.4434	65
62	.8957	43	.7683	50	.6212	57	.4811	64
63	.9175	42	.8016	49	.6597	56	.5189	63

Table A-7   Continued

$N_S = 8$

S_{Low}	$N_L = 8$	S_{Up}	$N_L = 9$	S_{Up}	$N_L = 10$	S_{Up}
36	.0001	100	.0000	108	.0000	116
37	.0002	99	.0001	107	.0000	115
38	.0003	98	.0002	106	.0001	114
39	.0005	97	.0003	105	.0002	113
40	.0009	96	.0005	104	.0003	112
41	.0015	95	.0008	103	.0004	111
42	.0023	94	.0012	102	.0007	110
43	.0035	93	.0019	101	.0010	109
44	.0052	92	.0028	100	.0015	108
45	.0074	91	.0039	99	.0022	107
46	.0103	90	.0056	98	.0031	106
47	.0141	89	.0076	97	.0043	105
48	.0190	88	.0103	96	.0058	104
49	.0249	87	.0137	95	.0078	103
50	.0325	86	.0180	94	.0103	102
51	.0415	85	.0232	93	.0133	101
52	.0524	84	.0296	92	.0171	100
53	.0652	83	.0372	91	.0217	99
54	.0803	82	.0464	90	.0273	98
55	.0974	81	.0570	89	.0338	97
56	.1172	80	.0694	88	.0416	96
57	.1393	79	.0836	87	.0506	95
58	.1641	78	.0998	86	.0610	94
59	.1911	77	.1179	85	.0729	93
60	.2209	76	.1383	84	.0864	92
61	.2527	75	.1606	83	.1015	91
62	.2869	74	.1852	82	.1185	90
63	.3227	73	.2117	81	.1371	89
64	.3605	72	.2404	80	.1577	88
65	.3992	71	.2707	79	.1800	87
66	.4392	70	.3029	78	.2041	86
67	.4796	69	.3365	77	.2299	85
68	.5204	68	.3715	76	.2574	84
69	.5608	67	.4074	75	.2863	83
70	.6008	66	.4442	74	.3167	82
71	.6395	65	.4813	73	.3482	81
72	.6773	64	.5187	72	.3809	80
73	.7131	63	.5558	71	.4143	79
74	.7473	62	.5926	70	.4484	78
75	.7791	61	.6285	69	.4827	77
76	.8089	60	.6635	68	.5173	76

Table A-7 Continued

$N_S = 9$

S_{Low}	$N_L = 9$	S_{Up}	$N_L = 10$	S_{Up}	S_{Low}	$N_L = 9$	S_{Up}	$N_L = 10$	S_{Up}
45	.0000	126	.0000	135	68	.0680	103	.0394	112
46	.0000	125	.0000	134	69	.0807	102	.0474	111
47	.0001	124	.0000	133	70	.0951	101	.0564	110
48	.0001	123	.0001	132	71	.1112	100	.0667	109
49	.0002	122	.0001	131	72	.1290	99	.0782	108
50	.0004	121	.0002	130	73	.1487	98	.0912	107
51	.0006	120	.0003	129	74	.1701	97	.1055	106
52	.0009	119	.0005	128	75	.1933	96	.1214	105
53	.0014	118	.0007	127	76	.2181	95	.1388	104
54	.0020	117	.0011	126	77	.2447	94	.1577	103
55	.0028	116	.0015	125	78	.2729	93	.1781	102
56	.0039	115	.0021	124	79	.3024	92	.2001	101
57	.0053	114	.0028	123	80	.3332	91	.2235	100
58	.0071	113	.0038	122	81	.3652	90	.2483	99
59	.0094	112	.0051	121	82	.3981	89	.2745	98
60	.0122	111	.0066	120	83	.4317	88	.3019	97
61	.0157	110	.0086	119	84	.4657	87	.3304	96
62	.0200	109	.0110	118	85	.5000	86	.3598	95
63	.0252	108	.0140	117	86	.5343	85	.3901	94
64	.0313	107	.0175	116	87	.5683	84	.4211	93
65	.0385	106	.0217	115	88	.6019	83	.4524	92
66	.0470	105	.0267	114	89	.6348	82	.4841	91
67	.0567	104	.0326	113	90	.6668	81	.5159	90

Table A-7 Continued

$N_S = 10$

S_{Low}	$N_L = 10$	S_{Up}	S_{Low}	$N_L = 10$	S_{Up}
55	.0000	155	81	.0376	129
56	.0000	154	82	.0446	128
57	.0000	153	83	.0526	127
58	.0000	152	84	.0615	126
59	.0001	151	85	.0716	125
60	.0001	150	86	.0827	124
61	.0002	149	87	.0952	123
62	.0002	148	88	.1088	122
63	.0004	147	89	.1237	121
64	.0005	146	90	.1399	120
65	.0008	145	91	.1575	119
66	.0010	144	92	.1763	118
67	.0014	143	93	.1965	117
68	.0019	142	94	.2179	116
69	.0026	141	95	.2406	115
70	.0034	140	96	.2644	114
71	.0045	139	97	.2894	113
72	.0057	138	98	.3153	112
73	.0073	137	99	.3421	111
74	.0093	136	100	.3697	110
75	.0116	135	101	.3980	109
76	.0144	134	102	.4267	108
77	.0177	133	103	.4559	107
78	.0216	132	104	.4853	106
79	.0262	131	105	.5147	105
80	.0315	130			

S	N=3	4	5	6	7	8	9
3	.6250						
4	.3750						
5	.2500	.5625					
6	.1250	.4375					
7		.3125					
8		.1875	.5000				
9		.1250	.4063				
10		.0625	.3125				
11			.2188	.5000			
12			.1563	.4219			
13			.0938	.3438			
14			.0625	.2813	.5313		
15			.0313	.2188	.4688		
16				.1563	.4063		
17				.1094	.3438		
18				.0781	.2891	.5273	
19				.0469	.2344	.4727	
20				.0313	.1875	.4219	
21				.0156	.1484	.3711	
22					.1094	.3203	
23					.0781	.2734	.5000
24					.0547	.2305	.4551
25					.0391	.1914	.4102
26					.0234	.1563	.3672
27					.0156	.1250	.3262
28					.0078	.0977	.2852
29						.0742	.2480
30						.0547	.2129
31						.0391	.1797
32						.0273	.1504
33						.0195	.1250
34						.0117	.1016
35						.0078	.0820
36						.0039	.0645
37							.0488
38							.0371
39							.0273
40							.0195
41							.0137
42							.0098
43							.0059
44							.0039
45							.0020

*Table entries for a given N is $P[S^+ \geq S]$, the probability that S^+ is greater than or equal to the sum S.

S	N						S	N		
	10	11	12	13	14	15		13	14	15
28	.5000						79	.0085	.0520	.1514
29	.4609						80	.0067	.0453	.1384
30	.4229						81	.0052	.0392	.1262
31	.3848						82	.0040	.0338	.1147
32	.3477						83	.0031	.0290	.1039
33	.3125	.5171					84	.0023	.0247	.0938
34	.2783	.4829					85	.0017	.0209	.0844
35	.2461	.4492					86	.0012	.0176	.0757
36	.2158	.4155					87	.0009	.0148	.0677
37	.1875	.3823					88	.0006	.0123	.0603
38	.1611	.3501					89	.0004	.0101	.0535
39	.1377	.3188	.5151				90	.0002	.0083	.0473
40	.1162	.2886	.4859				91	.0001	.0067	.0416
41	.0967	.2598	.4548				92		.0054	.0365
42	.0801	.2324	.4250				93		.0043	.0319
43	.0654	.2056	.3955				94		.0034	.0277
44	.0527	.1826	.3667				95		.0026	.0240
45	.0420	.1602	.3386				96		.0020	.0206
46	.0322	.1392	.3110	.5000			97		.0015	.0177
47	.0244	.1201	.2847	.4730			98		.0012	.0151
48	.0186	.1030	.2593	.4463			99		.0009	.0128
49	.0137	.0874	.2349	.4197			100		.0006	.0108
50	.0098	.0737	.2119	.3934			101		.0004	.0090
51	.0068	.0615	.1902	.3677			102		.0003	.0075
52	.0049	.0508	.1697	.3424			103		.0002	.0062
53	.0029	.0415	.1506	.3177	.5000		104		.0001	.0051
54	.0020	.0337	.1331	.2939	.4758		105		.0001	.0042
55	.0010	.0269	.1167	.2709	.4516		106			.0034
56		.0210	.1018	.2487	.4276		107			.0027
57		.0161	.0881	.2274	.4039		108			.0021
58		.0122	.0757	.2072	.3804		109			.0017
59		.0093	.0647	.1879	.3574		110			.0013
60		.0068	.0549	.1698	.3349	.5110	111			.0010
61		.0049	.0461	.1527	.3129	.4890	112			.0008
62		.0034	.0386	.1367	.2915	.4670	113			.0006
63		.0024	.0320	.1219	.2708	.4452	114			.0004
64		.0015	.0261	.1082	.2508	.4235	115			.0003
65		.0010	.0212	.0955	.2316	.4020	116			.0002
66		.0005	.0171	.0839	.2131	.3808	117			.0002
67			.0134	.0732	.1955	.3599	118			.0001
68			.0105	.0636	.1788	.3394	119			.0001
69			.0081	.0549	.1629	.3193	120			.0000+
70			.0061	.0471	.1479	.2997				
71			.0046	.0402	.1338	.2807				
72			.0034	.0341	.1206	.2622				
73			.0024	.0287	.1083	.2444				
74			.0017	.0239	.0969	.2271				
75			.0012	.0199	.0863	.2106				
76			.0007	.0164	.0765	.1947				
77			.0005	.0133	.0676	.1796				
78			.0002	.0107	.0594	.1651				

Appendix B

Reliability and Validity

Educators and psychologists who use standardized tests have devoted considerable thought and effort over the last 100 years to the quality of the information their measurements produce. The procedures, findings, and concerns that follow from this inquiry are typically confined to books with words like measurement or testing in the title, with the unfortunate consequence that researchers and students who are not directly concerned with the use of standardized tests on humans frequently overlook some important aspects of measurement theory as they apply to the broader arena of the behavioral and social sciences. This Appendix provides a brief overview of some of the major issues in evaluating the quality of data as they have arisen in the testing area. Some concerns that are more specific to the laboratory setting, were discussed in Chapters 9 and 10.

It is possible to state quite simply the two desirable properties that high-quality measurements should have: They should be accurate representations of the property being measured, and they should be easy to interpret correctly. Accuracy implies that it is possible to divide the measurement scale into units small enough for the scientist's purposes (precision of measurement) and that an observation receives the correct scale value (accuracy of category assignment). Ease of interpretation means that there is verification and general agreement among users and potential users of the scale about the information contained in the number that results from a measurement.

The physical sciences have been very successful in developing scales and measurement procedures that yield high-quality measurement. There is now universal agreement on a standard unit of distance, the meter (equal to 1,650,763.73 wavelengths of monochromatic orange light emitted by an atom of krypton-86 measured in a vacuum), that is defined with sufficient refinement and agreement that the accuracy and meaning of a physicist's measurement of this property is seldom in doubt. The same is true for such properties as time and temperature. The quality of measurement is so high in the physical sciences that there are usually only two reasons for a difference in measurements: human error in reading the scale or a real change or difference in the trait or entity being observed. (However, at the extreme limits of distance, measures of subatomic particles or intergalactic distances, even the physicist can encounter difficulty.)

Unfortunately, social and behavioral scientists have been unable to develop measurement procedures of this quality at any point in the continuum of phenomena of interest. Certainly, one of the reasons is that they have only been at the task for about 125 years. Measurement procedures for time and space have been under development for more than 20 times that long. Another reason is the number of assumed relationships between the measurement and the property. In the measurement of time, the interval between two ticks of the clock is defined as the unit of measurement and is equivalent to the property of interest without further extrapolation. (Actually, the interval of one second is now defined in terms of atomic events as well, hence the phrase atomic clock.) What is called the operational definition of the property, the procedure used to measure it, is equivalent to the property itself; in the social and behavioral sciences, this is seldom the case. For example, we generally use responses to a series of test items to measure something we call intelligence. But, getting a test item correct is not itself intelligence; it is assumed to reflect intelligent behavior, which is assumed to occur because there is intelligence around somewhere. (The somewhere is ordinarily assumed to be between the ears.)

A third reason for the lack of high-quality measurements in the social and behavioral sciences is that the entities themselves, usually humans, are constantly changing. It is very difficult to develop a high-quality scale on something that is undergoing change, because it is hard to know whether it is the object or the measuring operation that causes the instability. Consider, for example, trying to develop a procedure for measuring length when the only rulers you have are made of rubber and you have stretched the ruler more for one observation than for another.

All of these problems have led testing specialties to develop a vocabulary for measurement quality and a series of procedures to estimate the accuracy and usefulness of particular measurements. In the first section of this Appendix, issues related to accuracy and consistency of measurement are discussed under the general heading of reliability. The second section, validity, covers the problems of usefulness and interpretability from the philosophical-scientific side and from the practical-prediction side.

Reliability

The term reliability is usually used when we are concerned with the stability, consistency, or reproducibility of measurement. The question is whether a set of measurements of trait T on group G remains the same over time or over repeated measurements of the same trait. A measuring instrument is said to be reliable if it yields the same scale value for an individual each time the measurement is made, assuming it is reasonable to believe that the individual's status on the trait has not changed. The fundamental concepts in reliability partition observed variance into different components in much the same manner that we have already done in Chapter 8. We shall consider these partitions after having examined an example.

Consider some learnable and repeatable act such as throwing darts at a circular dart board. Let us assume that Charlie is throwing darts. As a budding scientist, Charlie reasons that it should be possible to measure a person's status on the trait of dart-throwing ability. His scientific training has led him to a firm belief in the value of controlled conditions for measurement, so he carefully sets up a measurement procedure. A standard board is used, as are standard-weight darts with properly sharpened points and combed feathers. The people to be measured stand in a fixed location, and temperature, humidity, and air movement are carefully controlled. Everything is perfect.

Enter Charlie with his first candidate, one Annie Oakley. A premeasurement interview has revealed that Annie has never thrown a dart before, but she is willing to help Charlie develop his measurement procedure. She takes her place and tosses a dart; bull's eye! Charlie can be heard mumbling to himself that it must be luck: Annie could not have a 20 on his scale of 0 to 20. This measurement must be wrong; a second measurement is taken (score of 8) and a third (17) and a fourth (0). Now Charlie is really confused. He has four different measurements of Annie's ability, each quite different from the others, with no way of telling which is the right one.

While this example is obviously contrived, it serves to point out a major factor of measurement in the social and behavioral sciences. Repeated measurements of the same trait on the same people often show a great deal of variability. Sometimes it is reasonable to attribute this variation to actual changes in the person over time. For example, one's expressed interest in eating or the time needed to fall asleep should vary throughout a single day; other characteristics such as moods can be expected to fluctuate from day to day. However, even simultaneous measurements of the behavioral characteristics of a person almost always show some differences, leading us to conclude that there is always some amount of variation that cannot be explained in this way. (As personal testimonial to the inherent variation in measurements, one of your authors [RMT] once tested a new electronic bathroom scale. During a 15-minute testing period the scale recorded weights that differed by as much as 11 pounds. Since it is quite unreasonable for a person's weight to change this much in such a short period, the scale was returned as defective, but lesser variations might have gone unnoticed even though they were not due to real changes in weight.)

Let us return to the dart board for a moment to take a look at the variation in performance. Remember that we are measuring dart-throwing ability and that each toss is a measurement of that ability. (We could even think of dart tosses as similar to test items on a personality or ability measure.) Suppose that Annie has been practicing diligently for several weeks now and has reached a fairly uniform level of performance. It would be reasonable to think of measuring her true level of ability; we could even say that the average or sum of her scores on ten darts was her true ability. With this in mind, we ask her to throw ten darts, and then, just to be sure, we have her throw ten more. Imagine our dismay when the first measurement and the second disagree.

Our dart example is not too unlike what seems to be going on in many measurement situations. It is reasonable to assume that the individual possesses a relatively stable amount of a quantifiable characteristic. Our experience with direct measurement of physical attributes suggests that we should obtain fairly consistent measurements. However, we find that repeated measurements of behavioral or social characteristics can give results showing much more than the degree of variability expected from errors in reading the instrument. For example, your knowledge of statistics could be assumed to be constant over a period of half an hour. But if we give you two 15-minute, ten-item quizzes on statistics, you probably will not get the same score on each. Is it that you changed, the instruments are measuring different things, or your behavior is just inconsistent? The measurement model has been devised to account for this variability in observations.

The *measurement model* is an outgrowth of what used to be called the *true-score model,* originally proposed by Charles Spearman in 1904. In its earlier form, this model held that each observation was composed of two parts, the individual's true level on the trait of interest, known as the true score (T_i for individual i) and some error of measurement that occurred in this particular measurement of this particular person (e_{ij}). A person's true score was assumed to be constant across all measurements of the same trait within a reasonable time span, and errors were assumed to be completely random chance events. Over a large number of measurements, the distribution of these errors was assumed to be normal in shape with a mean of zero.

The major difference between the true-score model and most current formulations of the measurement model is that both true score and error are seen as being partitionable into several sources of variance. The notions of true score and error have been replaced by consistent or between-persons variance and inconsistent or within-persons variance. An early formulation of this model (R. L. Thorndike, 1949) identified five major categories of variation; later contributors (cf. Cronbach et al., 1972) have suggested many more factors that need to be considered.

The particular sources of variance proposed under the measurement model would depend on the research context. There is the possibility of real differences in what is measured by two procedures or tests that seem to measure the same thing. There is the possibility of real differences between classrooms or schools that are due, not to characteristics of the children who are measured, but to physical or social differences in the school environment or the way the teacher administered the test. It is sometimes quite important to identify such sources of variance as these as different from either true scores or random errors of measurement. Our present discussion will be phrased in terms of the true-score model because it is the simplest approach to the problem and the principle of partitioning observed variance into parts remains the key concept in more complex versions. A more complete discussion of these concepts can be found in Feldt & Brennan (1989) or Thorndike & Thorndike (1994).

In the form of an equation, then, the true-score model says that the measurement of individual i on trait X at occasion j is

$$X_{ij} = T_i + e_{ij} \tag{B.1}$$

That is, an observed score X_{ij} is composed of two parts: the individual's true score T_i, which is assumed to be constant across measurement occasions, and an error of measurement, e_{ij}, which is a random or chance event that varies from one measurement to another. The model predicts that we can always expect to find some variations among repeated measurements of the same individual.

Consider our dart-throwing example again. It would be possible to measure Annie a very large number of times and get a frequency distribution of her scores. The mean of the distribution of observed scores is her true score (current authors use the term universe score instead of true score to indicate that they are talking about a least-squares estimate of a population mean), because the mean of the e_{ij} is assumed to be zero. The variance of observed scores around the true score can be found by subtracting the mean from each score, squaring, summing, and dividing by N to obtain

$$S^2_{x_{ij}} = S^2_{e_{ij}}$$

The variance in *observed scores of an individual* is entirely due to errors of measurement, according to the true-score model. The standard deviation of the distribution of an individual's observed scores is known as the *standard error of measurement.*

The standard error of measurement, sometimes abbreviated *SEM* and sometimes S_e, is a standard deviation. It is the standard deviation that we would get from the distribution of repeated measurements of a single individual on a single trait; it is the standard deviation of the distribution of errors of measurement.

In the real world, it is seldom possible to calculate the SEM directly because its definition requires a constant true score. Repeated measurements tend to change the person being measured, thereby confusing error of measurement with true changes. Even in a situation as simple as our dart-throwing example, our subject will probably show some changes over time. In more complex measurement situations such as the measurement of human ability, it is clear that we cannot retest people many times and expect them to remain unchanged. (One exception is the case of classical psychophysics, as developed by Weber and Fechner, in which repeated measurements of sensory discriminations were used to measure sensory processes. Similar procedures are used in modern signal detection research.)

We attempt to overcome the problem of nonrepeatability of measurements by making a small number of measurements, usually two, on a group of people and letting the large number of individuals substitute for the repeated measurements that we cannot make. For example, we may make two measurements of each of N people, in which case each person's score on first testing is expressed as

$$X_{1i} = T_i + e_{1i} \qquad \text{(B.2)}$$

The second score is the same way,

$$X_{2i} = T_i + e_{2i}$$

Applying the same steps to equation B.2 that we have used several times before, we determine that the total variance in observed scores X_{1i} is composed of two parts. First, subtract $\overline{X}_1$ from both sides and square the result. Then, by summing across the N individuals and dividing by N, we get three variances,

$$S^2_{x_1} = S^2_T + S^2_{e_1} \qquad \text{(B.3)}$$

The total variance in X-scores at time 1 is composed of variance due to differences among people in their true levels on the trait (S^2_T) and variance due to errors in measurement at time 1 ($S^2_{e_1}$). The same form of the equation also holds for scores at time 2; that is,

$$S^2_{x_2} = S^2_T + S^2_{e_1}$$

Unfortunately, we have no way of knowing at this stage what values to place on S^2_T and S^2_e. They must be smaller than S^2_x, but we do not know their relative magnitudes.

The solution to the problem is to compute the correlation of X_1 and X_2. First, let us simplify the notation for the problem by putting equation B.2 in the form of deviation scores

$$x_{1i} = t_i + e_{1i}$$
$$x_{2i} = t_i + e_{2i} \qquad \text{(B.4)}$$

Then we can express the correlation of x_1 with x_2 as

$$r_{x_1 x_2} = \frac{\sum_{i=1}^{N} x_{1i} x_{2i}}{N S_{x_1} S_{x_2}} = \frac{1}{N} \frac{\sum x_1 x_2}{S_{x_1} S_{x_2}} \qquad \text{(see 8.2)}$$

By substituting from equation B.4, we get

$$r_{x_1 x_2} = \frac{1}{N} \frac{\sum (t_i + e_{1i})(t_i + e_{2i})}{S_{x_1} S_{x_2}}$$

which, when we multiply the numerator out, yields

$$r_{x_1 x_2} = \frac{1}{N} \frac{\sum (t_i^2 + t_i e_{1i} + t_i e_{2i} + e_{1i} e_{2i})}{S_{x_1} S_{x_2}}$$

Distributing the summation sign and $1/N$ produces

$$r_{x_1 x_2} = \frac{\dfrac{\sum t_i^2}{N} + \dfrac{\sum t_i e_{1i}}{N} + \dfrac{\sum t_i e_{2i}}{N} + \dfrac{\sum e_{1i} e_{2i}}{N}}{S_{x_1} S_{x_2}} \qquad \text{(B.5)}$$

Now we must remember an assumption that we made and introduce another one. First, we assumed that the errors of measurement were chance events. If they are, then they should be uncorrelated with each other and with true scores because chance events are unrelated to anything (remember our assumption of independence in chapter 6). Noting that each of the terms in the numerator of equation B.5 is a variance or covariance (a covariance is a correlation multiplied by the standard deviations of the two variables; cov $(X, Y) = r_{xy} S_x S_y$), the last three terms must be zero because they each contain the correlation of errors of measurement either with the true score or with other errors of measurement. Therefore, equation B.5 becomes

$$r_{x_1 x_2} = \frac{\sum t_i^2 / N}{S_{x_1} S_{x_2}} = \frac{S_T^2}{S_{x_1} S_{x_2}} \qquad \text{(B.6)}$$

since $\sum t_i^2 / N$ is the same as the variance in true scores S_T^2. Our new assumption is that since the two tests really are measures of the same thing, they have equal standard deviations. If $S_{x_1} = S_{x_2}$, then $S_{x_1} S_{x_2} = S_x^2$, the observed variance of raw scores, which we can substitute into equation B.6 to get

$$r_{x_1 x_2} = \frac{S_T^2}{S_x^2} \qquad \text{(B.7)}$$

The correlation coefficient in equation B.7 is a special type of correlation known as a *reliability coefficient,* which we symbolize r_{xx}; it is the correlation between two measures of the same thing. Note carefully that we have made some assumptions in its development and that its interpretation is different from that of other correlations. First, we assumed that an observed score could be broken down into two unobservable parts. Second, true scores were assumed to be constant, identical from one testing to the other. In addition, errors of measurement were assumed to be chance events uncorrelated with anything, and the standard deviations of the two sets of measures were assumed to be equal.

A reliability coefficient is interpreted as the proportion of variance in observed scores that is due to true-score differences among people. In this respect, it is quite different from other correlations because it is a direct variance ratio. Remember that in Chapter 8 we found the square of the correlation coefficient to be a ratio of two variances (see equation 8.9). The key to the difference is that the reliability coefficient involves the unobservable true score that is constant in both measuring instruments rather than what is common to two different instruments.

As noted, we never observe true scores or errors of measurement directly; all we ever see are observed scores and their correlations, but with the aid of equations B.3 and B.7 we can get estimates or best guesses about S_T^2 and S_e^2. From equation B.7, we find that

$$S_T^2 = r_{xx} S_x^2$$

The variance in true scores is the observed score variance multiplied by the reliability. Substituting this result into equation B.3 and solving for S_e^2 yields

$$S_x^2 - S_x^2 r_{xx} = S_e^2$$

which is equivalent to

$$S_x^2 (1 - r_{xx}) = S_e^2$$

By taking the square root of both sides, we obtain

$$S_e = S_x \sqrt{1 - r_{xx}} \qquad \textbf{(B.8)}$$

as an expression of the standard error of measurement.

The important thing about equation B.8 is that it is in terms of observable quantities. Without an equation like this, it would not be possible to get a value for the standard error of measurement for most variables of interest because repeated testings are not possible. The reason that S_e, or SEM, is important is that it allows us to express our level of uncertainty about an individual's performance. It functions like any other standard deviation in that it can be used to form confidence intervals at whatever level we choose.

Perhaps an example will help clarify the situation. Let us assume that reading test X has a standard deviation of 20 and a reliability of .91. (A reliability of .91 is about what we might expect from a carefully developed and standardized test of this type. Most measurements in the behavioral and social sciences have reliabilities well below this level.) From the information given, we calculate

$$S_e = S_x \sqrt{1 - r_{xx}} = 20 \sqrt{1 - .91}$$
$$= 20 \sqrt{.09} = 20 \cdot (0.3) = 6$$

The standard error of measurement of this test is six points. This means that we would expect the standard deviation of observed scores around the true score for a person to be six points if we could collect those repeated measurements. Using the table of the normal distribution, we find that 95 percent of observed scores can be expected to fall in a band $11\frac{3}{4}$ points on either side of the true score. The 95 percent confidence interval is $23\frac{1}{2}$ score points wide! This means that there is about one chance in 20 that a person with a true ability of 100 would earn an observed score above 112 or below 88. If we are making decisions about people, it is well to remember the standard error of measurement because it gives an indication of how likely errors of a particular size are.

The standard error of measurement is an important piece of information to consider whenever indirect measurements are involved. It reflects our degree of uncertainty about the exact value of a measurement and should lead to caution in interpreting the results of a measurement. This caution should be greatest when dealing with the scores of single individuals rather than groups because the errors of measurement tend to cancel each other out when scores are averaged for a group.

The reliability coefficient and the standard error of measurement give complementary pieces of information in the same way that the correlation between two variables and the standard error of estimate do. The standard errors reflect the inconsistency of performance or inaccuracy of predictions. The reliability and correlation coefficients tell us the degree to which the test orders people in the same way on two measurements or to which the order of people is the same on two variables. Where concern is with the individual observation, the standard error is likely to be most useful, but where groups are being considered the reliability/correlation coefficient will often give the best information.

The problem of unreliability in our measurements is probably one of the major reasons that research in the behavioral and social sciences has depended heavily on group averages rather than the scores of individuals. If a measuring device yields highly reliable measurements, each individual will retain the same relative position in a group on first and second measurements. Few measurements accomplish this very satisfactorily for individuals, but averaging the scores for the group yields results that are quite stable or reliable. The larger the group over which the average is taken, the more reliable the result will be. Thus, we have tended to overcome the problem of unreliability or instability in our measurements by focusing attention on group averages. Unfortunately, this trend may often mask situations where we can say things that are true for a group but less than true for any individual in the group.

Validity

The reliability or unreliability of a set of measurements is an important issue when considering the quality of data. Unreliable data, data where all or most of the observed score variance is due to errors of measurement, are not likely to advance scientific understanding. Because of the random nature of errors of measurement in the model (random deviations around true scores), unreliably measured variables are unlikely to show relationships with other variables. Unreliable measurements will probably result in frustration but will probably not lead to serious misdirection.

A far more serious question in data quality is whether a measuring procedure actually measures the property of interest; this is the general issue in the validity of measurement. We can say for the purpose of starting our discussion that a measurement is valid if it in fact it measures the property or trait that we think it measures. The consequence of using an invalid measurement, then, can be a serious misdirection of the investigator's thinking and consequently lead to drawing erroneous conclusions.

We can think of the difference between the reliability and validity of data as the difference between random errors and an inherent lack of relationship. A measure that is unreliable will not have a high correlation with anything, not even itself. A reliable measure is consistent, it shows a high relationship with itself, but it still may not be related to anything else.

The key to validity is that a measurement has validity for some purpose; the measurement is related in some way to something other than itself. That something may be scholastic or job performance, voting behavior, mental or physical health, or learning a maze, for example. In each case the measure is valid for purpose X. It is an indicator of some underlying unobserved trait. Validity, therefore, is the crucial issue for a science that relies on indirect measurement.

Because there are a very large number of uses for measurements, there are a number of different types of validity. Each type has its own particular features and ways in which it is assessed, but all types of validity are concerned with the relation of a measurement to something other than itself. We describe three major types of validity. While there are many others that we do not cover, these three are sufficiently general and of widespread concern to cover most situations. (At the opposite end of the spectrum in terms of their presentation but not in their conclusions are writers such as Messick [1989] and Sheppard [1993] who argue that there is really only one type of validity. They view the types of validity described in what follows as different ways to gather evidence about a single unified concept. While their basic argument is valid, it is sometimes useful in the early stages of thinking about validity to make these distinctions.)

Empirical Validity.
Empirical validity covers those situations where there is a measured predictor variable or group of predictors and a measured criterion (outcome) variable or (in very rare cases) group of criterion variables. The empirical validity of the predictor is its correlation with the criterion. The numerical value of the correlation coefficient is considered to be the validity of the predictor; a high correlation is equated with high validity because it leads to accurate predictions. Our discussion of regression and correlation in Chapters 7 and 8 primarily relate to empirical validity.

It soon became obvious to researchers and test users that there was not just one type of empirical validity, but rather a continuum based on the time relationship between when the predictor measurement and the criterion measurement were taken. A rough but useful distinction has been drawn between *predictive validity* and *concurrent validity*. Both are empirical validities, but predictive validity involves collecting predictor information, waiting for a period of time, and then collecting criterion information. Concurrent validity refers to those situations where both measurements are made at essentially the same time.

Over the years, some other important restrictions on empirical validity have been identified. A predictive measure may have somewhat different levels of validity (magnitudes of correlation) for different but similar criteria: Test X may be a good predictor of performance as a plumber but not as an electrician. Also, some have argued that a test may be a good predictor in one company but not in another. Perhaps even more important has been the finding that measures may have different levels of validity for different identifiable groups of people. A test may have one level of validity for men and a different level of validity (higher or lower) for women; there may be different correlations and different least-squares regression lines for various subgroups within the population.

Such findings as these have led most psychologists and educators to reject the notion of a single validity for any measurement. We must now speak of a test, or other measurement procedure, as having a validity for a particular purpose (job, school achievement) with a particular group of people. It is important to make these qualifications because the empirical validity of a measure for an individual is directly related to the accuracy of our prediction for that person. The standard error of estimate increases as the empirical validity drops, so prediction may be more accurate for members of one group than another.

Messick (1989) has argued that we must also consider the social desirability of possible outcomes of test use as part of the test's validity. A test may predict quite accurately, according to Messick, but result in socially undesirable outcomes such as unequal access to jobs or education. The description of a test's validity and the decision to use it for a particular purpose imply accepting the social value of the outcomes.

Content Validity.
Educators have long been concerned with tests measuring such specific domains of content knowledge as principles of statistics or the facts of European history in the 14th century. To a lesser extent, other social and behavioral scientists may be concerned with well-defined content areas. The extent to which a measure covers a domain of content is known as its content validity.

Content validity is not an obvious concern of all measuring instruments; it is important only for those topics where it is possible to specify in considerable detail the elements of a domain. The clearest and most obvious examples come from education. For example, we might consider that our domain includes the 144 multiplication facts for the numbers 1–12. It is possible to state for any behavior whether it is in this domain or not.

What a test measures, its content, depends on more than the items that compose it. What a test measures is affected by how it is administered and the status of the persons to whom it is given. For example, a test of addition facts might be a measure of mathematics achievement for a group of second graders, but the same test given to high school students with a strict time limit becomes a measure of processing speed.

There are two fundamental features that a measure must have in order to be considered to possess content validity. First, each element of the measurement operation (usually a test item) must come from the domain in the sense of both content and cognitive process. If there are elements included that are not in the domain, they detract from the content validity of the instrument. Second, the elements must cover all aspects of the domain. To the extent that there are content or process subareas of the domain that are not represented by elements of the measurement, the instrument lacks content validity. For example, the multiplication fact $13 \times 13 = 169$ is not in the domain we defined, so including it in a measurement of that domain would detract from content validity. Likewise, failure to include any items covering the numerals 11 and 12 would constitute incomplete coverage of the domain.

As we noted, content validity is primarily a concern for educational achievement tests. However, there may be other areas where the idea of a domain of content is appropriate or can help in instrument construction. For example, some work in attitude measurement has utilized the basic idea of content validity. As knowledge expands in any area of research on behavior, the possibilities of specifying a domain of content or of behavioral manifestations of a trait of interest increase.

Construct Validity.
In the 1950s, psychologists and educators began to focus their attention on an important philosophical issue that had not been widely addressed by behavioral and social scientists. The issue was the place of measurement operations in science, and it arose from the work of P. E. Meehl and his associates (Cronbach and Meehl, 1955; MacCorquodale and Meehl, 1948). It is probably the most important link between thought and observation in the conduct of science.

Let us begin by considering what a science is. Recall that early in Chapter 1, we discussed collecting information; science involves collecting information. The conditions under which information is collected, for example, proper controls to rule out unwanted sources of variation, are known as the scientific method. The tightness of controls and other research conditions vary from one science to another, as do reliance on formal and agreed-upon measurement procedures. However, all sciences contain three basic elements: the real world of events to be observed; a set of more or less abstractly derived principles to explain the events in the real world; and a set of measurement operations that connects the real world with the abstract principles.

We can take an example from experimental psychology to illustrate the structure of a scientific inquiry. Suppose that we are interested in measuring and studying learning. Learning is an abstract principle that we use to explain real-world events. We might define it as a relatively permanent change in behavior or behavior potential that occurs as a result of experience. It is not, at the present stage of our knowledge, possible to observe learning directly; we cannot even prove that it exists in a formal sense. (As early as 1900 some theorists argued that learning involved changes in the nervous system, and recent research has found evidence of different chemical and structural properties in the brains of organisms that have had special learning experiences, but even this level of abstraction does not capture what most people mean by learning.) What we can do is observe behavior, manipulate the environment, and measure what happens. We put a hungry rat in a maze and record the amount of time taken and the number of errors made before the rat reaches food. Repeating this process, we record shorter times and fewer errors on later trials in the maze.

Note the three parts of a science at work here. We have the real world of rat behavior (substitute people or cockroaches, if you wish), the abstract principle of learning that we wish to use to explain a change in behavior, and the measurements of the real world (time and errors) that tie the two together. We develop our science by making predictions about the real world from the elements of our theory—our set of prior beliefs—about learning and by confirming those predictions with measurements.

So far, so good. The abstract principles that form the scientific theory are called *hypothetical constructs*. They are constructs (entities or processes) that are hypothesized to exist and to account for the events we observe. The problem that was identified by Cronbach and Meehl is that our measurement procedures may or may not actually involve the constructs of interest. They coined the term *construct validity* to refer to this issue. An instrument is construct valid if it is a measure of the hypothetical construct it is supposed to measure.

Let us take an example from education. Suppose we have a hypothetical construct that we label mathematics aptitude. Our theory of human performance says that aptitude precedes achievement; that is, the theory says that someone may possess an aptitude for mathematics without any experience that is manifest in actual performance of mathematical behavior. The aptitude could exist even if the opportunity for experience never arose (but it would not be manifest).

Now, consider a measure of mathematics aptitude. Since our construct does not require performance of mathematical behavior, our measuring instrument probably should avoid such behavior. Suppose that this leads us to devise a measure that requires a lot of reading and other verbal skills as well as whatever it is that is mathematics aptitude. We now measure an individual and find a low level of aptitude, but the low score may also be due to lack of reading ability, not lack of mathematics aptitude. Conversely, someone might get a relatively high score on our test if they had outstanding reading or verbal reasoning ability but relatively little mathematics aptitude. To the extent that the measurement procedure depends upon, or is affected by, things that are not part of the construct, it will not be a measure of that construct.

Construct validity is an essential quality for any measuring procedure to have if it is to be used for scientific purposes. In those situations where prediction of a criterion is the only thing that matters, we need not worry about what the instrument measures, only whether it correlates highly with the criterion. However, when scientific understanding is the objective, it is necessary for the measuring procedure actually to measure what we assume it measures. Otherwise, the causes or constructs that we infer to be operative may not be present at all.

A measure that possesses construct validity will probably satisfy other validity requirements as well. A construct-valid instrument will demonstrate predictive validity for criteria that are related to the construct. In fact, we infer that the measure has construct validity if it has the pattern of empirical relationships predicted by theory. If there is a domain of content relevant to the construct, the instrument should have content validity for that domain. Construct validity is an issue when the instrument is being used to measure a trait that is part of a scientific theory. Since we can only know about the construct through measurements that should be related to it, we infer simultaneously the existence of the construct and the validity of our measures of it; neither can be separated entirely from the other. Both grow and gain credibility to the extent that predictions made from them are confirmed. For example, confirmation of a prediction that faster running times in the maze and fewer errors should both occur on later trials in the maze adds an increment of credibility to the construct of learning and to changes in running time and number of errors made as measures of that construct.

SUMMARY

There are two primary qualities that measurement procedures should have in order to produce data of high quality; these are *reliability* and *validity*.

Reliability refers to the consistency or stability of measurements. In the behavioral and social sciences, tests and other measures of behavior tend to show variation in repeated measurements of the same individual. If it is reasonable to assume that the person's status on the trait has not changed, the variation in score is attributed to error of measurement. Because it is seldom possible to obtain many repeated measurements on a person, consistency of measurement is assessed by partitioning the variance in a group of people on two measurements into a portion that is between-persons variance, or true-score variance, and a portion that is within-persons, or error, variance. The reliability coefficient is the correlation between two measurements of the same thing and is the ratio of between-persons variance to total variance. The *standard error of measurement* is the square root of the error variance and can be used to obtain a confidence interval for a person's score.

Validity is the relationship of a measure to something other than itself. The *empirical validity* of a predictor is its correlation with a criterion. *Content validity* is the relationship between a test and a domain of content. A content-valid measure includes all aspects of the domain. *Construct validity* relates to the scientific value or appropriateness of a measure. Science consists of hypothetical constructs, the real world, and measurement operations that relate the constructs to the real world. To be construct valid, a measure must relate constructs to real-world observations.

REFERENCES

Cronbach, L. J., Gleser, G. C., Nanda, H., & Rajaratnam, N. (1972). *The dependability of behavioral measurements: Theory of generalizability for scores and profiles.* New York: Wiley.

Cronbach, L. J., & Meehl, P. E. (1955). Construct validity in psychological tests. *Psychological Bulletin, 52,* 281–302.

Feldt, L. S., & Brennan, R. L. (1989). Reliability. In R. L. Linn (Ed.), *Educational measurement* (3rd ed., pp. 105–146). New York: American Council on Education and Macmillan.

MacCorquodale, K., & Meehl, P. E. (1948). On the distinction between hypothetical constructs and intervening variables. *Psychological Review, 55,* 95–107.

Messick, S. (1989). *Validity.* In R. L. Linn (Ed.), *Educational measurement* (3rd ed., pp. 13–103). New York: American Council on Education and Macmillan.

Shepard, L. A. (1993). Evaluating test validity. In L. Darling-Hammond (Ed.), *Review of Educational Research* (Vol. 19, pp. 405–450). Washington, DC: American Educational Research Association.

Thorndike, R. L. (1949). *Personnel selection: Test and measurement techniques.* New York: Wiley.

Thorndike, R. L., & Thorndike, R. M. (1994). Reliability in educational and psychological measurement. In T. Husen & N. Postlethwaite (Eds.), *International Encyclopedia of Education* (2nd ed., pp. 4981–4995). New York: Pergamon Press.

Appendix C

Data for the Big Time University Anxiety Study

	id	lastnam	firstnam	age	sex	major	hsgpa	cgpa	nmathc	mathapt	mathanx	final	nhomwrk	homgrade	labsec
1	1	Alcott	Louisa	20	0	1	2.24	2.73	2	60	14	37	8	10.5	3
2	2	Alexander	Patty	23	0	2	2.85	2.44	3	56	13	47	10	9.9	3
3	3	Allen	Steven	21	1	2	1.68	1.13	3	46	18	19	10	7.3	1
4	4	Allport	Gordon	22	1	3	2.38	2.51	1	44	17	37	6	12.0	3
5	5	Alwood	Biffy	21	0	3	2.47	1.78	4	60	19	33	8	7.3	1
6	6	Bentler	Peter	19	1	2	3.06	2.94	4	61	17	48	8	12.2	3
7	7	Blivit	Cherie	22	0	2	2.25	1.86	2	36	7	20	4	8.7	3
8	8	Blowers	Mike	21	1	3	2.71	3.64	2	62	21	44	4	8.6	3
9	9	Blunt	Melanie	23	0	2	2.49	2.75	1	62	25	44	4	10.2	3
10	10	Borgatta	Steven	22	1	2	2.82	1.34	3	58	11	34	10	18.2	2
11	11	Borgia	Lucretia	22	0	2	2.61	2.40	3	63	20	38	9	7.1	2
12	12	Browne	Mike	23	1	2	1.44	1.41	0	35	16	14	7	11.9	1
13	13	Campbell	Don	21	1	1	2.37	1.33	0	33	25	4	5	13.9	3
14	14	Carter	Kim	23	1	3	2.50	2.83	3	50	21	33	8	6.5	2
15	15	Cattell	Jim	20	1	2	2.36	2.14	2	69	13	35	3	7.8	3
16	16	Cattell	Ray	21	1	1	3.00	2.24	2	56	13	37	10	2.5	2
17	17	Cummins	Dalia	23	0	2	2.43	1.22	0	38	27	8	6	8.4	3
18	18	Downing	Tom	20	1	2	2.37	3.32	4	57	24	37	7	7.3	1
19	19	Dyxter	Jammie	21	0	2	1.67	2.71	4	45	17	37	9	8.8	2
20	20	Eber	Herb	23	1	1	2.56	1.87	2	47	23	31	8	10.6	1
21	21	Ellington	Duke	24	1	2	1.44	1.52	1	45	21	22	5	4.0	2
22	22	Embretson	Sue	20	0	3	2.45	2.56	3	55	14	49	9	15.4	2
23	23	Foxx	George	19	1	1	1.15	1.44	2	48	19	28	6	13.6	3
24	24	Foxx	Red	20	1	2	1.95	2.09	2	53	32	34	4	10.9	2
25	25	Framm	Paul	21	1	2	4.00	3.31	5	77	10	61	10	16.0	1
26	26	Geldt	Silver	20	0	2	2.55	3.30	5	56	22	44	5	12.9	1
27	27	Good	Johnnie	21	1	2	2.74	2.97	5	79	7	55	10	13.6	2
28	28	Hakasami	Isoroku	19	1	3	2.84	1.63	1	40	25	19	4	7.2	3
29	29	Hamm	Shelia	20	0	1	3.72	3.59	3	74	13	50	6	9.1	1
30	30	Hammar	Ted	24	1	2	1.08	1.28	4	41	23	23	5	10.0	3
31	31	Helfgott	Lenny	21	1	3	3.68	1.83	3	69	23	48	9	8.1	1
32	32	Holder	Patty	20	0	1	3.44	1.79	4	56	21	34	9	10.9	1
33	33	House	Sam	22	1	3	1.53	1.04	1	41	20	13	6	4.7	3
34	34	Ishii	Kristi	22	0	3	2.73	3.66	3	64	13	65	9	12.7	1
35	35	Jackson	Phil	23	1	1	2.67	1.45	2	45	19	25	7	10.5	2
36	36	Johnson	Alex	19	1	3	3.48	1.85	2	67	21	38	10	10.7	3

	id	lastnam	firstnam	age	sex	major	hsgpa	cgpa	nmathc	mathapt	mathanx	final	nhomwrk	homgrade	labsec
37	37	Johnson	Sally	20	0	1	3.24	3.06	4	65	15	46	8	12.1	2
38	38	Kravitz	Lillie	21	0	1	2.34	1.70	1	51	17	25	8	8.6	3
39	39	Krell	Jack	22	1	2	2.16	1.79	3	51	18	34	5	8.6	3
40	40	LaLonde	Bobbie	21	0	3	1.74	1.83	1	31	23	7	3	11.4	2
41	41	Lamb	Spring	22	0	1	3.19	2.45	2	65	18	47	8	7.2	1
42	42	Lange	Harold	24	1	1	1.69	2.36	2	44	28	25	5	7.3	2
43	43	Lyons	Karl	23	1	2	2.35	1.53	3	43	16	20	3	10.5	3
44	44	McCallum	Rob	21	1	1	1.60	1.70	2	43	28	22	6	12.1	1
45	45	Montange	Sue	23	0	2	2.10	2.04	2	41	19	21	3	13.9	2
46	46	Montegue	Phil	22	1	1	3.35	2.96	1	61	13	45	6	9.8	2
47	47	Muse	Jim	22	1	1	1.04	1.88	3	37	23	25	6	6.0	1
48	48	Nees	Pat	22	1	3	1.67	3.94	3	72	17	43	9	3.3	2
49	49	Neuman	Joy	24	0	3	1.90	2.49	3	66	11	40	4	14.5	1
50	50	Nouse	Paula	20	0	1	1.34	1.51	4	54	18	29	7	16.2	3
51	51	Novick	Mel	20	1	1	3.64	3.42	3	61	12	55	8	4.3	3
52	52	Randell	Ronda	20	0	3	2.56	1.42	4	54	16	32	7	14.3	3
53	53	Robertson	Rob	19	1	3	2.10	2.17	3	50	22	28	10	16.3	3
54	54	Robson	Paul	21	1	2	4.00	4.00	4	71	5	59	7	7.6	1
55	55	Schmidt	Frank	21	1	2	3.25	2.61	3	64	22	41	6	14.5	1
56	56	Sloop	Ona	20	0	3	2.89	3.68	5	71	16	54	9	9.4	1
57	57	Smith	Ginger	23	0	2	1.60	1.70	3	45	25	29	4	12.2	2
58	58	Smith	Jenny	21	0	2	2.37	1.39	1	39	26	16	5	11.0	2
59	59	Squall	Peter	21	1	3	1.76	1.90	2	44	28	30	10	12.2	1
60	60	Squidly	Tom	20	1	3	2.15	2.47	3	47	15	30	8	7.6	3
61	61	Squires	Julie	23	0	1	1.66	1.08	2	42	24	18	7	9.5	3
62	62	Tattersal	Edwina	20	0	3	1.68	2.75	2	57	15	42	7	8.5	1
63	63	Tefft	Pam	22	0	2	1.90	1.19	2	42	25	14	6	10.9	2
64	64	Telcor	Jack	22	1	3	2.72	3.08	2	58	16	43	9	10.7	1
65	65	Washo	Marsha	20	0	1	3.09	2.61	4	58	17	39	4	10.0	2
66	66	Williams	John	22	1	1	3.71	3.55	3	64	11	56	10	12.8	1
67	67	Zyborg	Chip	19	1	3	2.12	2.59	4	57	21	39	10	13.5	2
68	68	Zylstra	Sonya	22	0	3	2.81	1.54	3	44	26	29	6	8.7	2

Appendix D

Data for the Searchers Study

	id	name	age	income	denom	member	involve	form	attitude
1	1	Abbott, Pete	42	44376	2	2	19	2	38
2	2	Alice, Peter	33	35481	1	2	19	1	41
3	3	Anderson, Earl	59	51246	2	1	30	3	60
4	4	Archer, Al	36	45485	2	2	22	2	28
5	5	Armour, Tommy	44	41318	2	1	20	1	45
6	6	Baltch, Karl	20	35424	2	1	10	3	24
7	7	Battle, Bill	24	41736	1	1	18	3	28
8	8	Brakkee, Rich	57	51701	2	1	17	2	48
9	9	Champagne, Sandy	49	50955	1	2	21	3	45
10	10	Chipman, Phil	46	51062	1	2	18	3	41
11	11	Chompsky, Moam	24	27731	1	1	16	3	25
12	12	Churchill, Win	46	47200	2	1	16	1	40
13	13	Coccus, Strepto	27	38986	1	1	21	2	39
14	14	Cummings, Terry	28	42387	1	2	19	1	34
15	15	Cyborg, B. A.	49	50211	1	1	14	2	55
16	16	DeCock, Gustav	21	42754	1	1	17	1	28
17	17	Ebby, Anson	35	36189	2	1	19	2	31
18	18	Fleetwood, Hugh	27	39989	2	2	10	3	29
19	19	Freud, Seigfried	40	47429	1	1	15	3	44
20	20	Gibbs, John	23	41229	1	2	14	2	39
21	21	Giblett, Turk	40	52993	2	2	25	1	40
22	22	Giddings, Verne	50	44219	1	1	24	1	41
23	23	Graybill, Fred	44	50404	1	2	16	1	45
24	24	Hillman, Art	35	45130	1	1	18	1	35
25	25	Hoquiam, Pat	27	40998	2	2	16	3	26
26	26	Jackson, Andy	45	55599	2	2	28	1	44
27	27	Kemp, Jack	25	39306	1	2	14	2	27
28	28	Kemp, Sean	37	42456	1	2	9	2	41
29	29	Klein, Ron	55	44506	1	1	13	2	48
30	30	Knudson, Cal	59	55273	2	2	24	3	55
31	31	Lincoln, Abe	30	30449	2	2	17	1	34
32	32	Marx, Groucho	31	34460	1	1	21	2	40
33	33	Marx, Harpo	47	50842	1	1	17	3	45
34	34	McCallum, Sandy	21	40833	1	2	6	3	27
35	35	Mendelsohn, Phil	41	49239	2	1	19	2	44

	id	name	age	income	denom	member	involve	form	attitude
36	36	Messer, Schmitt	29	37773	1	1	19	2	38
37	37	Nixon, Dick	39	44499	1	2	15	2	39
38	38	Nixon, Pat	46	42569	2	1	22	1	43
39	39	Porter, Seth	29	35234	2	2	13	3	32
40	40	Reynolds, Steve	22	29942	2	2	13	2	38
41	41	Schmeltz, Art	56	48573	2	2	17	1	46
42	42	Schwartz, Maury	45	45457	1	2	27	1	44
43	43	Simpson, Carl	39	46321	2	1	31	3	37
44	44	Sledge, Tom	48	52841	2	1	18	2	42
45	45	Smith, Bill	50	55169	1	2	18	1	42
46	46	Squim, Slim	53	48912	2	2	19	2	54
47	47	Stalone, Sam	54	45798	2	1	17	2	45
48	48	Stewart, Earl	43	50993	2	1	26	3	49
49	49	Stewart, Gordie	27	45562	2	1	12	3	33
50	50	Suzcek, Chris	30	34690	2	2	14	3	31
51	51	Telford, Pete	26	43818	1	2	14	3	32
52	52	Thwaites, Cyrus	31	41882	1	2	25	3	46
53	53	Towson, Rick	53	49461	2	1	16	1	49
54	54	Truman, Horton	45	49709	2	2	14	1	47
55	55	Van Donnette, Sly	54	58682	2	2	31	2	52
56	56	Wagner, Gus	48	42179	1	1	14	1	48
57	57	Warner, Vincent	52	57821	1	1	14	1	48
58	58	Whattis, Nom	56	58318	1	1	15	3	48
59	59	Wilson, Pete	34	53835	1	2	18	2	42
60	60	Yeltson, Boris	57	55616	2	1	22	1	45

Solutions to Odd-Numbered Exercises

CHAPTER 1

1. A population is all of the people or objects to which we wish to apply our conclusions from research. A sample is a subgroup of the population from which we usually collect information to allow us to arrive at conclusions.

3. a. The population is all of the registered voters in the state.
   b. The sample is the 852 voters who responded to the exit poll.
   c. The parameter is 45% support of the initiative.
   d. The statistic is 51% support of the initiative.
   e. The sample was not representative of the population. Perhaps the poll was taken only in certain precincts that had a slight bias in favor of the initiative (location). Furthermore, the poll could have been taken at a particular time of the day that was more conducive in some manner to voters who were sympathetic to the initiative (time). Perhaps some of the respondents lied about their vote in order to respond to some demand characteristic of the pollsters. Perhaps the pollster just *happened* to pick a sample purely by chance in which the proportion was 51%. There may be other possible reasons.

5. The first and most basic requirement is that each item of information must be accurate. Accuracy can be achieved by utilizing sources of information that are verifiable and knowable. In addition, accuracy involves a critical level of inference that assures the information that was obtained represents the characteristic that we wish to know about.

   The second requirement is that the information is representative. If we wish to arrive at conclusions that extend beyond the information that we have collected, the information that we have must adequately represent the population. The problems with sampling procedures usually revolve around sampling of people (Does our sample contain all of the relevant traits of people that are in the population to which we wish to generalize?), sampling of content (Do our measures effectively sample all of the content related to the behavior or trait of interest?), and sampling of times and locations (Have we taken into consideration that responses might vary as a function of when and where we have measured the behavior or trait?).

7. One of the basic tenets of psychology is that people differ. This difference is usually a function of different biological traits and different experiences. Within any population, a group of people may be highly similar on behaviors and traits that are important to the behavior or trait of interest. If we obtain a sample of these highly similar people then the information we obtain may not apply to the entire population of people but merely to the population of people who share those common behaviors or traits. Thus, our ability to generalize is limited. If we want to generalize to a larger group then our sample must be representative of that group.

   Usually in psychological and educational research we cannot sample all of the content related to the behavior or trait of interest. A good information-gathering procedure will reveal all aspects of the trait; however, we can seldom cover the topic of interest exhaustively. Thus, we must rely on an information-gathering procedure that contains a representative sample of the content related to the behavior or trait of interest.

Many of the behaviors of interest to psychologists and educators vary as a function of the time of day (e.g., alertness, hunger, physical and mental exhaustion). Since individuals vary throughout the day on the salience of each of these factors, it is important that researchers make observations at various times during the day so that the behavior or trait being observed or measured is representative of what would happen in the population.

Since people with certain traits are more likely to be at particular places, the location where we measure the behavior or trait may impact scores. For example, asking questions about economic issues might yield different responses if the questions were asked near an exclusive restaurant versus near a fast-food restaurant.

9. It is important to have measurement scales that are reliable so that the results of research studies can be replicated (reproduced) in the future. If results of research studies can be replicated, it strengthens the conclusions of those research studies.

   Measurement scales are valid if they measure the construct we believe they were intended to measure. If we use measurement scales that are not valid, our research conclusions will be erroneous.

11. The four types of scales are nominal, ordinal, interval, and ratio. Nominal scales are scales that use numbers only to indicate membership in a group or category. Ordinal scales are scales for which numbers indicate relative amount or ordering on some characteristic. Interval scales are scales that use numbers to indicate differential performance such that the difference between two consecutive integers has the same meaning anywhere on the scale and has an arbitrary zero point. Ratio scales are scales that use numbers to indicate differential performance such that the difference between two consecutive integers has the same meaning anywhere on the scale but the scale has an absolute zero point (i.e., the zero point means the absence of the measured trait).

   Answers will vary on the examples of scales. Some examples of nominal scales are gender, ethnicity, socioeconomic status, religious preference, marital status, etc. Some examples of ordinal scales are class rank, place finish in an art competition, career preferences, travel preferences, etc. Some examples of interval scales are intelligence measures, anxiety, stress, depression, reading comprehension, etc. Some examples of ratio scales are task completion time, number of correct answers on a test, age, number of hours of sleep, etc.

13. A dependent variable is a measure of behavior that is influenced by the independent variable. A manipulated independent variable is a variable that is actively controlled or assigned by the researcher. A status variable is a variable for which different categories of the participants already exist. A control variable is a variable that is held constant by the researcher so that it does not affect the participants' measured behavior in unequal ways. Examples will vary. The following is a sample response. Researchers were interested in measuring the effect of two methods of teaching reading (manipulated independent variable) on boys and girls (status variable of gender) on reading comprehension (dependent variable) after the participants were given 45 hours of reading instruction (control variable) at 9 A.M. each school day (control variable).

15. Answers will vary. Sample answers are given below. The variables that should be measured in this study are biological sex or gender, method of teaching algebra, algebra achievement test scores, and algebra attitude scores. For gender, students could merely list their gender on a demographic sheet that is attached to the other scales. For method of teaching algebra, the algebra classrooms from Gatekeeper High School and Lockedgate High School could be utilized since teachers at Gatekeeper were already trained to teach algebra using the metacognitive approach and teachers at Lockedgate were not. For the algebra achievement tests scores, a commercially available test should be selected so that all students take the same test. For measuring students' attitudes toward algebra, one of several math attitude scales could be used. Gender and teaching method are both nominal scale variables. Attitude toward algebra and achievement in algebra are both interval scale variables.

There are several questions related to this problem. First, is there a difference in change of attitude toward algebra as a function of the gender of the student? To answer this question, the gender of the students could be recorded when they complete a survey of their attitude toward math at the beginning of the school year. Students would also complete the same survey at the end of the school year, and the two attitude scores would then be paired. Second, is there a difference in change in algebra achievement as a function of gender? To answer this question, the gender of the students could be recorded when they complete an algebra achievement test at the beginning of the school year. They would also complete the algebra achievement test at the end of the school year and the two scores could be paired. Third, is there a difference in the change in attitude toward algebra as a function of the method of teaching algebra? At the beginning of the school year, students would be given an attitude toward math survey. Students would also be required to indicate what school they attended since the method of teaching algebra is indicated by the school the students attend. Students would then complete the math attitude survey at the end of the school year and the two scores would be paired. Fourth, is there a difference in the change in algebra achievement as a function of the method of teaching algebra? Students would complete the algebra achievement test at the beginning of the year. The school they attended would also be recorded since the teaching method is a function of the school the students attended. Students would also complete the algebra achievement test at the end of the school year and scores would be paired. Fifth, is there a difference in attitude toward algebra as a function of both the student's gender and the method of

teaching algebra? Again students would complete the attitude toward math scale at the beginning and the end of the school year. In addition, the students would be asked to indicate their gender and the school they attended since this is indicative of the algebra teaching approach. Finally, is there a difference in algebra achievement as a function of both the gender of the students and the method of teaching algebra? Students would take the algebra achievement test at the beginning and the end of the school year. In addition, the students would be asked to indicate their gender and the school they attended since this is indicative of the algebra teaching approach.

17. Answers will vary. The following is a sample answer. *Note:* This sample response does not include all of the possible responses that could be provided. The answer is abbreviated due to space considerations. Some of the variables that should be included in this study are culture (nominal scale); gender (nominal scale); individualistic and collectivistic self-esteem (interval scale); achievement motivation (interval scale); and independent and interdependent self-construal (interval scale). For both culture and gender, participants could be asked to indicate on a checklist their culture and their gender. For individualistic self-esteem, collectivistic self-esteem, independent self-construal, interdependent self-construal, and achievement motivation, existing scales in the professional literature could be used. The following represent some, but not all, of the questions that could be addressed. Is there a difference in individualistic self-esteem, collectivistic self-esteem, independent self-construal, interdependent self-construal, and achievement motivation as a function of both culture and gender? Is there a difference in individualistic self-esteem, collectivistic self-esteem, independent self-construal, interdependent self-construal, and achievement motivation as a function of gender? Is there a difference in individualistic self-esteem, collectivistic self-esteem, independent self-construal, interdependent self-construal, and achievement motivation as a function of culture? These questions could be answered by recruiting participants from at least two different cultures that differed on the basis of individualistic vs. collectivistic orientations (e.g., the United States and Japan). The culture and the gender of the participants would be recorded, and participants would be asked to complete a survey that contained measures of individualistic self-esteem, collectivistic self-esteem, independent self-construal, interdependent self-construal, and achievement motivation. This survey would be presented in the primary language of each culture.

CHAPTER 2

1. The raw frequency distribution is a listing of all possible scores between and including the lowest and highest scores. These scores are arranged from highest to lowest or from lowest to highest. The number of times each score occurs is then listed. In using this method only the identification of a particular score with an individual is lost.

The grouped frequency distribution is a listing of in-

tervals of scores. The frequency of scores within each interval is recorded. The midpoint of the interval is then used to represent all scores in this interval. Thus, there is a loss in precision of the score as well as identification of a particular score with an individual.

In a cumulative frequency distribution, all possible scores between and including the highest and lowest scores are listed in either increasing or decreasing order.

Beginning with the lowest score, the number of times that the score or a lower score appears in the distribution is recorded.

In a relative frequency distribution, all possible scores between and including the highest and lowest scores are listed in either increasing or decreasing order. Beginning with the lowest score, the proportion of total number of scores that are that score or lower is recorded.

3. a.

X	Tally	f	cf
20	//	2	19
19	///	3	17
18	////	4	14
17		0	10
16	///	3	10
15	//	2	7
14	////	4	5
13	/	1	1

b.

X	Tally	f	cf
49	/	1	62
48		0	61
47	/	1	61
46		0	60
45	/	1	60
44		0	59
43		0	59
42		0	59
41	//	2	59
40		0	57
39		0	57
38	/	1	57
37		0	56
36	/	1	56
35	/	1	55
34	//	2	54
33	///	3	52
32		0	49
31	///	3	49
30		0	46
29	///	3	46
28	////	4	43
27	/////	5	39
26		0	34
25	///	3	34
24	//	2	31
23	///	3	29
22	/////	5	26
21	/	1	21
20	//	2	20
19	//	2	18
18	///	3	16
17	/	1	13
16	///	3	12
15		0	9
14	/	1	9
13	//	2	8
12	//	2	6
11	//	2	4
10	//	2	2

c.

X	Tally	f	cf
32	/	1	10
31	/	1	9
30	//	2	8
29		0	6
28		0	6
27		0	6
26		0	6
25		0	6
24	/	1	6
23	/	1	5
22	/	1	4
21		0	3
20		0	3
19	/	1	3
18	/	1	2
17		0	1
16		0	1
15	/	1	1

5. a. The difference in the highest (63) and lowest (23) score is 40. Since we want between 10 and 20 intervals, if we divide 40 by 10 we would get an interval width of 4, and if we divide 40 by 20 we would get an interval width of 2. However, we want the interval width to be odd if possible. The only odd number between 2 and 4 is 3. Thus, we should select an interval width of 3.

b. The difference in the highest (96) and lowest (24) score is 72. Since we want an odd interval width, we should divide 72 by 3 which yields approximately 24 intervals, 72 by 5 which yields approximately 15 intervals, and 72 by 7 which yields approximately 10 intervals. Although interval widths of 5 and 7 will conform to the guidelines of having 10 to 20 intervals, the best selection is an interval width of 5.

c. The difference in the highest (115) and lowest (68) score is 47. If we want between 10 and 20 intervals, we should divide 47 by 10 and by 20. This results in interval widths of 5 and 3, respectively. When a choice exists, it is best to use the smaller interval width since it results in less loss of information. Thus the best choice for an interval width is 3.

d. The difference in the highest (48) and lowest (2) score is 46. If we want interval widths that are odd numbers, we could divide the difference by odd numbers beginning with 3 until the quotient is smaller than 10. Thus, 46 divided by 3 would yield approximately 16 intervals, and 46 divided by 5 would yield approximately 10 intervals. Although both of these are acceptable interval widths, the best interval width is 3.

7. For the grouped frequency distribution table, the grouped cumulative frequency distribution table, and the grouped relative frequency distribution table, I will use an interval width of 3, and the lowest score in each interval will be divisible by the interval width 3. (*Note:* Your answers will vary if you elect to have the midpoint or the highest score in each interval divisible by the interval width.)

Grouped Frequency Distribution Table

Score Interval	Midpoint X	Frequency f
78–80	79	2
75–77	76	7
72–74	73	3
69–71	70	17
66–68	67	13
63–65	64	9
60–62	61	9
57–59	58	4
54–56	55	5
51–53	52	2

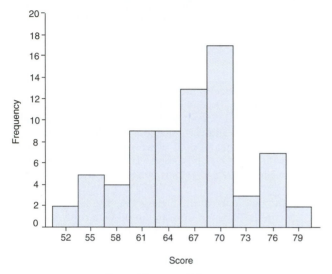

Grouped frequency histogram

Grouped Frequency Table for the Data Set from Exercise 3(b)

Score Interval	Midpoint X	Frequency f
48–50	49	1
45–47	46	2
42–44	43	0
39–41	40	2
36–38	37	2
33–35	34	6
30–32	31	3
27–29	28	12
24–26	25	5
21–23	22	9
18–20	19	7
15–17	16	4
12–14	13	5
9–11	10	4

11. a. Stem-and-Leaf Display

4	|	33
5	|	14455788
6	|	445799
7	|	12235678899
8	|	47
9	|	0177

Grouped Frequency Table for Interval Width = 3

Score Interval	Midpoint (X)	Frequency (f)
96–98	97	2
93–95	94	0
90–92	91	2
87–89	88	1
84–86	85	1
81–83	82	0
78–80	79	4
75–77	76	3
72–74	73	3
69–71	70	3
66–68	67	1
63–65	64	3
60–62	61	0
57–59	58	3
54–56	55	4
51–53	52	1
48–50	49	0
45–47	46	0
42–44	43	2

Grouped Frequency Table for Interval Width = 5

Score Interval	Midpoint (X)	Frequency (f)
95–99	97	2
90–94	92	2
85–89	87	1
80–84	82	1
75–79	77	7
70–74	72	4
65–69	67	4
60–64	62	2
55–59	57	5
50–54	52	3
45–49	47	0
40–44	42	2

9. The set of data from Exercise 3(a) could not be represented by a grouped frequency distribution since the difference in the highest score (20) and the lowest score (13) is only 7. Thus, with an interval width of 1 there are only 8 intervals. An interval width of greater than 1 would result in fewer than the required minimum number of intervals (10). Furthermore, there are only 19 scores in the set of data. Thus, there doesn't appear to be any gain in efficiency by grouping the data.

The set of data from Exercise 3(b) could be represented by a grouped frequency distribution since the difference in the highest score (49) and the lowest score (10) is 39. An interval width of 3 would yield approximately 14 intervals that would be within the acceptable guidelines. Furthermore, there are 62 scores which are not very manageable in terms of our ability to discern meaningful patterns in the data.

The set of data for Exercise 3(c) would not be appropriately represented by a grouped frequency distribution since the difference in the highest score (31) and the lowest score (15) is only 16 and this falls within the recommended number of intervals (10–20). Furthermore, it is relatively easy to manage the 10 scores in this set of data without summarization procedures.

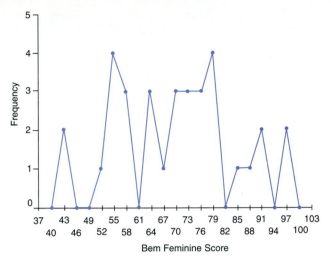

Grouped frequency polygon for interval width = 3

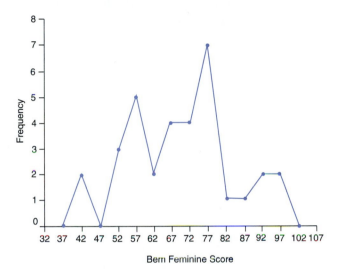

Grouped frequency polygon for interval width = 5

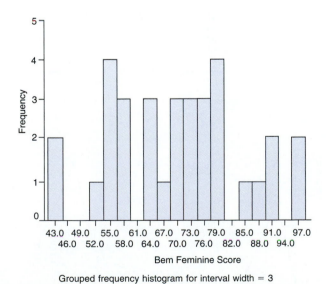

Grouped frequency histogram for interval width = 3

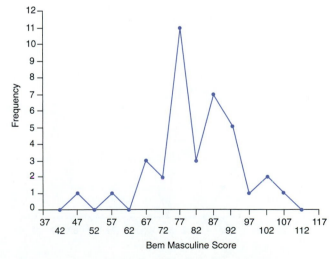

Grouped frequency histogram for interval width = 5

b. Stem-and-Leaf Display

```
 4  |  9
 5  |  7
 6  |  557
 7  |  1456667788999
 8  |  0006677889
 9  |  111226
10  |  235
```

Frequency Distribution Table for Interval Width = 5

Score Interval	Midpoint (X)	Frequency (f)
105–109	107	1
100–104	102	2
95–99	97	1
90–94	92	5
85–89	87	7
80–84	82	3
75–79	77	11
70–74	72	2
65–69	67	3
60–64	62	0
55–59	57	1
50–54	52	0
45–49	47	1

Grouped frequency polygon for interval width = 5

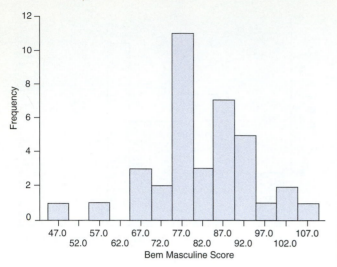

Grouped histogram for interval width = 5

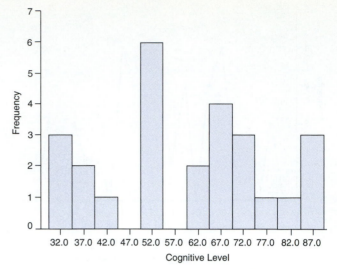

Grouped histogram for interval width = 5

c. Stem-and-Leaf Display

```
3  |  12378
4  |  2
5  |  023444
6  |  225678
7  |  2448
8  |  1556
```

d. Stem-and-Leaf Display

```
1  |  237
2  |
3  |  789
4  |  0559
5  |  2399
6  |  0135
7  |  0556
8  |  0689
9  |  349
```

Grouped Frequency Table for Interval Width = 5

Score Interval	Midpoint (X)	Frequency (f)
85–89	87	3
80–84	82	1
75–79	77	1
70–74	72	3
65–69	67	4
60–64	62	2
55–59	57	0
50–54	52	6
45–49	47	0
40–44	42	1
35–39	37	2
30–34	32	3

Grouped Frequency Table for Interval Width = 7

Score Interval	Midpoint (X)	Frequency (f)
98–104	101	1
91–97	94	2
84–90	87	3
77–83	80	1
70–76	73	4
63–69	66	2
56–62	59	4
49–55	52	3
42–48	45	2
35–41	38	4
28–34	31	0
21–27	24	0
14–20	17	1
7–13	10	2

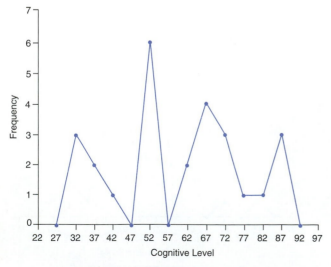

Grouped frequency polygon for interval width = 5

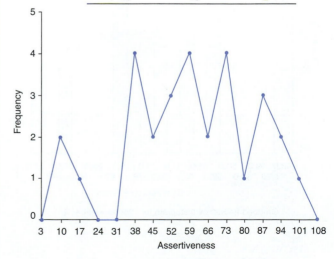

Grouped frequency polygon for interval width = 7

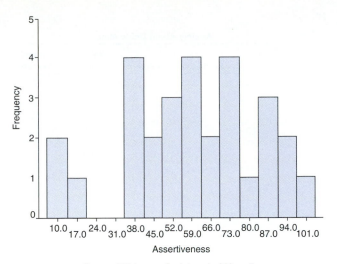

Grouped histogram for interval width = 7

13.

X	Tally	Frequency (f)
1	/	1
2	//	2
3	///// /	6
4	///// /	6
5	//	2
6	///// /////	10
7	///	3
8	///// ///// ///// //	17
9	///	3

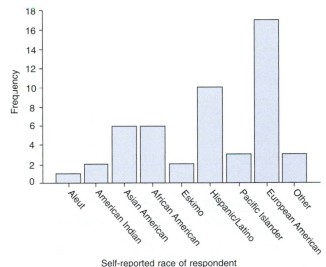

Self-reported race of respondent

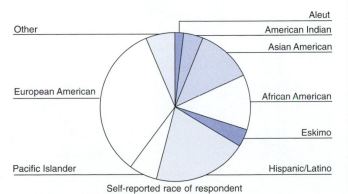

Self-reported race of respondent

From the graphs, the largest self-reported racial group in this sample was White/European American followed by Hispanic/Latino. There were approximately 1.5 times more participants who reported their racial group as White/European American than Hispanic/Latino. The next most frequently reported racial group was tied between Black/African American and Asian American. Compared with the population of the United States, members of White/European Americans were under-represented whereas the members of the other racial groups were over-represented in this sample.

15. a.

ITED 9th Grade Math Score

		Frequency	Percent	Valid Percent	Cumulative Percent
Valid	45	1	2.0	2.0	2.0
	52	1	2.0	2.0	4.1
	58	2	4.1	4.1	8.2
	59	1	2.0	2.0	10.2
	60	3	6.1	6.1	16.3
	61	2	4.1	4.1	20.4
	63	1	2.0	2.0	22.4
	65	2	4.1	4.1	26.5
	66	1	2.0	2.0	28.6
	67	1	2.0	2.0	30.6
	69	1	2.0	2.0	32.7
	70	1	2.0	2.0	34.7
	71	1	2.0	2.0	36.7
	72	1	2.0	2.0	38.8
	73	2	4.1	4.1	42.9
	75	1	2.0	2.0	44.9
	76	1	2.0	2.0	46.9
	78	1	2.0	2.0	49.0
	79	3	6.1	6.1	55.1
	80	2	4.1	4.1	59.2
	81	2	4.1	4.1	63.3
	83	2	4.1	4.1	67.3
	84	1	2.0	2.0	69.4
	85	3	6.1	6.1	75.5
	86	2	4.1	4.1	79.6
	88	1	2.0	2.0	81.6
	89	2	4.1	4.1	85.7
	90	1	2.0	2.0	87.8
	91	2	4.1	4.1	91.8
	95	1	2.0	2.0	93.9
	96	2	4.1	4.1	98.0
	97	1	2.0	2.0	100.0
	Total	49	100.0	100.0	

b.

Score Interval	Cumulative Midpoint X	Frequency f	Cumulative Frequency cf
96–98	97	3	49
93–95	94	1	46
90–92	91	3	45
87–89	88	3	42
84–86	85	6	39
81–83	82	4	33
78–80	79	6	29
75–77	76	2	23
72–74	73	3	21
69–71	70	3	18
66–68	67	2	15
63–65	64	3	13
60–62	61	5	10
57–59	58	3	5
54–56	55	0	2
51–53	52	1	2
48–50	49	0	1
45–47	46	1	1

Solutions to Odd-Numbered Exercises   479

c.

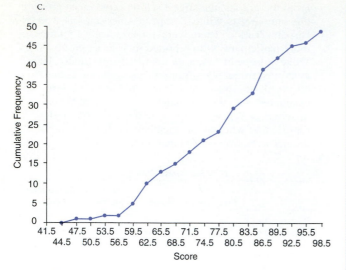

Grouped cumulative frequency curve for interval width = 3

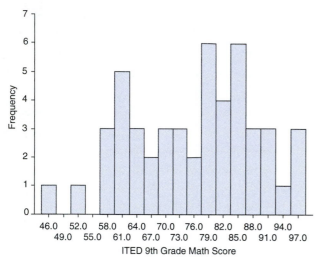

Grouped frequency histogram for interval width = 3

Anxiety About Doing Math

		Frequency	Percent	Valid Percent	Cumulative Percent
Valid	14	1	.5	.5	.5
	17	1	.5	.5	1.0
	18	2	1.0	1.0	2.0
	19	1	.5	.5	2.5
	20	1	.5	.5	3.0
	21	3	1.5	1.5	4.5
	22	2	1.0	1.0	5.5
	23	1	.5	.5	6.0
	25	1	.5	.5	6.5
	26	1	.5	.5	7.0
	27	3	1.5	1.5	8.5
	28	2	1.0	1.0	9.5
	29	5	2.5	2.5	12.0
	30	5	2.5	2.5	14.5
	31	3	1.5	1.5	16.0
	32	5	2.5	2.5	18.5
	33	7	3.5	3.5	22.0
	34	3	1.5	1.5	23.5
	35	6	3.0	3.0	26.5
	36	5	2.5	2.5	29.0
	37	6	3.0	3.0	32.0
	38	12	6.0	6.0	38.0
	39	10	5.0	5.0	43.0
	40	3	1.5	1.5	44.5
	41	4	2.0	2.0	46.5
	42	10	5.0	5.0	51.5
	43	4	2.0	2.0	53.5
	45	9	4.5	4.5	58.0
	46	11	5.5	5.5	63.5
	47	4	2.0	2.0	65.5
	48	10	5.0	5.0	70.5
	49	9	4.5	4.5	75.0
	50	7	3.5	3.5	78.5
	51	6	3.0	3.0	81.5
	52	5	2.5	2.5	84.0
	53	5	2.5	2.5	86.5
	54	4	2.0	2.0	88.5
	55	2	1.0	1.0	89.5
	56	6	3.0	3.0	92.5
	57	1	.5	.5	93.0
	58	3	1.5	1.5	94.5
	59	3	1.5	1.5	96.0
	60	8	4.0	4.0	100.0
	Total	200	100.0	100.0	

b.

Grouped Frequency Table, Interval Width = 3

Score Interval	Midpoint X	Frequency f	Cumulative Frequency cf
60–62	61	8	200
57–59	58	7	192
54–56	55	12	185
51–53	52	16	173
48–50	49	26	157
45–47	46	24	131
42–44	43	14	107
39–41	40	17	93
36–38	37	23	76
33–35	34	16	53
30–32	31	13	37
27–29	28	10	24
24–26	25	2	14
21–23	22	6	12
18–20	19	4	6
15–17	16	1	2
12–14	13	1	1

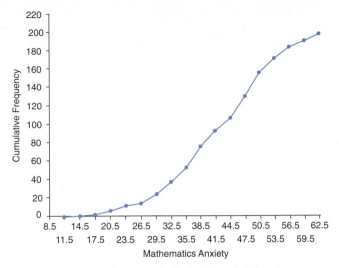

Cumulative frequency distribution, interval width = 3

c.

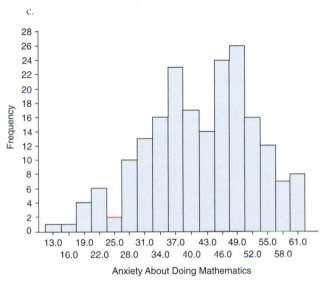

d.

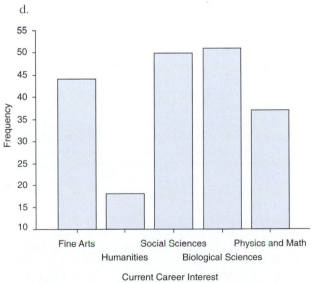

Current Career Interest

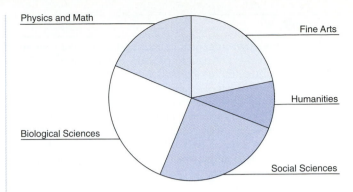

e.

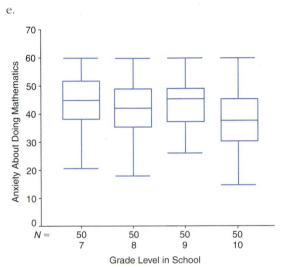

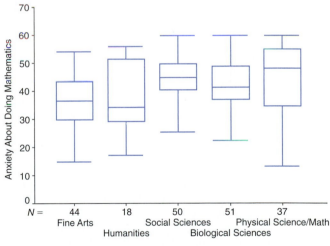

Current Career Interest

1. Researchers use numerical indices to describe the central tendency of a set of data to provide an efficient way of communicating the properties of the data. Although the complete set of data has a richness in terms of description, it would be a time-consuming task to try to discern patterns in the data or to compare sets of data. A measure of center allows a researcher to describe where a set of data is located on a scale by providing a single number (or sometimes a few numbers). This number is the most representative value of the set of data. By using the most representative value of each set of data, researchers can also compare sets of data more efficiently. The problem with this process is that the description does not include how spread out the scores in the set of data must be. It is possible for two sets of data to have the same measures of central tendency but to be very different in terms of the degree to which the numbers are spread out in the set of data. Thus, not only should measures of central tendency be used but also measures of spread or variability.

3. The mode is the most frequently occurring score or scores in a set of data. The mode is the only measure of central tendency that can also be used to characterize data that are not quantitative. In looking at a frequency distribution, it is very easy to find the mode. No computations are involved. It is also representative of the set of data in that it is the value that occurs more frequently than any other value. However, representativeness may be problematic since only the observations that are at the mode actually determine its value. A second problem is that there is not always a unique value for the mode. Several observations could occur with equal frequencies. In the extreme case of observations all occurring with equal frequency, it is possible that no mode exists at all. In addition, the mode may not be stable from group to group. Small variations in scores could lead to dramatic changes in the mode. In addition, because the mode itself is not derived mathematically from the scores themselves, it cannot be used in any further analysis that requires mathematical operations. Thus, information that is lost by summarizing the set of data by using the mode cannot be recovered in further analysis.

5. The mean is the score that results from summing all of the scores in the set of data and dividing by the number of scores. It represents the average of the scores and is most appropriately labeled the arithmetic mean. The mean is computed in such a way that it yields a single, unique value for a given set of data. Furthermore, every score in the set of data is used in computing the mean. Thus, in this sense it is more representative of all of the data than the median or the mode. Given the way it is computed, the mean is also very stable from group to group. Finally, since the mean is computed by using every score in the set of data, most of the techniques for analyzing the data further can be utilized. This allows the researcher to capture some information regarding the set of data that may be lost as a result of computing a value to summarize the data. The biggest disadvantage of the mean is that extreme scores called outliers can affect the value of the mean. This is a result of the fact

that every score is equally weighted in computing the mean.

7. a. *Bem sex role inventory feminine scores*. This distribution is multimodal. The frequency of occurrence of the scores is either one time or two times. Thus, the modes are the scores that occur twice: 43, 54, 55, 58, 64, 69, 72, 75, 78, 79, and 97.

   If the 33 scores are arranged in descending order (97, 97, 91, 90, 87, 84, 79, 79, 78, 78, 76, 75, 75, 73, 72, 72, 71, 69, 69, 67, 65, 64, 64, 58, 58, 57, 55, 55, 54, 54, 51, 43, 43), the middle score, the median, is the 17th score = 71. If the scores are placed into a cumulative frequency table, the following table results:

Score	Score Interval	f	cf
97	96.5–97.5	2	33
91	90.5–91.5	1	31
90	89.5–90.5	1	30
87	86.5–87.5	1	29
84	83.5–84.5	1	28
79	78.5–79.5	2	27
78	77.5–78.5	2	25
76	75.5–76.5	1	23
75	74.5–75.5	2	22
73	72.5–73.5	1	20
72	71.5–72.5	2	19
71	70.5–71.5	1	17
69	68.5–69.5	2	16
67	66.5–67.5	1	14
65	64.5–65.5	1	13
64	63.5–64.5	2	12
58	57.5–58.5	2	10
57	56.5–57.5	1	8
55	54.5–55.5	2	7
54	53.5–54.5	2	5
51	50.5–51.5	1	3
43	42.5–43.5	2	2

Use $Mdn = LL_j + \{[.5(N) - cf_{j-1}]/f_j\}$ to calculate the median. First calculate $.5(N) = .5(33) = 16.5$. The first score where the value in the cf is equal to or greater than 16.5 is the score that contains the median. This score is $j = 71$ from the table. The lower limit of this score is $LL_j = 70.5$. From the table $cf_{j-1} = 16$ is the value of the cumulative frequency of the interval that is immediately below the interval that contains the median, and $f_j = 1$ is the frequency of the scores in the interval that contains the median. Thus, $Mdn = 70.5 + \{[.5(33) - 16]/1\} = 70.5 + 0.5 = 71$.

For the mean, $\overline{X} = \Sigma f_j X_j / \Sigma f_j = [2(43) + 1(51) + 2(54) + 2(55) + 1(57) + 2(58) + 2(64) + 1(65) + 1(67) + 2(69) + 1(71) + 2(72) + 1(73) + 2(75) + 1(76) + 2(78) + 2(79) + 1(84) + 1(87) + 1(90) + 1(91) + 2(97)]/(2 + 1 + 2 + 2 + 1 + 2 + 2 + 1 + 1 + 2 + 1 + 2 + 1 + 2 + 1 + 2 + 1 + 2 + 2 + 1 + 1 + 2) = 2302/33 = 69.76$. The grouped frequency distributions from Chapter 2, Exercise 11(a) are as follows:

Interval Width = 3				Interval Width = 5			
Interval	*Midpt*	*f*	*cf*	*Interval*	*Midpt*	*f*	*cf*
96–98	97	2	33	95–99	97	2	33
93–95	94	0	31	90–94	92	2	31
90–92	91	2	31	85–89	87	1	29
87–89	88	1	29	80–84	82	1	28
84–86	85	1	28	75–79	77	7	27
81–83	82	0	27	70–74	72	4	20
78–80	79	4	27	65–69	67	4	16
75–77	76	3	23	60–64	62	2	12
72–74	73	3	20	55–59	57	5	10
69–71	70	3	17	50–54	52	3	5
66–68	67	1	14	45–49	47	0	2
63–65	64	3	13	40–44	42	2	2
60–62	61	0	10				
57–59	58	3	10				
54–56	55	4	7				
51–53	52	1	3				
48–50	49	0	2				
45–47	46	0	2				
42–44	43	2	2				

The mode would be determined by the midpoint of the most frequently occurring interval(s). Thus for an interval width of 3 the modes are 55 and 79, and for an interval width of 5 the mode is 77.

Use the formula $Mdn = LL_j + \{[.5(N) - cf_{j-1}]/f_j\}$ (J). For interval width 3 calculate $.5(N) = .5(33) = 16.5$. The interval where the value in the cf is equal to or greater than 16.5 is the interval that contains the median. This interval is 69–71 from the table. The lower limit of this interval is $LL_j = 68.5$. From the table $cf_{j-1} = 14$ is the value of the cumulative frequency of the interval that is immediately below the interval that contains the median, and $f_j = 3$ is the frequency of the scores in the interval that contains the median. Finally J is the width of the interval = 3. Thus, $Md = 68.5 + \{[.5(33) - 14]/3\}(3) = 71$. For interval width 5 calculate $.5(N) = .5(33) = 16.5$. The interval where the value in the cf is equal to or greater than 16.5 is the interval that contains the median. This interval is 70–74 from the table. The lower limit of this interval is $LL_j = 69.5$. From the table $cj_{j-1} = 16$ is the value of the cumulative frequency of the interval that is immediately below the interval that contains the median, and $f_j = 4$ is the frequency of the scores in the interval that contains the median. Finally J is the width of the interval = 5. Thus, $Mdn = 69.5 + \{[.5(33) - 16]/4\}(5) = 70.13$.

To calculate the mean using the cumulative frequency distribution table with interval width 3, use the formula $\bar{X} = \Sigma f_j X_j / \Sigma f_j = [2(43) + 0(46) + 0(49) + 1(52) + 4(55) + 3(58) + 0(61) + 3(64) + 1(67) + 3(70) + 3(73) + 3(76) + 4(79) + 0(82) + 1(85) + 1(88) + 2(91) + 0(94) + 2(97)]/(2 + 0 + 0 + 1 + 4 + 3 + 0 + 3 + 1 + 3 + 3 + 3 + 4 + 0 + 1 + 1 + 2 + 0 + 2) = 2313/33 = 70.09$. The mean for the scores with interval width 5 is $\bar{X} = \Sigma f_j X_j / \Sigma f_j = [2(42) + 0(47) + 3(52) + 5(57) + 2(62) + 4(67) + 4(72) + 7(77) + 1(82) + 1(87) + 2(92) + 2(97)]/(2 + 0 + 3 + 5 + 2 + 4 + 4 + 7 + 1 + 1 + 2 + 2) = 2291/33 = 69.42$.

b. *Bem sex role inventory masculine scores*. This distribution is multimodal. The frequency of occurrence of the scores is either one time, two times, or three times. Thus, the modes are the scores that occur three times: 76, 79, 80, and 91.

If the 37 scores are arranged in descending order (105, 103, 102, 96, 92, 92, 91, 91, 91, 89, 88, 88, 87, 87, 86, 86, 80, 80, 80, 79, 79, 79, 78, 78, 77, 77, 76, 76, 76, 75, 74, 71, 67, 65, 65, 57, 49), the middle score is the 19th = 80. If the scores are placed into a cumulative frequency table, the following table results:

Score	Score Interval	*f*	*cf*
105	104.5–105.5	1	37
103	102.5–103.5	1	36
102	101.5–102.5	1	35
96	95.5–96.5	1	34
92	91.5–92.5	2	33
91	90.5–91.5	3	31
89	88.5–89.5	1	28
88	87.5–88.5	2	27
87	86.5–87.5	2	25
86	85.5–86.5	2	23
80	79.5–80.5	3	21
79	78.5–79.5	3	18
78	77.5–78.5	2	15
77	76.5–77.5	2	13
76	75.5–76.5	3	11
75	74.5–75.5	1	8
74	73.5–74.5	1	7
71	70.5–71.5	1	6
67	66.5–67.5	1	5
65	64.5–65.5	2	4
57	56.5–57.5	1	2
49	48.5–49.5	1	1

Use $Mdn = LL_j + \{[.5(N) - cf_{j-1}]/f_j\}$ to calculate the median. First calculate $.5(N) = .5(37) = 18.5$. The first score where the value in the cf is equal to or greater than 18.5 is the score that contains the median. This score is $j = 80$ from the table. The lower limit of this score is $LL_j = 79.5$. From the table $cf_{j-1} = 18$ is the value of the cumulative frequency of the interval that is immediately below the interval that contains the median, and $f_j = 3$ is the frequency of the scores in the interval that contains the median. Thus, $Mdn = 79.5 + \{[.5(37) - 18]/3\} = 79.5 + 0.17 = 79.67$.

For the mean, $\bar{X} = \Sigma f_j X_j / \Sigma f_j = [1(49) + 1(57) + 2(65) + 1(67) + 1(71) + 1(74) + 1(75) + 3(76) + 2(77) + 2(78) + 3(79) + 3(80) + 2(86) + 2(87) + 2(88) + 1(89) + 3(91) + 2(92) + 1(96) + 1(102) + 1(103) + 1(105)]/(1 + 1 + 2 + 1 + 1 + 1 + 1 + 3 + 2 + 2 + 3 + 3 + 2 + 2 + 2 + 1 + 3 + 2 + 1 + 1 + 1 + 1) = 3012/37 = 81.41$. The grouped frequency distribution from Chapter 2, Exercise 11(b) is as follows:

Interval Width = 5			
Interval	**Midpt**	**f**	**cf**
105–109	107	1	37
100–104	102	2	36
95–99	97	1	34
90–94	92	5	33
85–89	87	7	28
80–84	82	3	21
75–79	77	11	18
70–74	72	2	7
65–69	67	3	5
60–64	62	0	2
55–59	57	1	2
50–54	52	0	1
45–49	47	1	1

The mode would be determined by the midpoint of the most frequently occurring interval(s). Thus the mode is 77.

Use the formula $Mdn = LL_j + \{[.5(N) - cf_{j-1}]/f_j\}$ (J). Calculate $.5(N) = .5(37) = 18.5$. The interval where the value in the cf is equal to or greater than 18.5 is the interval that contains the median. This interval is 80–84 from the table. The lower limit of this interval is $LL_j = 79.5$. From the table $cf_{j-1} = 18$ is the value of the cumulative frequency of the interval that is immediately below the interval that contains the median, and $f_j = 3$ is the frequency of the scores in the interval that contains the median. Finally J is the width of the interval = 5. Thus, $Mdn = 79.5 + \{[.5(37) - 18]/3\}(5) = 80.33$.

To calculate the mean using the cumulative frequency distribution table, use the formula $\overline{X} = \Sigma f_j X_j / \Sigma f_j = [1(47) + 0(52) + 1(57) + 0(62) + 3(67) + 2(72) + 11(77) + 3(82) + 7(87) + 5(92) + 1(97) + 2(102) + 1(107)]/(1 + 0 + 1 + 0 + 3 + 2 + 11 + 3 + 7 + 5 + 1 + 2 + 1) = 3019/37 = 81.59$.

c. *Cognitive level scores.* The mode is 54. It is the only score that occurs 3 times.

If the 26 scores are arranged in descending order (86, 85, 85, 81, 78, 74, 74, 72, 68, 67, 66, 65, 62, 62, 54, 54, 54, 53, 52, 50, 42, 38, 37, 33, 32, 31), the middle score is the average of the 13th and 14th scores = $(62 + 62)/2 = 62$. If the scores are placed into a cumulative frequency table, the following table results:

Score	Score Interval	f	cf
86	85.5–86.5	1	26
85	84.5–85.5	2	25
81	80.5–81.5	1	23
78	77.5–78.5	1	22
74	73.5–74.5	2	21
72	71.5–72.5	1	19
68	67.5–68.5	1	18
67	66.5–67.5	1	17
66	65.5–66.5	1	16
65	64.5–65.5	1	15
62	61.5–62.5	2	14
54	53.5–54.5	3	12
53	52.5–53.5	1	9
52	51.5–52.5	1	8
50	49.5–50.5	1	7
42	41.5–42.5	1	6
38	37.5–38.5	1	5
37	36.5–37.5	1	4
33	32.5–33.5	1	3
32	31.5–32.5	1	2
31	30.5–31.5	1	1

Use $Mdn = LL_j + \{[.5(N) - cf_{j-1}]/f\}$ to calculate the median. First calculate $.5(N) = .5(26) = 13$. The first score where the value in the cf is equal to or greater than 13 is the score that contains the median. This score is $j = 62$ from the table. The lower limit of this score is $LL_j = 61.5$. From the table cf_{j-1}

= 12 is the value of the cumulative frequency of the interval that is immediately below the interval that contains the median, and $f_j = 2$ is the frequency of the scores in the interval that contains the median. Thus, $Mdn = 61.5 + \{[.5(26) - 12]/2\} = 61.5 + 0.5 = 62$.

For the mean $\overline{X} = \Sigma f_j X_j / \Sigma f_j = [1(31) + 1(32) + 1(33) + 1(37) + 1(38) + 1(42) + 1(50) + 1(52) + 1(53) + 3(54) + 2(62) + 1(65) + 1(66) + 1(67) + 1(68) + 1(72) + 2(74) + 1(78) + 1(81) + 2(85) + 1(86)]/(1 + 1 + 1 + 1 + 1 + 1 + 1 + 1 + 1 + 3 + 2 + 1 + 1 + 1 + 1 + 1 + 2 + 1 + 1 + 2 + 1) = 1555/26 = 59.81$. The grouped frequency distribution from Chapter 2, Exercise 11(c) is

Interval Width = 5			
Interval	Midpt	f	cf
85–89	87	3	26
80–84	82	1	23
75–79	77	1	22
70–74	72	3	21
65–69	67	4	18
60–64	62	2	14
55–59	57	0	12
50–54	52	6	12
45–49	47	0	6
40–44	42	1	6
35–39	37	2	5
30–34	32	3	3

The mode would be determined by the midpoint of the most frequently occurring interval(s). Thus the mode is 52.

Use the formula $Mdn = LL_j + \{[.5(N) - cf_{j-1}]/f_j\}(J)$. Calculate $.5(N) = .5(26) = 13$. The interval where the value in the cf is equal to or greater than 13 is the interval that contains the median. This interval is 60–64 from the table. The lower limit of this interval is $LL_j = 59.5$. From the table $cf_{j-1} = 12$ is the value of the cumulative frequency of the interval that is immediately below the interval that contains the median, and $f_j = 2$ is the frequency of the scores in the interval that contains the median. Finally J is the width of the interval = 5. Thus, $Mdn = 59.5 + \{[.5(26) - 12]/2\}(5) = 62$.

To calculate the mean using the cumulative frequency distribution table, use the formula $\overline{X} = \Sigma f_j X_j / \Sigma f_j = [3(32) + 2(37) + 1(42) + 0(47) + 6(52) + 0(57) + 2(62) + 4(67) + 3(72) + 1(77) + 1(82) + 3(87)]/(3 + 2 + 1 + 0 + 6 + 0 + 2 + 4 + 3 + 1 + 1 + 3) = 1552/26 = 59.69$.

d. *Assertiveness scores.* Scores only occur 1 or 2 times in this data set. Three scores occur 2 times. These scores are the modes: 75, 59, 45.

If the 29 scores are arranged in descending order (99, 94, 93, 89, 88, 86, 80, 76, 75, 75, 70, 65, 63, 61, 60, 59, 59, 53, 52, 49, 45, 45, 40, 39, 38, 37, 17, 13, 12), the middle score is the 15th score = 60. If the scores are placed into a cumulative frequency table, the following table results:

Score	Score Interval	f	cf
99	98.5–99.5	1	29
94	93.5–94.5	1	28
93	92.5–93.5	1	27
89	88.5–89.5	1	26
88	87.5–88.5	1	25
86	85.5–86.5	1	24
80	79.5–80.5	1	23
76	75.5–76.5	1	22
75	74.5–75.5	2	21
70	69.5–70.5	1	19
65	64.5–65.5	1	18
63	62.5–63.5	1	17
61	60.5–61.5	1	16
60	59.5–60.5	1	15
59	58.5–59.5	2	14
53	52.5–53.5	1	12
52	51.5–52.5	1	11
49	48.5–49.5	1	10
45	44.5–45.5	2	9
40	39.5–40.5	1	7
39	38.5–39.5	1	6
38	37.5–38.5	1	5
37	36.5–37.5	1	4
17	16.5–17.5	1	3
13	12.5–13.5	1	2
12	11.5–12.5	1	1

Use $Mdn = LL_j + \{[.5(N) - cf_{j-1}]/f_j\}$ to calculate the median. First calculate $.5(N) = .5(29) = 14.5$. The first score where the value in the cf is equal to or greater than 14.5 is the score that contains the median. This score is $j = 60$ from the table. The lower limit of this score is $LL_j = 59.5$. From the table $cf_{j-1} = 14$ is the value of the cumulative frequency of the interval that is immediately below the interval that contains the median, and $f_j = 1$ is the frequency of the scores in the interval that contains the median. Thus, $Mdn = 59.5 + \{(.5(29) - 14]/1\} = 59.5 + 0.5 = 60$.

For the mean $\overline{X} = \Sigma f_j X_j / \Sigma f_j = [1(12) + 1(13) + 1(17) + 1(37) + 1(38) + 1(39) + 1(40) + 2(45) + 1(49) + 1(52) + 1(53) + 2(59) + 1(60) + 1(61) + 1(63) + 1(65) + 1(70) + 2(75) + 1(76) + 1(80) + 1(86) + 1(88) + 1(89) + 1(93) + 1(94) + 1(99)]/ (1 + 1 + 1 + 1 + 1 + 1 + 1 + 2 + 1 + 1 + 1 + 2 + 1 + 1 + 1 + 1 + 1 + 2 + 1 + 1 + 1 + 1 + 1 + 1 + 1 + 1) = 1719/29 = 59.72$. The grouped frequency distribution from Chapter 2, Exercise 11(d) is as follows:

Interval Width = 7

Interval	Midpt	f	cf
98–104	101	1	29
91–97	94	2	28
84–90	87	3	26
77–83	80	1	23
70–76	73	4	22
63–69	66	2	18
56–62	59	4	16
49–55	52	3	12
42–48	45	2	9
35–41	38	4	7
28–34	31	0	3
21–27	24	0	3
14–20	17	1	3
7–13	10	2	2

The mode would be determined by the midpoint of the most frequently occurring interval(s). Thus the modes are 38, 59, and 73 since the intervals that contain those values as midpoints all contain 4 scores.

Use the formula $Mdn = LL_j + \{[.5(N) - cf_{j-1}]/f_j\}(J)$. Calculate $.5(N) = .5(29) = 14.5$. The interval where the value in the cf is equal to or greater than 14.5 is the interval that contains the median. This interval is 56–62 from the table. The lower limit of this interval is $LL_j = 55.5$. From the table $cf_{j-1} = 12$ is the value of the cumulative frequency of the interval that is immediately below the interval that contains the median, and $f_j = 4$ is the frequency of the scores in the interval that contains the median. Finally J is the width of the interval = 7. Thus, $Mdn = 55.5 + \{[.5(29) - 12]/4\}(7) = 59.88$.

To calculate the mean using the cumulative frequency distribution, use the formula $\overline{X} = \Sigma f_j X_j / \Sigma f_j = [2(10) + 1(17) + 0(24) + 0(31) + 4(38) + 2(45) + 3(52) + 4(59) + 2(66) + 4(73) + 1(80) + 3(87) + 2(94) + 1(101)]/(2 + 1 + 0 + 0 + 4 + 2 + 3 + 4 + 2 + 4 + 1 + 3 + 2 + 1) = 1725/29 = 59.48$.

The values for the two different approaches to calculating measures of central tendency are different because summarizing information in a grouped format results in losing information about the individual data points. This loss in richness of the data can result in values from the grouped frequency table yielding different values compared to using the (ungrouped) frequency table.

9. A distribution of scores is symmetric when there is a line that can be drawn through the distribution such that every point on the distribution to the left of the line maps into one and only one point on the distribution to the right of the line. If the distribution is unimodal and symmetric, the mean, median, and mode are equal. If the distribution is multimodal and symmetric, the mean and the median are equal to each other and to the average of the modes.

11. A distribution of scores is negatively skewed when a majority of the scores lie in the upper end of the distribution. In a negatively skewed distribution, of the three measures of central tendency, the mean is the lowest, the median will be the next lowest, and the mode will be the highest score.

13. a. The anxiety scores are interval level scores. In addition, the distribution of scores is symmetric and unimodal. Thus, any measure of central tendency is appropriate here since they are all the same value.

b. The reinforcers are measures at the nominal level. The only measure of central tendency that is appropriate for nominal data is the mode.

c. The data in this case are rank-ordered. Thus, they are on an ordinal level of measurement. Since we want to declare an overall winner, the most appropriate statistic would involve some computation with the data at some level. Thus, the median is the most appropriate measure of central tendency.

   Note: The mean will actually work here since the average ranks are sufficient statistics for Thurstone scale values that transform ordinal scales to interval scales. If you are interested in learning more about this, we suggest that you enroll in a course in measurement.

d. The data in this case comprise a ratio level of measurement. However, there appear to be some extreme scores or outliers in the data. Thus, the median would be the most appropriate measure of central tendency.

e. The data here are really coded as "yes" and "no" responses. If the person showed an increase in galvanic skin response compared to baseline, a "yes" was recorded. If no such increase occurred, a "no" was recorded. Since assigning numbers to these categories is purely arbitrary and since the purpose of the study was to note the category in which there was the most frequent response, the most appropriate measure of central tendency is the mode.

15. a. *Socially-Oriented Achievement Motivation for Japanese Participants.*

Score Interval	Cumulative Midpoint X	Frequency f	Cumulative Frequency cf
135–143	139	1	94
126–134	130	3	93
117–125	121	2	90
108–116	112	5	88
99–107	103	17	83
90–98	94	16	66
81–89	85	12	50
72–80	76	13	38
63–71	67	7	25
54–62	58	6	18
45–53	49	8	12
36–44	40	4	4

Using the table above, the mode for Socially-Oriented Achievement Motivation for Japanese participants is 103.

The median is $80.5 + \{[.5(94) - 38]/12\}(9) = 87.25$.

The mean is $[4(40) + 8(49) + 6(58) + 7(67) + 13(76) + 12(85) + 16(94) + 17(103) + 5(112) + 2(121) + 3(130) + 1(139)]/(4 + 8 + 6 + 7 + 13 + 12 + 16 + 17 + 5 + 2 + 3 + 1) = 7963/94 = 84.71$.

b. *Socially-Oriented Achievement Motivation for American Participants.*

Score Interval	Cumulative Midpoint X	Frequency f	Cumulative Frequency cf
153–161	157	1	105
144–152	148	1	104
135–143	139	1	103
126–134	130	4	102
117–125	121	10	98
108–116	112	18	88
99–107	103	15	70
90–98	94	17	55
81–89	85	12	38
72–80	76	13	26
63–71	67	3	13
54–62	58	8	10
45–53	49	1	2
36–44	40	0	1
27–35	31	1	1

Using the table above, the mode for Socially-Oriented Achievement Motivation for American participants is 112.

The median is $89.5 + \{[.5(105) - 38]/17\}(9) = 97.18$.

The mean is $[1(31) + 0(40) + 1(49) + 8(58) + 3(67) + 13(76) + 12(85) + 17(94) + 15(103) +$

$18(112) + 10(121) + 4(130) + 1(139) + 1(148) + 1(157)]/(1 + 0 + 1 + 8 + 3 + 13 + 12 + 17 + 15 + 18 + 10 + 4 + 1 + 1 + 1) = 10086/105 = 96.06$.

c. *For Japanese Participants.*

Statistics: Socially-Oriented Achievement Motivation	
N Valid	94
Missing	0
Mean	85.05
Median	86.50
Mode	65[a]

[a]Multiple modes exist. The smallest value is shown.

Two modes exist for Japanese Participants' scores on Socially-Oriented Achievement Motivation, but the SPSS program only prints the smallest. The two modes are 65 and 102. The median is 86.5 and is calculated by the discrete score method. The mean is 85.05.

d. *For American Participants.*

Statistics: Socially-Oriented Achievement Motivation	
N Valid	105
Missing	0
Mean	95.72
Median	97.00
Mode	109

Using the table from the SPSS printout, the mode is 109, the median is 97, and the mean is 95.72 for Socially-Oriented Achievement Motivation for American participants.

e. For Japanese participants the mode for Socially-Oriented Achievement Motivation was 103 for the grouped frequency method. However, two modes, 65 and 102, existed when the raw score method was used. The median was 87.25 for the grouped frequency method and 86.5 using the raw score method. Finally, the mean was 84.71 using the grouped frequency method and 85.05 using the raw score method.

For American participants, the mode was 112 using the grouped frequency method and 109 using the raw score method. The median was 97.18 using the grouped frequency method and 97 using the raw score method. The mean was 96.06 using the grouped frequency method and 95.72 using the raw score method.

The values of the mean and median are reasonably close. They are not equivalent because the grouped frequency method uses approximations of scores with the lower interval or the midpoints of the intervals. Thus, the approximations are shown in this example. Since the mode is not calculated by the actual values it will vary more freely in the two methods as indicated in this example.

17. a. *Empathic concern scores.*

Score Interval	Midpoint X	Frequency f	Cumulative Frequency cf
48–50	49	4	120
45–47	46	9	116
42–44	43	7	107
39–41	40	14	100
36–38	37	15	86
33–35	34	12	71
30–32	31	17	59
27–29	28	15	42
24–26	25	16	27
21–23	22	10	11
18–20	19	1	1

Using the table above, the mode is the midpoint of the interval that contains the most frequently occurring scores. In this case the mode is 31.

The median is $32.5 + \{[.5(120) - 59]/12\}(3) = 32.75$.

The mean is $[1(19) + 10(22) + 16(25) + 15(28) + 17(31) + 12(34) + 15(37) + 14(40) + 7(43) + 9(46) + 4(49)]/(1 + 10 + 16 + 15 + 17 + 12 + 15 + 14 + 7 + 9 + 4) = 4020/120 = 33.50$.

b.

Statistics: Empathic Concern

N Valid	120
Missing	0
Mean	33.47
Median	33.00
Mode	31

For empathic concern the mode is 31, the median is 33, and the mean is 33.47 from the SPSS printout.

c. In both the raw score method and the grouped frequency method the mode was 31. The median was 33 in the raw score method and 32.75 in the grouped frequency method. The mean was 33.47 in the raw score method and 33.50 in the grouped frequency method. It is unusual to have the mode be equivalent using the two methods. The discrepancies in the measures of central tendency for the median and the mean result because the grouped frequency method uses approximations of scores with the lower interval or the midpoints of the intervals. Thus, the approximations are shown in this example.

19. a. For the distribution in Exercise 18(a), the mean is less than the median which in turn is less than the mode.

b. For the distribution in Exercise 18(b), the mean and the median are equal to each other and are equal to the average of the two modes.

c. For the distribution in Exercise 18(c), the mean and the median are equal. Since no mode exists, it cannot be compared to the mean and the median.

d. For the distribution in Exercise 18(d), the mean is greater than the median, which in turn is greater than the mode.

e. For the distribution in Exercise 18(e), the mean, median, and mode are all equal.

CHAPTER 4

1. Researchers usually collect data to observe them for patterns. These patterns will help researchers arrive at conclusions about the population. Usually the data are too extensive to allow for effective observation of patterns so researchers rely on procedures for summarizing the data. Usually these summarization procedures are concerned with the measures of central tendency or indices of location that provide representative values that characterize the sample or the population. Through other procedures researchers can then use the values from the sample to indicate the behaviors that might be expected in the population. However, measures of central tendency are not the only values that can be used to characterize a distribution of scores. In fact, it is possible that the measures of central tendency for two samples may be equal but the distribution of scores may look very different. Thus, conclusions that use only the measures of central tendency may be limited. A second way of characterizing the distribution of a sample and a population is to describe the spread or variability of the scores. Two samples or populations may have the same measures of center yet differ widely on measures of spread. One distribution may have the scores clustered very tightly around the measure of central tendency whereas the other may have scores spread very widely. These differences may tell us information about the characteristics of the group or the nature of the trait we are measuring. For example, if we are looking at two different age groups on some measure of problem-solving ability, and the two groups have the same measure of central tendency but a different measure of variability, we might use the age variable to explain why one group is more homogeneous or similar in its measured behavior. In addition, we might approach two different samples using different procedures such as two different counseling approaches. Although the two approaches may be rated equally effective, one approach may lead to more consistent ratings (i.e., less variability). Thus, if we are looking for consistency of behavior, we might favor that approach.

3. a. *Bem sex role inventory feminine scores.*

Raw Score Method: The range is calculated by subtracting the smallest score in the distribution from the largest score in the distribution and then adding 1. Thus, range = $(97 - 43) + 1 = 55$.

Grouped Frequency Table Method: The range is calculated by subtracting the midpoint of the lowest interval from the midpoint of the highest interval and then adding the interval width. Using the table below, range = $(97 - 43) + 3 = 57$.

Grouped Frequency Table for Interval Width = 3

Score Interval	Midpoint (X)	Frequency (f)
96–98	97	2
93–95	94	0
90–92	91	2
87–89	88	1
84–86	85	1
81–83	82	0
78–80	79	4
75–77	76	3
72–74	73	3
69–71	70	3
66–68	67	1
63–65	64	3
60–62	61	0
57–59	58	3
54–56	55	4
51–53	52	1
48–50	49	0
45–47	46	0
42–44	43	2

To demonstrate the effect of interval width, the range is also calculated for the Bem Feminine Scores from a grouped frequency distribution with an interval width of 5. Using the table below, range = $(97 - 42) + 5 = 60$.

Grouped Frequency Table for Interval Width = 5

Score Interval	Midpoint (X)	Frequency (f)
95–99	97	2
90–94	92	2
85–89	87	1
80–84	82	1
75–79	77	7
70–74	72	4
65–69	67	4
60–64	62	2
55–59	57	5
50–54	52	3
45–49	47	0
40–44	42	2

b. *Bem sex role inventory masculine scores.*

Raw Score Method: The range is calculated by subtracting the smallest score in the distribution from the largest score in the distribution and then adding 1. Thus, range = $(105 - 49) + 1 = 57$.

Grouped Frequency Table Method: The range is calculated by subtracting the midpoint of the lowest interval from the midpoint of the highest interval and then adding the interval width. Using the table below, range = $(107 - 47) + 5 = 65$.

Frequency Distribution Table for Interval Width = 5

Score Interval	Midpoint (X)	Frequency (f)
105–109	107	1
100–104	102	2
95–99	97	1
90–94	92	5
85–89	87	7
80–84	82	3
75–79	77	11
70–74	72	2
65–69	67	3
60–64	62	0
55–59	57	1
50–54	52	0
45–49	47	1

c. *Cognitive level.*

Raw Score Method: The range is calculated by subtracting the smallest score in the distribution from the largest score in the distribution and then adding 1. Thus, range = $(86 - 31) + 1 = 56$.

Grouped Frequency Table Method: The range is calculated by subtracting the midpoint of the lowest interval from the midpoint of the highest interval and then adding the interval width. Using the table below, range = $(87 - 32) + 5 = 60$.

Grouped Frequency Table for Interval Width = 5

Score Interval	Midpoint (X)	Frequency (f)
85–89	87	3
80–84	82	1
75–79	77	1
70–74	72	3
65–69	67	4
60–64	62	2
55–59	57	0
50–54	52	6
45–49	47	0
40–44	42	1
35–39	37	2
30–34	32	3

d. *Assertiveness.*

Raw Score Method: The range is calculated by subtracting the smallest score in the distribution from the largest score in the distribution and then adding 1. Thus, range = $(99 - 12) + 1 = 88$.

Grouped Frequency Table Method: The range is calculated by subtracting the midpoint of the lowest interval from the midpoint of the highest interval and then adding the interval width. Using the table below, range = $(101 - 10) + 7 = 98$.

Grouped Frequency Table for Interval Width = 7

Score Interval	Midpoint (X)	Frequency (f)
98–104	101	1
91–97	94	2
84–90	87	3
77–83	80	1
70–76	73	4
63–69	66	2
56–62	59	4
49–55	52	3
42–48	45	2
35–41	38	4
28–34	31	0
21–27	24	0
14–20	17	1
7–13	10	2

5. a. *Bem sex role inventory feminine scores.*

Score	Score Interval	f	cf
97	96.5–97.5	2	33
91	90.5–91.5	1	31
90	89.5–90.5	1	30
87	86.5–87.5	1	29
84	83.5–84.5	1	28
79	78.5–79.5	2	27
78	77.5–78.5	2	25
76	75.5–76.5	1	23
75	74.5–75.5	2	22
73	72.5–73.5	1	20
72	71.5–72.5	2	19
71	70.5–71.5	1	17
69	68.5–69.5	2	16
67	66.5–67.5	1	14
65	64.5–65.5	1	13
64	63.5–64.5	2	12
58	57.5–58.5	2	10
57	56.5–57.5	1	8
55	54.5–55.5	2	7
54	53.5–54.5	2	5
51	50.5–51.5	1	3
43	42.5–43.5	2	2

Use the following formulas to compute Q_1 and Q_3 for a cumulative frequency distribution:

$$Q_1 = LL_{25} + \{[.25(N) - cf_{25-1}]/f_{25}\}$$

$$Q_3 = LL_{75} + \{[.75(N) - cf_{75-1}]/f_{75}\}$$

$$Q_1 = 57.5 + \{[.25(33) - 8]/2\} = 57.62$$

$$Q_3 = 77.5 + \{[.75(33) - 23]/2\} = 78.38$$

Interquartile Range $= Q_3 - Q_1 = 78.38 - 57.62$
$= 20.76$

Semi-Interquartile Range $= (Q_3 - Q_1)/2 = 20.76/2$
$= 10.38$

Interval Width = 3

Interval	Midpt	f	cf
96–98	97	2	33
93–95	94	0	31
90–92	91	2	31
87–89	88	1	29
84–86	85	1	28
81–83	82	0	27
78–80	79	4	27
75–77	76	3	23
72–74	73	3	20
69–71	70	3	17
66–68	67	1	14
63–65	64	3	13
60–62	61	0	10
57–59	58	3	10
54–56	55	4	7
51–53	52	1	3
48–50	49	0	2
45–47	46	0	2
42–44	43	2	2

Interval Width = 5

Interval	Midpt	f	cf
95–99	97	2	33
90–94	92	2	31
85–89	87	1	29
80–84	82	1	28
75–79	77	7	27
70–74	72	4	20
65–69	67	4	16
60–64	62	2	12
55–59	57	5	10
50–54	52	3	5
45–49	47	0	2
40–44	42	2	2

Use the following formulas for a grouped frequency distribution:

$$Q_1 = LL_{25} + \{[.25(N) - cf_{25-1}]/f_{25}\}\,(J)$$

$$Q_3 = LL_{75} + \{[.75(N) - cf_{75-1}]/f_{75}\}\,(J)$$

For interval width $= 3$

$$Q_1 = 56.5 + \{[.25(33) - 7]/3\}(3) = 57.75$$

$$Q_3 = 77.5 + \{[.75(33) - 23]/4\}(3) = 78.81$$

Interquartile Range $= Q_3 - Q_1 = 78.81 - 57.75$
$= 21.06$

Semi-Interquartile Range $= (Q_3 - Q_1)/2 = 21.06/2$
$= 10.53$

For interval width $= 5$

$$Q_1 = 54.5 + \{[.25(33) - 5]/5\}\,(5) = 57.75$$

$$Q_3 = 74.5 + \{[.75(33) - 20]/7\}\,(5) = 77.89$$

Interquartile Range $= Q_3 - Q_1 = 77.89 - 57.75$
$= 20.14$

Semi-Interquartile Range $= (Q_3 - Q_1)/2 = 20.14/2$
$= 10.07$

b. *Bem sex role inventory masculine scores.*

Score	Score Interval	f	cf
105	104.5–105.5	1	37
103	102.5–103.5	1	36
102	101.5–102.5	1	35
96	95.5–96.5	1	34
92	91.5–92.5	2	33
91	90.5–91.5	3	31
89	88.5–89.5	1	28
88	87.5–88.5	2	27
87	86.5–87.5	2	25
86	85.5–86.5	2	23
80	79.5–80.5	3	21
79	78.5–79.5	3	18
78	77.5–78.5	2	15
77	76.5–77.5	2	13
76	75.5–76.5	3	11
75	74.5–75.5	1	8
74	73.5–74.5	1	7
71	70.5–71.5	1	6
67	66.5–67.5	1	5
65	64.5–65.5	2	4
57	56.5–57.5	1	2
49	48.5–49.5	1	1

Use the following formulas to compute Q_1 and Q_3 for a cumulative frequency distribution:

$$Q_1 = LL_{.25} + \{[.25(N) - cf_{.25-1}]/f_{.25}\}$$
$$Q_3 = LL_{.75} + \{[.75(N) - cf_{.75-1}]/f_{.75}\}$$
$$Q_1 = 75.5 + \{[.25(37) - 8]/3\} = 75.92$$
$$Q_3 = 88.5 + \{[.75(37) - 27]/1\} = 89.25$$

Interquartile Range $= Q_3 - Q_1 = 89.25 - 75.92$
$$= 13.33$$

Semi-Interquartile Range $= (Q_3 - Q_1)/2 = 13.33/2$
$$= 6.67$$

Interval Width = 5

Interval	Midpt	f	cf
105–109	107	1	37
100–104	102	2	36
95–99	97	1	34
90–94	92	5	33
85–89	87	7	28
80–84	82	3	21
75–79	77	11	18
70–74	72	2	7
65–69	67	3	5
60–64	62	0	2
55–59	57	1	2
50–54	52	0	1
45–49	47	1	1

Use the following formulas for a grouped frequency distribution:

$$Q_1 = LL_{.25} + \{[.25(N) - cf_{.25-1}]/f_{.25}\}\,(J)$$
$$Q_3 = LL_{.75} + \{[.75(N) - cf_{.75-1}]/f_{.75}\}\,(J)$$
$$Q_1 = 74.5 + \{[.25(37) - 7]/11\}\,(5) = 75.52$$
$$Q_3 = 84.5 + \{[.75(37) - 21]/7\}\,(5) = 89.32$$

Interquartile Range $= Q_3 - Q_1 = 89.32 - 75.52$
$$= 13.80$$

Semi-Interquartile Range $= (Q_3 - Q_1)/2 = 13.80/2$
$$= 6.90$$

c. *Cognitive level.*

Score	Score Interval	f	cf
86	85.5–86.5	1	26
85	84.5–85.5	2	25
81	80.5–81.5	1	23
78	77.5–78.5	1	22
74	73.5–74.5	2	21
72	71.5–72.5	1	19
68	67.5–68.5	1	18
67	66.5–67.5	1	17
66	65.5–66.5	1	16
65	64.5–65.5	1	15
62	61.5–62.5	2	14
54	53.5–54.5	3	12
53	52.5–53.5	1	9
52	51.5–52.5	1	8
50	49.5–50.5	1	7
42	41.5–42.5	1	6
38	37.5–38.5	1	5
37	36.5–37.5	1	4
33	32.5–33.5	1	3
32	31.5–32.5	1	2
31	30.5–31.5	1	1

Use the following formulas to compute Q_1 and Q_3 for a cumulative frequency distribution:

$$Q_1 = LL_{.25} + \{[.25(N) - cf_{.25-1}]/f_{.25}\}$$
$$Q_3 = LL_{.75} + \{[.75(N) - cf_{.75-1}]/f_{.75}\}$$
$$Q_1 = 49.5 + \{[.25(26) - 6]/1\} = 50.00$$
$$Q_3 = 73.5 + \{[.75(26) - 19]/2\} = 73.75$$

Interquartile Range $= Q_3 - Q_1 = 73.75 - 50.00$
$$= 23.75$$

Semi-Interquartile Range $= (Q_3 - Q_1)/2 = 23.75/2$
$$= 11.88$$

Interval Width = 5

Interval	Midpt	f	cf
85–89	87	3	26
80–84	82	1	23
75–79	77	1	22
70–74	72	3	21
65–69	67	4	18
60–64	62	2	14
55–59	57	0	12
50–54	52	6	12
45–49	47	0	6
40–44	42	1	6
35–39	37	2	5
30–34	32	3	3

Use the following formulas for a grouped frequency distribution:

$$Q_1 = LL_{.25} + \{[.25(N) - cf_{.25-1}]/f_{.25}\}\,(J)$$
$$Q_3 = LL_{.75} + \{[.75(N) - cf_{.75-1}]/f_{.75}\}\,(J)$$
$$Q_1 = 49.5 + \{[.25(26) - 6]/6\}(5) = 49.92$$
$$Q_3 = 69.5 + \{[.75(26) - 18]/3\}(5) = 72.00$$

Interquartile Range $= Q_3 - Q_1 = 72.00 - 49.92$
$$= 22.08$$

Semi-Interquartile Range $= (Q_3 - Q_1)/2 = 22.08/2$
$$= 11.04$$

d. *Assertiveness.*

Score	Score Interval	f	cf
99	98.5–99.5	1	29
94	93.5–94.5	1	28
93	92.5–93.5	1	27
89	88.5–89.5	1	26
88	87.5–88.5	1	25
86	85.5–86.5	1	24
80	79.5–80.5	1	23
76	75.5–76.5	1	22
75	74.5–75.5	2	21
70	69.5–70.5	1	19
65	64.5–65.5	1	18
63	62.5–63.5	1	17
61	60.5–61.5	1	16
60	59.5–60.5	1	15
59	58.5–59.5	2	14
53	52.5–53.5	1	12
52	51.5–52.5	1	11
49	48.5–49.5	1	10
45	44.5–45.5	2	9
40	39.5–40.5	1	7
39	38.5–39.5	1	6
38	37.5–38.5	1	5
37	36.5–37.5	1	4
17	16.5–17.5	1	3
13	12.5–13.5	1	2
12	11.5–12.5	1	1

Use the following formulas to compute Q_1 and Q_3 for a cumulative frequency distribution:

$$Q_1 = LL_{25} + \{[.25(N) - cf_{25-1}]/f_{25}\}$$
$$Q_3 = LL_{75} + \{[.75(N) - cf_{75-1}]/f_{75}\}$$
$$Q_1 = 44.5 + \{[.25(29) - 7]/2\} = 44.63$$
$$Q_3 = 75.5 + \{[.75(29) - 21]/1\} = 76.25$$

Interquartile Range $= Q_3 - Q_1 = 76.25 - 44.63$
$$= 31.62$$

Semi-Interquartile Range $= (Q_3 - Q_1)/2 = 31.62/2$
$$= 15.81$$

Interval Width = 7

Interval	Midpt	f	cf
98–104	101	1	29
91–97	94	2	28
84–90	87	3	26
77–83	80	1	23
70–76	73	4	22
63–69	66	2	18
56–62	59	4	16
49–55	52	3	12
42–48	45	2	9
35–41	38	4	7
28–34	31	0	3
21–27	24	0	3
14–20	17	1	3
7–13	10	2	2

Use the following formulas for a grouped frequency distribution:

$$Q_1 = LL_{25} + \{[.25(N) - cf_{25-1}]/f_{25}\}\,(J)$$
$$Q_3 = LL_{75} + \{[.75(N) - cf_{75-1}]/f_{75}\}\,(J)$$
$$Q_1 = 41.5 + \{[.25(29) - 7]/2\}(7) = 42.38$$
$$Q_3 = 69.5 + \{[.75(29) - 18]/4\}(7) = 76.06$$

Interquartile Range $= Q_3 - Q_1 = 76.06 - 42.38$
$$= 33.68$$

Semi-Interquartile Range $= (Q_3 - Q_1)/2 = 33.68/2$
$$= 16.84$$

7. Although the range of the two groups is the same, the groups are very different. The interquartile range of the second group is much smaller than the first group, as is the median. If these groups represent different treatment groups for rattlesnake phobias, the treatment given to the second group is much more successful if the groups were no different on initial heart rate in the presence of a rattlesnake.

9. a. *Bem sex role inventory feminine scores.*

Conceptual Formula (4.8)

X	$(X - \overline{X})^2$
77	$(77 - 69.76)^2 = 52.42$
58	$(58 - 69.76)^2 = 138.30$
64	$(64 - 69.76)^2 = 33.18$
73	$(73 - 69.76)^2 = 10.50$
87	$(87 - 69.76)^2 = 297.22$
71	$(71 - 69.76)^2 = 1.54$
91	$(91 - 69.76)^2 = 451.14$
64	$(64 - 69.76)^2 = 33.18$
72	$(72 - 69.76)^2 = 5.02$
84	$(84 - 69.76)^2 = 202.78$
65	$(65 - 69.76)^2 = 22.66$
78	$(78 - 69.76)^2 = 97.90$
72	$(72 - 69.76)^2 = 5.02$
75	$(75 - 69.76)^2 = 27.46$
55	$(55 - 69.76)^2 = 217.86$
78	$(78 - 69.76)^2 = 67.90$
43	$(43 - 69.76)^2 = 716.10$
54	$(54 - 69.76)^2 = 248.38$
55	$(55 - 69.76)^2 = 217.86$
79	$(79 - 69.76)^2 = 85.38$
69	$(69 - 69.76)^2 = 0.58$
76	$(76 - 69.76)^2 = 38.94$
79	$(79 - 69.76)^2 = 85.38$
97	$(97 - 69.76)^2 = 742.02$
67	$(67 - 69.76)^2 = 7.62$
69	$(69 - 69.76)^2 = 0.58$
90	$(90 - 69.76)^2 = 409.66$
54	$(54 - 69.76)^2 = 248.38$
57	$(57 - 69.76)^2 = 162.82$
58	$(58 - 69.76)^2 = 138.30$
97	$(97 - 69.76)^2 = 742.02$
43	$(43 - 69.76)^2 = 716.10$
51	$(51 - 69.76)^2 = 351.94$
$\Sigma X = 2302$	$\Sigma(X - \overline{X})^2 = 6546.06$

$$\overline{X} = 69.76$$

$$S^2 = \frac{\Sigma(X - \overline{X})^2}{N} = 6546.06/33 = 198.37$$

$$S = \sqrt{198.37} = 14.08$$

Computational Formula (4.12)

X	X²
77	5929
58	3364
64	4096
73	5329
87	7569
71	5041
91	8281
64	4096
72	5184
84	7056
65	4225
78	6084
72	5184
75	5625
55	3025
78	6084
43	1849
54	2916
55	3025
79	6241
69	4761
76	5776
79	6241
97	9409
67	4489
69	4761
90	8100
54	2916
57	3249
58	3364
97	9409
43	1849
51	2601

$$\Sigma X = 2302 \qquad \Sigma X^2 = 167128$$

$$S^2 = \frac{N\Sigma X^2 - (\Sigma X)^2}{N^2} = \frac{(33)(167128) - (2302)^2}{33^2}$$

$$= 198.37$$

$$S = \sqrt{198.37} = 14.08$$

b. *Bem sex role inventory masculine scores.*

Conceptual Formula (4.8)

X	$(X - \bar{X})^2$
75	$(75 - 81.41)^2 =$ 41.09
80	$(80 - 81.41)^2 =$ 1.99
103	$(103 - 81.41)^2 =$ 466.13
65	$(65 - 81.41)^2 =$ 269.29
79	$(79 - 81.41)^2 =$ 5.81
88	$(88 - 81.41)^2 =$ 43.43
71	$(71 - 81.41)^2 =$ 108.37
86	$(86 - 81.41)^2 =$ 21.07
77	$(77 - 81.41)^2 =$ 19.45
78	$(78 - 81.41)^2 =$ 11.63
49	$(49 - 81.41)^2 =$ 1050.41
86	$(86 - 81.41)^2 =$ 21.07
76	$(76 - 81.41)^2 =$ 29.27
96	$(96 - 81.41)^2 =$ 212.87
76	$(76 - 81.41)^2 =$ 29.27
92	$(92 - 81.41)^2 =$ 112.15
87	$(87 - 81.41)^2 =$ 31.25
87	$(87 - 81.41)^2 =$ 31.25
80	$(80 - 81.41)^2 =$ 1.99
91	$(91 - 81.41)^2 =$ 91.97
79	$(79 - 81.41)^2 =$ 5.81
92	$(92 - 81.41)^2 =$ 112.15
91	$(91 - 81.41)^2 =$ 91.97
65	$(65 - 81.41)^2 =$ 269.29
78	$(78 - 81.41)^2 =$ 11.63
91	$(91 - 81.41)^2 =$ 91.97
88	$(88 - 81.41)^2 =$ 43.43
89	$(89 - 81.41)^2 =$ 57.61
105	$(105 - 81.41)^2 =$ 556.49
76	$(76 - 81.41)^2 =$ 29.27
67	$(67 - 81.41)^2 =$ 207.65
80	$(80 - 81.41)^2 =$ 1.99
102	$(102 - 81.41)^2 =$ 423.95
79	$(79 - 81.41)^2 =$ 5.81
57	$(57 - 81.41)^2 =$ 595.85
77	$(77 - 81.41)^2 =$ 19.45
74	$(74 - 81.41)^2 =$ 54.91

$$\Sigma X = 3012 \qquad \Sigma(X - \bar{X})^2 = 5178.92$$

$$\bar{X} = 81.41$$

$$S^2 = \frac{\Sigma(X - \bar{X})^2}{N} = 5178.92/37 = 139.97$$

$$S = \sqrt{139.97} = 11.83$$

Computational Formula (4.12)

X	X²
75	5625
80	6400
103	10609
65	4225
79	6241
88	7744
71	5041
86	7396
77	5929
78	6084
49	2401
86	7396
76	5776
96	9216
76	5776
92	8464
87	7569
87	7569
80	6400
91	8281
79	6241
92	8464
91	8281
65	4225
78	6084
91	8281
88	7744
89	7921
105	11025
76	5776
67	4489
80	6400
102	10404
79	6241
57	3249
77	5929
74	5476

$\Sigma X = 3012 \quad \Sigma X^2 = 250372$

$$S^2 = \frac{N\Sigma X^2 - (\Sigma X)^2}{N^2} = \frac{(37)(250372) - (3012)^2}{37^2}$$

$$= 139.97$$

$$S = \sqrt{139.97} = 11.83$$

c. *Cognitive level.*

Conceptual Formula (4.8)

X	$(X - \overline{X})^2$
74	$(74 - 59.81)^2 = 201.36$
50	$(50 - 59.81)^2 = 96.24$
66	$(66 - 59.81)^2 = 38.32$
65	$(65 - 59.81)^2 = 26.94$
67	$(67 - 59.81)^2 = 51.70$
54	$(54 - 59.81)^2 = 33.76$
54	$(54 - 59.81)^2 = 33.76$
78	$(78 - 59.81)^2 = 330.88$
52	$(52 - 59.81)^2 = 61.00$
72	$(72 - 59.81)^2 = 148.60$
53	$(53 - 59.81)^2 = 46.38$
62	$(62 - 59.81)^2 = 4.80$
81	$(81 - 59.81)^2 = 449.02$
68	$(68 - 59.81)^2 = 67.08$
54	$(54 - 59.81)^2 = 33.76$
62	$(62 - 59.81)^2 = 4.80$
74	$(74 - 59.81)^2 = 201.36$
85	$(85 - 59.81)^2 = 634.54$
42	$(42 - 59.81)^2 = 317.20$
33	$(33 - 59.81)^2 = 718.78$
38	$(38 - 59.81)^2 = 475.68$
37	$(37 - 59.81)^2 = 520.30$
86	$(86 - 59.81)^2 = 685.92$
32	$(32 - 59.81)^2 = 733.40$
85	$(85 - 59.81)^2 = 634.54$
31	$(31 - 59.81)^2 = 830.02$

$\Sigma X = 1555 \qquad \Sigma(X - \overline{X})^2 = 7420.04$

$$\overline{X} = 59.81$$

$$S^2 = \frac{\Sigma(X - \overline{X})^2}{N} = 7420/26 = 285.38$$

$$S = \sqrt{285.38} = 16.89$$

Computational Formula (4.12)

X	X²
74	5476
50	2500
66	4356
65	4225
67	4489
54	2916
54	2916
78	6084
52	2704
72	5184
53	2809
62	3844
81	6561
68	4624
54	2916
62	3844
74	5476
85	7225
42	1764
33	1089
38	1444
37	1369
86	7396
32	1024
85	7225
31	961

$\Sigma X = 1555 \qquad \Sigma X^2 = 100421$

$$S^2 = \frac{N\Sigma X^2 - (\Sigma X)^2}{N^2} = \frac{(26)(100421) - (1555)^2}{26^2}$$

$$= 285.39$$

$$S = \sqrt{285.39} = 16.89$$

d. *Assertiveness.*

Conceptual Formula (4.8)

X	(X − X̄)²
75	$(75 - 59.72)^2 = 233.48$
37	$(37 - 59.72)^2 = 516.20$
13	$(13 - 59.72)^2 = 2182.76$
60	$(60 - 59.72)^2 = 0.08$
49	$(49 - 59.72)^2 = 114.92$
76	$(76 - 59.72)^2 = 265.04$
88	$(88 - 59.72)^2 = 799.76$
12	$(12 - 59.72)^2 = 2277.20$
70	$(70 - 59.72)^2 = 105.68$
89	$(89 - 59.72)^2 = 857.32$
61	$(61 - 59.72)^2 = 1.64$
53	$(53 - 59.72)^2 = 45.16$
93	$(93 - 59.72)^2 = 1107.56$
45	$(45 - 59.72)^2 = 216.68$
52	$(52 - 59.72)^2 = 59.60$
99	$(99 - 59.72)^2 = 1542.92$
17	$(17 - 59.72)^2 = 1825.00$
59	$(59 - 59.72)^2 = 0.52$
40	$(40 - 59.72)^2 = 388.88$
63	$(63 - 59.72)^2 = 10.76$
94	$(94 - 59.72)^2 = 1175.12$
59	$(59 - 59.72)^2 = 0.52$
86	$(86 - 59.72)^2 = 690.64$
75	$(75 - 59.72)^2 = 233.48$
39	$(39 - 59.72)^2 = 429.32$
80	$(80 - 59.72)^2 = 311.28$
45	$(45 - 59.72)^2 = 216.68$
65	$(65 - 59.72)^2 = 27.88$
38	$(38 - 59.72)^2 = 471.76$
$\Sigma X = 1732$	$\Sigma(X - \bar{X})^2 = 16207.79$

$$\bar{X} = 59.72$$

$$S^2 = \frac{\Sigma(X - \bar{X})^2}{N} = 16207.79/29 = 558.89$$

$$S = \sqrt{558.89} = 23.64$$

Computational Formula (4.12)

X	X²
75	5625
37	1369
13	169
60	3600
49	2401
76	5776
88	7744
12	144
70	4900
89	7921
61	3721
53	2809
93	8649
45	2025
52	2704
99	9801
17	289
59	3481
40	1600
63	3969
94	8836
59	3481
86	7396
75	5625
39	1521
80	6400
45	2025
65	4225
38	1444
$\Sigma X = 1732$	$\Sigma X^2 = 119650$

$$S^2 = \frac{N\Sigma X^2 - (\Sigma X)^2}{N^2} = \frac{(29)(119650) - (1732)^2}{29^2}$$

$$= 558.89$$

$$S = \sqrt{558.89} = 23.64$$

11. The sum of squares is merely the sum of all of the squared differences (deviations) of raw scores from the mean of the raw scores. Since the sum of the deviation scores will always be zero, we need to square the deviation scores before summing them to get a measure of distance without the confound of direction. Least squares means that the sum of the squared deviations of each raw score from the mean of the scores is the smallest value that can be obtained by subtracting a measure of central tendency, or any other value, from each raw score, squaring that difference, and then adding all of the squared difference scores.

13. a. The anxiety data are unimodal, symmetric, and continuous. In this case the most sensitive measure of variability would be either the variance or the standard deviation. Because most people do not have a good understanding of what a squared value means with respect to a data set, the standard deviation may be a better measure of variability than the variance.
    b. No measure of range is appropriate in this situation since the range is defined only for quantitative values.
    c. The data in this example are ordinal. Both the range and the interquartile range are appropriate for ordinal data. The range is sensitive to extreme cases. The interquartile range is a more stable measure of spread. In this situation it is probably best to report both the range and the interquartile range.
    d. Despite the fact that the data in this example are ratio level data, there are extreme scores that will affect the

measure of spread as reported by a standard deviation or a variance. As a result, the most appropriate measure of spread would be the interquartile range.

e. The data in this example are ratio level data. Since there is no mention of problems of symmetry or multimodality, the most appropriate measures of variability would be variance and standard deviation. However, since standard deviations are more familiar to readers, it is usually the measure of spread of choice.

15. A skewed distribution can be further classified into negatively skewed and positively skewed. In a negatively skewed distribution, the preponderance of scores occur at the higher end of the scale whereas in a positively skewed distribution the preponderance of scores occur at the lower end of the scale.

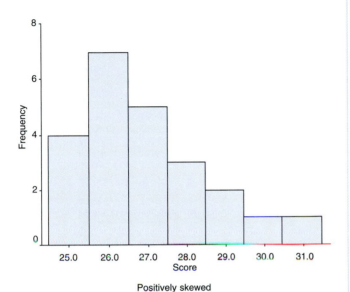

Positively skewed

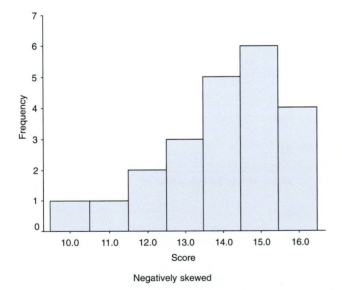

Negatively skewed

17. a. *Scores on a statistics quiz.*

X	X²	(X − X̄)³	(X − X̄)⁴
8	64	$(8-11)^3 = -27$	$(8-11)^4 = 81$
14	196	$(14-11)^3 = 27$	$(14-11)^4 = 81$
13	169	$(13-11)^3 = 8$	$(13-11)^4 = 16$
9	81	$(9-11)^3 = -8$	$(9-11)^4 = 16$
11	121	$(11-11)^3 = 0$	$(11-11)^4 = 0$
12	144	$(12-11)^3 = 1$	$(12-11)^4 = 1$
10	100	$(10-11)^3 = -1$	$(10-11)^4 = 1$
11	121	$(11-11)^3 = 0$	$(11-11)^4 = 0$
12	144	$(12-11)^3 = 1$	$(12-11)^4 = 1$
13	169	$(13-11)^3 = 8$	$(13-11)^4 = 16$
9	81	$(9-11)^3 = -8$	$(9-11)^4 = 16$
10	100	$(10-11)^3 = -1$	$(10-11)^4 = 1$
11	121	$(11-11)^3 = 0$	$(11-11)^4 = 0$
10	100	$(10-11)^3 = -1$	$(10-11)^4 = 1$
12	144	$(12-11)^3 = 1$	$(12-11)^4 = 1$
11	121	$(11-11)^3 = 0$	$(11-11)^4 = 0$

$\Sigma X = 176 \quad \Sigma X^2 = 1976 \quad \Sigma(X-\overline{X})^3 = 0 \quad \Sigma(X-\overline{X})^4 = 232$

$$S = \sqrt{\frac{N(\Sigma X^2) - (\Sigma X)^2}{N^2}} = \sqrt{\frac{16(1976) - (176)^2}{16^2}} = 1.58$$

$$Skew = \frac{\Sigma(X-\overline{X})^3}{N(S^3)} = \frac{0}{16(1.58^3)} = 0$$

$$K = \frac{\Sigma(X-\overline{X})^4}{N(S^4)} - 3 = \frac{232}{16(1.58^4)} - 3 = -0.67$$

Thus, the distribution is symmetric and platykurtic.

b. *Anxiety scores.*

X	X²	(X − X̄)³	(X − X̄)⁴
23	529	$(23-23)^3 = 0$	$(23-23)^4 = 0$
21	441	$(21-23)^3 = -8$	$(21-23)^4 = 16$
25	625	$(25-23)^3 = 8$	$(25-23)^4 = 16$
23	529	$(23-23)^3 = 0$	$(23-23)^4 = 0$
22	484	$(22-23)^3 = -1$	$(22-23)^4 = 1$
24	576	$(24-23)^3 = 1$	$(24-23)^4 = 1$
26	676	$(26-23)^3 = 27$	$(26-23)^4 = 81$
21	441	$(21-23)^3 = -8$	$(21-23)^4 = 16$
23	529	$(23-23)^3 = 0$	$(23-23)^4 = 0$
22	484	$(22-23)^3 = -1$	$(22-23)^4 = 1$
23	529	$(23-23)^3 = 0$	$(23-23)^4 = 0$
25	625	$(25-23)^3 = 8$	$(25-23)^4 = 16$
20	400	$(20-23)^3 = -27$	$(20-23)^4 = 81$
23	529	$(23-23)^3 = 0$	$(23-23)^4 = 0$
24	576	$(24-23)^3 = 1$	$(24-23)^4 = 1$
22	484	$(22-23)^3 = -1$	$(22-23)^4 = 1$
23	529	$(23-23)^3 = 0$	$(23-23)^4 = 0$
24	576	$(24-23)^3 = 1$	$(24-23)^4 = 1$
23	529	$(23-23)^3 = 0$	$(23-23)^4 = 0$
23	529	$(23-23)^3 = 0$	$(23-23)^4 = 0$

$\Sigma X = 460 \quad \Sigma X^2 = 10620 \quad \Sigma(X-\overline{X})^3 = 0 \quad \Sigma(X-\overline{X})^4 = 232$

$$S = \sqrt{\frac{N(\Sigma X^2) - (\Sigma X)^2}{N^2}} = \sqrt{\frac{20(10620) - (460)^2}{20^2}} = 1.41$$

$$Skew = \frac{\Sigma(X-\overline{X})^3}{N(S^3)} = \frac{0}{20(1.41^3)} = 0$$

$$K = \frac{\Sigma(X-\overline{X})^4}{N(S^4)} - 3 = \frac{232}{20(1.41^4)} - 3 = -0.07$$

Thus, the distribution is symmetric and mesokurtic.

c. *Creativity scores.*

X	X²	(X − X̄)³	(X − X̄)⁴
62	3844	$(62 - 63)^3 =$ −1	$(62 - 63)^4 =$ 1
68	4624	$(68 - 63)^3 =$ 125	$(68 - 63)^4 =$ 625
63	3969	$(63 - 63)^3 =$ 0	$(63 - 63)^4 =$ 0
60	3600	$(60 - 63)^3 =$ −27	$(60 - 63)^4 =$ 81
62	3844	$(62 - 63)^3 =$ −1	$(62 - 63)^4 =$ 1
65	4225	$(65 - 63)^3 =$ 8	$(65 - 63)^4 =$ 16
66	4356	$(66 - 63)^3 =$ 27	$(66 - 63)^4 =$ 81
61	3721	$(61 - 63)^3 =$ −8	$(61 - 63)^4 =$ 16
62	3844	$(62 - 63)^3 =$ −1	$(62 - 63)^4 =$ 1
63	3969	$(63 - 63)^3 =$ 0	$(63 - 63)^4 =$ 0
61	3721	$(61 - 63)^3 =$ −8	$(61 - 63)^4 =$ 16
67	4489	$(67 - 63)^3 =$ 64	$(67 - 63)^4 =$ 256
65	4225	$(65 - 63)^3 =$ 8	$(65 - 63)^4 =$ 16
64	4096	$(64 - 63)^3 =$ 1	$(64 - 63)^4 =$ 1
62	3844	$(62 - 63)^3 =$ −1	$(62 - 63)^4 =$ 1
60	3600	$(60 - 63)^3 =$ −27	$(60 - 63)^4 =$ 81
64	4096	$(64 - 63)^3 =$ 1	$(64 - 63)^4 =$ 1
62	3844	$(62 - 63)^3 =$ −1	$(62 - 63)^4 =$ 1
63	3969	$(63 - 63)^3 =$ 0	$(63 - 63)^4 =$ 0
62	3844	$(62 - 63)^3 =$ −1	$(62 - 63)^4 =$ 1
61	3721	$(61 - 63)^3 =$ −8	$(61 - 63)^4 =$ 16

$\Sigma X = 1323$ $\Sigma X^2 = 83445$ $\Sigma(X - \bar{X})^3 = 150$ $\Sigma(X - \bar{X})^4 = 1212$

$$S = \sqrt{\frac{N(\Sigma X^2) - (\Sigma X)^2}{N^2}} = \sqrt{\frac{21(83445) - (1323)^2}{21^2}} = 2.14$$

$$Skew = \frac{\Sigma(X - \bar{X})^3}{N(S^3)} = \frac{150}{21(2.14^3)} = 0.73$$

$$K = \frac{\Sigma(X - \bar{X})^4}{N(S^4)} - 3 = \frac{1212}{21(2.14^4)} - 3 = -0.25$$

Thus, the distribution is positively skewed and platykurtic.

d. *Science achievement scores: see Table A4-1.*

$$S = \sqrt{\frac{N(\Sigma X^2) - (\Sigma X)^2}{N^2}} = \sqrt{\frac{15(94279) - (1189)^2}{15^2}} = 1.44$$

$$Skew = \frac{\Sigma(X - \bar{X})^3}{N(S^3)} = \frac{-33.34}{15(1.44^3)} = -0.74$$

$$K = \frac{\Sigma(X - \bar{X})^4}{N(S^4)} - 3 = \frac{174.40}{15(1.44^4)} - 3 = -0.30$$

Thus, the distribution is negatively skewed and platykurtic.

19. a. *Bem sex role inventory feminine scores.* Pearson's index of skewness

$$P_s = \frac{3(\bar{X} - Mdn)}{S_x} = \frac{3(69.76 - 71)}{14.08} = -0.26$$

Thus, the distribution of scores is slightly negatively skewed.

	Skewness Index	
X	(X − X̄)³	
77	$(77 - 69.76)^3 =$	379.50
58	$(58 - 69.76)^3 =$	−1626.38
64	$(64 - 69.76)^3 =$	−191.10
73	$(73 - 69.76)^3 =$	34.01
87	$(87 - 69.76)^3 =$	5124.03
71	$(71 - 69.76)^3 =$	1.91
91	$(91 - 69.76)^3 =$	9582.16
64	$(64 - 69.76)^3 =$	−191.10
72	$(72 - 69.76)^3 =$	11.24
84	$(84 - 69.76)^3 =$	2887.55
65	$(65 - 69.76)^3 =$	−107.85
78	$(78 - 69.76)^3 =$	559.48
72	$(72 - 69.76)^3 =$	11.24
75	$(75 - 69.76)^3 =$	143.88
55	$(55 - 69.76)^3 =$	−3215.58
78	$(78 - 69.76)^3 =$	559.48
43	$(43 - 69.76)^3 =$	−19162.77
54	$(54 - 69.76)^3 =$	−3914.43
55	$(55 - 69.76)^3 =$	−3215.58
79	$(79 - 69.76)^3 =$	788.89
69	$(69 - 69.76)^3 =$	−0.44
76	$(76 - 69.76)^3 =$	242.97
79	$(79 - 69.76)^3 =$	788.89
97	$(97 - 69.76)^3 =$	20212.56
67	$(67 - 69.76)^3 =$	−21.02
69	$(69 - 69.76)^3 =$	−0.44
90	$(90 - 69.76)^3 =$	8291.47
54	$(54 - 69.76)^3 =$	−3914.43
57	$(57 - 69.76)^3 =$	−2077.55
58	$(58 - 69.76)^3 =$	−1626.38
97	$(97 - 69.76)^3 =$	20212.56
43	$(43 - 69.76)^3 =$	−19162.77
51	$(51 - 69.76)^3 =$	−6602.35

$\Sigma(X - \bar{X})^3 = 4801.63$

$$Skew = \frac{\Sigma(X - \bar{X})^3}{N(S^3)} = \frac{4801.63}{33(14.08^3)} = 0.05$$

Thus, the distribution of scores is slightly positively skewed but it is almost symmetric.

Table A4-1   Science Achievement Test Scores

X	X²	(X − X̄)³	(X − X̄)⁴
81	6561	$(81 - 79.27)^3 =$ 5.18	$(81 - 79.27)^4 =$ 8.96
80	6400	$(80 - 79.27)^3 =$ 0.39	$(80 - 79.27)^4 =$ 0.28
76	5776	$(76 - 79.27)^3 =$ −34.97	$(76 - 79.27)^4 =$ 114.34
80	6400	$(80 - 79.27)^3 =$ 0.39	$(80 - 79.27)^4 =$ 0.28
79	6241	$(79 - 79.27)^3 =$ −0.02	$(79 - 79.27)^4 =$ 0.01
78	6084	$(78 - 79.27)^3 =$ −2.05	$(78 - 79.27)^4 =$ 2.60
77	5929	$(77 - 79.27)^3 =$ −11.70	$(77 - 79.27)^4 =$ 26.55
81	6561	$(81 - 79.27)^3 =$ 5.18	$(81 - 79.27)^4 =$ 8.96
80	6400	$(80 - 79.27)^3 =$ 0.39	$(80 - 79.27)^4 =$ 0.28
79	6241	$(79 - 79.27)^3 =$ −0.02	$(79 - 79.27)^4 =$ 0.01
78	6084	$(78 - 79.27)^3 =$ −2.05	$(78 - 79.27)^4 =$ 2.60
80	6400	$(80 - 79.27)^3 =$ 0.39	$(80 - 79.27)^4 =$ 0.28
81	6561	$(81 - 79.27)^3 =$ 5.18	$(81 - 79.27)^4 =$ 8.96
79	6241	$(79 - 79.27)^3 =$ −0.02	$(79 - 79.27)^4 =$ 0.01
80	6400	$(80 - 79.27)^3 =$ 0.39	$(80 - 79.27)^4 =$ 0.28

$\Sigma X = 1189$ $\Sigma X^2 = 94279$ $\Sigma(X - \bar{X})^3 = -33.34$ $\Sigma(X - \bar{X})^4 = 174.40$

<table>
<thead>
<tr><th colspan="2" align="center">Kurtosis Index</th></tr>
<tr><th>X</th><th>$(X - \bar{X})^4$</th></tr>
</thead>
<tbody>
<tr><td>77</td><td>$(77 - 69.76)^4 =$ 2747.60</td></tr>
<tr><td>58</td><td>$(58 - 69.76)^4 =$ 19126.89</td></tr>
<tr><td>64</td><td>$(64 - 69.76)^4 =$ 1100.91</td></tr>
<tr><td>73</td><td>$(73 - 69.76)^4 =$ 110.25</td></tr>
<tr><td>87</td><td>$(87 - 69.76)^4 =$ 88339.73</td></tr>
<tr><td>71</td><td>$(71 - 69.76)^4 =$ 2.37</td></tr>
<tr><td>91</td><td>$(91 - 69.76)^4 =$ 203527.29</td></tr>
<tr><td>64</td><td>$(64 - 69.76)^4 =$ 1100.91</td></tr>
<tr><td>72</td><td>$(72 - 69.76)^4 =$ 25.20</td></tr>
<tr><td>84</td><td>$(84 - 69.76)^4 =$ 41119.73</td></tr>
<tr><td>65</td><td>$(65 - 69.76)^4 =$ 513.48</td></tr>
<tr><td>78</td><td>$(78 - 69.76)^4 =$ 9584.41</td></tr>
<tr><td>72</td><td>$(72 - 69.76)^4 =$ 25.02</td></tr>
<tr><td>75</td><td>$(75 - 69.76)^4 =$ 754.05</td></tr>
<tr><td>55</td><td>$(55 - 69.76)^4 =$ 47462.98</td></tr>
<tr><td>78</td><td>$(78 - 69.76)^4 =$ 4610.41</td></tr>
<tr><td>43</td><td>$(43 - 69.76)^4 =$ 512799.21</td></tr>
<tr><td>54</td><td>$(54 - 69.76)^4 =$ 61692.62</td></tr>
<tr><td>55</td><td>$(55 - 69.76)^4 =$ 47462.98</td></tr>
<tr><td>79</td><td>$(79 - 69.76)^4 =$ 7289.74</td></tr>
<tr><td>69</td><td>$(69 - 69.76)^4 =$ 0.34</td></tr>
<tr><td>76</td><td>$(76 - 69.76)^4 =$ 1516.32</td></tr>
<tr><td>79</td><td>$(79 - 69.76)^4 =$ 7289.74</td></tr>
<tr><td>97</td><td>$(97 - 69.76)^4 =$ 550593.68</td></tr>
<tr><td>67</td><td>$(67 - 69.76)^4 =$ 58.06</td></tr>
<tr><td>69</td><td>$(69 - 69.76)^4 =$ 0.34</td></tr>
<tr><td>90</td><td>$(90 - 69.76)^4 =$ 167821.31</td></tr>
<tr><td>54</td><td>$(54 - 69.76)^4 =$ 61692.62</td></tr>
<tr><td>57</td><td>$(57 - 69.76)^4 =$ 26510.35</td></tr>
<tr><td>58</td><td>$(58 - 69.76)^4 =$ 19126.89</td></tr>
<tr><td>97</td><td>$(97 - 69.76)^4 =$ 550593.68</td></tr>
<tr><td>43</td><td>$(43 - 69.76)^4 =$ 512799.21</td></tr>
<tr><td>51</td><td>$(51 - 69.76)^4 =$ 123861.76</td></tr>
</tbody>
</table>

$$\Sigma(X - \bar{X})^4 = 3066355.14$$

$$K = \frac{\Sigma(X - \bar{X})^4}{N(S^4)} - 3 = \frac{3066355.14}{33(14.08^4)} - 3 = -0.64$$

Thus, the distribution of scores is slightly platykurtic.

b. *Bem sex role inventory masculine scores.* Pearson's index of skewness

$$P_s = \frac{3(\bar{X} - Mdn)}{S_x} = \frac{3(81.41 - 79.67)}{11.83} = 0.44$$

Thus, the distribution of scores is positively skewed.

<table>
<thead>
<tr><th colspan="2" align="center">Skewness Index</th></tr>
<tr><th>X</th><th>$(X - \bar{X})^3$</th></tr>
</thead>
<tbody>
<tr><td>75</td><td>$(75 - 81.41)^3 =$ -263.37</td></tr>
<tr><td>80</td><td>$(80 - 81.41)^3 =$ -2.80</td></tr>
<tr><td>103</td><td>$(103 - 81.41)^3 =$ 10063.71</td></tr>
<tr><td>65</td><td>$(65 - 81.41)^3 =$ -4419.02</td></tr>
<tr><td>79</td><td>$(79 - 81.41)^3 =$ -14.00</td></tr>
<tr><td>88</td><td>$(88 - 81.41)^3 =$ 286.19</td></tr>
<tr><td>71</td><td>$(71 - 81.41)^3 =$ -1128.11</td></tr>
<tr><td>86</td><td>$(86 - 81.41)^3 =$ 96.70</td></tr>
<tr><td>77</td><td>$(77 - 81.41)^3 =$ -85.77</td></tr>
<tr><td>78</td><td>$(78 - 81.41)^3 =$ -39.65</td></tr>
<tr><td>49</td><td>$(49 - 81.41)^3 =$ -34043.73</td></tr>
<tr><td>86</td><td>$(86 - 81.41)^3 =$ 96.70</td></tr>
<tr><td>76</td><td>$(76 - 81.41)^3 =$ -158.34</td></tr>
<tr><td>96</td><td>$(96 - 81.41)^3 =$ 3105.75</td></tr>
<tr><td>76</td><td>$(75 - 81.41)^3 =$ -158.34</td></tr>
<tr><td>92</td><td>$(92 - 81.41)^3 =$ 1187.65</td></tr>
<tr><td>87</td><td>$(87 - 81.41)^3 =$ 174.68</td></tr>
<tr><td>87</td><td>$(87 - 81.41)^3 =$ 174.68</td></tr>
<tr><td>80</td><td>$(80 - 81.41)^3 =$ -2.80</td></tr>
<tr><td>91</td><td>$(91 - 81.41)^3 =$ 881.97</td></tr>
<tr><td>79</td><td>$(79 - 81.41)^3 =$ -14.00</td></tr>
<tr><td>92</td><td>$(92 - 81.41)^3 =$ 118.65</td></tr>
<tr><td>91</td><td>$(91 - 81.41)^3 =$ 881.97</td></tr>
<tr><td>65</td><td>$(65 - 81.41)^3 =$ -4419.02</td></tr>
<tr><td>78</td><td>$(78 - 81.41)^3 =$ -39.65</td></tr>
<tr><td>91</td><td>$(91 - 81.41)^3 =$ 881.97</td></tr>
<tr><td>88</td><td>$(88 - 81.41)^3 =$ 286.19</td></tr>
<tr><td>89</td><td>$(89 - 81.41)^3 =$ 437.25</td></tr>
<tr><td>105</td><td>$(105 - 81.41)^3 =$ 13127.55</td></tr>
<tr><td>76</td><td>$(76 - 81.41)^3 =$ -158.34</td></tr>
<tr><td>67</td><td>$(67 - 81.41)^3 =$ -2992.21</td></tr>
<tr><td>80</td><td>$(80 - 81.41)^3 =$ -2.80</td></tr>
<tr><td>102</td><td>$(102 - 81.41)^3 =$ 8729.09</td></tr>
<tr><td>79</td><td>$(79 - 81.41)^3 =$ -14.00</td></tr>
<tr><td>57</td><td>$(57 - 81.41)^3 =$ -14544.73</td></tr>
<tr><td>77</td><td>$(77 - 81.41)^3 =$ -85.77</td></tr>
<tr><td>74</td><td>$(74 - 81.41)^3 =$ -406.87</td></tr>
</tbody>
</table>

$$\Sigma(X - \bar{X})^3 = -21393.54$$

$$Skew = \frac{\Sigma(X - \bar{X})^3}{N(S^3)} = \frac{-21393.54}{37(11.83^3)} = -0.35$$

Thus, the distribution of scores is negatively skewed.

Kurtosis Index

X	$(X - \bar{X})^4$	
75	$(75 - 81.41)^4 =$	1688.23
80	$(80 - 81.41)^4 =$	3.95
103	$(103 - 81.41)^4 =$	217275.41
65	$(65 - 81.41)^4 =$	72516.08
79	$(79 - 81.41)^4 =$	33.73
88	$(88 - 81.41)^4 =$	1886.00
71	$(71 - 81.41)^4 =$	11743.65
86	$(86 - 81.41)^4 =$	443.86
77	$(77 - 81.41)^4 =$	378.23
78	$(78 - 81.41)^4 =$	135.21
49	$(49 - 81.41)^4 =$	1103357.20
86	$(86 - 81.41)^4 =$	443.86
76	$(76 - 81.41)^4 =$	856.62
96	$(96 - 81.41)^4 =$	45312.83
76	$(7 - 81.41)^4 =$	856.62
92	$(92 - 81.41)^4 =$	12577.20
87	$(87 - 81.41)^4 =$	976.44
87	$(87 - 81.41)^4 =$	976.44
80	$(80 - 81.41)^4 =$	3.95
91	$(91 - 81.41)^4 =$	8458.13
79	$(79 - 81.41)^4 =$	33.73
92	$(92 - 81.41)^4 =$	12577.20
91	$(91 - 81.41)^4 =$	8458.13
65	$(65 - 81.41)^4 =$	72516.08
78	$(78 - 81.41)^4 =$	135.21
91	$(91 - 81.41)^4 =$	8458.13
88	$(88 - 81.41)^4 =$	1886.00
89	$(89 - 81.41)^4 =$	3318.69
105	$(105 - 81.41)^4 =$	309679.01
76	$(76 - 81.41)^4 =$	856.62
67	$(67 - 81.41)^4 =$	43117.73
80	$(80 - 81.41)^4 =$	3.95
102	$(102 - 81.41)^4 =$	179732.00
79	$(79 - 81.41)^4 =$	33.73
57	$(57 - 81.41)^4 =$	355035.00
77	$(77 - 81.41)^4 =$	378.23
74	$(74 - 81.41)^4 =$	3014.90

$$\Sigma(X - \bar{X})^4 = 2479157.94$$

$$K = \frac{\Sigma(X - \bar{X})^4}{N(S^4)} - 3 = \frac{2479157.94}{37(11.83^4)} - 3 = 0.42$$

Thus, the distribution of scores is leptokurtic.

c. *Cognitive level scores.* Pearson's index of skewness

$$P_s = \frac{3(\bar{X} - Mdn)}{S_x} = \frac{3(59.81 - 62)}{16.89} = -0.39$$

Thus, the distribution of scores is negatively skewed.

Skewness Index

X	$(X - \bar{X})^3$	
74	$(74 - 59.81)^3 =$	2857.24
50	$(50 - 59.81)^3 =$	−944.08
66	$(66 - 59.81)^3 =$	237.18
65	$(65 - 59.81)^3 =$	139.80
67	$(67 - 59.81)^3 =$	371.69
54	$(54 - 59.81)^3 =$	−196.12
54	$(54 - 59.81)^3 =$	−196.12
78	$(78 - 59.81)^3 =$	6018.64
52	$(52 - 59.81)^3 =$	−476.38
72	$(72 - 59.81)^3 =$	1811.39
53	$(53 - 59.81)^3 =$	−315.82
62	$(62 - 59.81)^3 =$	10.50
81	$(81 - 59.81)^3 =$	9514.65
68	$(68 - 59.81)^3 =$	549.35
54	$(54 - 59.81)^3 =$	−196.12
62	$(62 - 59.81)^3 =$	10.50
74	$(74 - 59.81)^3 =$	2857.24
85	$(85 - 59.81)^3 =$	15983.96
42	$(42 - 59.81)^3 =$	−5649.26
33	$(33 - 59.81)^3 =$	−19270.39
38	$(38 - 59.81)^3 =$	−10374.50
37	$(37 - 59.81)^3 =$	−11868.95
86	$(86 - 59.81)^3 =$	17964.14
32	$(32 - 59.81)^3 =$	−21508.15
85	$(85 - 59.81)^3 =$	15983.96
31	$(31 - 59.81)^3 =$	−23912.76

$$\Sigma(X - \bar{X})^3 = -20597.39$$

$$Skew = \frac{\Sigma(X - \bar{X})^3}{N(S^3)} = \frac{-20597.39}{26(16.89^3)} = -0.16$$

Thus, the distribution of scores is negatively skewed.

Kurtosis Index

X	$(X - \bar{X})^4$	
74	$(74 - 59.81)^4 =$	40544.28
50	$(50 - 59.81)^4 =$	9261.39
66	$(66 - 59.81)^4 =$	1468.12
65	$(65 - 59.81)^4 =$	725.55
67	$(67 - 59.81)^4 =$	2672.49
54	$(54 - 59.81)^4 =$	1139.47
54	$(54 - 59.81)^4 =$	1139.47
78	$(78 - 59.81)^4 =$	109479.00
52	$(52 - 59.81)^4 =$	3720.52
72	$(72 - 59.81)^4 =$	22080.80
53	$(53 - 59.81)^4 =$	2150.74
62	$(62 - 59.81)^4 =$	23.00
81	$(81 - 59.81)^4 =$	201615.46
68	$(68 - 59.81)^4 =$	4499.20
54	$(54 - 59.81)^4 =$	1139.47
62	$(62 - 59.81)^4 =$	23.00
74	$(74 - 59.81)^4 =$	40544.28
85	$(85 - 59.81)^4 =$	402636.06
42	$(42 - 59.81)^4 =$	100613.37
33	$(33 - 59.81)^4 =$	516639.08
38	$(38 - 59.81)^4 =$	226267.75
37	$(37 - 59.81)^4 =$	270708.03
86	$(86 - 59.81)^4 =$	470480.90
32	$(32 - 59.81)^4 =$	598141.53
85	$(85 - 59.81)^4 =$	402636.06
31	$(31 - 59.81)^4 =$	688926.73

$$\Sigma(X - \bar{X})^4 = 4119275.80$$

$$K = \frac{\Sigma(X - \bar{X})^4}{N(S^4)} - 3 = \frac{4119275.80}{26(16.89^4)} - 3 = -1.05$$

Thus, the distribution of scores is platykurtic.

d. *Assertiveness scores.* Pearson's index of skewness

$$P_s = \frac{3(\overline{X} - Mdn)}{S_x} = \frac{3(59.72 - 60)}{23.64} = -0.04$$

Thus, the distribution of scores is slightly negatively skewed, but really close to being symmetric.

Skewness Index

X	$(X - \overline{X})^3$
75	$(75 - 59.72)^3 =$ 3567.55
37	$(37 - 59.72)^3 =$ −11728.03
13	$(13 - 59.72)^3 =$ −101978.47
60	$(60 - 59.72)^3 =$ 0.02
49	$(49 - 59.72)^3 =$ −1231.93
76	$(76 - 59.72)^3 =$ 4314.83
88	$(88 - 59.72)^3 =$ 22617.17
12	$(12 - 59.72)^3 =$ −108667.91
70	$(70 - 59.72)^3 =$ 1086.37
89	$(89 - 59.72)^3 =$ 25102.28
61	$(61 - 59.72)^3 =$ 2.10
53	$(53 - 59.72)^3 =$ −303.46
93	$(93 - 59.72)^3 =$ 36859.54
45	$(45 - 59.72)^3 =$ −3189.51
52	$(52 - 59.72)^3 =$ −460.10
99	$(99 - 59.72)^3 =$ 60605.83
17	$(17 - 59.72)^3 =$ −77963.93
59	$(59 - 59.72)^3 =$ −0.37
40	$(40 - 59.72)^3 =$ −7668.68
63	$(63 - 59.72)^3 =$ 35.29
94	$(94 - 59.72)^3 =$ 40283.06
59	$(59 - 59.72)^3 =$ −0.37
86	$(86 - 59.72)^3 =$ 18149.98
75	$(75 - 59.72)^3 =$ 3567.55
39	$(39 - 59.72)^3 =$ −8895.48
80	$(80 - 59.72)^3 =$ 8340.73
45	$(45 - 59.72)^3 =$ −3189.51
65	$(65 - 59.72)^3 =$ 147.20
38	$(38 - 59.72)^3 =$ −10246.59

$$\sum(X - \overline{X})^3 = -110844.84$$

$$Skew = \frac{\sum(X - \overline{X})^3}{N(S^3)} = \frac{-110844.84}{29(23.64^3)} = -0.29$$

Thus, the distribution of scores is negatively skewed.

Kurtosis Index

X	$(X - \overline{X})^4$
75	$(75 - 59.72)^4 =$ 54512.16
37	$(37 - 59.72)^4 =$ 266460.79
13	$(13 - 59.72)^4 =$ 4764434.20
60	$(60 - 59.72)^4 =$ 0.01
49	$(49 - 59.72)^4 =$ 13206.24
76	$(76 - 59.72)^4 =$ 70245.35
88	$(88 - 59.72)^4 =$ 639613.50
12	$(12 - 59.72)^4 =$ 5185633.60
70	$(70 - 59.72)^4 =$ 11167.92
89	$(89 - 59.72)^4 =$ 734994.84
61	$(61 - 59.72)^4 =$ 2.68
53	$(53 - 59.72)^4 =$ 2039.28
93	$(93 - 59.72)^4 =$ 1226686.60
45	$(45 - 59.72)^4 =$ 46949.53
52	$(52 - 59.72)^4 =$ 3551.97
99	$(99 - 59.72)^4 =$ 2380597.20
17	$(17 - 59.72)^4 =$ 3330619.20
59	$(59 - 59.72)^4 =$ 0.27
40	$(40 - 59.72)^4 =$ 151226.41
63	$(63 - 59.72)^4 =$ 115.74
94	$(94 - 59.72)^4 =$ 1380903.30
59	$(59 - 59.72)^4 =$ 0.27
86	$(86 - 59.72)^4 =$ 476981.40
75	$(75 - 59.72)^4 =$ 54512.16
39	$(39 - 59.72)^4 =$ 184314.29
80	$(80 - 59.72)^4 =$ 169149.92
45	$(45 - 59.72)^4 =$ 46949.53
65	$(65 - 59.72)^4 =$ 777.21
38	$(38 - 59.72)^4 =$ 222555.99

$$\sum(X - \overline{X})^4 = 21318199.00$$

$$K = \frac{\sum(X - \overline{X})^4}{N(S^4)} - 3 = \frac{21318199}{29(23.64^4)} - 3 = -0.65$$

Thus, the distribution of scores is platykurtic.

21. a. *See Table A4-2.*

b. For Japanese participants the Collectivism Scale scores were negatively skewed (−.225) and slightly leptokurtic (.091). The Individually-Oriented Achievement Motivation scores were slightly positively skewed (.027) and platykurtic (−.117). The Socially-Oriented Achievement Motivation scores were negatively skewed (−.119) and platykurtic (−.427). The Individualistic Self-Esteem scores were positively skewed (.275) and platykurtic (−.107). The Collectivistic Self-Esteem scores were positively skewed (.249) and leptokurtic (.493). The Independent Self-Construal scores

Table A4-2 Statistics

		Collectivism Scale Score	Individually-Oriented Achievement Motivation	Socially-Oriented Achievement Motivation	Individualistic Self-Esteem	Collectivistic Self-Esteem	Independent Self-Construal	Interdependent Self-Construal
N Valid		94	94	94	94	94	94	94
Missing		0	0	0	0	0	0	0
Std. Deviation		6.76	16.50	22.99	9.46	13.24	6.48	6.91
Variance		45.76	272.12	528.42	89.56	175.39	42.02	47.73
Skewness		−.225	.027	−.119	.275	−.775	−.033	−.492
Std. Error of Skewness		.249	.249	.249	.249	.249	.249	.249
Kurtosis		.091	−.117	−.427	−.107	1.119	−.105	.206
Std. Error of Kurtosis		.493	.493	.493	.493	.493	.493	.493
Range		33	81	104	43	75	31	35
Percentiles	25	36.75	113.00	70.75	37.00	68.00	56.00	47.75
	50	42.00	123.00	86.50	42.50	78.00	60.00	54.00
	75	46.00	135.00	102.25	49.00	85.00	64.00	57.00

		Confidence in Doing Math	Perception of Mother's Attitude Toward Math	Perception of Father's Attitude Toward Math	Success at Doing Math
N Valid		200	200	200	200
Missing		0	0	0	0
Std. Deviation		9.62	9.21	8.93	6.55
Variance		92.62	84.80	79.81	42.87
Skewness		−.442	−1.114	−1.252	−.445
Std. Error of Skewness		.172	.172	.172	.172
Kurtosis		−.161	2.878	3.236	.098
Std. Error of Kurtosis		.342	.342	.342	.342
Range		45	57	52	36
Percentiles	25	38.00	40.25	39.00	41.00
	50	46.00	48.00	46.00	46.00
	75	53.00	54.00	52.75	51.00

Table A4-4   Statistics

		Perception of Current Math Teacher's Attitude Toward Math	Perception of Math as a Male Domain	Perception of the Usefulness of Math	Anxiety About Doing Math	Perception of Math as a Fun Activity
N Valid		200	200	200	200	200
Missing		0	0	0	0	0
Std. Deviation		8.31	9.64	10.00	10.40	10.13
Variance		69.07	92.97	99.94	108.15	102.65
Skewness		−.083	−1.635	−1.038	−.312	−.136
Std. Error of Skewness		.172	.172	.172	.172	.172
Kurtosis		−.162	2.704	1.004	−.412	−.703
Std. Error of Kurtosis		.342	.342	.342	.342	.342
Range		44	48	48	46	46
Percentiles	25	36.00	47.00	42.00	35.00	32.25
	50	42.00	54.00	50.00	42.00	40.00
	75	47.00	57.75	56.00	49.75	48.00

were slightly negatively skewed (−.033) and platykurtic (−.105). The Interdependent Self-Construal scores were negatively skewed (−.492) and leptokurtic (.206).

23. a. *See Tables A4-3 and A4-4.*
   b. For confidence in doing math the distribution of scores was negatively skewed (−.442) and platykurtic (−.161). For perception of mother's attitude toward math, the distribution of scores was negatively skewed (−1.114) and leptokurtic (2.878). For perception of father's attitude toward math, the distribution of scores was negatively skewed (−1.252) and leptokurtic (3.236). For success at doing math, the distribution of scores was negatively skewed (−.445) and

slightly leptokurtic (.098). For perception of current math teacher's attitude toward math, the distribution of scores was slightly negatively skewed (−.083) and platykurtic (−.162). For perception of math as a male domain, the distribution of scores was negatively skewed (−1.635) and leptokurtic (2.704). For perception of the usefulness of math, the distribution of scores was negatively skewed (−1.038) and leptokurtic (1.004). For anxiety about doing math, the distribution of scores was negatively skewed (−.312) and platykurtic (−.412). For perception of math as a fun activity, the distribution of scores was negatively skewed (−.136) and platykurtic (−.703).

CHAPTER 5

1. Answers will vary on this exercise. Self-concept has a number of indicators such as the number of friends a person has, how well a person achieves in tasks, and how a person feels about himself or herself. These could be measured in several ways. One way that the number of friends a person has could be measured is to ask the person to list his or her friends. A second way that we could measure the number of friends a person has is to ask the people in the person's environment (e.g., school, clubs, and church) to list their friends and then count the number of times that the person is listed. Positive self-concept may also be indicated by achievement in school. Thus, the grade point average of a person could

be indicative of his or her self-concept. Positive self-concept might be measured by how a person feels about himself or herself. This might be measured by responses to statements such as "In general, I like myself," "I believe that I am a good person," etc.

3. A direct measure is a measure where the trait being measured is observable by a second person; for example, the amount of time it takes a person to complete a puzzle or whether the answer a person gives to a mathematics question is correct. An indirect measure is a measure of some behavior that can be observed by a second person and from which the existence of a trait can be inferred. Examples of indirect measures will vary in this ex-

ercise. The level of anxiety a person feels in a certain situation, a person's knowledge of science, and a person's memory skills are all indirect measures. The level of anxiety could be measured by giving the person a questionnaire in which she is asked to rate her level of anxiety in a number of situations such as speaking in front of people, taking a test, or asking another person out on a date. Another way of measuring anxiety is by measuring the galvanic skin response to a series of questions. To measure a person's knowledge of science, a science achievement test could be administered. To measure a person's memory, a list of nonsense words could be given and the person could be asked to recall them at some later time.

5. a. absolute measure
   b. relative measure
   c. relative measure
   d. absolute measure
   e. relative measure

7. a. All people who have obtained a Ph.D. and have 2,000 hours of supervised counseling.
   b. All people who have completed the educational requirements necessary to be a teacher.
   c. All people who have met the eligibility requirements for taking the LSAT.
   d. All people who are qualified to apply for a job in a state.
   e. All people who are entering high school.

9.

X	f	cf
29	1	50
28	2	49
27	5	47
26	0	42
25	4	42
24	6	38
23	4	32
22	4	28
21	5	24
20	3	19
19	3	16
18	3	13
17	4	10
16	4	6
15	2	2

If the upper 10% are highly stressed, the researcher must find the 90th percentile [100 − 10 = 90]. If the next 20% are moderately stressed, the researcher must find the 70th percentile [100 − (10 + 20) = 70]. If the next 40% are somewhat stressed, the researcher must find the 30th percentile [100 − (10 + 20 + 40) = 30]. If the next 20% are slightly stressed, the researcher must find the 10th percentile [100 − (10 + 20 + 40 + 20) = 10]. Finally, if the last 10% are not stressed, the researcher must find the 0th percentile [100 − (10 + 20 + 40 + 20 + 10) = 0] which will be the lower limit of the lowest score.

To find the percentiles use equation 5.1.

$$X_p = LL_j + \left(\frac{pN - cf_{j-1}}{f_j} \right)(J)$$

$$X_{.90} = 26.5 + \left(\frac{.90(50) - 42}{5} \right)(1)$$

= 27.10 is the 90th percentile
for highly stressed individuals.

$$X_{.70} = 23.5 + \left(\frac{.70(50) - 32}{6} \right)(1)$$

= 24.00 is the 70th percentile
for moderately stressed individuals.

$$X_{.30} = 18.5 + \left(\frac{.30(50) - 13}{3} \right)(1)$$

= 19.17 is the 30th percentile
for somewhat stressed individuals.

$$X_{.10} = 15.5 + \left(\frac{.10(50) - 2}{4} \right)(1)$$

= 16.25 is the 10th percentile
for slightly stressed individuals.

$X_{.00} = 14.5$ is the 0th percentile
for individuals who are not stressed.

11. To find percentile ranks of observed scores use equation 5.2 for the quiz scores.

$$PR_j = \left(\frac{cf_{j-1} + \frac{1}{2}f_j}{N} \right)$$

$$PR_{12} = \left(\frac{1 + \frac{1}{2}(0)}{19} \right) = .05 \text{ or a percentile rank of } 5$$

$$PR_{14} = \left(\frac{1 + \frac{1}{2}(2)}{19} \right) = .11 \text{ or a percentile rank of } 11$$

$$PR_{17} = \left(\frac{8 + \frac{1}{2}(0)}{19} \right) = .42 \text{ or a percentile rank of } 42$$

$$PR_{18} = \left(\frac{8 + \frac{1}{2}(5)}{19} \right) = .55 \text{ or a percentile rank of } 55$$

13. A percentile indicates the point in the distribution below which a certain specified proportion of the scores fall, whereas the percentile rank of a score is the proportion of the reference group that earned scores below the given score. Thus, percentiles are usually measured on a continuous scale, whereas percentile ranks are usually measured on a discrete scale of obtainable score values. Answers will vary with regard to situations where percentiles or percentile ranks would be more appropriate. For example, in situations where someone wants to know the scores that will divide the group into defined proportions, percentiles should be used. For example, if a person wants to assign grades such that 7% get A's, 23% get B's, 40% get C's, 23% get D's, and 7% get F's, then percentiles should be calculated. Suppose someone developed a scale such that 10% should be in the first category, 25% in the next category, 30% in the third category, 25% in the fourth category, and 10% in the final category (such as confidence as defined by some theory), then the percentiles should be calculated. Suppose that a school wants to identify individuals who scored in the lower 25% on a test of reading comprehension so that those students could receive additional instruction in reading. The students' percentile ranks on the reading comprehension test would need to be reported. In order to receive a university academic scholarship in English, an applicant must score in the upper 2% on the Verbal Test of the Scholastic Aptitude Test. Thus, scores should be reported as percentile ranks, with a person receiving a scholarship if she or he scored a percentile rank of 98 or higher.

15. Using the frequency table in Exercise 8 and the formula

$$\overline{X} = \frac{\Sigma f(X)}{N}$$

$$= \frac{2(20) + 4(19) + 5(18) + 3(16) + 2(15) + 2(14) + 1(11)}{19}$$

$$= \frac{323}{19} = 17$$

For variance and standard deviation, use the following formula to find the sum of squares: $\Sigma f(X)^2 = 2(20)^2 + 4(19)^2 + 5(18)^2 + 3(16)^2 + 2(15)^2 + 2(14)^2 + 1(11)^2 = 5595$.

   Use the following formula to calculate variance:

$$S^2 = \frac{N(\Sigma f(X)^2) - (\Sigma f(X))^2}{N^2} = \frac{19(5595) - (323)^2}{19^2} = 5.47$$

$$S = \sqrt{S^2} = \sqrt{5.47} = 2.34$$

a. Adding 5 to each score and using the formula

$$\overline{X} = \frac{\Sigma f(X)}{N}$$

$$= \frac{2(25) + 4(24) + 5(23) + 3(21) + 2(20) + 2(19) + 1(16)}{19}$$

$$= \frac{418}{19} = 22$$

For variance and standard deviation, use the following formula to find the sum of squares: $\Sigma f(X)^2 = 2(25)^2 + 4(24)^2 + 5(23)^2 + 3(21)^2 + 2(20)^2 + 2(19)^2 + 1(16)^2 = 9300$.

   Use the following formula to calculate variance:

$$S^2 = \frac{N(\Sigma f(X)^2) - (\Sigma f(X))^2}{N^2} = \frac{19(9300) - (418)^2}{19^2} = 5.47$$

$$S = \sqrt{S^2} = \sqrt{5.47} = 2.34$$

b. Subtracting 11 from each score and using the formula

$$\overline{X} = \frac{\Sigma f(X)}{N}$$

$$= \frac{2(29) + 4(8) + 5(7) + 3(5) + 2(4) + 2(3) + 1(0)}{19}$$

$$= \frac{114}{19} = 6$$

For variance and standard deviation, use the following formula to find the sum of squares: $\Sigma f(X)^2 = 2(9)^2 + 4(8)^2 + 5(7)^2 + 3(5)^2 + 2(4)^2 + 2(3)^2 + 1(0)^2 = 788$.

   Use the following formula to calculate variance:

$$S^2 = \frac{N(\Sigma f(X)^2) - (\Sigma f(X))^2}{N^2} = \frac{19(788) - (114)^2}{19^2} = 5.47$$

$$S = \sqrt{S^2} = \sqrt{5.47} = 2.34$$

c. Multiplying each score by 10 and using the formula

$$\overline{X} = \frac{\Sigma f(X)}{N}$$

$$= \frac{\begin{matrix} 2(200) + 4(190) + 5(180) + 3(160) \\ + 2(150) + 2(140) + 1(110) \end{matrix}}{19}$$

$$= \frac{3230}{19} = 170$$

For variance and standard deviation, use the following formula to find the sum of squares: $\Sigma f(X)^2 = 2(200)^2 + 4(190)^2 + 5(180)^2 + 3(160)^2 + 2(150)^2 + 2(140)^2 + 1(110)^2 = 559500$.

   Use the following formula to calculate variance:

$$S^2 = \frac{N(\Sigma f(X)^2) - (\Sigma f(X))^2}{N^2} = \frac{19(559500) - (3230)^2}{19^2}$$

$$= 547.37$$

$$S = \sqrt{S^2} = \sqrt{547.37} = 23.40$$

d. Dividing each score by 2 and using the formula

$$\overline{X} = \frac{\Sigma f(X)}{N}$$

$$= \frac{2(10) + 4(9.5) + 5(9) + 3(8) + 2(7.5) + 2(7) + 1(5.5)}{19}$$

$$= \frac{161.5}{19} = 8.5$$

For variance and standard deviation, use the following formula to find the sum of squares: $\Sigma f(X)^2 = 2(10)^2 + 4(9.5)^2 + 5(9)^2 + 3(8)^2 + 2(7.5)^2 + 2(7)^2 + 1(5.5)^2 = 1398.75$.

   Use the following formula to calculate variance:

$$S^2 = \frac{N(\Sigma f(X)^2) - (\Sigma f(X))^2}{N^2} = \frac{19(1398.75) - (161.5)^2}{19^2}$$

$$= 1.37$$

$$S = \sqrt{S^2} = \sqrt{1.37} = 1.17$$

e. If a constant is added to each score in a set of scores, the mean is increased by that constant, and the variance and standard deviation are unchanged. If a constant is subtracted from each score in a set of scores, the mean is decreased by that constant, and the variance and standard deviation are unchanged. If every score in a set of scores is multiplied by a constant, the new mean is the product of the constant and the original mean, the new variance is the product of the square of the constant and the original variance, and the new standard deviation is the product of the constant and the original standard deviation. If every score in a set of scores is divided by a constant, the new mean is the original mean divided by the constant, the new variance is the original variance divided by the square of the constant, and the new standard deviation is the original standard deviation divided by the constant.

f. To convert to standard scores, use the following formula:

$$Z_i = \frac{X_i - \overline{X}}{S_X}$$

$$Z_{20} = \frac{20 - 17}{2.34} = 1.28 \qquad Z_{19} = \frac{19 - 17}{2.34} = 0.85$$

$$Z_{18} = \frac{18 - 17}{2.34} = 0.43 \qquad Z_{16} = \frac{16 - 17}{2.34} = -0.43$$

$$Z_{15} = \frac{15 - 17}{2.34} = -0.85 \qquad Z_{14} = \frac{14 - 17}{2.34} = -1.28$$

$$Z_{11} = \frac{11 - 17}{2.34} = -2.56$$

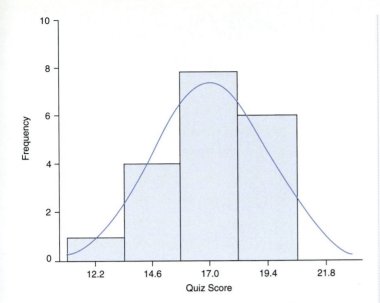

Histogram of the raw quiz scores with a normal curve

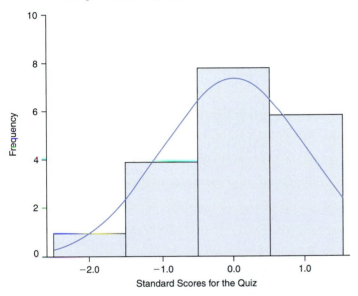

Histogram of the standard scores for the quiz with a normal curve

There is no change in the shape of the distribution of scores when they are converted to standard scores.

g. To convert the raw scores to IQ scores, the scores must first be converted to Z-scores and then the IQ scores using the following formula:

$$IQ_j = 100 + 15Z_j$$
$$IQ_{20} = 100 + 15(1.28) = 119.20$$

Since IQ scores are usually rounded to the nearest unit the score would be 119.

$$IQ_{19} = 100 + 15(0.85) = 112.75$$

Since IQ scores are usually rounded to the nearest unit the score would be 113.

$$IQ_{18} = 100 + 15(0.43) = 106.45$$

Since IQ scores are usually rounded to the nearest unit the score would be 106.

$$IQ_{16} = 100 + 15(-0.43) = 93.55$$

Since IQ scores are usually rounded to the nearest unit the score would be 94.

$$IQ_{15} = 100 + 15(-0.85) = 87.25$$

Since IQ scores are usually rounded to the nearest unit the score would be 87.

$$IQ_{14} = 100 + 15(-1.28) = 80.80$$

Since IQ scores are usually rounded to the nearest unit the score would be 81.

$$IQ_{11} = 100 + 15(-2.56) = 61.60$$

Since IQ scores are usually rounded to the nearest unit the score would be 62.

17. Irrespective of the distribution of the raw scores and the measures of center and variability, the mean of the set of Z-scores will always be zero (within rounding error), and the standard deviation will always be 1.0 (within rounding error). It should be noted that the shape of the distribution of Z-scores is the same as the shape of the distribution of the raw scores. Z-scores indicate the number of standard deviations that a person scores below or above the mean, but they preserve the shape of the original distribution. Thus, if we want to conduct any analysis in which we want to make normative comparisons but want the shape of the distribution preserved, we should use Z-scores since percentile ranks change the shape of the distribution.

19. Solving the equation $X' = 50X - 10$ for X, the following equation is obtained:

$$X = \frac{X' + 10}{50} = \frac{X'}{50} + \frac{10}{50} = \frac{X'}{50} + .2$$

Thus, $\overline{X} = \frac{\overline{X}'}{50} + .2 = \frac{468}{50} + .2 = 9.56$ since dividing a set of scores by a constant yields a mean that is the original mean divided by the constant and adding a constant to a set of scores results in a mean that is the sum of the original mean and the constant. $S = \frac{S'}{50} = \frac{125}{50} = 2.5$ since the standard deviation is unaffected by adding a constant to each score, but is the quotient of the original standard deviation and the constant when each score is divided by a constant.

CHAPTER 6

1. The three kinds of probability are subjective probability, theoretical probability, and empirical probability. Subjective probability is the guess that we make about an event occurring based on our past experiences. An exam-ple of a subjective probability is "What is the probability of your statistics teacher asking you a question when you have not read the assigned readings?" Theoretical probability is a branch of applied mathematics in which

the distribution of a chance event is studied in the abstract. Problems that involve theoretical probability are usually thought exercises and do not require the actual collection of data. An example of a theoretical probability is "What is the probability of obtaining a 3 on a single roll of a die?" Empirical probability is the probability of a future event happening based on careful recording of the frequency of occurrence of that event in the past. An example of an empirical probability is "Based on the voter turnout in the last 20 years during a presidential election, what is the probability that the voter turnout will exceed 60%?"

3.

Host	Player 1	Player 2	Winners
Red	Red	Red	Player 1 ($2) & Player 2 ($2)
Red	Red	Green	Host ($1) & Player 1 ($2)
Red	Red	Purple	Host ($1) & Player 1 ($2)
Red	Green	Red	Host ($1) & Player 2 ($2)
Red	Green	Green	Host ($2)
Red	Green	Purple	Host ($2)
Red	Purple	Red	Host ($1) & Player 2 ($2)
Red	Purple	Green	Host ($2)
Red	Purple	Purple	Host ($2)
Green	Red	Red	Host ($2)
Green	Red	Green	Host ($1) & Player 2 ($2)
Green	Red	Purple	Host ($2)
Green	Green	Red	Host ($1) & Player 1 ($2)
Green	Green	Green	Player 1 ($2) & Player 2 ($2)
Green	Green	Purple	Host ($1) & Player 1 ($2)
Green	Purple	Red	Host ($2)
Green	Purple	Green	Host ($1) & Player 2 ($2)
Green	Purple	Purple	Host ($2)
Purple	Red	Red	Host ($2)
Purple	Red	Green	Host ($2)
Purple	Red	Purple	Host ($1) & Player 2 ($2)
Purple	Green	Red	Host ($2)
Purple	Green	Green	Host ($2)
Purple	Green	Purple	Host ($1) & Player 2 ($2)
Purple	Purple	Red	Host ($1) & Player 1 ($2)
Purple	Purple	Green	Host ($1) & Player 1 ($2)
Purple	Purple	Purple	Player 1 ($2) & Player ($2)

Yes, the game is fair. Each participant has an equally likely chance of selecting a red, green, or purple bead. Furthermore, the total amount the host can win across the 27 possibilities is $36 whereas the combined total that the two players can win across the 27 possibilities is $36 ($18 each). Thus, neither the host nor the players have an advantage.

5. There are two laws associated with probability theory: the addition law and the multiplication law. If event A and event B are mutually exclusive, the addition law states that the probability of *either* event A *or* event B occurring is equal to the *sum* of their individual probabilities. If event A and event B are mutually exclusive, the multiplication law states that the probability of *both* event A *and* event B occurring is equal to the *product* of their individual probabilities.

7. For this exercise, there are a total of 20 red cards and 24 black cards.

a. $P_r = \left(\dfrac{20}{44}\right) = .45 \qquad P_b = \left(\dfrac{24}{44}\right) = .55$

b. $P_{rr} = (P_r)(P_r) = \left(\dfrac{20}{44}\right)\left(\dfrac{20}{44}\right) = .21$

$P_{bb} = (P_b)(P_b) = \left(\dfrac{24}{44}\right)\left(\dfrac{24}{44}\right) = .30$

c. $P_{two\ different\ colors} = P_{rb\ or\ br} = P_{rb} + P_{br} = (P_{r\ and\ b}) + (P_{b\ and\ r})$
$= [(P_r)(P_b)] + [(P_b)(P_r)]$
$= \left[\left(\dfrac{20}{44}\right)\left(\dfrac{24}{44}\right)\right] + \left[\left(\dfrac{24}{44}\right)\left(\dfrac{20}{44}\right)\right]$
$= .50$

d. $P_{rrr} = \left(\dfrac{20}{44}\right)\left(\dfrac{20}{44}\right)\left(\dfrac{20}{44}\right) = .09$

$P_{rrb} = \left(\dfrac{20}{44}\right)\left(\dfrac{20}{44}\right)\left(\dfrac{24}{44}\right) = .11$

$P_{rbr} = \left(\dfrac{20}{44}\right)\left(\dfrac{24}{44}\right)\left(\dfrac{20}{44}\right) = .11$

$P_{brr} = \left(\dfrac{24}{44}\right)\left(\dfrac{20}{44}\right)\left(\dfrac{20}{44}\right) = .11$

$P_{rbb} = \left(\dfrac{20}{44}\right)\left(\dfrac{24}{44}\right)\left(\dfrac{24}{44}\right) = .14$

$P_{brb} = \left(\dfrac{24}{44}\right)\left(\dfrac{20}{44}\right)\left(\dfrac{24}{44}\right) = .14$

$P_{bbr} = \left(\dfrac{24}{44}\right)\left(\dfrac{24}{44}\right)\left(\dfrac{20}{44}\right) = .14$

$P_{bbb} = \left(\dfrac{24}{44}\right)\left(\dfrac{24}{44}\right)\left(\dfrac{24}{44}\right) = .16$

The sum of all of these probabilities is 1.

9. a. $P_{circle} = \left(\dfrac{1}{6}\right) = .17$

b. $P_{red\ object} = \left(\dfrac{10}{11}\right) = .91$

c. $P_{blue\ square} = P_{blue\ and\ square} = (P_{blue})(P_{square})$
$= \left(\dfrac{1}{11}\right)\left(\dfrac{5}{6}\right) = .08$

d. $P_{red\ circle} = P_{red\ and\ circle} = (P_{red})(P_{circle}) = \left(\dfrac{10}{11}\right)\left(\dfrac{1}{6}\right) = .15$

e. No, because the events involved are not mutually exclusive. A square could be red or blue. Something that is blue could be a square or a circle. Thus, the events overlap.

11. a. Since there are two colors (red and black) in a standard deck of playing cards, the number of possible outcomes of 6 draws with replacement is $2^6 = 64$.

b. Since there are 13 different face values (ace, 2, 3, 4, 5, 6, 7, 8, 9, 10, jack, queen, king) in a standard deck of playing cards, the number of possible outcomes of 5 draws with replacement is $13^5 = 371{,}293$.

c. Since there are 4 different suits (diamonds, hearts, spades, clubs) in a standard deck of playing cards, the number of possible outcomes of 8 draws with replacement is $4^8 = 65{,}536$.

d. A die contains six faces with 6 different numbers. The total number of possible outcomes of 7 tosses of the die is $6^7 = 279,936$.

e. Since there are five different colors of beads that are equal in number, the total number of possible outcomes of 4 draws with replacement is $5^4 = 625$.

13. Formula 6.2 is

$$_NC_r = \frac{N!}{r!(N-r)!}$$

a. There are two outcomes for each event: a red card is drawn or a black card is drawn. The total number of possible outcomes for 6 draws is $N = 2^6 = 64$.

$$_6C_4 = \frac{6!}{4!(6-4)!} = \frac{6 \cdot 5 \cdot 4 \cdot 3 \cdot 2 \cdot 1}{(4 \cdot 3 \cdot 2 \cdot 1)(2 \cdot 1)} = 15$$

Thus, $P_{4 \text{ red cards out of 6 draws}} = \frac{_6C_4}{N} = \frac{15}{64} = .23$

b. There are two outcomes for each event: a head is tossed or a tail is tossed. The total number of possible outcomes for 8 tosses is $N = 2^8 = 256$.

$$_8C_2 = \frac{8!}{2!(8-2)!} = \frac{8 \cdot 7 \cdot 6 \cdot 5 \cdot 4 \cdot 3 \cdot 2 \cdot 1}{(2 \cdot 1)(6 \cdot 5 \cdot 4 \cdot 3 \cdot 2 \cdot 1)} = 28$$

Thus, $P_{2 \text{ beads out of 8 tosses}} = \frac{_8C_2}{N} = \frac{28}{256} = .11$

c. There are two outcomes for each event: a red bead is drawn or a blue bead is drawn. The total number of possible outcomes for 10 draws is $N = 2^{10} = 1024$.

$$_{10}C_0 = \frac{10!}{0!(10-0)!} = \frac{10!}{10!} = 1$$

Thus, $P_{0 \text{ red beads out of 10 draws}} = \frac{_{10}C_0}{N} = \frac{1}{1024} = .00098$

15. The total number of possible outcomes for tossing a die 5 times is $6^5 = 7,776$. From equation 6.2

$$_NC_r = \frac{N!}{r!(N-r)!} = \frac{5!}{3!(5-3)!}$$

There are $5^2 = 25$ ways of obtaining numbers to fill the remaining 2 values from among the 5 possible numbers. Thus,

$$P_{\text{exactly three 2s on 5 tosses of a die}} = \frac{10}{7776}(5^2) = .03215$$

Note: The same value can be obtained if we use equation 6.4

$$P_{3 \text{ of } 5} = \frac{5!}{3!(5-3)!}\left(\frac{1}{6}\right)^3\left(\frac{5}{6}\right)^2 = .03215$$

17. The normal distribution is the distribution of scores that seems to conform to naturally occurring phenomena such as height and weight. Note that the normal distribution is the limiting case of a binomial distribution with an infinite number of trials. The normal distribution is symmetric, mesokurtic, and unimodal.

19. The possible outcomes of the quiz scores are: 0, 1, 2, 3, 4, 5, 6, 7, 8, 9, 10. The probability of each outcome can be calculated using equation 6.4, $P_{r \text{ of } N} = \frac{N!}{r!(N-r)!} \times$ $(P^r)(Q^{N-r})$. Since there are four possible alternatives, the

probability of guessing the correct answer on a particular question is $P = .25$ whereas the probability of guessing an incorrect answer on a particular question is $Q = .75$. Thus,

$$P_{0 \text{ of } 10} = \frac{10!}{0!(10-0)!}(.25^0)(.75^{10-0}) = .0563$$

$$P_{1 \text{ of } 10} = \frac{10!}{1!(10-1)!}(.25^1)(.75^{10-1}) = .1877$$

$$P_{2 \text{ of } 10} = \frac{10!}{2!(10-2)!}(.25^2)(.75^{10-2}) = .2816$$

$$P_{3 \text{ of } 10} = \frac{10!}{3!(10-3)!}(.25^3)(.75^{10-3}) = .2503$$

$$P_{4 \text{ of } 10} = \frac{10!}{4!(10-4)!}(.25^4)(.75^{10-4}) = .1460$$

$$P_{5 \text{ of } 10} = \frac{10!}{5!(10-5)!}(.25^5)(.75^{10-5}) = .05840$$

$$P_{6 \text{ of } 10} = \frac{10!}{6!(10-6)!}(.25^6)(.75^{10-6}) = .01622$$

$$P_{7 \text{ of } 10} = \frac{10!}{7!(10-7)!}(.25^7)(.75^{10-7}) = .003090$$

$$P_{8 \text{ of } 10} = \frac{10!}{8!(10-8)!}(.25^8)(.75^{10-8}) = .0003862$$

$$P_{9 \text{ of } 10} = \frac{10!}{9!(10-9)!}(.25^9)(.75^{10-9}) = .0000286$$

$$P_{10 \text{ of } 10} = \frac{10!}{10!(10-10)!}(.25^{10})(.75^{10-10}) = .0000009$$

The most probable score on the quiz is 2 out of 10.

21. a. First convert 70 to a Z-score, $Z = \frac{70-100}{15} = -2.00$; then use Appendix Table A-1 to find the proportion of the area beyond 2.00 since the distribution is symmetric. This value is .0228, the probability of scoring less than 70 on the IQ test and receiving the designation of "educably mentally retarded."

b. First convert 90 to a Z-score, $Z = \frac{90-100}{15} = -0.67$; then use Appendix Table A-1 to find the proportion of the area between the mean and .67 since the distribution is symmetric. This value is .2486, the probability of scoring between the mean and 90 on the IQ test. Next convert 110 to a Z-score, $Z = \frac{110-100}{15} = .67$; then use Appendix Table A-1 to find the proportion of the area between the mean and .67. This value is .2486, the probability of scoring between the mean and 110 on the IQ test. The probability of scoring between 90 and 110 on the IQ test is the sum of the probability of scoring between the mean and 90 (.2486) and the mean and 110 (.2486). Thus, $P_{90 < IQ < 110} = .2486 + .2486 = .4972$, the probability of receiving the "normal" designation.

c. First convert 130 to a Z-score, $Z = \frac{130-100}{15} = 2.00$; then use Appendix Table A-1 to find the proportion of the area beyond 2.00. This value is .0228, the probability of scoring greater than 130 on the IQ test and receiving a designation of "gifted."

23. The formula $Z = \dfrac{X - \bar{X}}{S}$ and Appendix Table A-1 will be used to find percentile ranks.

a. $Z_{humanities} = \dfrac{77 - 81}{9} = -.44.$ Since Appendix Table A-1 has only positive Z values and since the distribution of Z values is symmetric, find the Z-score of .44 in Column 1 and read the value in Column 3. This value is .3300. Thus a score of 77 corresponds to a percentile rank of 33 for the humanities students.

$Z_{business\ administration} = \dfrac{77 - 69}{15} = .53.$ Since this value is positive, find the Z-score of .53 in Column 1, read the value in Column 2, and add .5 to this value since all scores below the mean comprise 50% of the distribution. This value is $.2019 + .5 = .7019$. Thus a score of 77 corresponds to a percentile rank of 70 if Jennifer had been a business administration student.

b. First, we must find the Z-score that corresponds to a percentile rank of 33 using Appendix Table A-1. Since the percentile rank is 33, we know the score is below the mean (percentile rank of 50). Thus, the Z-score will be negative. If we convert a percentile rank of 33 to a decimal, we obtain .3300. If we find the value closest to .3300 in Column 3 of Appendix Table A-1, we can read the Z-score value in Column 1. Thus, the Z-score that corresponds to a percentile rank of 33 is $-.44$. From this Z-score, we can find Martin's raw score. If we solve the formula for obtaining a Z-score for the raw score, we obtain the following equation:

$$X = (Z)(S) + \bar{X}$$

where $S = 15$ and $\bar{X} = 69$ for business administration students. Thus, Martin's raw score is

$$X = (-.44)(15) + 69 = 62.40$$

Since raw scores are usually recorded as whole numbers, we will round this score to 62. To determine what percentile rank Martin would have achieved with a score of 62 if he were a humanities student, we must first find the Z-score that corresponds to a 62 for humanities students. Thus,

$$Z_{humanities} = \dfrac{62 - 81}{9} = -2.11$$

Since Appendix Table A-1 has only positive Z values and since the distribution of Z values is symmetric,

find the Z-score of 2.11 in Column 1 and read the value in Column 3. This value is .0174. Thus a score of 62 corresponds to a percentile rank of 2 for Martin if he were a humanities student.

c. First, we must find the Z-score that corresponds to a percentile rank of 55 using Appendix Table A-1. Since the percentile rank is 55, we know the score is above the mean (percentile rank of 50). Thus, the Z-score will be positive. If we convert a percentile rank of 55 to a decimal, we obtain .5500. Since the score is above the mean, we must subtract .5 from .5500 to obtain a value that is presented in Appendix Table A-1. Thus, the value is .0500. If we find the value closest to .0500 in Column 2 of Appendix Table A-1, we can read the Z-score value in Column 1. Thus, the Z-score that corresponds to a percentile rank of 55 is .13. From this Z-score, we can find Alex's raw score. If we solve the formula for obtaining a Z-score for the raw score, we obtain the following formula:

$$X = (Z)(S) + \bar{X}$$

where $S = 15$ and $\bar{X} = 69$ for business administration students. Thus, Alex's raw score is

$$X = (.13)(15) + 69 = 70.95.$$

Since raw scores are usually reported as whole numbers, Alex's score will be rounded to 71.

First, we must find the Z-score that corresponds to a percentile rank of 43 using Appendix Table A-1. Since the percentile rank is 43, we know the score is below the mean (percentile rank of 50). Thus, the Z-score will be negative. If we convert a percentile rank of 43 to a decimal, we obtain .4300. If we find the value closest to .4300 in Column 3 of Appendix Table A-1, we can read the Z-score value in Column 1. Thus, the Z-score that corresponds to a percentile rank of 43 is $-.18$. From this Z-score, we can find Chris's raw score. If we solve the formula for obtaining a Z-score for the raw score, we obtain the following formula:

$$X = (Z)(S) + \bar{X}$$

where $S = 9$ and $\bar{X} = 81$ for humanities students. Thus, Chris's raw score is $X = (-.18)(9) + 81 = 79.38$. Since raw scores are usually reported as whole numbers, Chris's score will be rounded to 79. Thus Chris had a higher exam score than Alex.

CHAPTER 7

1. Answers will vary on this for the examples in this exercise. The following examples are provided to help illustrate each type of decision.

Selection decisions: Picking some individuals from a larger group for some defined purpose. Examples: (1) Based on the skills demonstrated in tryouts, a coach chooses the members of the varsity basketball team. (2) The counselors in a school choose the peer counselors from among the candidates who were nominated by the students.

Classification decisions: Assigning individuals to categories in which they are likely to make the greatest contribution to an organization. Examples: (1) The scores of a postal exam are used to determine to which job each qualified applicant will be assigned. (2) Volunteers are assigned specific duties associated with helping a town recover from a tornado (e.g., cooking, cleaning up debris, helping with childcare).

Placement decisions: Assigning individuals to a category that will maximize benefits to them. Examples:

(1) Students are assigned to a gifted program in English based on standardized test results. (2) Students are assigned to a remedial reading program based on their scores on a standardized reading test.

Personal decisions: Usually subjective choices that individuals make at various stages in their lives. Examples: (1) A student decides what residence hall to live in. (2) A person decides what type of car to buy.

3. a. criterion variable: tendency to procrastinate
   predictor variables: perfectionism, self-esteem, locus of control, and level of pessimism
   b. criterion variable: ability to achieve in algebra
   predictor variables: formal operational thinking, spatial ability, and understanding of basic mathematics concepts
   c. criterion variable: level of social phobia
   predictor variables: self-confidence, neuroticism, and embarrassability
   d. criterion variable: reading ability
   predictor variables: the number of books at home, the education level of the parents, the number of hours a parent reads to the child, and the child's vocabulary

5. a. Alex practiced the physical skill for one hour so we will use the mean and standard error of estimate of the one-hour practice group, 9 and 1.08, respectively. Since we are looking for a score above a criterion score, we must use the upper limit of the score 10, which is 10.5. Thus, we calculate the Z-score as

   $$Z_{10.5} = \frac{10.5 - 9}{1.08} = 1.39.$$

   We then look up this value in the column for the area beyond Z of Appendix Table A-1. The value from the table is .0823, which we round to .08. Thus, there is an 8 percent chance that the person will score above 10.
   b. Andrea practiced the physical skill for two hours so we will use the mean and standard error of estimate of the two-hour practice group, 15 and 1.08, respectively. Since we are looking for a score below a criterion score, we must use the lower limit of the score 13, which is 12.5. Thus, we calculate the Z-score as

   $Z_{12.5} = \frac{12.5 - 15}{1.08} = -2.31.$ We then look up this

   value in the column for the area beyond Z of Appendix Table A-1. The value from the table is .0104, which we round to .01. Thus, there is a 1 percent chance that the person will score below 13.
   c. Antionne practiced the physical skill for three hours so we will use the mean and standard error of estimate of the three-hour practice group, 18 and 1.08, respectively. Since we are looking for a score above a criterion score, we must also use the upper limit of the score 18, which is 18.5. Thus, we calculate the

   Z-score as $Z_{18.5} = \frac{18.5 - 18}{1.08} = .46.$ We then look up

   this value in the column for the area beyond Z of Appendix Table A-1. The value from the table is .3228, which we round to .32. Thus, there is a 32 percent chance that the person will score above 18.

7. a. If the slope of the regression line is 1.15 and the Y-intercept is 13.72, then the equation of the regres-

sion line is Science Final Exam Score = (1.15)(Attitude About Science) + 13.72.

b.

Student	X	Y	$\hat{Y} = 1.15X + 13.72$
1	50	70	1.15(50) + 13.72 = 71.22
2	40	60	1.15(40) + 13.72 = 59.72
3	40	65	1.15(40) + 13.72 = 59.72
4	60	85	1.15(60) + 13.72 = 82.72
5	65	85	1.15(65) + 13.72 = 88.47
6	58	78	1.15(58) + 13.72 = 80.42
7	48	68	1.15(48) + 13.72 = 68.92
8	50	73	1.15(50) + 13.72 = 71.22
9	60	84	1.15(60) + 13.72 = 82.72
10	60	82	1.15(60) + 13.72 = 82.72
11	55	80	1.15(55) + 13.72 = 76.97
12	65	90	1.15(65) + 13.72 = 88.47
13	70	95	1.15(70) + 13.72 = 94.22
14	50	68	1.15(50) + 13.72 = 71.22
15	55	74	1.15(55) + 13.72 = 76.97
16	62	88	1.15(62) + 13.72 = 85.02
17	55	78	1.15(55) + 13.72 = 76.97
18	68	94	1.15(68) + 13.72 = 91.92
19	45	64	1.15(45) + 13.72 = 65.47
20	55	70	1.15(55) + 13.72 = 76.97

c. Using the predicted scores from part (b)

Student	$(Y - \hat{Y})$	$(Y - \hat{Y})^2$
1	(70 − 71.22) = −1.22	$-1.22^2 =$ 1.4884
2	(60 − 59.72) = .28	$.28^2 =$ .0784
3	(65 − 59.72) = 5.28	$5.28^2 =$ 27.8784
4	(85 − 82.72) = 2.28	$2.28^2 =$ 5.1984
5	(85 − 88.47) = −3.47	$-3.47^2 =$ 12.0409
6	(78 − 80.42) = −2.42	$-2.42^2 =$ 5.8564
7	(68 − 68.92) = −.92	$-.92^2 =$ .8464
8	(73 − 71.22) = 1.78	$1.78^2 =$ 3.1684
9	(84 − 82.72) = 1.28	$1.28^2 =$ 1.6384
10	(82 − 82.72) = −.72	$-.72^2 =$ .5184
11	(80 − 76.97) = 3.03	$3.03^2 =$ 9.1809
12	(90 − 88.47) = 1.53	$1.53^2 =$ 2.3409
13	(95 − 94.22) = .78	$.78^2 =$ .6084
14	(68 − 71.22) = −3.22	$-3.22^2 =$ 10.3684
15	(74 − 76.97) = −2.97	$-2.97^2 =$ 8.8209
16	(88 − 85.02) = 2.98	$2.98^2 =$ 8.8804
17	(78 − 76.97) = 1.03	$1.03^2 =$ 1.0609
18	(94 − 91.92) = 2.08	$2.08^2 =$ 4.3264
19	(64 − 65.47) = −1.47	$-1.47^2 =$ 2.1609
20	(70 − 76.97) = −6.97	$-6.97^2 =$ 48.5809

$$\Sigma(Y - \hat{Y})^2 = 155.0415$$

$$S_{Y \cdot X} = \sqrt{\frac{\Sigma(Y - \hat{Y})^2}{N}} = \sqrt{\frac{155.0415}{20}} = 2.78$$

9. a. First calculate Randrick's predicted score using the formula $\hat{Y} = (4)(6) + 12 = 36$. Since the critical score is 34, we will use the lower real limit of 33.5 and convert to a Z-score. Since Randrick's predicted score is 36 and the standard error of estimate is 1.26, then $Z_{33.5} =$ $\frac{33.5 - 36}{1.26} = -1.98$. Using the normal curve table, we must add .5 to the area between the mean and Z, .4761, to obtain the desired probability. Thus, the probability that Randrick will obtain a grade of C− or higher is .9761.

b. Using the real limits of the raw scores, we find the Z of 3.5 or −3.5 to be $Z = \frac{\pm 3.5}{1.26} = \pm 2.78$. Using Appendix Table A-1, we find the probability associated with a score higher than a Z-score of 2.78 to be .0027. We find the same probability associated with a score lower than a Z-score of −2.78. Since the addition law of probability holds for this example, then the probability of scoring more than three points below or three points above the predicted score is .0027 + .0027 = .0054.

c. The probability associated with being 95% certain of a range of scores based on the predicted score can be obtained by considering a $2\frac{1}{2}$% chance of being too low and a $2\frac{1}{2}$% chance of being too high. Using Appendix Table A-1, these values are associated with Z-scores of −1.96 and +1.96. We convert these Z-scores to deviations by multiplying by the standard error of estimate (1.26) to get ±2.47. Thus, Randrick can be 95% confident that his score will lie between $(36 − 2.47) = 33.53$ and $(36 + 2.47) = 38.47$.

11. a.

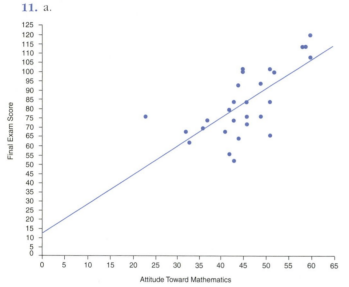

b.

Attitude Toward Mathematics (X)	X²	Final Exam Score (Y)	XY	$\hat{Y}$	$(Y − \hat{Y})^2$
59	3481	114	6726	105.13	78.74
58	3364	114	6612	103.56	108.94
45	2025	100	4500	83.23	281.08
52	2704	100	5200	94.18	33.87
46	2116	72	3312	84.80	163.79
45	2025	102	4590	83.23	352.14
60	3600	120	7200	106.69	177.16
36	1296	70	2520	69.16	.70
44	1936	64	2816	81.67	312.26
49	2401	76	3724	89.49	181.96
43	1849	84	3612	80.11	15.15
41	1681	68	2788	76.98	80.64
60	3600	120	7200	106.69	177.16
46	2116	76	3496	84.80	77.41
49	2401	94	4606	89.49	20.35
42	1764	80	3360	78.54	2.12
43	1849	52	2236	80.11	790.01
33	1089	62	2046	64.47	6.10
23	529	76	1748	48.83	738.04
42	1764	80	3360	78.54	2.12
46	2116	84	3884	84.80	.64

Attitude Toward Mathematics (X)	X²	Final Exam Score (Y)	XY	$\hat{Y}$	$(Y − \hat{Y})^2$
60	3600	108	6480	106.69	1.72
43	1849	74	3182	80.11	37.30
51	2601	84	4284	92.62	74.25
51	2601	66	3366	92.62	708.45
51	2601	102	5202	92.62	88.05
32	1024	68	2176	62.91	25.94
37	1369	74	2738	70.72	10.73
42	1764	56	2352	78.54	508.21
44	1936	93	4092	81.67	128.35

$\Sigma X = 1373$ $\Sigma X^2 = 65051$ $\Sigma Y = 2533$ $\Sigma XY = 119388$ $\Sigma(Y − \hat{Y})^2 = 5183.36$

Using equation 7.4,

$$B_{Y \cdot X} = \frac{N(\Sigma XY) − (\Sigma X)(\Sigma Y)}{N(\Sigma X^2) − (\Sigma X)^2}$$
$$= \frac{(30)(119388) − (1373)(2533)}{(30)(65051) − (1373)^2} = 1.56$$

In addition $\overline{X} = 45.77$ and $\overline{Y} = 84.43$. To find the Y-intercept, use equation 7.6, $A = \overline{Y} − B\overline{X} = 84.43 − (1.56)(45.77) = 13.03$. Thus, the regression equation is: (Final Exam Score) = 1.56(Attitude Toward Mathematics) + 13.03.

c.
$$\hat{Y}_{26} = (1.56)(26) + (13.03) = 53.59$$
$$\hat{Y}_{43} = (1.56)(43) + (13.03) = 80.11$$
$$\hat{Y}_{53} = (1.56)(53) + (13.03) = 95.71$$

d. First, we must calculate the standard error of estimate using the formula

$$S_{Y \cdot X} = \sqrt{\frac{\Sigma(Y − \hat{Y})^2}{N}} = \sqrt{\frac{5183.36}{30}} = 13.14$$

Given a score of 43 on the Attitude Toward Mathematics scale, the predicted Final Exam score is 80.11. Since the critical score is 70, we will use the upper real limit of 70.5 and convert to a Z-score. Since the predicted score is 80.11 and the standard error of estimate is 13.14, then $Z_{70.5} = \frac{70.5 − 80.11}{13.14} = −.73$. Using the normal curve table, we must use the column designating the area beyond Z, .2327, to obtain the desired probability.

e. Given a score of 26 on the Attitude Toward Mathematics scale, the predicted Final Exam score is 53.59. Since the critical score is 100, we will use the lower real limit of 99.5 and convert to a Z-score. Since the predicted score is 53.59 and the standard error of estimate is 13.14, then $Z_{99.5} = \frac{99.5 − 53.59}{13.14} = 3.49$. Using the normal curve table, we must use the column designating the area beyond Z to obtain the desired probability. However, since 3.49 is not in the table, use the lowest, closest score to 3.49, which is 3.45. Thus, the probability is .0003.

13. a.

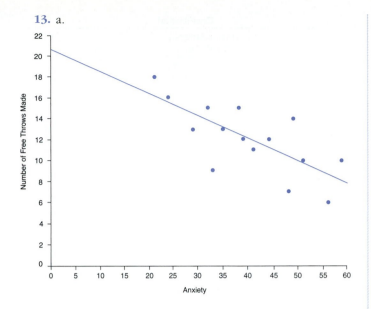

Anxiety (X)	X^2	Number of Free Throws Made (Y)	XY	$\hat{Y}$	$(Y - \hat{Y})^2$
56	3136	8	448	9.04	1.09
41	1681	11	451	11.99	.98
32	1024	15	480	13.76	1.54
51	2601	10	510	10.03	.00
24	576	16	384	15.33	.45
35	1225	13	455	13.17	.03
21	441	18	378	15.92	4.33
49	2401	14	686	10.42	12.82
44	1936	12	528	11.40	.36
33	1089	9	297	13.56	20.81
29	841	13	377	14.35	1.82
48	2304	7	336	10.62	13.07
59	3481	10	590	8.46	2.39
39	1521	12	468	12.38	.15
38	1444	15	570	12.58	5.86

$\Sigma X = 599 \quad \Sigma X^2 = 25701 \quad \Sigma Y = 183 \quad \Sigma XY = 6958 \quad \Sigma(Y - \hat{Y})^2 = 65.69$

b. Using equation 7.4,

$$B_{Y \cdot X} = \frac{N(\Sigma XY) - (\Sigma X)(\Sigma Y)}{N(\Sigma X^2) - (\Sigma X)^2}$$

$$= \frac{(15)(6958) - (599)(183)}{(15)(25701) - (599)^2} = -.20$$

In addition $\overline{X} = 39.93$ and $\overline{Y} = 12.20$. To find the Y-intercept, use equation 7.6, $A = \overline{Y} - B\overline{X} = 12.20 - (-.20)(39.93) = 20.19$. The regression equation is:

(Number of Free Throws Made) $= -.20$(Anxiety) $+ 20.19$

c. Number of Free Throws Made $= -.20(46) + 20.19$ $= 10.99$

d. To find the standard error of estimate, use the formula

$$S_{Y \cdot X} = \sqrt{\frac{\Sigma(Y - \hat{Y})^2}{N}} = \sqrt{\frac{65.69}{15}} = 2.09$$

e. The probability associated with being 95% certain of a range of scores based on the predicted score can be obtained by considering a $2\frac{1}{2}$% chance of being too low and a $2\frac{1}{2}$% chance of being too high. Using Appendix Table A-1, these values are associated with Z-scores of -1.96 and $+1.96$. We convert these Z-scores to deviations by multiplying by the standard error of estimate (2.09) to get ± 4.10. Thus, the 95 percent confidence interval for Julie's predicted free throws is $(10.99 - 4.10) = 6.89$ and $(10.99 + 4.10) = 15.09$.

15. a. Output from the SPSS analysis:

Regression

Variables Entered/Removed[b]

Model	Variables Entered	Variables Removed	Method
1	Post-instruction math attitude[a]		Enter

[a]All requested variables entered.
[b]Dependent Variable: Final algebra grade.

Model Summary

Model	R	R Square	Adjusted R Square	Std. Error of the Estimate
1	.739[a]	.546	.536	.5966

[a]Predictors: (Constant), Post-instruction math attitude.

ANOVA[b]

Model		Sum of Squares	df	Mean Square	F	Sig.
1	Regression	20.122	1	20.122	56.535	.000[a]
	Residual	16.728	47	.356		
	Total	36.850	48			

[a]Predictors: (Constant), Post-instruction math attitude.
[b]Dependent Variable: Final algebra grade.

Coefficients[a]

Model		Unstandardized Coefficients B	Unstandardized Coefficients Std. Error	Standardized Coefficients Beta	t	Sig.
1	(Constant)	4.367	.231		18.929	.000
	Post-instruction math attitude	-3.37E-02	.004	$-.739$	-7.519	.000

[a]Dependent Variable: Final algebra grade.

Thus, the regression equation is

(Final Algebra Grade)
$= -.0337$(Post-Instruction Math Attitude) $+ 4.367$

b. Output from the SPSS analysis:

Regression

Variables Entered/Removed[b]

Model	Variables Entered	Variables Removed	Method
1	Post-instruction math attitude[a]		Enter

[a]All requested variables entered.
[b]Dependent Variable: ITED 9th grade math score.

Model Summary

Model	R	R Square	Adjusted R Square	Std. Error of the Estimate
1	.789[a]	.622	.614	7.9572

[a]Predictors: (Constant), Post-instruction math attitude.

ANOVA[b]

Model		Sum of Squares	df	Mean Square	F	Sig.
1	Regression	4899.423	1	4899.423	77.379	.000[a]
	Residual	2975.924	47	63.318		
	Total	7875.347	48			

[a]Predictors: (Constant), Post-instruction math attitude.
[b]Dependent Variable: ITED 9th grade math score.

Coefficients[a]

Model		Unstandardized Coefficients B	Std. Error	Standardized Coefficients Beta	t	Sig.
1	(Constant)	100.971	3.077		32.812	.000
	Post-instruction math attitude	−.526	.060	−.789	−8.797	.000

[a]Dependent Variable: ITED 9th grade math score.

Thus, the regression equation is

ITED 9th Grade Math Score
= −.526(Post-Instruction Math Attitude) + 100.971

c. Output from the SPSS analysis:

Regression

Variables Entered/Removed[b]

Model	Variables Entered	Variables Removed	Method
1	Confidence in Doing Math[a]		Enter

[a]All requested variables entered.
[b]Dependent Variable: Final Grade in Math.

Model Summary

Model	R	R Square	Adjusted R Square	Std. Error of the Estimate
1	.590[a]	.348	.345	.896

[a]Predictors: (Constant), Confidence in Doing Math

ANOVA[b]

Model		Sum of Squares	df	Mean Square	F	Sig.
1	Regression	84.971	1	84.971	105.829	.000[a]
	Residual	158.977	198	.803		
	Total	243.949	199			

[a]Predictors: (Constant), Confidence in Doing Math.
[b]Dependent Variable: Final Grade in Math.

Coefficients[a]

Model		Unstandardized Coefficients B	Std. Error	Standardized Coefficients Beta	t	Sig.
1	(Constant)	−.551	.305		−1.807	.072
	Confidence in Doing Math	6.790E-02	.007	.590	10.287	.000

[a]Dependent Variable: Final Grade in Math.

Thus, the regression equation is

(Final Grade in Math)
= .0679 (Confidence in Doing Math) − .551

d. Output from the SPSS analysis:

Regression

Variables Entered/Removed[b]

Model	Variables Entered	Variables Removed	Method
1	Empathic Concern[a]		Enter

[a]All requested variables entered.
[b]Dependent Variable: Altruism.

Model Summary

Model	R	R Square	Adjusted R Square	Std. Error of the Estimate
1	.703[a]	.494	.490	4.84

[a]Predictors: (Constant), Empathic Concern.

ANOVA[b]

Model		Sum of Squares	df	Mean Square	F	Sig.
1	Regression	2706.255	1	2706.255	115.384	.000[a]
	Residual	2767.611	118	23.454		
	Total	5473.867	119			

[a]Predictors: (Constant), Empathic Concern.
[b]Dependent Variable: Altruism.

Coefficients[a]

Model		Unstandardized Coefficients B	Std. Error	Standardized Coefficients Beta	t	Sig.
1	(Constant)	15.354	1.969		7.797	.000
	Empathic Concern	.616	.057	.703	10.742	.000

[a]Dependent Variable: Altruism.

Thus, the regression equation is

(Altruism) = .616(Empathic Concern) + 15.354

CHAPTER 8

1.

Self-Esteem (X)	X²	Achievement (Y)	Y²	x	y	Z_X	Z_Y	XY	xy	Z_XZ_Y
25	625	52	2704	−7	−9	−0.78	−0.66	1300	63	0.51
34	1156	66	4356	2	5	0.22	0.36	2244	10	0.08
43	1849	78	6084	11	17	1.22	1.24	3354	187	1.51
21	441	48	2304	−11	−13	−1.22	−0.95	1008	143	1.16
48	2304	85	7225	16	24	1.77	1.75	4080	384	3.10
29	841	62	3844	−3	1	−0.33	0.07	1798	−3	−0.02
38	1444	72	5184	6	11	0.66	0.80	2736	66	0.53
41	1681	70	4900	9	9	1.00	0.66	2870	81	0.65
36	1296	66	4356	4	5	0.44	0.36	2376	20	0.16
18	324	40	1600	−14	−21	−1.55	−1.53	720	294	2.38
27	729	50	2500	−5	−11	−0.55	−0.80	1350	55	0.44
24	576	43	1849	−8	−18	−0.89	−1.31	1032	144	1.16
Σ 384	13266	732	46906	0	0	0	0	24868	1444	11.67

$$\overline{X} = \frac{\Sigma X}{N} = \frac{384}{12} = 32 \qquad \overline{Y} = \frac{\Sigma Y}{N} = \frac{732}{12} = 61$$

a. $r_{XY} = \dfrac{\Sigma Z_X Z_Y}{N} = \dfrac{11.67}{12} = .9725 \approx .97$

b. $S_X = \sqrt{\dfrac{N(\Sigma X^2) - (\Sigma X)^2}{N^2}} = \sqrt{\dfrac{(12)(13266) - (384)^2}{12^2}}$

   $= 9.03$

$S_Y = \sqrt{\dfrac{N(\Sigma Y^2) - (\Sigma Y)^2}{N^2}} = \sqrt{\dfrac{(12)(46906) - (732)^2}{12^2}}$

   $= 13.71$

$r_{XY} = \dfrac{\Sigma xy}{(N)(S_X)(S_Y)} = \dfrac{1444}{(12)(9.03)(13.71)} = .9720 \approx .97$

c. $r_{XY} = \dfrac{N(\Sigma XY) - (\Sigma X)(\Sigma Y)}{\sqrt{N(\Sigma X^2) - (\Sigma X)^2}\sqrt{N(\Sigma Y^2) - (\Sigma Y)^2}}$

   $= \dfrac{(12)(24868) - (384)(732)}{\sqrt{(12)(13266) - (384)^2}\sqrt{(12)(46906) - (732)^2}}$

   $= .9726 \approx .97$

3.

Attitude About Science (X)	X²	Science Final Exam Score (Y)	Y²	XY
50	2500	70	4900	3500
40	1600	60	3600	2400
40	1600	65	4225	2600
60	3600	85	7225	5100
65	4225	85	7225	5525
58	3364	78	6084	4524
48	2304	68	4624	3264
50	2500	73	5329	3650
60	3600	84	7056	5040
60	3600	82	6724	4920
55	3025	80	6400	4400
65	4225	90	8100	5850
70	4900	95	9025	6650
50	2500	68	4624	3400
55	3025	74	5476	4070
62	3844	88	7744	5456
55	3025	78	6084	4290
68	4624	94	8836	6392
45	2025	64	4096	2880
55	3025	70	4900	3850
Σ 1111	63111	1551	122277	87761

$$\overline{X} = \frac{\Sigma X}{N} = \frac{1111}{20} = 55.55 \qquad \overline{Y} = \frac{\Sigma Y}{N} = \frac{1551}{20} = 77.55$$

a. $r_{XY} = r_{YX} = \dfrac{N(\Sigma XY) - (\Sigma X)(\Sigma Y)}{\sqrt{N(\Sigma X^2) - (\Sigma X)^2}\sqrt{N(\Sigma Y^2) - (\Sigma Y)^2}}$

   $= \dfrac{(20)(87761) - (1111)(1551)}{\sqrt{(20)(63111) - (1111)^2}\sqrt{(20)(122277) - (1551)^2}}$

   $= .96$

This correlation means that there is a very strong, positive relationship between attitude about science and achievement in science as measured by the science final exam. Thus, lower scores on the attitude measure would imply lower scores on the achievement measure, or higher scores on the attitude measure would imply higher scores on the achievement measure.

$r_{XY}^2 = (.96)^2 = .92$

Thus, 92% of the variance in the measure of science achievement (science final exam score) is linearly related to the science attitude measure.

b. $S_{attitude} = S_X = \sqrt{\dfrac{N(\Sigma X^2) - (\Sigma X)^2}{N^2}}$

   $= \sqrt{\dfrac{(20)(63111) - (1111)^2}{20^2}} = 8.35$

$S_{exam} = S_Y = \sqrt{\dfrac{N(\Sigma Y^2) - (\Sigma Y)^2}{N^2}}$

   $= \sqrt{\dfrac{(20)(122277) - (1551)^2}{20^2}} = 9.99$

$B_{Y \cdot X} = r_{Y \cdot X}\dfrac{S_Y}{S_X} = (.96)\dfrac{9.99}{8.35} = 1.149$

$A = \overline{Y} - B\overline{X} = 77.55 - (1.149)(55.55) = 13.72$

c. $S_{Y \cdot X} = S_Y\sqrt{1 - r_{YX}^2} = 9.99\sqrt{1 - .96^2} = 2.80$

d. $Y = BX + A = (1.149)(52) + (13.72) = 73.47$

is the predicted raw score.

To obtain the predicted standard score, we must first convert an attitude score of 52 to a Z-score.

$Z_X = \dfrac{X - \overline{X}}{S_X} = \dfrac{52 - 55.55}{8.35} = -.425$

Solutions to Odd-Numbered Exercises 511

$\hat{Z}_Y = r_{XY}Z_X = (.96)(-.425) = -.408$ is the predicted standard score.

e. $B_{X \cdot Y} = r_{X \cdot Y} \dfrac{S_X}{S_Y} = (.96)\dfrac{8.35}{9.99} = .80$

$A = \overline{X} - B\overline{X} = 55.55 - (.80)(77.55) = -6.49$

f. $S_{X \cdot Y} = S_X\sqrt{1 - r_{XY}^2} = 8.35\sqrt{1 - .96^2} = 2.34$

5.

Neuroticism (X)	X²	Social Phobia (Y)	Y²	XY
6	36	11	121	66
4	16	12	144	48
19	361	22	484	418
8	64	10	100	80
5	25	12	144	60
9	81	17	289	153
9	81	15	225	135
6	36	13	169	78
15	225	15	225	225
14	196	23	529	322
10	100	15	225	150
11	121	17	289	187
20	400	26	676	520
19	361	18	324	342
20	400	26	676	520
16	256	22	484	352
9	81	14	196	126
6	36	14	196	84
17	289	20	400	340
14	196	23	529	322
Σ 237	3361	345	6425	4528

$\overline{X} = \dfrac{\Sigma X}{N} = \dfrac{237}{20} = 11.85 \qquad \overline{Y} = \dfrac{\Sigma Y}{N} = \dfrac{345}{20} = 17.25$

a. $r_{XY} = r_{YX} = \dfrac{N(\Sigma XY) - (\Sigma X)(\Sigma Y)}{\sqrt{N(\Sigma X^2) - (\Sigma X)^2}\sqrt{N(\Sigma Y^2) - (\Sigma Y)^2}}$

$= \dfrac{(20)(4528) - (237)(345)}{\sqrt{(20)(3361) - (237)^2}\sqrt{(20)(6425) - (345)^2}} = .86$

This correlation means that there is a very strong, positive relationship between neuroticism and social phobia. Thus, lower scores on the neuroticism measure would imply lower scores on the social phobia measure, or higher scores on the neuroticism measure would imply higher scores on the social phobia measure.

$r_{XY}^2 = (.86)^2 = .74$

Thus, 74% of the variance in the measure of social phobia is linearly related to the neuroticism measure.

b. $S_{neuroticism} = \sqrt{\dfrac{N(\Sigma X^2) - (\Sigma X)^2}{N^2}}$

$= \sqrt{\dfrac{(20)(3361) - (237)^2}{20^2}}$

$= 5.26$

$S_{social\ phobia} = \sqrt{\dfrac{N(\Sigma Y^2) - (\Sigma Y)^2}{N^2}}$

$= \sqrt{\dfrac{(20)(6425) - (345)^2}{20^2}}$

$= 4.87$

$B_{Y \cdot X} = r_{Y \cdot X}\dfrac{S_Y}{S_X} = (.86)\dfrac{4.87}{5.26} = .796$

$A = \overline{Y} - B\overline{X} = 17.25 - (.796)(11.85) = 7.82$

c. $S_{Y \cdot X} = S_Y\sqrt{1 - r_{YX}^2} = 4.87\sqrt{1 - .86^2} = 2.49$

d. $\hat{Y} = BX + A = (.796)(18) + 7.82 = 22.15$ is the predicted raw score.

To obtain the predicted standard score, we must first convert an attitude score of 18 to a Z-score.

$Z_X = \dfrac{X - \overline{X}}{S_X} = \dfrac{18 - 11.85}{5.26} = 1.17$

$\hat{Z}_Y = r_{XY}Z_X = (.86)(1.17) = 1.01$ is the predicted standard score.

e.

(X)	(Y)	(Y − Ȳ)²	Ŷ	(Ŷ − Ȳ)²	(Y − Ŷ)²
6	11	39.06	12.60	21.66	2.55
4	12	27.56	11.00	39.01	0.99
19	22	22.56	22.94	32.42	0.89
8	10	52.56	14.19	9.38	17.54
5	12	27.56	11.80	29.70	0.04
9	17	0.06	14.98	5.13	4.06
9	15	5.06	14.98	5.13	0.00
6	13	18.06	12.60	21.66	0.16
15	15	5.06	19.76	6.30	22.66
14	23	33.06	18.96	2.94	16.29
10	15	5.06	15.78	2.16	0.61
11	17	0.06	16.58	0.45	0.18
20	26	76.56	23.74	42.12	5.11
19	18	0.56	22.94	32.42	24.44
20	26	76.56	23.74	42.12	5.11
16	22	22.56	20.56	10.93	2.09
9	14	10.56	14.98	5.13	0.97
6	14	10.56	12.60	21.66	1.97
17	20	7.56	21.35	16.83	1.83
14	23	33.06	18.96	2.94	16.29
		$SS_T = 473.75$		$SS_P = 350.10$	$SS_E = 123.77$

Total Variance $= S_T^2 = \dfrac{\Sigma(Y - \overline{Y})^2}{N} = \dfrac{473.75}{20} = 23.6875$

Predictable Variance $= S_{\hat{Y}}^2 = \dfrac{\Sigma(\hat{Y} - \overline{Y})^2}{N}$

$= \dfrac{350.10}{20} = 17.505$

Variance Error of Estimate $= \dfrac{\Sigma(Y - \hat{Y})^2}{N}$

$= \dfrac{123.77}{20} = 6.1885$

f. $r_{YX}^2 = \dfrac{S_{\hat{Y}}^2}{S_Y^2} = \dfrac{17.505}{23.717} = .738 \approx .74$

which is the same value that was obtained in part (a).

7. a.

Correlations

		Post-Instruc-tion Math Attitude	ITED 9th Grade Math Score
Post-instruction math attitude	Pearson Correlation	1.000	−.789**
	Sig. (2-tailed)	.	.000
	N	49	49
ITED 9th grade math score	Pearson Correlation	−.789**	1.000
	Sig. (2-tailed)	.000	.
	N	49	49

**Correlation is significant at the 0.01 level (2-tailed).

b.

Correlations

		Confidence in Doing Math	Final Grade in Math
Confidence in Doing Math	Pearson Correlation	1.000	.590**
	Sig. (2-tailed)	.	.000
	N	200	200
Final Grade in Math	Pearson Correlation	.590**	1.000
	Sig. (2-tailed)	.000	.
	N	200	200

**Correlation is significant at the 0.01 level (2-tailed).

c.

Correlations

		Empathic Concern	Altruism
Empathic Concern	Pearson Correlation	1.000	.703**
	Sig. (2-tailed)	.	.000
	N	120	120
Altruism	Pearson Correlation	.703**	1.000
	Sig. (2-tailed)	.000	.
	N	120	120

**Correlation is significant at the 0.01 level (2-tailed).

9. a. −.40
   b. .00
   c. .75
   d. .30
   e. .85
   f. −.70

CHAPTER 9

1. A population is composed of all entities about which we want to make statements or draw conclusions. A sample is a subset of the population.

3. There are several properties of samples that allow for proper generalizations to the population. When this occurs we say we have a "good" sample. One of the most important properties of a good sample is that it is representative of the population. To be representative, the sample should have the same general characteristics that occur with the same relative frequency as in the population. Second, a sample is a good sample if it is unbiased. Since all samples contain some amount of sampling error, it is important that the sampling error does not arise as a result of errors in sampling which tend to be systematic. Finally, a sample is a good sample if the entities are selected by a random procedure.

5. a. cluster sample—a special form of stratified random sampling
   b. selected cases sample
   c. (simple) random sample
   d. convenience sample
   e. systematic random sample
   f. stratified random sample

7. The results of a study are confounded by another variable, when the researcher cannot determine whether one variable (e.g., the independent variable) or the other variable (the confounding variable) or a combination of the two produced the desired effect (e.g., the difference in the dependent variable).

Examples will vary. Two examples are provided. Suppose that a researcher wants to determine whether a client-centered counseling approach or a rational-emotive counseling approach results in higher levels of ratings of expertise of the counselor. Two counselors are chosen for the study, one who is trained in a client-centered approach and one who is trained in a rational-emotive approach. If there was a difference on ratings of expertise in this study, we could not determine if the difference was due to the approaches used by the two counselors or some personal quality of the counselor besides the counseling approach, such as the tone of voice or the physical attractiveness of the counselor. In the second example, suppose a group of second graders is tested on a physical skill that is taught in a particular way, and a group of third graders is tested on the same physical skill that is taught in a second way. The two groups are then measured on their ability to do the physical skill. In this case we cannot be certain that if there were a difference, the difference could be attributed to the teaching method or to the age of the participants.

9. a. The sampling population is all students in the school district who are transitioning from an elementary school to the sixth grade at a middle school.
   b. The treatment population is every sixth grader who is in a middle school for the first year after an elementary school in the school district and who is exposed to the Transitions program.
   c. The accessible population is all students in the school district who are transitioning from an elementary school to the sixth grade at a middle school.
   d. The target population is all sixth grade middle school students in the United States who attended fifth grade at an elementary school.
   e. The study is an experimental study because it involves random selection of participants and active manipulation of a variable. As a result, we can make in-

ferences to the population about the effectiveness of the Transitions program in causing a change in social, emotional, and cognitive factors that assure better adjustment for this particular school district. Any generalization beyond the school district is flawed because the population only included 6th graders in this school district.

11. a. The sampling population is all 5-year-olds in the three major metropolitan areas in the United States.
    b. One treatment population is all 5-year-olds in the three major metropolitan areas in the United States who suffered from malnutrition from birth to age 2. The second treatment population is all 5-year-olds in the three major metropolitan areas in the United States who did not suffer from malnutrition from birth to age 2.
    c. The accessible population is all 5-year-olds in the three major metropolitan areas in the United States.

d. The target population is all children who are age 5.
e. This is a correlational study since the conditions of malnutrition and non-malnutrition already existed prior to the study. The researcher is merely getting information from pre-existing groups. Since no manipulation is involved, no inferences regarding causation can be made. This study is correlational. However, relational inferences from this study to all children of age 5 should not be made since the sampling population is different from the target population. The researchers can make relational inferences to all 5-year-old children in the three major metropolitan areas in the United States.

13. a. convenience sample
    b. stratified random sample
    c. (simple) random sample
    d. selected cases sample
    e. systematic random sample

CHAPTER 10

1. Status variables allow the researcher to divide participants into pre-existing categories. These are qualities over which the researcher has no control in the sense that she or he must accept the categories to which the participant belongs. A manipulated variable also allows for participants to be divided into categories. However, for a manipulated variable, the researcher has control over who is assigned to the categories.

    Examples of status variables and manipulated variables will vary. Some examples of status variables are gender, socioeconomic status, grade in school, the cultural group with whom a person identifies, etc. Status variables can also be assigned according to scores on some measure. For example, we might give participants a measure of empathy. If participants scored below the mean they would be in the low empathy group. If they scored above the mean they would be in the high empathy group. The level of empathy isn't manipulated since that condition existed prior to the measure of the behavior. The researcher merely categorized people based on their score.

    Some examples of a manipulated variable would include the following: the color of the paper on which text information is written (e.g., white vs. gray vs. peach), the method that is used to teach students reading (e.g., phonics vs. whole language), the counseling approach that is used in a counseling session (directive vs. nondirective), etc.

3. a. The dependent variable is the score on the perspective-taking test.
   b. There are no manipulated variables in this research study.
   c. The status variables in this study are age (5 vs. 8) and gender (female vs. male).

5. a. The dependent variable in this research study is the number of uncontrollable anger episodes in a three-month period.
   b. The manipulated variable is the type of anger management class (cognitive-behavioral only vs. cognitive-behavioral with role play).
   c. There are no status variables in this research study.

7. Evaluation studies are usually conducted to determine if some manipulation has had the desired effect. If so, then usually some action is taken such that the manipulation that yielded the desired effect is implemented in some applied setting. The researcher who uses evaluation studies usually either studies every member of the population or has the ability to generalize to the entire population. Thus, there is little need for statistical analyses beyond descriptive statistics. Often evaluation studies are less controlled and less restricted than research studies. In addition, interpretation of the results of evaluation studies is somewhat casual or informal. Finally, the researcher in an evaluation study has a vested interest in demonstrating that something is "true." In contrast, research studies are conducted to demonstrate phenomena without regard to the practical utility of the findings. Thus, research studies typically do not focus on applying the knowledge in some practical setting. In addition, since the goal of researchers conducting research studies is to generalize the results to the widest possible population, more sophisticated statistical procedures may be required. In addition, more control of the research studies is required. Thus, the conclusions and generalizations that are drawn are more restricted. The rules and methods of research are under much tighter control for research studies.

9. Examples will vary. Evaluation examples should result in a study in which a recommendation is to be made regarding whether a policy, program, or existing practice should change. Usually this would involve testing the population and having a limited number of controls. For example, a company is interested in whether employees are more productive if they are given more frequent short breaks (e.g., two 15-minute breaks in the morning and two 15-minute breaks in the afternoon) rather than the traditional longer 30-minute breaks at midmorning and midafternoon. This new policy is implemented for a six-week trial period. Production for this six-week trial period is then compared to production for the previous six-week trial period. Based on this comparison, the company will decide whether to change the policy on work breaks.

The examples for a research study will also vary. However, the research study will involve more elements of control, will test a theory, and will involve an attempt at a greater generalization. An example might be the following. According to social theories, participants are more likely to help a victim if the participants are alone rather than in a group because the participant who is alone may think that she or he is the last chance the victim has for help whereas if the participant is in a group the responsibility for helping is diffused. In this study, participants are randomly assigned to either an individual setting or a group setting in which a victim is encountered. In the group setting, all of the other members are confederates (associates of the researcher) who obviously see the victim but make no attempt to help. The victim is always a male who has been apparently beaten (a good make-up job). The researcher receives a report from the victim of whether the participant made any attempt to help. (*Note:* This research study was actually conducted in the 1960s. However, in today's world, this research study is ethically questionable because informed consent could not be obtained from the participant prior to conducting the study because it would affect the outcome of the study. Since this study could be challenged from an ethical perspective, students are not recommended to conduct such a study.)

11. a. The dependent variables in this research study are students' attitude toward mathematics and students' achievement in algebra.
   b. For this research study, the manipulated variable is the method of teaching algebra (metacognitive vs. traditional).
   c. For this research study, there are no status variables.
   d. To rule out the rival hypothesis that period (or time) of day when algebra is taken could account for both attitudinal and achievement differences, both of the teaching methods were offered during the second period of the day.
      To rule out the rival hypothesis that the number of years of teaching experience could account for both attitudinal and achievement differences, the teachers of each method both had eight years of teaching experience.
      To rule out the rival hypothesis that differences in teacher effectiveness could account for both attitudinal and achievement differences, the teachers in both of the teaching methods had been cited for their outstanding teaching.
   e. Rival hypotheses will vary. One example of a rival hypothesis is provided for your consideration. At least one major rival hypothesis that could account for attitudinal and achievement differences is the teacher's enthusiasm or excitement about mathematics. Despite the fact that both teachers were cited for outstanding teaching, one of the teachers could be much more enthusiastic about mathematics. This enthusiasm could positively impact both the attitude and the achievement of the students in this teacher's class irrespective of the teaching method. If a method could be devised to measure the teachers' enthusiasm toward mathematics, then this could be used to select teachers who are equally enthusiastic.

13. Scientific research cannot be used to "prove" that something is true for a number of reasons. First, the instruments that we use to measure people on some behavior provide only approximations of reality. Thus, if the decisions are made based on these approximations there is always the possibility that those decisions or conclusions are wrong. Second, scientific research requires that we use inductive reasoning in which we collect a number of observations that lead to the proclamation of some general principle. The only valid conclusion that we can arrive at in this process is one that involves a negation—not all A behavior is consistent with B. This statement does not provide a universal truth about A and B. Finally, in scientific research, we can never control all of the variables in such a way that all rival hypotheses are eliminated. We can, at best, control the major variables that eliminate some of the major rival hypotheses.

15. a. The principle that is claimed is that higher salaries paid to professional athletes are responsible for the decline in interest in professional sports. There are three specific facts that are given. First, the salaries of professional baseball players have increased by 248% in the last 10 years. Second, attendance at baseball parks has decreased 15% in the last 10 years. Third, television viewing of baseball is down 18% in the last 10 years. The conclusion is that the salaries paid to professional athletes are responsible for the decline in interest in professional sports. This represents inductive reasoning. However, the conclusion is not valid from an inductive reasoning perspective since specific cases cannot be used to confirm a general principle. In addition, there is no evidence that these three statements are related to each other. They may, in fact, be independent of each other. Attendance at baseball parks or the reduction in television viewing of baseball may have decreased because there are more alternative forms of entertainment available that compete with baseball. There are other equally viable explanations for the decrease in both attendance and television viewing that are not related to salary increase.
   b. There are several facts asserted in this article. Jolene Carr is 68 years old. Jolene Carr died. Jolene Carr was on a diet and lost weight. Jolene Carr had a normal body mass index for her age group. The general principle is that women between the ages of 65 and 99 who have a normal body mass index and lose weight are four times more likely to die than women in this age category who maintained their weight or gained some weight. The conclusion is that a contributing factor in Jolene's death was that she lost weight when her body mass index was normal for her age. This reasoning is deductive.
   c. There are several specific facts in this article. First, Jose Garcia and Julie Chan both had dental implants. Second, Jose Garcia and Julie Chan both smoke. Third, the dental implants for both Jose Garcia and Julie Chan failed. The general principle and conclusion is that dental implants will fail if the people who have those implants smoke. This is inductive reasoning. However, it is invalid since specific facts cannot be used to demonstrate a general principle. The failure of the dental implants could be the result of many other "facts" that we do not know, such as both Jose and Julie ate foods that required a lot of chewing before they were advised to do so. Or both Jose and Julie were involved in accidents that resulted in exces-

sive force being applied to the mouth region that contained the dental implants.

17. a. The control variables in this study are the variables that are held constant for both groups (Ritalin vs. cognitive-behavioral approaches). In this regard, the control variables are that each child was given a one-month time frame to adjust to her or his treatment, that children were all observed at the same time of day, that children were all observed for 10 days before and 10 days after treatment, that measurement error was kept fairly low by utilizing two observers who had an interrater reliability of .92 which is very high, that all observations were made for one hour for each observation of each child, and that children were randomly assigned to treatment conditions. There was only one variable that was actively manipulated by the researcher—the type of treatment that was given to the hyperactive child. There were no status variables in this study as described. The research hypothesis is that cognitive-behavioral approaches to treating hyperactive children will be at least as effective in controlling inappropriate behavior as Ritalin. The alternative hypothesis is that cognitive-behavioral approaches to treating hyperactive children will be less effective in controlling inappropriate behavior than Ritalin. Effectiveness is measured as a decline in the average number of inappropriate responses from pre-treatment to post-treatment. There are several rival hypotheses that might explain any differences or lack of differences in the data. These rival hypotheses include the amount of time to learn or adjust to the new treatment, the length of the observation during any observation of a single participant, the number of observations during pre- and post-treatment, the bias of an observer who is recording behavior, and the time of day when the behavior is observed. These rival hypotheses can be eliminated by careful control of the variables as indicated above. Thus, the controls that were initiated were that each child was given a one-month time frame to adjust to her or his treatment, that children were all observed at the same time of day, that children were all observed for 10 days before and 10 days after treatment, that measurement error was kept fairly low by utilizing two observers who had an interrater reliability of .92 which is very high, that all observations were made for one hour for each observation of each child, and that children were randomly assigned to treatment conditions. There were other possible rival hypotheses that were not controlled. These include the hypothesis that gender might account for the results. In particular, if the proportion of boys to girls is different in the two treatments, this hypothesis might be plausible. To control for this, the researcher should assign children using a stratified random sample that will assure an equal proportion of boys and girls in each treatment. A second rival hypothesis that was not controlled was the activity that the child was supposed to engage in during the period of observation. Some activities may result in more inappropriate behaviors than other activities. Thus, to control for this, the children should be observed while they are supposed to be engaging in the same activities. Can you think of other rival hypotheses that were not controlled?

b. The control variables in this study are limited. There is only one major control established in the study. This control is that the instruction was given for two days in students' health education classes. It is considered a control variable because it is held constant for the two educational approaches. The manipulated variable in this study is the approach to teaching adolescents about smoking issues (factual vs. social skills/reinforcement). This is a manipulated variable because the researcher is in control of which adolescents get a particular approach. There are no status variables in this research study. The research hypothesis is that adolescents who are exposed to a social skills/reinforcement approach to teaching about cigarette smoking will express a more negative attitude toward smoking than adolescents who are exposed to factual information about cigarette smoking. The alternate hypotheses is that adolescents who are exposed to a social skills/reinforcement approach to teaching about cigarette smoking will express either equivalent or more positive attitudes about cigarette smoking than adolescents who are exposed to factual information about cigarette smoking. A rival hypothesis could be that the amount of time spent teaching adolescents about cigarette smoking will affect their attitudes toward smoking. In this regard, the researchers controlled for this by allowing two days of educational exposure for each method. There are a number of variables that were not controlled that may support rival hypotheses. An example of a rival hypothesis is that gender differences may account for differences in attitude toward cigarette smoking. To control for this rival hypothesis, a stratified random assignment procedure should have been invoked instead of merely grouping on the basis of period of the day as was done in this research study. Thus, the proportion of males and females in the class would be determined. The males would be randomly assigned to one of the teaching approaches and the females would be randomly assigned to one of the teaching approaches. This would assure that there would be a proportional gender representation in the two educational approaches. A second rival hypothesis that was uncontrolled is that students in one approach may already have a negative attitude toward smoking. Since we cannot be certain if the approach was responsible for any difference in attitude or if the pre-existing attitudes were responsible for any difference in attitude, the researcher should have given pre- and post-instruction measures of attitude. A difference score could then be calculated to determine the effect of the program on a change in attitude toward smoking. There are other rival hypotheses that can be identified and controlled. Can you provide some of these?

c. For this study, two variables were held constant or controlled to eliminate a rival hypothesis to explain the results. One control variable was that the reading levels of the two passages were both rated at the sixth grade level of difficulty. This is a control variable because it is held constant in the two research conditions involving lighting. The other control variable is that the participants were all adults. The manipulated variable in this study is the lighting condition (natural vs. fluorescent). The researcher was in charge of which participants got certain lighting conditions at certain times. The status variable in this research study was the dyslexia diagnosis (not dyslexic vs. dyslexic). This is a status variable because the condition existed before the research study began and the researcher could not manipulate it. The research hypothesis stated that dyslexic and nondyslexic adult readers would read at the same rate under natural lighting conditions. However, under fluorescent lighting conditions nondyslexic adult readers would read faster than dyslexic adult readers. An alternate hypothesis would be that dyslexic adult readers would read slower than nondyslexic adult readers under natural lighting conditions but at the same rate under fluorescent lighting conditions. There are several other alternate hypotheses in this case. One rival hypothesis is that these differences could be accounted for by differences in reading difficulty of the passages. This rival hypothesis was controlled in this study by equating the two reading passages on the basis of reading difficulty. They were both rated at the sixth grade level for reading difficulty. A second rival hypothesis is that educational experience may account for any

differences. Using adult readers only controlled for this rival hypothesis. There are several rival hypotheses that were not controlled in this research study. One rival hypothesis that was not controlled is that one of the passages could have been more familiar than the other passage even though they were written at the same level of reading difficulty. To control for this rival hypothesis, the researcher could have had half of the participants in each group read one passage first and then the other passage second while the reverse would be true of the other half of the participants. Another way to control for this rival hypothesis is to equate the two passages on familiarity. A second rival hypothesis is that the order in which the passages were read could have affected reading time. That is, readers may have read the second passage slower because they were exhausted from reading the first passage. To control for this rival hypothesis, half of the participants in each group could have read under natural light first and fluorescent light second while the reverse was true of the other half of the participants in each group. There are several other rival hypotheses for which there were no control measures. Can you provide these hypotheses as well as describe how the researcher could have controlled for them?

CHAPTER 11

1. a. A sampling distribution of the mean is a frequency distribution of the means of all samples of a given sample size from the same population.
   b. A sampling distribution of the median is a frequency distribution of the medians of all samples of a given sample size from the same population.
   c. A sampling distribution of the standard deviation is a frequency distribution of the standard deviations of all samples of a given sample size from the same population.
   d. A sampling distribution of the correlation is a frequency distribution of the correlation coefficients of all samples of a given sample size from the same population.

3. a. The standard error of the mean is the standard deviation of the distribution of means from all possible samples of a given sample size.
   b. The standard error of the variance is the standard deviation of the distribution of variances from all possible samples of a given sample size.
   c. The standard error of the correlation is the standard deviation of the distribution of correlations from all possible samples of a given sample size.

5. a. The expected mean of the sample is 500, the population mean.
   b. The standard error of the mean is

   $$\sigma_{\overline{X}} = \frac{\sigma}{\sqrt{N}} = \frac{100}{\sqrt{25}} = 20$$

   c. First convert 530 to a Z-score using equation 11.2.

   $$Z_{\overline{X}} = \frac{\overline{X} - \mu}{\sigma_{\overline{X}}} = \frac{530 - 500}{20} = 1.5$$

Looking this value up in Appendix Table A-1, we find that the proportion of the normal curve beyond $Z_{530} = 1.5$ is .0668. Thus, the probability of obtaining a sample with a mean that is greater than 530 is about .067 or about 67 in 1000. *Note:* To answer in this way, we must assume that the distribution of means is normal.

   d. First convert 450 to a Z-score using equation 11.2.

   $$Z_{\overline{X}} = \frac{\overline{X} - \mu}{\sigma_{\overline{X}}} = \frac{450 - 500}{20} = -2.5$$

Looking this value up in Appendix Table A-1, we find that the proportion of the normal curve beyond $Z_{450} = -2.5$ is .0062. Thus, the probability of obtaining a sample with a mean that is less than 450 is about .006 or about 6 in 1000.

   e. First convert 525 to a Z-score using equation 11.2.

   $$Z_{\overline{X}} = \frac{\overline{X} - \mu}{\sigma_{\overline{X}}} = \frac{525 - 500}{20} = 1.25$$

Looking this value up in Appendix Table A-1, we find that the proportion of the normal curve between the mean and $Z_{525} = 1.25$ is .3944. Then convert 475 to a Z-score using equation 11.2.

   $$Z_{\overline{X}} = \frac{\overline{X} - \mu}{\sigma_{\overline{X}}} = \frac{475 - 500}{20} = -1.25$$

Looking this value up in Appendix Table A-1, we find that the proportion of the normal curve between the mean and $Z_{475} = -1.25$ is .3944. Thus, the probability of obtaining a sample with a mean that is between 475 and 525 is about .3944 + .3944 = .7888 ≈ .789 or about 789 in 1000.

f. The Z-scores that are associated with a two-tailed test for a 95% confidence interval are ±1.96 according to Appendix Table A-1. Thus, the 95% confidence interval can be obtained by the formula $\overline{X} \pm (Z_{.95})(\sigma_{\overline{X}})$. Thus,

$$CI_{lower} = 515 - (1.96)(20) = 475.8$$
$$CI_{upper} = 515 + (1.96)(20) = 554.2$$

The interval from 475.8 to 554.2 is the 95% confidence interval for μ based on this sample.

g. The Z-scores that are associated with a two-tailed test for an 80% confidence interval are ±1.28 according to Appendix Table A-1. Thus, the 80% confidence interval can be obtained by the formula $\overline{X} \pm (Z_{.80})(\sigma_{\overline{X}})$. Thus,

$$CI_{lower} = 487 - (1.28)(20) = 461.4$$
$$CI_{upper} = 487 + (1.28)(20) = 512.6$$

The interval from 461.4 to 512.6 is the 80% confidence interval for μ based on this sample.

7. a. The expected mean of the sample is 80, the population mean.

b. The standard error of the mean is

$$\sigma_{\overline{X}} = \frac{\sigma}{\sqrt{N}} = \frac{12}{\sqrt{50}} = 1.70$$

c. First convert 84 to a Z-score using equation 11.2.

$$Z_{\overline{X}} = \frac{\overline{X} - \mu}{\sigma_{\overline{X}}} = \frac{84 - 80}{1.70} = 2.35$$

Looking this value up in Appendix Table A-1, we find that the proportion of the normal curve beyond $Z_{84} = 2.35$ is .0094. Thus, the probability of obtaining a sample with a mean that is greater than 84 is about .009 or about 9 in 1000.

d. First convert 75 to a Z-score using equation 11.2.

$$Z_{\overline{X}} = \frac{\overline{X} - \mu}{\sigma_{\overline{X}}} = \frac{75 - 80}{1.70} = -2.94$$

Looking this value up in Appendix Table A-1, we find that the proportion of the normal curve beyond $Z_{75} = -2.94$ is .0016. Thus, the probability of obtaining a sample with a mean that is less than 75 is about .002 or about 2 in 1000.

e. First convert 82 to a Z-score using equation 11.2.

$$Z_{\overline{X}} = \frac{\overline{X} - \mu}{\sigma_{\overline{X}}} = \frac{82 - 80}{1.70} = 1.18$$

Looking this value up in Appendix Table A-1, we find that the proportion of the normal curve between the mean and $Z_{82} = 1.18$ is .3810. Then convert 77 to a Z-score using equation 11.2.

$$Z_{\overline{X}} = \frac{\overline{X} - \mu}{\sigma_{\overline{X}}} = \frac{77 - 80}{1.70} = -1.76$$

Looking this value up in Appendix Table A-1, we find that the proportion of the normal curve between the

mean and $Z_{77} = -1.76$ is .4608. Thus, the probability of obtaining a sample with a mean that is between 77 and 82 is about .3810 + .4608 = .8418 ≈ .842 or about 842 in 1000.

f. The Z-scores that are associated with a two-tailed test for a 95% confidence interval are ±1.96 according to Appendix Table A-1. Thus, the 95% confidence interval can be obtained by the formula $\overline{X} \pm (Z_{.95})(\sigma_{\overline{X}})$. Thus,

$$CI_{lower} = 83 - (1.96)(1.70) = 79.67$$
$$CI_{upper} = 83 + (1.96)(1.70) = 86.33$$

The interval from 79.67 to 86.33 is the 95% confidence interval for μ based on this sample.

g. $\sigma_{\overline{X}} = \dfrac{\sigma}{\sqrt{N}} = \dfrac{12}{\sqrt{150}} = 0.98$

The Z-scores that are associated with a two-tailed test for a 95% confidence interval are ±1.96 according to Appendix Table A-1. Thus, the 95% confidence interval can be obtained by the formula $\overline{X} \pm (Z_{.95})(\sigma_{\overline{X}})$. Thus,

$$CI_{lower} = 83 - (1.96)(0.98) = 81.08$$
$$CI_{upper} = 83 + (1.96)(0.98) = 84.92$$

The interval from 81.08 to 84.92 is the 95% confidence interval for μ based on this sample. *Note:* This confidence interval does not contain the known population mean of 80.

9. a. To calculate variance, use the formula

$$S^2 = \frac{N(\Sigma X^2) - (\Sigma X)^2}{N^2}$$

$$\Sigma X = (47 + 57 + 44 + 58 + 49 + 49 + 51 + 54 + 50 + 51) = 510$$

$$\Sigma X^2 = (47^2 + 57^2 + 44^2 + 58^2 + 49^2 + 49^2 + 51^2 + 54^2 + 50^2 + 51^2) = 26{,}178$$

$$S^2 = \frac{N(\Sigma X^2) - (\Sigma X)^2}{N^2} = \frac{(10)(26{,}178) - (510)^2}{10^2} = 16.8$$

b. $\hat{S}^2 = \dfrac{N}{N-1} S^2 = \dfrac{10}{10-1}(16.8) = 18.67$

c. $df = N - 1 = 10 - 1 = 9$

d. Use the degrees of freedom from part (c) and Appendix Table A-2.
For a 95% confidence interval, the critical values of $t = \pm 2.262$.
For a 99% confidence interval, the critical values of $t = \pm 3.250$.
For a 99.9% confidence interval, the critical values of $t = \pm 4.781$.

11. Use the formula

$$\hat{S}^2 = \frac{N}{N-1}S^2 = \frac{28}{28-1}(169) = 175.26$$

13. Using the formula for the mean,

$$\overline{X} = \frac{\Sigma X}{N} = \frac{510}{10} = 51$$

Using the formula for the estimate of the standard error,

$$\hat{S}_{\overline{X}} = \frac{\hat{S}}{\sqrt{N}}$$

where

$$\hat{S} = \sqrt{\hat{S}^2} = \sqrt{18.67} = 4.32, \hat{S}_{\overline{X}} = \frac{\hat{S}}{\sqrt{N}} = \frac{4.32}{\sqrt{10}} = 1.37$$

From Exercise 9(d)
For a 95% confidence interval, the critical values of $t = \pm 2.262$.
For a 99% confidence interval, the critical values of $t = \pm 3.250$.
For a 99.9% confidence interval, the critical values of $t = \pm 4.781$.
The confidence interval can be obtained from the formula

$$CI = \overline{X} \pm (t_{CI})(\hat{S}_{\overline{X}})$$

a. The 95% confidence interval for the empathic concern scores contains the following limits:

$$CI_{lower} = 51 - (2.262)(1.37) = 47.90$$
$$CI_{upper} = 51 + (2.262)(1.37) = 54.10$$

The interval from 47.90 to 54.10 is the 95% confidence interval for μ based on this sample.

b. The 99% confidence interval for the empathic concern scores contains the following limits:

$$CI_{lower} = 51 - (3.25)(1.37) = 46.55$$
$$CI_{upper} = 51 + (3.25)(1.37) = 55.45$$

The interval from 46.55 to 55.45 is the 99% confidence interval for μ based on this sample.

c. The 99.9% confidence interval for the empathic concern scores contains the following limits:

$$CI_{lower} = 51 - (4.781)(1.37) = 44.45$$
$$CI_{upper} = 51 + (4.781)(1.37) = 57.55$$

The interval from 44.45 to 57.55 is the 99.9% confidence interval for μ based on this sample.

15. $\Sigma X = (28 + 21 + 19 + 33 + 10 + 15 + 18 + 21 + 32$
$+ 28 + 22 + 25 + 38 + 24 + 20 + 18 + 19 + 29$
$+ 31 + 30 + 13 + 18 + 21 + 26) = 559$

$\Sigma X^2 = (28^2 + 21^2 + 19^2 + 33^2 + 10^2 + 15^2 + 18^2 + 21^2$
$+ 32^2 + 28^2 + 22^2 + 25^2 + 38^2 + 24^2 + 20^2 + 18^2$
$+ 19^2 + 29^2 + 31^2 + 30^2 + 13^2 + 18^2 + 21^2 + 26^2)$
$= 14,099$

Calculate the mean $\overline{X} = \frac{\Sigma X}{N} = \frac{559}{24} = 23.29$

To calculate the unbiased estimate of the standard deviation, use the formula

$$\hat{S} = \sqrt{\frac{N(\Sigma X^2) - (\Sigma X)^2}{N(N-1)}} = \sqrt{\frac{(24)(14,099) - (559)^2}{(24)(24-1)}} = 6.85$$

$$\hat{S}_{\overline{X}} = \frac{\hat{S}}{\sqrt{N}} = \frac{6.85}{\sqrt{24}} = 1.40$$

$$df = N - 1 = 24 - 1 = 23$$

From Appendix Table A-2, the critical t values for a 95% confidence interval with 23 degrees of freedom are ± 2.069.

$$CI = \overline{X} \pm (t_{CI})(\hat{S}_{\overline{X}})$$

The 95% confidence interval for the anger scores contains the following limits:

$$CI_{lower} = 23.29 - (2.069)(1.40) = 20.39$$
$$CI_{upper} = 23.29 + (2.069)(1.40) = 26.19$$

The interval from 20.39 to 26.19 is the 95% confidence interval for μ based on this sample.

17. $\hat{S}_{\overline{X}} = \frac{\hat{S}}{\sqrt{N}} = \frac{6.28}{\sqrt{100}} = 0.63$

$$df = N - 1 = 100 - 1 = 99$$

From Appendix Table A-2, the critical t values for a 95% confidence interval with 99 degrees of freedom are not listed. Thus, you should use the critical t values for the next lowest degrees of freedom in the table (60) which yields critical t values of ± 2.000.

$$CI = \overline{X} \pm (t_{CI})(\hat{S}_{\overline{X}})$$

The 95% confidence interval for the self-esteem scores contains the following limits:

$$CI_{lower} = 52.45 - (2.000)(0.63) = 51.19$$
$$CI_{upper} = 52.45 + (2.000)(0.63) = 53.71$$

The interval from 51.19 to 53.71 is the 95% confidence interval for μ based on this sample.

19. The SPSS printout for the 90% confidence interval is as printed below. Note that the output has been edited to give only the descriptive information for the two variables.

Explore

Descriptives

			Statistic	Std. Error
Perspective Taking	Mean		42.83	.87
	90% Confidence Interval for Mean	Lower Bound	41.39	
		Upper Bound	44.26	
	5% Trimmed Mean		42.69	
	Median		42.00	
	Variance		90.381	
	Std. Deviation		9.51	
	Minimum		22	
	Maximum		64	
	Range		42	
	Interquartile Range		14.75	
	Skewness		.303	.221
	Kurtosis		−.524	.438
Empathic Concern	Mean		33.47	.71
	90% Confidence Interval for Mean	Lower Bound	32.29	
		Upper Bound	34.64	
	5% Trimmed Mean		33.37	
	Median		33.00	
	Variance		59.948	
	Std. Deviation		7.74	
	Minimum		19	
	Maximum		49	
	Range		30	
	Interquartile Range		12.00	
	Skewness		.187	.221
	Kurtosis		−.942	.438

The SPSS printout for the 99.9% confidence interval is as printed below. Note that the output has been edited to give only the descriptive information for the two variables.

Explore

Descriptives

			Statistic	Std. Error
Perspective Taking	Mean		42.83	.87
	99.9% Confidence Interval for Mean	Lower Bound	39.90	
		Upper Bound	45.75	
	5% Trimmed Mean		42.69	
	Median		42.00	
	Variance		90.381	
	Std. Deviation		9.51	
	Minimum		22	
	Maximum		64	
	Range		42	
	Interquartile Range		14.75	
	Skewness		.303	.221
	Kurtosis		−.524	.438
Empathic Concern	Mean		33.47	.71
	99.9% Confidence Interval for Mean	Lower Bound	31.08	
		Upper Bound	35.85	
	5% Trimmed Mean		33.37	
	Median		33.00	
	Variance		59.948	
	Std. Deviation		7.74	
	Minimum		19	
	Maximum		49	
	Range		30	
	Interquartile Range		12.00	
	Skewness		.187	.221
	Kurtosis		−.942	.438

CHAPTER 12

1. a. μ represents the population mean number of sessions to extinction of lever-pressing behavior

H_0: $\mu_{variable\ schedule\ of\ reinforcement} \leq \mu_{constant\ schedule\ of\ reinforcement}$

H_1: $\mu_{variable\ schedule\ of\ reinforcement} > \mu_{constant\ schedule\ of\ reinforcement}$

b. μ represents the population mean self-esteem score

H_0: $\mu_{upper\ income\ level} \leq \mu_{lower\ income\ level}$

H_1: $\mu_{upper\ income\ level} > \mu_{lower\ income\ level}$

c. μ represents the population mean moral reasoning score

H_0: $\mu_{eighth\ graders\ from\ rural\ schools} = \mu_{eighth\ graders\ from\ metropolitan\ schools}$

H_1: $\mu_{eighth\ graders\ from\ rural\ schools} \neq \mu_{eighth\ graders\ from\ metropolitan\ schools}$

d. μ represents the population mean job satisfaction rating

H_0: $\mu_{entry\ level\ employees} = \mu_{middle\ management}$

H_1: $\mu_{entry\ level\ employees} \neq \mu_{middle\ management}$

e. μ represents the population mean intelligence score

H_0: $\mu_{children\ raised\ in\ conditions\ of\ emotional\ deprivation} \geq 100$

H_1: $\mu_{children\ raised\ in\ conditions\ of\ emotional\ deprivation} < 100$

f. μ represents the population mean reading comprehension score

H_0: $\mu_{children\ taught\ reading\ by\ phonics} \leq \mu_{children\ taught\ reading\ by\ whole\ language}$

H_1: $\mu_{children\ taught\ reading\ by\ phonics} > \mu_{children\ taught\ reading\ by\ whole\ language}$

g. μ represents the population mean depression score

H_0: $\mu_{older\ adults\ who\ exercise} \geq \mu_{older\ adults\ who\ do\ not\ exercise}$

H_1: $\mu_{older\ adults\ who\ exercise} < \mu_{older\ adults\ who\ do\ not\ exercise}$

h. μ represents the population mean spatial rotation score

H_0: $\mu_{men} \leq \mu_{women}$

H_1: $\mu_{men} > \mu_{women}$

i. μ represents the population mean resting heart rate

H_0: $\mu_{marathon\ runners} \geq \mu_{sprinters}$

H_1: $\mu_{marathon\ runners} < \mu_{sprinters}$

j. μ represents the population mean body temperature

H_0: $\mu_{malnourished\ children} \geq 98.6°$

H_1: $\mu_{malnourished\ children} < 98.6°$

3. Answers will vary. The following is a sample answer. The null hypothesis is a statement that our sample came from a particular population that we specify. In other words, the null hypothesis specifies the characteristics of the sampling distribution. The alternate hypothesis is a statement that the sample did not come from the population specified in the null hypothesis. In other words, it is the negation of the null hypothesis. In most cases in psychology and education, the alternate hypothesis is the statement of what we predict will happen based on the knowledge/theories that we have about the phenomenon under investigation.

5. When we test the null hypothesis, we make several assumptions. These assumptions are listed below:

a. The measure of behavior that we are using is valid. That is, it measures the behavior or construct that we think it measures. Furthermore, the way that we have measured these behaviors or constructs is consistent with the behaviors or constructs as specified in the theory that supports our research.

b. The measure of the behavior or construct is accurate and reliable. This merely means that there is a high degree of consistency in the measure of the behavior or construct that we are using.

c. We have designed and conducted our study in such a way that all confounding variables have been controlled or held constant in such a way that plausible rival hypotheses are ruled out.

d. The sample that we have used has been selected randomly from the population and is representative of the population.

e. The conditions regarding the population that are specified in the null hypothesis are true.

7. a. For H_0: $\mu \leq 100$, $\sigma = 15$, $N = 25$, $\alpha = .05$, true $\mu = 110$

i. Since this is a one-tailed test, the critical value for $\alpha = .05$ can be determined from Appendix Table A-1. We are looking for the Z-score whose area beyond the mean is .0500. This value is $Z = +1.645$ since the null hypothesis is a one-tailed test where μ is *less than or equal to* some constant if we are working with the Z-distribution. The critical value is calculated by the formula $CV = (Z)(\sigma_{\bar{x}}) + \mu$ if we are working with the raw score distribution. Thus, we must first calculate

$$\sigma_{\bar{x}} = \frac{\sigma}{\sqrt{N}} = \frac{15}{\sqrt{25}} = 3$$

Thus, $CV = (1.645)(3) + 100 = 104.94$.

ii. To calculate β, we must first calculate the standard error, $\sigma_{\bar{x}} = \sigma/\sqrt{N} = 15/\sqrt{25} = 3$. We must also calculate the critical value in the raw score distribution. The critical value is calculated by the formula $CV = (Z)(\sigma_{\bar{x}}) + \mu$ if we are working with the raw score distribution. Thus, $CV = (1.645)(3) + 100 = 104.94$. Next determine the Z-score of 104.94 in the alternate distribution with a mean of 110 and standard error of 3, $Z = (104.94 - 110)/3 = -1.69$. If we look up this value in Appendix Table A-1 for the area beyond Z, we find .0455. Thus, $\beta = .0455$ since the critical value is to the left of the mean of the population of the alternate distribution. Thus, we have approximately a 4.5%

chance of failing to detect that $\mu > 100$ if the true value of $\mu = 110$.

iii. Power $= 1 - \beta = 1 - .0455 = .9545$.

b. For H_0: $\mu > 500$, $\sigma = 100$, $N = 50$, $\alpha = .01$, true $\mu = 450$

i. Since this is a one-tailed test, the critical value for $\alpha = .01$ can be determined from Appendix Table A-1. We are looking for the Z-score whose area beyond the mean is .0100. This value is $Z = -2.326$ since the null hypothesis is a one-tailed test where μ is *greater than or equal to* some constant if we are working with the Z-distribution. The critical value is calculated by the formula $CV = (Z)(\sigma_{\bar{x}}) + \mu$ if we are working with the raw score distribution. Thus, we must first calculate

$$\sigma_{\bar{x}} = \frac{\sigma}{\sqrt{N}} = \frac{100}{\sqrt{50}} = 14.14$$

Thus, $CV = (-2.326)(14.14) + 500 = 467.11$.

ii. To calculate β, we must first calculate the standard error,

$$\sigma_{\bar{x}} = \frac{\sigma}{\sqrt{N}} = \frac{100}{\sqrt{50}} = 14.14$$

We must also calculate the critical value in the raw score distribution. The critical value is calculated by the formula $CV = (Z)(\sigma_{\bar{x}}) + \mu$ if we are working with the raw score distribution. Thus, $CV = (-2.326)(14.14) + 500 = 467.11$. Next determine the Z-score of 467.11 in the alternate distribution with a mean of 450 and standard error of 14.14, $Z = (467.11 - 450)/14.14 = 1.21$. If we look up this value in Appendix Table A-1 for the area beyond Z, we find .1131. Thus, $\beta = .11$. Thus, we have approximately an 11% chance of failing to detect that $\mu < 500$ if the true value of $\mu = 450$.

iii. Power $= 1 - \beta = 1 - .11 = .89$.

c. For H_0: $\mu = 25$, $\sigma = 3$, $N = 15$, $\alpha = .05$, true $\mu = 25.8$

i. Since this is a two-tailed test, the critical value for $\alpha = .05$ must be divided in half (.025) to use Appendix Table A-1. We are looking for the Z-score whose area beyond the mean is .0250. This value is $Z = +1.96$ since the null hypothesis is a two-tailed test. The critical values are calculated by the formula $CV = \pm(Z)(\sigma_{\bar{x}}) + \mu$ if we are working with the raw score distribution. Thus, we must first calculate

$$\sigma_{\bar{x}} = \frac{\sigma}{\sqrt{N}} = \frac{3}{\sqrt{15}} = 0.77$$

Thus, $CV_{upper} = (1.96)(0.77) + 25 = 26.51$ and $CV_{lower} = -(1.96)(0.77) + 25 = 23.49$.

ii. To calculate β, we must first calculate the standard error, $\sigma_{\bar{x}} = \sigma/\sqrt{N} = 3/\sqrt{15} = 0.77$. We must also calculate the critical values in the raw score distribution. The critical values are calculated by the formula $CV = (Z)(\sigma_{\bar{x}}) + \mu$ if we are working with the raw score distribution. Thus, $CV_{upper} = (1.96)(0.77) + 25 = 26.51$ and $CV_{lower} = -(1.96)(0.77) + 25 = 23.49$. Next determine the Z-score of 26.51 in the alternate distribution with a mean of 25.8 and standard error of 0.77, $Z = (26.51 - 25.8)/0.77 = 0.92$. If we look up this value in Appendix Table A-1 for the area between Z and the mean, we find .3212. Thus, $\beta = .5 + .32 = .82$. Thus, we have approximately an 82% chance of failing to detect that $\mu \neq 25$ if the true value of $\mu = 25.8$.

iii. Power $= 1 - \beta = 1 - .82 = .18$.

9. a. $\sigma_{\bar{x}} = \dfrac{\sigma}{\sqrt{N}} = \dfrac{15}{\sqrt{9}} = 5$

The critical Z-values for a two-tailed test with $\alpha = .05$ for the null hypothesis $\mu = 100$ are ± 1.96 according to Appendix Table A-1. For the raw score distribution $CV = (Z)(\sigma_{\bar{x}}) + \mu$. Thus, $CV_{upper} = (1.96)(5) + 100 = 109.8$ and $CV_{lower} = -(1.96)(5) + 100 = 90.2$.

There are two critical regions (CR). $CR_{lower} \leq 90.2$ and $CR_{upper} \geq 109.8$ for a test of the null hypothesis with $\alpha = .05$.

b. $\sigma_{\bar{x}} = \dfrac{\sigma}{\sqrt{N}} = \dfrac{100}{\sqrt{25}} = 20$

The critical Z-value for a one-tailed test with $\alpha = .01$ for the null hypothesis $\mu \leq 500$ is 2.326 according to Appendix Table A-1. For the raw score distribution $CV = (Z)(\sigma_{\bar{x}}) + \mu$. Thus, $CV = (2.326)(20) + 500 = 546.5$.

There is one critical region (CR). $CR \geq 546.5$ for a test of the null hypothesis with $\alpha = .01$.

c. $\sigma_{\bar{x}} = \dfrac{\sigma}{\sqrt{N}} = \dfrac{1.33}{\sqrt{30}} = 0.24$

The critical Z-values for a two-tailed test with $\alpha = .01$ for the null hypothesis $\mu = 16$ are ± 2.576 according to Appendix Table A-1. For the raw score distribution $CV = \pm(Z)(\sigma_{\bar{x}}) + \mu$. Thus, $CV_{upper} = (2.576)(0.24) + 16 = 16.62$ and $CV_{lower} = -(2.576)(0.24) + 16 = 15.38$.

There are two critical regions (CR). $CR_{lower} \leq 15.38$ and $CR_{upper} \geq 16.62$ for a test of the null hypothesis with $\alpha = .01$.

d. $\sigma_{\bar{x}} = \dfrac{\sigma}{\sqrt{N}} = \dfrac{1.75}{\sqrt{20}} = 0.39$

The critical Z-value for a one-tailed test with $\alpha = .05$ for the null hypothesis $\mu \geq 18.8$ is -1.645 according to Appendix Table A-1. For the raw score distribution $CV = (Z)(\sigma_{\bar{x}}) + \mu$. Thus, $CV = (-1.645)(0.39) + 18.8 = 18.16$.

There is one critical region (CR). $CR \leq 18.16$ for a test of the null hypothesis with $\alpha = .05$.

e. $\sigma_{\bar{x}} = \dfrac{\sigma}{\sqrt{N}} = \dfrac{9.73}{\sqrt{80}} = 1.09$

The critical Z-values for a two-tailed test with $\alpha = .001$ for the null hypothesis $\mu = 63.2$ are ± 3.30 according to Appendix Table A-1. For the raw score distribution $CV = +(Z)(\sigma_{\bar{x}}) + \mu$. Thus, $CV_{upper} = $

$(3.30)(1.09) + 63.2 = 66.80$ and $CV_{lower} = -(3.30)(1.09) + 63.2 = 59.60$.

There are two critical regions (CR). $CR_{lower} \leq 59.60$ and $CR_{upper} \geq 66.80$ for a test of the null hypothesis with $\alpha = .001$.

f. $\sigma_{\bar{x}} = \dfrac{\sigma}{\sqrt{N}} = \dfrac{15}{\sqrt{36}} = 2.5$

The critical Z-value for a one-tailed test with $\alpha = .05$ for the null hypothesis $\mu \leq 100$ is 1.645 according to Appendix Table A-1. For the raw score distribution $CV = (Z)(\sigma_{\bar{x}}) + \mu$. Thus, $CV = (1.645)(2.5) + 100 = 104.11$.

There is one critical region (CR). $CR \geq 104.11$ for a test of the null hypothesis with $\alpha = .05$.

11. Suppose $\mu = 80$ and $\sigma = 9$ for the Creativity Test

a. $\sigma_{\bar{x}} = \dfrac{\sigma}{\sqrt{N}} = \dfrac{9}{\sqrt{36}} = 1.5$

For $\alpha = .05$
$CV = +(Z)(\sigma_{\bar{x}}) + \mu$ with $Z = 1.96$ for a two-tailed test with $\alpha = .05$. Thus, $CV_{upper} = (1.96)(1.5) + 80 = 82.94$ and $CV_{lower} = -(1.96)(1.5) + 80 = 77.06$. Next determine the Z-score of 82.94 in the true distribution with a mean of 84 and standard error of the mean of 1.5, $Z = (82.94 - 84)/1.5 = -0.71$. If we look up this value in Appendix Table A-1 for the area beyond Z, we find .2389. Thus, $\beta = .24$ and power $= 1 - \beta = 1 - .24 = .76$.
For $\alpha = .01$
$CV = \pm(Z)(\sigma_{\bar{x}}) + \mu$ with $Z = 2.576$ for a two-tailed test with $\alpha = .01$.
Thus, $CV_{upper} = (2.576)(1.5) + 80 = 83.86$ and $CV_{lower} = -(2.576)(1.5) + 80 = 76.14$. Next determine the Z-score of 83.86 in the true distribution with a mean of 84 and standard error of the mean of 1.5, $Z = (83.86 - 84)/1.5 = -0.09$. If we look up this value in Appendix Table A-1 for the area beyond Z, we find .4641. Thus, $\beta = .46$ and power $= 1 - \beta = 1 - .46 = .54$.

From the calculations above, the consequences for power and Type II error of testing the hypothesis at $\alpha = .01$ instead of $\alpha = .05$ if the treatment mean is 84 are that the probability of committing a Type II error increases from .24 to .46, and the power of the test decreases from .76 to .54. Thus by lowering α, we have increased the probability of making a Type II error and decreased the power of the test.

b. For $N = 49$, a one-tailed test at $\alpha = .01$, $\sigma_{\bar{x}} = \sigma/\sqrt{N} = 9/\sqrt{49} = 1.29$ and $CV = (Z)(\sigma_{\bar{x}}) + \mu$ with $Z = 2.326$ with $\alpha = .01$.
Thus, $CV = (2.326)(1.29) + 80 = 83.00$. Next determine the Z-score of 83.00 in the true distribution with a mean of 84 and standard error of the mean of 1.29, $Z = (83.00 - 84)/1.29 = -0.78$. If we look up this value in Appendix Table A-1 for the area beyond Z, we find .2177. Thus, $\beta = .22$ and power $= 1 - \beta = 1 - .22 = .78$.

For $N = 64$, a one-tailed test at $\alpha = .01$, $\sigma_{\bar{x}} = \sigma/\sqrt{N} = 9/\sqrt{64} = 1.13$. $CV = (Z)(\sigma_{\bar{x}}) + \mu$ with $Z = 2.326$ for a one-tailed test with $\alpha = .01$. Thus, $CV = (2.326)(1.13) + 80 = 82.63$. Next determine the Z-

score of 82.63 in the true distribution with a mean of 84 and standard error of the mean of 1.13, $Z = (82.63 - 84)/1.13 = -1.21$. If we look up this value in Appendix Table A-1 for the area beyond Z, we find .1131. Thus, $\beta = .11$ and power $= 1 - \beta = 1 - .11 = .89$.

From the calculations above, the consequences for power, and Type II error of testing the hypothesis at $\alpha = .01$ with a sample size of 64 instead of 49 if the treatment population mean is 84 are that the probability of committing a Type II error decreases from .22 to .11, and the power of the test increases from .78 to .89. Thus by increasing N, we have decreased the power of the test.

c. For $N = 25$, $\alpha = .05$, $\sigma_{\bar{x}} = \sigma/\sqrt{N} = 9/\sqrt{25} = 1.8$. For a two-tailed test: $CV = \pm(Z)(\sigma_{\bar{x}}) + \mu$ with $Z = 1.96$ for a two-tailed test with $\alpha = .05$. Thus, $CV_{upper} = (1.96)(1.8) + 80 = 83.53$ and $CV_{lower} = -(1.96)(1.8) + 80 = 76.47$. Next determine the Z-score of 83.53 in the true distribution with a mean of 84 and standard error of the mean of 1.8, $Z = (83.53 - 84)/1.8 = -0.26$. If we look up this value in Appendix Table A-1 for the area beyond Z, we find .3974. Thus, $\beta = .40$ and power $= 1 - \beta = 1 - .40 = .60$.
For a one-tailed test: $CV = (Z)(\sigma_{\bar{x}}) + \mu$ with $Z = 1.645$ for a one-tailed test with $\alpha = .05$. Thus, $CV = (1.645)(1.8) + 80 = 82.96$. Next determine the Z-score of 82.96 in the true distribution with a mean of 84 and standard error of the mean of 1.8, $Z = (82.96 - $

$84)/1.8 = -0.58$. If we look up this value in Appendix Table A-1 for the area beyond Z, we find .2810. Thus, $\beta = .28$ and power $= 1 - \beta = 1 - .28 = .72$.

From the calculations above, the consequences for power, and Type II error of testing the hypothesis at $\alpha = .05$ with a directional test of the null hypothesis instead of a nondirectional test for a constant treatment population mean and sample size are that the probability of committing a Type II error decreases from .40 to .28, and the power of the test increases from .60 to .72. Thus by using a directional test instead of a nondirectional test, we have decreased the probability of making a Type II error and increased the power of the test.

d. If a researcher wants to reduce the probability of making a Type I error, he or she can merely reduce the α of the test. However, by reducing the α, the researcher correspondingly increases the probability of committing a Type II error and decreases the power of the test.

If the researcher wants to decrease the probability of making a Type II error, he or she may do so by raising the α of the test (This is not recommended), increasing the sample size, or conducting a directional test.

If the researcher wants to increase the power of the test, he or she may do so by raising the α of the test (This is not recommended), increasing the sample size, or conducting a directional test.

CHAPTER 13

1. In testing a hypothesis, we obtain a test statistic from the descriptive statistics that are computed from a sampling process. We establish two hypotheses based on the research question we hope to answer. The null hypothesis is a symbolic statement that is the opposite of our research prediction (the alternate hypothesis) and is the hypothesis we are testing. A frequency sampling distribution based on the null hypothesis is known. Thus, we can determine the probability of any score in the distribution. We establish a value or values (the critical values) that divide the frequency sampling distribution into a confidence interval and critical region(s). If the test statistic falls within the confidence interval, the probability of this occurring is high given the conditions specified in the null hypothesis. If the test statistic falls in this region we fail to reject the null hypothesis. If the test statistic falls within the critical region(s), the probability of this occurring under the conditions specified in the null hypothesis is extremely low. Thus, we choose to reject the null hypothesis.

3. a. H_0: $\mu \leq 80$
   H_1: $\mu > 80$
   b. $\Sigma X = 78 + 84 + 88 + 83 + 89 + 79 + 80 + 81 + 86 + 87 + 82 + 80 + 76 + 79 + 80 + 79 + 83 + 86 + 82 + 81 + 84 + 80 + 83 + 81 + 79 = 2050$

   $\bar{X} = \dfrac{\Sigma X}{N} = \dfrac{2050}{25} = 82.00$

   $\mu = 80$, $\sigma = 10$, $N = 25$, Thus, $\sigma_{\bar{x}} = \dfrac{\sigma}{\sqrt{N}} = \dfrac{10}{\sqrt{25}} = 2.00$

Test statistic:

$$Z_{\bar{x}} = \frac{\bar{X} - K}{\sigma_{\bar{x}}} = \frac{82.00 - 80}{2.00} = 1.00$$

Critical value: for a one-tailed test with $\alpha = .01$, $Z = 2.326$. Thus, we fail to reject H_0: $\mu \leq 80$ at $\alpha = .01$
c. Based on these results, it appears that the average creativity level of the children in the educational program was not significantly enhanced.

5. a. H_0: $\mu_{female} \leq \mu_{male}$
   H_1: $\mu_{female} > \mu_{male}$

b. $\Sigma X_{female} = 510 + 530 + 520 + 540 + 450 + 600 + 620 + 480 + 510 + 530 + 540 + 550 + 520 + 560 + 500 + 540 + 570 + 530 + 520 + 540 = 10,660$

$\bar{X}_{female} = \dfrac{\Sigma X_{female}}{N_{female}} = \dfrac{10,660}{20} = 533$

$\Sigma X_{male} = 490 + 560 + 510 + 500 + 580 + 440 + 510 + 470 + 520 + 540 + 510 + 490 + 480 + 520 + 540 + 510 + 500 + 520 + 510 + 500 = 10,200$

$\bar{X}_{male} = \dfrac{\Sigma X_{male}}{N_{male}} = \dfrac{10,200}{20} = 510$

$\sigma_{\bar{X}_{female} - \bar{X}_{male}} = \sqrt{\dfrac{\sigma^2_{female}}{N_{female}} + \dfrac{\sigma^2_{male}}{N_{male}}} = \sqrt{\dfrac{90^2}{20} + \dfrac{110^2}{20}} = 31.78$

Test statistic:

$$Z_{\overline{X}_{female}-\overline{X}_{male}} = \frac{(\overline{X}_{female} - \overline{X}_{male}) - (\mu_{female} - \mu_{male})}{\sigma_{\overline{X}_{female}-\overline{X}_{male}}}$$

$$= \frac{(533 - 510) - 0}{31.78} = 0.72$$

Critical value: for a one-tailed test with $\alpha = .05$, $Z = 1.645$. Thus, we fail to reject H_0: $\mu_{female} \leq \mu_{male}$ at $\alpha = .05$

c. Based on these results, it appears that the mean Verbal Ability Test score of the 14-year-old females in this community does not differ significantly from the mean Verbal Ability Test score of the 14-year-old males in the community.

7. a. H_0: $\mu_{male} \leq \mu_{female}$
   H_1: $\mu_{male} > \mu_{female}$
   b. $\Sigma X_{male} = 42 + 36 + 46 + 52 + 44 + 40 + 36 + 45 + 43 + 43 + 47 + 40 + 45 + 40 = 599$

$$\overline{X}_{male} = \frac{\Sigma X_{male}}{N_{male}} = \frac{599}{14} = 42.79$$

$\Sigma X_{female} = 43 + 40 + 39 + 40 + 40 + 28 + 47 + 30 + 32 + 27 + 43 + 34 + 50 + 36 = 529$

$$\overline{X}_{female} = \frac{\Sigma X_{female}}{N_{female}} = \frac{529}{14} = 37.79$$

$$\sigma_{\overline{X}_{male}-\overline{X}_{female}} = \sqrt{\frac{\sigma^2_{male}}{N_{male}} + \frac{\sigma^2_{female}}{N_{female}}} = \sqrt{\frac{4.5^2}{14} + \frac{6.9^2}{14}} = 2.20$$

Test statistic:

$$Z_{\overline{X}_{male}-\overline{X}_{female}} = \frac{(\overline{X}_{male} - \overline{X}_{female}) - (\mu_{male} - \mu_{female})}{\sigma_{\overline{X}_{male}-\overline{X}_{female}}}$$

$$= \frac{(42.79 - 37.79) - 0}{2.20} = 2.27$$

Critical value: for a one-tailed test with $\alpha = .05$, $Z = 1.645$. Thus, we reject H_0: $\mu_{male} \leq \mu_{female}$ and accept H_1: $\mu_{male} > \mu_{female}$ at $\alpha = .05$

c. Based on these results, it appears that the mean spatial reasoning score of males is higher than the mean spatial reasoning score of females.

9. a. H_0: $\mu \geq 16$
   H_1: $\mu < 16$
   b. $\Sigma X = 18 + 14 + 13 + 17 + 16 + 15 + 17 + 18 + 13 + 12 + 15 + 16 + 15 + 16 + 14 + 15 = 244$

$$\overline{X} = \frac{\Sigma X}{N} = \frac{244}{16} = 15.25$$

$\Sigma X^2 = 18^2 + 14^2 + 13^2 + 17^2 + 16^2 + 15^2 + 17^2 + 18^2 + 13^2 + 12^2 + 15^2 + 16^2 + 15^2 + 16^2 + 14^2 + 15^2 = 3768$

$$\hat{S} = \sqrt{\frac{(N)(\Sigma X^2) - (\Sigma X)^2}{(N)(N-1)}} = \sqrt{\frac{(16)(3768) - (244)^2}{(16)(16-1)}} = 1.77$$

$$\hat{S}_{\overline{X}} = \frac{\hat{S}}{\sqrt{N}} = \frac{1.77}{\sqrt{16}} = .44$$

Test statistic:

$$t_{df} = \frac{\overline{X} - K}{\hat{S}_{\overline{X}}} = \frac{15.25 - 16}{.44} = -1.70$$

Critical value: for a one-tailed test with $\alpha = .01$, $t(15) = -2.602$. Thus, we fail to reject H_0: $\mu \geq 16$ at $\alpha = .01$

c. Based on these results, it appears that on average there are not significantly fewer than 16 chips in the advertised cookie.

11. a. H_0: $\mu \geq 14.3$
    H_1: $\mu < 14.3$
    b. $\Sigma X = 12 + 15 + 11 + 10 + 9 + 15 + 8 + 11 + 10 + 13 + 14 + 11 + 12 + 16 + 12 + 11 + 6 + 19 + 21 + 14 = 250$

$$\overline{X} = \frac{\Sigma X}{N} = \frac{250}{20} = 12.5$$

$\Sigma X^2 = 12^2 + 15^2 + 11^2 + 10^2 + 9^2 + 15^2 + 8^2 + 11^2 + 10^2 + 13^2 + 14^2 + 11^2 + 12^2 + 16^2 + 12^2 + 11^2 + 6^2 + 19^2 + 21^2 + 14^2 = 3366$

$$\hat{S} = \sqrt{\frac{(N)(\Sigma X^2) - (\Sigma X)^2}{(N)(N-1)}} = \sqrt{\frac{(20)(3366) - (250)^2}{(20)(20-1)}} = 3.56$$

$$\hat{S}_{\overline{X}} = \frac{\hat{S}}{\sqrt{N}} = \frac{3.56}{\sqrt{20}} = 0.80$$

Test statistic:

$$t_{df} = \frac{\overline{X} - K}{\hat{S}_{\overline{X}}} = \frac{12.5 - 14.3}{0.80} = -2.25$$

Critical value: for a one-tailed test with $\alpha = .05$, $t(19) = -1.729$. Thus, we reject H_0: $\mu \geq 14.3$ and accept H_1: $\mu < 14.3$ at $\alpha = .05$

c. Based on these results, it appears that the course was effective in decreasing the mean number of cigarettes adolescents smoked per day.

13. a. H_0: $\mu_{re} = \mu_{cc}$
    H_1: $\mu_{re} \neq \mu_{cc}$
    b. $\Sigma X_{re} = 56 + 42 + 38 + 41 + 42 + 25 + 66 + 38 + 55 + 61 = 464$

$$\overline{X}_{re} = \frac{\Sigma X_{re}}{N_{re}} = \frac{464}{10} = 46.4$$

$\Sigma X^2_{re} = 56^2 + 42^2 + 38^2 + 41^2 + 42^2 + 25^2 + 66^2 + 38^2 + 55^2 + 61^2 = 22,960$

$$\hat{S}_{re} = \sqrt{\frac{(N_{re})(\Sigma X^2_{re}) - (\Sigma X_{re})^2}{(N_{re})(N_{re}-1)}} = \sqrt{\frac{(10)(22,960) - (464)^2}{(10)(10-1)}} = 12.61$$

$\Sigma X_{cc} = 51 + 45 + 43 + 40 + 48 + 22 + 61 + 51 + 64 + 56 = 481$

$$\overline{X}_{cc} = \frac{\Sigma X_{cc}}{N_{cc}} = \frac{481}{10} = 48.1$$

$\Sigma X^2_{cc} = 51^2 + 45^2 + 43^2 + 40^2 + 48^2 + 22^2 + 61^2 + 51^2 + 64^2 + 56^2 = 24,417$

$$\hat{S}_{cc} = \sqrt{\frac{(N_{cc})(\Sigma X^2_{cc}) - (\Sigma X_{cc})^2}{(N_{cc})(N_{cc}-1)}} = \sqrt{\frac{(10)(24,417) - (481)^2}{(10)(10-1)}} = 11.93$$

$$\hat{S}^2_p = \frac{(N_{re} - 1)(\hat{S}^2_{re}) + (N_{cc} - 1)(\hat{S}^2_{cc})}{(N_{re} + N_{cc} - 2)}$$

$$= \frac{(10 - 1)(12.61^2) + (10 - 1)(11.93^2)}{(10 + 10 - 2)} = 150.67$$

$$\hat{S}_{\overline{X}_{re}-\overline{X}_{cc}} = \sqrt{\frac{\hat{S}^2_p}{N_{re}} + \frac{\hat{S}^2_p}{N_{cc}}} = \sqrt{\frac{150.67}{10} + \frac{150.67}{10}} = 5.49$$

Test statistic:

$$t_{df} = \frac{(\overline{X}_{re} - \overline{X}_{cc}) - (\mu_{re} - \mu_{cc})}{\hat{S}_{\overline{X}_{re}-\overline{X}_{cc}}} = \frac{(46.4 - 48.1) - 0}{5.49} = -0.31$$

Critical value: for a two-tailed test with $\alpha = .05$, $t(18) = \pm 2.101$. Thus, we fail to reject H_0: $\mu_{re} = \mu_{cc}$ at $\alpha = .05$

c. Based on these results, it appears that the mean counseling expertise rating of counselors who used a rational-emotive approach to counseling did not differ significantly from the mean counseling expertise rating of counselors who used a client-centered approach.

d. $r_{pb}^2 = \dfrac{t^2}{t^2 + df} = \dfrac{-0.31^2}{-0.31^2 + 18} = .005$

This means that $\frac{1}{2}$ of 1% of the variability in the ratings of counselor expertise can be attributed to the counseling approach of the counselor who was rated. Thus, the effect was almost nonexistent which is consistent with the results of the test of the null hypothesis.

$d = \dfrac{|\overline{X}_{re} - \overline{X}_{cc}|}{\sigma} = \dfrac{|46.4 - 48.1|}{\sqrt{150.67}} = .14$

This reflects the ratio of the differences in treatment means from the samples in relation to the variability of the scores of the participants in the research study.

e. Lower $CI_{\overline{X}_{re} - \overline{X}_{cc}} = -(t)(\hat{S}_{\overline{X}_{re} - \overline{X}_{cc}}) + (\overline{X}_{re} - \overline{X}_{cc})$
$= -(2.101)(5.49) + (46.4 - 48.1)$
$= -13.23$

Upper $CI_{\overline{X}_{re} - \overline{X}_{cc}} = (t)(\hat{S}_{\overline{X}_{re} - \overline{X}_{cc}}) + (\overline{X}_{re} - \overline{X}_{cc})$
$= (2.101)(5.49) + (46.4 - 48.1)$
$= 9.83$

Thus $-13.23 \leq CI \leq 9.83$.
$H_0: \mu_{re} = \mu_{cc}$ can be rewritten as $H_0: \mu_{re} - \mu_{cc} = 0$. Since the difference in the population means is 0 and this falls within the CI, the null value could be the population parameter.

15. a. $H_0: \mu_{bb} \leq \mu_{bo}$
$H_1: \mu_{bb} > \mu_{bo}$

b. $\Sigma X_{bb} = 5 + 6 + 7 + 6 + 5 + 3 + 6 + 7 + 4 + 6$
$= 55$

$\overline{X}_{bb} = \dfrac{\Sigma X_{bb}}{N_{bb}} = \dfrac{55}{10} = 5.5$

$\Sigma X_{bb}^2 = 5^2 + 6^2 + 7^2 + 6^2 + 5^2 + 3^2 + 6^2 + 7^2 + 4^2 + 6^2 = 317$

$\hat{S}_{bb} = \sqrt{\dfrac{(N_{bb})(\Sigma X_{bb}^2) - (\Sigma X_{bb})^2}{(N_{bb})(N_{bb} - 1)}} = \sqrt{\dfrac{(10)(317) - (55)^2}{(10)(10 - 1)}} = 1.27$

$\Sigma X_{bo} = 4 + 3 + 6 + 4 + 5 + 2 + 4 + 5 + 7 + 4 = 44$

$\overline{X}_{bo} = \dfrac{\Sigma X_{bo}}{N_{bo}} = \dfrac{44}{10} = 4.4$

$\Sigma X_{bo}^2 = 4^2 + 3^2 + 6^2 + 4^2 + 5^2 + 2^2 + 4^2 + 5^2 + 7^2 + 4^2 = 212$

$\hat{S}_{bo} = \sqrt{\dfrac{(N_{bo})(\Sigma X_{bo}^2) - (\Sigma X_{bo})^2}{(N_{bo})(N_{bo} - 1)}} = \sqrt{\dfrac{(10)(212) - (44)^2}{(10)(10 - 1)}} = 1.43$

$\hat{S}_p^2 = \dfrac{(N_{bb} - 1)(\hat{S}_{bb}^2) + (N_{bo} - 1)(\hat{S}_{bo}^2)}{(N_{bb} + N_{bo} - 2)}$

$= \dfrac{(10 - 1)(1.27^2) + (10 - 1)(1.43^2)}{(10 + 10 - 2)} = 1.83$

$\hat{S}_{\overline{X}_{bb} - \overline{X}_{bo}} = \sqrt{\dfrac{\hat{S}_p^2}{N_{bb}} + \dfrac{\hat{S}_p^2}{N_{bo}}} = \sqrt{\dfrac{1.83}{10} + \dfrac{1.83}{10}} = 0.60$

Test statistic:

$t_{df} = \dfrac{(\overline{X}_{bb} - \overline{X}_{bo}) - (\mu_{bb} - \mu_{bo})}{\hat{S}_{\overline{X}_{bb} - \overline{X}_{bo}}} = \dfrac{(5.5 - 4.4) - 0}{0.60} = 1.83$

Critical value: for a one-tailed test with $\alpha = .05$, $t(18) = 1.734$. Thus, we reject $H_0: \mu_{bb} \leq \mu_{bo}$ and accept $H_1: \mu_{bb} > \mu_{bo}$ at $\alpha = .05$

c. Based on these results, it appears that horror preceded by humor leads to higher levels of experienced horror on average than viewing horror only.

d. $r_{pb}^2 = \dfrac{t^2}{t^2 + df} = \dfrac{1.83^2}{1.83^2 + 18} = .157$

This means that 15.7% of the variability in experienced terror can be attributed to the horror treatment conditions. This effect appears rather small.

$d = \dfrac{|\overline{X}_{bb} - \overline{X}_{bo}|}{\sigma} = \dfrac{|5.5 - 4.4|}{\sqrt{1.83}} = .81$

This reflects the ratio of the differences in treatment means from the samples in relation to the variability of the experienced horror scores of the participants in the research study.

e. Since this is a one-tailed test there is only one critical value
$CI_{\overline{X}_{bb} - \overline{X}_{bo}} = -(t)(\hat{S}_{\overline{X}_{bb} - \overline{X}_{bo}}) + (\overline{X}_{bb} - \overline{X}_{bo})$
$= (-1.734)(0.60) + (5.5 - 4.4)$
$= 0.06$

Thus $0.06 \leq CI < +\infty$.
$H_0: \mu_{bb} \leq \mu_{bo}$ can be rewritten as $H_0: \mu_{bb} - \mu_{bo} \leq 0$. Since the difference in the population means is 0 or smaller and this does not fall within the CI, the null value is unlikely to be the population parameter.

17. The advantage of using a matched-pairs or repeated-measures design over an independent-samples design is that error variance is reduced in most situations. As a result, the power of the test is increased, thereby reducing the probability of making a Type II error. In addition it is possible to reject the null hypothesis at a given level of α with a smaller difference in the means.

19. a. $H_0: \mu_p \leq \mu_r$
$H_1: \mu_p > \mu_r$

b.

Using the Direct Difference Method

Paraphrase (X_p)	Reread (X_r)	Difference (D)	Difference² (D^2)
15	11	$15 - 11 =$ 4	16
10	12	$10 - 12 = -2$	4
13	9	$13 - 9 =$ 4	16
14	14	$14 - 14 =$ 0	0
14	12	$14 - 12 =$ 2	4
15	14	$15 - 14 =$ 1	1
14	10	$14 - 10 =$ 4	16
15	11	$15 - 11 =$ 4	16
9	8	$9 - 8 =$ 1	1
14	11	$14 - 11 =$ 3	9
17	16	$17 - 16 =$ 1	1
26	22	$26 - 22 =$ 4	16
15	18	$15 - 18 = -3$	9
11	15	$11 - 15 = -4$	16
11	10	$11 - 10 =$ 1	1
8	8	$8 - 8 =$ 0	0
14	12	$14 - 12 =$ 2	4
17	15	$17 - 15 =$ 2	4
8	6	$8 - 6 =$ 2	4
11	11	$11 - 11 =$ 0	0
$\Sigma X_p = 271$	$\Sigma X_r = 245$	$\Sigma D = 26$	$\Sigma D^2 = 138$

$$\overline{D} = \frac{\Sigma D}{N} = \frac{26}{20} = 1.3$$

$$\hat{S}_D = \sqrt{\frac{(N)(\Sigma D^2) - (\Sigma D)^2}{(N)(N-1)}} = \sqrt{\frac{(20)(138) - (26)^2}{(20)(20-1)}} = 2.34$$

$$\hat{S}_{\overline{D}} = \frac{\hat{S}_D}{\sqrt{N}} = \frac{2.34}{\sqrt{20}} = 0.52$$

Test statistic:

$$t_{df} = \frac{\overline{D} - K}{\hat{S}_{\overline{D}}} = \frac{1.3 - 0}{0.52} = 2.50$$

Critical value: for a one-tailed test with $\alpha = .01$, $t(19) = 2.539$. Thus, we fail to reject H_0: $\mu_p \le \mu_r$ at $\alpha = .01$

c. Based on these results, it appears that recall of information is not significantly enhanced if that information is paraphrased vs. reread.

21. a. Given that $d = .80$ and $\sigma^2 = 8$ ($\sigma = 2.83$), then $\overline{X}_1 - \overline{X}_2$ can be calculated from the formula $d = (\overline{X}_1 - \overline{X}_2)/\sigma$. Thus $\overline{X}_1 - \overline{X}_2 = (d)(\sigma) = (.8)(2.83) = 2.26$. Select a sample size of 10. Then

$$\sigma_{\overline{X}_1 - \overline{X}_2} = \sqrt{\frac{\sigma^2}{N_1} + \frac{\sigma^2}{N_2}} = \sqrt{\frac{8}{10} + \frac{8}{10}} = 1.26$$

The critical values of Z for $\alpha = .05$ are ± 1.960. Computing the critical value of $\overline{X}_1 - \overline{X}_2$ from the null distribution we get $(\overline{X}_1 - \overline{X}_2)_{critical} = \mu + Z_{critical}\sigma_{\overline{X}_1 - \overline{X}_2} = 0 + (1.960)(1.26) = 2.47$

$$Z = \frac{(\overline{X}_1 - \overline{X}_2) - (\mu_1 - \mu_2)}{\sigma_{\overline{X}_1 - \overline{X}_2}} = \frac{2.47 - 2.26}{1.26} = .17$$

The probability of obtaining a Z-value of 0.17 or higher is .4325, lower than the power of .60 that we want to achieve. Thus, we need to increase the sample size. Increase the sample size to 20. Then

$$\sigma_{\overline{X}_1 - \overline{X}_2} = \sqrt{\frac{\sigma^2}{N_1} + \frac{\sigma^2}{N_2}} = \sqrt{\frac{8}{20} + \frac{8}{20}} = 0.89$$

The critical values of Z for $\alpha = .05$ are ± 1.960. Computing the critical value of $\overline{X}_1 - \overline{X}_2$ from the null distribution we get $(\overline{X}_1 - \overline{X}_2)_{critical} = \mu + Z_{critical}\sigma_{\overline{X}_1 - \overline{X}_2} = 0 + (1.960)(0.89) = 1.74$

$$Z = \frac{(\overline{X}_1 - \overline{X}_2) - (\mu_1 - \mu_2)}{\sigma_{\overline{X}_1 - \overline{X}_2}} = \frac{1.74 - 2.26}{0.89} = -.58$$

The probability of obtaining a Z-value of -0.58 or higher is .7190, higher than the power of .60 that we want to achieve. Thus, we need a sample size between 10 and 20. Select a sample size of 15. Then

$$\sigma_{\overline{X}_1 - \overline{X}_2} = \sqrt{\frac{\sigma^2}{N_1} + \frac{\sigma^2}{N_2}} = \sqrt{\frac{8}{15} + \frac{8}{15}} = 1.03$$

The critical values of Z for $\alpha = .05$ are ± 1.960. Computing the critical value of $\overline{X}_1 - \overline{X}_2$ from the null distribution we get $(\overline{X}_1 - \overline{X}_2)_{critical} = \mu + Z_{critical}\sigma_{\overline{X}_1 - \overline{X}_2} = 0 + (1.960)(1.03) = 2.02$

$$Z = \frac{(\overline{X}_1 - \overline{X}_2) - (\mu_1 - \mu_2)}{\sigma_{\overline{X}_1 - \overline{X}_2}} = \frac{2.02 - 2.26}{1.03} = -0.23$$

The probability of obtaining a Z-value of -0.23 or higher is .5910, slightly lower than the power of .60 that we want to achieve. Thus, we would need about 16 participants per group to achieve a power of .60.

b. Given that $d = 1.80$ and $\sigma^2 = 15$ ($\sigma = 3.87$), then $\overline{X}_1 - \overline{X}_2$ can be calculated from the formula $d = (\overline{X}_1 - \overline{X}_2)/\sigma$. Thus $\overline{X}_1 - \overline{X}_2 = (d)(\sigma) = (1.80)(3.87) = 6.97$. Select a sample size of 10. Then

$$\sigma_{\overline{X}_1 - \overline{X}_2} = \sqrt{\frac{\sigma^2}{N_1} + \frac{\sigma^2}{N_2}} = \sqrt{\frac{15}{10} + \frac{15}{10}} = 1.73$$

The critical values of Z for $\alpha = .01$ are ± 2.576. Computing the critical value of $\overline{X}_1 - \overline{X}_2$ from the null distribution we get $(\overline{X}_1 - \overline{X}_2)_{critical} = \mu + Z_{critical}\sigma_{\overline{X}_1 - \overline{X}_2} = 0 + (2.576)(1.73) = 4.45$

$$Z = \frac{(\overline{X}_1 - \overline{X}_2) - (\mu_1 - \mu_2)}{\sigma_{\overline{X}_1 - \overline{X}_2}} = \frac{4.45 - 6.97}{1.73} = -1.46$$

The probability of obtaining a Z-value of -1.46 or higher is .9279, higher than the power of 75 that we want to achieve. Lower the sample size to 5. Then

$$\sigma_{\overline{X}_1 - \overline{X}_2} = \sqrt{\frac{\sigma^2}{N_1} + \frac{\sigma^2}{N_2}} = \sqrt{\frac{15}{5} + \frac{15}{5}} = 2.45$$

The critical values of Z for $\alpha = .01$ are ± 2.576. Computing the critical value of $\overline{X}_1 - \overline{X}_2$ from the null distribution we get $(\overline{X}_1 - \overline{X}_2)_{critical} = \mu + Z_{critical}\sigma_{\overline{X}_1 - \overline{X}_2} = 0 + (2.576)(2.45) = 6.31$.

$$Z = \frac{(\overline{X}_1 - \overline{X}_2) - (\mu_1 - \mu_2)}{\sigma_{\overline{X}_1 - \overline{X}_2}} = \frac{6.31 - 6.97}{2.45} = -0.27$$

The probability of obtaining a Z-value of -0.27 or higher is .6064, lower than the power of .75 that we want to achieve. Increase the sample size to 7. Then,

$$\sigma_{\overline{X}_1 - \overline{X}_2} = \sqrt{\frac{\sigma^2}{N_1} + \frac{\sigma^2}{N_2}} = \sqrt{\frac{15}{7} + \frac{15}{7}} = 2.07$$

The critical values of Z for $\alpha = .01$ are ± 2.576. Computing the critical value of $\overline{X}_1 - \overline{X}_2$ from the null distribution we get $(\overline{X}_1 - \overline{X}_2)_{critical} = \mu + Z_{critical}\sigma_{\overline{X}_1 - \overline{X}_2} = 0 + (2.576)(2.07) = 5.33$.

$$Z = \frac{(\overline{X}_1 - \overline{X}_2) - (\mu_1 - \mu_2)}{\sigma_{\overline{X}_1 - \overline{X}_2}} = \frac{5.33 - 6.97}{2.07} = -0.79$$

The probability of obtaining a Z-value of -0.79 or higher is .7852, slightly higher than the power of .75 that we want to achieve. Thus, we need about 7 participants per group to achieve a power of .75.

23. a. The test of the difference in independent self-construal of Japanese vs. Americans yielded $t(197) = .809$ with a probability [Sig. (2-tailed)] of .419. Since this probability is larger than $\alpha = .05$, we would not conclude that there was a difference in the independent self-construal of Japanese ($M = 59.88$, $SD = 6.48$) vs. Americans ($M = 59.08$, $SD = 7.47$).

b. The test of the difference in interdependent self-construal of Japanese vs. Americans yielded $t(197) = -1.277$ with a probability [Sig. (2-tailed)] of .203. Since this probability is larger than $\alpha = .05$, we would not conclude that there was a difference in the interdependent self-construal of Japanese ($M = 52.61$, $SD = 6.91$) vs. Americans ($M = 53.91$, $SD = 7.47$).

c. The test of the difference in the individualistic self-esteem of Japanese vs. Americans yielded $t(197) = -7.75$ with a probability [Sig. (2-tailed)] of .000 (The probability is not really .000. It is merely so small that rounding to three decimal places yields .000. Usually researchers report this as $p < .001$). Since this probability is smaller than $\alpha = .05$, we would conclude that Japanese ($M = 43.72$, $SD = 9.46$) scored lower than Americans ($M = 54.83$, $SD = 10.62$) on the measure of individualistic self-esteem.

d. The test of the difference in the collectivistic self-esteem of Japanese vs. Americans yielded $t(197) = -4.79$ with a probability [Sig. (2-tailed)] of .000 (The probability is not really .000. It is merely so small that rounding to three decimal places yields .000. Usually researchers report this as $p < .001$). Since this probability is smaller than $\alpha = .05$, we would conclude that Japanese ($M = 75.83$, $SD = 13.24$) scored lower than Americans ($M = 85.04$, $SD = 13.79$) on the measure of collectivistic self-esteem.

e. The test of the difference in the individually-oriented achievement motivation of Japanese vs. Americans

yielded $t(197) = -6.72$ with a probability [Sig. (2-tailed)] of .000 (The probability is not really .000. It is merely so small that rounding to three decimal places yields .000. Usually researchers report this as $p < .001$). Since this probability is smaller than $\alpha = .05$, we would conclude that Japanese ($M = 123.28$, $SD = 16.50$) scored lower than Americans ($M = 140.35$, $SD = 19.06$) on the measure of individually-oriented achievement motivation.

f. The test of the difference in the socially-oriented achievement motivation of Japanese vs. Americans yielded $t(197) = -3.34$ with a probability [Sig. (2-tailed)] of .001. Since this probability is smaller than $\alpha = .05$, we would conclude that Japanese ($M = 85.05$, $SD = 22.99$) scored lower than Americans ($M = 95.72$, $SD = 22.04$) on the measure of socially-oriented achievement motivation.

CHAPTER 14

1. a. For $df = 1$, with 90 percent (.90 from the top of the table) of the distribution to the right of the critical value, the critical value is $\chi^2 = .02$.

   b. First 5 percent to the *left* of the distribution must be translated to 95 percent to the *right* of the distribution. For $df = 2$, with 95 percent (.95 from the top of the table) of the distribution to the right (5 percent to the left) of the critical value, the critical value is $\chi^2 = .10$.

   c. For $df = 3$, with 5 percent (.05 from the top of the table) of the distribution to the right of the critical value, the critical value is $\chi^2 = 7.8$.

   d. First if the value cuts off 30 percent of the bottom, then 70 percent of the top is to the right of the value. Thus, for $df = 10$ with 70 percent (.70 from the top of the table) of the distribution to the right of the critical value, the critical value is $\chi^2 = 7.3$.

   e. If we are looking for the values between which the middle 80 percent of the distribution lies, we must use the values that correspond to 10 percent to the left (or 90 percent to the right) and 10 percent to the right. Thus, for $df = 6$, the value that cuts off the distribution with 90 percent of the scores to the right (10 percent to the left) is 2.2 and the value that cuts off the distribution with 10 percent of the scores to the right is 10.6.

   f. If we are looking for the values between which the middle 98 percent of the distribution lies, we must use the values that correspond to 1 percent to the left (or 99 percent to the right) and 1 percent to the right. Thus, for $df = 5$, the value that cuts off the distribution with 99 percent of the scores to the right (1 percent to the left) is .55 and the value that cuts off the distribution with 1 percent of the scores to the right is 15.1.

3. For $df = 29$ we must use columns .95 and .05 to calculate the 90% confidence interval. Thus, $\chi^2 = 17.7$ and 42.6 for these values, respectively.

$$CI_L = \frac{(N-1)\hat{S}^2}{\chi^2_{.05}} = \frac{(30-1)(300)}{42.6} = 204.2$$

$$CI_U = \frac{(N-1)\hat{S}^2}{\chi^2_{.95}} = \frac{(30-1)(300)}{17.7} = 491.5$$

$$204.2 \leq \sigma^2 \leq 491.5$$

Thus, we can be 90% certain that our sample was drawn from a population with a variance somewhere between 204.2 and 491.5.

For $df = 29$ we must use columns .99 and .01 to calculate the 98% confidence interval. Thus, $\chi^2 = 14.3$ and 49.6 for these values, respectively.

$$CI_L = \frac{(N-1)\hat{S}^2}{\chi^2_{.01}} = \frac{(30-1)(300)}{49.6} = 175.4$$

$$CI_U = \frac{(N-1)\hat{S}^2}{\chi^2_{.99}} = \frac{(30-1)(300)}{14.3} = 608.4$$

$$175.4 \leq \sigma^2 \leq 608.4$$

Thus, we can be 98% certain that our sample was drawn from a population with a variance somewhere between 175.4 and 608.4.

5. We must first do some calculations.

$$\Sigma X = 88 + 95 + 100 + 98 + 88 + 80 + 110 + 82 + 84 + 99 + 92 + 90 + 88 + 86 = 1280$$

$$\Sigma X^2 = 88^2 + 95^2 + 100^2 + 98^2 + 88^2 + 80^2 + 110^2 + 82^2 + 84^2 + 99^2 + 92^2 + 90^2 + 88^2 + 86^2 = 117{,}902$$

$$\hat{S}^2 = \frac{N(\Sigma X^2) - (\Sigma X)^2}{N(N-1)} = \frac{(14)(117{,}902) - (1280)^2}{(14)(14-1)} = 67.19$$

$H_0: \sigma^2 = 81$
$H_1: \sigma^2 \neq 81$ for $\alpha = .10$

$$\chi^2 = \frac{(N-1)\hat{S}^2}{\sigma^2} = \frac{(14-1)(67.19)}{81} = 10.78$$

Using Appendix Table A-3 for $df = 13$, $\chi^2 = 5.9$ for $p = .95$ and $\chi^2 = 22.4$ for $p = .05$. Since our value lies between the table values, we fail to reject $H_0: \sigma^2 = 81$ for $\alpha = .10$. Thus, creative individuals might be no different from the general population on their variability on the test of creativity.

7. a. For $df_1 = 3$, $df_2 = 40$, $\alpha = .05$, $F = 2.84$
   b. For $df_1 = 10$, $df_2 = 60$, $\alpha = .01$, $F = 2.63$
   c. For $df_1 = 20$, $df_2 = 120$, $\alpha = .001$, $F = 2.53$

9. a. First the variance estimates of each sample must be calculated.

$$\Sigma X_j = 66 + 68 + 67 + 52 + 57 + 54 + 53 + 59 + 63 + 69 = 608$$

$$\Sigma X_j^2 = 66^2 + 68^2 + 67^2 + 52^2 + 57^2 + 54^2 + 53^2 + 59^2 + 63^2 + 69^2 = 37{,}358$$

$$\hat{S}_J^2 = \frac{N_J(\Sigma X_J^2) - (\Sigma X_J)^2}{N_J(N_J - 1)} = \frac{(10)(37,358) - (608)^2}{(10)(10 - 1)} = 43.51$$

$$\Sigma X_{US} = 65 + 78 + 81 + 64 + 74 + 77 + 66 + 59 + 79$$
$$+ 66 = 709$$

$$\Sigma X_{US}^2 = 65^2 + 78^2 + 81^2 + 64^2 + 74^2 + 77^2 + 66^2 + 59^2$$
$$+ 79^2 + 66^2 = 50,805$$

$$\hat{S}_{US}^2 = \frac{N_{US}(\Sigma X_{US}^2) - (\Sigma X_{US})^2}{N_{US}(N_{US} - 1)} = \frac{(10)(50,805) - (709)^2}{(10)(10 - 1)}$$
$$= 59.66$$

Thus, according to Appendix Table A-4, the critical F for 9 and 9 degrees of freedom for $\alpha = .05$ is 3.18. But from our data $F = 59.66/43.51 = 1.37$. Since this value is smaller than the critical value, we cannot reject the hypothesis that the variances in the two populations are equal.

b. First the variance estimates of each sample must be calculated.

$$\Sigma X_F = 54 + 58 + 49 + 56 + 55 + 62 + 66 + 54 + 58$$
$$+ 60 + 70 + 50 = 692$$

$$\Sigma X_F^2 = 54^2 + 58^2 + 49^2 + 56^2 + 55^2 + 62^2 + 66^2 + 54^2$$
$$+ 58^2 + 60^2 + 70^2 + 50^2 = 40,322$$

$$\hat{S}_F^2 = \frac{N_F(\Sigma X_F^2) - (\Sigma X_F)^2}{N_F(N_F - 1)} = \frac{(12)(40,322) - (692)^2}{(12)(12 - 1)}$$
$$= 37.88$$

$$\Sigma X_M = 48 + 42 + 56 + 44 + 38 + 59 + 62 + 66 + 44$$
$$+ 48 + 56 + 75 = 638$$

$$\Sigma X_M^2 = 48^2 + 42^2 + 56^2 + 44^2 + 38^2 + 59^2 + 62^2 + 66^2$$
$$+ 44^2 + 48^2 + 56^2 + 75^2 = 35,266$$

$$\hat{S}_M^2 = \frac{N_M(\Sigma X_M^2) - (\Sigma X_M)^2}{N_M(N_M - 1)} = \frac{(12)(35,266) - (638)^2}{(12)(12 - 1)}$$
$$= 122.33$$

Thus, according to Appendix Table A-4, the critical F for 10 and 11 degrees of freedom for $\alpha = .05$ is 2.85 [Since $df_1 = 11$ is not in the table we must use $df_1 = 10$ as the best estimate]. But from our data $F = 122.33/37.88 = 3.23$. Since this value is larger than the critical value, we reject the hypothesis that the variances in the two populations are equal.

11. a. First we must calculate the correlation for the sample data.

$$\Sigma X_{IND} = 42 + 55 + 44 + 49 + 59 + 26 + 32 + 30 + 44$$
$$+ 58 = 439$$

$$\Sigma X_{IND}^2 = 42^2 + 55^2 + 44^2 + 49^2 + 59^2 + 26^2 + 32^2 + 30^2$$
$$+ 44^2 + 58^2 = 20,507$$

$$\Sigma X_{INT} = 38 + 26 + 58 + 26 + 54 + 28 + 55 + 44 + 45$$
$$+ 38 = 412$$

$$\Sigma X_{INT}^2 = 38^2 + 26^2 + 58^2 + 26^2 + 54^2 + 28^2 + 55^2 + 44^2$$
$$+ 45^2 + 38^2 = 18,290$$

$$\Sigma (X_{IND})(X_{INT}) = (42)(38) + (55)(26) + (44)(58)$$
$$+ (49)(26) + (59)(54) + (26)(28)$$
$$+ (32)(55) + (30)(44) + (44)(45)$$
$$+ (58)(38) = 18,030$$

$r_{X_{IND}X_{INT}}$

$$= \frac{N(\Sigma (X_{IND})(X_{INT})) - (\Sigma X_{IND})(\Sigma X_{INT})}{[\sqrt{N(\Sigma X_{IND}^2) - (\Sigma X_{IND})^2}][\sqrt{N(\Sigma X_{INT}^2) - (\Sigma X_{INT})^2}]}$$

$$= \frac{(10)(18,030) - (439)(412)}{[\sqrt{(10)(20,507) - (439)^2}][\sqrt{(10)(18,290) - (412)^2}]}$$

$$= -.045$$

$H_0: \rho = 0$
$H_1: \rho \neq 0$ at $\alpha = .05$

$$t_{N-2} = \frac{r - \rho}{\sqrt{\dfrac{1 - r^2}{N - 2}}} = \frac{-.045 - 0}{\sqrt{\dfrac{1 - (-.045)^2}{10 - 2}}} = -0.13$$

For $df = N - 2 = 10 - 2 = 8$ and $\alpha = .05$ and for a two-tailed test the critical $t(8) = \pm2.306$. Since -0.13 is between -2.306 and $+2.306$, we cannot reject the hypothesis that the correlation in the population is zero.

b. First we must calculate the correlation for the sample data.

$$\Sigma X_{emp} = 32 + 48 + 55 + 65 + 54 + 36 + 66 + 44 + 48$$
$$= 448$$

$$\Sigma X_{emp}^2 = 32^2 + 48^2 + 55^2 + 65^2 + 54^2 + 36^2 + 66^2 + 44^2$$
$$+ 48^2 = 23,386$$

$$\Sigma X_{agg} = 64 + 53 + 46 + 49 + 40 + 56 + 32 + 51 + 48$$
$$= 439$$

$$\Sigma X_{agg}^2 = 64^2 + 53^2 + 46^2 + 49^2 + 40^2 + 56^2 + 32^2 + 51^2$$
$$+ 48^2 = 22,087$$

$$\Sigma (X_{emp})(X_{agg}) = (32)(64) + (48)(53) + (55)(46)$$
$$+ (65)(49) + (54)(40) + (36)(56)$$
$$+ (66)(32) + (44)(51) + (48)(48)$$
$$= 21,143$$

$r_{X_{emp}X_{agg}}$

$$= \frac{N(\Sigma (X_{emp})(X_{agg})) - (\Sigma X_{emp})(\Sigma X_{agg})}{[\sqrt{N(\Sigma X_{emp}^2) - (\Sigma X_{emp})^2}][\sqrt{N(\Sigma X_{agg}^2) - (\Sigma X_{agg})^2}]}$$

$$= \frac{(9)(21,143) - (448)(439)}{[\sqrt{(9)(23,386) - (448)^2}][\sqrt{(9)(22,087) - (439)^2}]}$$

$$= -.83$$

$H_0: \rho = 0$
$H_1: \rho \neq 0$ at $\alpha = .05$

$$t_{N-2} = \frac{r - \rho}{\sqrt{\dfrac{1 - r^2}{N - 2}}} = \frac{-.83 - 0}{\sqrt{\dfrac{1 - (-.83)^2}{9 - 2}}} = -3.94$$

For $df = N - 2 = 9 - 2 = 7$ and $\alpha = .05$ and for a two-tailed test the critical $t(7) = \pm2.365$. Since -3.94 is outside of the area of -2.365 and $+2.365$, we can reject the hypothesis that the correlation in the population is zero and accept that the correlation in the population is different from zero.

c. First we must calculate the correlation for the sample data.

$$\Sigma X_{sby} = 23 + 33 + 55 + 48 + 42 + 38 + 24 + 51 + 41$$
$$+ 44 + 56 + 54 = 509$$

$$\Sigma X_{sby}^2 = 23^2 + 33^2 + 55^2 + 48^2 + 42^2 + 38^2 + 24^2 + 51^2$$
$$+ 41^2 + 44^2 + 56^2 + 54^2 = 23,001$$

$\sum X_{dep} = 31 + 37 + 49 + 42 + 40 + 39 + 29 + 50 + 43$
$\qquad + 45 + 52 + 54 = 511$

$\sum X_{dep}^2 = 31^2 + 37^2 + 49^2 + 42^2 + 40^2 + 39^2 + 29^2 + 50^2$
$\qquad + 43^2 + 45^2 + 52^2 + 54^2 = 22{,}451$

$\sum (X_{sby})(X_{dep}) = (23)(31) + (33)(37) + (55)(49)$
$\qquad + (48)(42) + (42)(40) + (38)(39)$
$\qquad + (24)(29) + (51)(50) + (41)(43)$
$\qquad + (44)(45) + (56)(52) + (54)(54)$
$\qquad = 22{,}624$

$r_{X_{sby} X_{dep}}$

$= \dfrac{N(\sum (X_{sby})(\sum X_{dep})) - (\sum X_{sby})(\sum X_{dep})}{[\sqrt{N(\sum X_{sby}^2) - (\sum X_{sby})^2}][\sqrt{N(\sum X_{dep}^2) - (\sum X_{dep})^2}]}$

$= \dfrac{(12)(22{,}624) - (509)(511)}{[\sqrt{(12)(23{,}001) - (509)^2}][\sqrt{(12)(22{,}451) - (511)^2}]}$

$= .96$

$H_0: \rho = 0$
$H_1: \rho \neq 0$ at $\alpha = .05$

$t_{N-2} = \dfrac{r - \rho}{\sqrt{\dfrac{1 - r^2}{N - 2}}} = \dfrac{.96 - 0}{\sqrt{\dfrac{1 - (.96)^2}{12 - 2}}} = 10.84$

For $df = N - 2 = 12 - 2 = 10$ and $\alpha = .05$ and for a two-tailed test the critical $t(10) = \pm 2.228$. Since 10.84 is outside of the area of -2.228 and $+2.228$, we can reject the hypothesis that the correlation in the population is zero and accept that the correlation in the population is different from zero.

13. a. If $Z_F = 0.56$ then $r = .51$
    b. If $Z_F = 0.15$ then $r = .15$
    c. If $Z_F = 0.89$ then $r = .71$
    d. If $Z_F = 1.83$ then $r = .95$
    e. If $Z_F = 0.32$ then $r = .31$

15. a. A sample of size $N = 36$ is drawn and a sample correlation of $r = .48$ is computed. First, we must use Appendix Table A-5 to find $Z_F = 0.523$ for a correlation of .48. For $\alpha = .05$ and a two-tailed test, $Z = \pm 1.96$. In addition,

$\sigma_{Z_F} = \dfrac{1}{\sqrt{N - 3}} = \dfrac{1}{\sqrt{36 - 3}} = 0.17$

$CI_L = (-Z_\alpha)(\sigma_{Z_F}) + Z_{F_r} = (-1.96)(0.17) + 0.523 = 0.19$

$CI_U = (Z_\alpha)(\sigma_{Z_F}) + Z_{F_r} = (1.96)(0.17) + 0.523 = 0.86$

Translating these Z-scores back into correlations using Appendix Table A-6, we obtain a confidence interval of .19 to .70. Thus, there is a 95% probability that the correlation in the population is between .19 and .70.

b. A sample of size $N = 100$ is drawn and a sample correlation of $r = -.66$ is computed. First, we must use Appendix Table A-5 to find $Z_F = -0.793$ for a correlation of $-.66$. For $\alpha = .05$ and a two-tailed test, $Z = \pm 1.96$. In addition,

$\sigma_{Z_F} = \dfrac{1}{\sqrt{N - 3}} = \dfrac{1}{\sqrt{100 - 3}} = 0.10$

$CI_L = (-Z_\alpha)(\sigma_{Z_F}) + Z_{F_r} = (-1.96)(0.10) + -0.793$
$\qquad = -0.99$

$CI_U = (Z_\alpha)(\sigma_{Z_F}) + Z_{F_r} = (1.96)(0.10) + -0.793 = -0.60$

Translating these Z-scores back into correlations using Appendix Table A-6, we obtain a confidence interval of $-.76$ to $-.54$. Thus, there is a 95% probability that the correlation in the population is between $-.76$ and $-.54$.

c. A sample of size $N = 75$ is drawn and a sample correlation of $r = .78$ is computed. First, we must use Appendix Table A-5 to find $Z_F = 1.045$ for a correlation of .78. For $\alpha = .01$ and a two-tailed test, $Z = \pm 2.576$. In addition,

$\sigma_{Z_F} = \dfrac{1}{\sqrt{N - 3}} = \dfrac{1}{\sqrt{75 - 3}} = 0.12$

$CI_L = (-Z_\alpha)(\sigma_{Z_F}) + Z_{F_r} = (-2.576)(0.12) + 1.045 = 0.74$

$CI_U = (Z_\alpha)(\sigma_{Z_F}) + Z_{F_r} = (2.576)(0.12) + 1.045 = 1.35$

Translating these Z-scores back into correlations using Appendix Table A-6, we obtain a confidence interval of .63 to .87. Thus, there is a 99% probability that the correlation in the population is between .63 and .87.

d. A sample of size $N = 50$ is drawn and a sample correlation of $r = -.22$ is computed. First, we must use Appendix Table A-5 to find $Z_F = -0.224$ for a correlation of $-.22$. For $\alpha = .01$ and a two-tailed test, $Z = \pm 2.576$. In addition,

$\sigma_{Z_F} = \dfrac{1}{\sqrt{N - 3}} = \dfrac{1}{\sqrt{50 - 3}} = 0.15$

$CI_L = (-Z_\alpha)(\sigma_{Z_F}) + Z_{F_r} = (-2.576)(0.15) + -0.224$
$\qquad = -0.61$

$CI_U = (Z_\alpha)(\sigma_{Z_F}) + Z_{F_r} = (2.576)(0.15) + -0.224 = 0.16$

Translating these Z-scores back into correlations using Appendix Table A-6, we obtain a confidence interval of $-.54$ to $+.16$. Thus, there is a 99% probability that the correlation in the population is between $-.54$ and $+.16$.

17. a. $r_w = .68$, $N_w = 40$; $r_m = .39$, $N_m - 35$. First transform each correlation to a Z-score.

$Z_{F_w} = 0.829$ and $Z_{F_m} = 0.412$

Next, calculate the standard error of the difference between two Z_F's.

$\sigma_{Z_{F_w} - Z_{F_m}} = \sqrt{\dfrac{1}{N_w - 3} + \dfrac{1}{N_m - 3}} = \sqrt{\dfrac{1}{40 - 3} + \dfrac{1}{35 - 3}}$
$\qquad = 0.24$

$H_0: \rho_w \leq \rho_m$
$H_1: \rho_w > \rho_m$

Calculate the Z ratio.

$Z = \dfrac{(Z_{F_{r_w}} - Z_{F_{r_m}}) - (Z_{F_{\rho_w}} - Z_{F_{\rho_m}})}{\sigma_{Z_{F_w} - Z_{F_m}}} = \dfrac{(0.829 - 0.412) - (0)}{0.24}$

$\qquad = 1.74$

For a one-tailed test with $\alpha = .05$, $Z = 1.645$. Since 1.74 is greater than 1.645, we reject the null hypothesis $H_0: \rho_w \leq \rho_m$.

b. $r_i = .35$, $N_i = 28$; $r_e = .18$, $N_e = 22$. First transform each correlation to a Z-score.

$Z_{F_i} = 0.366$ and $Z_{F_e} = 0.182$

Next, calculate the standard error of the difference between two Z_F's.

$$\sigma_{Z_{F_i}-Z_{F_e}} = \sqrt{\frac{1}{N_i-3}+\frac{1}{N_e-3}} = \sqrt{\frac{1}{28-3}+\frac{1}{22-3}}$$
$$= 0.30$$

$H_0: \rho_i \leq \rho_e$
$H_1: \rho_i > \rho_e$

Calculate the Z ratio.

$$Z = \frac{(Z_{F_{r_i}} - Z_{F_{r_e}}) - (Z_{F_{\rho_i}} - Z_{F_{\rho_e}})}{\sigma_{Z_{F_i}-Z_{F_e}}} = \frac{(0.366 - 0.182) - (0)}{0.30}$$
$$= 0.61$$

For a one-tailed test with $\alpha = .05$, $Z = 1.645$. Since 0.61 is less than 1.645, we fail to reject the null hypothesis $H_0: \rho_i \leq \rho_e$.

c. $r_e = -.54$, $N_e = 34$; $r_y = -.28$, $N_y = 27$. First transform each correlation to a Z-score.

$$Z_{F_e} = -0.604 \text{ and } Z_{F_y} = -0.288$$

Next, calculate the standard error of the difference between two Z_F's.

$$\sigma_{Z_{F_e}-Z_{F_y}} = \sqrt{\frac{1}{N_e-3}+\frac{1}{N_y-3}} = \sqrt{\frac{1}{34-3}+\frac{1}{27-3}}$$
$$= 0.27$$

$H_0: \rho_e \geq \rho_y$
$H_1: \rho_e < \rho_y$

Calculate the Z ratio.

$$Z = \frac{(Z_{F_{r_e}} - Z_{F_{r_y}}) - (Z_{F_{\rho_e}} - Z_{F_{\rho_y}})}{\sigma_{Z_{F_e}-Z_{F_y}}}$$
$$= \frac{(-0.604 - -0.288) - (0)}{0.27}$$
$$= -1.17$$

For a one-tailed test with $\alpha = .05$, $Z = -1.645$. Since -1.17 is not less than -1.645, we fail to reject the null hypothesis $H_0: \rho_e \geq \rho_y$.

19. a. $r_{bosagg} = .45$; $r_{angagg} = .58$; $r_{bosang} = .36$; $N = 100$
To test $H_0: \rho_{bosagg} = \rho_{angagg}$, we must use the formula

$$Z = \frac{\sqrt{N}(r_{bosagg} - r_{angagg})}{\sqrt{(1 - r_{bosagg}^2)^2 + (1 - r_{angagg}^2)^2 - 2r_{bosang}^3 - (2r_{bosang} - r_{bosagg}r_{angagg})(1 - r_{bosagg}^2 - r_{angagg}^2 - r_{bosang}^2)}}$$

$$= \frac{\sqrt{100}(.45 - .58)}{\sqrt{(1 - .45^2)^2 + (1 - .58^2)^2 - 2(.36)^3 - (2(.36) - (.45)(.58))(1 - .45^2 - .58^2 - .36^2)}}$$

$$= -1.42$$

For a two-tailed test with $\alpha = .05$, $Z = \pm 1.96$. Since -1.42 is between -1.96 and $+1.96$, we fail to reject the null hypothesis that the correlations are equal in the population.

b. $r_{shse} = -.59$; $r_{prse} = -.38$; $r_{prsb} = .43$; $N = 75$
To test $H_0: \rho_{shse} = \rho_{prse}$, we must use the formula

$$Z = \frac{\sqrt{N}(r_{shse} - r_{prse})}{\sqrt{(1 - r_{shse}^2)^2 + (1 - r_{prse}^2)^2 - 2r_{prsb}^3 - (2r_{prsb} - r_{shse}r_{prse})(1 - r_{shse}^2 - r_{prse}^2 - r_{prsb}^2)}}$$

$$= \frac{\sqrt{75}((-.59) - (-.38))}{\sqrt{(1 - (-.59)^2)^2 + (1 - (-.38)^2)^2 - 2(.43)^3 - (2(.43) - (-.59)(-.38))(1 - (-.59)^2 - (-.38)^2 - .43^2)}}$$

$$= -2.04$$

For a two-tailed test with $\alpha = .05$, $Z = \pm 1.96$. Since -2.04 is outside -1.96 and $+1.96$, we reject the null hypothesis that the correlations are equal in the population.

CHAPTER 15

Table A15-1 Expected Cell Frequencies

$$E_{rc} = \frac{N_r N_c}{N_T}$$

Adolescent	Antisocial Risk-taking			Row Marginal
	High	Moderate	Low	
Native American Living Off	$\frac{(196)(263)}{849} = 60.72$	$\frac{(196)(329)}{849} = 75.95$	$\frac{(196)(257)}{849} = 59.33$	196
Native American Living On	$\frac{(205)(263)}{849} = 63.50$	$\frac{(205)(329)}{849} = 79.44$	$\frac{(205)(257)}{849} = 62.06$	205
Black	$\frac{(223)(263)}{849} = 69.08$	$\frac{(223)(329)}{849} = 86.42$	$\frac{(223)(257)}{849} = 67.50$	223
White	$\frac{(225)(263)}{849} = 69.70$	$\frac{(225)(329)}{849} = 87.19$	$\frac{(225)(257)}{849} = 68.11$	225
Column Marginal	263	329	257	849

1. a.

Observed Frequencies

Class	No Further Vocational Education Beyond High School	Vocational Technical Education	College Education	Row Marginal
Senior	5	7	10	22
Freshman	9	5	6	20
Column Marginal	14	12	16	42

b.

Expected Cell Frequencies

$$E_{rc} = \frac{N_r N_c}{N_T}$$

Class	No Further Education Beyond High School	Vocational Technical Education	College Education	Row Marginal
Senior	$\frac{(22)(14)}{42} = 7.33$	$\frac{(22)(12)}{42} = 6.29$	$\frac{(22)(16)}{42} = 8.38$	22
Freshman	$\frac{(20)(14)}{42} = 6.67$	$\frac{(12)(20)}{42} = 5.71$	$\frac{(20)(16)}{42} = 7.62$	20
Column Marginal	14	12	16	42

c. $\chi^2 = \sum_{r=1}^{R}\sum_{c=1}^{C}\frac{(O_{rc} - E_{rc})^2}{E_{rc}} = \frac{(5 - 7.33)^2}{7.33} + \frac{(7 - 6.29)^2}{6.29}$

$+ \frac{(10 - 8.38)^2}{8.38} + \frac{(9 - 6.67)^2}{6.67} + \frac{(5 - 5.71)^2}{5.71}$

$+ \frac{(6 - 7.62)^2}{7.62} = 2.38$

Using Appendix Table A-3 for $df = (R-1)(C-1) = (2-1)(3-1) = 2$ for $\alpha = .05$, $\chi^2 = 6.0$.

d. Since the observed χ^2 value (2.38) does not exceed the critical χ^2 value (6.0), we can conclude that the career choices as reflected in education beyond high school do not change significantly from the freshman year to the senior year.

3. a.

Observed Frequencies

Adolescent	Antisocial Risk-taking			Row Marginal
	High	Moderate	Low	
Native American Living Off	98	57	41	196
Native American Living On	54	85	66	205
Black	53	91	79	223
White	58	96	71	225
Column Marginal	263	329	257	849

b. See Table A15-1 above.

c. $\chi^2 = \sum_{r=1}^{R}\sum_{c=1}^{C}\frac{(O_{rc} - E_{rc})^2}{E_{rc}} = \frac{(98 - 60.72)^2}{60.72} + \frac{(57 - 75.95)^2}{75.95}$

$+ \frac{(41 - 59.33)^2}{59.33} + \frac{(54 - 63.50)^2}{63.50} + \frac{(85 - 79.44)^2}{79.44}$

$+ \frac{(66 - 62.06)^2}{62.06} + \frac{(53 - 69.08)^2}{69.08} + \frac{(91 - 86.42)^2}{86.42}$

$+ \frac{(79 - 67.50)^2}{67.50} + \frac{(58 - 69.70)^2}{69.70} + \frac{(96 - 87.19)^2}{87.19}$

$+ \frac{(71 - 68.11)^2}{68.11} = 44.25$

Using Appendix Table A-3 for $df = (R-1)(C-1) = (4-1)(3-1) = 6$ for $\alpha = .05$, $\chi^2 = 12.6$.

d. Since the observed χ^2 value (44.25) exceeds the critical χ^2 value (12.6), we can conclude that there is a difference in the degree of antisocial risk-taking behavior for the four identified adolescent groups. If we examine the values in the contingency table, it appears that the adolescent group that is at the highest risk is the Native American who lives off the reservation. All other adolescent groups appear to be highly similar in terms of level of antisocial risk-taking behavior.

5.

Counseling Center Usage Pattern	Observed Frequency	Expected Frequency	Discrepancy (O − E)	χ^2
Considerably More Women	189	$\left(\frac{1}{3}\right)(486) = 162$	27	4.50
An Equal Number of Men and Women	158	$\left(\frac{1}{3}\right)(486) = 162$	−4	0.10
Considerably More Men	139	$\left(\frac{1}{3}\right)(486) = 162$	−23	3.27

$$\chi^2 = \sum_{r=1}^{R} \frac{(O_r - E_r)^2}{E_r} = \frac{(189 - 162)^2}{162} + \frac{(158 - 162)^2}{162}$$

$$+ \frac{(139 - 162)^2}{162} = 4.5 + 0.1 + 3.27 = 7.87$$

Using Appendix Table A-3 for $df = (R - 1) = (3 - 1) = 2$ for $\alpha = .05$, $\chi^2 = 6.0$. Thus, the frequencies are different from the expectations. It seems that the counseling center is viewed by more people as serving considerably more women and considerably fewer men.

7.

Age Group of Drivers		Yes	No	TOTAL
Adolescents	Observed	24	26	50
	Expected	19.5	30.5	50.0
Young Adults	Observed	21	29	50
	Expected	19.5	30.5	50.0
Middle-Aged Adults	Observed	18	32	50
	Expected	19.5	30.5	50.0
Elderly Adults	Observed	15	35	50
	Expected	19.5	30.5	50.0
TOTAL		78	122	200

(header: Exceeded the Mean Number of Errors?)

$$\chi^2 = \sum_{r=1}^{R} \sum_{c=1}^{C} \frac{(O_{rc} - E_{rc})^2}{E_{rc}} = \frac{(26 - 30.5)^2}{30.5} + \frac{(29 - 30.5)^2}{30.5}$$

$$+ \frac{(32 - 30.5)^2}{30.5} + \frac{(35 - 30.5)^2}{30.5} + \frac{(24 - 19.5)^2}{19.5}$$

$$+ \frac{(21 - 19.5)^2}{19.5} + \frac{(18 - 19.5)^2}{19.5} + \frac{(15 - 19.5)^2}{19.5} = 0.6$$

$$+ 0.07 + 0.07 + 0.66 + 1.04 + 0.12 + 0.12$$

$$+ 1.04 = 3.78$$

Using Appendix Table A-3 for $df = (4 - 1)(2 - 1) = 3$ for $\alpha = .05$, $\chi^2 = 7.8$. Thus, it appears there is not a significant difference in driving error rates of adolescents, young adults, middle-aged adults, and elderly adults.

9.

	Status Before Education	
Status After Education	Driving After Consuming Alcohol	No Driving After Consuming Alcohol
No Driving After Consuming Alcohol	42	275
Driving After Consuming Alcohol	114	28

$$\chi^2 = \frac{(A - D)^2}{A + D} = \frac{(42 - 28)^2}{42 + 28} = 2.80$$

Using Appendix Table A-3 for $df = 1$ for $\alpha = .05$, $\chi^2 = 3.8$. Thus, the educational program does not seem to have decreased driving after consuming alcohol significantly more than it increased driving after consuming alcohol.

11.

Empathy 24, 24, 27, 33, 34, 38, 44, 47, 48, 52, 55, 62
Rank 1.5, 1.5, 3, 4, 5, 6, 7, 8, 9, 10, 11, 12
Aggression 25, 26, 31, 35, 36, 38, 40, 42, 42, 43, 47, 51
Rank 1, 2, 3, 4, 5, 6, 7, 8.5, 8.5, 10, 11, 12
Note: You will get the same correlation coefficient if you rank the scores in the reverse order from the order above.

Empathy	r_E	Aggression	r_A	$d^2 = (r_E - r_A)^2$
24	1.5	47	11	$(1.5 - 11)^2 = 90.25$
48	9	42	8.5	$(9 - 8.5)^2 = 0.25$
33	4	51	12	$(4 - 12)^2 = 64.00$
62	12	43	10	$(12 - 10)^2 = 4.00$
24	1.5	35	4	$(1.5 - 4)^2 = 6.25$
27	3	40	7	$(3 - 7)^2 = 16.00$
44	7	36	5	$(7 - 5)^2 = 4.00$
38	6	38	6	$(6 - 6)^2 = 0.00$
55	11	25	1	$(11 - 1)^2 = 100.00$
52	10	26	2	$(10 - 2)^2 = 64.00$
47	8	31	3	$(8 - 3)^2 = 25.00$
34	5	42	8.5	$(5 - 8.5)^2 = 12.25$
				$\sum d^2 = 386.00$

$$\rho_S = 1 - \frac{6\sum d^2}{N(N^2 - 1)} = 1 - \frac{6(386)}{(12)(12^2 - 1)} = -.35$$

$$t_{(N-2)} = \frac{\rho_S - \rho}{\sqrt{\dfrac{(1 - \rho_S^2)}{N - 2}}} = \frac{-.35 - 0}{\sqrt{\dfrac{(1 - (-.35)^2)}{12 - 2}}} = -1.18$$

Using Appendix Table A-2 for $df = N - 2 = 12 - 2 = 10$ and $\alpha = .05$ for a two-tailed test, the t-values are ± 2.228. Since -1.18 is between -2.228 and $+2.228$, we fail to reject the null hypothesis that the correlation is zero.

13.

Procrastination 29, 30, 31, 36, 40, 47, 48, 49, 57, 64
Rank 1, 2, 3, 4, 5, 6, 7, 8, 9, 10
Fear of Negative 26, 31, 31, 35, 37, 51, 52, 56, 58, 58
Evaluation
Rank 1, 2.5, 2.5, 4, 5, 6, 7, 8, 9.5, 9.5

Procrastination	r_P	Fear of Negative Evaluation	r_{FNE}	$d^2 = (r_P - r_{FNE})^2$
47	6	56	8	$(6 - 8)^2 = 4$
30	2	37	5	$(2 - 5)^2 = 9$
36	4	31	2.5	$(4 - 2.5)^2 = 2.25$

64	10	58	9.5	$(10 - 9.5)^2 = 0.25$
49	8	52	7	$(8 - 7)^2 = 1$
31	3	31	2.5	$(3 - 2.5)^2 = 0.25$
48	7	51	6	$(7 - 6)^2 = 1$
40	5	35	4	$(5 - 4)^2 = 1$
57	9	58	9.5	$(9 - 9.5)^2 = 0.25$
29	1	26	1	$(1 - 1)^2 = 0$
				$\Sigma d^2 = 19$

$$\rho_S = 1 - \frac{6\Sigma d^2}{N(N^2 - 1)} = 1 - \frac{6(19)}{(10)(10^2 - 1)} = .88$$

$$t_{(N-2)} = \frac{\rho_S - \rho}{\sqrt{\frac{(1 - \rho_S^2)}{N - 2}}} = \frac{.88 - 0}{\sqrt{\frac{1 - (.88)^2}{10 - 2}}} = 5.24$$

Using Appendix Table A-2 for $df = N - 2 = 10 - 2 = 8$ and $\alpha = .05$ for a one-tailed test, the t-value is 1.860. Since 5.24 is larger than 1.86 we reject the null hypothesis that the correlation is zero or smaller and accept that the correlation is positive.

15.

Score 11, 14, 14, 15, 16, 17, 18, 18, 23, 25
Rank 1, 2.5, 2.5, 4, 5, 6, 7.5, 7.5, 9, 10

Score on the Dart Game		Score on the Dart Game	
Physical Practice	r_P	Mental Practice	r_M
23	9	17	6
18	7.5	15	4
25	10	11	1
16	5	14	2.5
18	7.5		
14	2.5		
	$\Sigma r_P = 41.5$		$W_S = \Sigma r_M = 13.5$

$N_S = 4$ and $N_L = 6$

From Appendix Table A-7, for the table $N_S = 4$ and using the column $N_L = 6$ and the row 13.5 in S_{LOW} we find a probability of .0333 for row 13 and .0571 for row 14. We must double these for a two-tailed test. These values will then be .0666 and .1142. Despite what value we choose it is still larger than $\alpha = .05$. Thus, we fail to reject the null hypothesis that physical practice of the movements of throwing a dart will lead to the same level of dart-throwing skill as mental practice.

17. Score 0, 1, 2, 2, 3, 3, 4, 4, 4, 4, 5, 5, 5
Rank 1, 2, 3.5, 3.5, 5.5, 5.5, 8.5, 8.5, 8.5, 8.5, 12, 12, 12

Score 6, 6, 6, 7, 8, 8, 9, 9, 10, 11
Rank 15, 15, 15, 17, 18.5, 18.5, 20.5, 20.5, 22, 23

Errors		Errors	
Caffeine	r_C	No Caffeine	r_N
4	8.5	3	5.5
5	12	4	8.5
9	20.5	6	15
6	15	5	12
8	18.5	0	1
4	8.5	2	3.5
11	23	4	8.5
8	18.5	1	2
3	5.5	7	17
6	15	5	12
9	20.5	2	3.5
10	22		
	$\Sigma r_C = 187.5$		$W_S = \Sigma r_N = 88.5$

$N_S = 11 \qquad N_L = 12 \qquad N_T = 11 + 12 = 23$

$$\mu_{W_S} = \frac{N_S(N_T + 1)}{2} = \frac{(11)(23 + 1)}{2} = 132$$

$$\sigma_{W_S} = \sqrt{\frac{N_S(N_L)(N_T + 1)}{12}} = \sqrt{\frac{(11)(12)(23 + 1)}{12}} = 16.25$$

$$Z = \frac{W_S + .5 - \mu_{W_S}}{\sigma_{W_S}} = \frac{88.5 + .5 - 132}{16.25} = -2.65$$

Since the critical value for a one-tailed test with $\alpha = .05$ is -1.645 in the Z-distribution we would reject the null hypothesis that the number of errors committed by rats given no caffeine will be greater than or equal to the number of errors committed by rats that were given caffeine. Thus, giving rats concentrated doses of caffeine appears to increase the number of errors that they commit.

19.

Control	Injected	d	$|d|$	Rank	Signed Rank
48	36	$48 - 36 = 12$	12	4	4
57	48	$57 - 48 = 9$	9	3	3
35	36	$35 - 36 = -1$	1	1	−1
63	41	$63 - 41 = 22$	22	5	5
42	47	$42 - 47 = -5$	5	2	−2

Since our research hypothesis is that rats in the control condition will have more lever presses than rats who were given scopolamine, we sum the positive ranks $(4 + 3 + 5 = 12)$. From Appendix Table A-8, use $N = 5$ (the number of pairs) for a one-tailed test to find the probability associated with a sum of positive ranks of 12. The probability is .1563. Since this is not smaller than our $\alpha = .05$ we fail to reject the null hypothesis that the control rats have fewer or an equal number of lever presses than the rats who were given scopolamine. Thus, scopolamine does not seem to significantly deter rats from lever pressing to receive food.

21.

Aircraft Noise	White Noise	d	$|d|$	Rank	Signed Rank
46	41	$(46 - 41) = 5$	5	5.5	5.5
58	42	$(58 - 42) = 16$	16	16	16
61	55	$(61 - 55) = 6$	6	7.5	7.5
65	57	$(65 - 57) = 8$	8	10	10
59	52	$(59 - 52) = 7$	7	9	9
34	37	$(34 - 37) = -3$	3	4	−4
72	54	$(72 - 54) = 18$	18	17	17
69	57	$(69 - 57) = 12$	12	12.5	12.5
45	47	$(45 - 47) = -2$	2	2.5	−2.5
58	44	$(58 - 44) = 14$	14	15	15
64	51	$(64 - 51) = 13$	13	14	14
65	55	$(65 - 55) = 10$	10	11	11
66	64	$(66 - 64) = 2$	2	2.5	2.5
47	53	$(47 - 53) = -6$	6	7.5	−7.5
49	50	$(49 - 50) = -1$	1	1	−1
43	48	$(43 - 48) = -5$	5	5.5	−5.5
63	51	$(63 - 51) = 12$	12	12.5	12.5

Since the research hypothesis was that participants exposed to aircraft noise would have different white cell counts than participants who were exposed to white noise, you can sum either the positive or the negative ranks. Summing the positive ranks yields (5.5 + 16 + 7.5 + 10 + 9 + 17 + 12.5 + 15 + 14 + 11 + 2.5 + 12.5 = 132.5).

$$\mu_s = \frac{N(N+1)}{4} = \frac{(17)(17+1)}{4} = 76.5$$

$$\sigma_s = \sqrt{\frac{N(N+1)(2N+1)}{24}} = \sqrt{\frac{(17)(17+1)(34+1)}{24}}$$

$$= 21.12$$

$$Z = \frac{S - \mu_s}{\sigma_s} = \frac{132.5 - 76.5}{21.12} = 2.65$$

Since the critical value for a two-tailed test with $\alpha = .05$ are ± 1.96 in the Z-distribution we would reject the null hypothesis ($1.96 < 2.65$). Thus, constant aircraft noise seems to increase the white cell count more than white noise.

23. The following is the SPSS output:

Crosstabs

Case Processing Summary

	Cases					
	Valid		Missing		Total	
	N	Percent	N	Percent	N	Percent
Current Career Interest *Gender of Participant	200	100.0%	0	.0%	200	100.0%

Current Career Interest *Gender of Participant Crosstabulation

			Gender of Participant		
			Male	Female	Total
Current Career Interest	Fine Arts	Count	19	25	44
		Expected Count	22.0	22.0	44.0
	Humanities	Count	6	12	18
		Expected Count	9.0	9.0	18.0
	Social Sciences	Count	22	28	50
		Expected Count	25.0	25.0	50.0
	Biological Sciences	Count	26	25	51
		Expected Count	25.5	25.5	51.0
	Physical Sciences and Math	Count	27	10	37
		Expected Count	18.5	18.5	37.0
Total		Count	100	100	200
		Expected Count	100.0	100.0	200.0

Chi-Square Tests

	Value	df	Asymp. Sig. (2-sided)
Pearson Chi-Square	11.369[a]	4	.023
Likelihood Ratio	11.713	4	.020
Linear-by-Linear Association	7.145	1	.008
N of Valid Cases	200		

[a] 0 cells (.0%) have expected count less than 5. The minimum expected count is 9.00.

From the results in the analysis above, there is a gender difference in career interests of 7th to 10th grade students. It appears that the largest difference is in physical sciences and math where males are more likely to have a career interest than females. The significance of the test was .023 which is smaller than the traditional $\alpha = .05$.

25. The following is the SPSS output:

Nonparametric Correlations

Correlations

			Post-instruction math attitude	Final algebra grade
Spearman's rho	Post-instruction math attitude	Correlation Coefficient	1.000	−.784**
		Sig. (1-tailed)	.	.000
		N	49	49
	Final Algebra grade	Correlation Coefficient	−.784**	1.000
		Sig. (1-tailed)	.000	.
		N	49	49

**Correlation is significant at the .01 level (1-tailed).

From the results above, there was a significant negative correlation ($-.784$) between a negative attitude toward math and the student's final grade in algebra. This is consistent with the researcher's prediction. The significance of the test was .000 which is smaller than the traditional $\alpha = .05$.

CHAPTER 16

1. The first reason it is necessary or desirable to include more than two groups in a single study is that many things that are of interest to psychologists and educators have more than two groups involved. Selecting only two groups limits our ability to generalize our findings since it does not represent the way things occur in the world outside of the laboratory. We may also reach incorrect conclusions about the phenomenon being studied since we are not looking at all of the possible groups or conditions.

A second reason for studying more than two groups at the same time is economy. We can examine several different groups or conditions in a single study that would require numerous studies to investigate in pairs. For example, if we have three conditions, it would take three separate studies to examine the relationship in a pairwise fashion. Furthermore, if we have 30 participants per condition in our study with 3 conditions, we would need a total of 90 participants. However, if we have 30 participants in 3 separate studies in which we investigate relationships between these same variables in a pairwise fashion, we would need 60 participants per study for a total of 180 participants. Thus, we also economize on participants.

3. Partitioning the variance is a process of separating the total variance into predictable and unpredictable portions. The total variability is related to how much an individual score differs from the mean of all of the scores. This variability is related to a predictable portion which is how much the means of each group or condition differ from the overall mean and an unpredictable portion which is how much the individual scores within a group differ from the mean of that group. We can always construct these differences for each participant. However, if we try to sum them across participants the sum is always zero, so we must square each deviation score before we sum

them. This leads to three sums of squares which are sum of squares total, sum of squares between treatment, and sum of squares within each treatment (error). If we divide each of these sums of squares by the number of scores comprising them, we will obtain three variances: total variance (how much each individual score varies from the overall mean), between-treatment variance (how much each treatment mean varies from the overall mean), and within-treatment variance (how much each individual score differs from its respective treatment mean).

5. A planned comparison is a test between two means or two averages of means. Planned comparisons are conducted when the researcher knows before she or he collects data what comparisons should be made based on theoretical considerations. In order for this process to be tenable the theories should be necessarily strongly supported in the research literature.

7. a. Left Hemispheric Preference = L; Integrated Hemispheric Preference = I; Right Hemispheric Preference = R

X_L	$(X_L - \overline{X}_L)^2$	X_I	$(X_I - \overline{X}_I)^2$	X_R	$(X_R - \overline{X}_R)^2$
238	278.89	194	161.29	176	501.76
206	234.09	207	0.09	185	179.56
224	7.29	213	39.69	210	134.56
225	13.69	226	372.49	189	88.36
229	59.29	208	1.69	211	158.76
195	691.69	195	136.89	186	153.76
213	68.89	191	246.49	196	5.76
244	515.29	221	204.49	216	309.76
191	918.09	203	13.69	205	43.56
248	712.89	209	5.29	210	134.56
2213	3,500.10	2067	1,182.10	1984	1,710.40

$$\overline{X}_L = \frac{\sum X_L}{N_L} = \frac{2213}{10} = 221.3 \qquad \overline{X}_I = \frac{\sum X_I}{N_I} = \frac{2067}{10} = 206.7$$

$$\overline{X}_R = \frac{\sum X_R}{N_R} = \frac{1984}{10} = 198.4$$

$$\overline{X}_T = \frac{\sum X_T}{N_T} = \frac{\sum X_L + \sum X_I + \sum X_R}{N_L + N_I + N_R}$$

$$= \frac{2213 + 2067 + 1984}{10 + 10 + 10} = 208.8$$

$$SS_W = \sum_j (\sum_i [\overline{X}_{ij} - \overline{X}_j]^2) = \sum_L (X_L - \overline{X}_L)^2 + \sum_I (X_I - \overline{X}_I)^2 + \sum_R (X_R - \overline{X}_R)^2$$
$$= 3,500.10 + 1,182.10 + 1,710.40 = 6392.60$$

$$SS_B = \sum_j (N[\overline{X}_j - \overline{X}_T]^2) = (10)(221.3 - 208.8)^2 + (10)(206.7 - 208.8)^2 + (10)(198.4 - 208.8)^2 = 2688.20$$

X_L	$(X_L - \overline{X}_T)^2$	X_I	$(X_I - \overline{X}_T)^2$	X_R	$(X_R - \overline{X}_T)^2$
238	852.64	194	219.04	176	1075.84
206	7.84	207	3.24	185	566.44
224	231.04	213	17.64	210	1.44
225	262.44	226	295.84	189	392.04
229	408.04	208	0.64	211	4.84
195	190.44	195	190.44	186	519.84
213	17.64	191	316.84	196	163.84
244	1239.04	221	148.84	216	51.84
191	316.84	203	33.64	205	14.44
248	1536.64	209	0.04	210	1.44
2213	5,062.60	2067	1,226.20	1984	2,792.00

Thus, $SS_T = \sum_j \sum_i (X_{ij} - \overline{X}_T)^2 = \sum_i (X_L - \overline{X}_T)^2 + \sum_i (X_I - \overline{X}_T)^2 + \sum_i (X_R - \overline{X}_T)^2 = 5,062.60 + 1,226.20 + 2,792.00 = 9080.80$

b.

X_L	X_L^2	X_I	X_I^2	X_R	X_R^2
238	56,664	194	37,636	176	30,976
206	42,436	207	42,849	185	34,225
224	50,176	213	45,369	210	44,100
225	50,625	226	51,076	189	35,721
229	52,441	208	43,264	211	44,521
195	38,025	195	38,025	186	34,596
213	45,369	191	36,481	196	38,416
244	59,536	221	48,841	216	46,656
191	36,481	203	41,209	205	42,025
248	61,504	209	43,681	210	44,100
2213	493,237	2067	428,431	1984	395,336

$$1 \rightarrow \frac{(\sum_j \sum_i X_{ij})^2}{JN} = \frac{(2213 + 2067 + 1984)^2}{(3)(10)} = 1,307,923.20$$

$$J \rightarrow \sum_j \frac{(\sum_i X_{ij})^2}{N_j} = \frac{(2213)^2}{10} + \frac{(2067)^2}{10} + \frac{(1984)^2}{10}$$

$$= 1,310,611.40$$

$$JN \rightarrow \sum_j \sum_i X_{ij}^2 = 493,237 + 428,431 + 395,336 = 1,317,004$$

$$SS_T = \sum_j \sum_i X_{ij}^2 - \frac{(\sum_j \sum_i X_{ij})^2}{JN} = 1,317,004 - 1,307,923.2$$

$$= 9080.80$$

$$SS_B = \sum_j \frac{(\sum_i X_{ij})^2}{N_j} - \frac{(\sum_j \sum_i X_{ij})^2}{JN} = 1,310,611.4 - 1,307,923.2$$

$$= 2688.20$$

$$SS_W = \sum_j \sum_i X_{ij}^2 - \sum_j \frac{(\sum_i X_{ij})^2}{N_j} = 1,317,004 - 1,310,611.4$$

$$= 6392.60$$

c. $df_T = JN - 1 = (3)(10) - 1 = 29$

$df_B = J - 1 = 3 - 1 = 2$

$df_W = JN - J = (3)(10) - 3 = 27$

d. $MS_T = \dfrac{SS_T}{df_T} = \dfrac{9080.8}{29} = 313.13$

$MS_B = \dfrac{SS_B}{df_B} = \dfrac{2688.2}{2} = 1344.10$

$MS_W = \dfrac{SS_W}{df_W} = \dfrac{6392.6}{27} = 236.76$

$F = \dfrac{MS_B}{MS_W} = \dfrac{1344.1}{236.76} = 5.68$

e.

Source	SS	df	MS	F
Between groups	2688.20	2	1344.10	5.68
Within groups	6392.60	27	236.76	
Total	9080.80	29		

f. $S_{X_L} = \sqrt{\dfrac{\sum (X_L - \overline{X}_L)^2}{N_L}} = \sqrt{\dfrac{3,500.1}{10}} = 18.71$

$S_{X_I} = \sqrt{\dfrac{\sum (X_I - \overline{X}_I)^2}{N_I}} = \sqrt{\dfrac{1,182.1}{10}} = 10.87$

$S_{X_R} = \sqrt{\dfrac{\sum (X_R - \overline{X}_R)^2}{N_R}} = \sqrt{\dfrac{1,710.4}{10}} = 13.08$

g. From Appendix Table A-4 with $df_B = 2$ and $df_W = 27$ for $\alpha = .05$, the critical $F = 3.35$.

h. Since our obtained $F = 5.68$ is greater than the critical $F = 3.35$, we reject the null hypothesis of equal treatment means in the population. Thus, at least one of the means is different in the population.

i. i. $C = \overline{X}_L - \overline{X}_R = 221.3 - 198.4 = 22.9$

$$\hat{S}_C = \sqrt{MS_W\left[\left(\frac{1}{N_L}\right) + \left(\frac{1}{N_R}\right)\right]}$$

$$= \sqrt{(236.76)\left[\left(\frac{1}{10}\right) + \left(\frac{1}{10}\right)\right]}$$

$$= 6.88$$

$$t = \left(\frac{C - K}{\hat{S}_C}\right) = \frac{22.9 - 0}{6.88} = 3.33$$

From Appendix Table A-2 for a two-tailed test with $\alpha = .05$ and $df = 27$, the critical values are ± 2.052. Thus, we can conclude that it took people with a left hemispheric preference a different number of moves to solve the Tower of Hanoi problem than people with a right hemispheric preference.

ii. $C = \overline{X}_L - \overline{X}_I = 221.3 - 206.7 = 14.6$

$$\hat{S}_C = \sqrt{MS_W\left[\left(\frac{1}{N_L}\right) + \left(\frac{1}{N_I}\right)\right]}$$

$$= \sqrt{(236.76)\left[\left(\frac{1}{10}\right) + \left(\frac{1}{10}\right)\right]}$$

$$= 6.88$$

$$t = \left(\frac{C - K}{\hat{S}_C}\right) = \frac{14.6 - 0}{6.88} = 2.12$$

From Appendix Table A-2 for a two-tailed test with $\alpha = .05$ and $df = 27$, the critical values are ± 2.052. Thus, we can conclude that it took people with a left hemispheric preference a different number of moves to solve the Tower of Hanoi problem than people with an integrated hemispheric preference.

iii. $C = \overline{X}_I - \overline{X}_R = 206.7 - 198.4 = 8.3$

$$\hat{S}_C = \sqrt{MS_W\left[\left(\frac{1}{N_I}\right) + \left(\frac{1}{N_R}\right)\right]}$$

$$= \sqrt{(236.76)\left[\left(\frac{1}{10}\right) + \left(\frac{1}{10}\right)\right]}$$

$$= 6.88$$

$$t = \left(\frac{C - K}{\hat{S}_C}\right) = \frac{8.3 - 0}{6.88} = 1.21$$

From Appendix Table A-2 for a two-tailed test with $\alpha = .05$ and $df = 27$, the critical values are ± 2.052. Thus, we can conclude that people with an integrated hemispheric preference did not differ significantly on the number of moves to solve the Tower of Hanoi problem from people with a right hemispheric preference.

iv. $C = \dfrac{\overline{X}_I + \overline{X}_R}{2} - \overline{X}_L = \dfrac{206.7 + 198.4}{2} - 221.3$

$$= -18.75$$

$$\hat{S}_C = \sqrt{MS_W\left[\left(\frac{1}{N_I + N_R}\right) + \left(\frac{1}{N_L}\right)\right]}$$

$$= \sqrt{(236.76)\left[\left(\frac{1}{10 + 10}\right) + \left(\frac{1}{10}\right)\right]}$$

$$= 5.96$$

$$t = \left(\frac{C - K}{\hat{S}_C}\right) = \frac{18.75 - 0}{5.96} = -3.15$$

From Appendix Table A-2 for a two-tailed test with $\alpha = .05$ and $df = 27$, the critical values are ± 2.052. Thus, we can conclude that participants who have either an integrated or right hemispheric preference differed on the number of moves required to solve the Tower of Hanoi problem from people with a left hemispheric preference.

v. $C = \overline{X}_I - \dfrac{\overline{X}_L + \overline{X}_R}{2} = 206.7 - \dfrac{221.3 + 198.4}{2}$

$$= -3.15$$

$$\hat{S}_C = \sqrt{MS_W\left[\left(\frac{1}{N_I}\right) + \left(\frac{1}{N_L + N_R}\right)\right]}$$

$$= \sqrt{(236.76)\left[\left(\frac{1}{10}\right) + \left(\frac{1}{10 + 10}\right)\right]} = 5.96$$

$$t = \left(\frac{C - K}{\hat{S}_C}\right) = \frac{3.15 - 0}{5.96} = -0.53$$

From Appendix Table A-2 for a two-tailed test with $\alpha = .05$ and $df = 27$, the critical values are ± 2.052. Thus, participants who have an integrated hemispheric preference did not differ significantly on the number of moves required to solve the Tower of Hanoi problem from people with a left or right hemispheric preference.

j. i. $C = \overline{X}_L - \overline{X}_R = 221.3 - 198.4 = 22.9$

$$\hat{S}_C = \sqrt{MS_W\left[\left(\frac{1}{N_L}\right) + \left(\frac{1}{N_R}\right)\right]}$$

$$= \sqrt{(236.76)\left[\left(\frac{1}{10}\right) + \left(\frac{1}{10}\right)\right]}$$

$$= 6.88$$

$$t = \left(\frac{C - K}{\hat{S}_C}\right) = \frac{22.9 - 0}{6.88} = 3.33$$

To obtain the corrected critical value

$$t_{corrected} = \sqrt{(df_B)(F_{(df_B, df_W)})} = \sqrt{(2)(3.35)} = 2.59$$

Since $3.33 > 2.59$, we can reject H_0 and conclude that people with a left hemispheric preference took a different number of moves to solve the Tower of Hanoi problem than people with a right hemispheric preference.

ii. $C = \overline{X}_L - \overline{X}_I = 221.3 - 206.7 = 14.6$

$$\hat{S}_C = \sqrt{MS_W\left[\left(\frac{1}{N_L}\right) + \left(\frac{1}{N_I}\right)\right]}$$

$$= \sqrt{(236.76)\left[\left(\frac{1}{10}\right) + \left(\frac{1}{10}\right)\right]}$$

$$= 6.88$$

$$t = \left(\frac{C - K}{\hat{S}_C}\right) = \frac{14.6 - 0}{6.88} = 2.12$$

To obtain the corrected critical value

$$t_{corrected} = \sqrt{(df_B)(F_{(df_B, df_W)})} = \sqrt{(2)(3.35)} = 2.59$$

Since $2.12 < 2.59$, we fail to reject H_0 and conclude that people with a left hemispheric preference did not differ significantly on the number of moves to solve the Tower of Hanoi problem from people with an integrated hemispheric preference.

iii. $C = \overline{X}_I - \overline{X}_R = 206.7 - 198.4 = 8.3$

$$\hat{S}_C = \sqrt{MS_W\left[\left(\frac{1}{N_I}\right) + \left(\frac{1}{N_R}\right)\right]}$$

$$= \sqrt{(236.76)\left[\left(\frac{1}{10}\right) + \left(\frac{1}{10}\right)\right]}$$

$$= 6.88$$

$$t = \left(\frac{C - K}{\hat{S}_C}\right) = \frac{8.3 - 0}{6.88} = 1.21$$

To obtain the corrected critical value

$$t_{corrected} = \sqrt{(df_B)(F_{(df_B, df_W)})} = \sqrt{(2)(3.35)} = 2.59$$

Since $1.21 < 2.59$, we fail to reject H_0 and conclude that people with an integrated hemispheric preference did not differ significantly on the number of moves to solve the Tower of Hanoi problem from people with a right hemispheric preference.

k. The results of 7i(i) and 7j(i), and 7i(iii) and 7j(iii) are consistent. However, the results of 7i(ii) and 7j(ii) are not the same. This difference occurred because an adjustment was made in the computational algorithms for the Scheffé test to account for familywise error whereas no such adjustment was made in the planned comparisons.

l.

$$\eta^2 = \frac{SS_B}{SS_T} = \frac{2688.2}{9080.8} = .30$$

This means that 30% of the variance in the number of moves it took to solve the Tower of Hanoi problem could be related to the hemispheric preference of the people completing the task in the sample.

$$\hat{\omega}^2 = \frac{SS_B - (J - 1)MS_W}{SS_T + MS_W} = \frac{2688.2 - (3 - 1)(236.76)}{9080.8 + 236.76} = .24$$

This estimate of the proportion of variability in the number of moves it took to solve the Tower of Hanoi problem that is accounted for by hemispheric preference is .24. This is considered a moderately large effect.

$$\hat{f} = \sqrt{\frac{\left(\frac{J-1}{JN}\right)(MS_B - MS_W)}{MS_W}}$$

$$= \sqrt{\frac{\left(\frac{3-1}{(3)(10)}\right)(1344.1 - 236.76)}{236.76}} = .56$$

This is considered to be a moderately large effect of the hemispheric preference on the number of moves it takes to solve the Tower of Hanoi problem.

Note: The two effect size indices $\hat{f}$ and $\hat{\omega}^2$ are related in the following way:

$$\hat{f} = \sqrt{\frac{\hat{\omega}^2}{1 - \hat{\omega}^2}}$$

This fact makes computation easier once we know the value of either $\hat{f}$ or $\hat{\omega}^2$.

m. For left and integrated hemispheric preference:

$$d = \frac{\overline{X}_L - \overline{X}_I}{\sqrt{MS_W}} = \frac{221.3 - 206.7}{\sqrt{236.76}} = 0.95$$

This is a moderate effect.

For left and right hemispheric preference:

$$d = \frac{\overline{X}_L - \overline{X}_R}{\sqrt{MS_W}} = \frac{221.3 - 198.4}{\sqrt{236.76}} = 1.49$$

This is a moderately large effect.

For integrated and right hemispheric preference:

$$d = \frac{\overline{X}_I - \overline{X}_R}{\sqrt{MS_W}} = \frac{206.7 - 198.4}{\sqrt{236.76}} = 0.54$$

This is a somewhat small effect.

9. The comparisons involve six distinct pairs of sample means as follows:

$\overline{X}_1$ and $\overline{X}_2$; $\overline{X}_1$ and $\overline{X}_3$; $\overline{X}_1$ and $\overline{X}_4$; $\overline{X}_2$ and $\overline{X}_3$; $\overline{X}_2$ and $\overline{X}_4$; $\overline{X}_3$ and $\overline{X}_4$

The familywise error rate is

$$\alpha_{FW} < 1 - (1 - \alpha)^c = 1 - (1 - .05)^6 = .26$$

Using the Bonferroni method, the α for each comparison should be adjusted to $\frac{.05}{c} = \frac{.05}{6} = .008$ in order to keep familywise error at .05.

11. a. $C = \dfrac{\overline{X}_P + \overline{X}_M + \overline{X}_{CO}}{3} - \overline{X}_C = \dfrac{21.5 + 15 + 27.75}{3} - 14.67$

$$= 6.75$$

$$\hat{S}_C = \sqrt{MS_W\left[\left(\frac{1}{N_P + N_M + N_{CO}}\right) + \left(\frac{1}{N_C}\right)\right]}$$

$$= \sqrt{(21.36)\left[\left(\frac{1}{12 + 12 + 12}\right) + \left(\frac{1}{12}\right)\right]} = 1.54$$

$$t = \left(\frac{C - K}{\hat{S}_C}\right) = \frac{6.75 - 0}{1.54} = 4.38$$

From Appendix Table A-2 for a one-tailed test with $\alpha = .05$ and $df = 44$, the critical value is 1.684. (Use $df = 40$ since $df = 44$ is not in the table.) Thus, individuals who engaged in some form of practice of the motion involved in throwing the dart scored higher on a dart game than individuals who did not practice.

b. $C = \overline{X}_P - \overline{X}_M = 21.5 - 15 = 6.5$

$$\hat{S}_C = \sqrt{MS_W\left[\left(\frac{1}{N_P}\right) + \left(\frac{1}{N_M}\right)\right]}$$

$$= \sqrt{(21.36)\left[\left(\frac{1}{12}\right)+\left(\frac{1}{12}\right)\right]} = 1.89$$

$$t = \left(\frac{C-K}{\hat{S}_C}\right) = \frac{6.5-0}{1.89} = 3.44$$

From Appendix Table A-2 for a two-tailed test with $\alpha = .05$ and $df = 44$, the critical values are ± 2.021. Since $3.44 > 2.021$, we can conclude that physical practice of the motion involved in throwing darts leads to better performance on a dart game than mental practice of the motion involved in throwing darts.

c. $C = \dfrac{\bar{X}_P + \bar{X}_M}{2} - \bar{X}_{CO} = \dfrac{21.5 + 15}{2} - 27.75 = -9.5$

$$\hat{S}_C = \sqrt{MS_W\left[\left(\frac{1}{N_P + N_M}\right)+\left(\frac{1}{N_{CO}}\right)\right]}$$

$$= \sqrt{(21.36)\left[\left(\frac{1}{12+12}\right)+\left(\frac{1}{12}\right)\right]} = 1.63$$

$$t = \left(\frac{C-K}{\hat{S}_C}\right) = \frac{-9.5-0}{1.63} = -5.83$$

From Appendix Table A-2 for a one-tailed test with $\alpha = .05$ and $df = 44$, the critical value is -1.684. Therefore, individuals who engaged in some form of single practice of the motion involved in throwing the dart scored lower on a dart game than individuals who engaged in combined practice.

13. a. $C = \bar{X}_S - \bar{X}_I = 8.5 - 12 = -3.5$

$$\hat{S}_C = \sqrt{MS_W\left[\left(\frac{1}{N_S}\right)+\left(\frac{1}{N_I}\right)\right]}$$

$$= \sqrt{(6.67)\left[\left(\frac{1}{8}\right)+\left(\frac{1}{8}\right)\right]} = 1.29$$

$$t = \left(\frac{C-K}{\hat{S}_C}\right) = \frac{-3.5-0}{1.29} = -2.71$$

To obtain the corrected critical value

$$t_{corrected} = \sqrt{(df_B)(F_{(df_B, df_W)})} = \sqrt{(2)(5.78)} = 3.4$$

Since -2.71 is not smaller than -3.4, we can conclude that recall for shallow and intermediate levels of processing is not significantly different.

b. $C = \bar{X}_S - \bar{X}_D = 8.5 - 13.5 = -5$

$$\hat{S}_C = \sqrt{MS_W\left[\left(\frac{1}{N_S}\right)+\left(\frac{1}{N_D}\right)\right]}$$

$$= \sqrt{(6.67)\left[\left(\frac{1}{8}\right)+\left(\frac{1}{8}\right)\right]} = 1.29$$

$$t = \left(\frac{C-K}{\hat{S}_C}\right) = \frac{-5-0}{1.29} = -3.88$$

To obtain the corrected critical value

$$t_{corrected} = \sqrt{(df_B)(F_{(df_B, df_W)})} = \sqrt{(2)(5.78)} = 3.4$$

Since $-3.88 < -3.4$, we can conclude that participants who read the passage at a shallow level of pro-

cessing recalled less information about the passage than participants who read the information at a deep level.

c. $C = \bar{X}_I - \bar{X}_D = 12 - 13.5 = -1.5$

$$\hat{S}_C = \sqrt{MS_W\left[\left(\frac{1}{N_I}\right)+\left(\frac{1}{N_D}\right)\right]}$$

$$= \sqrt{(6.67)\left[\left(\frac{1}{8}\right)+\left(\frac{1}{8}\right)\right]} = 1.29$$

$$t = \left(\frac{C-K}{\hat{S}_C}\right) = \frac{-1.5-0}{1.29} = -1.16$$

To obtain the corrected critical value

$$t_{corrected} = \sqrt{(df_B)(F_{(df_B, df_W)})} = \sqrt{(2)(5.78)} = 3.4$$

Since -1.16 is not smaller than -3.4, we can conclude that people who processed information at a deep level did not recall significantly more information than people who processed information at an intermediate level.

Based on the Scheffé post hoc comparisons, the levels of processing theory was only partially supported. As indicated by the theory, individuals who processed information at a deep level recalled more information than individuals who processed information at a shallow level. However, inconsistent with the theory, individuals who processed the information at an intermediate level did not differ significantly in their level of recall from individuals who processed the information at a shallow level or at a deep level.

15. a. $C = \bar{X}_T - \bar{X}_S = 17.13 - 17.88 = -0.75$

$$\hat{S}_C = \sqrt{MS_W\left[\left(\frac{1}{N_T}\right)+\left(\frac{1}{N_S}\right)\right]} = \sqrt{(16.41)\left[\left(\frac{1}{8}\right)+\left(\frac{1}{8}\right)\right]}$$
$$= 2.03$$

$$t = \left(\frac{C-K}{\hat{S}_C}\right) = \frac{-0.75-0}{2.03} = -0.37$$

To obtain the corrected critical value

$$t_{corrected} = \sqrt{(df_B)(F_{(df_B, df_W)})} = \sqrt{(2)(3.47)} = 2.63$$

Since -0.37 is not smaller than -2.63, we can conclude that the traditional analytic presentation and the social context presentation did not lead to a significant difference in problem-solving scores.

b. $C = \bar{X}_T - \bar{X}_C = 17.13 - 22.13 = -5.00$

$$\hat{S}_C = \sqrt{MS_W\left[\left(\frac{1}{N_T}\right)+\left(\frac{1}{N_C}\right)\right]} = \sqrt{(16.41)\left[\left(\frac{1}{8}\right)+\left(\frac{1}{8}\right)\right]}$$
$$= 2.03$$

$$t = \left(\frac{C-K}{\hat{S}_C}\right) = \frac{-5.00-0}{2.03} = -2.46$$

To obtain the corrected critical value

$$t_{corrected} = \sqrt{(df_B)(F_{(df_B, df_W)})} = \sqrt{(2)(3.47)} = 2.63$$

Since -2.46 is not smaller than -2.63, we can conclude that the traditional analytic presentation and the culturally relevant presentation did not lead to a significant difference in problem-solving scores.

c. $C = \overline{X}_S - \overline{X}_C = 17.88 - 22.13 = -4.25$

$$\hat{S}_C = \sqrt{MS_W\left[\left(\frac{1}{N_S}\right) + \left(\frac{1}{N_C}\right)\right]} = \sqrt{(16.41)\left[\left(\frac{1}{8}\right) + \left(\frac{1}{8}\right)\right]}$$
$$= 2.03$$

$$t = \left(\frac{C - K}{\hat{S}_C}\right) = \frac{-4.25 - 0}{2.03} = -2.09$$

To obtain the corrected critical value

$$t_{corrected} = \sqrt{(df_B)(F_{(df_B, df_W)})} = \sqrt{(2)(3.47)} = 2.63$$

Since -2.09 is not smaller than -2.63, we can conclude that the culturally relevant presentation and the social context presentation did not lead to a significant difference in problem-solving scores.

From the results of the overall F-test we concluded that among the types of presentations at least one mean was different from the others. However in conducting Scheffé post hoc comparisons for each pair of means no differences were found. These contradictory results can be explained by the fact that a corrected critical value was calculated for the Scheffé post hoc comparisons to adjust for familywise error. In making these corrections the net effect is to test the pairwise comparisons at some α less than .05.

17. Since the overall F-test was not significant it is not appropriate to conduct Scheffé post hoc comparisons.

19. The following is the SPSS output:

Oneway

Descriptives

Perception of Math as a Fun Activity

	N	Mean	Std. Deviation	Std. Error	95% Confidence Interval for Mean Lower Bound	95% Confidence Interval for Mean Upper Bound	Minimum	Maximum
7	50	43.38	11.41	1.61	40.14	46.62	14	60
8	50	38.28	9.31	1.32	35.63	40.93	22	60
9	50	41.32	7.66	1.08	39.14	43.50	22	56
10	50	35.22	10.12	1.43	32.34	38.10	19	58
Total	200	39.55	10.13	.72	38.14	40.96	14	60

ANOVA

Perception of Math as a Fun Activity

	Sum of Squares	df	Mean Square	F	Sig.
Between Groups	1908.180	3	636.060	6.732	.000
Within Groups	18519.320	196	94.486		
Total	20427.500	199			

Post Hoc Tests

Multiple Comparisons

Dependent Variable: Perception of Math as a Fun Activity
Scheffé

(I) Grade Level in School	(J) Grade Level in School	Mean Difference (I-J)	Std. Error	Sig.	95% Confidence Interval Lower Bound	95% Confidence Interval Upper Bound
7	8	5.10	1.94	.079	−.38	10.58
	9	2.06	1.94	.772	−3.42	7.54
	10	8.16*	1.94	.001	2.68	13.64
8	7	−5.10	1.94	.079	−10.58	.38
	9	−3.04	1.94	.487	−8.52	2.44
	10	3.06	1.94	.481	−2.42	8.54
9	7	−2.06	1.94	.772	−7.54	3.42
	8	3.04	1.94	.487	−2.44	8.52
	10	6.10*	1.94	.022	.62	11.58
10	7	−8.16*	1.94	.001	−13.64	−2.68
	8	−3.06	1.94	.481	−8.54	2.42
	9	−6.10*	1.94	.022	−11.58	−.62

*The mean difference is significant at the .05 level.

Homogeneous Subsets

Perception of Math as a Fun Activity

Scheffé[a]

Grade Level in School	N	Subset for alpha = .05	
		1	2
10	50	35.22	
8	50	38.28	38.28
9	50		41.32
7	50		43.38
Sig.		.481	.079

Means for groups in homogeneous subsets are displayed.
[a]Uses Harmonic Mean Sample Size = 50.000.

Overall, there is a difference among 7th, 8th, 9th, and 10th grade students in their perceptions of mathematics as a fun activity. In particular, 10th grade students were less likely to view mathematics as a fun activity than 7th and 9th grade students. Seventh, 8th, and 9th grade students did not differ significantly in their perceptions of mathematics as a fun activity. In addition, 8th and 10th grade students did not differ significantly in their perceptions of mathematics as a fun activity.

21. The following is the SPSS output:

Oneway

ANOVA

Perception of Math as a Fun Activity

	Sum of Squares	df	Mean Square	F	Sig.
Between Groups	1908.180	3	636.060	6.732	.000
Within Groups	18519.320	196	94.486		
Total	20427.500	199			

Contrast Coefficients

	Grade Level in School			
Contrast	7	8	9	10
1	1	1	−1	−1
2	1	1	1	−3
3	1	1	−2	0
4	1	−1	0	0

Contrast Tests

		Contrast	Value of Contrast	Std. Error	t	df	Sig. (2-tailed)
Perception of Math as a Fun Activity	Assume equal variances	1	5.12	2.75	1.862	196	.064
		2	17.32	4.76	3.637	196	.000
		3	−.98	3.37	−.291	196	.771
		4	5.10	1.94	2.623	196	.009
	Does not assume equal variances	1	5.12	2.75	1.862	182.344	.064
		2	17.32	4.89	3.538	80.008	.001
		3	−.98	3.00	−.326	125.600	.745
		4	5.10	2.08	2.449	94.226	.016

The familywise error rate is $\alpha_{FW} \le 1 - (1 - .05)^4 = .19$. If we want our familywise error rate to be .05, then we should test each comparison at $\alpha = .05/4 = .0125$. If we apply $\alpha = .0125$ to the printout, the results suggest the following about the predictions.

1. Since the column labeled "Sig. (2-tailed)" gives a value of .064 > .0125, the first planned comparison is not significant. Thus, the prediction that middle school students will have a different perception of mathematics as a fun activity than high school students is not supported.

2. Since the column labeled "Sig. (2-tailed)" gives a value of .000 < .0125, the second planned comparison is significant. Thus, the prediction that senior high school students will have a different perception of mathematics as a fun activity than junior high school students is supported. [NOTE: The two-tailed probability is not really zero as reported above. The value is so small that when we round to three decimal places it yields .000.]

3. Since the column labeled "Sig. (2-tailed)" gives a value of .771 > .0125, the third planned comparison is not significant. Thus, the prediction that 9th grade students will have a different perception of mathematics as a fun activity than middle school students is not supported.

4. Since the column labeled "Sig. (2-tailed)" gives a value of .009 < .0125, the fourth planned comparison is significant. Thus, the prediction that 8th grade students will have a different perception of mathematics as a fun activity than 7th grade students is supported.

Index